Nanoscale Zerovalent Iron Particles for Environmental Restoration

Tanapon Phenrat • Gregory V. Lowry
Editors

Nanoscale Zerovalent Iron Particles for Environmental Restoration

From Fundamental Science to Field Scale Engineering Applications

Springer

Editors
Tanapon Phenrat
Department of Civil Engineering
Environmental Engineering Program
Naresuan University
Phitsanulok, Thailand

Center of Excellence for
Sustainability of Health
Environment and Industry (SHEI)
Faculty of Engineering
Naresuan University
Phitsanulok, Thailand

Gregory V. Lowry
Department of Civil and Environmental
Engineering
Canrnegie Mellon University
Pittsburgh, PA, USA

ISBN 978-3-319-95338-0 ISBN 978-3-319-95340-3 (eBook)
https://doi.org/10.1007/978-3-319-95340-3

Library of Congress Control Number: 2018964898

This Springer imprint is published by the registered company Springer International Publishing AG part of Springer Nature.
The registered company address is: Gewerbestrasse 11, 6330 Cham, Switzerland

Preface

Soil and groundwater contamination remains a serious threat to humans and to the ecosystem. Sources of contamination include accidental spills or routine improper hazardous waste management from various industrial activities, for example, mining, petrochemical, refinery, electronic, petrol station (underground storage tanks), treatment, storage, and disposal facility industries. Hundreds of thousands of contaminated sites have been identified in the USA. Similarly, in Europe, in 2011, approximately 1.2 million potentially contaminated sites and 130,000 contaminated sites have been identified. In 2014, the Chinese National Soil Survey found that 16.1% of all soil and 19.4% of cultivated land were contaminated by hazardous organic and inorganic chemicals. This amounted to roughly 25 million ha. of contaminated soil and 3.5 million ha. of farmland that was too contaminated to safely allow agriculture. Even a small country like Thailand has more than 12 identified contaminated sites and 40 potentially contaminated sites throughout the country.

Tremendous advances have been made regarding regulations, tools for site characterization, risk assessment, and remediation technology. However, the cost of remediation remains high. The US EPA (2004) estimated that cleaning up 294,000 sites from 2004 to 2033 would cost about US$209 billion. Similarly, in Europe, annual national expenditures for managing contaminated sites are, on average, EUR 10.7 per capita (an average of 0.041% of the national gross domestic product [GDP]). In China, in 2016, the State Council of China issued a Soil Pollution Prevention and Control Action Plan to restore 90% of contaminated farmland soil to be safe for human use by 2020 and 95% by 2030. Thailand's government is spending US$14.44 million to clean up the lead-contaminated sediment at Klity Creek, Kanchanaburi Province. This is the first court-ordered remedial action in Thailand. Over the next several decades, remediation research and business will be needed to address legacy and emerging pollutants (e.g., PFOS) in countries across worldwide.

Nanoscale zerovalent iron (NZVI) has proven to be versatile and has the potential to compliment and overcome critical limitations of many of the standard remediation technologies. The small size allows zerovalent iron (ZVI) typically used to build a permeable reactive barrier, a passive technique, to become an active technique that can be delivered into deep or difficult to reach source zones to speed up remediation. Over two decades of active NZVI research have provided a large body of knowledge about the uses, benefits, and limitations of NZVI. The number of peer-reviewed papers and citations related to NZVI has increased quadratically every year. In 2017, there were 262 peer-reviewed NZVI journal papers and 8094 citations. There are 77 documented pilot- and full-scale demonstration sites where NZVI was used as part of the remediation technology, primarily for chlorinated solvents. Significant progress has been made in understanding how environmental conditions affect reactivity, how to emplace NZVI in situ, and how to leverage the complex, coupled interactions between NZVI and bioremediation. Remediation performance expectations are slowly becoming clear, depending on the type of NZVI, the type of contaminant, site conditions, and injection methods. Thus, it is the right time for a book that is solely devoted to advancing NZVI research and its field application.

This book presents the past, the present, and the future of NZVI research from fundamental science to field-scale engineering applications. The 16 chapters are based on more than two decades of laboratory research and development and field-scaled demonstrations of NZVI implementation. The first chapter presents a brief history of NZVI research and applications over the last two decades, while the other 15 chapters in this book provide detailed contents on the progress of NZVI, which is essential for researchers and practitioners aiming to study or to deploy the technology. The chapters review well-established aspects of NZVI, including NZVI synthesis, characterization (Chap. 2), and reactivity toward organic contaminants (Chap. 3), metals, and metalloids (Chap. 4), fundamental and modeling aspects of NZVI modification for subsurface delivery including theoretical explanations of rapid agglomeration and poor dispersion stability of NZVI and surface engineering approaches to overcome these limitations (Chap. 5), and modeling tools for field-scale applications (Chap. 6). Chapters 7 and 8 provide details of experiences and best practices for field-scale applications of NZVI from real practitioners. Chapters 9, 10, 11, 12, 13, and 14 present recent advances and novel combined remedies using NZVI, including sulfidation of NZVI (Chap. 9), NZVI and bioremediation (Chap. 10), thermally enhanced NZVI remediation using electromagnetic induction (Chap. 11), pre-magnetization and application of a weak magnetic field (Chap. 12), vadose-zone applications of NZVI with electromagnetic induction (Chap. 13), and aerosol-based iron-carbon nanocomposites for synergistic adsorption and reduction of Cr (VI) (Chap. 14). Risk and environmental implication of NZVI is also discussed in depth in Chap. 15. Finally, Chap. 16 concludes with the state of the knowledge and future needs for NZVI applications in subsurface remediation.

The authors of each chapter are leading researchers and/or practitioners in NZVI technology. The intended audience includes remediation professionals, policymakers, practicing engineers, researchers, and students seeking to understand and to improve NZVI performance in remediation. This volume should serve as a comprehensive technical resource for remediation professionals and researchers to support effective application and development of NZVI-based technologies.

Phitsanulok, Thailand Tanapon Phenrat
Pittsburgh, PA, USA Gregory V. Lowry

Contents

About the Editors

Tanapon Phenrat is an assistant professor and a principal investigator of Research Unit for Integrated Natural Resources Remediation and Reclamation (IN3R), Department of Civil Engineering, Faculty of Engineering, Naresuan University, Phitsanulok. Also, he serves as the deputy director of Center of Excellence for Sustainability of Health, Environment, and Industry (SHEI), Faculty of Engineering, Naresuan University, Phitsanulok, Thailand. He has 13-year experience in conducting research and professional work in the field of remediation of contaminated sites. He has published 36 peer-reviewed research articles in international journals including several highly cited research articles in nanoscale zerovalent iron applications. He is also a practitioner in remedial engineering in Thailand and has worked collaboratively as a consultant for affected communities, government agencies, and potentially responsible contaminated site owners. He investigated more than 40 contaminated sites and successfully remediated 6 major contaminated sites using soil vapor extraction, in situ chemical oxidation, colloidal carbon injection, colloidal iron oxide injection, magnetic-assisted soil washing with zerovalent iron, and phytoremediation.

He is serving as an official tripartite member of committees to resolve the conflict between gold mining in Phichit Province and affected communities as appointed by the Minister of Industry as well as an official tripartite member of Klity Creek Restoration appointed by the Pollution Control Department. He also served as an expert witness for the case of arsenic contaminated soil and sediment at Loei's gold mining as requested by the National Human Rights Commission of Thailand. Dr. Phenrat was awarded with 16 research awards (7 international awards) in the field of environmental remediation. Dr. Phenrat earned his B.Eng in Civil Engineering from Kasetsart University, M.S. in Hazardous Waste Management from Chularlongkorn University, and Ph.D. in Civil and Environmental Engineering from Carnegie Mellon University.

Gregory V. Lowry is the Walter J. Blenko, Sr., professor of Civil and Environmental Engineering at Carnegie Mellon University. He is the deputy director of the

NSF/EPA Center for Environmental Implications of Nanotechnology (CEINT). He is on the editorial board for *Environmental Science: Nano* and *Nature: Scientific Data*. He served as the founding associate editor of *Environmental Science: Nano* for 5 years (2013–2018). Dr. Lowry earned his B.S. in Chemical Engineering from the University of California in Davis, M.S. from the University of Wisconsin in Madison, and Ph.D. in Civil and Environmental Engineering from Stanford University. His current research and teaching focus on safely harnessing the unique properties of engineered nanomaterials for applications in remediation, water treatment, and food production. He has served on the National Academy of Science Committee on "Science Breakthroughs 2030: A Strategy for Food and Agricultural Research," and the National Research Council Committee to develop a research strategy for environmental health and safety aspects of engineered nanomaterials. Dr. Lowry is a Fellow of the American Association for the Advancement of Science (AAAS) and has received awards from the ASCE (Walter L. Huber Civil Engineering Research Prize), AEESP (Malcolm Pirnie/AEESP Frontiers in Research Award), and ACS (Best Feature Article in ES&T for 2012). He was recognized as a highly cited researcher (top 1%) in the area of environment and ecology by Thomson Reuters in 2014 and 2015. Dr. Lowry has served as PI or co-PI on grants from the National Science Foundation, US Department of Defense, US Department of Energy, and US Environmental Protection Agency, and industry.

About the Authors

Ling Ai received her B.S. degree in Environmental Engineering from Sichuan Agricultural University in 2013 and master's degree in Environmental Engineering from the Dalian University of Technology in 2016. Currently, she is working in the Tianjin YXT Science and Technology Co., Ltd., as an environmental engineer, focusing on the engineering design for VOCs and wastewater treatments.

Peyman Babakhani is currently near completion of a dual Ph.D. program at the University of Liverpool and National Tsing Hua University. His research interests include modeling the fate and transport of nanoparticles in the environment as well as the remediation of groundwater contaminants using nanoparticles. Peyman is currently working on the MT3DM-USGS code to incorporate various transport mechanisms of nanoparticles in porous media. To date, he has incorporated several mechanisms in the model including the aggregation of nanoparticles. He has also proposed a method based on artificial neural network to face the challenge of predictivity of the numerical models used for groundwater simulation. This method enables the use of nanoparticle transport models without the need for fitting the model to the observation data.

Michael Borda is a strategist, implementer, change agent, and connector. Currently the director of Strategic Initiatives at the University of Pennsylvania, Borda is maximizing the impact of Penn's investment in innovation and diversifying Penn's research portfolio. Previously, Dr. Borda was a senior consultant with the Golder Associates and a leader in the development and application of nanoscale zerovalent iron (NZVI) remediation technologies. Dr. Borda has published over 30 peer-reviewed articles, book chapters, and encyclopedia contributions in the environmental chemistry and microbiology fields and holds US and international patents for an innovative remediation technology. Michael has presented research and projects at conferences and workshops around the globe. In 2013 he was selected by the *Engineering News-Record* as one of the top 20 engineers under the age of 40, and in 2017 he was awarded a Model of Excellence at Penn.

Julian Bosch managing director and co-founder of the Intrapore GmbH, Germany, is an environmental scientist and holds a Ph.D. in Geomicrobiology. As a postdoc and EU project leader at the Institute of Groundwater Ecology of the Helmholtz Zentrum München, he has more than 10 years of experience in the development and application of nano- and microparticles for groundwater remediation.

Jonathan Bridge is an environmental geoscientist based at Sheffield Hallam University, UK, where he teaches geography and environmental science, specializing in water quality and water management. He completed his Ph.D. and postdoctoral research in the Groundwater Protection and Restoration Group and the Cell-Mineral Research Centre at the University of Sheffield where he led a work on time-lapse fluorescence imaging of colloid transport through porous media and contributed to the development of agent-based modeling approaches for simulation of fungal weathering in soils. He has applied both experimental and theoretical approaches to the nuclear industry through work on sludge filtration processes in rapid sand bed filters and is currently supporting research on the integration of coupled aggregation-sedimentation models with lab- and field-scale numerical simulations of nanoparticle transport and fate.

Johannes Bruns is a geologist with a Ph.D. in soil mechanics and geotechnics. He started his professional career as a state regulator in Germany, working for 5 years on research projects for soil and groundwater remediation. Afterward he worked for 10 years as department manager at two large-scale industry enterprises in Germany, performing brownfield soil and groundwater remediation. Then he worked for 19 years as director and senior partner for Golder Associates GmbH on international remediation projects in Europe, Asia, and South America. His professional activities include remediation of soil and groundwater contamination, applying innovative site investigation and remediation technologies such as nanoscale zerovalent iron (NZVI) and nanocomposite materials. The experience gathered with particle application for environmental remediation has been published by Dr. Bruns on diverse conferences and in articles.

Alexis Wells Carpenter is principal investigator at AxNano, a small business entity with over 30 years' experience developing early-stage technologies to address specific market needs. Dr. Carpenter's expertise is in the design and application of multifunctional materials for use in environmental and biomedical markets. Dr. Carpenter has a Ph.D. in Inorganic Chemistry from the University North Carolina at Chapel Hill where her dissertation focused on manipulating particle formulations for controlled drug delivery and antimicrobial applications, under the advisement of Professor Mark Schoenfisch. She subsequently worked with Dr. Mark Wiesner in the Superfund Research Center and Center for Environmental Implications of Nanotechnology at Duke University, developing zerovalent iron technologies for in situ sediment remediation and enhanced wastewater treatment. AxNano is a sub-entity underneath Triad Growth Partners, LLC (www.triadgrowthpartners.com), which connects entrepreneurs and scientists to launch or enhance existing technologies and build profitable businesses. AxNano creates advanced technologies for the environmental space, and it is the producer of the RemRx™ line of remediation products, which currently includes RemRx™ CRP, a controlled release polymeric system for in situ chemical oxidation, and RemRx™ CSI, a zerovalent iron-based composite for in situ chemical reduction.

Ruey-an Doong was the former dean of the College of Nuclear Science at National Tsing Hua University in Taiwan before joining the Institute of Environmental Engineering, National Chiao Tung University, also in Taiwan, as a distinguished professor. Prof. Doong received his B.S. in Environmental Engineering from National Chung Hsing University and his Ph.D. degree in Environmental Engineering from National Taiwan University. He is internationally known for his research in environmental nanotechnology, energy materials, and biosensors and nanosensing devices and was invited as a speaker at the Gordon Research Conference on Environmental Nanotechnology in 2015. Prof. Doong has published over 150 papers with 6000+ citations and h-index of 46. He has received many awards and honors for his contribution in research and teaching in the field of environmental engineering and science. He received the International Honorary Member Award from the American Academy of Environmental Engineers and Scientists in 2016 for his worthy and sustained contributions to the advancement of environmental protection and his eminence in the field of environmental engineering and sciences. He is also a recipient of the prestigious national award of Outstanding Research Award from the Ministry of Science and Technology in 2008 and 2015 in addition to the Most Cited Paper Award from the *Journal of Chromatography* in 2006.

Jan Filip received his master's and Ph.D. degrees in Mineralogy at Masaryk University in Brno, Czech Republic (2002 and 2008, respectively). His dissertation focused on the crystal chemistry of borosilicate minerals and OH defects in nominally anhydrous minerals. In 2005, he joined the research group of Prof. Radek Zbořil and currently leads the group "Environmental nanotechnologies" of the Regional Centre of Advanced Technologies and Materials, Palacký University in Olomouc, Czech Republic. He underwent several foreign stays (e.g., University of Vienna, Austria; Swedish Museum of Natural History, Stockholm, Sweden) and is actively engaged in national and international collaboration projects. His research interests cover development, optimization, and application of zerovalent iron nanoparticles, ferrates, and iron oxide nanoparticles in technologies of water treatment, identification of nanoparticles in the environment, and application of X-ray-based techniques (X-ray diffraction, X-ray fluorescence, and X-ray photoelectron spectroscopy) in material science. He has coauthored over 90 papers in refereed journals and 5 patents.

Anett Georgi is a senior scientist at the Helmholtz Center for Environmental Research – UFZ in Leipzig, Germany. She is leading the "Environmental catalysis – Oxidation processes" Group within the Department of Environmental Engineering.

She completed a Ph.D. in Chemistry within the field of chemical fate and impact in the environment. Her current research is focused on sorptive, catalytic, and other advanced oxidation processes for removal of organic contaminants from water. This includes material and process development for in situ nanoremediation, such as adsorptive and reactive colloidal particles using activated carbon or zeolite supports (Trap-Ox®). She has experience in the study of surface chemistry, colloidal properties, and transport of nano- and microscale particles. Finding good scientific solutions for challenging environmental problems is a motivation for her work.

Florin Gheorghiu has 30 years of remediation experience developed while working at the Golder Associates Inc. and currently as a principal at Applied Testing & Geosciences, LLC. Mr. Gheorghiu is an expert in the field of environmental restoration, contaminant hydrogeology, and numerical and solute transport modeling for unconsolidated and fractured bedrock sites. Mr. Gheorghiu international experience includes environmental projects conducted in 15 countries from North America, Europe, Asia, and South America. Mr. Gheorghiu was the project director and technical manager for the design and implementation of a large bedrock remedial system in Pennsylvania that received the Outstanding Groundwater Remediation Project Award from the National Groundwater Association. Mr. Gheorghiu directed or worked as a member of a team to design and implement numerous pilot test and full-scale nanoscale zerovalent iron (NZVI) remedies including the design and implementation of the very first NZVI pilot test carried out in fractured bedrock. The results and lessons learned from these studies have been summarized by Mr. Gheorghiu in professional publications and presented at industry conferences.

Khara Grieger is currently an environmental research scientist and expert within risk and decision support at RTI International's Risk Analysis Program within the Center for Health and Environmental Modelling. She is an internationally recognized expert in risk assessment, decision support, uncertainty, and governance related to emerging technologies, including engineered nanomaterials. To date, she has published over 30 peer-reviewed articles on nanomaterial and nanotechnology risks in the past decade and currently leads RTI's nanomaterial risk research efforts. She is a co-PI on a joint research project with North Carolina State University focused on sustainable innovation of nanomaterials for use in water treatment systems and has served as RTI's representative to the US Technical Advisory Group ISO/TC 229 Nanotechnologies (2014–2018). She also has over 5 years of experience working with US Federal agencies on diverse projects related to risk analysis, risk ranking, and decision support and currently is a university scholar within the Risk and Resilience Program at Duke University.

Xiaohong Guan is a full professor working in the Department of Municipal Engineering, Tongji University. Her research interests are developing novel physical and chemical technologies for water and wastewater treatment and exploring the involved mechanisms. She pioneered the research on weak magnetic field-enhanced zerovalent iron technology and bisulfite-activated permanganate oxidation process.

She has authored over 100 SCI-cited papers including 18 in *Water Research* and 16 in *Environmental Science & Technology*. She has been awarded several Best Poster Awards at international conferences, excellent doctor student supervisor for Tongji University, Environmental Science & Technology's Excellence in Review Award, and Shouchuang Water Star Award.

Claudia Gunsch is an associate professor in the Development of Civil and Environmental Engineering at Duke University. Her research has focused on characterizing and engineering environmental microbiomes. Current research projects focus on exploring ballast water microbiomes, exploring correlations between microbial adaptation and evolution stemming from their exposure to contaminants, characterizing the fate of genetically modified crop transgenes, developing genetic bioaugmentation technologies for improving the bioremediation efficacy of recalcitrant contaminants, and developing innovative water treatment technologies for industrial and developing world applications.

Jiawei He received his B.S. degree in Environmental Engineering from Shenyang University of Technology in 2013. In 2016, he obtained his master's degree in Environmental Sciences and Engineering from the Dalian University of Technology. Currently, he is working in the Panjin Municipal Environmental Protection Bureau.

Rune Hjorth is an environmental scientist with international experience at the Technical University of Denmark. For almost a decade, he has worked with human and environmental risk assessments in various contexts, including evaluations of engineered nanomaterials. He is also interested in pursuing more societal topics in connection with risk and enjoys working in the interplay between science, policy, uncertainty, and innovation. He was recently awarded DTU's Ph.D. Thesis of the Year (2017).

Danail Hristozov is the Head of Research at Greendecision Srl, a spin-off company of University Ca' Foscari of Venice (UNIVE) in Italy. There he performs integrative research across the areas of risk assessment and governance of nano(bio)materials used in medicine and consumer products as part of the large-scale EU H2020 projects GRACIOUS, BIORIMA, REFINE and caLIBRAte. In addition, Danail has been a senior research scientist at the Department of Environmental Sciences, Informatics and Statistics of UNIVE, where he acted as the Principal Investigator of the large-scale EU FP7 project SUN. Over the last years Danail contributed to several other EU projects dealing with the assessment and management of risks posed by chemical and emerging technologies (e.g. MARINA, NANOforART, ENPRA, ITS-NANO, GLOCOM, ECONANOSOPRB) also as research staff of the private sector companies Venice Research Consortium and Veneto Nanotech. In his early years as a researcher, Danail was employed at the Chair of Industrial Sustainability of the Brandenburg University of Technology in Germany.

David Jassby is an associate professor at the Civil and Environmental Engineering Department at the University of California, Los Angeles. Previously he was an assistant professor in the Department of Chemical and Environmental Engineering at the University of California, Riverside (2012–2017). Dr. Jassby received his B.S. in Biology from the Hebrew University of Jerusalem (2002); an M.S. in Civil and Environmental Engineering from the University of California, Davis (2004); and a Ph.D. in Civil and Environmental Engineering from Duke University (2011). Dr. Jassby is a recipient of the National Science Foundation CAREER Award and the ACS Petroleum Research Fund Doctoral New Investigator Award. Dr. Jassby's research interests include water treatment and desalination, membrane separation processes, membrane material fabrication, electrochemistry, and environmental applications of nanotechnology.

Eakalak Khan is a professor in Civil and Environmental Engineering and Construction Department and the director of Water Resources Research Program at the University of Nevada, Las Vegas, USA. From 2002 to 2017, he was a professor in Civil and Environmental Engineering Department at North Dakota State University (NDSU), USA. He also served as the chair of Civil Engineering Department at the NDSU from 2010 to 2013. Prior to the NDSU, he was an assistant professor in the Department of Civil Engineering at Polytechnic University, New York, USA, from 1999 to 2002. He received his Bachelor of Engineering in Environmental Engineering from Chiang Mai University, Thailand, in 1990; M.S. in Agricultural Engineering from the University of Hawaii, USA, in 1993; and Ph.D. in Civil Engineering from the University of California, Los Angeles (UCLA), USA, in 1997. In 1998, he was a postdoctoral research associate at the Institute of Environment, UCLA. Eakalak has published more than 100 refereed journal articles. He was awarded a CAREER grant from NSF in 2005. His honors include the NDSU Odney Award for Excellence in Teaching in 2008; Researcher of the Year, College of Engineering, NDSU, in 2005; and Water B. Booth Endowed Distinguished Professorship, NDSU, in 2017.

Frederick Klaessig is currently with the Pennsylvania Bio Nano Systems, a consulting firm focusing on reference materials used in investigating chromatographic effects at the nanoscale. In prior years, he was the US technical director for Evonik Degussa's Aerosil & Silanes and later the business director for the Aerosil Business Line. His assignments ranged from commercial overview (product management, production, sales) to technical responsibilities involving customer support, new product introduction, liaison with the R&D Department in Germany, and for US regulatory matters. Fumed silica, titania, and related materials are utilized in many fields for reinforcement, rheology control, abrasion, and UV absorption. In recent years, the great interest in nanotechnology has raised safety and registration concerns. These issues, both everyday technology and EHS, led to involvement in ASTM (E56), ISO (TC229), and industry organizations. Fred received a B.S. in Chemistry from the University of California, Berkeley, and a Ph. D. in Physical

Chemistry from the Rensselaer Polytechnic Institute. His earlier industrial experiences were with the Bio-Rad Laboratories as a quality control chemist and various R&D management positions at Betz Laboratories, now a division of Suez, where his responsibilities involved scale, corrosion, and microbiological control in many chemical industrial processes.

Chris M. Kocur is currently a postdoctoral scholar at the Oregon Health & Science University in Portland, Oregon, where he investigates core concepts of redox processes in all phases of site management. His work is funded through an NSERC Postdoctoral Fellowship (Canada) and is developing emerging tools to estimate and understand abiotic contaminant degradation. Chris received his Ph.D. at Western University in London, Ontario, in Environmental Engineering with a focus on field-scale application of nanoscale zerovalent iron technologies. His suite of studies in *Environmental Science & Technology* explores nanoparticle mobility and characterization in environmental systems, abiotic and biotic degradation of chlorinated compounds, and the synergistic interaction of microbial communities with NZVI on remediation sites. Chris has received numerous awards as a young researcher, including the national graduate scholarship from the Canadian Foundation for Geotechnique, in part, for his dedication to service through various professional societies. Chris' work has been funded through the Walkerton Clean Water Centre and Industrial Postgraduate Scholarship (NSERC-IPS) in collaboration with CH2M Canada Ltd.

Jan Kolařík received his master's degrees in Material Chemistry at Palacký University in Olomouc, Czech Republic, in 2012. In the same year, he joined the Ph.D. studies at the Department of Physical Chemistry and also joined as a junior researcher at the Regional Centre of Advanced Technologies and Materials in "Environmental group." His research is focused on studies of kinetics and mechanisms of removal of inorganic pollutants, mainly heavy metals and metalloids by zerovalent iron and ferrate-based materials. He underwent foreign stay in the Florida Institute of Technology and Texas A&M University, USA.

Panaya Kotchaplai received her bachelor's degree with first-class honor in Biochemistry from Chulalongkorn University, Thailand, in 2009 and her Ph.D. in Environmental Management from Chulalongkorn University, Thailand, in 2017 under the Royal Golden Jubilee Scholarship provided by the Thailand Research Fund. During her Ph.D. study under the supervision of Assoc. Prof. Alisa Vangnai and Prof. Eakalak Khan, she focused on the effects of nanoscale zerovalent iron (NZVI) which is a promising catalyst for environmental remediation on bacteria as well as bacterial adaptation. She is currently a postdoctoral researcher in Biocatalyst and Environmental Biotechnology Research Unit, Department of Biochemistry, Chulalongkorn University.

Amy E. Landis is now the first Presidential Faculty Fellow for Access, Attainment, and Diversity at the Colorado School of Mines. Previously, she was a full professor at Clemson University from 2015 to 2017 as the Thomas F. Hash ’69 Endowed Chair in Sustainable Development. She served as director of Clemson’s Institute for Sustainability, which brings together interdisciplinary research, education, and business. Landis was an associate professor at the Arizona State University’s School of Sustainable Engineering and the Built Environment from 2012 to 2015. During her tenure at ASU, she served as the director of research for the Center for Earth Systems Engineering and Management, senior sustainability scientist for the Global Institute of Sustainability, Lincoln Fellow of Sustainable Development and Ethics for the Lincoln Center for Applied Ethics, and Tooker professor of STEM Education for the Ira A. Fulton Schools of Engineering. Landis began her career as an assistant professor at the University of Pittsburgh after earning her Ph.D. in 2007 from the University of Illinois at Chicago under the supervision of Thomas L. Theis. Landis has developed a research program in sustainable engineering of bioproducts. Her research ranges from design of systems based on industrial ecology and by-product synergies, life cycle, and sustainability assessments of biopolymers and biofuels and design and analysis of sustainable solutions for health care.

Thi Song Thao Le is currently a full-time Ph.D. student in Environmental Engineering at the University of New South Wales, Australia. She completed her B.S. in Environmental Technology at the Ho Chi Minh City University of Science, Vietnam, in 2013. She was awarded a fully funded scholarship to complete her Master of Engineering at Naresuan University, Thailand, in 2015. Her major research interests include Fenton’s reaction, nanoscale zerovalent iron, and TiO_2 photocatalysis and fluorinated surfactants (PFASs). She is now working on the interfacial adsorption and the fate and transport of PFASs in the water.

Emilie Lefèvre obtained her Ph.D. in 2007 in Aquatic Microbial Ecology at the University of Clermont-Ferrand II, France. Her doctoral research focused on the trophic roles of heterotrophic flagellates and zoosporic fungi in lacustrine ecosystems. As a postdoctoral fellow in the Department of Biological Sciences of the University of Alabama, USA, she continued her research on the taxonomic and functional diversity of aquatic zoosporic fungi through the development of molecular tools. Then, during her postdoc in the Department of Chemical and Biological Engineering at Colorado State University, she explored the microbial processes driving acid mine drainage bioremediation. Subsequently, she moved to the Biology and Civil and Environmental Engineering Departments at Duke University, where her research included the exploration of the biogeography of foliar fungal endophytes, the impact of carbonaceous materials and NZVI on anaerobic sludge microbial communities degrading flame retardants, and the roles and diversity of plant-associated bacteria degrading hydrocarbons. Her research interest is at the intersection of microbial ecology and environmental engineering, and she is currently a Max Planck Tandem Group Leader and professor at the Universidad Nacional de Colombia, Bogotá, where she is conducting research on the taxonomic and metabolic diversity of the endo-microbiome of medicinal plants for therapeutic purposes.

Jinxiang Li is a postdoc in Xiaohong Guan's research group, College of Environmental Science and Technology, Tongji University, Shanghai, China. His research interests are improving the reactivity and electron selectivity of zerovalent iron toward various metal(loid)s by multiple physicochemical technologies.

Liping Liang is a full-time teacher working in the Department of Environmental Science and Engineering, Shaoxing University. She is a master supervisor. Her research interests are developing novel physical and chemical technologies for water and wastewater treatment and exploring the involved mechanisms. She pioneered the research on weak magnetic field-enhanced zerovalent iron technology and sulfidation of zerovalent iron for water decontamination and oil/water separation. She has authored over 20 SCI-cited papers including 2 in *Water Research* and 2 in *Environmental Science & Technology.*

Igor Linkov leads the Risk and Decision Science Team at the US Army Corps of Engineers and is adjunct professor at Carnegie Mellon University. Dr. Linkov has managed multiple risk assessments and risk management projects utilizing advanced method of risk assessment, life cycle assessment, and decision analysis. He currently supports basic and applied research projects on evaluation of the use of nanotechnology, synthetic biology, and additive manufacturing in military and civilian settings. He leads several efforts on the use of data and decision fusion, including the use of weight of evidence techniques for assessing adverse outcome pathways. He supports development of computational tools for exposure and toxicity assessment for several agencies, including US EPE, FDrugA, and Consumer Product Safety Commission. Dr. Linkov has published widely on environmental policy, modeling, and risk analysis, including 20 books and over 300 peer-reviewed papers and book chapters. Dr. Linkov is elected fellow with the Society for Risk Analysis and American Association for the Advancement of Science. He is Army representative in the National Nanotechnology Initiative.

Yuan Long received his B.S. degree in Environmental Engineering from Anhui Jianzhu University in 2014 and master's degree in Environmental Engineering from the Dalian University of Technology in 2017. Currently, he is working in Shougang Group as a researcher.

Katrin Mackenzie is an environmental chemist, senior scientist, and group leader at the Helmholtz Centre for Environmental Research – UFZ in Leipzig, Germany. She comes originally from the fields of heterogeneous catalysis and organometallic chemistry which she bridged in her Ph.D. work and is now dedicated to material design for groundwater treatment. After postdoctoral research stations at the University of Oxford (UK) and the State University of New York in Binghamton (USA), she started research in the environmental sector where reduction processes and dehalogenation of pollutants wove like a thread through her further career. She is now leading the "Environmental catalysis – Reduction processes" Group within the

Department of Environmental Engineering. Her main research areas are iron-based reactions, sorption-coupled reaction, colloids and catalytic hydrodehalogenation, and design of reactive materials for environmental applications.

Bhanuphong Naknakorn is an environmental engineer at SCG Chemicals, Thailand, who received his M.S. degree from Carnegie Mellon University, USA. During his study, he worked on the effectiveness of NZVI on dechlorination of 1,2-dichloroethane which is the main raw material of PVC manufacturing. Besides environmental and occupational health management in petrochemical factories, his responsibility also mainly focuses on groundwater contamination and remediation which he has been experiencing for at least 10 years.

Denis M. O'Carroll is a future fellow and associate professor at the University of New South Wales (UNSW). Before taking up his position as the director of the Water Research Center at UNSW in Sydney, NSW, Australia, he was an associate professor in Civil and Environmental Engineering at Western University. He received his Ph.D. and completed postdoc work at the University of Michigan, before being awarded NSERC Postdoc funding to work at the University of Toronto. He has been the recipient of several Excellence in Teaching Awards, as well as a Green Award for his cultivation of environmentally sustainable practices. Denis has ongoing research projects developing nanometals for contaminated site remediation, investigating the environmental fate and ecotoxicity of nanoparticles released from commercial products, improving the understanding of the fate of nonaqueous phase liquids in the subsurface and developing climate change mitigation measures in urban areas. He is an associate editor for the *Water Resources Research*, the *Vadose Zone Journal*, and the *Journal of Contaminant Hydrology*.

Eleni Petala received her M.Sc. degree in "Materials Chemistry and Technology" from the Departments of Chemistry and Materials Science & Engineering of the University of Ioannina, Greece. In 2017 she received her Ph.D. from the Department of Materials Science and Engineering, University of Ioannina, in collaboration with the Regional Centre of Advanced Technologies and Materials (RCPTM) in Olomouc, Czech Republic. Her Ph.D. thesis was focused on synthesis, characterization, and evaluation of various composites based on nanoscale zerovalent iron particles with high remediation ability toward a wide range of pollutants, especially focused on hexavalent chromium removal mechanisms. She has an expertise on modification routes of materials, such as by metal and elemental doping, surface coating, and deposition of nanoparticles on support materials and hybrid composites, in order to enhance their properties and widen their performance activities. During her work as a scientific researcher in RCPTM, she has been involved in projects where she could exploit the modification, reactivity, and applicability of NZVI. She is related to four publications and one patent and has fulfilled a scientific internship in the Korea Basic Science Institute, while she is keeping an active international collaborating character.

Martin Petr received his master's degrees in Nanotechnology at Palacký University in Olomouc, Czech Republic (2013). In the same year, he started his Ph.D. studies (Applied Physics) at the Department of Experimental Physics, Palacký University, in Olomouc, Czech Republic. In 2014 he became a "junior scientist" in the group "Environmental nanotechnologies" of the Regional Centre of Advanced Technologies and Materials, Palacký University, in Olomouc, Czech Republic. He underwent several foreign stays (University of Leicester, Leicester, United Kingdom; Polytechnic University of Catalonia, Barcelona, Spain; Italian National Research Council, Rome, Italy) and is actively engaged in national and international collaboration projects. His research interests cover development, optimization, and application of carbon-based (iron-based) nanomaterials in technologies of energy storage and application of X-ray photoelectron spectroscopy and ultraviolet photoelectron spectroscopy in material science. He has coauthored around 20 papers in refereed journals, and his h-index is 9.

Brent E. Sleep is currently chair and professor of Civil and Mineral Engineering at the University of Toronto. His work covers a broad range of remediation techniques including bioremediation, thermal remediation, and the applications of chemical oxidants and nanoscale zerovalent iron. Receiving his Ph.D. from the University of Waterloo, he has built a body of work that is internationally recognized in the study and modeling of contaminated soil, groundwater, and fractured rock. He has served on the US NRC/NAS Committee on Source Removal of Contaminants in the Subsurface, is an associate editor of Advances in Water Resources, and was appointed editor in chief of the *Journal of Contaminant Hydrology*. Brent is the lead researcher and principal investigator of the Remediation Education Network (RENEW) and is funded by an NSERC Collaborative Research and Training Experience grant.

Kullapa Soratana is an assistant professor at the School of Logistics and Supply Chain, Naresuan University, Phitsanulok, Thailand. She received her B.Sc. and M. Sc. in Environmental and Hazardous Waste Management from Chulalongkorn University and received her Ph.D. in Sustainability and Green Design from the Swanson School of Engineering, University of Pittsburgh. Her expertise is in sustainability, particularly on environmental sustainability and life cycle assessment. Her research focuses are on circular economy, sustainable supply chain, sustainable tourism, and life cycle assessment.

Yiming Su is a lecturer at Civil and Environmental Engineering Department at the University of California, Los Angeles, where he teaches environmental applications and implications of nanotechnology. Dr. Su received his Ph.D. degree from Tongji University (2015) in China. After 1-year postdoctoral research at Tongji University (2016), Dr. Su joined Dr. David Jassby's group as a postdoc at the University of California, Riverside (2017). Dr. Su's research focuses on the synthesis and characterization of nano-iron-based material, heavy metal pollution treatment/remediation by nano-iron-based material, implications of nano-iron-based material, and emerging contaminant degradation.

Yuankui Sun is a research associate professor working in the Institute for Advanced Study, Tongji University. He received his Ph.D. in Municipal Engineering from Tongji University. In the following years, he worked as a postdoctoral fellow at Xi'an University of Architecture and Technology. After spending 2 years in Xi'an, he joined Tongji University in September 2017. His research focuses on water and wastewater treatment processes using iron-based materials.

Ondřej Šráček is a professor of Hydrogeology and Environmental Geochemistry at Palacký University in Olomouc, Czech Republic. He received his B.Sc. degree in Geological Engineering from the Technical University of Mining, Ostrava, Czech Republic; M.Sc. degree from the University of Waterloo, Waterloo, Canada; and Ph. D. degree in the Université Laval, Québec City, Canada. He worked and lectured in several countries including Canada, Sweden, Mexico, Argentina, Bolivia, Brazil, Botswana, Zambia, Namibia, Bangladesh, India, and Taiwan. Currently, he is involved in the project of environmental impact of mining evaluation in the sub-Saharan Africa. His principal research interests are environmental arsenic, mine drainage, geochemical and reactive transport modeling, and natural attenuation of contaminants.

Alisa S. Vangnai is an associate professor in Biochemistry, Department of Biochemistry, Faculty of Science, Chulalongkorn University, Bangkok, Thailand. She received her B.S. and M.S. in Biotechnology from Chulalongkorn University in 1992 and 1995, respectively. In 2002, she received her Ph.D. in Biochemistry from Oregon State University, Oregon, USA. Her research focus involves bacterial characterization including stress responses and their modification for bioremediation of aromatic pollutants and bioconversion of biomass. Currently, she is the head of the Biocatalyst and Environmental Biotechnology Research Unit, Department of Biochemistry, Chulalongkorn University, Bangkok, Thailand.

Yiyan Wang received his B.S. degree in Environmental Sciences from Huaibei Normal University in 2015. Currently, he is a master student under the supervision of Prof. Jingjing Zhan in the School of Food and Environment, Dalian University of Technology. His research interest includes the transport and reactivity of NZVI particles for in situ remediation of organic contaminants in soils and groundwater.

Chaoliang Wei received his B.S. degree in 2001 and master's degree in 2004 from the Dalian University of Technology. Currently, he is a Ph.D. student in the same university. His research focuses on the design and synthesis of novel environmental catalysts.

Fern Wickson is a senior scientist and research leader at the GenØk Centre for Biosafety in Tromsø, Norway, where she coordinates the transdisciplinary collaborative for Responsible and Sustainable BioTechnoscience (RootS). Committed to a politics of ecological care, her work integrates environmental science, policy, and philosophy in pursuit of good governance of new and emerging technologies. With a Ph.D. in Biology and Political Science, she performs transdisciplinary research that aims to advance (a) the theory and practice of ecological ethics, (b) the development of responsible research and innovation, and (c) the creation of sustainable socio-ecological futures. Beyond her research, Fern is an expert member of the Norwegian Biotechnology Advisory Board and a past president of the international Society for the Study of New and Emerging Technologies (S.NET). She was also a member of the expert working group of the Intergovernmental Panel on Biodiversity and Ecosystem Services (IPBES) on the diverse conceptualization of values in nature. In her spare time, Fern enjoys hiking, snowboarding, and trying to grow her own food. She is also a yoga and meditation teacher and runs her own studio in the north of Norway called *The Peaceful Wild*.

Radek Zbořil is the general director of the Regional Centre of Advanced Technologies and Materials at Palacký University in Olomouc, Czech Republic. After his Ph.D. study (2000), he underwent several short-term stays, e.g., at the University of Delaware and University of Tokyo. From 2010, he is a full professor at Palacký University. His research is focused on nanomaterials for antimicrobial technologies, energy applications, biomedicine, magnetic devices, catalysis, and water treatment. He has coauthored several books and book chapters and more than 445 papers in peer-reviewed journals. His publications have received almost 17,000 citations (h-index = 58). He is also a coauthor of several European and US patents and technologies implemented on the market of water treatment (EP2164656), biomedicine (US 9505027), and food processing (EP 2873329). He is a member of the Learned Society of the Czech Republic, of the board of the Technology Agency of the Czech Republic, and of the editorial boards of various journals published by Elsevier, Nature, or Wiley. He was awarded by the Prize of Minister of Education of the Czech Republic (2011 and 1996).

Jingjing Zhan received his B.S. degree in Biochemical Engineering from Anhui Polytechnic University in 1999 and his M.S. degree in Organic Chemistry from the Dalian University of Technology in 2004. He obtained his Ph.D. in Chemical and Biomolecular Engineering in 2010 under the supervision of Prof. Vijay T. John at Tulane University. From 2010 to 2012, he worked in the same group as a research scientist. In 2013, he joined the faculty in the School of Food and Environment at the Dalian University of Technology as a full professor. His main research interests focus on energy and environment, environmental remediation, environmental materials, environmental nanotechnology, and so on.

Yalei Zhang is the professor of Environmental Science and Engineering at Tongji University, where he also serves as dean of the Institute of Engineering and Industry and executive vice dean of the Institute of New Rural Development. Dr. Zhang received his Ph.D. degree in Environmental Engineering from Tongji University in 1999. His research interests include wastewater treatment and principles and methods for the resource recovery and energy regeneration from wastewater, environmental fate of contaminants and treatment of water and wastewater by advanced environmental functional materials (NZVI and the like), the regulation mechanism of microalgae cultivation using wastewater and the resource recovery from microalgae biomass, and integrated equipment development for rural drinking water treatment. He has published to date 203 refereed articles in professional journals and 6 books. Additionally, he has held more than 80 patents for the wastewater treatment techniques. Yalei Zhang was awarded the 2016 National Science Fund for Distinguished Young Scholars of China, the 2015 Science and Technology Youth Innovation Award of Ho Leung Ho Lee Foundation, and the National Award for Technological Invention (the second prize) in 2013. Professor Zhang is also the committee member of Heavy Metal Pollution Prevention Professional Committee of Chinese Society of Environmental Sciences.

Chapter 1
Nanoscale Zerovalent Iron (NZVI) for Environmental Decontamination: A Brief History of 20 Years of Research and Field-Scale Application

Tanapon Phenrat, Gregory V. Lowry, and Peyman Babakhani

Abstract Environmental contamination continues to pose a serious threat to human health and the ecosystem. Over the next several decades, remediation research and business will be actively restoring both legacy and newly spilled sites in many countries worldwide. This chapter critically reviews the 20-year progress (1997–2017) in nanoscale zerovalent iron (NZVI) research and development from laboratory testing to pilot- and field-scale demonstrations. Several major areas of NZVI research, including (1) NZVI synthesis and reactivity, (2) aggregation, (3) transport in porous media, (4) polymer modification including carboxymethyl cellulose (CMC), (5) toxicity, (6) sulfidation, and (7) use of electromagnetic fields to enhance remediation, are discussed. Additionally, we summarize important aspects of pilot- and field-scale NZVI applications from 27 peer-reviewed articles and credible reports including the types of contaminants and NZVI used; delivery techniques; injection concentration, rates, and durations; hydrogeological conditions of the sites; pre-operations (before NZVI application); unexpected phenomena (such as clogging) during or after NZVI application; and performance monitoring

T. Phenrat (✉)
Department of Civil Engineering, Environmental Engineering Program, Naresuan University, Pitsanulok, Thailand

Center of Excellence for Sustainability of Health, Environment and Industry (SHEI), Faculty of Engineering, Naresuan University, Phitsanulok, Thailand

G. V. Lowry
Center for Environmental Implications of Nanotechnology (CEINT)
Carnegie Mellon University, Pittsburgh, PA, USA

Department of Civil & Environmental Engineering, Carnegie Mellon University, Pittsburgh, PA, USA

P. Babakhani
Department of Biomedical Engineering and Environmental Sciences
National Tsing Hua University, Hsinchu City, Taiwan

Center for Engineering Sustainability, School of Engineering, University of Liverpool, Liverpool, UK

T. Phenrat, G. V. Lowry (eds.), *Nanoscale Zerovalent Iron Particles for Environmental Restoration*, https://doi.org/10.1007/978-3-319-95340-3_1

including the radius of influence, treatment efficiency, and rebound. Finally, this chapter links the past, present, and future of NZVI research and application to the remaining 15 chapters of this book.

Keywords Nanoscale zerovalent iron · Historical perspectives · Research and development · Field-scaled implementation · Timeline

1.1 International Status of Soil and Groundwater Contamination and Remediation

Environmental contamination with hazardous substances continues to pose a serious threat to human and ecological health worldwide. Sources of contamination are broad and include accidental spills or routine improper hazardous waste management from various industrial activities, for example, the mining, petrochemical, refinery, electronic, petrol station (underground storage tanks), treatment, storage, and disposal facility industries. Contamination of soil and groundwater is pervasive throughout the world. As of 2018, 1566 contaminated sites have been listed in the National Priority List (NPL) in the USA, and only 381 have been sufficiently restored to meet acceptable risk and have been announced for site deletion (US.EPA 2018b). In 2011, approximately 1.2 million potentially contaminated sites and 130,000 contaminated sites have been identified in the European Economic Area (European Environment Agency 2014). Many of those sites contain chlorinated organic contaminants that are amenable to treatment by NZVI (Panagos et al. 2013). In 2014, the Chinese government published a national soil survey report showing that 16.1% of all soil and 19.4% of cultivated land were contaminated by both organic and inorganic chemical contaminants. The total area of contaminated soil was roughly 25 million ha., while 3.5 million ha. of farmland was so contaminated that no agriculture should be allowed on it (The Economist 2017). Cadmium and arsenic were found in 40% of the affected lands, likely leading to abnormally high levels of liver, lung, esophageal, and gastric cancers for the people living near them (Lu et al. 2015). Thailand has significant contamination problems from past mining activities (e.g., Phichit's gold mine) (Rujivanarom 2018). Metal contamination and hazardous organic contaminants, such as phenol and total petroleum hydrocarbons, due to illegal dumping, contaminate shallow groundwater used for consumption by affected villagers (Phenrat et al. 2017). For other Asian countries, such as Myanmar and Vietnam, contaminated sites and their effects on the quality of life have been increasingly reported (Rujivanarom 2017; Phenrat et al. 2018). Thus, concern regarding the health threats from groundwater and soil contamination in those countries may follow the same path as China and Thailand.

The costs to remediate these sites will be large. The USEPA (2004) estimated that, from 2004 to 2033, the cost for remediation of US contaminated sites may be as high as US$209 billion for as many as 294,000 sites. As of 2018, 58,340 sites in EU have been remediated (European Environment Agency 2014). Annual national expenditures for managing contaminated sites are, on average, EUR 10.7 per capita (an average of 0.041% of the national gross domestic product [GDP]). In 2016, the State Council of China issued a Soil Pollution Prevention and Control Action Plan. The plan was to restore 90% of contaminated farmland soil to be safe for human use by 2020 and 95% by 2030, as specified in China's 13th Five-Year Plan (Li et al. 2015). The government estimated that, with 3.33 million ha. of contaminated farmland already identified, the total budget remediation could reach as much as US$ 157 billion (based on 2018 USD) (Reuters 2017).

However, remediation also yields benefits to society from the redevelopment of contaminated sites and cost savings from avoiding health risks from hazardous substance exposure (US.EPA 2011). For example, an annual national benefit from redevelopment at Superfund sites was estimated to be US$54.8 billion for 2017 alone (US.EPA 2018a). Kiel and Zabel (2001) estimated that the benefit from cleaning up two Superfund sites in Woburn, Massachusetts, ranges from US$72 to US$122 million (1992 USD), which is greater than the present estimated costs for cleaning up these sites (Kiel and Zabel 2001). Remediation, especially in Asian countries where remediation services markets are just emerging, will continue to be an important business. This gives a plenty of opportunity to NZVI research and development.

1.2 Nanoscale Zerovalent Iron: A Brief History of Research and Development

Although more than 59 remediation technologies from 14 types of treatment processes are available in various stages of development and application (FRTR 2016) (see examples in Table 1.1), innovative in situ remediation technologies like NZVI are needed to overcome critical limitations of existing techniques. The small size allows zerovalent iron (ZVI) typically used to build a permeable reactive barrier, a passive technique (Table 1.1), to become an active technique that can be delivered into deep or difficult to reach contaminant source areas to speed up remediation. Figure 1.1 illustrates interesting statistics on the research and development (R&D) in the NZVI field of study based on the Web of Science (WOS) database from 2001 to 2018.

Notably, the number of peer-reviewed papers and citations in this field has increased quadratically every year until now. In 2017, there were 262 peer-reviewed journal papers and 8094 citations related to NZVI research. Figure 1.2 illustrates a timeline of NZVI R&D over the last two decades (from 1997 to 2017). Based on our perspective, we divide the NZVI R&D into several major topics, including (1) NZVI

Table 1.1 Examples of available remediation technologies

Technology	Description/definition	Application	Limitation
Steam-enhanced extraction	Combination of steam injection and vacuum extraction Enhancement of soil vapor and groundwater extraction Injection using pulsed mode with continuous vacuum extraction until steam breakthrough	Recovery of organic compounds from unsaturated and saturated subsurface Most effective in coarse-grained aquifers	Limited applicability in silt or clay
Thermal conduction heating	Applies heat to the soil using a combination of thermal conduction and vacuum Heats soil to temperatures above 100 °C Configured for treatment of the surface (thermal blankets) or subsurface (thermal wells)	High boiling-point contaminants Not dependent on permeability of soil	Limited to unsaturated zones May require extensive dewatering or hydraulic control, if applied in saturated zones
Surfactant-enhanced aquifer remediation (SEAR)	Involves injecting and extracting surfactant solution to solubilize and/or mobilize dense nonaqueous phase liquids (DNAPLs) Solubilized DNAPL travels within microemulsions that are readily transported through the aquifer Includes hydraulic control to direct the flow of injectate and prevent uncontrolled DNAPL migration Extracted fluids are treated above ground to separate contaminants from the surfactant waste stream	Applicable to treatment of volatile and semi-volatile compounds in the saturated zone May work beneath structures that are inaccessible using other techniques Recovery and recycling of surfactant may be necessary to be cost-effective	The DNAPL zone must be carefully characterized prior to SEAR activities to develop a cost-effective design Low-permeability zones may not be treated effectively May result in uncontrolled DNAPL migration
Enhanced reductive bioremediation	Injection of organic substrate into the subsurface releases hydrogen (electron donor) upon fermentation Applied by direct-push injection or backfill auguring Commercial electron	Applicable to hydrophobic adsorbed contaminants and dissolved-phase contaminants May be applicable to moderate levels of residual DNAPL	May not be cost-effective in highly aerobic zones May not be applicable at DNAPL sources (unfavorable conditions for microbial communities)

(continued)

Table 1.1 (continued)

Technology	Description/definition	Application	Limitation
	donors (e.g., sodium lactate, hydrogen release compound, molasses, methanol, and vegetable oil) are available Enhances reductive bio-attenuation process Facilitates desorption of sorbed materials via development of a steep concentration gradient and release of biosurfactant		
Permanganate	Permanganate injected into the subsurface Delivery methods include soil mixing, direct injection, and oxidant recirculation Results in direct oxidation of organic contaminants	Oxidizes many CVOCs	Subsurface heterogeneities can cause nonuniform distribution of oxidants Limited by mass transfer limitations and/or poor contact with oxidant May cause increase in dissolved manganese levels Effective subsurface porosity may be reduced due to precipitation of MnO_2 May need more than one application of oxidant to remediate rebound effects
Permeable reactive (or sorption) barrier	Emplacement of reactive or sorptive media in the subsurface is designed to intercept a contaminant plume to transform the contaminants into benign forms to attain remediation concentration goals downgradient of the barrier	Treats many contaminants Cost-effective remediation with low maintenance and operational cost Good for corrective action since the site can be used during remediation	Lengthy time for cleanup and monitoring Passive treatment cannot address the DNAPL source zone to speed up remediation Long-term reduction of permeability of the permeable reactive barrier Installation is difficult and expensive

synthesis and reactivity, (2) aggregation, (3) transport in a porous medium, (4) polymer modification, (5) toxicity, (6) sulfidation, and (7) use of electromagnetic fields to enhance remediation. NZVI synthesis and reactivity studies started with the first introduction of NZVI and remain very active since there remains a need to develop better NZVI-based materials for remediation. Other major topics emerged as

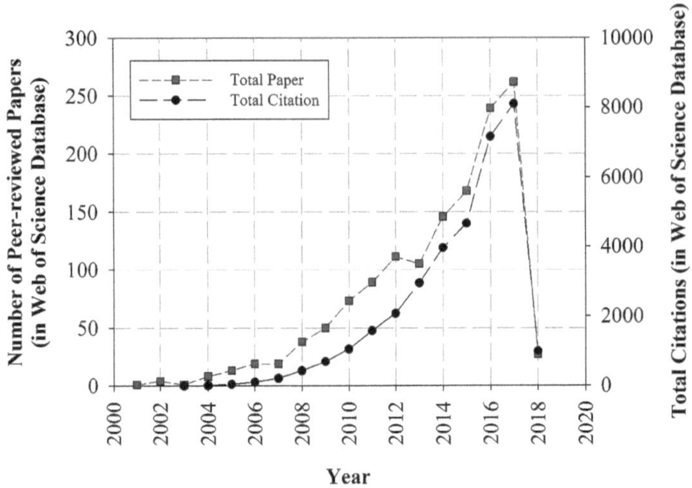

Fig. 1.1 Total peer-reviewed papers and total citations of NZVI research per year from 2001 to 2018 (as of February 2018) based on the Web of Science database

researchers increasingly understand NZVI applications and limitations. Each major topic has seminal papers, which revealed the importance of the topic and called for attention from the research community to perform further investigations. The timeline in Fig. 1.2 shows such seminal papers for each major field of study selected with respect to the year of publication and total citations from WOS.

The timeline illustrates the brief history of NZVI R&D from the authors' perspective. The application of ZVI in the form of iron filings or micron-sized particles in a permeable reactive barrier started in 1990 (Reynolds et al. 1900; Gillham and O'Hannesin 1994). The synthesis of nanoscale iron particles from the borohydride reduction of Fe(II) and Fe(III) salts was reported more than 50 years ago (Oppegard et al. 1961), but the application of NZVI to degrade environmental contaminants was first introduced in 1997 by Wang and Zhang. NZVI is an excellent electron donor (Eq. 1.1) capable of reductive transformation of contaminants with $E_H{}^0$ greater than −0.447 V (Table 1.2). The transformation involves reductive dechlorination of chlorinated organics (trichloroethylene [TCE], tetrachloroethylene [PCE], and vinyl chloride [VC]) and reduction and immobilization of metals. For the metals with an $E_H{}^0$ lower than −0.41 V, detoxification by NZVI is mainly by sorption/surface complexation (Li and Zhang 2007).

$$Fe^{2+} + 2e^- \rightarrow Fe^0 \quad E_H{}^0 = -0.447 \text{ V} \tag{1.1}$$

The small size of NZVI results in an increasing fraction of atoms at the surface, excess surface energy, and high surface to volume ratio (Wang and Zhang 1997). These properties lead to higher contaminant degradation/immobilization rates per mass of the remediation agents compared to bulk materials. Additionally, the small size of NZVI offers the potential for injection into the subsurface for in situ

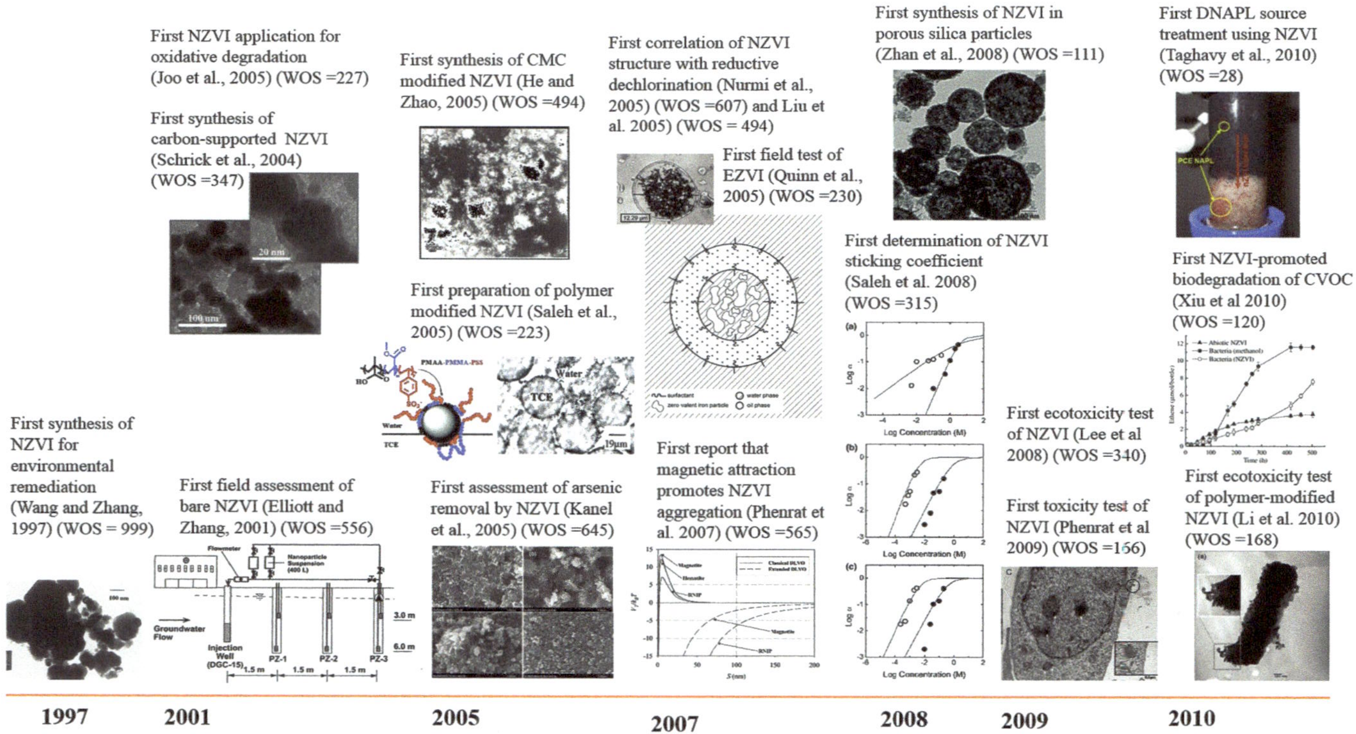

Fig. 1.2 Timeline of NZVI research and development (1997–2016) with the number of citations from the Web of Science (WOS) as of January 2019. (Adapted with permission from Wang and Zhang (1997), Elliott and Zhang (2001), Schrick et al. (2004), He and Zhao (2005), Joo et al. (2005), Kanel et al. (2005), Liu et al. (2005), Nurmi et al. (2005), Quinn et al. (2005), Saleh et al. (2005, 2008), Phenrat et al. (2007, 2009c, 2010a, 2016), Lee et al. (2008), Zhan et al. (2008), Bennett et al. (2010), Grieger et al. (2010), Li et al. (2010), Taghavy et al. (2010), Xiu et al. (2010), Ding et al. (2013), Fan et al. (2013), Krol et al. (2013), Liang et al. (2014), Babakhani et al. (2015), Busch et al. (2015), Kocur et al. (2015); copyright (1997, 2001, 2004, 2007, 2008, 2009, 2010, 2013, 2014, 2015, 2016) American Chemical Society; copyright (2010) Elsevier; copyright (2015) Springer Nature)

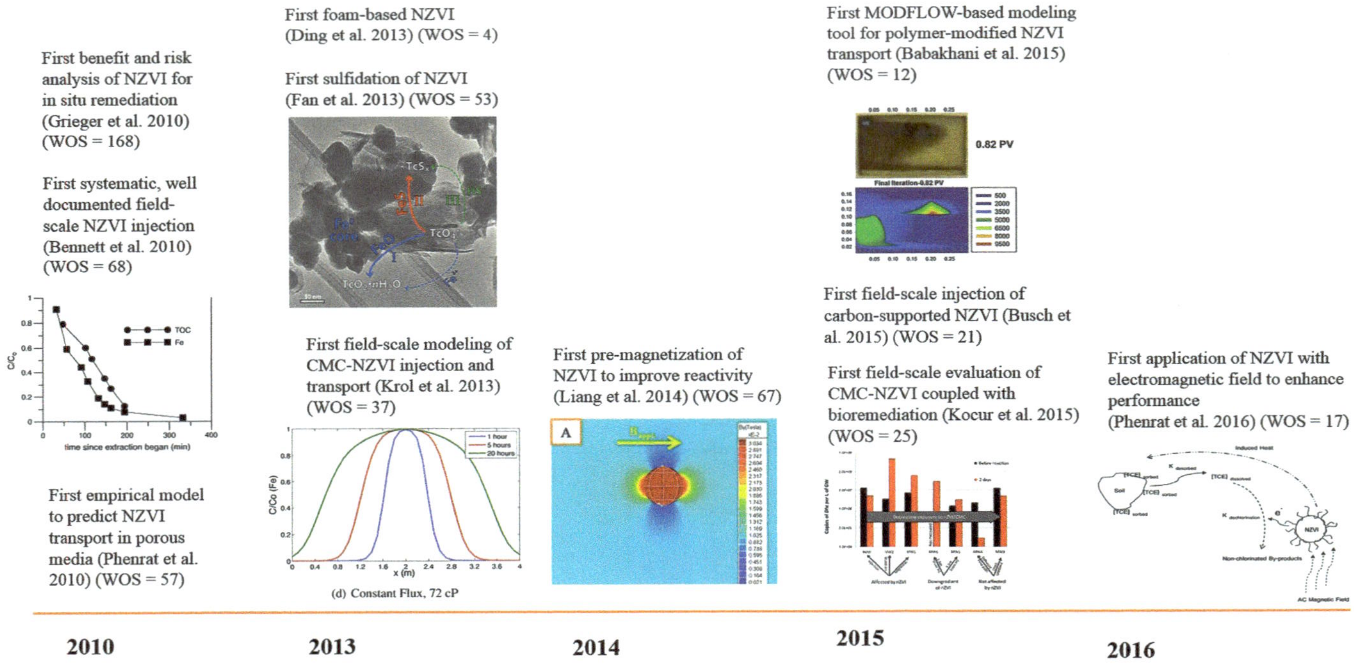

Fig. 1.2 (continued)

Table 1.2 Contaminants of concern amendable by NZVI and their standard redox potentials (E^0) in aqueous solution at 25 °C (Bard et al. 1985; O'Carroll et al. 2013)

Aqueous solution	Half reactions	E^0 (V)
Chromium (Cr)	$CrO_4^{2-} + 8H^+ + 3e^- \leftrightarrow Cr^{3+} + 4H_2O$	1.51
Chromium (Cr)	$Cr_2O_7^{2-} + 14H^+ + 6e^- \leftrightarrow 2Cr^{3+} + 7H_2O$	1.36
Platinum (Pt)	$Pt^{2+} + 2e^- \leftrightarrow Pt$	1.19
Palladium (P)	$Pd^{2+} + 2e^- \leftrightarrow Pd$	0.92
Mercury (Hg)	$Hg^{2+} + 2e^- \leftrightarrow Hg$	0.86
Silver (Ag)	$Ag^+ + e^- \leftrightarrow Ag$	0.80
Arsenic (As^V)	$H_3AsO_4 + 2H^+ + 2e^- \leftrightarrow HAsO_2 + 4H_2O$	0.56
Copper (Cu)	$Cu^{2+} + 2e^- \leftrightarrow Cu$	0.34
Uranium (U)	$UO_2^{2+} + 4H^+ + 2e^- \leftrightarrow U^{4+} + 2H_2O$	0.27
Arsenic (As^{III})	$H_3AsO_3 + 3H^+ + 3e^- \leftrightarrow As + 3H_2O$	0.24
Copper $(Cu)^+$	$Cu^{2+} + e^- \leftrightarrow Cu^+$	0.16
Lead (Pb)	$Pb^{2+} + 2e^- \leftrightarrow Pb$	−0.13
Nickel (Ni)	$Ni^{2+} + 2e^- \leftrightarrow Ni$	−0.25
Cadmium (Cd)	$Cd^{2+} + 2e^- \leftrightarrow Cd$	−0.40
Iron (Fe)	$Fe^{2+} + 2e^- \leftrightarrow Fe$	−0.44
Zinc (Zn)	$Zn^{2+} + 2e^- \leftrightarrow Zn$	−0.76
Barium (Ba)	$Ba^{2+} + 2e^- \leftrightarrow Ba$	−2.92
1,2-Dichloroethane	$ClH_2C{-}CH_2Cl + 2e^- \leftrightarrow H_2C{=}CH_2 + 2Cl^-$	0.74
Carbon tetrachloride (CT)	$CCl_4 + H^+ + 2e^- \leftrightarrow CHCl_3 + Cl^-$	0.67
Tetrachloroethylene (PCE)	$Cl_2C{=}CCl_2 + H^+ + 2e^- \leftrightarrow Cl_2C{=}CHCl + Cl^-$	0.57
Trichloroethylene (TCE)	$Cl_2C{=}CHCl + H^+ + 2e^- \leftrightarrow Cl_2C{=}CH_2 + Cl^-$	0.53
Vinyl chloride (VC)	$ClHC{=}CH_2 + H^+ + 2e^- \leftrightarrow H_2C{=}CH_2 + Cl^-$	0.45
1,1-Dichloroethane (1,1-DCE)	$Cl_2C{=}CH_2 + H^+ + 2e^- \leftrightarrow ClHC{=}CH_2 + Cl^-$	0.42

remediation (Elliott and Zhang 2001; Schrick et al. 2004). Thus, after several relevant papers on synthesis and reactivity evaluation of bimetallic NZVI and supported NZVI from 1997 to 2001 by Zhang's group and Mallouk's group (Wang and Zhang 1997; Zhang et al. 1998; Lien and Zhang 1999, 2001; Ponder et al. 2000, 2001), NZVI research rapidly moved to its first field-scale trial in 2001 (Elliott and Zhang 2001). In this demonstration, the TCE reduction efficiency was reported to be up to 96% over a 4-week monitoring period with the highest values observed at the NZVI injection well (Elliott and Zhang 2001).

This first field trial used uncoated bimetallic NZVI, so a low concentration of the NZVI dispersion (0.75 and 1.5 g/L) was used to reduce subsurface pore clogging by NZVI due to aggregation (Elliott and Zhang 2001). Not long after that first field trial, researchers realized the high reactivity and small size alone were insufficient to make NZVI a good in situ remediation agent. The NZVI must also be readily dispersible in water such that it can be delivered at a relatively high particle concentration (3–20 g/L for effective source-zone removal) through a water-saturated porous medium to the contaminated area and must be able to target specific contaminants such as dense

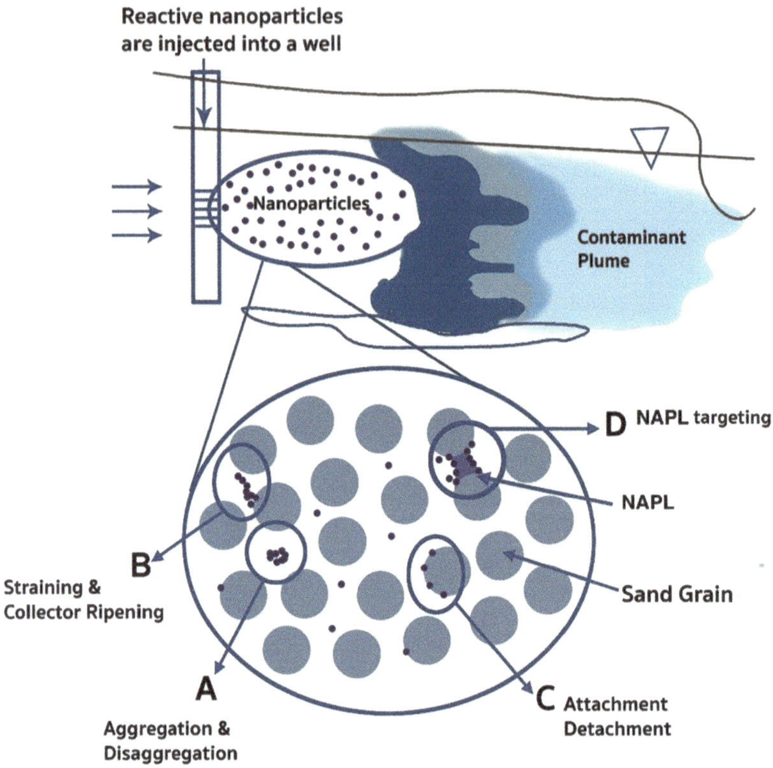

Fig. 1.3 Ideal conceptual mode of delivering NZVI for in situ nonaqueous phase liquid (NAPL) source-zone remediation. Particle mobility and contaminant targetability are needed for effective remediation

nonaqueous phase liquid (DNAPL; Fig. 1.3). Researchers realized that the mobility of uncoated NZVI in saturated porous media is very limited (i.e., practical transport distances of only a few centimeters or less for uncoated, unsupported particles; Schrick et al. 2004; Phenrat et al. 2007; Saleh et al. 2007). They hypothesized that the limited mobility may be due to either NZVI deposition or aggregation or both (Schrick et al. 2004; Phenrat et al. 2007; Saleh et al. 2007).

Thus, from 2004 on, research was aimed at developing approaches to enhance NZVI mobility and deliverability in porous media (Fig. 1.2). Mallouk's group proposed the first synthesis of carbon-supported NZVI in 2004 (Schrick et al. 2004), while Lowry and Tilton's group proposed the use of triblock copolymers and homopolymers as surface modifiers for pre-synthesized NZVI in 2005 (Saleh et al. 2005; Phenrat et al. 2008, 2009b). This polymeric surface modification provides not only electrosteric repulsion to decrease agglomeration and deposition (Phenrat et al. 2009a, 2010b; Phenrat and Lowry 2009) but also functional groups to target the nonaqueous phase liquid (NAPL) source zone (Saleh et al. 2005; Phenrat et al. 2011).

Additionally, in 2005, Zhao's group proposed a one-pot synthesis of carboxymethyl cellulose (CMC)-modified NZVI to enhance NZVI mobility in porous media (He and Zhao 2005), while Quinn et al. (2005) developed emulsified zerovalent iron (EZVI) to deliver NZVI into the NAPL source zone. In addition to abiotic degradation by NZVI, biodegradation enhanced by the presence of oil and surfactant in the EZVI emulsion was hypothesized in its first field trial (Quinn et al. 2005). Later, NZVI entrapped in porous silica particles was first synthesized by John's group (Zhan et al. 2008; Zheng et al. 2008). The importance of the magnetic properties of NZVI on its rapid agglomeration and limited mobility in porous media was revealed by Phenrat et al. (2007). After these works, NZVI agglomeration and transport in porous media became a significant area of NZVI R&D. Moreover, polymeric surface modification, especially CMC modification, became a very popular approach to prepare NZVI for both laboratory study and field-scale application.

Around 2006, the potential for undesirable effects of nanotechnologies became a concern in the scientific community (Wiesner et al. 2006). Several research groups studied environmental fate and transport and the toxicity of various nanoparticles used in commercial products. Unavoidably, for NZVI, which is directly applied to the environment, there was a need to understand the mobility of polymer-modified NZVI at low particle concentrations in porous media to assess the potential for unintended off-site migration and ecological or human toxicity and risk. The first paper quantifying NZVI mobility (at a low particle concentration) using the clean-bed filtration theory via determination of the sticking coefficient was published in 2008 (Saleh et al. 2008). They revealed that electrosteric stabilization of polymer-modified NZVI at a low particle concentration (30 mg/L) provides resistance to NZVI particle attachment to aquifer materials, resulting in potential transport distances of tens to hundreds of meters in unconsolidated sandy aquifers at the injection velocities used for NZVI emplacement. At low particle concentration, agglomeration of the polymer-modified NZVI is minimal, and classical filtration theory can successfully model NZVI transport behavior.

Similarly, the first papers on NZVI toxicity to bacteria and mammalian cells were published in 2008 and 2009 (Lee et al. 2008; Phenrat et al. 2009c). Lee et al. (2008) reported that NZVI rapidly inactivated *Escherichia coli* via induced oxidative stress using dissolved Fe^{2+} and intracellular oxygen or hydrogen peroxide. They also noted that the bactericidal effect of uncoated NZVI was unique and was not observed in other types of iron-based particles or dissolved Fe(II) and Fe(III) species. For mammalian cells, Phenrat et al. (2009c) found that, while fresh NZVI produced morphological evidence of mitochondrial swelling and apoptosis, aged NZVI and surface-modified NZVI were much less toxic to mammalian cells, presumably due to the reduction of redox activity and particle exposure to the cells due to particle aging and surface modification, respectively. Environmental trade-offs between benefits and risks due to in situ remediation using NZVI were first assessed in 2010 (Grieger et al. 2010). They concluded that, based on the current state-of-the-art knowledge, there are no significant grounds to indicate that NZVI poses significant environmental risk. However, they also noted that evaluations of the most serious criteria for environmental concern (i.e., persistence, bioaccumulation, and toxicity) were largely unknown at that time due to a lack of suitable data.

Meanwhile, research on NZVI applications progressed for both reactivity and in situ emplacement. The first studies using NZVI for DNAPL source-zone treatment and a combined remedy of NZVI and bioremediation were published in 2010 (Taghavy et al. 2010; Xiu et al. 2010). Taghavy et al. (2010) found that delivering NZVI (60 g/L) into a PCE DNAPL (5% residual in sand) source zone resulted in partial conversion (30%) of PCE to ethane. Placement of NZVI within and downgradient from the DNAPL source zone enhanced the transformation efficiency of PCE and also reduced the effluent concentrations of PCE. The efficiency of DNAPL source-zone treatment was a strong function of flow rate and size of the source zone.

Studies assessing the viability of combining NZVI with bioremediation or natural attenuation emerged around 2010. Xiu et al. (2010) reported that NZVI can contribute a biostimulatory effect associated with H_2 production during its anaerobic corrosion and an inhibitory effect upon contact with cell surfaces. While methanogens were significantly biostimulated by NZVI, TCE-dechlorinating bacteria were initially inhibited by NZVI but later recovered following the partial oxidation and presumable passivation of the NZVI. Kirschling et al. (2010) found that NZVI addition stimulated both methanogens and sulfate reducers, but did not negatively affect total bacteria (Kirschling et al. 2010). A field-scale demonstration of this combined remedy concept was not evaluated until 2015 (Kocur et al. 2015). They found that the abundance of *Dehalococcoides* spp. increased throughout the NZVI/CMC-affected area relative to preinjection abundance, as well as other evidence that indicated biostimulation had occurred in the injection area.

Because delivering NZVI in situ to create a reactive treatment zone at a specific location is needed for effective in situ remediation, several research groups developed modeling tools for the design of NZVI injection at a high concentration (3–20 g/L) in the subsurface. In 2008, Kanel et al. applied SEAWAT to model polyacrylic acid-modified NZVI in a 2D flow cell (Kanel et al. 2008). In 2009, Phenrat et al. (2009a) clearly demonstrated that deep-bed filtration theory provides poor estimates of transport and deposition of concentrated polyelectrolyte-modified NZVI dispersions because particles agglomerate during transport, which violates a fundamental assumption of the theory. In 2010, Phenrat et al. proposed two empirical correlations for estimating the deposition and transport of concentrated polyelectrolyte-modified NZVI dispersions in saturated porous media. The first correlation determines the apparent stable agglomerate size formed during NZVI transport in porous media for a fixed hydrogeochemical condition. The second correlation estimates the attachment efficiency (sticking coefficient) of the stable agglomerates. Both correlations are described using dimensionless numbers derived from parameters affecting deposition and agglomeration in porous media. The exponents for the dimensionless numbers are determined from experimental data (Phenrat et al. 2010a). They also suggested that these empirical correlations when coupled with flow-field simulations in 2D or 3D can be an effective tool for designing NZVI delivery systems (see Chap. 6, where these correlations are coupled with the COMSOL-based flow model for predicting the emplacement and transport of polymer-modified NZVI in 2D heterogeneous porous media). In 2010, Bennett et al. conducted the first single-well push-pull tests to evaluate the in situ transport of CMC-stabilized NZVI particles in saturated sediments. The first field-scale modeling of CMC-NZVI injection and transport was conducted in 2013 (Krol et al. 2013). More recently, Babakhani et al. modified standard tools for groundwater modeling like MODFLOW and M3TD to include advection, dispersion, attachment, detachment, and agglomeration of polymer-modified NZVI (Babakhani et al. 2015). This approach, owing to the use of standard MODFLOW modules, provides NZVI modeling tools for remediation practitioners and researchers who are already familiar with MODFLOW.

Modifications to NZVI were introduced to the scientific community. For example, in 2013, sulfidation of NZVI was found to be a viable approach to enhance the reactive longevity of NZVI through inhibiting the reaction between Fe^0 and H_2O but maintaining significant reactivity with contaminants like TCE (Fan et al. 2013, 2016). Sulfidation of NZVI appears to be a very promising solution for nonselective oxidation of NZVI, a critical obstacle for effective in situ remediation using NZVI (see Chap. 9).

Other methods have been proposed to increase reactivity of NZVI. Based on an innovative cancer treatment via hyperthermia using magnetic nanoparticles, Phenrat et al. (2016) proposed a novel combined remediation technique for enhanced treatment of source-zone chlorinated volatile organic compounds (CVOCs) and CVOC-contaminated soil using NZVI and an electromagnetic field (EMF; Phenrat et al. 2016). A low-frequency (LF) EMF (180 kHz) is emitted to induce NZVI to generate heat within the contaminant source zone (Phenrat and Kumloet 2016). This heat will enhance the dissolution (mass transfer) of contaminants and their degradation by reductive dechlorination using NZVI. This novel application of EMF with NZVI enhances the treatment rate by 58 times (Phenrat and Kumloet 2016).

In addition to the saturated zone, researchers proposed to use NZVI for vadose-zone treatment. Ding et al. (2013) first reported the synthesis of a foam-based NZVI for vadose remediation to overcome the technical difficulty of dispersing water-based NZVI into the vadose zone, which may cause unintended CVOC dissolution and migration to underlying aquifers. Although the vadose zone may not have enough water or sufficient CVOC dissolution into pore water for substantial reductive dechlorination by NZVI, using foam-based NZVI with an LF EMF to heat the soil can serve as a combined remedy with soil vapor extraction (SVE; i.e., a thermal-enhanced SVE for vadose-zone restoration). Foam-based NZVI carrying 41 g/L of NZVI in the liquid phase of the foam generated heat to raise the temperature to 77 °C in 15 min under an applied LF EMF (150 kHz and 13 A). This enhanced TCE volatilization from TCE-DNAPL in unsaturated sand by 40 $\pm$ 6-fold compared to reactors without LF-EMF application (Chap. 13). Although still in development, these novel NZVI applications have potential for field applications.

1.3 Nanoscale Zerovalent Iron: A Brief History of Field-Scale Applications

Since 2001, NZVI has been used in pilot- and field-scale applications (Elliott and Zhang 2001). In 2009, NZVI were used to remediate 44 CVOC-contaminated sites (Karn et al. 2009). In 2013, the number of NZVI-treated sites reached 59 (Bardos et al. 2015). As of 2017, based on our survey, the number of NZVI sites has reached 77, all of which are CVOC-contaminated sites except one arsenic-contaminated site. These sites are in the USA, Canada, Europe, and Taiwan. Tables 1.3, 1.4, and 1.5 systematically summarize various important aspects of pilot- and field-scale NZVI applications from 27 peer-reviewed articles and other credible reports.

The referenced literature includes the most recent pilot- and field-scale studies conducted by the NanoRem project (2018) in 2017. Several important findings can be extracted from 17 years of field applications of NZVI. Table 1.3 summarizes the types of NZVI used, delivery techniques, COCs, volume/weight of injected particles, particle concentration, and injection rates and durations. Table 1.4 illustrates hydrogeological conditions of the sites, pre-operations (before NZVI application), and remarkable phenomena (such as clogging and ripening) during or after NZVI application. Table 1.5 summarizes the monitoring systems and performance of NZVI, including the reported radius of influence (ROI) of NZVI injection, treatment efficiency, and any rebound observed at the site. In chorological order (2001–2017), the tables depict the brief history of NZVI field-scale applications. While many relevant details are included in the tables, we briefly summarize some common features for each aspect of pilot- and field-scale applications. Chapters 7 and 8 provide complete details of experiences and best practices for field-scale applications of NZVI from practitioners. Furthermore, Chap. 16 concludes with the state of the knowledge and future needs for NZVI applications in subsurface remediation.

1.3.1 Types of NZVI Used and Contaminants Addressed

Uncoated NZVI and bimetallic NZVI were first used for a pilot test in 2001 (Elliott and Zhang 2001) and 2005 (Gavaskar et al. 2005), followed by an EZVI test in 2005 and 2006 (Quinn et al. 2005; O'Hara et al. 2006). After that, various kinds of polymer-modified NZVI (especially CMC-modified) have been intensively evaluated for pilot-scale applications from 2006 (Henn and Waddill 2006) to the present (Kocur et al. 2014, 2015, 2016; Bitsch et al. 2017; Stejskal et al. 2017).

Supported NZVI, such as CMC-modified Carbo-Iron colloids (CIC), have been used since 2015 (Busch et al. 2015). The NZVI was applied to treat CVOCs for every site except in Nitrastur, Spain, where NZVI was used for arsenic immobilization (Otaegi and Cagigal 2017). For CVOC sites, NZVI was primarily applied near source zones, some of which had as much as 85 g/kg and 83 g/L CVOC contamination in the soil and groundwater, respectively. The amount of NZVI injected for pilot- and field-scale implementation ranges from 0.34 to 1000 kg.

Table 1.3 Summary of nanoscale zerovalent iron (NZVI) types used, delivery techniques, contaminants of concern (COCs), volume/weight of injected particles, particle concentration, and injection rates and durations from 17 years of pilot- and field-scale applications (2001–2017)

	Reference	Particle type	Delivery technique	Contaminants of concern	Volume/weight of injected particles and particle concentration	Injection duration	Injection rate
1	Elliott and Zhang (2001)	Uncoated NZVI (Fe/Pd)	Gravity fed (single-injection well)	TCE and other CVOCs	Two injections: 0.89 m^3 of 1.5 g/L (1.34 kg NZVI) and 0.45 m^3 of 0.75 g/L (0.34 kg NZVI)	6.3 and 4 h	3.7–7.5 L/min
2	Quinn et al. (2005)	EZVI	Pressure pulse technology (PPT) with eight injection wells: three in diameter, two intervals (5.3–6.8 m and 6.8–7.9 m below ground surface (bgs))	TCE (DNAPL)	2.54 m^3 of EZVI; injection concentration is not available, but EZVI contained 44.3% water, 37.2% oil, 1.5% surfactant, and 17% iron by weight	5 days	–
3	O'Hara et al. (2006) based on Quinn et al. (2005)	EZVI	PPT was used at eight injection points. Four injection technologies were investigated: (a) pneumatic fracturing, (b) hydraulic fracturing, (c) pressure pulsing, and (d) direct-push injection	TCE (DNAPL)	0.66 m^3 of EZVI containing 349 kg of NZVI along with 6.16 m^3 of site groundwater	4 days	–
4	Gavaskar et al. (2005)	Bimetallic NZVI (BNP)	Direct-push (DP) technology (Geoprobe) at ten locations	PCE, TCE, 1,1,1-TCA, and degradation products *cis*-DCE and VC	136 kg of BNP was mixed with 68 m^3 of water yielding a BNP concentration of 2 g/L for injection. For each injection point, 9 kg of BNP was mixed with 4.5 m^3 of water	–	–

(continued)

Table 1.3 (continued)

	Reference	Particle type	Delivery technique	Contaminants of concern	Volume/weight of injected particles and particle concentration	Injection duration	Injection rate
5	Henn and Waddill (2006)	Palladium-catalyzed and polymer-coated NZVI	Two injection techniques were used 1. DP technology and gravity feed at ten locations 2. A closed-loop recirculation system of four injection and three extraction wells. Two recirculation cycles were performed	CVOCs, including PCE (4.4 g/kg), TCE (60.1 g/kg), 1,1-dichloroethene (1,1-DCE), 1,1,1-trichloroethane (TCA) (25.3 g/kg), and *cis*-1,2- DCE. Maximum dissolved-phase VOC concentration: 83 g/L	First recirculation: 136.1 kg of NZVI suspension; recirculated water + NZVI was ~13 m^3 ≈ 2 pore volumes Second recirculation: ~15.9 m^3 ≈ 2.5 pore volumes Direct push and gravity feed: 10 g/L; first recirculation: initiated at 2 g/L, then increased to 4.5 g/L; second recirculation: 4.5 g/L	First recirculation: 23 h Second recirculation: 21.5 h (12 days after first injection)	First recirculation: 12.1 L/min Second recirculation: 11.4–17.0 L/min
6	He et al. (2010)	1. CMC-modified NZVI with 0.1 wt% Pd and CMC of 0.1 wt% or 0.6 wt% prepared onsite	Two injections 1. Gravity fed through a single well (screened 13–15 m bgs) 2. Pressure injection (<34.5 kPa)	CVOCs, including PCE (1.2–12.0 mg/L), TCE (1.6–23.8 mg/L), *cis*-DCE (8.5–20 mg/L), *trans*-DCE, vinyl chloride (VC) (1.1–2.2 mg/L), and PCB1242 (6.9–97.4 μg/L)	1. 0.57 m^3 and 114 g of NZVI; particle concentration: 0.2 g/L 2. 0.587 m^3 and 568 g NZVI; particle concentration: 1 g/L	1. 3.7 h 2. After 1 month, 1.8 h	1. 2.54 L/min 2. 5.07 L/min
7	Bennett et al. (2010)	CMC-modified NZVI	Push-pull well ($d = 10$ cm) using a multilevel packer to prevent vertical movement	TCE and PCE	Batch#1: 0.12 m^3 of 0.69 g/L NZVI stabilized by 0.8 wt% CMC	7 days	#1: 3.4 L/min #2: 1.2 L/min #3: 2.6 L/min

			of the fluid between slotted intervals		Batch#2: 0.11 m^3 of 0.21 g/L NZVI +0.4 wt% CMC Batch#3: 0.33 m^3 of 0.34 g/L BNP (Pd/Fe: 0.1 wt%) and ~0.27 wt% CMC		
8	Wei et al. (2010)	1. NZVI from Lehigh Nanotech 2. Onsite synthesized NZVI using borohydride	Gravity feed using three injection wells (18 m deep with 15 m screen)	CVOCs	1. 2.25 m^3 containing 40 kg NZVI with palladium (1 wt %) 2. 8.5 m^3 containing 20 kg NZVI	Duration of project: 6 months	20 L/min
9	Edmiston et al. (2011)	Iron-Osorb®: a composite of Pd-modified NZVI and swellable organosilica Osorb®, which is organophilic; 2–10 wt% NZVI and 0.02–0.1 wt% palladium acetate	Three phases 1. Single-injection point: Geoprobe® pressure-activated 4-port probe with a 69-bar grout pump; injection depth, 18.3–21.4 m (note: unsuccessful injection) 2. Single-injection point: Geoprobe; DP; a bottom-up injection method; more concentrated slurry than Phase 1 (note: unsuccessful injection) 3. 11 injection points: DP with a piston pump at 52 bar pressure; higher pressure injection; wider bore and multiple injection points designed to inject thicker Iron-Osorb® slurry (note: successful injection)	TCE concentrations 0.5–1 mg/L	1. Carrier fluid: 0.45 m^3 2. Carrier fluid: 0.64 m^3 3. Carrier fluid: 0.57 m^3; 52 kg dry Iron-Osorb®	NA	45.4 L/min

(continued)

Table 1.3 (continued)

	Reference	Particle type	Delivery technique	Contaminants of concern	Volume/weight of injected particles and particle concentration	Injection duration	Injection rate
10	Wei et al. (2012)	NZVI coated with Pd and synthesized onsite. Two injection rounds 1. Commercial polyacrylic acid-modified NZVI from Lehigh Nanotech, LLC 2. Onsite synthesized NZVI stabilized with a nonionic surfactant, industrial-grade coconut fatty acid diethanolamide ($C_{11}H_{23}CO–N(CH_2CH_2OH)_2$)	Two rounds of gravity-fed injection through three injection wells (18 m deep). Circulation of groundwater (16.5 m^3) from the pumping well to injection well for 29 h	VC (primary); 1,2-dichloroethane (1,2-DCA); 1,1-DCA; 1,1-DCE; *cis*-1,2-DCE; and TCE	40 kg NZVI + 400 g palladium 1 m^3 of onsite synthesized NZVI stabilized with nonionic surfactant and 7.5 m^3 of NZVI (almost 20 kg NZVI + 100 g palladium) Round 1: 17 g/L iron Round 2: 2.5 g/L iron	NA	NA
11	Su et al. (2013)	EZVI: Toda RNIP-10DS (10% wt) + tap water (51%) + corn oil (38%) + surfactant (span 85) (1%)	Two sites 1. Pneumatic injection using nitrogen gas as the carrier 2. Direct injection using DP rig (Geoprobe)	1. CVOCs (38 kg) dominated by PCE and its chlorinated daughter products 2. CVOCs (0.155 kg)	1. 2.18 m^3 of EZVI containing 225 kg of Toda RNIP, 850 kg of corn oil, 22.5 kg of surfactant 2. 0.57 m^3 of EZVI	4 days	1. 21–107 L/min 2. 3–12 L/min
12	Mueller et al. (2012)	NZVI (70 nm particle size, stabilized by polycarboxylic acid)	Fe^0 suspension was introduced by sleeve-pipe injection (via plastic tubes with small holes at regular	PCE (1–2 tons)	1 ton of NZVI and 2 tons of micron-sized ZVI	NA	~90 g/L with 30 g/L NZVI and 60 g/L 2.5 μm iron

			intervals) to a 16–22 m depth. Ten wells were situated in the area with an injection radius of 2 m each				
13	Mueller et al. (2012)	NANOFER	A DP method was used for 82 injection wells under 0.8 MPa pressure	PCE, TCE, and DCE up to 70 mg/L	300 kg for each injection	NA	NA
14	Johnson et al. (2013)	CMC-modified NZVI	Three parallel sets of injection and extraction wells were used. The center injection well was used for the CMC/NZVI injection and the adjacent ones for a CMC-only solution (0.4 m^3 of 4.8 g/L CMC). Each well: ~2.5 cm i.d. using DP equipment. 0.75-cm-long screen (slot size 0.5 mm). Six extraction wells were used to minimize localized drawdown	Assessing injection and distribution in controlled pilot scale study	Total Fe ~900 mg/L and total CMC ~4.8 g/L. 0.4 m^3 in total	NA	Injection: 6 L/min and extraction: 3 L/min for each extraction well
15	Jordan et al. (2013)	NanoFe™ (PARS Environmental). NZVI injected in suspension with food-grade inorganic material	A series of permanent wells including two vertical wells and four 45°-oriented wells were used to inject NZVI. Screen length: 3.0–6.1 m. To minimize daylighting or fracturing, an injection pressure of less than 34 kPa was applied	PCE in soil: 2.7 g/kg and in shallow groundwater: 41 mg/L	726 kg; loading rate: 0.001 kg of iron per kg of aquifer material Injection concentration: 21.5 g/L	Continuous injection over 2 weeks	NA

(continued)

Table 1.3 (continued)

	Reference	Particle type	Delivery technique	Contaminants of concern	Volume/weight of injected particles and particle concentration	Injection duration	Injection rate
16	Köber et al. (2014)	Flake-like NZVI (85% Fe^0) coated by polyethylene glycol	DP injection into a depth of 10–12 m bgs; using pressure-activated injection probes at 5–6 bar pressure at 14 injection points at 4 depths in an omnidirectional setup	PCE with aqueous concentration of 20–50 mg/L (10–14 m bgs)	280 kg; concentration, 10 g/L	NA	13 L/min
17	Lacina et al. (2015)	NANOFER STAR (87 wt% of Fe^0)	Flow-dosing device	CVOCs (PCE, TCE, DCE, and VC) at maximum concentration of 79.9 mg/L in hot spots	100 kg (2 g/L)	5 h for 5 m^3 of NZVI suspension at 2 g/L	16.7 L/min
18	Busch et al. (2015)	CMC-modified Carbo-Iron colloids (CIC)	Injection and extraction wells at 5.3 m apart. The well screen of the injection well was at 3–6 m depth below the surface	Benzene (20 mg/L), MTBE (3.9 mg/L), ammonium (55 mg/L), and other contaminants	1.2 kg of CMC-modified CIC; 850 L of CMC-modified CIC	NA	8.33 L/min
19	Mackenzie et al. (2016)	CMC-modified CIC	Two injection events First injection test and second main injection. Injection was performed at a depth of 6 m bgs. The suspension was divided over 18 wells	1.1 tons of CVOCs, mostly PCE	First injection test: 20 kg CMC-modified CIC (10 g/L of CIC and 2 g/L of CMC) (total of 0.85 m^3 CIC) Second main injection: 110 kg CMC-modified CIC (15 g/L of CIC and 1.5 g/L of CMC); the CMC solution was used for pre- and postinjection	NA	8.33 L/min

20	Kocur et al. (2014, 2015, 2016)	CMC-modified NZVI	Constant-head gravity injection: a local gradient was also produced by circulating water from a well located 6 m downgradient of the injection well INJ1 injection into the injection well (diameter, 5.08 cm; screened, 2.5–3 m bgs) INJ2 injection into MW-2 (screened 1.5–3 m bgs) INJ3 injection into MW3 INJ4 injection into MW7	TCE and PCE	Total 760 L INJ1 0.375 kg INJ2 0.225 kg INJ3 0.075 kg INJ4 0.075 kg Concentration: 1 g/L	INJ1: 24 h (starting from time: 0) INJ2: 12 h (starting from time: 12) INJ3: 5 h (starting from time: 16) INJ4: 3 h (starting from time: 21)	INJ1 initial rate, 1.0 L/min; final rate, 0.08 L/min INJ2 initial rate, 1.8 L/min; final rate, 0.08 L/min INJ3 initial rate, 1.0 L/min; final rate, 0.08 L/min INJ4 initial rate, 0.4 L/min; final rate, 0.08 L/min
21	Stejskal et al. (2017)	NANOFER 25 S (modified biodegradable organic and inorganic stabilizers) and NANOFER STAR (i.e., STAR)	Five DP probes with an applied fluid pressure of 1–12 bars. Injection depth: 5–10.5 m bgs	CVOCs, mainly TCE, PCE, *cis*-DCE, tetrachloromethane (PCM), and trichloromethane (TCM). The highest CVOC concentration was 30 mg/L at 6.5 m bgs	First injection of NANOFER 25 S: 1000 kg of a 20% suspension of NANOFER 25 S was injected at a concentration of ~2.2 g/L; 91 m^3 of suspension was injected Second injection of NANOFER STAR: 300 kg of NZVI were injected (1500 kg of 20% suspension) at six injection points	NA	NA

(continued)

Table 1.3 (continued)

	Reference	Particle type	Delivery technique	Contaminants of concern	Volume/weight of injected particles and particle concentration	Injection duration	Injection rate
22	Bitsch et al. (2017)	NZVI in monoethylene glycol (FerMEG12)	Five injection wells at five bar	CVOCs including PCE, TCE, and hexachloroethane; up to 20 g/kg	NZVI concentrations of 10 g/L; 100 kg iron (in 10 m^3 of water) were injected	NA	50 L/min
23	Laszlo and Szabo (2017)	Carbo-Iron®	Three stations of DP technology. Injection depth at each injection point, 13 m bgs; injection pressure, 0.5–5 bar	PCE (95%) and TCE (15 kg CVOC) Maximum concentration: 15–20 mg/L	Carbo-Iron®, 176.8 kg; CMC, 21.2 kg; oxygen-free tap water, 12.35 m^3 Carbo-Iron® concentration, ~10–15 g/L; CMC concentration, ~1.5 g/L	NA	Injection-flow rate: 20–30 L/min
24	Cohen and Weisbrod (2017)	Carbo-Iron®	NA	NA	6 kg of Carbo-Iron® particles, 4.8 kg CMC stabilizer (Carbo-Iron®: CMC 1:0.8); 0.16 m^3 of local groundwater (purged by N_2)	NA	NA
25	Otaegi and Cagigal (2017)	NANOFER STAR	Three injection points	The maximum As concentration was 4300 mg/kg, while As (mostly As(V)) in groundwater is 5.5 mg/L	250 kg of NZVI	NA	Injection rate: 10–17 L/min

Table 1.4 Summary of hydrogeological conditions of the sites, pre-operations (before nanoscale zerovalent iron [NZVI] application), and remarkable phenomena (such as clogging and ripening) during or after NZVI application from 17 years of pilot- and field-scale applications (2001–2017)

	Reference	Pre-operation	Remarkable phenomena (clogging, ripening, etc.)	Geology	Description of media	Migration velocity	Site name and place	Hydraulic conductivity
1	Elliott and Zhang (2001)	Recirculation of groundwater was performed for 2 days	Minimal clogging of the well. No deterioration of aquifer permeability	NA	Saturated thickness was 6 m. Groundwater table: 1.5–2.1 m bgs	Apparent velocity of plume: 0.8 m/day (exceeded the natural seepage velocity of 0.3 m/day)	Manufacturing facility in Trenton, NJ	0.2 cm/s before and after treatment
2	Quinn et al. (2005)	Recirculation of groundwater was performed at 0.23 m^3/h for 3 weeks prior to injection of the EZVI	No substantial change in observed hydraulic conductivity following EZVI treatment	Aquifer extended from the water table to 13.7 m bgs, including an upper sand layer, a middle fine-grained unit, and a lower sand unit	4.57 × 2.90 m	NA	NASA's Launch Complex 34 (LC34)	0.015 cm/s before treatment, 0.013 cm/s after treatment
3	O'Hara et al. (2006) with Quinn et al. (2005)	–	Upward migration of a proportion of injected EZVI was observed. No change in hydraulic conductivity was shown by the slug test	Aquifer comprising an upper sand unit, middle fine-grained unit, and lower sand unit. Most DNAPL exists 5.3–7.9 m bgs within the upper sand unit	1.5 × 2.9 × 7.9 m Depth interval 4.9–5.8 m bgs	NA	NASA's Launch Complex 34 (LC34), Cape Canaveral Air Force Station, Cape Canaveral, FL, USA	NA
4	Gavaskar et al. (2005)	NA	NA	0.5–5.9% gravel, 86–94% sand, and 5–9% clay. Total organic carbon levels: 40–800 mg/kg	Two plume areas (787 and 404 m^2). Vertical extent, 21.33 m below the groundwater table; hot spot, 13.7–18.3 m below the groundwater table	NA	Naval Air Engineering Station Lakehurst, NJ	NA
5	Henn and Waddill (2006)	Catalyzed hydrogen peroxide injection was previously	No clogging. However, hydraulic conductivity dropped	Unconfined aquifer composed of clayey sand and silty very	Depth of injection: 2.3–7.0 m bgs	NA	Naval Air Station in Jacksonville, FL, USA.	Average hydraulic conductivity for all wells

(continued)

Table 1.4 (continued)

	Reference	Pre-operation	Remarkable phenomena (clogging, ripening, etc.)	Geology	Description of media	Migration velocity	Site name and place	Hydraulic conductivity
		implemented but unsuccessful. Before adding NZVI, recirculation was conducted for 3 h, causing an initial increase in CVOC concentrations due to increased dissolution of the sorbed CVOCs and DNAPL	~45% after injection. The reduction in hydraulic conductivity for different wells varied: −25% to +45%	fine- to fine-grained sand underlain by a low-permeability clay aquitard			Used for Navy operations since 1940	dropped 1.3–0.7 m/day, a 45% reduction
6	He et al. (2010)	Monitoring groundwater quality for about two decades before the test	NA	4.6 m deep, unconfined aquifer made of mainly stiff silty clay, yet with sand and gravel near the injection point	4.57 × 3.05 m	0.067 m/day; hydraulic gradient: ~0.012 m/m	A former manufacturing facility, California	1.98×10^{-3} cm/s; total porosity: 0.3
7	Bennett et al. (2010)	A pilot test injecting 2000 kg carbohydrate (dilute molasses) at 2 g/L, injecting 699 L of tap water and extracting 1440 L during the pull stage at 5.3 L/min	NA	Saturated sediments, predominately silt and clays; confined zone at a depth of 3 m; for depths 4.1 and 9.5 m: $D_{10} = 0.37$ and 0.22 mm; $D_{60} = 1.9$ and 2.3 mm	NA	After injection: minimum for #1–#3, 5.8, 8.5, 4.6 m/day; max for #1–#3, 24.9, 8.0, 31.8 m/day Before injection: 0.3–1.0 m/day	Aerospace facility near San Francisco Bay located in Palo Alto, CA	105 m/day
8	Wei et al. (2010)	NA	More accumulation of NZVI in the upper layer than in the lower layer since much of the	An unconfined aquifer medium-to-coarse sand and some silt	10 × 20 m and 4–8 m bgs	Linear (real, pore) groundwater velocity: 0.285 m/day	An active industrial complex including petrochemical plants	0.275 cm/s, transmissivity, 4349 m^2/day;

			NZVI first seeped through channels in the unsaturated zone due to injection via gravity				and vinyl chloride monomer manufacturing plants in Kaohsiung, Taiwan	specific yield, 0.18
9	Edmiston et al. (2011)	Iron injection followed by a "chaser" fluid containing water and citric acid; injection of citric acid/water solution 6 months later; 18.9 L of 0.2% surfactant water followed by injection of 45.4 L of Iron-Osorb® (containing 2.5 kg of material followed by 45.4 L water at a depth of 18.9 m, repeating the procedure at a depth of 20.4 m)	A change in flow condition and back pressure was observed when Iron-Osorb® was injected in a column experiment; injection in freezing condition using thick fluid caused clogging; thus, the Osorb® dispersion was diluted with surfactant as a carrier solution and constantly stirred	61 m of permeable sand and gravel outwash overlaid by a thin-layer alluvial deposit; depth to groundwater: 12.2–13.4 m bgs	NA	Groundwater velocity: ~0.06 m/day	A 6.2 acre industrial property, Coshocton County, Ohio	2×10^{-3} to 2×10^{-2} cm/s
10	Wei et al. (2012)	Before Round 2, 1000 L of 5 g/L nonionic surfactant solution was injected to condition the aquifer	NA	NA	NA	NA	NA	NA
11	Su et al. (2013)	NA	1. 121 L of injected EZVI daylighted. Mobilization of DNAPL from contaminated to non-contaminated sites. No change in flow direction. No	Two side-by-side treatment plots; depth of contamination: ~5.8 m bgs	1. 4.57 × 3.05 × 1.83–5.49 m bgs 2. 2.44 × 2.44 × 1.83–3.66 m bgs	Groundwater velocity: 0.046–0.055 m/day	Parris Island, SC, USA	4.66 m/day

(continued)

Table 1.4 (continued)

	Reference	Pre-operation	Remarkable phenomena (clogging, ripening, etc.)	Geology	Description of media	Migration velocity	Site name and place	Hydraulic conductivity
			flow-rate decrease after EZVI injection 2. 19 L of injected EZVI daylighted					
12	Mueller et al. (2012)	For 14 years, groundwater was treated by a combination of pump and treat and soil vapor extraction, removing around 5 tons of PCE at a cost of more than €1 million	NA	Sandy gravel	Several square kilometers to a depth of 20 m	NA	Bornheim (Rhein-Sieg-Kreis, Germany)	NA
13	Mueller et al. (2012)	NA	NA	NA	120 × 60 m at a depth of 3–10 m	NA	Hořice, Czech Republic	Less than 10^{-6} m/s
14	Johnson et al. (2013)	A fluorescein tracer test was conducted to evaluate the performance of the injection/extraction system	Reversible flow change caused by CMC/NZVI injection is due to the formation of hydrogen bubbles during the injection process, which could cause a temporary diversion of flow from the areas where NZVI is concentrated. The rate of oxidation of NZVI by oxygen is much more rapid than by other potential oxidants	The presence of partially confining low-permeability layers above and below the injection and extraction wells allowed a steep hydraulic gradient to be maintained	10 × 10 × 2.4 m deep	According to a tracer test, transport velocity between the injection and extraction wells was ~34.6 m/day	NA	NA

15	Jordan et al. (2013)	NA	Injection in some of the wells led to day-lighting. Well fouling was observed in three wells. No change in groundwater flow suggested that no fouling of the aquifer took place	Saprolite was 4.6 m bgs and partially weathered rock aquifer. Groundwater was at a depth of ~1.5 m bgs	NA	Groundwater velocity in saprolite: <0.0042 m/day	Dry cleaner located in Chapel Hill, North Carolina	Porosity, 35%; hydraulic conductivity for the shallow aquifer was 0.04 m/day and for shallow bedrock was 1.4 m/day
16	Köber et al. (2014)	NA	No significant sign of clogging; wide particle size distribution due to agglomeration during synthesis (200–4500 nm); oxygen can cause agglomeration and precipitation over 18 h	Quaternary sand layers	NA	NA	Former dry-cleaning facility, Breite St. in Braunschweig, Germany	$5–10 \times 10^{-5}$ m/s; porosity, 20%
17	Lacina et al. (2015)	Previous monitoring started approximately 4 months before application	NA	NA	400 m^2	NA	Metal fabrication industrial facility in the Czech Republic	NA
18	Busch et al. (2015)	Tracer test was performed before NZVI injection	NA	From the top to a depth of 2.9 m, sand and silt layers were identified, followed by sand layers to 3.9 m and a gravel layer to 6 m, followed by another silt layer	NA	2 m/h	City of Leuna, Saxony-Anhalt, Germany	NA
19	Mackenzie et al. (2016)	NA	No significant signs of clogging	A 25–30-m-thick interbedded sequence of quaternary mixed sands and clayey silt layers in the first	Contaminated plume: 100 × 120 m at 6 m depth	6–30 cm/day	NA	2.1×10^{-4} m/s; porosities: 15% and 30%

(continued)

Table 1.4 (continued)

	Reference	Pre-operation	Remarkable phenomena (clogging, ripening, etc.)	Geology	Description of media	Migration velocity	Site name and place	Hydraulic conductivity
				aquifer; below this system, an aquiclude of silty-clayey sediments of very low permeability follows. Groundwater table: 4–6 m bgs				
20	Kocur et al. (2014, 2015, 2016)	A tracer test was carried out by gravity injecting 0.15 m^3 of 6 g/L NaBr solution for 2.75 h	The flow field was highly affected by heterogeneity; the initial decrease and final increase in hydraulic conductivity and injection-well performance were ascribed to the formation of hydrogen gas temporarily clogging the pores near the wells	A highly complex fill of mixed sand and clay units capped by an engineered clay layer (to a depth of 1.5 m)	Distance from the injection well to the hydraulic control sump, 6 m; distance between the two side MWs, 2 m	Hydraulic gradient before injection, 0.002; during injection, 0.055; pore velocity, 0.018–0.864 m/day; tracer migration velocity, 4.58 m/day	North Utility Corridor, Sarnia, Ontario	Prior to injection: 1.2–4.6 × 10^{-6} m/s After injection: 5.3–10.2 × 10^{-6} m/day (two- to sixfold increase); porosity, 0.32
21	Stejskal et al. (2017)	NA	NA	Quaternary sand and gravel with a thickness of approximately 10 m underlain by a clay aquitard; groundwater table: 4 m bgs	10 × 20 m	Groundwater velocity: 0.2 m/day	Spolchemie I, Usti nad Labem, Czech Republic	NA
22	Bitsch et al. (2017)	A hydraulic barrier was installed in the plume in 2001. Further migration of	NA	Average water table: 13 m bgs; 60–80% gravel (0.5–5 cm) in a silty-sandy matrix	Plume length: 600 m	NA	Solvay, Switzerland	NA

		contaminants from the primary source to the area outside the plant is prevented by an underground impermeable wall constructed in 2008		intersected by a 1-m-thick layer with gravel and stones and sand (1%)				
23	Laszlo and Szabo (2017)	NA	NA	NA	12–13 m bgs	NA	Balassagyarmat, Hungary	NA
24	Cohen and Weisbrod (2017)	A tracer test was carried out using two fluorescent tracers	NA	Fractured chalk	NA	NA	NA	NA
25	Otaegi and Cagigal (2017)	NA	NA	Layer of quaternary alluvial deposits about 3 m thick, of sandy gravel with boulders that define the locally semi-confined aquifer of the site. Beneath the aquifer, there is less permeable, very hard carboniferous shale and sandstone	20 ha; 228,904 m^3 Site covered by a layer (177,000 m^2) of construction debris, slag from furnaces, waste from coal washing, and partially burned pyrite and pyrite ash, with variable thickness 2–9 m	Average flow, 55 m^3/s, and average velocity, 0.5 m/day	Nitrastur, Spain	NA

Table 1.5 Summary of monitoring systems and performance of nanoscale zerovalent iron (NZVI), including radius of influence (ROI), treatment efficiency, and rebound from 17 years of pilot- and field-scale applications (2001–2017)

	Reference	Monitoring system	Results
1	Elliott and Zhang (2001)	Three piezometers spaced 1.5 m apart (downgradient of the injection well)	Trichloroethylene (TCE) reduction efficiencies were up to 96% over 4 weeks of monitoring, with the highest observed at the injection well and adjacent piezometers. A sharp decrease of oxidation potential and an increase in pH were also observed
2	Quinn et al. (2005)	Soil and groundwater samples were analyzed for volatile organic compounds (VOCs) to evaluate the change in mass, concentration, and mass flux of VOCs. A series of four multilevel monitoring wells with five separate sample intervals and one fully screened well were installed	A reduction greater than 80% in TCE concentration in soil was observed for 67% of soil samples within 90 days of injection; a 57–100% reduction in TCE concentration in groundwater was observed. At all depths targeted with emulsified zerovalent iron (EZVI), groundwater samples showed a significant increase in concentration of *cis*-dichloroethene (DCE), vinyl chloride (VC), and ethane, which is evidence of dechlorination, including biodegradation
3	O'Hara et al. (2006) with Quinn et al. (2005)	Same as Row 2	Reductions of 80% and 60–100% of TCE were obtained in the soil and aqueous phases, respectively. The EZVI appeared to decrease TCE concentrations more than NZVI alone. A slight decrease in ORP and DO was observed after the NZVI injection. The pH remained stable (~neutral) during the demonstration. The TCE substantially decreased in the groundwater samples collected 18 months after the EZVI injection. This was attributed to biodegradation resulting from the presence of oil and surfactant in EZVI. Pneumatic fracturing and direct push (DP) were the best injection technologies. These methods did not show loss of EZVI from above or below the targeted area and provided a better distribution of EZVI, compared to pressure pulse technology (PPT)
4	Gavaskar et al. (2005)	Fourteen monitoring wells were used for monitoring downgradient of the injection locations. A monitoring well was screened 15.2–21.3 m below grade, which matches the target treatment-zone depth where BNP injections occurred. Seven sampling events were conducted	An apparent increase in VOC concentrations was observed in approximately 50% of the monitoring wells in the first sampling round performed 1 week after the BNP injections. Later, the two primary contaminants of concern, TCE and DCE, showed an average decrease of 79% and 83%, respectively. The average decrease in VOC concentrations was 74%. The highest concentration of chloride (evidence of dechlorination) was 31.5 mg/L in well RWI-1 during the last sampling event. The average concentration in the other wells throughout the monitoring period was approximately 7 mg/L. The ORP levels increased in most of the wells or remained relatively the same after injection but decreased (as expected) in three monitoring wells. Thus, this may indicate that not enough NZVI was injected to create the strongly reducing conditions necessary for abiotic reduction of CVOCs. The acidic conditions are likely responsible for the inability to lower the ORP. The average pH

			after injection (pH = 4) was lower than during the baseline sampling event (pH = 5) in the other sampling rounds
5	Henn and Waddill (2006)	Change in CVOCs, DO, ORP, and pH and visual observation of the iron through seven monitoring wells	The ORP dropped from a range of 100 to −100 mV to a range of −200 to −550 mV immediately after injection. The DO decreased from 1 mg/L to less than 0.2 mg/L in many wells. Observation of gray and black water in some but not all extraction wells indicated heterogeneous iron migration. Dissolved iron concentrations increased by almost 1 order of magnitude in the source wells and less in other wells over the first 5 weeks. Sulfate was decreased by 1 order of magnitude in the source well over the first 5 weeks. The hydrogen concentration doubled over the first 40 days after the second iron injection. Based on soil samples, a total contaminant reduction of 8–92% was obtained. Although the contaminant concentrations in the soil samples were decreased substantially, concentrations in the groundwater remained or rebounded. The reasons were suggested to be the possible existence of a DNAPL source zone near the wells, insufficient injected NZVI, inadequate contact between NZVI and DNAPL, lack of proper distribution of NZVI due to heterogeneity, and/or too-soon passivation of NZVI
6	He et al. (2010)	One monitoring well was 1.5 m up-gradient from the injection well (MW-3), and two monitoring wells were 1.5 and 3.0 m downgradient (MW-1 and MW-2, respectively). MW-1 and MW-3 were screened 13.7–15.2 m bgs, while MW-2 was screened 12.2–13.7 m bgs. A multiparameter probe with a flow-through cell was used for sample measurement	During injection 1, ~37.4% of injected Fe was observed in the first monitoring well (1.5 m), while ~3% of injected Fe was in MW-2 (3 m). During injection 2, 84% and <3% of Fe were observed in MW-1 and MW-2, respectively. Higher mobility of the particles was observed for higher pressure injections. The ORP of the downgradient monitoring wells, MW-1 and MW-2, reduced from the baseline value of −63 mV to −355 mV and −73 mV to −179 mV, respectively, 2 h after injection and remained low (~−190 mV) for 6 days in both monitoring wells. The ORP returned to preinjection levels after 8–10 days. The degradation of PCE and TCE were greatest in the first week. Chlorinated ethene concentrations returned to the preinjection levels ~2 weeks after injection. Despite the NZVI being exhausted after 2 weeks, there was evidence that is stimulated long-term biodegradation of the CVOCs
7	Bennett et al. (2010)	Samples from the extraction pump	The NZVI was mobile but lost mobility after 13 h. Rapid degradation of chlorinated ethenes was observed. The destroyed mass of chlorinated ethenes was low since the injected solution pushed the chlorinated ethenes away
8	Wei et al. (2010)	Thirteen nested multilevel monitoring wells in the downstream direction of each injection well	Degradation of VC was 50–99%. A decrease in ORP after NZVI injection was around fourfold. At least 3 m of effective travel distance was obtained for NZVI based on suspended solid and total solid analysis

(continued)

Table 1.5 (continued)

	Reference	Monitoring system	Results
9	Edmiston et al. (2011)	1. Sampling points, 6; monitoring over 4 months 2. Sampling points, 8; monitoring over 3 months 3. Sampling points, 8; monitoring over 6 months	1. Unsuccessful delivery: The groundwater cleanup criterion (TCE = 5 μg/L) was not met, presumably due to insufficient injected NZVI 2. Unsuccessful delivery, like Phase 1: Only a 20–30% reduction in TCE was observed. No sign of injected slurry 30 cm around the injection well. Yet, the material was found at a low concentration at 1–1.2 m from the injection point 3. Significant TCE reduction (>50%) was observed in most wells. The groundwater cleanup criteria (TCE = 5 μg/L) was met after 60 days. The only partially dechlorinated product was 1–3 μg/L *cis*-DCE. The low levels of contaminant in water persisted up to 120 days. Although the material did not migrate with the groundwater, the revised method of injection (improved from Phases 1 and 2) brought about a wider distribution of the injected slurry
10	Wei et al. (2012)	Thirteen nested multilevel monitoring wells. Each monitoring well was embedded with three separate wells, each with 3 m screens at nearly 6, 12, and 18 m deep	The NZVI was trapped in the upper layer and did not penetrate to lower layers. The travel distance of NZVI was up to 3 m, but no more than 5 m. The limited NZVI transport was attributed to agglomeration observed in the scanning electron microscope images of soil samples. A strong decrease of ORP (−400 mV) was observed at the injection point indicating favorable reaction conditions. The onsite synthesized NZVI performed better than the commercial NZVI due to its higher and fresher surface area. Evidence of 1,2-DCA reduction suggested some biostimulation had also occurred
11	Su et al. (2013)	Monitoring for 2.5 years with six fully screened (5.08 cm diameter) and seven multilevel (1.27 cm diameter) monitoring wells screened at seven depth intervals. Soil cores were obtained before injection	Significant reduction in concentration and mass flux (>85%) of PCE and TCE with a corresponding increase in degradation products (ethene) in downgradient wells was evident. The reduction in total CVOC mass was 86%. Biological reductive dechlorination caused by the corn oil took place following abiotic dechlorination by NZVI. Pneumatic injection showed a travel distance of up to 2.1 m and direct injection showed a travel distance of up to 0.89 m. Changes in geochemical parameters due to EZVI injection over time were observed. Mobilization of DNAPL in a fine sand at Darcy velocity >0.04 m/day was in accordance with the previous laboratory column study
12	Mueller et al. (2012)	Two-year monitoring	For the Bornheim site (Rhein-Sieg-Kreis, Germany), approx. 90% reduction of CVOCs was observed with no increase in the daughter products (e.g., TCE and DCE). Two years after injection, no rebound has been observed, and there is still a trend toward declining contaminant concentration. However, the success of the remediation cannot be unambiguously attributed to NZVI since both nanoscale and microscale ZVI were used

13	Mueller et al. (2012)		For the Hořice site (Czech Republic), reductions of 60–75% (>90% in the diffuse plume) of the original contaminant concentration were accomplished
14	Johnson et al. (2013)	For the hydraulic gradient: a pair of wells installed ~0.5 m from the injection and extraction wells (i.e., 1.5 m apart) (2.5 cm i.d., 15 cm screened intervals) equipped with submersible pressure transducer	A hydraulically constrained flow path between injection and extraction wells was necessary to obtain an NZVI transport of 2.5 m. Fresh NZVI transported only 1 m at 2% of the initial concentration. A decline in "reduction demand" of the aquifer led to a decrease in the oxidation of the NZVI. A sudden decline of ORP to 200–400 mV during and after the breakthrough of NZVI occurred in all cases. Neither ORP data nor total iron measurement nor pH could individually provide sufficient information about the transport of NZVI
15	Jordan et al. (2013)	A network of seven monitoring wells installed at 1.2–12.2 m from the injection points. A potassium bromide tracer (231 g) was used to investigate the hydrodynamics. The zone of influence was characterized by collecting ORP and pH data	Strong reducing conditions with very low ORP (−728 mV) initially, rebounding over 2 years but remaining negative in all injection wells. The pH followed the same trends, high (>10) after injection, with a slow decline over 2 years. Nearly complete removal of PCE was accomplished within 1 month, yielding *cis*-DCE (high concentration = 65 mg/L for 12 months) and ethane as major by-products. Detection of acetylene confirmed the presence of the β-elimination pathway. Vinyl chloride and 1,2-DCE concentrations dropped after 2.5 years. The radius of influence was reported to be at least 2.44 m. However, this was defined as areas where pH and ORP are influenced, rather than physical measurements of NZVI. Injections led to clogging of some wells and daylighting of NZVI in others. Lower concentrations of 10–15 g/L offered greater migration of NZVI with less fouling risk. Continuing injection with water free of NZVI after the NZVI injection also enhanced the iron distribution and lessened the risk of fouling of the wells
16	Köber et al. (2014)	Five novel NZVI in situ monitoring systems. DP-liner soil sampling was used to investigate the radius of influence (ROI) of NZVI. Monitoring was performed for 1 year	A transport distance of at least 190 cm for Fe^0 concentrations of 6–160 g/kg (sand) was achieved. A Fe^0 concentration of 1 g/kg (sand) to 100 g/kg (sand) was observed in 80% of the contaminated area. Iron flakes remained reactive for more than 1 year. The PCE concentration decreased from 20–30 mg/L to 0.1–2 mg/L (in source and plume). Mechanical mobilization of PCE-NAPL from the nearby sources could be the reason for the initial increase of the PCE concentration right after the NZVI injection
17	Lacina et al. (2015)	Monitoring wells by monitoring every 1–2 weeks over a period of 6 months	Over 6 months of monitoring, no significant change of pH was observed, while ORP substantially dropped and rebounded after 6 months. Within a month after NZVI injection, PCE and TCE decreased by 80% and gradually rebounded after 5 months. The DCE and VC were gradually decreased. Overall, a decrease across all selected contaminants was observed to be approximately 40–50%

(continued)

Table 1.5 (continued)

	Reference	Monitoring system	Results
18	Busch et al. (2015)	An observation well was installed in the middle between the injection and extraction wells. Groundwater samples were taken regularly every hour using an automatic sampling system. Additional samples were manually taken at 30-min intervals	Breakthrough of NZVI (CIC) was observed visually. About 12% of the amount introduced was found in an extraction well 5 m from the injection point based on total particle and iron concentration. No significant change of pH was observed. These data suggest that the Carbo-Iron has potential for greater mobility than unsupported NZVI
19	Mackenzie et al. (2016)	Monitoring wells	For the first injection test, during the first 2.5 months, a substantial decline of PCE in a monitoring well 4 m away was observed. By-products such as DCEs were initially found in very low concentrations but disappeared again, but no VC was observed. After 10 weeks, the PCE concentration steadily increased again but stayed well below the initial value. Considerable ethane and ethene production was still going on much longer than the expected CIC reactivity, suggesting biostimulation. The second injection reduced the PCE concentration from 19 mg/L to 1.5 μg/L after 200 days
20	Kocur et al. (2014, 2015, 2016)	Nine monitoring wells, 3 of which were also used for injection. They cover distances of 1, 2.5, and 3.5 m downgradient of the injection well	Total Fe breakthrough of 71% of the injected amount was observed at 1 m from the injection well. Stimulated growth of *Dehalococcoides* due to CMC-modified NZVI injection was evident. Over 2 years after NZVI injection (i.e., long after the depletion of the NZVI reduction capacity), enhanced dechlorination of chlorinated ethenes to nontoxic ethene was still observed, suggesting enhanced bioremediation. This is confirmed by an order-of-magnitude increase of *Dehalococcoides* and vinyl chloride reductase (vcrA) genes, monitored using qPCR, in the area affected by CMC-modified NZVI
21	Stejskal et al. (2017)	Fifteen sensors for in situ measurement of magnetic susceptibility and six multilevel monitoring wells (with micropumps); lithium chloride (LiCl) was injected as a tracer with the NZVI to make it easier to monitor the migration of the NZVI	For NANOFER 25S (first injection): A significant increase of iron concentration was observed in the monitoring wells in the contaminated zone. A significant decrease in ORP in the groundwater (from +400 to −300 mV) was observed, but this lasted only 20–50 days. After 112 days, the ORP value stabilized around +100 mV. The PCE concentration decreased (and ethene and ethane increased) immediately after NZVI injection for 277 days (i.e., the concentration decreased from 35–40 to 10 mg/L; 72–80% long-term PCE reduction). After this, the PCE concentration rebounded. A similar trend was observed for PCM (i.e., 50–60% reduction followed by a rebound in contaminant concentration) For NANOFER STAR (second injection): The second injection of NANOFER STAR caused a significant decrease in ORP at all monitoring points (from 200 to −100 mV). A significant decrease of total chlorinated ethene concentration was

			observed immediately after injection of NANOFER STAR. This decrease has lasted for at least 250 days. The PCE and TCE decreased more than 95% with no rebound, but DCE initially decreased but then showed some rebound, presumably due to PCE and TCE degradation
22	Bitsch et al. (2017)	Multilevel monitoring wells (with micropumps)	A travel distance for the FerMEG12 particles of at least 2 m was confirmed by visual observation of black water in two monitoring wells. In the source area, oxygen and nitrate have nearly completely been reduced by NZVI. Three months after injection, the concentration of contaminants at the sampling level in the injection zone was reduced by 49–89% for PCE, 81–97% for TCE, and 97–98% for hexachloroethane. A slight rebound was not observed until 1 year after injection. A high concentration of ethane was observed as a dechlorination by-product. Noticeably, the injection of high volumes of FerMEG12 suspension probably caused a (lateral) spread of the source and increased the dissolution of contaminants into the groundwater
23	Laszlo and Szabo (2017)	Eight continuous multilevel tubing monitoring wells	Carbo-Iron® colloids (CIC) were observed only 0.5 m from the injection point. A significant reduction of PCE and enhanced microbiological degradation/chemical reduction in the monitoring wells closest to the injection points were detected, but ethene and ethane concentrations, as indicators of abiotic reduction, were only detected in low concentrations right after injection. A small increase of TCE, *cis*-DCE, and VC concentration was detected compared to the change detected in PCE concentration. Poor abiotic degradation of CVOCs suggested that portions of Carbo-Iron® bypassed the target zone due to preferential pathways or blockage and thus did not participate in the contaminant reduction
24	Cohen and Weisbrod (2017)	Monitoring wells	Transport potential in a fractured rock matrix is high and can be efficiently manipulated by changing stabilizer concentration, particle properties, and flow velocity
25	Otaegi and Cagigal (2017)	Monitoring wells	The NZVI effectively reduced As into metalloid arsenic As (0) during the first 24 days. 24–180 days after injection, most NZVI was oxidized and precipitated, forming a reactive zone that can sorb dissolved As. Desorption of As 60–120 days after injection was also observed in some monitoring wells, probably related to different local hydrogeological settings and/or to some degree of NZVI corrosion. Thus, long-term monitoring is essential to evaluate the extent of partial desorption of As

1.3.2 Injection Methods

The NZVI has been injected by gravity feed (Elliott and Zhang 2001), but more recently the direct-push technique appears to be popular because they can enhance NZVI delivery and distribution (Gavaskar et al. 2005; Edmiston et al. 2011; Köber et al. 2014; Stejskal et al. 2017). Additionally, a closed-loop recirculation of NZVI was used for NZVI emplacement (Henn and Waddill 2006; Wei et al. 2012). This enhanced distribution of NZVI in comparison with a typical gravity feed or single direct-push injection. Gravity feed can deliver NZVI at a rate of 1–20 L/min, while direct push can deliver NZVI at a rate of 3–50 L/min. Pneumatic injection can inject at a rate of 21–107 L/min, while a closed-loop recirculation can deliver NZVI at a rate of 8–17 L/min. Uncoated NZVI was injected at concentrations ranging from 0.75 to 2 g/L, while polymer-modified NZVI and polymer-modified CIC were injected at concentrations ranging from 0.2 to 17 g/L and 10 to 15 g/L, respectively.

The NZVI has been applied mostly to gravel and sandy aquifers. Nevertheless, NZVI has been applied to clayey sand and clayey silt or even saprolite and weathered rock sites. Interestingly, NZVI delivery appeared to be very effective in a fracture rock site (Cohen and Weisbrod 2017). As shown in Table 1.4, NZVI has been tested in sites with hydraulic conductivity ranging from 0.04 to 105 m/day.

1.3.3 Performance

After NZVI injection, a sharp decrease of oxidation-reduction potential (ORP) and an increase in pH were typically observed. The ROI values of polymer-modified NZVI ranged from 1 to 5 m, that is, 1–2.5 m for CMC-modified NZVI; 3–5 m for NZVI stabilized with a nonionic surfactant, industrial-grade coconut fatty acid diethanolamide ($C_{11}H_{23}CO{-}N(CH_2CH_2OH)_2$; Wei et al. 2010, 2012); 1.9 m for flake-like NZVI coated by polyethylene glycol; and 2 m for NZVI in monoethylene glycol (FerMEG12; Bitsch et al. 2017). Alternatively, the ROI values for EZVI and CMC-modified CIC were 0.89–2.1 m (Su et al. 2013) and 5 m (Busch et al. 2015), respectively. Preferential pathways or aquifer blockage can cause a flow bypass, resulting in poor NZVI transport and emplacement (i.e., the ROI for CMC-modified CIC = 0.5 m because of the flow bypass; Laszlo and Szabo 2017). It should be noted that the ROI is defined differently in the different pilot studies. Standardizing how the ROI is defined would benefit future efforts to clearly delineate expectations for emplacement in different media.

The NZVI appeared to effectively address the goal of contaminant mass reduction. In general, the higher the injected NZVI mass, the greater the mass of the destroyed CVOCs, given that the delivery is effective. As shown in Table 1.5, in situ dechlorination using NZVI can decrease CVOC concentrations in soil and groundwater by 8–92% and 40–99%, respectively. Yet, only two cases of NZVI applications (Edmiston et al. 2011; Mackenzie et al. 2016) were able to achieve the cleanup level of CVOCs in groundwater (5 μg/L) after 60 and 200 days, respectively.

Importantly, most NZVI applications show a limited abiotic reactive lifetime, as evident from the rebound of CVOC concentration. The abiotic reactive lifetime was from 2 weeks (He et al. 2010) to 1 year (Köber et al. 2014); the average abiotic reactive lifetime was 211 ± 134 days, depending on subsurface conditions and the injected mass of NZVI. However, NZVI stabilized by polycarboxylic acid when used with micron-sized ZVI showed no rebound even after 2 years (Mueller et al. 2012). In some cases, NANOFER STAR (Stejskal et al. 2017) and NZVI in monoethylene glycol (FerMEG12; Bitsch et al. 2017) did not show any rebound for more than 250 days and 1 year, respectively.

NZVI can stimulate biotic dechlorination of CVOCs. This was first observed for a site remediated by EZVI (O'Hara et al. 2006). Later, the same effect was observed for CMC-modified NZVI (He et al. 2010), CMC-modified CIC (Mackenzie et al. 2016), and NANOSTAR (Stejskal et al. 2017). Nevertheless, the long-term biostimulation was not systematically studied until 2014 (Kocur et al. 2014, 2015, 2016). Initially, abiotic NZVI reactions degraded inhibitory and less biodegradable compounds like chloroform and consumed dissolved oxygen (DO) and formed H_2 from water, creating favorable geochemical conditions for anaerobic microbial species. Later, CMC fermentation and slow release of biomass stimulated dehalogenator growth (e.g., *Dehalococcoides*) and biotic degradation of chlorinated ethenes.

Besides CVOCs, a field-scale application of NZVI showed reduction of dissolved arsenic, As (III) and As (V), to metalloid arsenic As (0) (solid) after 24 days. After 24–180 days, most of the NZVI was oxidized and precipitated, forming a reactive zone that can adsorb dissolved As. However, desorption of As species 60–120 days after injection was observed in some monitoring wells. Thus, long-term monitoring is essential to evaluating the extent of the partial desorption of As (Otaegi and Cagigal 2017).

1.4 Past, Present, and Future of NZVI in This Book

This chapter presents a brief history of NZVI research and applications over the last two decades. The other 15 chapters in this book elaborate on the progress of NZVI research and practice. The chapters cover fundamental aspects of NZVI, including NZVI synthesis, characterization (Chap. 2), reactivity toward organic contaminants (Chap. 3), and metals and metalloids (Chap. 4). Fundamental aspects of NZVI modification for subsurface delivery are presented in Chaps. 5 and 6. Chapter 5 explains rapid agglomeration and poor dispersion stability of NZVI and presents surface engineering approaches to overcome these limitations. Chapter 6 reviews modeling tools and emphasizes their potential use for designing field-scale applications. Chapters 7 and 8 provide details of experiences and best practices for field-scale applications of NZVI from real practitioners, which provide more detail than the summary of pilot- and field-scale NZVI applications reviewed in this chapter. Chapters 9–14 present recent advances and novel combined remedies using NZVI, including sulfidation of NZVI (Chap. 9), NZVI and bioremediation (Chap. 10), thermally enhanced NZVI remediation using electromagnetic induction (Chap. 11), pre-magnetization and application of a weak magnetic field (Chap. 12), vadose-zone applications of NZVI with electromagnetic induction (Chap. 13), and aerosol-based iron-carbon nanocomposites for adsorption and reduction of Cr (VI) (Chap. 14). Risk and environmental implications of NZVI are discussed in depth in Chap. 15. Finally, Chap. 16 concludes with the state of the knowledge and future needs for NZVI applications in subsurface remediation.

All the chapters are written by experts that are developing or applying NZVI in the field. Most of the authors have presented the scientific community with one or more seminal papers mentioned in Fig. 1.2. In this book, they elaborate on their contributions in NZVI research from laboratory-scale development to field-scale applications.

References

Babakhani, P., Fagerlund, F., Shamsai, A., Lowry, G. V., & Phenrat, T. (2015). Modified MODFLOW-based model for simulating the agglomeration and transport of polymer-modified Fe nanoparticles in saturated porous media. *Environmental Science and Pollution Research International, 25*(8), 7180–7199.

Bard, A. J., Parsons, R., & Jordan, J. (1985). *Standard potentials in aqueous solution*. New York: CRC Press.

Bardos, P., Bone, B., Daly, P., Elliott, D., Jones, S., Lowry, G. V., & Merly, C. (2015). A risk/benefit appraisal for the application of nano-scale zero valent iron (nZVI) for the remediation of contaminated sites. *Taking nanotechnological remediation processes from lab scale to end user applications for the restoration of a clean environment*. European Union Seventh Framework Programme (FP7/2007-2013), Project Nr.: 309517, p. 76.

Bennett, P., He, F., Zhao, D., Aiken, B., & Feldman, L. (2010). In situ testing of metallic iron nanoparticle mobility and reactivity in a shallow granular aquifer. *Journal of Contaminant Hydrology, 116*, 35–46.

Bitsch, R., Matz, P., Kvapil, P., & Klaas, N. (2017). NanoRem Pilot Site – Solvay, Switzerland: Nanoscale zero-valent iron remediation of chlorinated solvents. *CL:AIRE's NanoRem bulletins*. CL:AIRE, UK, pp. 1–6.

Busch, J., Meißner, T., Potthoff, A., Bleyl, S., Georgi, A., Mackenzie, K., Trabitzsch, R., Werban, U., & Oswald, S. E. (2015). A field investigation on transport of carbon-supported nanoscale zero-valent iron (nZVI) in groundwater. *Journal of Contaminant Hydrology, 181*, 59–68.

Cohen, M., & Weisbrod, N. (2017). NanoRem Pilot Site – Neot Hovav, Israel: Transport of iron nanoparticles in fractured chalk. *CL:AIRE's NanoRem bulletin*. CL:AIRE, UK, pp. 1–4.

Ding, Y., Liu, B., Shen, X., Zhong, L., & Li, X. (2013). Foam-assisted delivery of nanoscale zero valent iron in porous media. *Journal of Environmental Engineering, 139*, 1206–1212.

Edmiston, P. L., Osborne, C., Reinbold, K. P., Pickett, D. C., & Underwood, L. A. (2011). Pilot scale testing composite swellable organosilica nanoscale zero-valent iron—Iron-Osorb®—for in situ remediation of trichloroethylene. *Remediation Journal, 22*, 105–123.

Elliott, D. W., & Zhang, W. X. (2001). Field assessment of nanoscale bimetallic particles for groundwater treatment. *Environmental Science & Technology, 15*, 4922–4926.

European Environment Agency. (2014). *Progress in management of contaminated sites*. Copenhagen, Denmark, p. 37.

Fan, D., Anitori, R. P., Tebo, B. M., Tratnyek, P. G., Lezama Pacheco, J. S., Kukkadapu, R. K., Engelhard, M. H., Bowden, M. E., Kovarik, L., & Arey, B. W. (2013). Reductive sequestration of pertechnetate ($^{99}TcO_4^-$) by nano zero-valent iron (nZVI) transformed by abiotic sulfide. *Environmental Science & Technology, 47*, 5302–5310.

Fan, D., O'Brien Johnson, G., Tratnyek, P. G., & Johnson, R. L. (2016). Sulfidation of nano zerovalent iron (nZVI) for improved selectivity during in-situ chemical reduction (ISCR). *Environmental Science & Technology, 50*, 9558–9565.

FRTR. (2016). *Remediation technologies screening matrix and reference guide, version 4.0*. Washington, DC: FRTR.

Gavaskar, A., Tatar, L., & Condit, W. (2005). *Cost and performance report: Nanoscale zero-valent iron technologies for source remediation*. Port Hueneme, CA: Naval Facilities Engineering Service Center.

Gillham, R. W., & O'Hannesin, S. F. (1994). Enhanced degradation of halogenated aliphatics by zero-valent iron. *Groundwater, 32*, 958–967.

Grieger, K. D., Fjordbøge, A., Hartmann, N. B., Eriksson, E., Bjerg, P. L., & Baun, A. (2010). Environmental benefits and risks of zero-valent iron nanoparticles (nZVI) for in situ remediation: Risk mitigation or trade-off? *Journal of Contaminant Hydrology, 118*, 165–183.

He, F., & Zhao, D. (2005). Preparation and characterization of a new class of starch-stabilized bimetallic nanoparticles for degradation of chlorinated hydrocarbons in water. *Environmental Science & Technology, 39*, 3314–3320.

He, F., Zhao, D., & Paul, C. (2010). Field assessment of carboxymethyl cellulose stabilized iron nanoparticles for in situ destruction of chlorinated solvents in source zones. *Water Research, 44*, 2360–2370.

Henn, K. W., & Waddill, D. W. (2006). Utilization of nanoscale zero-valent iron for source remediation – a case study. *Remediation Journal, 16*, 57–77.

Johnson, R. L., Nurmi, J. T., O'Brien Johnson, G. S., Fan, D., O'Brien Johnson, R. L., Shi, Z., Salter-Blanc, A. L., Tratnyek, P. G., & Lowry, G. V. (2013). Field-scale transport and transformation of carboxymethylcellulose-stabilized nano zero-valent iron. *Environmental Science & Technology, 47*, 1573–1580.

Joo, S. H., Feitz, A. J., Sedlak, D. L., & Waite, T. D. (2005). Quantification of the oxidizing capacity of nanoparticulate zero-valent iron. *Environmental Science & Technology, 39*, 1263–1268.

Jordan, M., Shetty, N., Zenker, M. J., & Brownfield, C. (2013). Remediation of a former dry cleaner using nanoscale zero valent iron. *Remediation Journal, 24*, 31–48.

Kanel, S. R., Manning, B., Charlet, L., & Choi, H. (2005). Removal of arsenic(III) from groundwater by nanoscale zero-valent iron. *Environmental Science & Technology, 39*, 1291–1298.

Kanel, S. R., Goswami, R. R., Clement, T. P., Barnett, M. O., & Zhao, D. (2008). Two dimensional transport characteristics of surface stabilized zero-valent iron nanoparticles in porous media. *Environmental Science & Technology, 42*, 896–900.

Karn, B., Kuiken, T., & Otto, M. (2009). Nanotechnology and in situ remediation: A review of the benefits and potential risks. *Environmental Health Perspectives, 117*, 1823–1831.

Kiel, K., & Zabel, J. (2001). Estimating the economic benefits of cleaning up superfund sites: The case of Woburn, Massachusetts. *The Journal of Real Estate Finance and Economics, 22*, 163–184.

Kirschling, T., Gregory, K., Minkley, N., Lowry, G., & Tilton, R. (2010). Impact of nanoscale zero valent iron on geochemistry and microbial populations. *Environmental Science & Technology, 44*, 3474–3480.

Köber, R., Hollert, H., Hornbruch, G., Jekel, M., Kamptner, A., Klaas, N., Maes, H., Mangold, K.-M., Martac, E., Matheis, A., Paar, H., Schäffer, A., Schell, H., Schiwy, A., Schmidt, K. R., Strutz, T. J., Thümmler, S., Tiehm, A., & Braun, J. (2014). Nanoscale zero-valent iron flakes for groundwater treatment. *Environment and Earth Science, 72*, 3339–3352.

Kocur, C., Chowdhury, A. I., Sakulchaicharoen, N., Boparai, H., Weber, K. P., Sharma, P., Krol, M. M., Austrins, L. M., Peace, C., Sleep, B. E., & O'Carroll, D. M. (2014). Characterization of nZVI mobility in a field scale test. *Environmental Science & Technology, 48*, 2862–2869.

Kocur, C. M., Lomheim, L., Boparai, H. K., Chowdhury, A. I., Weber, K. P., Austrins, L. M., Edwards, E. A., Sleep, B. E., & O'Carroll, D. M. (2015). Contributions of abiotic and biotic dechlorination following carboxymethyl cellulose stabilized nanoscale zero valent iron injection. *Environmental Science & Technology, 49*, 8648–8656.

Kocur, C. M. D., Lomheim, L., Molenda, O., Weber, K. P., Austrins, L. M., Sleep, B. E., Boparai, H. K., Edwards, E. A., & O'Carroll, D. M. (2016). Long-term field study of microbial community and dechlorinating activity following carboxymethyl cellulose-stabilized nanoscale zero-valent iron injection. *Environmental Science & Technology, 50*, 7658–7670.

Krol, M. M., Oleniuk, A. J., Kocur, C. M., Sleep, B. E., Bennett, P., Xiong, Z., & O'Carroll, D. M. (2013). A field-validated model for in situ transport of polymer-stabilized nZVI and implications for subsurface injection. *Environmental Science & Technology, 47*, 7332–7340.

Lacina, P., Dvorak, V., Vodickova, E., Barson, P., Kalivoda, J., & Goold, S. (2015). The application of nano-sized zero-valent iron for in situ remediation of chlorinated ethylenes in groundwater: A field case study. *Water Environment Research, 87*, 326–333.

Laszlo, T., & Szabo, M. (2017). NanoRem Pilot Site – Balassagyarmat, Hungary: In situ groundwater remediation using Carbo-Iron® nanoparticles. *CL:AIRE's NanoRem bulletin*. CL:AIRE, UK, pp. 1–6.

Lee, C., Kim, J. Y., Lee, W. I., Nelson, K. L., Yoon, J., & Sedlak, D. L. (2008). Bactericidal effect of zero-valent iron nanoparticles on Escherichia coli. *Environmental Science & Technology, 42*, 4927–4933.

Li, X.-Q., & Zhang, W.-X. (2007). Sequestration of metal cations with zerovalent iron nanoparticles a study with high resolution X-ray photoelectron spectroscopy (HR-XPS). *Journal of Physical Chemistry C, 111*, 6939–6946.

Li, Z., Greden, K., Alvarez, P. J., Gregory, K. B., & Lowry, G. V. (2010). Adsorbed polymer and NOM limits adhesion and toxicity of nano scale zerovalent iron to E. coli. *Environmental Science & Technology, 44*, 3462–3467.

Li, X. N., Jiao, W. T., Xiao, R. B., Chen, W. P., & Chang, A. C. (2015). Soil pollution and site remediation policies in China: A review. *Environmental Reviews, 23*, 1–12.

Liang, L., Guan, X., Shi, Z., Li, J., Wu, Y., & Tratnyek, P. (2014). Coupled effects of aging and weak magnetic fields on sequestration of selenite by zero-valent iron. *Environmental Science & Technology, 48*, 6326–6334.

Lien, H.-L., & Zhang, W. X. (1999). Transformation of chlorinated methanes by nanoscale Iron particles. *Journal of Environmental Engineering, 125*, 1042–1047.

Lien, H.-L., & Zhang, W. X. (2001). Nanoscale iron particles for complete reduction of chlorinated ethenes. *Colloids and Surfaces A, 191*, 97–105.

Liu, Y., Majetich, S. A., Tilton, R. D., Sholl, D. S., & Lowry, G. V. (2005). TCE dechlorination rates, pathways, and efficiency of nanoscale iron particles with different properties. *Environmental Science & Technology, 39*, 1338–1345.

Lu, Y., Song, S., Wang, R., Liu, Z., Meng, J., Sweetman, A. J., Jenkins, A., Ferrier, R. C., Li, H., Luo, W., & Wang, T. (2015). Impacts of soil and water pollution on food safety and health risks in China. *Environment International, 77*, 5–15.

Mackenzie, K., Bleyl, S., Kopinke, F. D., Doose, H., & Bruns, J. (2016). Carbo-Iron as improvement of the nanoiron technology: Fromlaboratory design to the field test. *The Science of the Total Environment, 563–564*, 641–648.

Mueller, N. C., Braun, J., Bruns, J., Černík, M., Rissing, P., Rickerby, D., & Nowack, B. (2012). Application of nanoscale zero valent iron (NZVI) for groundwater remediation in Europe. *Environmental Science and Pollution Research International, 19*, 550–558.

Nurmi, J. T., Tratnyek, P. G., Sarathy, V., Baer, D. R., Amonette, J. E., Pecher, K., Wang, C., Linehan, J. C., Matson, D. W., Penn, R. L., & Driessen, M. D. (2005). Characterization and properties of metallic iron nanoparticles: Spectroscopy, electrochemistry, and kinetics. *Environmental Science & Technology, 39*, 1221–1230.

NanoRem (2018). Taking Nanotechnological Remediation Processes from Lab Scale to End User Applications for the Restoration of a Clean Environment. NanoRem. http://www.nanorem.eu/

O'Carroll, D., Sleep, B., Krol, M., Boparai, H., & Kocur, C. (2013). Nanoscale zero valent iron and bimetallic particles for contaminated site remediation. *Advances in Water Resources, 51*, 104–122.

O'Hara, S., Krug, T., Quinn, J., Clausen, C., & Geiger, C. (2006). Field and laboratory evaluation of the treatment of DNAPL source zones using emulsified zero-valent iron. *Remediation Journal, 16*, 35–56.

Oppegard, A. L., Darnell, F. J., & Miller, H. C. (1961). Magnetic properties of single-domain iron and iron-cobalt particles prepared by borohydride reduction. *Journal of Applied Physics, 32*, S184.

Otaegi, N., & Cagigal, E. (2017). NanoRem Pilot Site – Nitrastur, Spain: Remediation of arsenic in groundwater using nanoscale zero-valent iron. *CL:AIRE's NanoRem bulletin*. CL:AIRE, UK, pp. 1–6.

Panagos, P., Liedekerke, M. V., Yigini, Y., & Montanarella, L. (2013). Contaminated sites in Europe: Review of the current situation based on data collected through a European network. *Journal of Environmental and Public Health, 2013*, 158764.

Phenrat, T., & Kumloet, I. (2016). Electromagnetic induction of nanoscale zerovalent iron particles accelerates the degradation of chlorinated dense non-aqueous phase liquid: Proof of concept. *Water Research, 107*, 19–28.

Phenrat, T., & Lowry, G. V. (2009). Physicochemistry of polyelectrolyte coatings that increase stability, mobility, and contaminant specificity of reactive nanoparticles used for groundwater remediation. In N. Savage, M. Diallo, J. Duncan, A. Street, & R. Sustich (Eds.), *Nanotechnology applications: Solutions for improving water quality* (pp. 249–267). New York: William Andrew Publisher.

Phenrat, T., Saleh, N., Sirk, K., Tilton, R. D., & Lowry, G. V. (2007). Aggregation and sedimentation of aqueous nanoscale zerovalent iron dispersions. *Environmental Science & Technology, 41*, 284–290.

Phenrat, T., Saleh, N., Sirk, K., Kim, H.-J., Tilton, R. D., & Lowry, G. V. (2008). Stabilization of aqueous nanoscale zerovalent iron dispersions by anionic polyelectrolytes: Adsorbed anionic polyelectrolyte layer properties and their effect on aggregation and sedimentation. *Journal of Nanoparticle Research, 10*, 795–814.

Phenrat, T., Kim, H.-J., Fagerlund, F., Illangasekare, T., Tilton, R. D., & Lowry, G. V. (2009a). Particle size distribution, concentration, and magnetic attraction affect transport of polymer-modified Fe^0 nanoparticles in sand columns. *Environmental Science & Technology, 43*, 5079–5085.

Phenrat, T., Liu, Y., Tilton, R. D., & Lowry, G. V. (2009b). Adsorbed polyelectrolyte coatings decrease Fe^0 nanoparticle reactivity with TCE in water: Conceptual model and mechanisms. *Environmental Science & Technology, 43*, 1507–1514.

Phenrat, T., Long, T. C., Lowry, G. V., & Veronesi, B. (2009c). Partial oxidation ("aging") and surface modification decrease the toxicity of nanosized zerovalent iron. *Environmental Science & Technology, 43*, 195–200.

Phenrat, T., Kim, H.-J., Fagerlund, F., Illangasekare, T., & Lowry, G. V. (2010a). Empirical correlations to estimate agglomerate size and deposition during injection of a polyelectrolyte-modified Fe0 nanoparticle at high particle concentration in saturated sand. *Journal of Contaminant Hydrology, 118*, 152–164.

Phenrat, T., Song, J. E., Cisneros, C. M., Schoenfelder, D. P., Tilton, R. D., & Lowry, G. V. (2010b). Estimating attachment of nano- and submicrometer-particles coated with organic macromolecules in porous media: Development of an empirical model. *Environmental Science & Technology, 44*, 4531–4538.

Phenrat, T., Fagerlund, F., Illanagasekare, T., Lowry, G. V., & Tilton, R. D. (2011). Polymer-modified Fe0 nanoparticles target entrapped NAPL in two dimensional porous media: Effect of particle concentration, NAPL saturation, and injection strategy. *Environmental Science & Technology, 45*, 6102–6109.

Phenrat, T., Thongboot, T., & Lowry, G. V. (2016). Electromagnetic induction of zerovalent iron (ZVI) powder and nanoscale zerovalent iron (NZVI) particles enhances dechlorination of trichloroethylene in contaminated groundwater and soil: Proof of concept. *Environmental Science & Technology, 50*, 872–880.

Phenrat, T., Teeratitayangkul, P., Prasertsung, I., Parichatprecha, R., Jitsangiam, P., Chomchalow, N., & Wichai, S. (2017). Vetiver plantlets in aerated system degrade phenol in illegally dumped industrial wastewater by phytochemical and rhizomicrobial degradation. *Environmental Science and Pollution Research, 24*, 13235–13246.

Phenrat, T., Ba, Q. T., Piaowan, K., Thongboot, T., Le, S. T., & Sawasdee, T. (2018). Arsenic residue in residential area after cleanup of pesticide illegal dumping sources in Thanh Hoa province, Central Vietnam. *Environmental Forensics, 19*, 66–78.

Ponder, S. M., Darab, J. G., & Mallouk, T. E. (2000). Remediation of Cr(VI) and Pb(II) aqueous solutions using supported, nanoscale zero-valent Iron. *Environmental Science & Technology, 34*, 2564–2569.

Ponder, S. M., Darab, J. G., Bucher, J., Caulder, D., Craig, I., Davis, L., Edelstein, N., Lukens, W., Nitsche, H., Rao, L., Shuh, D. K., & Mallouk, T. E. (2001). Surface chemistry and electrochemistry of supported zerovalent iron nanoparticles in the remediation of aqueous metal contaminants. *Chemistry of Materials, 13*, 479–486.

Quinn, J., Geiger, C., Clausen, C., Brooks, K., Coon, C., O'Hara, S., Krug, T., Major, D., Yoon, W. S., Gavaskar, A., & Holdsworth, T. (2005). Field demonstration of DNAPL dehalogenation using emulsified zero-valent iron. *Environmental Engineering Science, 39*, 1309–1318.

Reuters. (2017). *China needs patience to fight costly war against soil pollution: Government*. Reuters.

Reynolds, G. W., Hoff, J. T., & Gillham, R. W. (1900). Sampling bias caused by materials used to monitor halocarbons in groundwater. *Environmental Science & Technology, 24*, 135–142.

Rujivanarom, P. (2017). *Thai mine "destroyed Myanmar water sources"*. The Nation, Bangkok: The Nation.

Rujivanarom, P. (2018). Residents demand help after report concludes mine hazard. *The Nation*, Bangkok. http://www.nationmultimedia.com/detail/national/30339973.

Saleh, N., Phenrat, T., Sirk, K., Dufour, B., Ok, J., Sarbu, T., Matyjaszewski, K., Tilton, R. D., & Lowry, G. V. (2005). Adsorbed triblock copolymers deliver reactive iron nanoparticles to the oil/water interface. *Nano Letters, 5*, 2489–2494.

Saleh, N., Sirk, K., Liu, Y., Phenrat, T., Dufour, B., Matyjaszewski, K., Tilton, R. D., & Lowry, G. V. (2007). Surface modifications enhance nanoiron transport and NAPL targeting in saturated porous media. *Environmental Engineering Science, 24*, 45–57.

Saleh, N., Kim, H. J., Phenrat, T., Matyjaszewski, K., Tilton, R. D., & Lowry, G. V. (2008). Ionic strength and composition affect the mobility of surface-modified Fe0 nanoparticles in water-saturated sand columns. *Environmental Science & Technology, 42*, 3349–3355.

Schrick, B., Hydutsky, B. W., Blough, J. L., & Mallouk, T. E. (2004). Delivery vehicles for zerovalent metal nanoparticles in soil and groundwater. *Chemistry of Materials, 16*, 2187–2193.

Stejskal, V., Lederer, T., Kvapil, P., Slunsky, J., & Skácelová, P. (2017). NanoRem Pilot Site – Spolchemie I, Czech Republic: Nanoscale zero-valent iron remediation of chlorinated hydrocarbons. *NanoRem bulletin*. CL:AIRE, UK, pp. 1–8.

Su, C., Puls, R. W., Krug, T. A., Watling, M. T., O'Hara, S. K., Quinn, J. W., & Ruiz, N. E. (2013). Travel distance and transformation of injected emulsified zerovalent iron nanoparticles in the subsurface during two and half years. *Water Research, 47*, 4095–4106.

Taghavy, A., Costanza, J., Pennell, K. D., & Abrio, L. M. (2010). Effectiveness of nanoscale zero-valent iron for treatment of a PCE–DNAPL source zone. *Journal of Contaminant Hydrology, 118*, 128–142.

The Economist. (2017). The most neglected threat to public health in China is toxic soil. *The Economist*.

US.EPA. (2011). *Beneficial effects of the superfund program* (p. 25). Washington, DC: Office of Superfund Remediation and Technology Innovation.

US.EPA. (2018a). *Redevelopment economics at superfund sites*. Washington, DC: US.EPA.

US.EPA. (2018b). *Superfund: National priorities list (NPL)*. Washington, DC: US.EPA.

USEPA. (2004). Cleaning Up the Nation's Waste Sites: Markets and Technology Trends, EPA 542-R-04-015, September 2004, Cincinnati, OH, 338 page.

Wang, C. B., & Zhang, W. X. (1997). Synthesizing nanoscale iron particles for rapid and complete dechlorination of TCE and PCBs. *Environmental Science & Technology, 31*, 2154–2156.

Wei, Y.-T., Wu, S.-C., Chou, C.-M., Che, C.-H., Tsai, S.-M., & Lien, H.-L. (2010). Influence of nanoscale zero-valent iron on geochemical properties of groundwater and vinyl chloride degradation: A field case study. *Water Research, 44*, 131–140.

Wei, Y. T., Wu, S. C., Yang, S. W., Che, C.-H., Lien, H.-L., & Huang, D. H. (2012). Biodegradable surfactant stabilized nanoscale zero-valent iron for in situ treatment of vinyl chloride and 1,2-dichloroethane. *Journal of Hazardous Materials, 211–212*, 373–380.

Wiesner, M. R., Lowry, G. V., Alvarez, P., Dionysiou, D., & Biswas, P. (2006). Assessing the risks of manufactured nanomaterials. *Environmental Science & Technology, 40*, 4336–4345.

Xiu, Z. M., Jin, Z. H., Li, T. L., Mahendra, S., Lowry, G. V., & Alvarez, P. J. (2010). Effects of nano-scale zero-valent iron particles on a mixed culture dechlorinating trichloroethylene. *Bioresource Technology, 101*, 1141–1146.

Zhan, J., Zheng, T., Piringer, G., Day, C., McPherson, G. L., Lu, Y., Papadopoulos, K., & John, V. T. (2008). Transport characteristics of nanoscale functional zerovalent iron/silica composites for in situ remediation of trichloroethylene. *Environmental Science & Technology, 42*, 8871–8876.

Zhang, W.-X., Wang, C.-B., & Lien, H.-L. (1998). Treatment of chlorinated organic comtaminants with nanoscale bimetallic particles. *Catalysis Today, 40*, 387–395.

Zheng, T., Zhan, J., He, J., Day, C., Lu, Y., McPherson, G. L., Piringer, G., & John, V. T. (2008). Reactivity characteristics of nanoscale zerovalent iron--silica composites for trichloroethylene remediation. *Environmental Science & Technology, 42*, 4494–4499.

Chapter 2
NZVI Synthesis and Characterization

Katrin Mackenzie and Anett Georgi

Abstract This chapter provides an overview of NZVI types used to date for environmental restoration. The particle types are introduced systematically from bare NZVI to the manifold modifications leading to NZVI-containing composites or emulsions. Properties of these NZVI types which are important for the intended use as water treatment reagent and methods for their characterization are compiled. For each of the main NZVI groups – bare and bimetallic NZVI, polymer-modified NZVI, supported NZVI and emulsified NZVI, approved synthesis strategies and resulting NZVI properties are described.

Keywords Nanoscale zerovalent iron · Synthesis · Characterization

2.1 Overview of NZVI Types, Important Properties, and Characterization Techniques

2.1.1 Overview of NZVI Types

Initiated by the first studies on the preparation of NZVI and its application for reductive dechlorination of chlorinated groundwater contaminants in the last decade of the previous century (Wang and Zhang 1997; Lien and Zhang 1999) and in the first decade of this century (Elliott and Zhang 2001; Cao et al. 2003; Liu et al. 2005a, b; Nurmi et al. 2005), the great potential of NZVI for groundwater remediation was soon recognized by the scientific community as well as by practitioners. At the same time, it also became clear that there is significant room for improvement of NZVI properties with respect to applicability, as revealed by detailed lab research and also field trials. This obvious gap boosted the creativity and efforts of scientists and engineers and finally led to a wide variety of NZVI-

K. Mackenzie (✉) · A. Georgi
Helmholtz Centre for Environmental Research – UFZ, Leipzig, Germany
e-mail: katrin.mackenzie@ufz.de; anett.georgi@ufz.de

T. Phenrat, G. V. Lowry (eds.), *Nanoscale Zerovalent Iron Particles for Environmental Restoration*, https://doi.org/10.1007/978-3-319-95340-3_2

based particles being investigated in research labs all over the world; a number of them are already being produced at large scales and applied in the field. NZVI modifications were mainly intended for improvements regarding one or several of the following key features of NZVI: (i) range of treatable contaminants, (ii) colloidal properties and suitability for subsurface delivery, (iii) lifetime, (iv) efficiency of Fe^0 utilization, and (v) potential for combination of various modes of contaminant removal, such as sorption and reaction.

This chapter gives an overview of various NZVI types (Fig. 2.2) and modifications together with the underlying intentions for performance improvements. Details regarding preparation procedures, characteristics, and performance of specific particle types are dealt with in the individual subchapters.

Bare NZVI

NZVI which was not intentionally given a shell (other than a self-developed iron oxide layer) will be discussed here as bare NZVI. Depending on the synthesis method, we can distinguish between several types of bare NZVI. Often, synthesis procedures are subdivided into top-down and bottom-up approaches. The top-down approach starts from micrometer- to millimeter-sized iron filings, which are ball-milled to fine, nano-sized particles. This approach has been applied for commercial production by Golder Assoc., USA. The bottom-up approach, which is most frequently applied for NZVI synthesis in the laboratory but also at commercial scale, starts from either dissolved iron salts, nano-sized iron oxides, or iron-containing molecules (e.g., $Fe(CO)_5$); and the various methods typically consist of a sequence of physical and chemical treatment steps (for details and references, see Sect. 2.2).

NZVI synthesized by reduction of dissolved Fe(II) or Fe(III) with dissolved borohydride (BH_4^-) is the most studied Fe^0 nanoparticles to date (often denoted as Fe^{BH}), mainly due to the fact that the synthesis can easily be performed in almost any laboratory. However, large-scale commercial production of NZVI usually involves gas-phase reduction of nanoscale iron oxides by H_2, followed by inertization in order to prevent pyrophoric reactions of fresh Fe^0 surfaces upon air-contact (often denoted as Fe^{H2}). Most field-scale applications in the USA used reactive nanoscale iron particles (RNIP) formerly commercially available from Toda Kogyo Corporation, Onoda, Japan (Toda Patent). In Europe, NANO IRON s.r.o. (Rajhrad, Czech Republic) has been the supplier of NZVI (Nanofer products) for various field tests (Mueller et al. 2012). Figure 2.1 gives an impression of the preparation and injection of NZVI slurry.

Bimetallics

NZVI's reactivity and the treatable contaminant spectrum can be increased by the addition of a catalytically active noble metal to form bimetallic NZVI particles (here, bimetallics; see Fig. 2.2b). The intimate contact with the second metal, which can be Pd, Ag, Ni, Co, or Cu, enhances the H_2 formation by Fe^0 corrosion (galvanic corrosion). The accelerated hydrogen generation can be combined with the catalytic hydrogenation properties of a noble metal to drive the reduction of

Fig. 2.1 NZVI slurry ready for injection. (Photo with permission from Kocur, Sleep and O'Carroll)

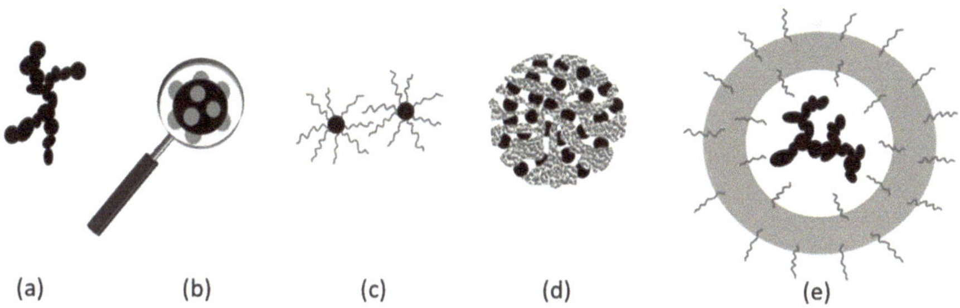

Fig. 2.2 Illustration of NZVI types: (**a**) bare NZVI, (**b**) bimetallics, (**c**) polymer-modified NZVI, (**d**) supported ZVI, (**e**) emulsified NZVI

chlorinated organic pollutants at the bimetallic interface, also for pollutants not amenable to reduction by Fe^0 alone (Grittini et al. 1995; Fang and Al-Abed 2008; Xu and Zhang 2000; Schrick et al. 2002). Compounds treatable by bimetallics include not only halogenated aromatic compounds (e.g., chlorobenzenes, polychlorinated biphenyls) but also heavy metals, oxyanions, and nitro- and azo-compounds (summarized in Liu et al. (2014) and in Chap. 3). The disadvantages of bimetallics are the expected short lifetime due to fast Fe^0 corrosion and the possible poisoning of the noble metal by groundwater constituents (e.g., sulfur compounds other than sulfate). Furthermore, environmental risks due to the use of potentially toxic metals have to be considered.

Polymer-Modified NZVI

Surface modification of NZVI by organic polymers and polyelectrolytes is a widely used strategy to improve the colloidal properties of NZVI, which, due to its high surface energy and magnetic properties, is prone to fast agglomeration. By virtue of their charge and/or steric confinements, organic coatings can increase repulsive forces between NZVI particles themselves and toward solid surfaces, thus enhancing suspension stability and subsurface transport of NZVI. In practice, there is no strict dividing line between bare and polymer-modified NZVI, since such modifications also include simple physical and thus reversible adsorption of dissolved polymers or surfactants added to aqueous suspensions. A wide variety of (semi)natural (e.g., carboxymethyl cellulose, humic acids, starch, xanthan, and guar gum) and synthetic products (e.g., poly(acrylic acid), poly(styrene sulfonate), polyaspartate) is applied (for references see Sect. 2.3). The formation of covalent bonds between the polymer and surface functional groups of the particle is an effective method for anchoring the coating material to the particle surface. In addition, particularly tight and hydrophobic coatings such as polysiloxanes (Krajangpan et al. 2012) or hydrophobic blocks of triblock copolymers (Saleh et al. 2005) are thought to protect the iron from oxidation and to improve attraction by DNAPL phases. However, surface coating of NZVI can also have negative effects on its reactivity (Saleh et al. 2007). Another downside specifically for synthetic polymeric coatings is that they may not be desirable in the subsurface; the eco-compatibility of modified particles needs to be proven. Zhao et al. (2016) provide a broad overview of stabilization methods for NZVI particles used for soil and groundwater remediation. Chapters 5 and 6 in this book are devoted to polymer-modified NZVI.

Emulsified NZVI

Emulsification of NZVI particles has been applied for improving direct treatment of DNAPL source zones. Emulsified NZVI or ENZVI is actually a water-in-oil-in-water emulsion prepared from NZVI particles, surfactants, and biodegradable oil (Quinn et al. 2005). ENZVI has two important features: (i) an outer hydrophobic surface, which improves adhesion to and penetration into DNAPL phases, and (ii) an inner aqueous phase, surrounding the NZVI particles and serving as reaction medium (Quinn et al. 2005; Su et al. 2012, 2013). Organic contaminants can reach this interior phase by diffusion. The impact of inorganic ions in the groundwater on Fe^0 is reduced. Injection and delivery of ENZVI is an engineering challenge and still remains an obstacle for its wide application (Mueller et al. 2012).

Supported ZVI

Supporting and dispersing iron nanoparticles on various types of porous solid materials is a strategy to overcome the undesirable fast agglomeration of NZVI in aqueous suspension. Silica (Zheng et al. 2008), organic polymers (Schrick et al. 2004), activated carbon (Mackenzie et al. 2012), carbon microspheres (Sunkara et al. 2010), and other carbon materials (Hoch et al. 2008; Zhang et al. 2006) have been used for composite formation. The composites can be further optimized with respect to (i) particle size and density for optimal subsurface transport, (ii) improved surface properties (charge and/or hydrophobicity) for suspension stability and/or subsurface delivery, and (iii) synergistic effects due to sorption properties (mainly in the case of carbon-based composites).

Particle Types Used in the Field

In 2014 Bardos et al. (2014) summarized in total 70 pilot or field site applications of NZVI and related products worldwide. Mueller et al. (2012) reviewed pilot tests of NZVI in Europe and compared the situation with NZVI applications in the USA. The greatest difference between Europe and the USA is the type of NZVI particle used. All remedial actions in Europe were carried out with standard NZVI (RNIP, NANOFER, or Fe^{BH}), while in the USA only 60% used standard NZVI. Until 2009, about 30% of the NZVI remediations in the USA used bimetallic NZVI and about 10% were carried out with ENZVI (Karn et al. 2009). Field tests in the USA have also been conducted with other types of supported NZVI (e.g., silica-supported NZVI) which are summarized in US EPA (2011). In Europe, there is no company working with ENZVI. The main reason is concerned about the injectability of an emulsified suspension (Mueller et al. 2012). In addition, no field application of bimetallic NZVI has been carried out in Europe so far, mainly due to concerns regarding the toxicity of the catalysts (e.g., Ni is considered to be a priority hazardous substance under the EU Water Framework Directive; Schrick et al. 2002). Novel NZVI types applied in recently conducted pilot tests in Europe include supported (Carbo-Iron, Mackenzie et al. 2016) and milled (Köber et al. 2014) NZVI. See more case studies of field-scale NZVI application in Chaps. 7 and 8. The most recent adaptation of NZVI to be tested is sulfidized NZVI (s-NZVI). This material holds signficant promise becuase it appears to limit the reaction of the NZVI with water, while maintaing or enhancing reactivity with the target contaminants. This material is discussed separately in Chap. 9.

2.1.2 Important Properties of NZVI

Core-Shell Structure, Aging, and Handling

Iron is the fourth most abundant element in the earth's crust. The most important minerals for iron production are magnetite and hematite. Iron exists in a wide range of oxidation states from −2 to +6, although +2 and +3 are the most common. While for the large-scale production of bulk iron from iron ore (e.g., for steel production), carbothermal reduction and reduction with CO are used, large-scale production of NZVI is mainly done by high-temperature reduction with hydrogen using nano-sized iron oxides.

Primary NZVI particles typically have a particle size in the range of 20–100 nm and specific surface area (determined by the BET method) of 20–40 m^2/g (Li et al. 2006a, b; Nurmi et al. 2005; Liu et al. 2005b; Eglal and Ramamurthy 2014). In this respect, Fe^{BH} and Fe^{H2} are rather similar. Nevertheless, various NZVI types have been shown to have significant differences in reactivity, which can be ascribed to structural differences in the various iron types (Liu et al. 2005a, b; Nurmi et al. 2005). One major difference lies in the crystallinity of the iron core: Fe^{H2} has a highly crystalline α-Fe^0 core whereas Fe^{BH} contains highly disordered Fe^0

Table 2.1 Comparison of Fe^{BH} and Fe^{H2} (based on two commercial products: RNIP from Toda, Japan, and Nanofer, from NANO IRON, Czech Republic) according to data reported by Liu et al. (2005a, b) (for Fe^{BH} and RNIP), Eglal and Ramamurthy (2014), and Soukupova et al. (2015) (for Nanofer)

	Fe^{BH}	RNIP	Nanofer
Primary particle size (nm)	20–40	40–60	20–100
BET surface area (m^2/g)	36	23	27
Fe^0 core crystallinity	Highly disordered	Highly crystalline	Highly crystalline
B content (wt%, bulk) (atom%)	5 (18)	0	0

crystallites of very small size (about 1 nm) (Liu et al. 2005a; Nurmi et al. 2005; Eglal and Ramamurthy 2014). At this scale, the distinction between amorphous and polycrystalline becomes ambiguous (Liu et al. 2005a).

Furthermore, impurities of elements other than iron can be important for reactivity: Fe^{BH} is the only NZVI type which contains a high B content, while a specificity of RNIP is its content of sulfur species resulting from the use of sulfuric acid in the preparation procedure (reduced sulfur identified in RNIP supplied as slurry (Nurmi et al. 2005) and iron sulfate in dry RNIP powder (Reinsch et al. 2010)). Similarities and differences reported for Fe^{BH} and Fe^{H2} are shown in Table 2.1.

The chemistry of iron nanoparticles is dominated by iron's extreme reactivity with oxidizing agents. This is a particular concern with air. Finely divided iron has long been known to be pyrophoric, something that should always be kept in mind when handling iron nanoparticles. Pre-oxidizing the nanoparticles is a strategy to control the reaction. By exposing iron nanoparticles to a low partial pressure of air, small amounts of dilute, air-free water, a carboxylic acid, or other weak oxidizers can provide the surface with a passivating oxide that allows the nanoparticles to be handled, at least briefly, in air with little additional oxidation (Huber 2005). The shell of iron oxide formed by these procedures is strongly adherent and acts as an oxidation barrier, preventing the sudden evolution of heat that occurs when a pristine iron surface meets air (Huber 2005).

Nevertheless, Fe^0 will be subject to corrosion as soon as it is in contact with water, even in the absence of oxygen. The rate of anaerobic corrosion is strongly pH-dependent. In a closed environment, corrosion is self-inhibiting because hydroxide ions are produced, leading to an increase in pH and consequently a reduction in the corrosion rate.

$$Fe^0 + 2H_2O \rightarrow Fe^{2+} + H_2 + 2OH^- \tag{2.1}$$

In the presence of dissolved oxygen, iron oxidation is much faster. Thus, the presence of oxygen is undesirable when NZVI is applied for contaminant *reduction*. However, the production of hydrogen peroxide and dissolved iron as a result of iron oxidation under aerobic conditions finally leads to hydroxyl radicals. This process can be utilized for *oxidation* of contaminants which are not amenable to reductive degradation (Keenan and Sedlak 2008) (see also Chap. 3).

The formation of an iron-oxide shell around freshly produced iron is unavoidable but in some cases (e.g., Toda RNIP) has been intentionally forced in order to form passivating or protective phases which impart stability to the particles in aqueous suspensions. Eventually, exposure of Fe^0 to water leads to the formation of an oxidic shell, which consists mainly of iron oxide for iron synthesized by reduction with hydrogen and of iron and boron oxides for borohydride-derived particles, where oxidation of the iron core progresses with time (Li et al. 2006a).

The core-shell structure, consisting of an inner Fe^0 core and a mixed-valence iron oxide shell, is therefore a common feature of all NZVI types; this has been demonstrated for NZVI produced via borohydride reduction as well as for RNIP and Nanofer products synthesized by gas-phase reduction of nanoscale iron oxides (Liu et al. 2005b; Nurmi et al. 2005; Sarathy et al. 2008; Eglal and Ramamurthy 2014). TEM images illustrating the core-shell structure for both NZVI types are shown later in the text (Figs. 2.10 and 2.11, Sect. 2.2). The mixed-valence iron oxide shell is largely insoluble under neutral pH conditions.

The type of iron oxide forming the shell is dependent on the specific synthesis conditions for NZVI and can easily be affected by storage and even sample preparation for analysis. Synthesis and handling under strictly anaerobic conditions favors the formation of a crystalline magnetite shell, although iron oxides which are rather amorphous (Ramos et al. 2009) and/or more rich in Fe(III) (Sarathy et al. 2008) have also been reported. Additional minor components of other mineral phases (e.g., iron sulfate for RNIP provided as dry product) have been reported as well (Reinsch et al. 2010). Surface properties of NZVI after immersion in water are relevant not only for interaction with contaminants but also for its colloidal properties, such as agglomeration and mobility in saturated porous media. The various iron oxides differ significantly with respect to surface charge as a function of pH. Magnetite, maghemite, and goethite have IEP values of ≤ 7, 9, and 7.5 to 9, respectively. Therefore, particles may be positively or negatively charged at $pH = 7$ with varying background electrolytes, depending on the species present at the surface (Kosmulski 2014).

The actual Fe^0 content at the time of application is one of the key properties of NZVI particles. Highest Fe^0 contents were reported for freshly synthesized Fe^{BH} (up to about 90%; Liu et al. 2005a, b), whereas Fe^0 content of commercial products, especially when supplied in aqueous suspensions, is generally lower, which is not only due to the fact that these products are designed to be more inert for safety reasons but also due to ongoing aging which cannot be fully prevented during shipping and storage times. Liu and Lowry (2006) studied aging of commercial RNIP slurries (300 g L^{-1}, pH > 10.6, anaerobic) and reported that within 19 days Fe^0 content decreased from the initial content of 60–70% (according to manufacturer) to 48%.

In order to prevent undesired aging and corrosion of NZVI, strict precautions have to be taken. NZVI powders and suspensions are best handled under strictly anaerobic conditions, i.e., in a glove box with inert and dry atmosphere, or at least using Schlenk techniques and/or purging of suspensions with inert gas (argon or nitrogen). NZVI is best stored as dry product in an airtight container under dry inert gas and at low temperature. Storage of NZVI in suspension requires at least

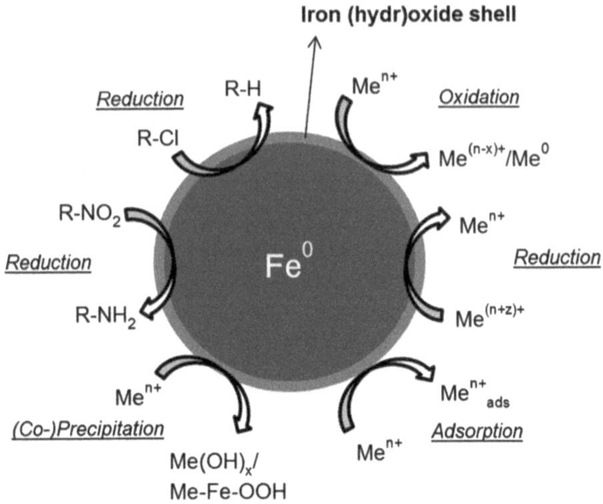

Fig. 2.3 Possible mechanisms of NZVI interaction with contaminants

anaerobic conditions but is always associated with iron corrosion. For converting suspended NZVI into a dry product, flash drying (i.e., rinsing with organic solvents during suction filtration with drying under inert gas) or freeze-drying under vacuum is applied (e.g., Miehr et al. 2004; Liu and Lowry 2006; Wang et al. 2010).

Chemical Reactivity

NZVI has been shown to react with chlorinated organic compounds, nitroaromatics, heavy metals, and various inorganic anions, as will be discussed in detail in Chaps. 3 and 4. Possible mechanisms are illustrated in Fig. 2.3.

The dechlorination of chlorinated organics such as TCE has been most extensively studied and can be described by the following general formulas:

$$Fe^0 \rightarrow Fe^{2+} + 2e^- \quad (2.2)$$

$$TCE + n\,e^- + (n - 3)H^+ \rightarrow \text{products} + 3\,Cl^- \quad (2.3)$$

whereby n depends on the types of products formed, which are mainly ethane, ethene, and acetylene but to some extent even build-up products with $n_C > 2$. The product selectivity depends on the type of NZVI applied. TCE reduction with Fe^{BH} shifts selectivity to more saturated C2 products (ethane), whereas with Fe^{H2} more unsaturated products are formed (acetylene, Liu et al. 2005a). At the same time, NZVI is consumed by the competing processes of reaction with target pollutants and corrosion, i.e., H_2 formation (Eq. 2.1). Thus, the electron efficiency of Fe^0, i.e., the percentage of electrons utilized for contaminant reduction and H_2 formation, is an important characteristic of NZVI-based reactions (Liu et al. 2005b).

Reactivity toward target contaminants, corrosion, and aging characteristics are influenced by intrinsic NZVI properties and external conditions (solution chemistry), as will be discussed in detail in Chap. 3. The crystallinity of the Fe^0 core has been identified as one of the key properties for NZVI reactivity, whereby the highly disordered structure in fresh Fe^{BH} is thought to be responsible for its initial high reactivity and its unique ability to activate and utilize H_2 in contaminant reduction (Liu et al. 2005a). At the same time, near-surface composition of NZVI is equally important, since the reduction reactions promoted by NZVI are surface-mediated.

When considering NZVI properties, one should keep in mind that under environmental conditions, NZVI is a steadily changing system. Aging of NZVI is a complex process which involves changes in chemical, mineralogical, and morphological composition as well as microscopic changes. Oxidation of iron generally leads to the formation of Fe(II)/(III) iron oxides, whereby magnetite is often considered as being the major oxide phase formed under anaerobic conditions (Nurmi et al. 2005; Liu and Lowry 2006; Reinsch et al. 2010). However, not only the pathway of formation of magnetite and/or other oxides seems again to depend on intrinsic factors of the specific NZVI type but also extrinsic factors (type of oxidants present, solution conditions), whereby even within one and the same reaction batch, a significant variability in morphological changes among individual particles was observed (Yan et al. 2012) (see Fig. 2.4). Mechanistic perceptions and experimentally observed changes during NZVI aging include (i) increasing thickness of oxide layer with no net change in particle size (Nurmi et al. 2005; Liu and Lowry 2006), (ii) decreasing particle size due to a shrinking iron core covered with a dynamic magnetite shell of constant thickness (Sohn et al. 2006), (iii) conversion of NZVI into hollowed-out particles (Yan et al. 2012), and (iv) partial dissolution of iron and formation of external precipitates with distinct morphology (Liu et al. 2005a; Nurmi et al. 2005; Yan et al. 2012).

Magnetic Properties and Agglomeration

Fe^0 and magnetite, as main components of NZVI, are ferro- and ferrimagnetic materials, respectively. Though very small, NZVI colloids are still ferromagnetic and contain a positive and negative pole (i.e., they are bipolar). Thus, each NZVI particle behaves as a single-domain magnetic particle with an intrinsic permanent magnetic dipole moment $\mu = (4\,\pi/3)r^3 M_s$, even in the absence of an applied external magnetic field (McCurrie 1994), where r is the particle radius and M_s is the saturation magnetization.

Magnetic properties of NZVI can be favorably used for particle detection in sediment cores or even in the field (Buchau et al. 2010). On the other hand, magnetic attraction is the dominant driving force for agglomeration of NZVI, which negatively affects not only suspension stability (Phenrat et al. 2007) but also transport in saturated porous media (Phenrat et al. 2009). The saturation magnetization of RNIP (with 14% Fe^0) was determined as 570 kA/m, which lies between that of magnetite (330 kA/m) and Fe^0 (1226 kA/m), consistent with its Fe^0/Fe_3O_4 core-shell structure and the low Fe^0 content (Phenrat et al. 2007).

Magnetic attractive forces cause RNIP particles to aggregate quickly to micrometer-size aggregates when suspended in water, even at low concentrations. Subsequently, these aggregates assemble themselves into fractal, chain-like clusters,

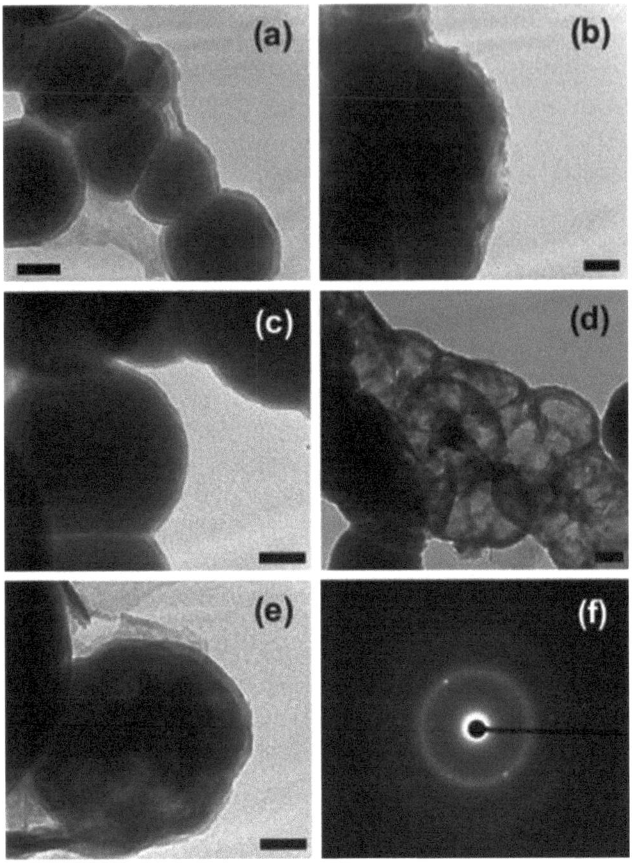

Fig. 2.4 Bright-field TEM images of NZVI reacted with As(III) (5 g/L of NZVI and 100 mg/L of As(III)) for (**a**, **b**) 24 h and (**c**–**e**) 146 days (Yan et al. 2012). Scale bar corresponds to 20 nm. (**f**) is the electron diffraction pattern of a particle shown in (**c**). After 24 h, core-shell structure remains largely unaltered (**a**), but surface is roughened by formation of additional granular iron oxide deposits at outer surface (**b**). After 146 d, images show various distinct morphologies existing in one sample: particles still having a core-shell structure (**c**), hollowed-out oxide shell spheres (**d**), and particles with intermediate structure (**e**), i.e., with tiny internal cavities, additional iron mineral platelets are observed in (**e**). (Reprinted with permission from Yan et al. (2012). Copyright (2012) American Chemical Society)

and under certain conditions even gelation occurs (Phenrat et al. 2007). Formation of chain-like aggregates was also observed for Fe^{BH} (Nurmi et al. 2005).

Stabilization of NZVI by polymer coatings can at least to some extent increase suspension stability (see Chap. 5), although in some cases at the expense of particle reactivity (Phenrat et al. 2009). Fixation of iron nanoclusters on larger particle supports is a more robust means of NZVI dispersion and can lead to very stable suspensions (Schrick et al. 2004; Mackenzie et al. 2012; Zhan et al. 2008).

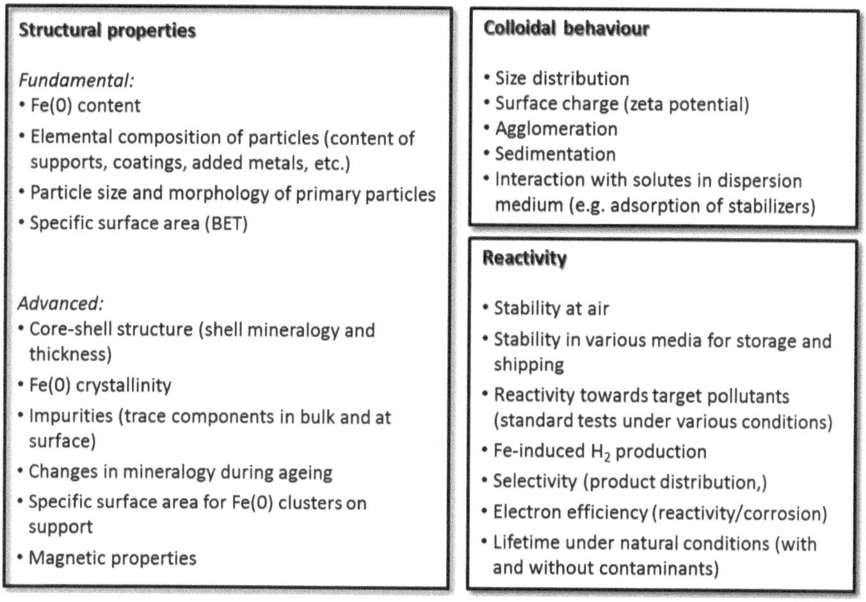

Fig. 2.5 Overview of relevant physical-chemical NZVI properties

As magnetic dipole moment increases with particle radius to the power of three, the larger particles within the size distribution are more important in initiating agglomeration. Formation of aggregates further increases magnetic attractive forces toward neighboring nanoparticles. Thus, synthesis of small particles with narrow size distribution (Wang et al. 2009) and early effective stabilization (pre-synthesis) are effective ways to decrease agglomeration tendency.

Figure 2.5 summarizes again the most important physical-chemical properties of NZVI, which, due to their relevance for NZVI subsurface delivery, reactivity, and fate, are frequently subject to investigation.

2.1.3 Methods for NZVI Characterization: Basic Principles, Advantages, and Limitations

A large variety of characterization techniques has been and is currently applied in order to elucidate not only structural and morphological properties of NZVI and its modifications but also elucidate correlations with reactivity, suspension stability, and particle transport (Table 2.2).

Morphology and Size Distribution of Primary Particles

The first questions usually asked about nanoparticles are concerned with aggregation state, size, and morphology; the direct visualization of particles by *scanning or*

Table 2.2 Overview of discussed analytical methods suitable for NZVI size characterization

Method	Size range	Concentration range	Polydisperse samples
SEM	10 nm ... > 1 μm	Not in suspension	Single particle rather than ensemble analysis
TEM	1 nm ... > 1 μm	Not in suspension	Single particle rather than ensemble analysis
DLS	3 nm ... 3 μm	ppm	Limited applicability
Laser diffraction	100 nm ... some mm	ppm	Suitable
Acoustic/electro-acoustic spectroscopy	5 nm ... 1 mm	ppt	Suitable
NTA	10 nm (samples with high refractive index) ... 1 μm	ppm	Suitable

transmission electron microscopy (*SEM and TEM*, respectively) is an ideal tool for answering such questions. In both methods, a beam of high-energy electrons is directed onto the sample, and signals resulting from electron-sample interactions are analyzed.

Signals of electron-sample interactions utilized in *SEM* include secondary electrons (that produce SEM images), backscattered electrons (BSE), diffracted backscattered electrons (EBSD that are used to determine crystal structures and orientations of minerals), and photons (characteristic X-rays that are used for elemental analysis). For the latter mode, which can give localized elemental information, SEM instruments can be equipped with EDS or EDX detectors (see below). With the help of software packages, particle size distribution and particle shape can be analyzed for a high number of particles in a single image. For conventional SEM analyses, the sample must be dry and conductive (if necessary, coated with a conductive material such as C or Au) and is usually transferred to high vacuum, although low vacuum or environmental SEM instruments also exist.

Imaging methods in *TEM* utilize the information contained in the electron waves exiting from the sample to form an image, with bright-field imaging being the most common mode. In this mode, the contrast formation is based on absorption of electrons in the sample. Thicker regions or regions with heavy atoms will appear dark, while regions with no sample in the beam path will appear bright. High-resolution transmission electron microscopy (HRTEM) can give information about crystal structure. For TEM analyses, samples have to be prepared as very thin slices (100–200 nm). If NPs are sufficiently small, they can simply be placed on support grids. Samples must be dry before transferring them into high vacuum, but the presence of a conducting layer such as in SEM is not required.

Specific Surface Area

In order to obtain information about specific surface area, N_2 adsorption/desorption at low temperatures is measured in surface area analyzers and data are evaluated according to the BET theory. In addition to BET surface area, micro- and mesopore volume of samples with intraparticle and/or interparticle porosity can be determined from N_2 adsorption/desorption isotherms. However, in the case of Fe particles supported on porous supports such as activated carbon or porous silica, differentiation between the actual surface area of the Fe clusters and the large surface area provided by the support remains an unsolved problem, which prevents, e.g., calculation of (Fe)-surface-area-normalized rate constants.

Size Distribution in Suspension, Surface Charge, Agglomeration, and Sedimentation Behavior

The most common methods for analyzing particle-size distribution in suspension are based on the interaction of particles with visible light. *Dynamic light scattering (DLS)* analyzes the fluctuations in the intensity of laser light scattered by the particles. As these fluctuations result from changing interferences caused by particle motions, the basic information obtained from DLS is the diffusion coefficient of the particles, which can be related to a particle radius. Data analysis is straightforward in the case of monodisperse samples but requires complicated algorithms in order to calculate size distribution in the case of polydisperse mixtures. In addition, small particles can be discriminated against in the presence of significantly larger particles, as the primary information is the intensity-weighted or z-average of the particle size distribution. In addition, suspensions must be highly diluted in order to exclude short-term interparticle interactions (i.e., multiple scattering). Thus, DLS is not suitable for obtaining information on real particle size distributions in concentrated suspensions. DLS is frequently applied after sonication of dilute NZVI suspensions in order to determine the intrinsic size distribution of primary particles and (unbreakable) sintered aggregates (e.g., Phenrat et al. 2009). In various studies, DLS measurements have been applied in order to follow agglomeration of NZVI particles with various coatings in dilute suspensions by measuring the increase in hydrodynamic radius over time (Phenrat et al. 2008; Tiraferri et al. 2008).

Nanoparticle tracking analysis (NTA) is another method which analyzes particle size based on Brownian motion. However, in this case individual particles (visualized by their spots of scattered light) are traced, thus eliminating to a certain extent the discrimination of small particles. The observable size range reaches down to 10 nm for materials with high refractive index and up to about 1 μm until Brownian motion becomes too slow and/or sedimentation movement is dominating. Raychoudhury et al. (2012) applied NTA for tracing the increase in hydrodynamic radius of CMC-stabilized NZVI over time due to agglomeration.

Composites and Fe-on-support particles as well as milled iron particles often have a larger size than NZVI, reaching into the μm range, and thus require different methods for size analysis. *Laser diffraction analysis* utilizes diffraction patterns of a laser beam passed through aerosols (prepared from dry powders) or suspensions and is applicable for particles from the nm to mm range. Particle size is reported as a volume-equivalent sphere diameter. However, as particle size falls below 1 μm, results obtained by simplified algorithms (i.e., without knowledge of the refractive index of the material) must be considered with caution.

Interesting alternatives for sufficiently concentrated suspensions (≥1 wt% solids) are *acoustic and electroacoustic spectroscopy*, which are based on measurements of ultrasound attenuation in the sample at various frequencies. The method is applicable in a very wide size range, from 5 nm to 1000 μm. The ultrasound attenuation spectrum is defined by the sample's properties and serves as the basis for calculation of its particle size distribution. Acoustic spectrometers have been applied for particle size analysis of NZVI suspensions in a concentration range of 1 to 10 g L^{-1} (Sun et al. 2006; Comba and Sethi 2009).

Surface charge The most common method for determining surface charge is *zeta potential measurement* which is often provided as an additional feature of DLS instruments (see above). Laser Doppler velocimetry is applied in order to measure the velocity of particles after applying an electric field across the suspension. This velocity is proportional to the magnitude of the zeta potential, which is the potential at the slipping plane separating mobile fluid from fluid that remains attached to the particle surface. This slipping plane lies within the electrical double layer (Fig. 2.6).

Zeta potential can be also determined using *electroacoustic spectrometers* by measuring the colloid vibration current that results from displacement of the electrical double layer of charged particles under the influence of ultrasound.

A zeta potential of ±30 mV is often used as an approximate threshold for stability of a particle suspension against agglomeration. Zeta potential is highly

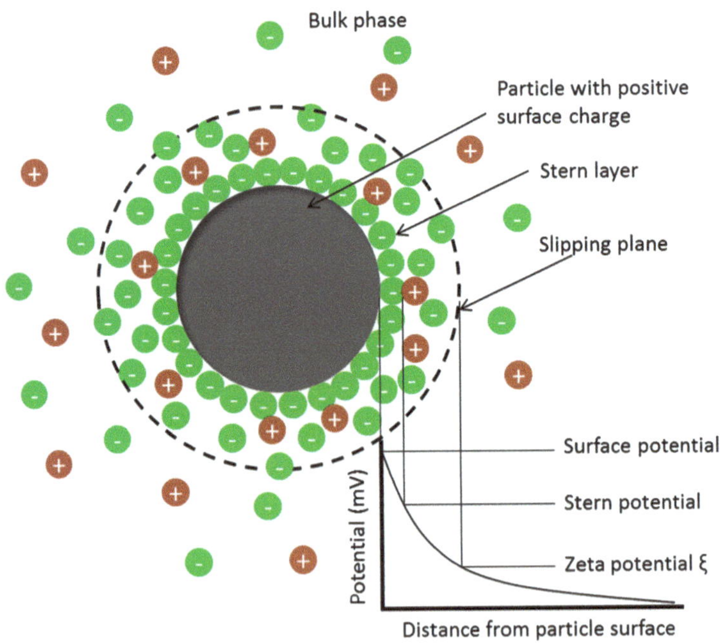

Fig. 2.6 Illustration of electric double layer and zeta potential of a nanoparticle

sensitive to the properties of the suspension medium, such as pH and ionic strength. The isoelectric point (IEP) is the pH where zeta potential is zero and particles are immobile in the electric field. The IEP of Fe^{BH} has been found to be at a slightly alkaline pH (pH = 7.8 (Kanel et al. 2005) and pH = 8.3 (Sun et al. 2006)), whereas for RNIP, values between 6.3 and 7 have been reported (Phenrat et al. 2008; Tiraferri et al. 2008). It should be noted that the zeta potential is a modeled value based on the measured electrophoretic mobility (EPM). The ionic strength and pH both affect the magnitude of the EPM, and hence zeta potenital, so these parameters should be controlled during measurement and reported along with the result. The presence of polymer coatings on NZVI affects the types of models used to calculate zeta potential from EPM, and the zeta potential of a polymer-coated or polyelectorlyte-coated NZVI particle should technically be reported as the "apparent" zeta potential. A guidance document for making robust and reporducible EPM measurements and calculations of zeta potential is available (Lowry et al. 2016).

Agglomeration/sedimentation Sedimentation processes can be followed by methods with varying degrees of automation. Given that a suitable method for determination of particle concentration is available (total Fe content or content of another characteristic component, e.g., carbon in case of C-supported Fe particles), manual sampling at one or various heights below the water table of a quiet suspension, repeated after certain time intervals, can be applied. Less labor-intensive options are the measurement of absorbance (in the high-wavelength range of visible light, i.e., >600 or 800 nm) in simple UV/Vis spectrometers or the measurement of turbidity over time. In these cases, the sample is placed in cuvettes, and the distance of observation below the water table is determined by the optical path (applied, e.g., in Phenrat et al. 2007; Tiraferri et al. 2008). The Turbiscan instrument analyzes backscattering and transmission along the height of a suspension sample. Thus, it provides automated and objective information on time and space dependence of both signals, which for a given dispersion depend only on concentration and particle size. Turbiscan analysis was applied in order to follow agglomeration and sedimentation processes in concentrated NZVI suspensions (Comba and Sethi 2009).

Bulk Elemental Composition and Crystallography

The simplest method for determination of the *Fe^0 content* of NZVI is based on H_2 formation by *acid digestion*. In the absence of gas analysis instruments for quantification of gas-phase H_2 concentrations, even a volumetric measurement can be performed if sufficient particle mass is available.

X-ray fluorescence (XRF) The emission of characteristic "secondary" (or fluorescent) X-rays from a material that has been excited by high-energy X-rays or gamma rays is a phenomenon which is widely used for elemental analysis. Most instruments detect elements with atomic number ≥ 11 and in a very wide concentration range from ppm to high %. Nondestructive analysis with little sample preparation is possible.

Analysis of total (organic) carbon (TC/TOC) and elemental analysis Measurement of CO_2 produced upon high-temperature catalytic combustion in TC analyzers is an important method for characterizing the content of organic or carbonaceous coatings and supports in modified NZVI particles. Differentiation between organic and inorganic carbon can be achieved by acid pretreatment of the sample, whereby inorganic carbon is removed as CO_2. Elemental analyzers allow quantification of the nonmetal components C, H, N, and S (sometimes also P and halogens), whereby high-temperature combustion and gas analysis after catalytic treatment of the combustion gases are the key steps of analysis.

X-ray diffraction (XRD) X-ray diffraction is a strong method to investigate the bulk structure of NZVI and its modifications based on atomic structure (i.e., interatomic distances and unit cell parameters). In particular, information about crystallinity and polymorph type (various crystal structures for minerals of the same chemical composition) can be obtained. XRD has been applied in order to determine the qualitative and quantitative contribution of various mineral phases in NZVI (e.g., Nurmi et al. 2005). However, XRD is an indirect method, so it is, e.g., not possible to distinguish between Fe_3O_4 and γ-Fe_2O_3 since both minerals possess an inverse spinel structure. In this case, XRD results must be combined with other analysis techniques (e.g., Moessbauer spectroscopy). The presence of bimetallic particles, as opposed to a mixture of monometallic particles, can also be demonstrated by XRD, since the diffraction pattern of the physical mixtures consists of overlapping lines of the two individual monometallic nanoparticles and is clearly different from that of the bimetallic nanoparticles. Information on crystallite size (S) can be obtained from the peak width of XRD spectra via the Scherrer equation, $S = \lambda/\omega \cos \Theta$, where λ is the wavelength of the beam, Θ is the diffraction angle, and ω is the width of the peak at half-maximum. Peak shape can give indications on whether the NZVI crystallite size is mono- or polydisperse (Nurmi et al. 2005). However, for nanoparticles or structural components consisting of less than hundreds of atoms, the acquisition of structural information by means of XRD may be difficult. The same applies to amorphous structures, which are in general XRD-silent. XRD has also been applied in order to estimate the Fe^0 content in NZVI, as an alternative to acid digestion methods (e.g., Nurmi et al. 2005; Sarathy et al. 2008). However, exact quantification in XRD requires identification of all mineral phases present in a sample, which might be possible for fresh NZVI but becomes increasingly difficult with progressive aging and formation of new mineral phases.

Extended X-ray absorption (XAS) XAS is one of the most powerful techniques for probing the local atomic structure. It can be applied irrespective of the physical state, also including amorphous materials and samples in solution. XAS gives average information about all crystallographic sites in a mineral. XAS is only able to determine surface modifications if the particles are small enough, i.e., have a sufficiently high surface-to-volume ratio (15–20% surface atoms, Rose et al. 2007). In all XAS methods, the X-ray beam energy is scanned through a certain range around the binding energy of the core-shell electrons of the atoms under consideration. The position and shape of the absorption edge reflects the excitation energy of the inner-shell electrons, whereby fine structures are affected by neighboring atoms around the excited atom. *XANES (X-ray absorption near edge structure) and EXAFS (extended X-ray absorption fine structure)* are characterization methods which focus on different parts of the X-ray absorption spectrum. *EXAFS* provides information on coordination number, chemical species, and symmetry, whereas *XANES* yields information on chemical bonds and symmetry and is more sensitive to site geometry and oxidation state of the target atom. EXAFS has been applied in order to distinguish between maghemite and magnetite as oxide-layer components of NZVI with quantitative information obtained on the basis of model compound spectra (Reinsch et al. 2010). Care must be taken with respect to chemical (redox) changes which can be induced in XANES when using X-rays from a synchrotron source.

Energy-dispersive X-ray microanalysis (EDX, EDAX, EDS) One of the most revealing analytical methods for the composition of NZVI and its modifications is energy-dispersive X-ray spectroscopy (EDX), which is usually coupled with a scanning or transmission electron microscope. Each element in the chosen nanoparticle emits X-rays at characteristic energies after electron beam irradiation; their intensity is proportional to the concentration of each element in the particle. Sampling depth is up to 1–2 μm, whereas the lateral resolution is about 1 nm when coupled with TEM and about 1 μm with SEM analysis. By this means, the spatial distribution of catalytic dopants (e.g., Pd) or impurities in the structure of NZVI can be visualized.

NMR spectroscopy NMR utilizes transitions in energetic states of nuclei with intrinsic magnetic moments in a strong magnetic field which are induced by radiofrequency irradiation. NMR can provide structural information on a range of different elements (those having isotopes with odd numbers of protons and/or neutrons) as well as their coordination. Thus mineral composition can be identified without the necessity of long-range order as in XRD. NMR spectroscopy of metal isotopes is a powerful technique for understanding the electronic environment of metal atoms in metallic particles by virtue of the NMR shifts caused by free electrons (so-called Knight shifts). The sensitivity depends on the natural abundances of the isotopes used. NMR is in most cases not suitable for trace analysis.

Moessbauer spectroscopy Moessbauer spectroscopy is an element-specific method which is based on the interaction of γ-rays with the nuclei of the atoms of a solid, which can be modulated by hyperfine interactions with the surrounding electrons. The location and intensity of the absorption peaks in the Moessbauer spectrum can give information about the coordination number, valency, spin state, and site distortion of the atoms under consideration and (if present) about the magnetic properties of the compound. With respect to NZVI characterization, Moessbauer spectroscopy is applied for determining not only the Fe(II)/Fe(III) ratio but also the Fe^0 content as well as the nature and content of the various (crystalline and amorphous) oxide phases (Kharisov et al. 2012). Limitations exist with respect to sensitivity; elements must be present at concentrations of at least 1–2 wt% (Rose et al. 2007). Most instruments require homogeneous dry sample powder.

Surface Elemental Characterization and Crystallography

X-ray photoelectron spectroscopy (XPS). Quantitative XPS analysis is a powerful tool for the elucidation of the surface composition, whereby dry samples are required. It is sensitive to the outer 3–5 nm of the sample. Nurmi et al. (2005) deduced, from the finding that Fe^0 in NZVI particles was detectable by XPS, that the oxide shell must be less than a few nanometers thick. Typical surface impurities of NZVI, such as sulfur in RNIP and boron in Fe^{BH}, have been detected by XPS (Nurmi et al. 2005). In the case of polymer- or surfactant-stabilized metal nanoparticles, XPS cannot work well if the dry samples required for XPS are prepared by evacuation of the dispersion of nanoparticles, because the stabilizer formed a thick organic coating on the nanoparticle surface. In the case of bimetallic nanoparticles, quantitative analysis by XPS can provide information on the types of elements present in the surface region.

2.2 Synthesis and Properties of Bare NZVI and Bimetallic NZVI

Synthesis Strategies

There are numerous synthesis routes for nanoscale zero-valent iron (NZVI) particles, of which only an incomplete selection can be presented here. When categorizing the synthesis methods for NZVI, the distinction between top-down and bottom-up methods (Fig. 2.7) is usually made (Yan et al. 2013); classification based on the

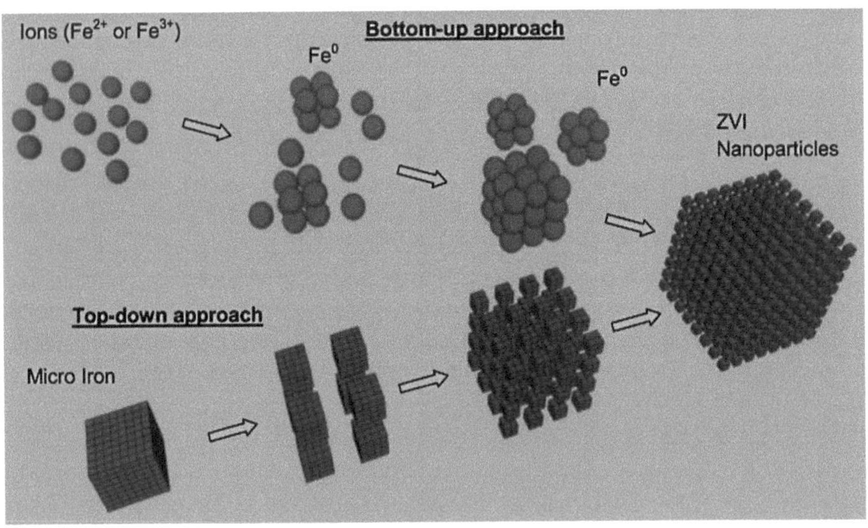

Fig. 2.7 Classification of NZVI production routes in bottom-up and top-down approaches. (Adapted from Li et al. (2009). Reprinted with permission from Li et al. (2009). Copyright (2009) Royal Society of Chemistry)

process type leads to the distinction between "physical," "physicochemical," "chemical," and "biological" methods (Kharisov et al. 2012). Independent of the synthesis route, iron metal or other iron-containing substances are precursors of the target particles, so that a classification according to source material type would also be logical. However, for production of NZVI intended for remediation purposes, gas-phase and liquid-phase reduction processes are most common.

Iron Metal as Precursor

Starting from iron metal, breakdown of the original metal body by high-energy ball milling, laser ablation, vaporization of metallic iron (thermal, laser-, arc-discharge-, or plasma-induced) in an inert atmosphere (He or Ar) followed by condensation/precipitation of iron clusters, and especially annealing lead to pure NZVI with α-Fe as the dominant Fe phase (Scott et al. 2010). Coprecipitation with other metals offers a synthesis route for generating bi- or heterometallic nanoparticles (Kharisov et al. 2012). Closely related to the above-described formation of NZVI via gas-phase-transported vapors is the metal organic chemical vapor deposition (MOCVD) technique whereby *thermally labile iron precursors* such as iron pentacarbonyl ($Fe(CO)_5$) are pyrolyzed, forming pure uniform NZVI particles (Choi et al. 2001). Thermal decomposition of $Fe(CO)_5$ can also be carried out in organic solvents by means of a sonolysis technique, such as application of high-intensity ultrasonic waves (Suslick et al. 1996). NZVI particles produced from clean metal or via the described pyrolytic routes are of high purity and have a narrow size distribution. However, this has to be paid for with considerable energy consumption and, when $Fe(CO)_5$ is used, with the handling of a highly toxic precursor.

Synthesis by H_2 Reduction

The NZVI type which has most often been applied in the field and used as reference material in the literature is the *reactive nanoscale iron particles (RNIPs)* supplied by Toda Kogyo, Inc., Onoda, Japan (more recently by Toda America Inc., Schaumburg, IL). RNIP has been produced from nano-goethite or nano-hematite. The oxidic iron precursor nanoparticles are commonly synthesized from iron scrap metal by dissolution in dilute H_2SO_4 and precipitation by pH adjustment. This makes their production pathway not only economically but also ecologically feasible. The precursor particles are treated in H_2 atmosphere at elevated temperatures (<600 °C); Eqs. 2.4 to 2.6 show the sequential reduction of hematite which is produced from goethite by heat dehydration in the first step in the temperature range between 250 and 350 °C. Often RNIP is referred to as Fe^{H2}, named after its reducing agent.

$$3\,Fe_2O_3 + H_2 \rightarrow 2Fe_3O_4 + H_2O \tag{2.4}$$

$$Fe_3O_4 + H_2 \rightarrow 3FeO + H_2O \tag{2.5}$$

$$FeO + H_2 \rightarrow Fe + H_2O \tag{2.6}$$

The thermal reduction process in hydrogen atmosphere forms the initially pyrophoric nanoparticles. The α-Fe formed is cooled and quenched with deionized water.

Fine pulverization (and breaking of particles aggregates) may be carried out by milling or applying mixing devices with shear forces. The product is usually supplied in water (Uegami et al. 2003). Freshly produced RNIPs show the classical Fe_3O_4-on-Fe^0 core-shell structure and have a Fe content of no less than 65% by weight, a particle size in the range of 50–300 nm, and a specific surface area between 7 and 55 m^2/g (Uegami et al. 2003). For the batch charges described more thoroughly in the literature, particle sizes between 20 and 70 nm and specific surface areas (N_2-BET) between 15 and 29 m^2/g have been reported by the groups studying the material (Phenrat et al. 2008; Saleh et al. 2007; Nurmi et al. 2005).

Reduction in Aqueous Solution

Because of its position in the electrochemical series, metallic iron is more difficult to generate in aqueous solution than, e.g., noble-metal catalyst nanoparticles such as Pd or Pt, which can be prepared simply by reduction with H_2. Classical wet reduction of ferrous (Fe^{2+}) and ferric (Fe^{3+}) salts requires borohydride, forming borohydride-iron or Fe^{BH} (sometimes also called Fe^B). Elliot et al. differentiate between Type I Fe^{BH} and Type II Fe^{BH}. The basic procedure published by Wang and Zhang (Type I Fe^{BH}, also called chloride method) is the reduction of ferric iron chloride by sodium borohydride at ambient temperature (Eq. 2.7) to produce NZVI with a particle size distribution such that >90% of particle mass is in the size range between 1 and 100 nm (Wang and Zhang 1997; Elliott et al. 2012).

$$4Fe^{3+}{}_{(aq)} + 3BH_{4(aq)}^{-} + 9H_2O \rightarrow 4Fe^0{}_{(s)} + 3H_2BO_{3(aq)}^{-} + 12H^+{}_{(aq)} + 6H_{2(g)} \quad (2.7)$$

When starting from an Fe(II)-salt, less borohydride is necessary for the reduction to zero-valent iron. With Type II Fe^{BH} starting from iron(II) sulfate (also called the sulfate method), additionally upscaled production of Fe^{BH} is possible.

$$2Fe^{2+}{}_{(aq)} + BH_4{}^-{}_{(aq)} + 3H_2O \rightarrow 2Fe^0{}_{(s)} + H_2BO_3{}^-{}_{(aq)} + 4H^+{}_{(aq)} + 2H_{2(g)} \quad (2.8)$$

The stoichiometric excess of borohydride used for Type II Fe^{BH} synthesis is lower than for Type I Fe^{BH} (3.6–5.6, respectively). Numerous modifications of the borohydride method have been described where the precursor salt, the solvent, and the borohydride type were varied (Capek 2004; Liu et al. 2005a, b; Ponder et al. 2000; He and Zhao 2007). When carrying out borohydride reduction in the presence of supports, pathways are also open to generate NZVI structures on various supports. This will be described in more detail in Sect. 2.4 (e.g., Ponder et al. 2000). Yan et al. (2013) stated that the borohydride method is easy to apply and especially useful for synthesis of NZVI with various desired physical properties since the particle generation in the presence of surfactants or polyelectrolytes allows control of the size, size distribution, and core-shell morphology. However, some disadvantages remain to be considered. The borohydride method is not able to produce pure NZVI: the mainly amorphous Fe^{BH} always contains boron impurities. In addition, Fe^{BH} is quite expensive (over $200 per kg NZVI) due to the need for the rather costly reducing agent $NaBH_4$ (Yan et al. 2013). Moreover, large-scale production suffers from the considerable hydrogen generation and the large amount of wastewater produced during the separation steps.

As an alternative to borohydride, dithionite can be used to reduce Fe(II) and produce NZVI under conditions of high pH and in the absence of oxygen (Eq. 2.9) (Sun et al. 2008).

$$Fe^{2+} + S_2O_4^{2-} + 4OH^- \rightarrow Fe^0_{(s)} + 2SO_3^{2-} + 2H_2O \quad (2.9)$$

The authors point out that the dithionite-NZVI can be much more cheaply produced than borohydride-NZVI, in addition avoiding the high amount of hydrogen generated by the borohydride method. The reactivity of dithionite-NZVI toward TCE is reported to be equivalent to that of borohydride-NZVI. In contrast to the spherical borohydride-NZVI, dithionite-NZVI forms small, thin, black, platelet-like crystals (Fig. 2.8b) that appear to be a mixture of very small Fe^0 particles.

Recently, techniques for obtaining nanoparticles using naturally occurring reagents, such as plant extracts, have also been further developed. The key active reagents in plant extracts are believed to be polyphenols present, for example, in tea leaves, wine and winery waste, and red grape pomace (Kharissova et al. 2013). The biomaterials are extracted by a water-alcohol mixture at elevated temperature, and the polyphenols released serve as mild green-reducing agents. Polydispersed colloids of nanoscale zero-valent iron (NZVI) particles are reported to have been prepared using tea (sp. *Camellia sinensis*) polyphenols (Hoag et al. 2009) and those extracted from a large variety of tree leaves (Machado et al. 2013). Advantages of the biogenic reducing agents are seen not only in their simple, cost-effective, and relatively reproducible application but also in that the NZVI results in comparatively more stable particle suspensions due to the chelating effect of the plant extract polyphenols (Wang 2013).

Bimetallics

There are several synthesis options to produce bimetallics:

(i) Doping of a ready-prepared NZVI particle with noble metal salt and reduction of the more precious metal on the iron surface according to the example shown for Pd deposition on NZVI (Eq. 2.10) (Wang and Zhang 1997).

$$Pd^{2+} + Fe^0 \rightarrow Pd^0 + Fe^{2+} \quad (2.10)$$

In a similar analogue way, bimetallics in the compositions Fe/Pt, Fe/Ag, Fe/Ni, Fe/Co, and Fe/Cu can be prepared (Xu and Zhang 2000; Zhou et al. 2010). This forms clusters, or even coatings, according to the core-shell type at the outer surface of the NZVI particles (Fig. 2.9a, b).

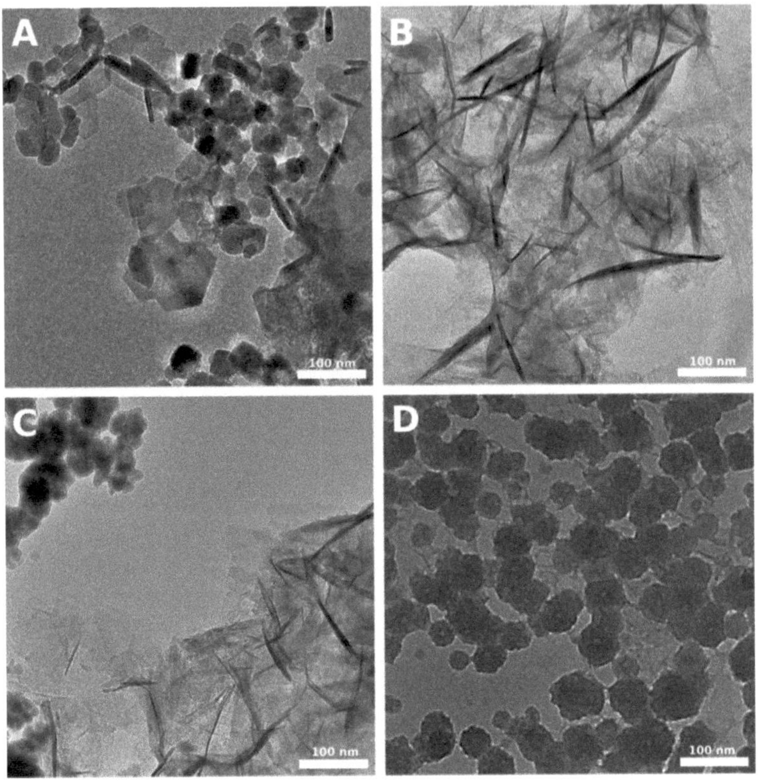

Fig. 2.8 TEM images of freshly synthesized NZVI samples: (**a**) and (**b**) NZVI were synthesized using ferrous sulfate and ferric chloride as precursors and sodium dithionite as reducing agent, respectively, while (**c**) and (**d**) were formed by using ferrous sulfate and ferric chloride as precursors and using sodium borohydride as reducing agent. (Reprinted with permission from Rónavári et al. (2016). Copyright (2016) Elsevier)

(ii) Coprecipitation of iron simultaneously with other metals offers a synthesis route to generate bi- or heterometallic alloyed nanoparticles (Fig. 2.9c) (Kharisov et al. 2012). Reduction of and coprecipitation of bimetals using borohydride in aqueous solution lead to bimetallics (Toshima and Yonezawa 1998). In order to adjust particle size, bimetallic nanoparticles can also be generated in the presence of a suspension stabilizer, such as CMC.

The formation of true nano-alloys can be achieved by co-deposition using two or more thermally labile precursors (via MOCVD, CVC, or solvent-mediated thermal destruction). MOCVD of single bimetallic precursors (such as mixed carbonyls of the type $[NEt_4][FeCo_3\text{-}(CO)_{12}]$) leads to defined iron nano-alloys such as $FeCo_3$, FePt, $FeNi_4$, and Fe_4Pt (Robinson et al. 2009).

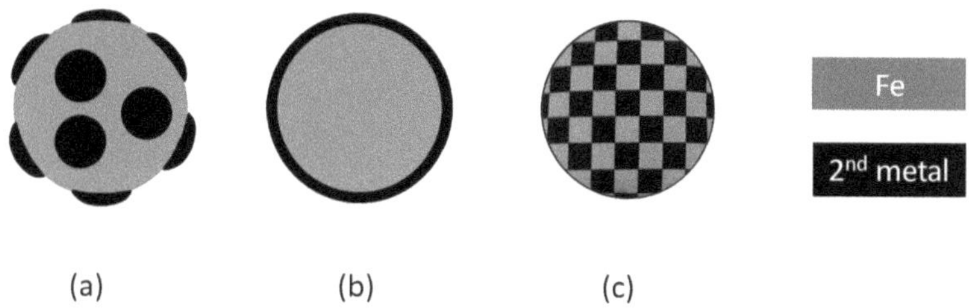

Fig. 2.9 Simplified scheme of various bimetallic types: (**a**) clustered dopant metal on iron, (**b**) full core-shell structure, (**c**) bimetallic alloy

(iii) Ball milling of both metals simultaneously forms bimetallics whereby composition and distribution of the metals can be controlled by milling intensity and time and by the precursor stoichiometry (Xu et al. 2012).

Bimetallics are composed of two different metals, where the atomic distribution may have a significant influence on their catalytic performance. However, it is seen still as a great challenge to understand the nucleation and growth mechanism of bimetallic nanocrystals (Liu et al. 2012). Figure 2.9 gives a simplified picture of bimetal structures. Exchange of metal atoms by segregation and alloying will form intermediate structures between a, b, and c.

Palladium (with <1 wt%) is the most often used metal dopant for iron bimetallics used for environmental application (Wang and Zhang 1997; Elliott and Zhang 2001; Lien and Zhang 2007). Palladium's ability to catalyze efficiently the hydrodechlorination of not only aliphatic but also aromatic chlorohydrocarbons immensely extends the spectrum of treatable pollutants. In addition, reaction rates of Pd-catalyzed dechlorination can extend that of iron by several orders of magnitude. Less dead-end products are expected than with pure NZVI. When using Pd, the changed reaction pathway almost completely avoids the formation of the toxic dehalogenation products VC, 1,1-dichloroethylene (1,1-DCE), cis-dichloroethylene (cis-DCE), and trans-dichloroethylene (trans-DCE), which are formed only in trace amounts (Schrick et al. 2002). However, Pd decreases the Fe lifetime due to higher corrosion rates and is subject to catalyst poisoning (Zhu and Lim 2007; Korte et al. 2000). Moreover, the reactive lifetime of Pd catalyst under in situ conditions may be limited, bringing into question the efficacy of bimetalics.

Properties

Maintaining pristine iron surfaces is difficult. NZVI particles rapidly develop a typical core-shell structure since it is almost unavoidable that an oxidic shell forms around the reactive iron core, whereby the composition of the shell can vary depending on the particle's immediate environment but mainly consists of mixed valent iron oxides (Fe^{2+} and Fe^{3+}) (Figs. 2.10b and 2.11b). RNIP particles are a typical example where the shell is intentionally formed. The pyrophoric properties of the freshly produced particles are especially prominent for the hydrothermal synthesis pathway. Controlled quenching of this "excessive reactivity" by partial oxidation

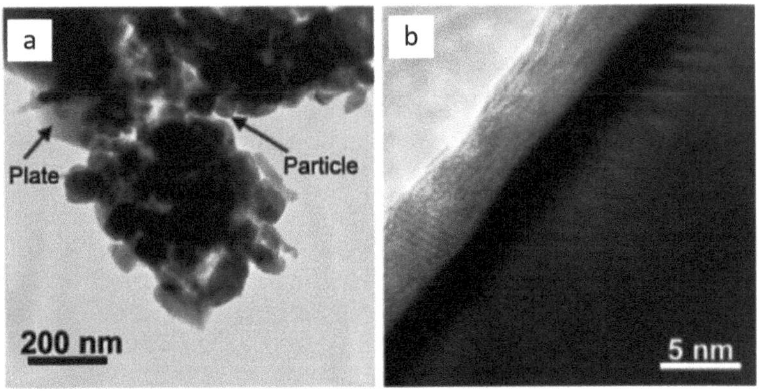

Fig. 2.10 TEMs of Fe^{H2}. (**a**) Low-magnification image of as-received sample. (**b**) Higher resolution image of oxide shell around metal particle (Nurmi et al. 2005). (Reprinted with permission from Nurmi et al. (2005). Copyright (2005) American Chemical Society)

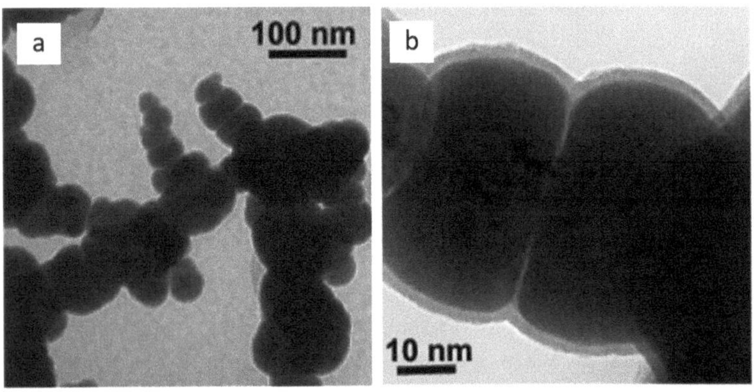

Fig. 2.11 TEMs of Fe^{BH}. (**a**) Low-magnification image of as-received sample. (**b**) Higher resolution image of metal particle (Nurmi et al. 2005). (Reprinted with permission from (Nurmi et al. 2005). Copyright (2005) American Chemical Society)

in water or an oxygen-containing gas phase forms a magnetite shell around the α-Fe core. It is believed that the oxidation of the iron core is the driver for the NZVI reactivity (Liu et al. 2005a, b; Nurmi et al. 2005).

The particle structure, composition, and reactivity parameters have been analyzed by Nurmi et al. and Liu et al. (2005b) for the two most widely used NZVI particle types (Nurmi et al. 2005; Liu et al. 2005b). In this study, RNIP (Fe^{H2}) and NZVI generated by reductive precipitation with borohydride (Fe^{BH}) were compared. Main differences were not only the initial Fe^0 content (for Fe^{H2} $x_{Fe0} \approx 70$ wt% and for Fe^{BH} $x_{Fe0} \approx 97$ wt%) but also the mean crystallite size for Fe^0 (α-phase), which was found to be one order of magnitude larger for Fe^{H2} (for Fe^{H2} $d_{Fe\ crystal} \approx$ 40 nm and for Fe^{BH} $d_{Fe\ crystal} < 1.5$ nm). Fe^{BH} consists of agglomerates of such small crystallites, forming 20–80 nm metallic Fe particles. The oxidic shell around Fe^{BH} contains a high amount of oxidized boron. In the magnetite shell of Fe^{H2}, reduced sulfur remained from the production process (originating from the scrap metal dissolution with H_2SO_4), and some other elements, e.g., Na and Ca, were found by X-ray photoelectron spectroscopy (XPS) (Nurmi et al. 2005). The impurities in NZVI particles, such as boron or sulfur, but also oxide and hydroxide, which are incorporated during synthesis, usually increase the amorphous iron content and presumably have an influence on the reactivity. Sulfur is claimed by the producer to increase the reactivity and boron is believed to be the reason for selectivity changes (Uegami et al. 2003; Liu et al. 2005b).

Due to their magnetic properties, NZVI particles strongly tend to form agglomerates. TEM data given by Nurmi et al. (2005) show this phenomenon for Fe^{H2} and Fe^{BH} (Figs. 2.10a and 2.11). Fe^{H2} is composed of aggregates of facet-like ordered platelets and smaller irregular particles, whereas the more spherical Fe^{BH} particles aggregate in a chain-like structure. Li et al. see in addition to particle size the core-shell structure, the composition, the specific surface area, and the magnetic properties as further important characteristics of NZVI particles (Li et al. 2006a). However, the latter is usually just seen as a nuisance when it comes to environmental application, since the magnetic attractive forces are largely responsible for the high agglomeration tendency of NZVI particles. Later in this chapter, various means of particle "construction" are discussed which counteract this behavior.

Obviously, NZVI is able to exhibit higher rates of reaction with contaminants compared to its larger-scale pendants, simply due to its larger specific surface area or higher surface-to-volume ratio. Specific surface areas are determined from N_2 physisorption measurements using dry particles (BET analysis). One could assume that dry particle aggregates vary in their specific surface areas from well-dispersed particles in suspension, a problem which can be expected to become more pronounced as particle size decreases and more contact area is available. Microscale iron particles with a mean diameter of <100 mesh (<150 μm) (fisher electrolytic iron) and typical Fe^{BH} (d_{mean} = 30–50 nm) differ in their N_2 BET-surface areas by more than two orders of magnitude: 0.18 m^2/g and 36.5 m^2/g, respectively (Lowry and Johnson 2004).

2.3 Synthesis and Properties of Polymer-Modified NZVI

A great variety of surface modifications of bare NZVI particles have been conducted which all followed the goal of overcoming the high tendency toward self-aggregation (due to NZVI's high surface energy and magnetic attractive forces). Surface coatings are utilized to lower subsurface transport limitations or to improve the source treatment abilities. They can provide protection, separation distance, and charge (positive or negative) to the particles. The materials used for coating NZVI can be divided into three classes: polyelectrolytes, surfactants, and polymers (Lowry

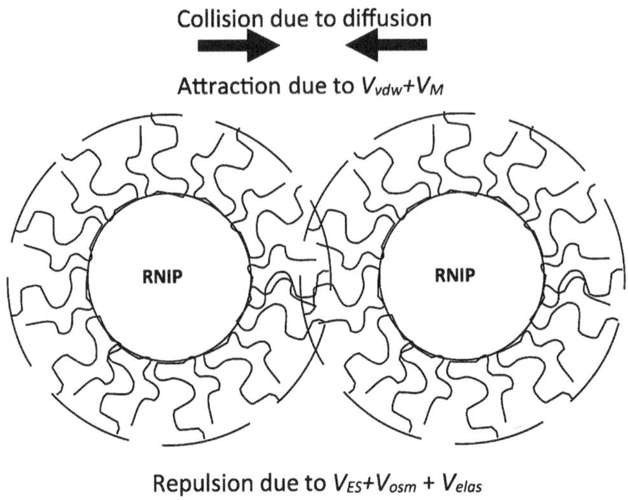

Fig. 2.12 Effects of surface coating. (Reprinted with permission from Phenrat et al. (2008). Copyright (2008) Springer Nature)

2007). For each of the coating material classes, both natural and synthetic varieties are available. In most cases, the coating strategies aim at hindrance of direct particle-particle and particle-surface interactions in order to combat agglomeration and particle interception. Some approaches of particle modification aim at targeted delivery to source zone areas and will be briefly discussed later in this section. Stabilization of NZVI structures in composites with *solid* polymer support particles (e.g., poly(vinyl alcohol), ferragels, alginate) is discussed in detail in Sect. 2.4.

Due to their small size, *surfactants* can solely offer electrostatic repulsive or attractive forces (charge-dependent) for transport enhancement, whereas *polyelectrolytes* are functionalized macromolecules and can provide electrosteric repulsion on account of their charge and steric demand (Fig. 2.12). Adsorption of surfactants is reversible and underlies a sorption/desorption equilibrium, whereas coatings of polyelectrolyte and nonionic *polymers* usually provide a longer-lasting modification.

There are several strategies to provide *polymer* surface coating of particles after their preparation. In the presence of dissolved polymeric suspension stabilizers, particle surfaces can be polymer-coated simply by (i) *adsorptive surface coverage*. (ii) The *formation of covalent bonds* between the polymer and surface functional groups of the particle is a method to anchor the coating material better to the particle surface. These coatings show essentially irreversible attachment and, thus, should be more suitable as long-lived delivery systems for groundwater remediation. Covalent bonding of a polymer to the particle surface can be also achieved by utilization of constituents of the *particle surface as polymerization aid*. Particle-aided polymerization is chosen to construct bonded polymer coatings directly at the surface with the aim of irreversible attachment.

(i) *Adsorptive surface coverage*

Dissolved biopolymers, such as carboxymethyl cellulose (CMC) (Raychoudhury et al. 2010; Krol et al. 2013; Johnson et al. 2013), xanthan gum (Dalla Vecchia et al. 2009), guar gum (Tiraferri et al. 2008), or polyelectrolytes, e.g., polystyrene sulfonate and polyaspartate (Phenrat et al. 2008), and polyelectrolyte mixtures (Hydutsky et al. 2007), are subject to surface adsorption, forming thin polymer coats. Naturally, sorption/desorption equilibria apply. Thus, the adsorption is reversible when solution conditions change, e.g., by mixing and dilution of the injected suspension with groundwater. The modification of the NZVI surface properties is achieved by increasing not only the surface charge (electrostatic stabilization) but also the steric demand, e.g., of long hydrophilically functionalized chains, which lead to prevention of particle agglomeration (steric stabilization).

In most studies, pre-grafted particles were brought into contact with low-concentrated polymer solution (in the order of milligrams per liter). However, particle synthesis in the presence of a polymer solution also leads to coated particles (post-grafting). A study of the four polymers carboxymethyl cellulose (CMC), poly(styrenesulfonate) (PSS), poly(acrylic acid) (PAA), and polyacrylamide (PAM) in pre- or post-grafting approaches has shown that, depending on the synthesis conditions, the particle size can be adjusted and thus has a major influence on particle reactivity and transport properties of NZVI (Cirtiu et al. 2011). When increasing the concentrations of the polymers to several grams per liter, the polymer not only adsorbs to the particles' surface, but can also modify the viscous properties of the suspension, which can lead to a dramatic decrease in Brownian motion (viscous stabilization). This limits the frequency of collisions among particles and thus the particle aggregation and sedimentation (Cantrell et al. 1997; Li et al. 2015; Gastone et al. 2014).

However, post-grafting of polymer-coated particles can only be achieved for bottom-up processes. In aqueous solution, this is only possible for processes using $NaBH_4$ as reducing agent (e.g., as in Cirtiu et al. 2011). In addition, the thermal decomposition of $Fe(CO)_5$ in a CVD process in the presence of monomers (e.g., styrene) allows particle formation and polymerization at the same time (Yang et al. 2003).

(ii) *Formation of covalent bonds*

By formation of strong covalent bonds between a polymer chain and the surface atoms, polymer layers can irreversibly bind to particle surfaces. However, it has been demonstrated that poly(ethylene oxide) brushes covalently bound to engineered nanoparticles are bioavailable and can support the growth of microorganisms (Kirschling et al. 2011). There are two approaches typically employed for covalent bonding of polymers to NPs. One is a "grafting to" method, involving the reaction of reactive functional groups in the polymer with complementary active groups on the particle surface. In general, the grafting density of "grafted to" polymer chains is limited because of the steric hindrance effect. Another method is a "grafting from" approach, involving the use of active species on the surface to initiate polymerization (particle-aided polymerization) and usually resulting in polymers with high grafting density. Blacha et al. summarize methods for covalent polymer bonding to surfaces (Blacha et al. 2011). Atom transfer radical polymerization (ATRP), ring-opening metathesis polymerization (ROMP), and ionic polymerization are reviewed. Compounds reactive to iron(oxide) phases, such as halogenosilanes, alkoxysilanes, thiols, and disulfides, can chemisorb to NZVI surfaces to form the covalent linkage between the organic building block (polymer) and the surface. This formation of self-assembled monolayers (SAM) is a spontaneous process. Strong coordinative interactions were also reported for poly(sodium acrylate) and poly(sodium vinylphosphonate) on iron oxide surfaces (Golas et al. 2010). Type, molecular weight distribution, and charge determine the stabilizing effect for magnetite nanoparticles for use under various pH and ionic strength conditions. Golas et al. recommend to use strong polyanions, such as sulfonated poly (hydroxyethylmethacrylate) and poly(styrene sulfonate), under neutral to acidic conditions whereas polyanions capable of forming coordinative bonds with the iron oxide surface, such as poly(sodium vinyl phosphonate) and poly(sodium acrylate), should be used under basic conditions. The behavior found for nanomagnetite particles should be applicable to commercial NZVI as well.

Particle Modification Aimed at Targeted Delivery to Source Zone Areas
Delivery approaches designed to target regions containing entrapped and pooled DNAPL are important innovations for source remediation using reactive iron particles. In order to accomplish effective delivery of NZVI to source zones, a delivery system is required that allows for dispersibility and colloidal stability of individual nanoparticles in water but at the same time provides an affinity for the water/contaminant interface. Due to the hydrophilic surface properties of NZVI, the source-zone compatibility of NZVI is rather poor. For this reason, organic polymer coatings for improving source-zone targeting have received considerable interest.

Saleh et al. (2005, 2007) showed that amphiphilic triblock copolymers with an A-B-C triblock architecture enhanced colloidal stability and increased affinity of the coated NZVI toward a water/organic interface. The triblock copolymers, poly (methacrylic acid)-b-(methyl methacrylate)-b-(styrene sulfonate) (PMAA-PMMA-PSS) (see Fig. 2.13a), contain both hydrophilic and hydrophobic blocks to promote nonpolar interactions. They were produced using atom transfer radical polymerization (ATRP) in conjunction with a post-polymerization ester-hydrolysis step and a post-polymerization sulfonation step. The mechanism of action for triblock copolymers in two-phase systems is based on the amphiphilic nature of the polymer construction: in water, the hydrophobic PMMA block collapses, while the hydrophilic PSS fully extends into the water bulk phase, providing electrosteric protection (Fig. 2.13b). Conversely at the NAPL/water interface, the hydrophobic PMMA block conveys hydrophobic affinity to the organic phase, thus anchoring the particle at the interface (see Fig. 2.13c). The triblock copolymer-coated particles do not enter the NAPL due to the weak solubility of the PSS block in the organic phase (Saleh et al. 2005).

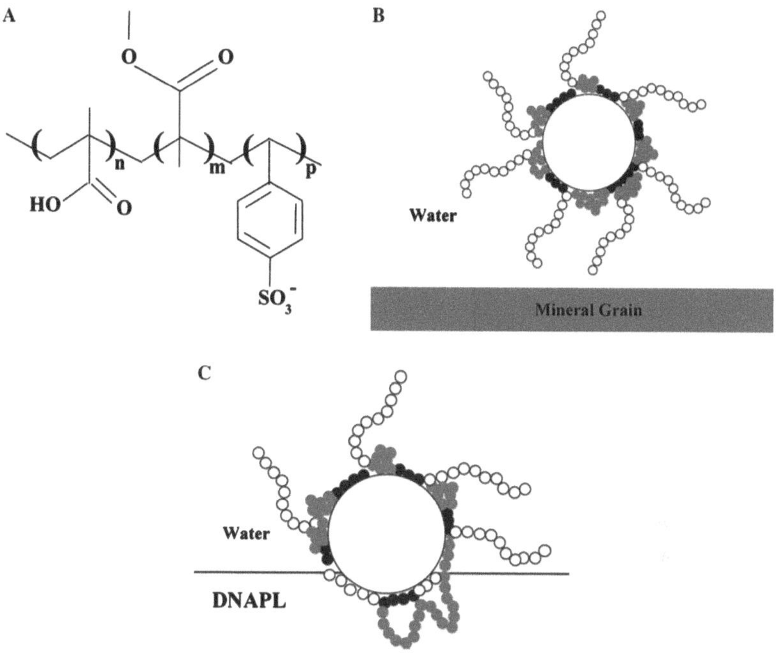

Fig. 2.13 Structure of amphiphilic triblock copolymers (**a**), leading to electrosteric stabilization in water (**b**) and anchoring at the DNAPL/water interface, with poly(methyl methacrylate) anchoring block (black circles in **b** and **c**), hydrophobic poly(methyl methacrylate) block (gray circles in **b** and **c**), and polystyrene sulfonate polyelectrolyte block (open circles in **b** and **c**); see text for further explanation. (From Saleh et al. (2007))

A similar concept was followed by Krajangpan et al. with functionalized amphiphilic polysiloxanes (Krajangpan et al. 2008, 2012). The synthesis of amphiphilic polysiloxane graft copolymers (APGC) is achieved through hydrosilylation of hydride-functioning polysiloxanes by polyethylene glycol (PEG) and tertiary butyl acrylate (Krajangpan et al. 2012). The PEG grafts are water-soluble and allow for colloidal stability and dispersibility in aqueous mediums. The polysiloxane backbone polymer imparts a hydrophobic barrier to the NZVI particles which protects them from excessive oxidation. It also causes an affinity of the coated NZVI for the water/contaminant interface and allows for permeation of the contaminants to the NZVI surface (Krajangpan et al. 2008, 2012). Treatment of NZVI with APGCs was found to enhance nanoparticle colloidal stability in water; the magnitude of the enhancement was a function of APGC chemical composition.

Characterization of Particle Coatings

Particle diameter, surface properties, coating thickness, and the influence on the colloidal stability or wettability by various solvents are of interest.

- The particle size measurement can in principle be carried out as for uncoated particles. However, in some cases (i.e., where Brownian movement plays a role, as for DLS and nanoparticle tracking analysis – NTA), the viscosity of the dispersant fluid could markedly alter the result and has to be taken into consideration.
- For the analysis of the mass of polymeric coating material attached to the particles, their carbon content is a good basis for measurement. After particle separation from the suspension (without disturbance of the polymer-particle sorption equilibrium), thermogravimetric (TGA) or elemental analyses (EA) allow quantification of the attached coating material.
- The influence of the coating material on the surface properties may reflect in a changed zeta potential. The measurement can be carried out similarly to that for bare particles with the caveats about the impacts of coatings as described above.
- Characterization of the coating can be conducted by direct visualization using atomic force microscopy or scanning tunneling microscopy. Indirect evidence of organic layer formation is collected based on a signal from specific surface functional groups, as in the case of electrochemical (voltammetry and impedance measurements) and spectroscopic (Raman, UV-Vis, and IR) methods (Blacha et al. 2011).

The great variety of NZVI modification approaches shows that each can provide advantages but none is a universal remedy. They all have their assets and drawbacks – hardly surprising considering the great range of different remediation requirements and situations found in the field. Besides improvements of particle delivery, interference with NZVI reactivity toward target contaminants and also toxicological aspects of synthetic polymer amendments need to be considered. The uncertainties surrounding the effects of surface coatings on the fate, transport, and potential toxicity of engineered nanomaterials makes this a rich area for current and future research on nanotechnology (Lowry 2007).

2.4 Synthesis and Properties of NZVI on Supports

The combination of NZVI or bimetals with support materials, focusing on environmental applications, mostly aims at improved transport and/or reactivity properties of the composites compared to bare NZVI. Bringing iron onto support materials is done in order to finely disperse the metal, to sterically hold the single iron nanoclusters apart from each other (spacer effect), and to generate composite particles which have properties from both the iron and the support material. Embedding NZVI particles in a porous support such as silica (Fig. 2.14a, c, or d) also reduces their magnetic attraction due to spatial separation of the iron structures. In these cases, where the iron is "hidden," particle surface properties are dominated by the support material. Covering the iron surface may also protect the iron to a certain extent from excessive oxidation.

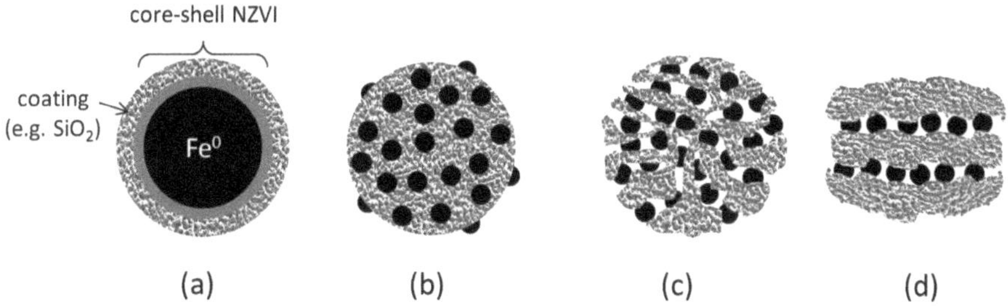

Fig. 2.14 Various types of NZVI-support combinations (**a** = coated NZVI (view enlarged), **b** = nanoiron clusters attached to the outer support surface, **c** = nanoiron clusters within porous supports, **d** = intercalated nanoiron clusters)

The combination of Fe^0 nanoparticles with other materials has been reported for several applications, whereby Fe^0 has been entrapped or surface-supported. NZVI can be:

- Coated with solids or polymers (see also Sect. 2.3) (Fig. 2.14a)
- Supported on the outer surface of solid particles (Fig. 2.14b)
- Integrated into other materials as composites or intercalates (Fig. 2.14c, d, respectively)

Coating shells may enhance the lifetime of the metal but also provide a spacer function between the iron nanoparticles, which prohibits their direct contact. This in turn lowers the tendency for agglomeration and improves the suspension stability. The materials reported for *embedding* of single particles are polymers (Saleh et al. 2005, 2007; Phenrat et al. 2011), metal oxides such as SiO_2 (Li et al. 2012; Wan et al. 2013), and increasingly carbon materials (Hoch et al. 2008). The coating of iron with polymers is discussed previously; here we will concentrate on iron-mineral and iron-carbon composites as well as iron on solid polymer particles.

The *transport and sedimentation behavior* can be influenced by the combination with a second material due to:

- Changed surface charge
- Larger size of primary particles
- Lower particle effective density
- Reduced agglomeration tendency due to separating matrix (shallower secondary minimum)

The *reaction* of NZVI in composites can be influenced by:

- Iron surface coverage
- Electronic effects of the second material on iron
- Possible charge transfer abilities of the carrier
- Hydrophobicity of the carrier and contaminant enrichment
- Diffusive contaminant delivery to iron surface

For enhanced suspension stability and subsurface mobility, the combination of iron with a porous low-density support offers the possibility to "design" optimized NZVI particles. Some of the multiple factors determining subsurface transport and stability of NPs can be actively influenced for engineered NPs. The surface properties of each material forming the composite contribute to the properties of the whole particles. A coating, for instance, predominantly determines the surface properties. Negative surface charge, as provided, e.g., by SiO_2, carbon materials, or accordingly functionalized polymers, can help to overcome iron's electrostatic attraction to the sediment grain during the aquifer passage. The coating and the carrier therefore have protecting and transport-improving functions. Schrick et al. (2004) showed, in a study with various carrier materials and sediments, that transport and reactivity can vary according to the compatibility between support and sediment properties.

Particle density and accordingly particle size also influence the sedimentation velocity (Stokes equation) and the subsurface transport (filtration) behavior of the composite particles. The effective density of a porous particle ($\rho_{p,w}$) such as composites of NZVI on a porous support in water (i.e., under conditions of fully water-filled pores) can be estimated according to

$$\rho_{p,w} = \frac{\rho_w + \frac{1}{V_{p,pore}}}{\frac{1}{\rho_{p,s} \times V_{p,pore}} + 1} \tag{2.11}$$

where $V_{p,pore}$ is the specific pore volume (mL g^{-1}) of the particles as determined from gas (e.g., N_2) adsorption/desorption measurements, ρ_w is the density of water, and $\rho_{p,s}$ is the true density of the particle's solid fraction. The latter can be determined by gas pycnometry. Alternatively, if the true density of all components including the porous support is known, $\rho_{p,s}$ for the solid composite fraction (consisting, e.g., of NZVI, magnetite, and a porous solid carrier) is estimated according to

$$\rho_{p,s} = \frac{1}{\sum \frac{x_i}{\rho_i}} = \frac{1}{\frac{x_{Fe(0)}}{\rho_{Fe(0)}} + \frac{x_{Fe3O4}}{\rho_{Fe3O4}} + \frac{x_{carrier}}{\rho_{carrier}}} \tag{2.12}$$

from the mass fractions (x_i, related to dry particle mass) and densities of the individual components ρ_i.

Synthesis Strategies

The synthesis pathways for these material combinations are manifold, and we see a seamless transition between them. Often similar strategies are followed as for the synthesis of pure iron nanoparticles (Sect. 2.2). For generation of coated NZVI particles, the bare core iron is usually synthesized first and can be subsequently coated (Fig. 2.15). Metal oxide coating (e.g., SiO_2, TiO_2, ZrO_2, Al_2O_3) can be applied not only by sol-gel polycondensation or impregnation techniques (Yuan et al. 2010) but also by arc discharge (Fernandez-Pacheco et al. 2006). For these techniques and for NZVI generated in the presence of coating material, the composites appear as core-shell structured particles.

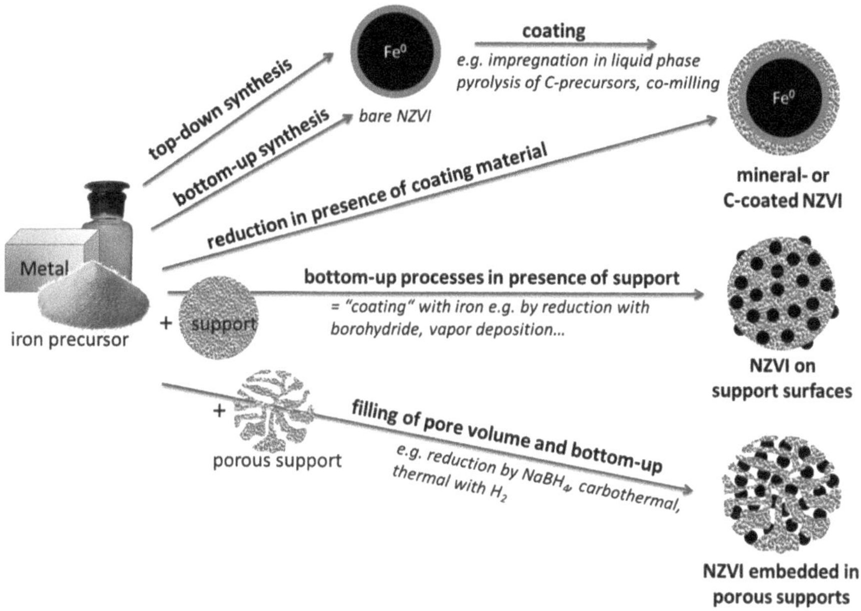

Fig. 2.15 Principal synthesis pathways for iron-mineral, iron-carbon, and iron-polymer composites

Generation of surface-supported NZVI or embedded iron in a porous carrier requires bottom-up processes employing iron precursors which precipitate at the outer surface or are introduced into the pore volume of the carrier. Reduction by borohydride or H_2 at elevated temperatures leads to NZVI structures. Also chemical vapor deposition (e.g., using $Fe(CO)_5$) is a means to introduce iron structures to inner and outer carrier surfaces (Chen et al. 1987). These synthesis strategies are applicable for all kinds of solid supports.

Metal Oxides as Coatings and Supports

Especially for inorganic contaminants, where hydrophobic enrichment does not play a role, high-surface-area metal oxides or natural minerals may be beneficial as support materials for iron. Immobilization of NZVI on solid particles, such as silica (Ponder and Mallouk 2004; Dror et al. 2012; Martínez-Baez et al. 2015), or clay minerals, such as bentonite (Zhang et al. 2011), montmorillonite (Jia and Wang 2013; Bhowmick et al. 2014), or kaolin (Wang et al. 2016), has been reported.

Metal-oxide carriers are commonly regarded as inert support materials. However, Boudart et al. (1975) found that, for instance, magnesium oxide is not merely an inert carrier for metallic iron, but interacts with the iron in a manner essential for the ultimate production and stabilization of small metallic particles. The results of Cr (VI) removal experiments using clay-supported NZVI revealed that the removal rates were strongly related to NZVI surface dispersity (Zhang et al. 2013). ZVI intercalated in the interlayers of montmorillonite is more stable than that located on the external surface, which can be attributed to the protective effect of the clay layers

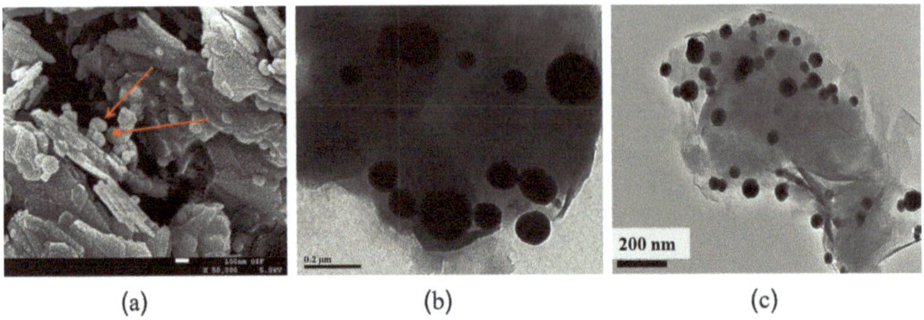

Fig. 2.16 Examples for Fe-clay composites: (**a**) SEM images of kaolin-supported NZVI, red arrows = spherical Fe structures within kaolin layers (Wang et al. 2016), (**b**) TEM of bentonite-supported NZVI (Zhang et al. 2011), and (**c**) TEM of montmorillonite-supported NZVI (Zhang et al. 2013). (Reprinted with permission from Zhang et al. (2011, 2013). Copyright (2011 and 2013) Elsevier)

against metal oxidation (Jia and Wang 2013). Also in the case of pillared bentonite, the carrier was found to play a significant role in enhanced reactivity and stability of NZVI; the authors state that this finding may shed new light on design and fabrication of supported NZVI for environmental remediation (Li et al. 2012). The composite's Cr(VI) removal efficiency was much higher than that for the sum of NZVI reduction and bentonite adsorption.

The composite particles formed by supporting NZVI are usually considerably larger than NZVI itself, determined by the carrier particle size (Fig. 2.16). Particle sizes ≈ 1 μm seem beneficial according to Tufenkji and Elimelech (2004) for this type of material, being quite similar to natural sediment particles.

Zheng et al. (2008) showed attachment of ZVI nanoclusters to porous submicron silica particles using an aerosol-assisted process and subsequent reduction with $NaBH_4$ (Fig. 2.17a) or H_2 (Fig. 2.17c). The figure clearly shows the spatial separation of the NZVI structures by the support matrices inhibiting the direct interaction (agglomeration) of NZVI particles.

In contrast, coating NZVI with metal oxides offers a means to generate individual NPs where the thickness of the coating determines on the spacer distance between the iron species. Figure 2.18 shows an example for silica-coated NZVI particles.

Carbon Materials as Coating and Supports

The combination of sorption-active carbon materials such as activated carbon (AC) and iron as reactive metal is a known method to prolong the lifetime of carbon adsorption beds by continuous AC regeneration (Tseng et al. 2011). However, for in situ groundwater treatment, the combination of NZVI with carbonaceous materials offers features additional in comparison to iron: the sorptive enrichment of hydrophobic contaminants and a higher affinity of the composites to organic solvent phases (Simon 2015). With carbon supports, especially with high-surface-area sorption-active carbon materials, the enrichment of organic halogenated pollutants and their degradation by iron can be coupled. The

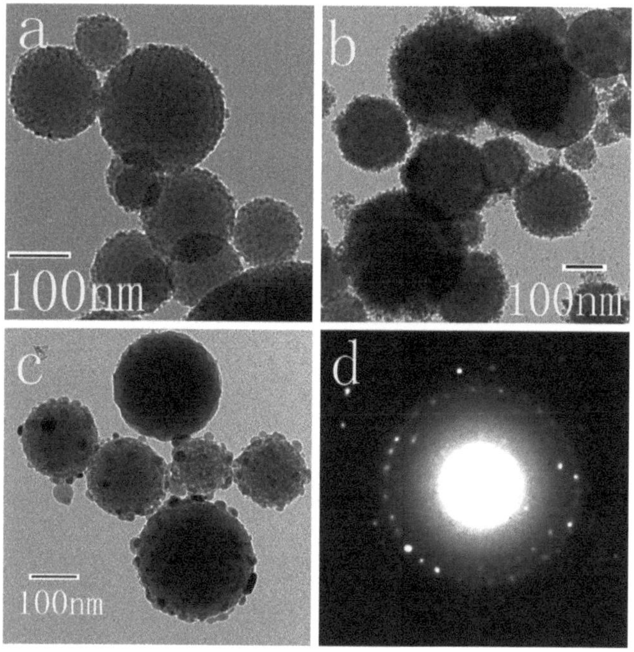

Fig. 2.17 TEM images of (**a**) Fe(B)/silica reduced by $NaBH_4$, (**b**) Fe(B)/ethyl-silica reduced by $NaBH_4$, (**c**) Fe(H)/silica reduced by H_2, and (**d**) electron diffraction pattern of Fe(H)/silica, illustrating polycrystallinity in the sample (Zheng et al. 2008). (Reprinted with permission from Zheng et al. (2008). Copyright (2008) American Chemical Society)

Fig. 2.18 Silica-coated NZVI (Li et al. 2012). (Reprinted with permission from Li et al. (2012). Copyright (2012) Elsevier)

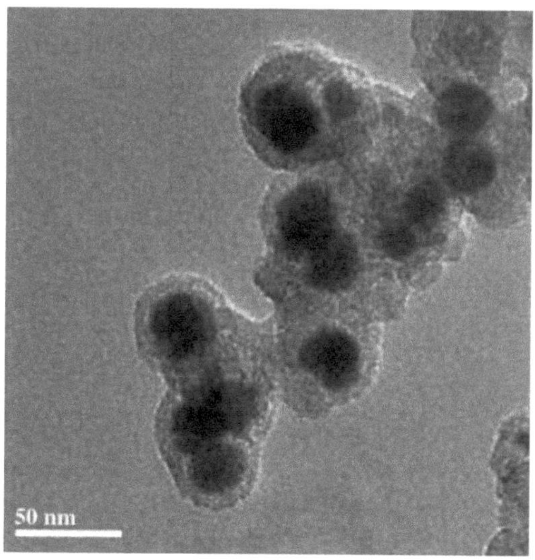

Table 2.3 Examples for in situ compatible NZVI-C composites

Carbon carrier	Description	Reference
Graphite	Graphite-encapsulated Fe^0 Laboratory-tested for treatment of lead or Cr(VI)-contaminated water	Cao et al. (2007)
Carbon black	Fe^0on 80 m^2/g carbon black from Fe salt impregnation and carbothermal reduction Fe^0 content 12...35 wt% Laboratory-tested for Cr(VI)	Hoch et al. (2008)
Activated carbon	Named "Carbo-Iron" Carrier colloidal AC (≈ 1 μm) impregnated with iron salt and reduced either by H_2 (500 °C) or carbothermally (700 °C) Fe^0 content 20...30 wt% Field-tested for PCE	Mackenzie et al. (2008, 2012, 2016) Bleyl et al. (2012)
Carbon microspheres from sucrose carbonization	0.1–3 μm hard carbon microspheres (≈ 300 m^2/g) with Fe^0 at the outer surface. Microsphere synthesis from hydrothermal treatment of sugar and carbonization, Fe introduction from Fe salts and reduction by $NaBH_4$. Fe^0 content ≈ 15 wt% Laboratory-tested for TCE	Sunkara et al. (2010, 2011, 2015) Zhan et al. (2009, 2011)
Hydrophilic carbon	Carrier: carbon black hydrophilized by treatment with sulfanilic acid (impregnated with iron salt and borohydride reduction) Fe^0 content ≈ 20 wt% Showed improved transport in laboratory tests	Schrick et al. (2004)
Graphene	Fe on high-surface graphene or Fe-oxide/graphene composites Lab-tested for Cr(VI) or Pd(II) removal	Jabeen et al. (2013) Lv et al. (2014)

combination of iron and various carbon materials has attracted interest in many ways during recent years (Hoch et al. 2008; Mackenzie et al. 2012; Pereira et al. 2010; Zhan et al. 2011; Ling et al. 2012) (see Table 2.3). Some of the preparation approaches (Fig. 2.15) stem from the usual synthesis methods used for bare NZVI and are conducted in the presence of a carbon support material (Bystrzejewski 2011; Wei et al. 2011a, b; Nadagouda and Lytle 2011). Others use organic coating materials and obtain a C/Fe material through thermal degradation of the coating (Kong et al. 2011). Pyrolytic deposition of carbon on NZVI (similar to CNT generation) also leads to C-coated NZVI (Dumitrache et al. 2004). By reduction of iron oxide in the presence of carbon at 1200 °C, CNTs and NZVI particles coated by graphite nanolayers can be generated (Tokoro et al. 2004).

The idea to use composites of carbon and iron for remediation goes back to the 2004 paper of Schrick and coworkers who used anionic, hydrophilized carbon black as support for iron (Schrick et al. 2004). The particles showed markedly improved transport performance through soil- and sand-packed columns compared to bare unsupported iron. As the main reason for the improvement, the negative surface charge of the Fe/C particles was discussed.

In the group of Mallouk, spheric particles with carbon-black-supported NZVI were generated by carbothermal reduction (Hoch et al. 2008). Iron salts were impregnated to the carbon support ($\approx$ 80 m^2/g) and carbothermally reduced in the temperature range between 600 and 800 °C under inert gas atmosphere. The iron (20–100 nm diameter) is attached mainly at the outer surface but gains improved properties for the reduction of Cr(VI) by the combination of the materials.

In a similar approach, monodisperse carbon particles ($\approx$ 300 m^2/g), which were prepared by hydrothermal dehydration and successive pyrolytic carbonization of sugar, were used as supports for iron (Zhan et al. 2009). Further development of the two-step process led to a semicontinuous aerosol-based synthesis method (Zhan et al. 2011). Reduction of the iron precursor was achieved by $NaBH_4$ but alternatively also by a carbothermal approach (Sunkara et al. 2010, 2011). TEM analysis of the Fe/C showed that the zero-valent iron was evenly attached to the outer surface of the carbon spheres (Fig. 2.19).

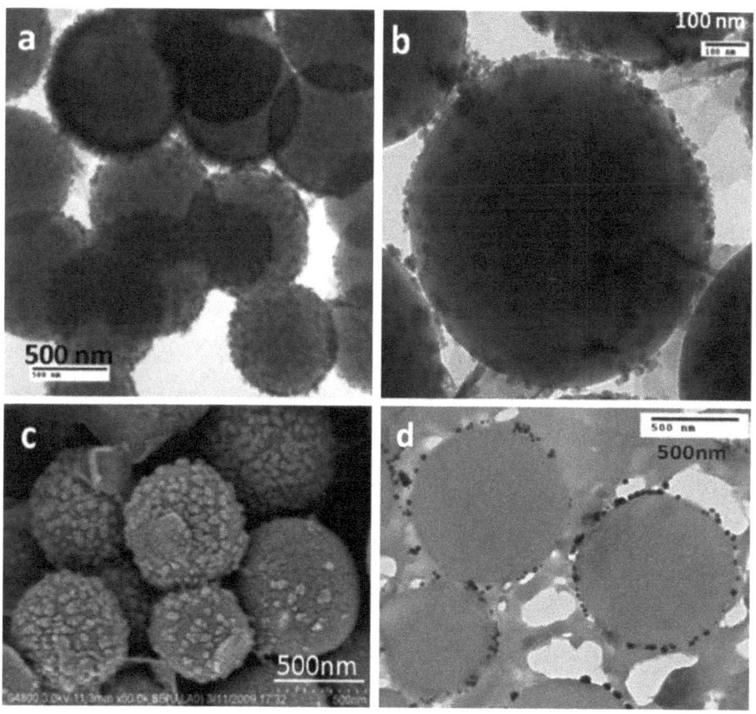

Fig. 2.19 (**a**) TEM of (CMC + NZVI)/carbon particles. (**b**) Higher-resolution TEM image of a single particle showing the distribution of NZVI (**a–c** Zhan et al. 2009). (**c**) SEM of (CMC + ZVI)/carbon particles. (**d**) TEM of a cut-section of the iron supported on carbon microspheres (Zhan et al. 2011). (Reprinted with permission from Zhan et al. (2011). Copyright (2011) American Chemical Society)

Stevenson et al. (1989) showed that iron can be finely deposited on graphite, carbon black, or alumina from iron carbonyls with post-reductive treatment in H_2 atmosphere. Graphite, graphene, and graphene-oxide or pyrolytically deposited carbon were utilized as alternative carbon sources to generate other C-Fe composites, which to our knowledge have not yet achieved field applicability (Cao et al. 2007; Jabeen et al. 2013; Lv et al. 2014).

For an efficient Fe-C system to be used for generation of a subsurface treatment zone, it is expedient to use a low-cost, highly porous carbon material, such as colloidal activated carbon, and combine it with metallic iron in order to follow the strategy of:

- Adsorption and enrichment of the contaminants
- Increase of their retention time in the treatment zone
- Degradation of the contaminants in the sorbed state

In principle RPI's BOS 100® follows this strategy but has a granular grain size of 12–40 mesh size (0.42–1.68 mm), which requires forced pressure-injection (Simon 2015) and therefore does not fall under the material specifications discussed here.

In a product named Carbo-Iron®, colloidal activated carbon (> 1000 m^2/g) with NZVI species generated in the inner pore volume (Fig. 2.20) assures the dominance of the carbon properties for particle transport. The Carbo-Iron composite has been synthesized with 20–30 wt% Fe^0 and optimized in its properties for subsurface utilization (Mackenzie et al. 2008, 2012, 2016).

In accordance with filtration theory, Carbo-Iron particles, which have a mean size of about 1 μm and an effective particle density in water of 1.7–2.0 g/cm^3 (according to Eq. 2.12), were found to be much more mobile than bare NZVI. In D.II column tests with relatively highly concentrated slurries (up to 30 g/L), transport lengths for Carbo-Iron were found to be in the meter range. For injection velocities of 10 m/d,

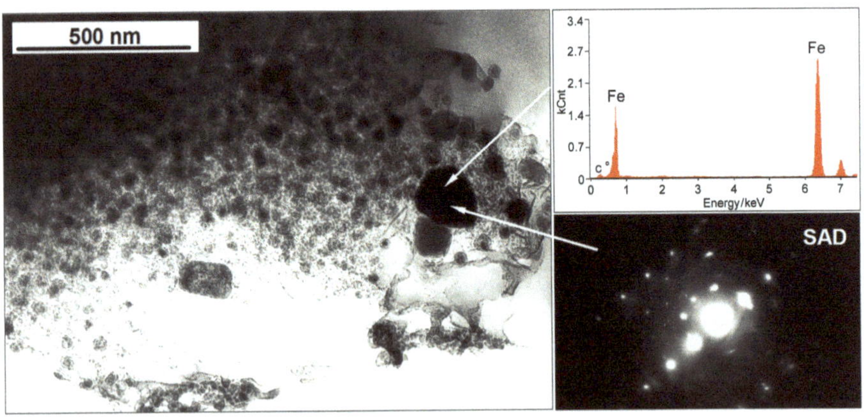

Fig. 2.20 TEM bright-field image of a Carbo-Iron cut (selected area diffraction pattern of the Fe structures are indicated by arrows) (measurements: G. Wagner, Leipzig University) (Mackenzie et al. 2012)

transport lengths (as L_T values) were derived within which a proportion of 50% particles ($L_{T,50}$) and a proportion of 99.9% ($L_{T,99.9}$) are deposited. In other words, a transport length which 50% or 0.1% of the particles pass through are $L_{T,50} = 1.25$ m and $L_{T,99.9} = 12$ m, respectively (Bleyl et al. 2015). By this means, broader treatment zones can be generated.

The product is registered in Germany as Carbo-Iron®. Two alternative synthesis pathways have been used for Carbo-Iron production: in H_2-containing atmosphere at about 500 °C and the carbothermal reduction (about 700 °C, inert gas) method (Bleyl et al. 2012). The utilization of the carbothermal reduction method is a specific feature for the C-Fe systems. It has the charm of using the support itself or its gasification products (mainly CO) as reducing agent. This avoids costly and unhealthy additional reducing agents such as sodium borohydride and is easy to scale-up. For all the processes of early iron smelting, charcoal was used as both heat source and reducing agent. The reduction processes taking place in the temperature zone from 400 to 900 °C are the relevant ones for carbothermal synthesis of Fe/C systems (having slightly lowered temperature windows there) and are summarized in Eqs. 2.13 to 2.15. At the higher end of this temperature region, carbon can directly act as reducing agent forming metallic iron from wustite ($FeO + C \rightarrow Fe + CO$) (Wiberg et al. 2001):

$$3\,Fe_2O_3 + CO \rightarrow 2Fe_3O_4 + CO_2 \quad (2.13)$$

$$Fe_3O_4 + CO \rightarrow 3FeO + CO_2 \quad (2.14)$$

$$FeO + CO \rightarrow Fe + CO_2 \quad (2.15)$$

Recent findings show that carbon having electron conducting properties, such as AC or graphite, can function as a mediator of redox equivalents between iron and the sorbate in contact – even when not unified in a composite structure (Tang et al. 2011). A mechanistic study revealed the transfer of both electrons and H* species from the metallic iron to the AC particles (Kopinke et al. 2016). This effect can be expected to be even more pronounced when iron is built into the AC structure, as is true for Carbo-Iron. It means not only the iron(oxide) surface but also the carbon surface can be the place of chemical reaction in these composite materials.

Fe-C systems seem to have a marked impact on product selectivity for reduction of chloroethenes. The prolonged retention of the pollutant (and its degradation intermediates) in the sorbed state appears to prevent detachment of partially dechlorinated daughter products, which helps to prevent DCE and VC generation. This finding was made not only with the Fe-C microsphere system under laboratory conditions for TCE and PCE dechlorination (Sunkara et al. 2015) but also in the field study of Carbo-Iron as recently reported (Mackenzie et al. 2016).

The hydrophobic nature of carbon and the large inner surface area of AC help to attract organic pollutants and to increase the particle affinity to organic phases (Fig. 2.21). This may be helpful for subsurface source attack.

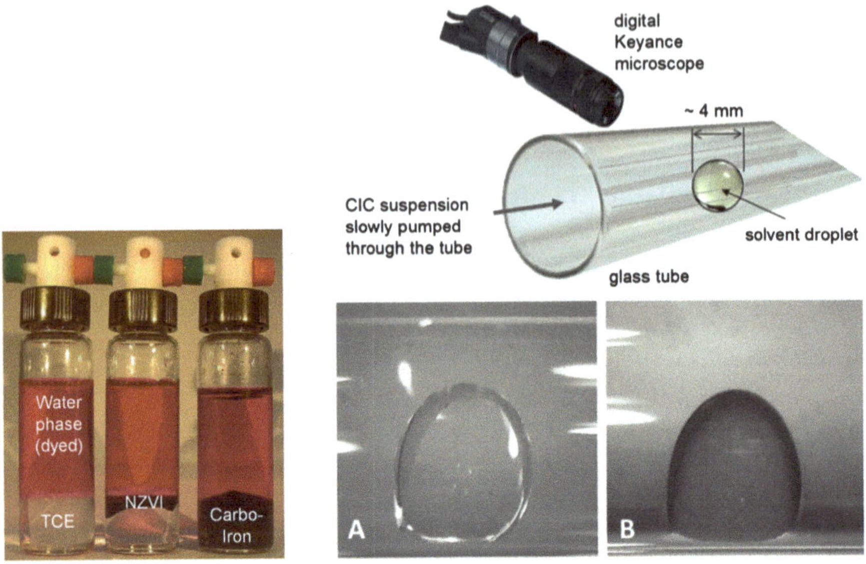

Fig. 2.21 Affinity of Fe/C to organic solvent phases. Left: Comparison of partitioning characteristics of NZVI and Carbo-Iron in a shaken two-phase water-TCE system. Right: Images of Carbo-Iron attraction to a TCE droplet exposed (undisturbed) to a low-concentrated Carbo-Iron suspension, taken with a digital microscope (Mackenzie et al. 2012). A: Fresh droplet of TCE. B: Enrichment of Carbo-Iron at the surface of the TCE droplet in water after 30 min of particle contact and rinsing with freshwater ($d_{\text{droplet}} = 4$ mm, $c_{\text{Carbo-Iron}} = 10$ mg/L, $\nu_{\text{suspension}} = 0.3$ cm/min)

Polymer Supports

In this chapter we distinguish between polymer-modified NZVI, i.e., particles with a polymer or macromolecule as surface coating (Sect. 2.3), and polymer-supported particles, i.e., systems where iron is deposited on polymeric solid support materials. There are manifold examples of polymer-NZVI composites which have been studied for environmental purpose (Zhao et al. 2011). This shows the great potential of polymer-based approaches. However, composites with NZVI located on or within organic polymer supports were mainly studied with the aim of obtaining NZVI-based materials with increased particle size (diameters of some tens of μm to a few mm) which are immobile and thus suitable for reactive barrier or on-site applications (Ponder et al. 2000, 2001; Bezbaruah et al. 2009, 2011). At the same time, dispersion and fixation on the polymer support prevents NZVI aggregation and ensures high specific surface area and reactivity of NZVI. Bai et al. (2009) prepared composites of NZVI and poly(vinyl alcohol) (PVA) with a particle diameter of 600–700 μm or 10–12 μm by means of an inverse suspension cross-linked method. Ponder et al. (Ponder et al. 2000, 2001) used nonporous hydrophobic polymer resin (20–30 μm) to produce a "ferragel" composite with NZVI. In addition, composites with biopolymers have been suggested, such as NZVI and Pd/NZVI on 2–3 mm spherical chitosan particles (Kustov et al. 2011) or NZVI encapsulated or entrapped in

Ca-alginate beads or capsules with diameters of a few mm (Bezbaruah et al. 2009, 2011). All the examples presented here are not regarded as injectable due to their large particle size.

2.5 Synthesis and Properties of Emulsified NZVI

Pure nanoiron exhibits rather hydrophilic surface properties. Thus, for its targeted delivery to contaminant source zones (i.e., DNAPL), emulsified zero-valent iron (EZVI) was designed and for the first time successfully applied in the field at NASA's Launch Complex 34 in 2005 (Quinn et al. 2005). EZVI is one of the few in situ approaches so far which proved to treat the DNAPL source (Su et al. 2012). EZVI is an emulsion of iron particles, food-grade surfactants, and biodegradable vegetable oil in water. The gathered iron particles (from commonly used NZVI or microparticle sources) form an iron-particle-containing core within a water droplet surrounded by an oil-liquid membrane (micelle structure) separating it from the water bulk phase (Fig. 2.22). Thus, EZVI is a water-in-oil-in-water formulation, making the iron miscible to the DNAPL. The average droplet diameter was reported to be in the range of 15 μm. In the classical EZVI application, reactive nanoiron particles (RNIPs) (from Toda America) were used. The mass ratios of the emulsion components were 17% RNIP, 44.3% water, 37.2% oil, and 1.5% surfactant (Quinn et al. 2005).

The outer membrane of the droplet has hydrophobic properties, which increases the affinity of the iron to the residual organic contaminant phase, and thus helps contacting and even partitioning into it. The organic contaminants easily diffuse

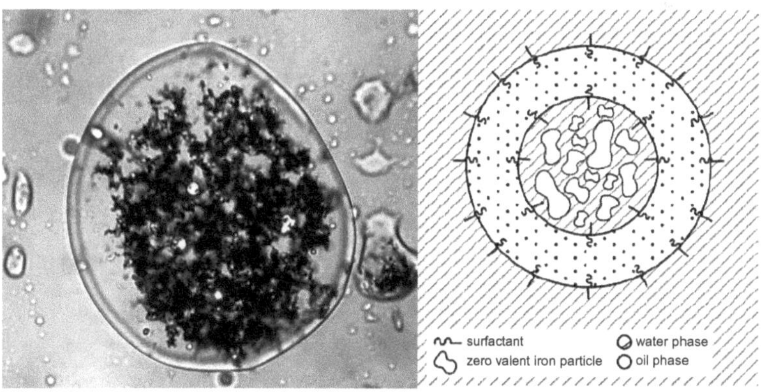

Fig. 2.22 Emulsified zero-valent iron. Left: Micrograph of a droplet formed in nano-iron emulsion (from http://www.nasa.gov/offices/oct/home/tech_life_ezvi.html). Right: Schematic picture of an EZVI droplet showing the oil-liquid membrane surrounding the ZVI particles in water (Quinn et al. 2005). (Reprinted with permission from Quinn et al. (2005). Copyright (2005) American Chemical Society)

into the droplet and are reduced within the interior's aqueous phase by the zero-valent iron particles. At the same time, particles within the emulsion droplets are shielded from other iron-consuming groundwater constituents such as reducible inorganics. Unlike other reductive technologies, EZVI is capable of working in sites with high concentrations of dissolved oxygen or under saline conditions. The injected RNIP iron is transformed from α-Fe^0 to magnetite, then to lepidocrocite, and finally to goethite. Ferrihydrite has been proposed to be an intermediate mineral during transformation of lepidocrocite to goethite (Su et al. 2012). EZVI has an average viscosity above 1942 centipoise (cp) and a specific gravity of approximately 1.1 g/cm^3 (Quinn et al. 2005). Possible problems of the product concern mainly the transport: the metal distribution is not as uniform as intended, the emulsions migrate upward, travel distances are poor, and, due to the high viscosity, higher injection pressures are required (Berge and Ramsburg 2009). It was found that the travel distance of EZVI (made from 10% of Toda iron (RNIP-10DS), 51% of water, 38% of corn oil, and 1% of a nonionic surfactant) was strongly dependent on the injection method. By means of pneumatic injection, EZVI was transported up to 2.1 m from the injection points; by direct injection a travel distance of 0.89 m was reached (Su et al. 2013). Due to the interfacial tension reduction and the high injection pressure necessary with EZVI, careful hydraulic design is advised in order to avoid unintended DNAPL mobilization (Pennell et al. 1996). Successful delivery of NZVI still remains a challenge.

In a closely related alternative approach, iron transport in columns packed with medium and fine sands was tested for nanoiron encapsulated within soybean oil droplets in water. These oil-in-water emulsions, generated by a phase-inversion procedure, formed droplet diameters between 1 and 2 μm. The iron transport efficiency in the water-saturated porous media did clearly benefit from the prolonged kinetic emulsion stability, the emulsion densities (which were very close to that of water, i.e., 0.996–1.00 g/mL at 22 °C), and the viscosity (which could be reduced by a factor of >20 compared to the aforementioned approach). The column experiments showed only little emulsion retention at a (groundwater-near) Darcy velocity of 0.4 m/day. However, the packed bed length was only 10–12 cm (Berge and Ramsburg 2009).

Emulsion properties such as their kinetic stability, droplet size, density, and viscosity ultimately determine the ability of the enclosed iron to reach the source zone. However, the concurrence of improved transport and targeted delivery by phase partitioning, and thus mass delivery to the target zone, still needs to be proved. Results from column tests using EZVI-type formulations with iron-oleate droplets and trichloroethylene source, which were charge-modified by various additives, indicated that the surface charge and the lipophilic character of the outer droplet skin play an important role in the optimization of the interplay between transport, targeted delivery, and particle capture by the source (Wang and Acosta 2013). After evaluation of US EPA studies, Cook summarizes that biodegradation, enhanced by the presence of oil and surfactant in the EZVI emulsion, has contributed to TCE reduction. Surfactants can be ionic or nonionic, but ZVI modified by nonionic surfactants showed less positive synergistic effects than ZVI modified by cationic surfactants applied on a carbon tetrachloride source (Cook 2009).

Applying electric fields can help to enhance electrokinetically the transport of charged oil-in-water droplets of encapsulated NZVI in porous media (studied for an electric potential gradient of 1 V/cm) (Yang and Chang 2011). After EZVI injection campaigns, the expected direct abiotic dechlorination by Fe^0 is initially observed, but in contrast to other ZVI applications, dramatic changes in biogeochemical parameters of groundwater can be found. Significant increases in dissolved sulfide, volatile fatty acids, and total organic carbon become apparent, which indicates a dominant biological reductive dechlorination. EZVI's organic emulsifiers – vegetable oil and surfactant components – enable the material to serve as a long-term electron donor to anaerobic biodegradation (Su et al. 2012).

References

Bai, X., Ye, Z. F., Qu, Y. Z., Li, Y. F., & Wang, Z. Y. (2009). Immobilization of nanoscale Fe-0 in and on PVA microspheres for nitrobenzene reduction. *Journal of Hazardous Materials, 172* (2–3), 1357–1364.

Bardos, P., Bone, B., Daly, P., Elliott, D., Jones, S., Lowry, G. V., & Merly, C. (2014). *A risk/ benefit appraisal for the application of nano-scale zero valent iron (nZVI) for the remediation of contaminated sites*. Available via: http://www.nanorem.eu/Displaynews.aspx?ID=525

Berge, N. D., & Ramsburg, C. A. (2009). Oil-in-water emulsions for encapsulated delivery of reactive iron particles. *Environmental Science & Technology, 43*(13), 5060–5066.

Bezbaruah, A. N., Krajangpan, S., Chisholm, B. J., Khan, E., & Bermudez, J. J. E. (2009). Entrapment of iron nanoparticles in calcium alginate beads for groundwater remediation applications. *Journal of Hazardous Materials, 166*(2–3), 1339–1343.

Bezbaruah, A. N., Shanbhogue, S. S., Simsek, S., & Khan, E. (2011). Encapsulation of iron nanoparticles in alginate biopolymer for trichloroethylene remediation. *Journal of Nanoparticle Research, 13*(12), 6673–6681.

Bhowmick, S., Chakraborty, S., Mondal, P., Van Renterghem, W., Van den Berghe, S., Roman-Ross, G., Chatterjee, D., & Iglesias, M. (2014). Montmorillonite-supported nanoscale zero-valent iron for removal of arsenic from aqueous solution: Kinetics and mechanism. *Chemical Engineering Journal, 243*, 14–23.

Blacha, A., Krukiewicz, K., & Zak, J. (2011). The covalent grafting of polymers to the solid surface. *CHEMIK, 65*(1), 11–19.

Bleyl, S., Kopinke, F.-D., & Mackenzie, K. (2012). Carbo-Iron®-synthesis and stabilization of $Fe^{(0)}$ -doped colloidal activated carbon for in situ groundwater treatment. *Chemical Engineering Journal, 191*, 588–595.

Bleyl, S., Mackenzie, K., Georgi, A., & Kopinke, F. –D. (2015). Nanoiron and Carbo-Iron® particle transport in aquifer sediments – Targeted deposition. In: *Conference proceedings, AquaConSoil 2015*. Copenhagen.

Boudart, M., Delbouille, A., Dumesic, J. A., Khammouma, S., & Topsoe, H. (1975). Surface, catalytic and magnetic-properties of small Iron particles. 1. Preparation and characterization of samples. *Journal of Catalysis, 37*(3), 486–502.

Buchau, A., Rucker, W. M., de Boer, C. V., & Klaas, N. (2010). Inductive detection and concentration measurement of nano sized zero valent iron in the subsurface. *IET Science, Measurement and Technology, 4*(6), 289–297.

Bystrzejewski, M. (2011). Synthesis of carbon-encapsulated iron nanoparticles via solid state reduction of iron oxide nanoparticles. *Journal of Solid State Chemistry, 184*(6), 1492–1498.
Cantrell, K. J., Kaplan, D. I., & Gilmore, T. J. (1997). Injection of colloidal Fe-0 particles in sand with shear-thinning fluids. *Journal of Environmental Engineering, ASCE, 123*(8), 786–791.
Cao, J. S., Elliott, D., & Zhang, W. X. (2003). Nanoscale iron particles for perchlorate reduction. *Abstracts of Papers of the American Chemical Society, 225*, U972–U972.
Cao, H., Li, R., Gui, Q. J., Wang, X. H., & Bin, X. B. (2007). Characteristics and microstructure of graphite encapsulated iron nanoparticles. *Journal of Wuhan University of Technology, 22*(2), 214–217.
Capek, I. (2004). Preparation of metal nanoparticles in water-in-oil (w/o) microemulsions. *Advances in Colloid and Interface Science, 110*(1–2), 49–74.
Chen, A. A., Vannice, M. A., & Phillips, J. (1987). Effect of support pretreatments on carbon-supported Fe particles. *The Journal of Physical Chemistry, 91*(24), 6257–6269.
Choi, C. J., Dong, X. L., & Kim, B. K. (2001). Characterization of Fe and Co nanoparticles synthesized by chemical vapor condensation. *Scripta Materialia, 44*(8–9), 2225–2229.
Cirtiu, C. M., Raychoudhury, T., Ghoshal, S., & Moores, A. (2011). Systematic comparison of the size, surface characteristics and colloidal stability of zero valent iron nanoparticles pre- and post-grafted with common polymers. *Colloids and Surfaces A, 390*(1–3), 95–104.
Comba, S., & Sethi, R. (2009). Stabilization of highly concentrated suspensions of iron nanoparticles using shear-thinning gels of xanthan gum. *Water Research, 43*(15), 3717–3726.
Cook, S. M. (2009). *Assessing the use and application of zero-valent iron nanoparticle technology for remediation at contaminated sites.* Jackson State University. https://clu-in.org/download/studentpapers/zero-valent-iron-cook.pdf
Dalla Vecchia, E., Luna, M., & Sethi, R. (2009). Transport in porous media of highly concentrated iron micro- and nanoparticles in the presence of xanthan gum. *Environmental Science & Technology, 43*(23), 8942–8947.
Dror, I., Jacov, O. M., Cortis, A., & Berkowitz, B. (2012). Catalytic transformation of persistent contaminants using a new composite material based on nanosized zero-valent iron. *ACS Applied Materials & Interfaces, 4*(7), 3416–3423.
Dumitrache, F., Morjan, I., Alexandrescu, R., Morjan, R. E., Voicu, I., Sandu, I., Soare, I., Ploscaru, M., Fleaca, C., Ciupina, V., Prodan, G., Rand, B., Brydson, R., & Woodword, A. (2004). Nearly monodispersed carbon coated iron nanoparticles for the catalytic growth of nanotubes/nanofibres. *Diamond and Related Materials, 13*(2), 362–370.
Eglal, M. M., & Ramamurthy, A. S. (2014). Nanofer ZVI: Morphology, particle characteristics, kinetics, and applications. *Journal of Nanomaterials, 29*, 1–11.
Elliott, D. W., & Zhang, W. X. (2001). Field assessment of nanoscale biometallic particles for groundwater treatment. *Environmental Science & Technology, 35*(24), 4922–4926.
Elliott, D. W., Lien, H.-L., & Zhang, W.-X. (2012). Nanoscale zero-valent iron (nZVI)for site remediation. In G. E. Fryxell & G. Cao (Eds.), *Environmental applications of nanomaterials: Synthesis, sorbents and sensors*. World Scientific. https://doi.org/10.1142/9781860948572_0002
Fang, Y. X., & Al-Abed, S. R. (2008). Dechlorination kinetics of monochlorobiphenyls by Fe/Pd: Effects of solvent, temperature, and PCB concentration. *Applied Catalysis B: Environmental, 78*(3–4), 371–380.
Fernandez-Pacheco, R., Arruebo, M., Marquina, C., Ibarra, R., Arbiol, J., & Santamaria, J. (2006). Highly magnetic silica-coated iron nanoparticles prepared by the arc-discharge method. *Nanotechnology, 17*(5), 1188–1192.
Gastone, F., Tosco, T., & Sethi, R. (2014). Guar gum solutions for improved delivery of iron particles in porous media (part 1): Porous medium rheology and guar gum-induced clogging. *Journal of Contaminant Hydrology, 166*, 23–33.
Golas, P., Matyjaszewski, K., Lowry, G. V., & Tilton, R. D. (2010). Comparative study of polymeric stabilizers for magnetite nanoparticles using ATRP. *Langmuir, 26*(22), 16890–16900.

Grittini, C., Malcomson, M., Fernando, Q., & Korte, N. (1995). Rapid dechlorination of polychlorinated-biphenyls on the surface of a Pd/Fe bimetallic system. *Environmental Science & Technology, 29*(11), 2898–2900.

He, F., & Zhao, D. Y. (2007). Manipulating the size and dispersibility of zerovalent iron nanoparticles by use of carboxymethyl cellulose stabilizers. *Environmental Science & Technology, 41*(17), 6216–6221.

Hoag, G. E., Collins, J. B., Holcomb, J. L., Hoag, J. R., Nadagouda, M. N., & Varma, R. S. (2009). Degradation of bromothymol blue by 'greener' nano-scale zero-valent iron synthesized using tea polyphenols. *Journal of Materials Chemistry, 19*(45), 8671–8677.

Hoch, L. B., Mack, E. J., Hydutsky, B. W., Hershman, J. M., Skluzacek, I. M., & Mallouk, T. E. (2008). Carbothermal synthesis of carbon-supported nanoscale zero-valent iron particles for the remediation of hexavalent chromium. *Environmental Science & Technology, 42*(7), 2600–2605.

Huber, D. L. (2005). Synthesis, properties, and applications of iron nanoparticles. *Small, 1*(5), 482–501.

Hydutsky, B. W., Mack, E. J., Beckerman, B. B., Skluzacek, J. M., & Mallouk, T. E. (2007). Optimization of nano- and microiron transport through sand columns using polyelectrolyte mixtures. *Environmental Science & Technology, 41*(18), 6418–6424.

Jabeen, H., Kemp, K. C., & Chandra, V. (2013). Synthesis of nano zerovalent iron nanoparticles – Graphene composite for the treatment of lead contaminated water. *Journal of Environmental Management, 130*, 429–435.

Jia, H. Z., & Wang, C. Y. (2013). Comparative studies on montmorillonite-supported zero-valent iron nanoparticles produced by different methods: Reactivity and stability. *Environmental Technology, 34*(1), 25–33.

Johnson, R. L., Nurmi, J. T., Johnson, G. S. O., Fan, D. M., Johnson, R. L. O., Shi, Z. Q., Salter-Blanc, A. J., Tratnyek, P. G., & Lowry, G. V. (2013). Field-scale transport and transformation of carboxymethylcellulose-stabilized nano zero-valent Iron. *Environmental Science & Technology, 47*(3), 1573–1580.

Kanel, S. R., Manning, B., Charlet, L., & Choi, H. (2005). Removal of arsenic(III) from groundwater by nanoscale zero-valent iron. *Environmental Science & Technology, 39*(5), 1291–1298.

Karn, B., Kuiken, T., & Otto, M. (2009). Nanotechnology and in situ remediation: A review of the benefits and potential risks. *Environmental Health Perspectives, 117*(12), 1823–1831.

Keenan, C. R., & Sedlak, D. L. (2008). Factors affecting the yield of oxidants from the reaction of manoparticulate zero-valent iron and oxygen. *Environmental Science & Technology, 42*(4), 1262–1267.

Kharisov, B. I., Dias, H. V. R., Kharissova, O. V., Jimenez-Perez, V. M., Perez, B. O., & Flores, B. M. (2012). Iron-containing nanomaterials: Synthesis, properties, and environmental applications. *RSC Advances, 2*(25), 9325–9358.

Kharissova, O. V., Dias, H. V. R., Kharisov, B. I., Perez, B. O., & Perez, V. M. J. (2013). The greener synthesis of nanoparticles. *Trends in Biotechnology, 31*(4), 240–248.

Kirschling, T. L., Golas, P. L., Unrine, J. M., Matyjaszewski, K., Gregory, K. B., Lowry, G. V., & Tilton, R. D. (2011). Microbial bioavailability of covalently bound polymer coatings on model engineered nanomaterials. *Environmental Science & Technology, 45*(12), 5253–5259.

Köber, R., Hollert, H., Hornbruch, G., Jekel, M., Kamptner, A., Klaas, N., Maes, H., Mangold, K. M., Martac, E., Matheis, A., Paar, H., Schaffer, A., Schell, H., Schiwy, A., Schmidt, K. R., Strutz, T. J., Thummler, S., Tiehm, A., & Braun, J. (2014). Nanoscale zero-valent iron flakes for groundwater treatment. *Environment and Earth Science, 72*(9), 3339–3352.

Kong, Q. S., Guo, C. X., Wang, B. B., Ji, Q., & Xia, Y. Z. (2011). A facile preparation of carbon-supported nanoscale zero-valent Iron fibers. *Materials Science Forum, 688*, 349–352.

Kopinke, F. D., Speichert, G., Mackenzie, K., & Hey-Hawkins, E. (2016). Reductive dechlorination in water: Interplay of sorption and reactivity. *Applied Catalysis B: Environmental, 181*, 747–753.

Korte, N. E., Zutman, J. L., Schlosser, R. M., Liang, L., Gu, B., & Fernando, Q. (2000). Field application of palladized iron for the dechlorination of trichloroethene. *Waste Management, 20* (8), 687–694.
Kosmulski, M. (2014). The pH dependent surface charging and points of zero charge. VI. Update. *Journal of Colloid and Interface Science, 426*, 209–212.
Krajangpan, S., Jarabek, L., Jepperson, J., Chisholm, B., & Bezbaruah, A. (2008). Polymer modified iron nanoparticles for environmental remediation. *Polymer Preprints, 49*, 921–922.
Krajangpan, S., Kalita, H., Chisholm, B. J., & Bezbaruah, A. N. (2012). Iron nanoparticles coated with amphiphilic polysiloxane graft copolymers: Dispersibility and contaminant treatability. *Environmental Science & Technology, 46*(18), 10130–10136.
Krol, M. M., Oleniuk, A. J., Kocur, C. M., Sleep, B. E., Bennett, P., Xiong, Z., & O'Carroll, D. M. (2013). A field-validated model for in situ transport of polymer-stabilized nZVI and implications for subsurface injection. *Environmental Science & Technology, 47*(13), 7332–7340.
Kustov, L. M., Finashina, E. D., Shuvalova, E. V., Tkachenko, O. P., & Kirichenko, O. A. (2011). Pd-Fe nanoparticles stabilized by chitosan derivatives for perchloroethene dechlorination. *Environment International, 37*(6), 1044–1052.
Li, L., Fan, M. H., Brown, R. C., Van Leeuwen, J. H., Wang, J. J., Wang, W. H., Song, Y. H., & Zhang, P. Y. (2006a). Synthesis, properties, and environmental applications of nanoscale iron-based materials: A review. *Critical Reviews in Environmental Science and Technology, 36*(5), 405–431.
Li, X. Q., Elliott, D. W., & Zhang, W. X. (2006b). Zero-valent iron nanoparticles for abatement of environmental pollutants: Materials and engineering aspects. *Critical Reviews in Solid State and Materials Sciences, 31*(4), 111–122.
Li, S. L., Yan, W. L., & Zhang, W. X. (2009). Solvent-free production of nanoscale zero-valent iron (nZVI) with precision milling. *Green Chemistry, 11*(10), 1618–1626.
Li, Y. C., Jin, Z. H., & Li, T. L. (2012). A novel and simple method to synthesize SiO2-coated Fe nanocomposites with enhanced Cr (VI) removal under various experimental conditions. *Desalination, 288*, 118–125.
Li, J., Bhattacharjee, S., & Ghoshal, S. (2015). The effects of viscosity of carboxymethyl cellulose on aggregation and transport of nanoscale zerovalent iron. *Colloids and Surfaces A, 481*, 451–459.
Lien, H. L., & Zhang, W. X. (1999). Transformation of chlorinated methanes by nanoscale iron particles. *Journal of Environmental Engineering, 125*(11), 1042–1047.
Lien, H. L., & Zhang, W. X. (2007). Nanoscale Pd/Fe bimetallic particles: Catalytic effects of palladium on hydrodechlorination. *Applied Catalysis B: Environmental, 77*(1–2), 110–116.
Ling, X. F., Li, J. S., Zhu, W., Zhu, Y. Y., Sun, X. Y., Shen, J. Y., Han, W. Q., & Wang, L. J. (2012). Synthesis of nanoscale zero-valent iron/ordered mesoporous carbon for adsorption and synergistic reduction of nitrobenzene. *Chemosphere, 87*(6), 655–660.
Liu, Y. Q., & Lowry, G. V. (2006). Effect of particle age (Fe-o content) and solution pH on NZVI reactivity: H-2 evolution and TCE dechlorination. *Environmental Science & Technology, 40* (19), 6085–6090.
Liu, Y. Q., Choi, H., Dionysiou, D., & Lowry, G. V. (2005a). Trichloroethene hydrodechlorination in water by highly disordered monometallic nanoiron. *Chemistry of Materials, 17*(21), 5315–5322.
Liu, Y. Q., Majetich, S. A., Tilton, R. D., Sholl, D. S., & Lowry, G. V. (2005b). TCE dechlorination rates, pathways, and efficiency of nanoscale iron particles with different properties. *Environmental Science & Technology, 39*(5), 1338–1345.
Liu, X. W., Wang, D. S., & Li, Y. D. (2012). Synthesis and catalytic properties of bimetallic nanomaterials with various architectures. *Nano Today, 7*(5), 448–466.
Liu, W. J., Qian, T. T., & Jiang, H. (2014). Bimetallic Fe nanoparticles: Recent advances in synthesis and application in catalytic elimination of environmental pollutants. *Chemical Engineering Journal, 236*, 448–463.
Lowry, G. V. (2007). Nanomaterials for groundwater remediation. In M. R. Wiesner & J.-Y. Bottero (Eds.), *Environmental nanotechnology*. New York: The McGraw-Hill Companies.

Lowry, G. V., & Johnson, K. M. (2004). Congener-specific dechlorination of dissolved PCBs by microscale and nanoscale zerovalent iron in a water/methanol solution. *Environmental Science & Technology, 38*(19), 5208–5216.

Lowry, G. V., Hill, R., Harper, S., Rawle, A. F., Hendren, C. O., Klaessig, F., Nobbmann, U., Syare, P., & Rumble, J. (2016). Guidance for measuring, interpreting, and reporting zeta potential measurements for environmental nanotechnology and Nanotoxicology. *Environmental Science: Nano, 3*, 953–965. https://doi.org/10.1039/C6EN00136J.

Lv, X. S., Xue, X. Q., Jiang, G. M., Wu, D. L., Sheng, T. T., Zhou, H. Y., & Xu, X. H. (2014). Nanoscale zero-valent iron (nZVI) assembled on magnetic Fe3O4/graphene for chromium (VI) removal from aqueous solution. *Journal of Colloid and Interface Science, 417*, 51–59.

Machado, S., Pinto, S. L., Grosso, J. P., Nouws, H. P. A., Albergaria, J. T., & Delerue-Matos, C. (2013). Green production of zero-valent iron nanoparticles using tree leaf extracts. *The Science of the Total Environment, 445*, 1–8.

Mackenzie, K., Schierz, A., Georgi, A., & Kopinke, F. D. (2008). Colloidal activated carbon and carbo-iron – Novel materials for in-situ groundwater treatment. *Global NEST Journal, 10*(1), 54–61.

Mackenzie, K., Bleyl, S., Georgi, A., & Kopinke, F. D. (2012). Carbo-Iron – An Fe/AC composite – As alternative to nano-iron for groundwater treatment. *Water Research, 46*(12), 3817–3826.

Mackenzie, K., Bleyl, S., Kopinke, F.-D., Doose, H., & Bruns, J. (2016). Carbo-Iron as improvement of the nanoiron technology: From laboratory design to the field test. *Science Total Environment*. https://doi.org/10.1016/j.scitotenv.2015.07.107.

Martinez-Baez, E., Dominguez, J., Ortega-Pijeira, M. S., Tejeda-Mazola, Y., Borroto, J., & Rivera-Denis, A. (2015). Synthesis and evaluation of ferragels as prospective solid Tc-99m radiotracers. *Journal of Radioanalytical and Nuclear Chemistry, 304*(1), 267–272.

McCurrie, R. A. (1994). *Ferromagnetic materials*. London: Academic Press.

Miehr, R., Tratnyek, P. G., Bandstra, J. Z., Scherer, M. M., Alowitz, M. J., & Bylaska, E. J. (2004). Diversity of contaminant reduction reactions by zerovalent iron: Role of the reductate. *Environmental Science & Technology, 38*(1), 139–147.

Mueller, N. C., Braun, J., Bruns, J., Cernik, M., Rissing, P., Rickerby, D., & Nowack, B. (2012). Application of nanoscale zero valent iron (NZVI) for groundwater remediation in Europe. *Environemental Science and Pollution Research, 19*(2), 550–558.

Nadagouda, M. N., & Lytle, D. A. (2011). Microwave-assisted combustion synthesis of nano iron oxide/iron-coated activated carbon, anthracite, cellulose fiber, and silica, with arsenic adsorption studies. *Journal of Nanotechnology, 972486*, 1–8.

Nurmi, J. T., Tratnyek, P. G., Sarathy, V., Baer, D. R., Amonette, J. E., Pecher, K., Wang, C. M., Linehan, J. C., Matson, D. W., Penn, R. L., & Driessen, M. D. (2005). Characterization and properties of metallic iron nanoparticles: Spectroscopy, electrochemistry, and kinetics. *Environmental Science & Technology, 39*(5), 1221–1230.

Pennell, K. D., Pope, G. A., & Abriola, L. M. (1996). Influence of viscous and buoyancy forces on the mobilization of residual tetrachloroethylene during surfactant flushing. *Environmental Science & Technology, 30*(4), 1328–1335.

Pereira, M. C., Coelho, F. S., Nascentes, C. C., Fabris, J. D., Araujo, M. H., Sapag, K., Oliveira, L. C. A., & Lago, R. M. (2010). Use of activated carbon as a reactive support to produce highly active-regenerable Fe-based reduction system for environmental remediation. *Chemosphere, 81* (1), 7–12.

Phenrat, T., Saleh, N., Sirk, K., Tilton, R. D., & Lowry, G. V. (2007). Aggregation and sedimentation of aqueous nanoscale zerovalent iron dispersions. *Environmental Science & Technology, 41*(1), 284–290.

Phenrat, T., Saleh, N., Sirk, K., Kim, H. J., Tilton, R. D., & Lowry, G. V. (2008). Stabilization of aqueous nanoscale zerovalent iron dispersions by anionic polyelectrolytes: Adsorbed anionic polyelectrolyte layer properties and their effect on aggregation and sedimentation. *Journal of Nanoparticle Research, 10*(5), 795–814.

Phenrat, T., Kim, H. J., Fagerlund, F., Illangasekare, T., Tilton, R. D., & Lowry, G. V. (2009). Particle size distribution, concentration, and magnetic attraction affect transport of polymer-modified Fe-0 nanoparticles in sand columns. *Environmental Science & Technology, 43*(13), 5079–5085.

Phenrat, T., Fagerlund, F., Illangasekare, T., Lowry, G. V., & Tilton, R. D. (2011). Polymer-modified Fe-0 nanoparticles target entrapped NAPL in two dimensional porous media: Effect of particle concentration, NAPL saturation, and injection strategy. *Environmental Science & Technology, 45*(14), 6102–6109.

Ponder, S. M., & Mallouk, T. F. (2004). *Powerful reductant for decontamination of groundwater and surface streams.* U.S. Patent No. 6,689,485.

Ponder, S. M., Darab, J. G., & Mallouk, T. E. (2000). Remediation of Cr(VI) and Pb(II) aqueous solutions using supported, nanoscale zero-valent iron. *Environmental Science & Technology, 34* (12), 2564–2569.

Ponder, S. M., Darab, J. G., Bucher, J., Caulder, D., Craig, I., Davis, L., Edelstein, N., Lukens, W., Nitsche, H., Rao, L. F., Shuh, D. K., & Mallouk, T. E. (2001). Surface chemistry and electrochemistry of supported zerovalent iron nanoparticles in the remediation of aqueous metal contaminants. *Chemistry of Materials, 13*(2), 479–486.

Quinn, J., Geiger, C., Clausen, C., Brooks, K., Coon, C., O'Hara, S., Krug, T., Major, D., Yoon, W. S., Gavaskar, A., & Holdsworth, T. (2005). Field demonstration of DNAPL dehalogenation using emulsified zero-valent iron. *Environmental Science & Technology, 39* (5), 1309–1318.

Ramos, M. A. V., Yan, W., Li, X. Q., Koel, B. E., & Zhang, W. X. (2009). Simultaneous oxidation and reduction of arsenic by zero-valent iron nanoparticles: Understanding the significance of the core-shell structure. *Journal of Physical Chemistry C, 113*(33), 14591–14594.

Raychoudhury, T., Naja, G., & Ghoshal, S. (2010). Assessment of transport of two polyelectrolyte-stabilized zero-valent iron nanoparticles in porous media. *Journal of Contaminant Hydrology, 118*(3–4), 143–151.

Raychoudhury, T., Tufenkji, N., & Ghoshal, S. (2012). Aggregation and deposition kinetics of carboxymethyl cellulose-modified zero-valent iron nanoparticles in porous media. *Water Research, 46*(6), 1735–1744.

Reinsch, B. C., Forsberg, B., Penn, R. L., Kim, C. S., & Lowry, G. V. (2010). Chemical transformations during aging of zerovalent iron nanoparticles in the presence of common groundwater dissolved constituents. *Environmental Science & Technology, 44*(9), 3455–3461.

Robinson, I., Zacchini, S., Tung, L. D., Maenosono, S., & Thanh, N. T. K. (2009). Synthesis and characterization of magnetic nanoalloys from bimetallic carbonyl clusters. *Chemistry of Materials, 21*(13), 3021–3026.

Rónavári, A., Balázs, M., Tolmacsov, P., Molnár, C., Kiss, I., Kukovecz, Á., & Kónya, Z. (2016). Impact of the morphology and reactivity of nanoscale zero-valent iron (NZVI) on dechlorinating bacteria. *Water Research, 95*, 165–173.

Rose, J., Thill, A., & Brant, J. (2007). Methods for structural and chemical characterization of nanomaterials. In M. R. Wiesner & J.-Y. Bottero (Eds.), *Environmental nanotechnology: Applications and impacts of nanomaterials* (pp. 105–154). New York: McGraw-Hill.

Saleh, N., Phenrat, T., Sirk, K., Dufour, B., Ok, J., Sarbu, T., Matyiaszewski, K., Tilton, R. D., & Lowry, G. V. (2005). Adsorbed triblock copolymers deliver reactive iron nanoparticles to the oil/water interface. *Nano Letters, 5*(12), 2489–2494.

Saleh, N., Sirk, K., Liu, Y. Q., Phenrat, T., Dufour, B., Matyjaszewski, K., Tilton, R. D., & Lowry, G. V. (2007). Surface modifications enhance nanoiron transport and NAPL targeting in saturated porous media. *Environmental Engineering Science, 24*(1), 45–57.

Sarathy, V., Tratnyek, P. G., Nurmi, J. T., Baer, D. R., Amonette, J. E., Chun, C. L., Penn, R. L., & Reardon, E. J. (2008). Aging of iron nanoparticles in aqueous solution: Effects on structure and reactivity. *Journal of Physical Chemistry C, 112*(7), 2286–2293.

Schrick, B., Blough, J. L., Jones, A. D., & Mallouk, T. E. (2002). Hydrodechlorination of trichloroethylene to hydrocarbons using bimetallic nickel-iron nanoparticles. *Chemistry of Materials, 14*(12), 5140–5147.

Schrick, B., Hydutsky, B. W., Blough, J. L., & Mallouk, T. E. (2004). Delivery vehicles for zerovalent metal nanoparticles in soil and groundwater. *Chemistry of Materials, 16*(11), 2187–2193.

Scott, T. B., Dickinson, M., Crane, R. A., Riba, O., Hughes, G. M., & Allen, G. C. (2010). The effects of vacuum annealing on the structure and surface chemistry of iron nanoparticles. *Journal of Nanoparticle Research, 12*(5), 1765–1775.

Simon, J. A. (2015). Editor's perspective-an in situ revelation: First retard migration, then treat. *Remediation Journal, 25*(2), 1–7.

Sohn, K., Kang, S. W., Ahn, S., Woo, M., & Yang, S. K. (2006). Fe(0) nanoparticles for nitrate reduction: Stability, reactivity, and transformation. *Environmental Science & Technology, 40* (17), 5514–5519.

Soukupova, J., Zboril, R., Medrik, I., Filip, J., Safarova, K., Ledl, R., Mashlan, M., Nosek, J., & Cernik, M. (2015). Highly concentrated, reactive and stable dispersion of zero-valent iron nanoparticles: Direct surface modification and site application. *Chemical Engineering Journal, 262*, 813–822.

Stevenson, S. A., Goddard, S. A., Arai, M., & Dumesic, J. A. (1989). Effects of preparation variables on particle-size and morphology for carbon-supported and alumina-supported metallic iron samples. *The Journal of Physical Chemistry, 93*(5), 2058–2065.

Su, C. M., Puls, R. W., Krug, T. A., Watling, M. T., O'Hara, S. K., Quinn, J. W., & Ruiz, N. E. (2012). A two and half-year-performance evaluation of a field test on treatment of source zone tetrachloroethene and its chlorinated daughter products using emulsified zero valent iron nanoparticles. *Water Research, 46*(16), 5071–5084.

Su, C. M., Puls, R. W., Krug, T. A., Watling, M. T., O'Hara, S. K., Quinn, J. W., & Ruiz, N. E. (2013). Travel distance and transformation of injected emulsified zerovalent iron nanoparticles in the subsurface during two and half years. *Water Research, 47*(12), 4095–4106.

Sun, Y. P., Li, X. Q., Cao, J. S., Zhang, W. X., & Wang, H. P. (2006). Characterization of zero-valent iron nanoparticles. *Advances in Colloid and Interface Science, 120*(1–3), 47–56.

Sun, Q., Feitz, A. J., Guan, J., & Waite, T. D. (2008). Comparison of the reactivity of nanosized zero-valent iron (nZVI) particles produced by borohydride and dithionite reduction of iron salts. *Nano, 3*(5), 341–349.

Sunkara, B., Zhan, J. J., He, J. B., McPherson, G. L., Piringer, G., & John, V. T. (2010). Nanoscale zerovalent iron supported on uniform carbon microspheres for the in situ remediation of chlorinated hydrocarbons. *ACS Applied Materials & Interfaces, 2*(10), 2854–2862.

Sunkara, B., Zhan, J. J., Kolesnichenko, I., Wang, Y. Q., He, J. B., Holland, J. E., McPherson, G. L., & John, V. T. (2011). Modifying metal nanoparticle placement on carbon supports using an aerosol-based process, with application to the environmental remediation of chlorinated hydrocarbons. *Langmuir, 27*(12), 7854–7859.

Sunkara, B., Su, Y., Zhan, J. J., He, J. B., Mcpherson, G. L., & John, V. T. (2015). Iron-carbon composite microspheres prepared through a facile aerosol-based process for the simultaneous adsorption and reduction of chlorinated hydrocarbons. *Frontiers of Environmental Science & Engineering, 9*(5), 939–947.

Suslick, K. S., Fang, M. M., & Hyeon, T. (1996). Sonochemical synthesis of iron colloids. *Journal of the American Chemical Society, 118*(47), 11960–11961.

Tang, H., Zhu, D. Q., Li, T. L., Kong, H. N., & Chen, W. (2011). Reductive dechlorination of activated carbon-adsorbed trichloroethylene by zero-valent iron: Carbon as electron shuttle. *Journal of Environmental Quality, 40*(6), 1878–1885.

Tiraferri, A., Chen, K. L., Sethi, R., & Elimelech, M. (2008). Reduced aggregation and sedimentation of zero-valent iron nanoparticles in the presence of guar gum. *Journal of Colloid and Interface Science, 324*(1–2), 71–79.

Tokoro, H., Fujii, S., & Oku, T. (2004). Iron nanoparticles coated with graphite nanolayers and carbon nanotubes. *Diamond and Related Materials, 13*(4–8), 1270–1273.

Toshima, N., & Yonezawa, T. (1998). Bimetallic nanoparticles – novel materials for chemical and physical applications. *New Journal of Chemistry, 22*(11), 1179–1201.

Tseng, H. H., Su, J. G., & Liang, C. J. (2011). Synthesis of granular activated carbon/zero valent iron composites for simultaneous adsorption/dechlorination of trichloroethylene. *Journal of Hazardous Materials, 192*(2), 500–506.

Tufenkji, N., & Elimelech, M. (2004). Correlation equation for predicting single-collector efficiency in physicochemical filtration in saturated porous media. *Environmental Science & Technology, 38*(2), 529–536.
U.S. EPA. (2011). *Selected sites using or testing nanoparticles for remediation*. https://clu-in.org/download/remed/nano-site-list.pdf
Uegami, M., Kawano, J., Okita, T., Fujii, Y., Okinaka, K., & Kakuyua, K. (2003). *Iron particles for purifying contaminated soil or ground water. Process for producing the iron particles, purifying agent comprising the iron particles, process for producing the purifying agent and method of purifying contaminated soil or ground water*. Toda Kogyo Corp., US Patent Application 2003/0217974 A1.
Wan, J. J., Wan, J. Q., Ma, Y. W., Huang, M. Z., Wang, Y., & Ren, R. (2013). Reactivity characteristics of SiO2-coated zero-valent iron nanoparticles for 2,4-dichlorophenol degradation. *Chemical Engineering Journal, 221*, 300–307.
Wang, Z. Q. (2013). Iron complex nanoparticles synthesized by eucalyptus leaves. *ACS Sustainable Chemistry & Engineering, 1*(12), 1551–1554.
Wang, Z. H., & Acosta, E. (2013). Formulation design for target delivery of iron nanoparticles to TCE zones. *Journal of Contaminant Hydrology, 155*, 9–19.
Wang, C. B., & Zhang, W. X. (1997). Synthesizing nanoscale iron particles for rapid and complete dechlorination of TCE and PCBs. *Environmental Science & Technology, 31*(7), 2154–2156.
Wang, Q. L., Kanel, S. R., Park, H., Ryu, A., & Choi, H. (2009). Controllable synthesis, characterization, and magnetic properties of nanoscale zerovalent iron with specific high Brunauer-Emmett-Teller surface area. *Journal of Nanoparticle Research, 11*(3), 749–755.
Wang, Q., Lee, S., & Choi, H. (2010). Aging study on the structure of Fe-0-nanoparticles: Stabilization, characterization, and reactivity. *Journal of Physical Chemistry C, 114*(5), 2027–2033.
Wang, C., Xu, Z., Ding, G., Wang, X., Zhao, M., Ho, S. S. H., & Li, Y. (2016). Comprehensive study on the removal of chromate from aqueous solution by synthesized kaolin supported nanoscale zero-valent iron. *Desalination and Water Treatment 57*(11), 5065–5078.
Wei, Z. Q., Liu, L. G., Yang, H., Zhang, C. R., & Feng, W. J. (2011a). Characterization of carbon encapsulated Fe-nanoparticles prepared by confined arc plasma. *Transactions of Nonferrous Metals Society of China, 21*(9), 2026–2030.
Wei, Z. Q., Wang, X. Y., & Yang, H. (2011b). Preparation of carbon-encapsulated Fe core-shell nanostructures by confined arc plasma. *Materials Science Forum, 688*, 245–249.
Wiberg, N., Holleman, A. F., & Wiberg, E. E. (2001). *Holleman-Wiberg's inorganic chemistry*. New York: Academic Press.
Xu, Y., & Zhang, W. X. (2000). Subcolloidal Fe/Ag particles for reductive dehalogenation of chlorinated benzenes. *Industrial and Engineering Chemistry Research, 39*(7), 2238–2244.
Xu, F. Y., Deng, S. B., Xu, J., Zhang, W., Wu, M., Wang, B., Huang, J., & Yu, G. (2012). Highly active and stable Ni-Fe bimetal prepared by ball milling for catalytic hydrodechlorination of 4-chlorophenol. *Environmental Science & Technology, 46*(8), 4576–4582.
Yan, W. L., Ramos, M. A. V., Koel, B. E., & Zhang, W. X. (2012). As(III) sequestration by iron nanoparticles: Study of solid-phase redox transformations with X-ray photoelectron spectroscopy. *Journal of Physical Chemistry C, 116*(9), 5303–5311.
Yan, W. L., Lien, H. L., Koel, B. E., & Zhang, W. X. (2013). Iron nanoparticles for environmental clean-up: Recent developments and future outlook. *Environmental Science: Processes & Impacts, 15*(1), 63–77.
Yang, G. C. C., & Chang, Y. I. (2011). Integration of emulsified nanoiron injection with the electrokinetic process for remediation of trichloroethylene in saturated soil. *Separation and Purification Technology, 79*(2), 278–284.
Yang, N. L., Desai, A., Mahajan, D., & Rafailovich, M. H. (2003). Synthesis and characterization of nano-sized iron particles on a polystyrene support as potential Fischer-Tropsch catalysts. *Abstracts of Papers of the American Chemical Society, 226*, U565–U565.

Yuan, M. L., Tao, J. H., Yan, G. J., Tan, M. Y., & Qiu, G. Z. (2010). Preparation and characterization of Fe/SiO2 core/shell nanocomposites. *Transactions of Nonferrous Metals Society of China, 20*(4), 632–636.

Zhan, J. J., Zheng, T. H., Piringer, G., Day, C., McPherson, G. L., Lu, Y. F., Papadopoulos, K., & John, V. T. (2008). Transport characteristics of nanoscale functional zerovalent iron/silica composites for in situ remediation of trichloroethylene. *Environmental Science & Technology, 42*(23), 8871–8876.

Zhan, J. J., Sunkara, B., Le, L., John, V. T., He, J. B., McPherson, G. L., Piringer, G., & Lu, Y. F. (2009). Multifunctional colloidal particles for in situ remediation of chlorinated hydrocarbons. *Environmental Science & Technology, 43*(22), 8616–8621.

Zhan, J. J., Kolesnichenko, I., Sunkara, B., He, J. B., McPherson, G. L., Piringer, G., & John, V. T. (2011). Multifunctional iron-carbon nanocomposites through an aerosol-based process for the in situ remediation of chlorinated hydrocarbons. *Environmental Science & Technology, 45*(5), 1949–1954.

Zhang, H., Jin, Z. H., Han, L., & Qin, C. H. (2006). Synthesis of nanoscale zero-valent iron supported on exfoliated graphite for removal of nitrate. *Transactions of Nonferrous Metals Society of China, 16*, S345–S349.

Zhang, Y., Li, Y. M., & Zheng, X. M. (2011). Removal of atrazine by nanoscale zero valent iron supported on organobentonite. *The Science of the Total Environment, 409*(3), 625–630.

Zhang, Y. Y., Jiang, H., Zhang, Y., & Xie, J. F. (2013). The dispersity-dependent interaction between montmorillonite supported nZVI and Cr(VI) in aqueous solution. *Chemical Engineering Journal, 229*, 412–419.

Zhao, X., Lv, L., Pan, B. C., Zhang, W. M., Zhang, S. J., & Zhang, Q. X. (2011). Polymer-supported nanocomposites for environmental application: A review. *Chemical Engineering Journal, 170*(2–3), 381–394.

Zhao, X., Liu, W., Cai, Z., Han, B., Qian, T., & Zhao, D. (2016). An overview of preparation and applications of stabilized zero-valent iron nanoparticles for soil and groundwater remediation. *Water Research, 100*, 245–266.

Zheng, T. H., Zhan, J. J., He, J. B., Day, C., Lu, Y. F., Mcpherson, G. L., Piringer, G., & John, V. T. (2008). Reactivity characteristics of nanoscale zerovalent iron-silica composites for trichloroethylene remediation. *Environmental Science & Technology, 42*(12), 4494–4499.

Zhou, T., Li, Y. Z., & Lim, T. T. (2010). Catalytic hydrodechlorination of chlorophenols by Pd/Fe nanoparticles: Comparisons with other bimetallic systems, kinetics and mechanism. *Separation and Purification Technology, 76*(2), 206–214.

Zhu, B. W., & Lim, T. T. (2007). Catalytic reduction of chlorobenzenes with Pd/Fe nanoparticles: Reactive sites, catalyst stability, particle aging, and regeneration. *Environmental Science & Technology, 41*(21), 7523–7529.

Chapter 3
Chemical Reduction and Oxidation of Organic Contaminants by Nanoscale Zerovalent Iron

Tanapon Phenrat, Thi Song Thao Le, Bhanuphong Naknakorn, and Gregory V. Lowry

Abstract This chapter critically reviews the kinetics of NZVI used to reductively and oxidatively transform various kinds of priority organic contaminants including chlorinated ethenes, chlorinated ethanes, chlorinated and aromatic nitro hydrocarbons, chlorinated biphenyls, halogenated bisphenol A, explosives, dyes, and pesticides. All kinds of NZVI, including bare, bimetallic, polymer-modified, and supported NZVI, are reviewed. A total of 102 datasets of laboratory-scale experiments over 20 years (1997–2017) of NZVI research are evaluated to extract state-of-the-art understanding. This chapter elaborates not only reductive transformation pathways of priority organic contaminants but also two factors governing NZVI reactivity: intrinsic properties of the materials and environmental conditions where NZVI particles are applied. These include the particle crystallinity and chemical composition (noble metal), the effect of polymeric surface modification, and the effects of sorptive support, aging effects, pH, anionic and cationic solutes, natural organic matter, aquifer material, contaminant concentration, and the presence of

T. Phenrat (✉) · T. S. T. Le
Department of Civil Engineering, Environmental Engineering Program, Naresuan University, Phitsanulok, Thailand

Center of Excellence for Sustainability of Health, Environment and Industry (SHEI), Faculty of Engineering, Naresuan University, Phitsanulok, Thailand

B. Naknakorn
Thai Plastic and Chemicals PCL, Bangkok, Thailand

Center for Environmental Implications of Nanotechnology (CEINT)
Carnegie Mellon University, Pittsburgh, PA, USA

G. V. Lowry
Center for Environmental Implications of Nanotechnology (CEINT)
Carnegie Mellon University, Pittsburgh, PA, USA

Department of Civil & Environmental Engineering, Carnegie Mellon University, Pittsburgh, PA, USA

T. Phenrat, G. V. Lowry (eds.), *Nanoscale Zerovalent Iron Particles for Environmental Restoration*, https://doi.org/10.1007/978-3-319-95340-3_3

dense nonaqueous phase liquid. Similarly, various factors affecting oxidative degradation of contaminants of concern using NZVI-induced Fenton's reaction are reviewed.

Keywords Nanoscale zerovalent iron · Reductive dechlorination · Fenton reaction · Chlorinated organics · Degradation kinetics · Degradation mechanisms

3.1 Reductive Transformation of Priority Organic Contaminants Using Nanoscale Zerovalent Iron

Nanoscale zerovalent iron (NZVI) is a well-known remediation agent due to its high reactivity to reductively detoxify a variety of contaminants, including metals (Chap. 4) and halogenated organics, which is the focus of this chapter. Table 3.1 provides an intensive review of the reductive transformation of priority organic contaminants, including chlorinated ethene, chlorinated ethane, chlorinated and aromatic nitro hydrocarbon, chlorinated biphenyl, halogenated bisphenol A, explosives, dye, and pesticides, using various kinds of NZVI, including bare, bimetallic, polymer-modified, and supported NZVI. This involves 102 datasets of laboratory-scaled experiments over 20 years (1997–2017) of NZVI research. In addition to presenting a database of organic contaminants, several interesting trends can be determined from Table 3.1, some of which will be discussed later in this chapter.

Although the idea of using NZVI for in situ subsurface remediation started in 1997 (Wang and Zhang 1997), bulk ZVI has long been recognized as an electron donor and has been used in the form of iron filings to build a permeable reactive barrier since 1994 (Reynolds et al. 1990; Gillham and O'Hannesin 1994). Undeniably, NZVI is much more reactive than its ZVI counterpart on the mass basis comparison. Nevertheless, it is still debatable if this is mainly due to a much larger specific surface area of NZVI in comparison to iron filings or micron-sized ZVI or if it has something to do with the "nano-effect," which may result in a greater density of reactive surface sites or surface sites of higher intrinsic reactivity.

The small size of nanoparticles causes an exponential increase in the number of atoms localized at the surface. These atoms are characterized by excess surface energy and are thermodynamically unstable (Ghosh 2015). This results in crystallographic changes including lattice contraction or deformation, the appearance of defects, rearrangements of the surface atoms, or changes in the morphology of nanoparticles (Jiang et al. 2001). This may yield quantum size effects, which cause changes in the Fermi level and band gap, leading to increases in intrinsic reactivity with decreasing particle size (Ghosh 2015).

Nevertheless, Tratnyek's group revealed that, for tetrachloromethane (or carbon tetrachloride (CT)), no "nano-effect" contributing to the greater intrinsic reactivity of the surface sites or the greater abundance of reactive sites on the surface was observed. This is because, when comparing surface-area normalized rate constants (k_{SA}) against mass normalized rate constants (k_M) for the reductive degradation of

Table 3.1 Summary of batch experiments for reductive treatment of organic contaminants using various NZVI (1997–2017)

No.	Contaminant (initial concentration)	Type of particle	Dispersant or supported material	Experimental condition	Rate constant (pseudo-first order)	Contaminant removal (%)	References
Chlorinated ethene							
1	Trichloroethylene (TCE) (7 mg/L)	Fe^{BH}	–	Batch experiment using 10 g/L bare and PV3A stabilized NZVI in deionized (DI) water	$k_{obs} = 4.11$/h per 10 g/L of NZVI	98% removal after 3 h	Sun et al. (2007)
2	TCE (7 mg/L)	Fe^{BH}	Polyvinyl alcohol-co-vinyl acetate-co-itaconic acid (PV3A) (5–10% by weight of NZVI)		$k_{obs} = 1.88$/h per 10 g/L of NZVI	98% removal after 3 h	Sun et al. (2007)
3	TCE (25 mg/L)	Commercial NZVI (100 nm)	–	Batch experiment using 0.1 g/L NZVI in DI water	$k_{obs} = 0.034$/h for 0.1 g/L NZVI	<10% removal in 120 min	He and Zhao (2005)
4	TCE (25 mg/L)	Commercial NZVI (100 nm)	Starch (either 0.2% (w/w) for 0.1 g/L Fe or 0.8% (w/w) for 1 g/L Fe solution)		$k_{obs} = 0.11$/h for 0.1 g/L NZVI or $k_{SA} = 0.02$ L/(h·m^2)	20% removal in 120 min	He and Zhao (2005)
5	TCE (20 mg/L)	Pd/Fe^{BH}	–	Batch experiment using 20 g/L NZVI in DI water	$k_{SA} = 0.1$ L/(h·m^2)	100% removal in 0.25 h	Wang and Zhang (1997)
6	TCE (25 mg/L)	Commercial NZVI with Pd (0.1% wt) (100 nm, Argonide, Sanford, FL)	–	Batch experiment using 1 g/L NZVI in DI water	$k_{obs} = 0.0016$/h for 1 g/L NZVI	<20% removal in 120 min	He and Zhao (2005)
7	TCE (4.4–290 mg/L)	RNIP (Fe^{H2})	–	Batch experiment in DI water at 0.35 g/L to 1.9 g/L NZVI	$k_{SA} = 4 \times 10^{-4}$ L/(h·m^2) for iron limited condition (0.35 g/L) and $k_{SA} = 3 \times 10^{-3}$	75% removal in 8 days	(Liu et al. 2005b)

(continued)

Table 3.1 (continued)

No.	Contaminant (initial concentration)	Type of particle	Dispersant or supported material	Experimental condition	Rate constant (pseudo-first order)	Contaminant removal (%)	References
					L/(h·m^2) for iron excess (1.9 g/L)		
8	TCE (4.4–290 mg/L)	Fe^{BH}	–		$k_{SA} = 2 \times 10^{-3}$ L/(h·m^2) for iron limited condition (0.35 g/L) and $k_{SA} = 1.4 \times 10^{-2}$ L/(h·m^2) for iron excess (1.9 g/L)	70–80% removal in less than 2 days	Liu et al. (2005b)
9	TCE (5.25 mg/L)	RNIP (Fe^{H2})	Poly(styrene sulfonate) (MW = 70 K) (adsorbed mass = 2.06 mg/m^2 of NZVI)	Batch experiment in 1 mM NaHCO3 at 2.8 g/L NZVI	$k_{SA} = 0.73 \pm 0.03 \times 10^{-3}$ L/(h·m^2)	90% removal in 160 h	Phenrat et al. (2009b)
10	TCE (5.25 mg/L)	RNIP (Fe^{H2})	Poly(styrene sulfonate) (MW = 1 M) (adsorbed mass = 1.92 mg/m^2 NZVI)		$k_{SA} = 0.44 \pm 0.04 \times 10^{-3}$L/(h·m^2)	–	Phenrat et al. (2009b)
11	TCE (5.25 mg/L)	RNIP (Fe^{H2})	Carboxymethylcellulose (MW = 90 K) (adsorbed mass = 0.98 mg/m^2 of NZVI)		$k_{SA} = 0.31 \pm 0.02 \times 10^{-3}$ L/(h·m^2)	–	Phenrat et al. (2009b)
12	TCE (5.25 mg/L)	RNIP (Fe^{H2})	Carboxymethylcellulose (MW = 700 K) (adsorbed mass = 2.03 mg/m^2 of NZVI)		$k_{SA} = 0.38 \pm 0.03 \times 10^{-3}$ L/(h·m^2)	–	Phenrat et al. (2009b)
13	TCE (5.25 mg/L)	RNIP (Fe^{H2})	Polyaspartate (MW = 2.5 K) (adsorbed		$k_{SA} = 0.15 \pm 0.01 \times 10^{-3}$ L/(h·m^2)	–	Phenrat et al. (2009b)

			mass = 2.29 mg/m^2 of NZVI)				
14	TCE (5.25 mg/L)	RNIP (Fe^{H2})	Polyaspartate (MW = 10 K) (adsorbed mass = 2.29 mg/m^2 of NZVI)		$k_{SA} = 0.14 \pm 0.01 \times 10^{-3}$ L/(h·m^2)	–	Phenrat et al. (2009b)
15	TCE (50 mg/L)	Pd/Fe^{BH} (Pd loading 0.1% wt)	–	Batch experiment in DI with iron dose of 0.1 g/L	k_{obs} = 0.44/h. per 0.1 g/L of Pd/Fe	40% removal after 120 min	He et al. (2007)
16	TCE (50 mg/L)	Pd/Fe^{BH} (Pd loading 0.1% wt)	CMC (0.2% (w/w))	Batch experiment using 0.1 g/L NZVI in DI water	k_{obs} = 7.4/h. per 0.1 g/L of Pd/Fe	Complete reduction after 40 min	He et al. (2007)
17	TCE (25 mg/L)	Pd/Fe^{BH} (Pd loading 0.1% wt)	–	Batch experiment using 0.1 g/L NZVI in DI water	k_{obs} = 0.0125/h for 0.1 g/L NZVI	80% removal	He and Zhao (2005)
18	TCE (25 mg/L)	Pd/Fe^{BH} (Pd loading 0.1% wt)	Starch (either 0.2% (w/w) for 0.1 g/L Fe or 0.8% (w/w) for 1 g/L Fe solution)	Batch experiment in DI with iron dose of 0.1 g/L	k_{obs} = 3.7/h for 0.1 g/L NZVI or k_{SA} = 0.67 L/(h·m^2)	98% removal	He and Zhao (2005)
19	TCE (30 mg/L)	Fe^{H2} (using H_2 at 500 °C)	Activated carbon (d_{50} = 0.8 μm) (NZVI loading = 11%-20% wt)	Batch experiment at 1 g/L of NZVI and pH = 8.5 in DI water with 0.2 M NH_4Cl/NH_3- or $NaHCO_3$ buffer	k_{obs} = 0.9 to 4.2 × 10^{-2}/h or k_{SA} = 0.6 to 2.7 × 10^{-3} L/(h·m^2)	90% removal in 100 h	Mackenzie et al. (2012)

(continued)

Table 3.1 (continued)

No.	Contaminant (initial concentration)	Type of particle	Dispersant or supported material	Experimental condition	Rate constant (pseudo-first order)	Contaminant removal (%)	References
20	Tetrachloroethylene (PCE) (37 mg/L)	RNIP (Fe^{H2})	–	Batch experiment in 5 mM NaHCO3 and 10 g/L NZVI	$k_{SA} = 7.6 \times 10^{-4}$ L/(h·m^2)	Complete removal in 2.5 days	Fagerlund et al. (2012)
21	PCE (4.8 mg/L)	MRNIP2 (Fe^{H2})	Poly(maleic acid-co-olefin) (MW = 12 K)	Batch experiment in 5 mM NaHCO3 and various NZVI concentrations from 1 to 20 g/L	k_{SA} from 2×10^{-4} to 4×10^{-4} L/(h·m^2)	Complete removal in 50 h	(Kim et al. 2017)
22	PCE (20 mg/L)	Pd/Fe^{BH} (Pd loading 0.1% wt)	–	Batch experiment in DI water at 5 g/L NZVI	$k_{SA} = 12.2 \pm 0.36 \times 10^{-3}$ L/(h·m^2)	Complete removal in 1.5 h	Lien and Zhang (2001)
23	TCE (20 mg/L)	Pd/Fe^{BH} (Pd loading 0.1% wt)	–		$k_{SA} = 18.2 \pm 1.18 \times 10^{-3}$ L/(h·m^2)	Complete removal in 1.5 h	Lien and Zhang (2001)
24	tran-1,2-Dichloroethylene (t-DCE) (20 mg/L)	Pd/Fe^{BH} (Pd loading 0.1% wt)	–		$k_{SA} = 15.1 \pm 2.08 \times 10^{-3}$ L/(h·m^2)	Complete removal in 1.5 h	Lien and Zhang (2001)
25	cis-1,2-Dichloroethylene (c-DCE) (20 mg/L)	Pd/Fe^{BH} (Pd loading 0.1% wt)	–		$k_{SA} = 17.6 \pm 1.34 \times 10^{-3}$ L/(h·m^2)	Complete removal in 1.5 h	(Lien and Zhang 2001)
26	1,1-Dichloroethylene (1,1-DCE) (20 mg/L)	Pd/Fe^{BH} (Pd loading 0.1% wt)	–		$k_{SA} = 11.5 \pm 1.25 \times 10^{-3}$ L/(h·m^2)	Complete removal in 1.5 h	Lien and Zhang (2001)

Chlorinated ethane							
27	Hexachloroethane (HCA) (23.6 mg/L)	Fe^{BH}	–	Batch experiment in DI water in NZVI concentration of 0.08 g/L at the initial pH 7.2 to 7.8	$k_{obs} = 1.73 \pm 0.12$/h. or $k_{SA} = 7.7 \pm 0.7 \times 10^{-1}$ L/(h·m^2)	>90% removal in 1.3 h	Song and Carraway (2005)
28	Pentachloroethane (PCA) (19.19 mg/L)	Fe^{BH}	–		$k_{obs} = 1.79 \pm 0.09$/h or $k_{SA} = 7.96 \pm 0.63 \times 10^{-1}$ L/(h·m^2)	>90% removal in 1.3 h	Song and Carraway (2005)
29	1,1,2,2-Tetrachloroethane (1,1,1,2-TeCA) (18.3 mg/L)	Fe^{BH}	–		$k_{obs} = 1.21 \pm 0.10$/h or $k_{SA} = 5.38 \pm 0.56 \times 10^{-1}$ L/(h·m^2)	–	Song and Carraway (2005)
30	1,1,2,2-TeCA (18.3 mg/L)	Fe^{BH}	–		$k_{obs} = 6.82 \pm 0.85 \times 10^{-2}$/h or $k_{SA} = 3.03 \pm 0.42 \times 10^{-2}$ L/(h·m^2)	90% removal in 55 h	Song and Carraway (2005)
31	1,1,1-Trichloroethane (1,1,1-TCA) (13.4 mg/L)	Fe^{BH}	–		$k_{obs} = 3.4 \pm 0.2 \times 10^{-1}$/h or $k_{SA} = 1.51 \pm 0.13 \times 10^{-1}$ L/(h·m^2)	80% removal in 5 h	Song and Carraway (2005)
32	1,1,2-Trichloroethane (1,1,2-TCA) (13.4 mg/L)	Fe^{BH}	–	Batch experiment in DI water in NZVI concentration of 0.08 g/L at the initial pH 7.2 to 7.8	$k_{obs} = 5.2 \pm 0.71 \times 10^{-3}$/h or $k_{SA} = 2.31 \pm 0.34 \times 10^{-3}$ L/(h·m^2)	50% removal in 110 h	Song and Carraway (2005)
33	1,1-Dichloroethane (1,1-DCA) (9.9 mg/L)	Fe^{BH}	–		$k_{obs} = 2.41 \pm 0.52 \times 10^{-4}$/h or $k_{SA} = 1.99 \pm 0.44 \times 10^{-5}$ L/(h·m^2)	–	Song and Carraway (2005)
34	1,2-Dichloroethane (1,2-DCA) (9.9 mg/L)	Fe^{BH}	–		$k_{obs} < 5 \times 10^{-5}$/h or $k_{SA} = 4 \times 10^{-6}$ L/(h·m^2)	–	Song and Carraway (2005)
35	HCA (20–30 mg/L)	Pd/Fe^{BH} (Pd loading 0.5% wt)	–	Batch experiment in DI	$k_{obs} = 3.43$/h		

(continued)

Table 3.1 (continued)

No.	Contaminant (initial concentration)	Type of particle	Dispersant or supported material	Experimental condition	Rate constant (pseudo-first order)	Contaminant removal (%)	References
				water in NZVI concentration of 5 g/L		100% removal in 1.7 h	Lien and Zhang (2005)
36	PCA	Pd/Fe^{BH} (Pd loading 0.5% wt)	–		$k_{obs} = 4.46/h$	–	(Lien and Zhang 2005)
37	1,1,1,2-TeCA	Pd/Fe^{BH} (Pd loading 0.5% wt)	–		$k_{obs} = 3.60/h$	–	Lien and Zhang (2005)
38	1,1,1,2-TeCA	Pd/Fe^{BH} (Pd loading 0.5% wt)	–		$k_{obs} = 1.51/h$	–	Lien and Zhang (2005)
39	1,1,1-TCA	Pd/Fe^{BH} (Pd loading 0.5% wt)	–		$k_{obs} = 0.93/h$	100% removal in 7 h	Lien and Zhang (2005)
Chlorinated methane							
40	Tetrachloromethane (or carbon tetrachloride (CT) (0.6 mg/L)	RNIP (Fe^{H2}) (surface area = 29 m^2/g)	–	Batch experiment in DI water in various NZVI concentrations	$k_{SA} = 1.2 \times 10^{-2}$ to 1.5×10^{-1} L/(h·m^2)	–	Nurmi et al. (2005)
41	CT (0.6 mg/L)	Fe^{BH} (surface area = 33.5 m^2/g)	–		$k_{SA} = 4.2 \times 10^{-2}$ to 3.0×10^{-1} L/(h·m^2)	–	Nurmi et al. (2005)
42	CT (0.6 mg/L)	Fe^{El} (surface area = 0.12 m^2/g)	–		k_{SA} = 0.12–2.4 L/(h·m^2)	–	Nurmi et al. (2005)

43	CT (30 mg/L)	Fe^{BH}	–	Batch experiment in DI water (pH 7.5) at 0.16 g/L NZVI	$k_{obs} = 5.02 \pm 0.44$/h $k_{SA} = 1.12 \pm 0.11$ L/(h·m^2)	100% removal in 1.5 h	Song and Carraway (2006)
44	Trichloromethane (or chloroform (CF) (11.9 mg/L)	Fe^{BH}	–		$k_{obs} = 1.89 \pm 0.18 \times 10^{-1}$/h $k_{SA} = 4.20 \pm 0.47 \times 10^{-2}$ L/(h·m^2)	80% removal in 10 h	Song and Carraway (2006)
45	CT (15.9 mg/L)	Fe^{BH}	–	Batch experiment in DI water at 12.5 g/L NZVI	$k_{SA} = 5.31 \times 10^{-4}$ L/(h·m^2)	100% removal in 20 h	Lien and Zhang (1999)
46	CT (15.9 mg/L)	Pd/Fe^{BH} (Pd loading is 0.06% wt)	–		$k_{SA} = 9.0 \times 10^{-3}$ L/(h·m^2)	100% removal in 1 h	Lien and Zhang (1999)
47	CF (14.8 mg/L)	Fe^{BH}	–		$k_{SA} = 8.41 \times 10^{-5}$ L/(h·m^2)	> 95% removal in 100 h	Lien and Zhang (1999)
48	CF (14.8 mg/L)	Pd/Fe^{BH} (Pd loading is 0.06% wt)	–		$k_{SA} = 6.5 \times 10^{-3}$ L/(h·m^2)	>95% removal in 1 h	Lien and Zhang (1999)
Chlorinated and aromatic nitro hydrocarbon							
49	Hexachlorobenzene (HCB)	Fe^{BH}	–	Batch experiment in DI with NZVI of 133 g/L at pH = 8	$k_{obs} = 0.14$/h	60% removal in 24 h	Shih et al. (2009)
50	HCB	Pd/Fe^{BH} (Pd loading is 0.02% wt)	–	Batch experiment in DI with NZVI of 133 g/L at pH = 8	$k_{obs} = 0.23$/h	70% removal in 24 h	Shih et al. (2009)
51	HCB (0.2 mg/L)	Cu/Fe (cu loading = 5% wt)	–	Batch experiment in DI with	$k_{obs} = 0.056 \pm 0.008$/h	98% removal in 3 h	Zhu et al. (2010)

(continued)

Table 3.1 (continued)

No.	Contaminant (initial concentration)	Type of particle	Dispersant or supported material	Experimental condition	Rate constant (pseudo-first order)	Contaminant removal (%)	References
52	HCB (0.2 mg/L)	Fe	–	NZVI of 40 g/L at pH = 4	$k_{obs} = 0.012 \pm 0.003$/h	88% removal in 192 h	Zhu et al. (2010)
53	Nitrobenzene (NB) (20 mg/L)	Fe^{H2}	–	Batch experiment at 1 g/L of NZVI	$k_{obs} = 9.30 \pm 2.13$/h for NZVI	56% removal for NZVI	Zhang et al. (2013)
54	NB (20 mg/L)	Fe^{H2}	Ordered mesoporous materials, including platelet shape and short mesochannels (SBA-15_S) and conventional fibrous SBA-15 (SBA-15_L)	Batch experiment at 1 g/L of NZVI	$k_{obs} = 29.40 \pm 2.52$/h for NZVI/SBA-15_S and $k_{obs} = 37.50 \pm 4.18$/h for NZVI/SBA-15_L	94% removal for NZVI/SBA-15_S and 83.9% removal for NZVI/SBA-15_L in 12 h	Zhang et al. (2013)
55	1, 2, 4-trichloro-benzene (1,2,4-TCB) (25 mg/L)	Cu/Fe^{BH} (cu loading is 0.5% wt)	Carboxymethyl cellulose (MW = 90 K) (0.2–0.4% wt)	Batch experiment in DI water at 1 g/L NZVI	$k_{obs} = 0.06$–0.09/h	75–88% removal in 25 h	Cao et al. (2011)
56	1,2,4-TCB (19.2 mg/L)	Pd/Fe^{BH} (Pd loading is 0.1% wt)	SDS (1 critical micelle concentration (CMC))	Batch experiment in DI water at 0.71 g/L NZVI	$k_{obs} = 6.06 \pm 1.14$/h	100% removal in 60 min	Zhu et al. (2008)
57	1,2,4-TCB (19.2 mg/L)	Pd/Fe^{BH} (Pd loading is 0.1% wt)	NPE (1 CMC)		$k_{obs} = 9.18 \pm 1.86$/h	100% removal in 60 min	Zhu et al. (2008)
58	1,2,4-TCB (19.2 mg/L)	Pd/Fe^{BH} (Pd loading is 0.1% wt)	TX-100 (1 CMC)		$k_{obs} = 5.64 \pm 1.44$/h	100% removal in 60 min	Zhu et al. (2008)
59	1,2,4-TCB (19.2 mg/L)	Pd/Fe^{BH} (Pd loading is 0.1% wt)	CTAB (1 CMC)		$k_{obs} = 13.9\ 8 \pm 1.62$/h	100% removal in 60 min	Zhu et al. (2008)

60	1,2,4-TCB (19.2 mg/L)	Pd/Fe^{BH} (Pd loading is 0.1% wt)	DCP (1 CMC)	Batch experiment in DI water at 0.71 g/L NZVI	$k_{obs} = 0.9 \pm 0.48$/h	100% removal in 60 min	Zhu et al. (2008)
61	1,2,4-TCB (19.2 mg/L)	Pd/Fe^{BH} (Pd loading is 0.1% wt)	Natural organic matter (NOM) (10 mg/L)		$k_{obs} = 3.54 \pm 0.78$/h	30% removal in 60 min	Zhu et al. (2008)
62	1,2,4-TCB (19.2 mg/L)	Pd/Fe^{BH} (Pd loading is 0.1% wt)	NOM (50 mg/L)		$k_{obs} = 0.54 \pm 0.42$/h	–	Zhu et al. (2008)
63	1,2,4-TCB (19.2 mg/L)	Pd/Fe^{BH} (Pd loading is 0.1% wt)	NOM (200 mg/L)		$k_{obs} = 0.18 \pm 0.06$/h	–	Zhu et al. (2008)
64	1,2,4-TCB (30.8 mg/L)	Pd/Fe^{BH} (Pd loading is 0.1–1% wt)	Chitosan (3.33 g/L)	Batch experiment in DI water at 1.65 g/L NZVI	$k_{obs} = 5.87$/h	100% removal in 60 min	Zhu et al. (2006)
65	1,2,4-TCB (30.8 mg/L)	Pd/Fe^{BH} (Pd loading is 0.1–1% wt)	Silica (3.33 g/L)		$k_{obs} = 1.86$/h	95% removal in 120 min	Zhu et al. (2006)
66	Para-nitrochlorobenzene (p-NCB) (40 mg/L)	Ni/Fe^{BH} (Ni loading is 2% wt)	–	Batch experiment in DI water at 6 g/L NZVI at pH =7	$k_{obs} = 2.39$/h	100% removal in 300 min	Xu et al. (2009)
67	Pentachlorophenol (PCP) (0.075 mM)	Fe^{BH} ($Fe^0 = 2.95\%$ on the weight basis of clay)	2% Smectite clay in the suspension	Batch experiment in DI water at pH = 9	$k_{obs} = 0.001$/h	20% removal in 14 days	Jia and Wang (2015)
68	PCP (0.075 mM)	Pd/Fe^{BH} (Fe^0:Pd^0 = 2.95%:0.065% on the weight basis of clay)	2% Smectite clay in the suspension	Batch experiment in DI water at pH = 9	$k_{obs} = 0.487$/h	95% removal in 6 h	(Jia and Wang 2015)
69	2,3,4,5-TeCP and 2,3,4,6-TeCP and 2,3,5,6-TeCP (0.075 mM) (TeCP = Tetrachlorophenol)	Pd/Fe^{BH} (Fe^0:Pd^0 = 2.95%:0.065% on the weight basis of clay)	2% Smectite clay in the suspension	Batch experiment in DI water at pH = 9	$k_{obs} = 0.502–0.586$/h	–	Jia and Wang (2015)
70		Pd/Fe^{BH} (Fe^0:Pd^0 = 2.95%:0.065% on	2% Smectite clay in the suspension		$k_{obs} = 0.968–1.064$/h	–	

(continued)

Table 3.1 (continued)

No.	Contaminant (initial concentration)	Type of particle	Dispersant or supported material	Experimental condition	Rate constant (pseudo-first order)	Contaminant removal (%)	References
	2,3,4-TCP and 2,3,6-TCP and 2,4,6-TCP (0.075 mM) (TCP is Trichlorophenol)	the weight basis of clay)		Batch experiment in DI water at pH = 9			Jia and Wang (2015)
71	3,4,5-TCP and 2,4,5-TCP and 2,3,5-TCP (0.075 mM)	Pd/Fe^{BH} (Fe^0:Pd^0 = 2.95%:0.065% on the weight basis of clay)	2% Smectite clay in the suspension	Batch experiment in DI water at pH = 9	k_{obs} = 1.300–1.459/h	–	Jia and Wang (2015)
72	2,6-Dichlorophenol (2,6-DCP) (0.075 mM)	Pd/Fe^{BH} (Fe^0:Pd^0 = 2.95%:0.065% on the weight basis of clay)	2% Smectite clay in the suspension	Batch experiment in DI water at pH = 9	k_{obs} = 1.912/h	–	Jia and Wang (2015)
73	3,5-Dichlorophenol (3,5-DCP) (0.075 mM)	Pd/Fe^{BH} (Fe^0:Pd^0 = 2.95%:0.065% on the weight basis of clay)	2% Smectite clay in the suspension	Batch experiment in DI water at pH = 9	k_{obs} = 2.139/h	–	Jia and Wang (2015)
74	2,3-DCP and 2,4-DCP and 3,4-DCP (0.075 mM)	Pd/Fe^{BH} (Fe^0:Pd^0 = 2.95%:0.065% on the weight basis of clay)	2% Smectite clay in the suspension	Batch experiment in DI water at pH = 9	k_{obs} = 2.234–2.351/h	–	Jia and Wang (2015)
75	2,5-DCP (0.075 mM)	Pd/Fe^{BH} (Fe^0:Pd^0 = 2.95%:0.065% on the weight basis of clay)	2% Smectite clay in the suspension	Batch experiment in DI water at pH = 9	k_{obs} = 3.265/h	–	Jia and Wang (2015)
76	4-CP and 2-CP (0.075 mM) (CP is chlorophenol)	Pd/Fe^{BH} (Fe^0:Pd^0 = 2.95%:0.065% on the weight basis of clay)	2% Smectite clay in the suspension	Batch experiment in DI water at pH = 9	k_{obs} = 8.606–9.603/h	–	Jia and Wang (2015)

77	3-CP (0.075 mM)	Pd/Fe^{BH} (Fe^0:Pd^0 = 2.95%:0.065% on the weight basis of clay)	2% Smectite clay in the suspension	Batch experiment in DI water at pH = 9	$k_{obs} = 15.63/h$	–	Jia and Wang (2015)
Chlorinated biphenyl							
78	Polychlorinated biphenyls (PCBs) (2.5 mg/L)	Pd/Fe^{BH} (Pd loading 0.1% wt)	–	Batch experiment using 0.1 g/L NZVI in DI water	$k_{obs} = 2.9 \times 10^{-3}/h$ for 0.1 g/L NZVI	20% removal in 100 h	He and Zhao (2005)
79	PCBs (2.5 mg/L)	Pd/Fe^{BH} (Pd loading 0.1% wt)	Starch (either 0.2% (w/w) for 0.1 g/L Fe or 0.8% (w/w) for 1 g/L Fe solution)	Batch experiment using 0.1 g/L NZVI in DI water	$k_{obs} = 0.017/h$ for 0.1 g/L NZVI	80% removal in 100 h	He and Zhao (2005)
80	2-Chlorobiphenyl (4 mg/L)	Pd/Fe^{BH} (0.68% Pd to 14.4% Fe)	Granular activated carbon (GAC)	Batch experiment in DI water at 100 g/L of GAC embedded with Pd/FeBH	Not determined due to complex sorption and dechlorination processes	1.16% removal after 2 days. The residual was adsorbed on the surface of GAC	Choi et al. (2008)
81	2-Chlorobiphenyl (4 mg/L)	Fe^{BH}	Granular activated carbon (GAC)		Not determined due to complex sorption and dechlorination processes	90% removal after 2 days. The residual was adsorbed on the surface of GAC	Choi et al. (2008)
Halogenated bisphenol A (flame retardant)							
82	Tetrabromobisphenol A (10 mg/L)	Ni/Fe^{BH} (Ni loading = 0.5% wt)	–	Batch experiment in DI water at 3 g/L NZVI and	$k_{obs} = 2.7/h$ or $k_{SA} = 3.3 \times 10^{-2}$ L/(h m^2)	Successfully degraded within 2 h	Li et al. (2016)

(continued)

Table 3.1 (continued)

No.	Contaminant (initial concentration)	Type of particle	Dispersant or supported material	Experimental condition	Rate constant (pseudo-first order)	Contaminant removal (%)	References
				pH 5–6 (optimal)			
83	Tetrabromobisphenol A (16.30 mg/L)	Pd/Fe^{BH} (Pd loading = 0.022%)	–	Batch experiment in DI water at 4 g/L NZVI and pH 4.2	$k_{obs} = 138/h$	99% removal efficiency in 45 min	Huang et al. (2013)
84	Tetrachlorobisphenol A (10 mg/L)	Ni/Fe^{BH} (Ni loading = 0.5% wt)	–	Batch experiment in DI water at 2 g/L NZVI and pH 7	$k_{obs} = 0.57 \pm 0.02/h$	95% removal in 360 min	Li et al. (2017)
85	Tetrachlorobisphenol A (10 mg/L)	Fe^{BH}	–		$k_{obs} = 0.03/h$	12% removal in 360 min	Li et al. (2017)
86	Tetrabromobisphenol A) (10 mg/L)	Ni/Fe^{BH} (Ni loading = 0.5% wt)	–		$k_{obs} = 2.24 \pm 0.17/h$	95% removal in 120 min	Li et al. (2017)
87	Tetrabromobisphenol A) (10 mg/L)	Fe^{BH}	–		$k_{obs} = 0.08/h$	15% in 120 min	Li et al. (2017)
Explosives							
88	2,4,6-trinitrotoluene (TNT) (80 mg/L)	NZVI (powder)	–	Batch experiment in DI water at 5 g/L NZVI and pH 4	$k_{obs} = 0.31/h$	Completely degraded within 3 h	Zhang et al. (2010)
89	1,3,5-Trinitro-1,3,5-triazinane (RDX) (0.18 mg/L)	Fe^{BH}	–	Batch experiment in DI water at 0.3 g/L NZVI and pH 5	$k_{obs} = 1.14 \pm 0.12/h$	75% removal in 60 min	Naja et al. (2008)
90	RDX (0.18 mg/L)	Fe^{BH}	Poly(acrylic acid) (PAA) (61 g/L, degree of polymerization = 100)	Batch experiment in DI	$k_{obs} = 4.92 \pm 0.12/h$	100% removal in 60 min	Naja et al. (2008)

91	RDX (0.18 mg/L)	Fe^{BH}	Carboxymethyl cellulose (CMC) (0.5% wt, degree of polymerization = 100)	water at 0.3 g/L NZVI and pH 5	$k_{obs} = 48.96 \pm 4.02$/h	100% removal in 5 min	Naja et al. (2008)
Dye, pesticides, and others							
92	Acid blue 113 dye azo dye solution	Fe^{BH}	Resin (NZVI loads of 4.9–50.8 mg/g)	Batch experiment with 20 g/L of resin imbedded NZVI and 50–600 mg/L acid blue 113	$k_{obs} = 8.22–45.36$/h by NZVI loads of 4.9–50.8 mg/g	Almost 100% dye removal and 12.6% TOC removal after 30 min	Shu et al. (2010)
93	Methyl orange (MO) dye	Fe^{BH}	Bentonite	Batch experiment at 1 g/L of bentonite-supported NZVI and 100–800 mg/L of MO	$k_{obs} = 2.04$/h at MO concentration of 800 mg/L to 6.48/h at 100 mg/L MO	79.46% removal after 10 min and Nearly 100% after 30 min	Chen et al. (2011b)
94	Methylene blue (MB) dye	Fe^{BH}	Palygorskite	Batch experiment at 1 g/L of palygorskite-supported NZVI 100 mg/L MB at a natural pH	–	89.36% in 30–60 min	Frost et al. (2010)
95	MO (60 mg/L)	Fe^{BH}	Biochar NZVI/BC (1:3, 1:5, or 1:7 by wt)	Batch experiment in DI water at 0.6–0.9 g/L biochar	$k_{obs} = 28.56–44.58$/h	93.3–98.5% removal in 25 min	Han et al. (2015)

(continued)

Table 3.1 (continued)

No.	Contaminant (initial concentration)	Type of particle	Dispersant or supported material	Experimental condition	Rate constant (pseudo-first order)	Contaminant removal (%)	References
				supported NZVI at pH 4.06–9.13			
96	2,4-dichloro-phenoxyacetic acid (10 mg/L)	Pd/Fe^{BH} (Pd loading = 0.8% wt)	–	Batch experiment in DI water at 0.6 g/L NZVI and pH 6	$k_{obs} = 8.42/h$	25% within 180 min	Zhou et al. (2011)
97	2,4-dichloro-phenoxyacetic acid (10 mg/L)	Pd/Fe^{BH} (Pd loading = 0.8% wt)	Carboxymethyl cellulose (CMC) (CMC/Fe, (wt/wt)) = 2/1 to 10/1)		$k_{obs} = 5.03/h$, 3.70/h, 8.91/h for CMC/Fe = 2/1, 10/1, and 5/1, respectively	45%, 80%, and 97% within 180 min for CMC/Fe = 2/1, 10/1, and 5/1, respectively	Zhou et al. (2011)
98	Alachlor [2-chloro-20,60-diethyl-N-(methoxymethyl) acetanilide] (10–40 mg/L)	Fe^{BH}	–	Batch experiment in DI water at 4 g/L NZVI and pH 7–8.5	$k_{SA} = 38.5 \times 10^{-5}$ L/(h·m^2)	92%–96% after 72 h and completely reduced after 5 days	Thompson et al. (2010)
99	Benzoquinone (21–210 mg/L)	RNIP (Fe^{H2}) (surface area = 29 m^2/g)	–	Batch experiment in DI water in various NZVI concentrations	$k_{SA} = 1.8 \times 10^{-2}$ to 1.2×10^{-1} L/(h·m^2)	–	Nurmi et al. (2005)
100	Benzoquinone (21–210 mg/L)	Fe^{BH} (surface area = 33.5 m^2/g)	–	Batch experiment in DI water in	$k_{SA} = 0.6–3$ L/(h·m^2)	–	Nurmi et al. (2005)

				various NZVI concentrations			
101	Benzoquinone (21–210 mg/L)	Fe^{El} (surface area = 0.12 m^2/g)	–	Batch experiment in DI water in various NZVI concentrations	$k_{SA} = 6 \times 10^{-2}$ to 3.6 L/(h·m^2)	–	Nurmi et al. (2005)
102	γ-Hexachlorocyclohexane (7.5 mg/L)	Fe^{BH}	–	Batch experiment in DI water in various NZVI concentrations (from 0.015 to 0.39 g/L)	$k_{SA} = 1.5$ to 7.20×10^{-2} L/(h m^2)	100% removal in 24 h at NZVI concentration of 0.1–0.39 mg/L	Elliott et al. (2009)

Note: Fe^{BH} NZVI synthesized by borohydride reduction (see Chap. 2)
Fe^{H2} NZVI synthesized by H2 reduction (see Chap. 2)
RNIP and MRNIP2 Fe^{H2} distributed by Toda, Japan
k_{obs} observed rate constant
k_{SA} surface-area normalized rate constant

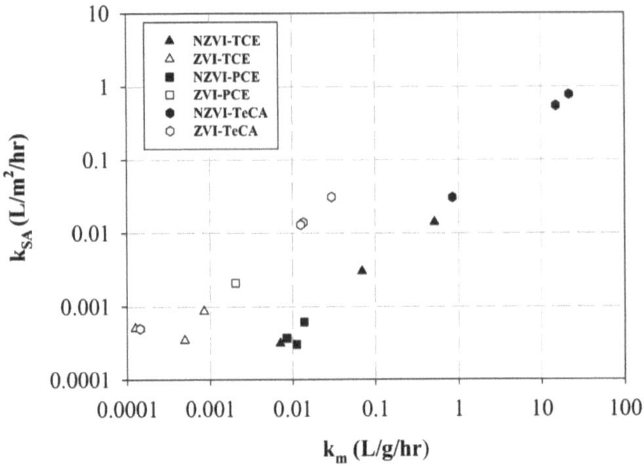

Fig. 3.1 Comparison of surface-area normalized rate constants (k_{SA}) against mass normalized rate constants (k_M) for the degradation of TCE, PCE, and TeCA by NZVI and bulk ZVI. (Data from Johnson et al. (1996), Liu et al. (2005b), Song and Carraway (2005), Amir and Lee (2011), Phenrat et al. (2016), Hepure Technology Inc. (2017), Kim et al. (2017))

CT by NZVI and bulk ZVI, the results show that NZVI gives larger k_M but similar k_{SA} values (Nurmi et al. 2005; Tratnyek and Johnson 2006). Nevertheless, when doing the same analysis for other contaminants, including trichloroethylene (TCE), tetrachloroethylene (PCE), and tetrachloroethane (TeCA) (Fig. 3.1), we found that a significant increase in k_{SA} for NZVI in comparison to ZVN was observed for TeCA and TCE but not for PCE. This suggests a potential "nano-effect," but it seems to be, at least in part, governed by the interaction between the contaminant and NZVI surface.

Regardless of its particle size and probable nano-effect, the fundamental chemistry of NZVI in an aqueous environment from a reaction viewpoint is the same as ZVI. For environmental remediation purposes, Fe^0 can be oxidized by halogenated organics (as electron acceptors) as long as such organics have an $E_H{}^0$ greater than −0.447 V. As a result of the electron transfer, in most cases, NZVI transforms such organic contaminants to more environmentally benign byproducts (Eqs. 3.1 and 3.2 using TCE as an example) (Liu et al. 2005a, b). In the meantime, Fe^0 can also react with water (or H^+) to produce H_2 gas (Eq. 3.3), which is a competing reaction to the dechlorination reaction and is strongly controlled by the availability of H^+ (i.e., pH).

$$Fe^{2+} + 2e^- \rightarrow Fe^0, \quad E_H{}^0 = -0.447 \text{ V} \tag{3.1}$$

$$\text{TCE} + ne^- + (n-3)H^+ \rightarrow \text{product} + 3Cl^- \tag{3.2}$$

$$H^+ + e^- \rightarrow H^* \rightarrow \frac{1}{2}H_2 \tag{3.3}$$

Although the oxidation of Fe^0 to Fe^{2+} (Eq. 3.1) is usually assumed, in environmentally relevant applications (i.e., groundwater at a natural pH), the transformation of Fe^0 to the film of iron oxide, such as magnetite (Fe_3O_4) (Eq. 3.4) and maghemite (Fe_2O_3), is often observed (Liu et al. 2005b; Reinsch et al. 2010). Thermodynamically, the Fe^0/Fe_3O_4 couple is more favorable than the Fe^0/Fe^{2+}couple at a pH above 6.1. The oxidation of Fe^0 to Fe_3O_4 provides an additional 2/3 mole of electrons per mole of Fe^0 oxidized in comparison to the oxidation to Fe^{2+} (Liu et al. 2005b). However, the formation of iron oxide film can passivate NZVI and adversely affect its reactivity and reactive life time (Liu et al. 2005b; Reinsch et al. 2010).

$$Fe_3O_4 + 8H^+ + 8e^- \rightarrow 3Fe^0 + 4H_2O \tag{3.4}$$

As shown in Eqs. 3.1, 3.2, 3.3, and 3.4, electrons from the NZVI oxidation can be used to reduce contaminants or to produce H_2. For bimetallic NZVI (i.e., NZVI modified by noble metals, such as Pd, Pt, and Rh; (Wang and Zhang 1997; Zhang et al. 1998) and highly disordered monometallic NZVI (Liu et al. 2005a); see Sect. 3.1.3.1), H_2 can be activated and used for dechlorination via hydrodechlorination (Eq. 3.5).

$$H_2(g) \leftrightarrow 2H_{ad} \leftrightarrow 2H^+_{(aq)} + 2e^- \tag{3.5}$$

This explains the database in Table 3.1, in that bimetallic NZVI is typically 10–50 times more reactive than bare NZVI for the same contaminant, considering the enhanced pseudo-first-order reaction rate constant. The cost-effectiveness of in situ remediation using NZVI can be substantially affected by how electrons are utilized. In addition, the ability to utilize H_2 for hydrodechlorination and the characteristics of the iron oxide film formed on the surface of NZVI influence its lifetime, reactivity, and thus treatment efficiency. For NZVI, these phenomena are controlled by (1) intrinsic properties of the materials and (2) environmental conditions where NZVI particles are applied. These factors will be discussed later in this chapter, but we will first discuss reductive degradation pathways of various kinds of contaminants in aqueous environments in Sect. 3.1.2.

3.1.1 Reductive Transformation Pathways

3.1.1.1 Chlorinated Methane

While chlorinated methane, such as tetrachloromethane (or carbon tetrachloride (CT)) and trichloromethane (or chloroform (CF)), are mendable by NZVI, dichloromethane (DCM) shows negligible reactivity toward NZVI, giving the reactivity order of

CT > chloroform (CF) >> DCM (Song and Carraway 2006); Table 3.1). Nevertheless, the reductive transformation of CT using NZVI (both Fe^{BH} and Fe^{H2}) cannot completely dechlorinate CT to methane. Instead, it produces toxic byproducts including CF (18.4–39.3% of the initial concentration of CT) and dichloromethane (0–2.4% of the initial concentration of CT) as well as unknown byproducts (23.3–42.5% of the initial concentration of CT) presumably methane, carbon monoxide, and formate, while CT retains 15.8–58.2% of its initial concentration after 27 h (see Table 3.1) (Nurmi et al. 2005; Song and Carraway 2006). Because the dechlorination reactions are heterogeneous, the rates depend greatly on the interaction between chlorinated organics and the NZVI surface (iron oxide). For dechlorination of highly chlorinated organics such as CT, CF, and DCM, stepwise electron transfer takes place. The weak sorption of chlorinated methane on the NZVI surface may result in a short residence time and desorption of incompletely dechlorinated compounds. Interestingly, while Fe^{BH} was reported to perform indirect dechlorination of chlorinated ethene using reactive hydrogen species (Liu et al. 2005a), for chlorinated methane, the dechlorination occurs via a direct electron transfer reduction mechanism, rather than the indirect mechanism (Song and Carraway 2006).

Two possible dechlorination pathways include hydrogenolysis and α elimination. Hydrogenolysis is a two-step electron transfer reaction initiated by direct dissociative electron transfer, in which the first electron addition results in dissociation of the molecule to a trichloromethyl free radical ($\bullet CCl_3$) and chloride (Mccormick and Adriaens 2004). The free radical can subsequently form CF through the direct abstraction of hydrogen (Fig. 3.2a) or could be further reduced to a trichloromethyl carbanion ($:CCl_3$), which then forms CF via the rapid addition of H^+ from the solution (Mccormick and Adriaens 2004; Song and Carraway 2006) (Fig. 3.2a). This is supposed to be a prevalent pathway of CT degradation by NZVI, given that CF is a major byproduct of the degradation. Similarly, hydrogenolysis is supposed to be a major degradation pathway for CF as well, given that DCM is the major byproduct (Fig. 3.2b) (Song and Carraway 2006).

On the other hand, CH_4 was also observed along with DCM in the reduction of CT and CF. Since DCM is a rather inert intermediate, the sequential hydrogenolysis of DCM to CH_4 is unlikely. Consequently, methane formation must be caused by a parallel CT degradation pathway, bypassing CF and DCM as an intermediate. This can happen via concerted reductive elimination steps (two-electron transfer (α elimination) and four-electron transfer, respectively) to form dichloromethylene, methylene, and eventually methane (Fig. 3.2c). Notedly, α elimination is a dichloro-elimination process involving a two-electron transfer to the molecule and the elimination of two chlorine atoms. The reaction products are less-saturated aliphatic hydrocarbons and two chloride ions.

Notedly, α elimination is the elimination of chlorine atoms from one carbon atom, while β elimination occurs when chlorine atoms are removed from two different carbons. Reductive β elimination is known to be a preferential pathway for compounds possessing (α, β) pairs of chlorine atoms (Arnold and Roberts 2000), while hydrogenolysis or reductive α elimination is the primary transformation pathway for compounds possessing only α chlorines (Mccormick and Adriaens 2004; Song and Carraway 2005).

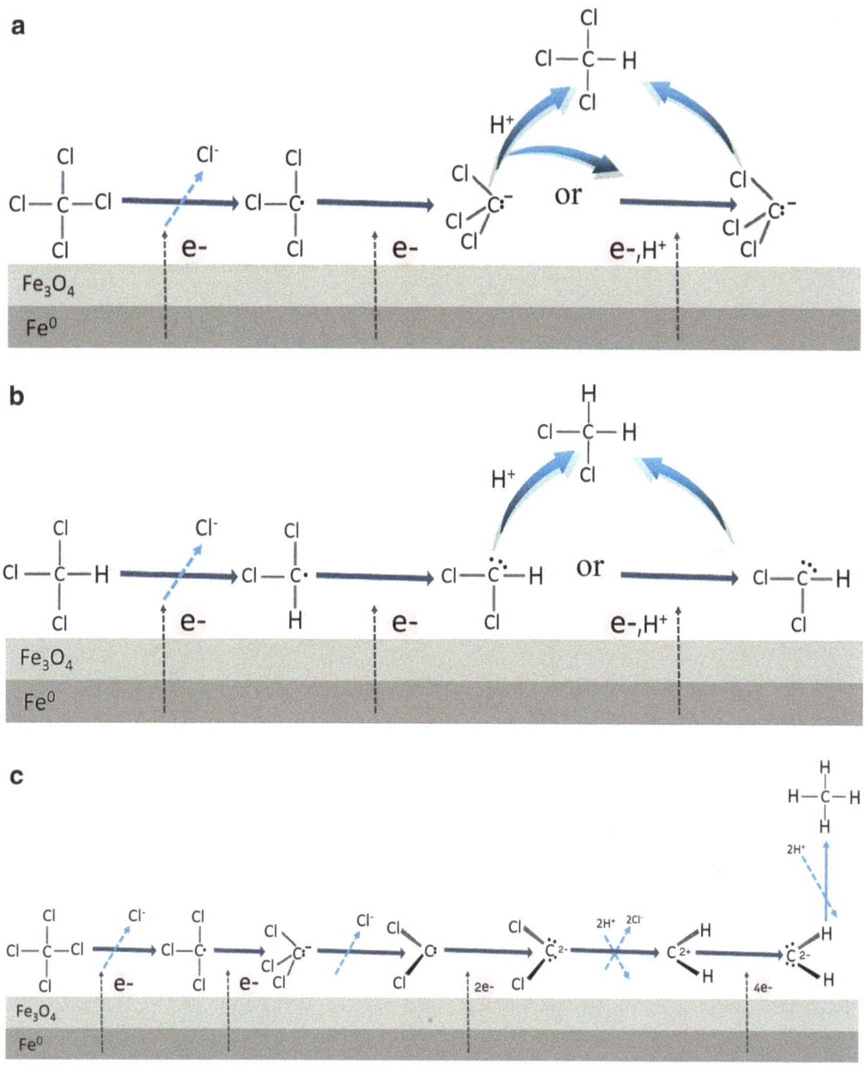

Fig. 3.2 (**a**) CT degradation pathway forming CF via trichloromethyl radical or carbanion intermediates, (**b**) CF degradation pathway forming DCM via dichloromethyl radical or carbanion intermediates, and (**c**) speculated methane (CH_4) formation pathway from the dichlorocarbene intermediate to CH_4 involving multi-electron transfers. (Adapted from Mccormick and Adriaens (2004), Song and Carraway (2006))

3.1.1.2 Chlorinated Ethene and Ethane

The NZVI can completely dechlorinate all chlorinated ethene and ethane (except 1,2-DCA) to non-chlorinated intermediates, such as acetylene, and byproducts including ethene and ethane. Unlike chlorinated methane, the main degradation pathway of chlorinated ethenes via NZVI is dichloro-elimination (β elimination)

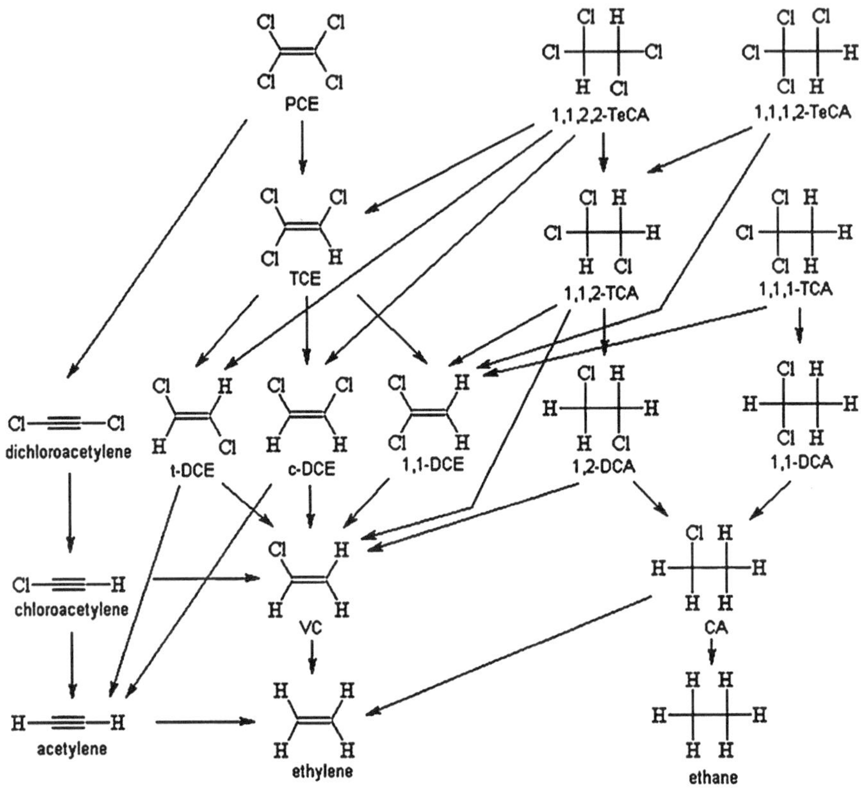

Fig. 3.3 Dechlorination pathway, ethene and ethane (Tobiszewski and Namieśnik 2012). (Reprinted with permission from Tobiszewski and Namieśnik (2012). Copyright (2012) Springer Nature)

followed by hydrogenolysis (Fig. 3.3). Similarly, the main degradation pathways of chlorinated ethane, such as HCA, PCA, and 1,1,1,2-TeCA, via NZVI is β elimination followed by hydrogenolysis (Fig. 3.3). Nevertheless, for 1,1,2,2-TeCA and both 1,1,1-TCA and 1,1,2-TCA, dehydrohalogenation becomes equally important, if not more important, than β elimination, especially at high a pH (Song and Carraway 2005).

Fundamentally, for chlorinated ethene, β elimination is preferable over hydrogenolysis because thermodynamic reduction potentials for two-electron reduction of chlorinated ethene and ethane via β elimination are more favorable than hydrogenolysis. This is demonstrated by an example of the half reaction of PCE transformation via β elimination (Eq. 3.6) and hydrogenolysis (Eq. 3.7) (Totten and Roberts 2010). From a kinetic viewpoint, the predominance of β elimination over hydrogenolysis is also partially due to the rapidity of intermolecular rearrangements (leading to β elimination) compared to bimolecular collisions with H^+ (leading to hydrogenolysis) at moderately basic pHs (Arnold and Roberts 2000).

$$CCl_2 = CCl_2 + 2e^- \rightarrow CCl \equiv CCl + 2Cl^- \quad E_H^0 = 0.631 \text{ V} \tag{3.6}$$

$$CCl_2 = CCl_2 + 2e^- + H^+ \rightarrow CCl_2 = CHCl + Cl^- \quad E_H^0 = 0.592 \text{ V} \tag{3.7}$$

Additionally, Fe^{BH} was reported to perform indirect dechlorination of chlorinated ethene using reactive hydrogen species (Liu et al. 2005a). This will be discussed in more detail in the next section.

Since the rate-limiting step in the reduction of chlorinated ethene and ethane using ZVI is supposed to be the transfer of a single electron and the formation of an alkyl radical (Johnson et al. 1996; Arnold and Roberts 2000), Scherer et al. (1998) proposed linear free energy relationships (LFERs) capable of explaining or predicting the rates of dehalogenation by ZVI. They showed that, at the same number of chlorine atoms, dechlorination rate constants of chlorinated ethanes are typically higher than chlorinated ethene. For an internal comparison among chlorinated ethene or among chlorinated ethane, the dechlorination rate constants tend to increase with increasing chlorination of the compounds. As for NZVI, the reactivity of chlorinated ethanes summarized in Table 3.1 reasonably agrees with this trend, in that HCA > PCA > 1,1,1,2-TeCA >1,1,1-TCA > 1,1,2,2-TeCA >1,1,2-TCA > 1,1-DCA. Moreover, the k_{SA} values of these chlorinated ethanes linearly correlate with the one-electron reduction potential (E1) and the lowest unoccupied molecular orbital (LUMO) energy of chlorinated ethanes used in the LFERs (Song and Carraway 2005). In contrast, chlorinated ethenes do not follow the proposed trend. As shown in Fig. 3.1 and Table 3.1, the k_{SA} of PCE is lower for TCE for bare Fe^{H2} (RNIP) (Liu et al. 2007; Fagerlund et al. 2012). A similar finding was also reported for ZVI for all the series of chlorinated ethene (i.e., k_{SA} of PCE < TCE < DCEs < VC) (Arnold and Roberts 2000). Based on this finding, Arnold and Roberts (2000) hypothesized that for chlorinated ethene, the transfer of a single electron is not the rate-limiting step but rather the formation of a di-σ-bonded intermediate (Fig. 3.4). This may also explain the findings on NZVI for TCE and PCE.

Noticeably, NZVI reduction is known to be incapable of completely treating 1,2-DCA (Maes et al. 2006; Deng and Hu 2007; Su et al. 2012a). The pseudo-first-order rate constants of 1,2-DCA using NZVI at the concentration of 0.4 g/L and at neutral pH were less than 5×10^{-5}/h (Table 3.1) (Song and Carraway 2005). The poor dechlorination efficacy of 1,2-DCA using NZVI may be due to the relatively high C–Cl bond strength (87.2 kcal/mol) (Cioslowski et al. 1997) compared to the higher chlorinated ethanes. For example, the bond strength of HCA and 1,1,1-TCA was 68.83 and 73.6 kcal/mol, respectively, while the k_{SA} of HCA was four times greater than that of 1,1,1-TCA using Pd/Fe^0 nanoparticles under the same experimental conditions (Lien and Zhang 2005).

Arnold and Roberts (2000) implied that the slow dechlorination rate of 1,2-DCA by ZVI is explainable by the slow formation of a di-σ-bonded intermediate. Recently, sulfidation of NZVI using dithionite was found to degrade more than 90% of 1,2-DCA, but over the course of a year, pseudo-first-order rate constants ranged from 3.8×10^{-3} to 7.8×10^{-3}/day (Garcia et al. 2016); for more details about NZVI sulfidation, see Chap. 9). Even with this novel NZVI modification, the

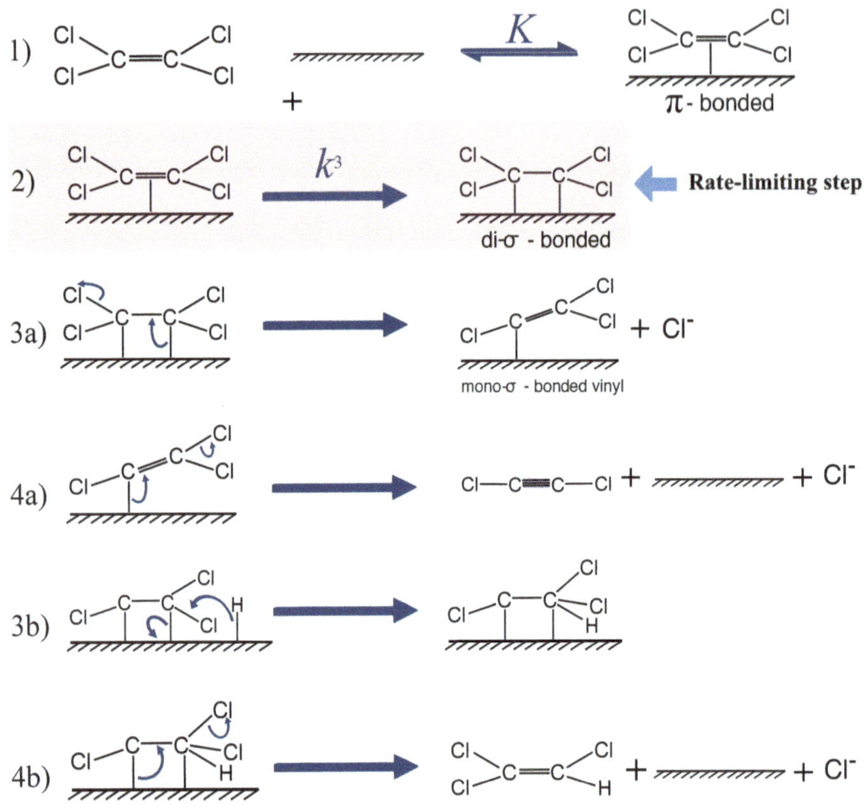

Fig. 3.4 Hypothesized rate-limiting step of the PCE dechlorination by ZVI (i.e., formation of a di-σ-bonded intermediate). After this step, the reaction may proceed via Steps 3a and 4a to give the reductive β-elimination product dichloroacetylene or via Steps 3b and 4b to give the hydrogenolysis product TCE (Arnold and Roberts 2000). (Adapted with permission from Arnold and Roberts (2000). Copyright (2000) American Chemical Society)

reductive dechlorination rate constants are still relatively low and may not be practical for source zone treatment of 1,2-DCA. In contrast, the Fenton process has shown its full potential to degrade 1,2-DCA. The NZVI-induced Fenton's reaction will be discussed more in the latter part of this chapter.

3.1.1.3 Chlorinated and Aromatic Nitro Hydrocarbon

Chlorinated aromatic hydrocarbon, such as chlorophenols, can be detoxified by NZVI to yield a much less hazardous byproduct, such as phenol. Nevertheless, the dechlorination rate constant of pentachlorophenol (PCP) by bare NZVI is relatively low ($K_{obs} = 0.001$/h using smectite-templated Fe^0 containing 13.3 g/L of smectite and 7 mM Fe^0). Thus, to improve the treatability, a Pd catalyst (0.8 mM Pd^0) was added to the nanocomposite, yielding the dechlorinate rate constant of 0.487/h.

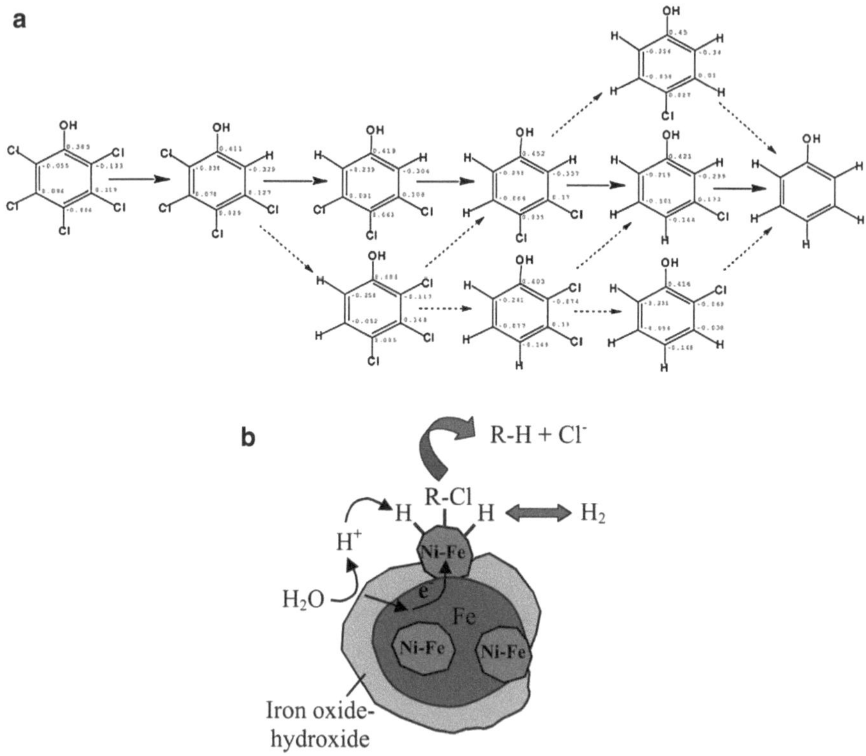

Fig. 3.5 (**a**) Primary dechlorination pathways of chlorinated phenols by smectite-templated Pd^0/Fe^0. Solid arrows refer to the major reaction route, and broken arrows refer to the minor reaction pathway (Jia and Wang 2015) and (**b**) hydrodechlorination driven by Pd^0/Fe^0 nanoparticles (Schrick et al. 2002). (Reprinted with permission from Schrick et al. (2002) and Jia and Wang (2015). Copyright (2015) Elsevier and Copyright (2002) American Chemical Society)

Figure 3.5a illustrates dechlorination pathways of chlorinated phenols by smectite-templated Pd^0/Fe^0 (Jia and Wang 2015).

Hydrodechlorination (Eq. 3.5) is the main dechlorination pathway for such a catalytic NZVI. Protons in the aqueous solution acquire electrons from NZVI, forming hydrogen radical on the Pd surfaces (Fig. 3.5b). In this process, the active H atom formed by Pd^0/Fe^0 attacks the chlorinated phenols via electrophilic H addition to the π system of the benzene ring (formation of H-aromatic complexes) and then removes the chlorine atom by two-electron transfer. Since this process is promoted by electrophilicity of the chlorinated aromatic hydrocarbon, the more chlorinated the phenols, the lower the dechlorination rate with Pd^0/Fe^0 particles because electron-withdrawing groups, such as Cl, that are attached to the aromatic ring reduces the electron density associated with the ring carbons, decreasing the probability of complexation between active H and the corresponding aromatic ring (Jia and Wang 2015). The selectivity of dechlorination of chlorinated phenols is mainly due to charges and steric hindrance governed by the interaction of substituent Cl atoms and individual aromatic carbon.

Figure 3.5a also shows the calculated charge on each carbon of chlorinated phenols in red. Substituent Cl at the ortho-position is preferentially dechlorinated from the phenol ring due to more negative charges associated with the ortho-position C, which is attributed to the influence of the adjacent OH group. The H in the formed H-aromatic ring complexes is prone to interact with more negatively charged carbon, leading to the scission of the CCl bond (Jia and Wang 2015).

Steric hindrance is another important factor governing the dechlorinate rate of chlorinated phenols. For example, for those chlorinated phenols not containing ortho-Cl, Cl located at the meta-position can be dechlorinated radially due to less steric hindrance. Consequently, as summarized in Table 3.1, the dechlorination rate constants of chlorinated phenols follow the order: PCP < 2,3,4,5-TeCP ~ 2,3,4,6-TeCP ~ 2,3,5,6-TeCP <2,3,4-TCP~2,3,6-TCP ~ 2,4,6-TCP < 3,4,5-TCP ~ 2,4,5-TCP ~ 2,3,5-TCP < 2,6-DCP < 3,5-DCP < 2,3.DCP ~ 2,4-DCP ~ 3,4-DCP < 2,5-DCP < 4-CP ~ 2-CP < 3.CP (Jia and Wang 2015).

Similarly, NZVI and bimetallic NZVI can sequentially dechlorinate chlorobenzenes yielding PeCB, TeCBs, TCBs, and DCBs as byproducts. Unlike chlorophenols, no selectivity of dechlorination product formation was observed for the case of HCB (Zhu et al. 2010). Noticeably, HCB dechlorination is not complete (i.e., no substantial formation of benzene was detected). Interestingly, for HCB dechlorination, Cu appears to be a better catalyst than Pd (Shih et al. 2009; Zhu et al. 2010). Furthermore, nitroaromatic compounds (NACs), such as para-nitrochlorobenzene (p-NCB), are treatable by Ni/Fe^0 nanoparticles. In addition, Ni plays the role of catalyst, decreasing the activation energy of hydrogenolysis of C–Cl bonds, while Fe^0 supplies electrons and reduces H^+ to produce hydrogen gas. Complete transformation (100% removal efficiency) is achieved using Ni/Fe^0 (2.0% Ni) at the concentration of 6 g/L after 300 min. The p-NCB is first adsorbed by NZVI and then quickly reduced to p-chloroaniline (p-CAN) and eventually to aniline (AN) via hydrogenolysis (Xu et al. 2009).

3.1.1.4 Halogenated Bisphenol a (Flame Retardant)

Bimetallic NZVI can completely dechlorinate halogenated bisphenol A, especially at a slightly acidic pH (pH 5–6). Two dehalogenation mechanisms take place simultaneously for different degrees at different pH ranges. Using debromination of tetrabromobisphenol (TBBPA) by Ni/Fe^{H2} nanoparticles as an example, at pH ranging from 3 to 9, sequential dehalogenation via hydrogenolysis takes place as shown in pathway (a) in Fig. 3.6. This hydrogenolysis yields brominated intermediates, such as tri-, di-, and mono-BBPA, as well as a small amount of bisphenol A (BPA; 7.2–7.6%) (Li et al. 2016, 2017). Nevertheless, at a slightly acidic pH (pH 5–6), concerted hydrogenolysis (pathway (b) in Fig. 3.6) is dominant, yielding BPA as a major byproduct (90.7–93.3%) with a small amount of tri-, di, and mono-BBPA (Li et al. 2016, 2017).

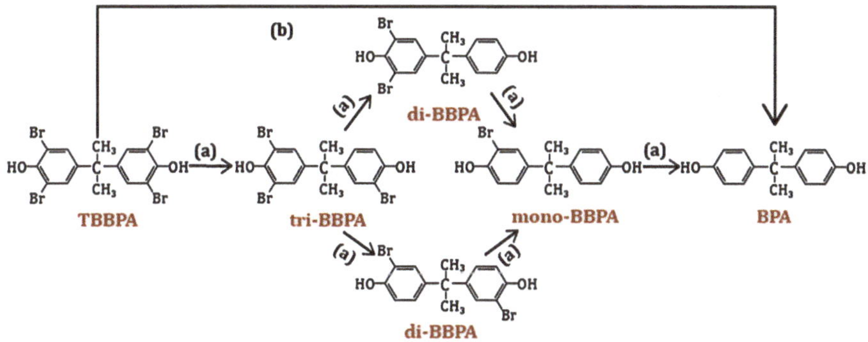

Fig. 3.6 Proposed primary debromination pathways of TBBPA by Ni/NZVI at different initial pHs (a) pH 3.0, 5.0, 6.0, 7.0, and 9.0 and (b) pH 5.0 and 6.0 (Li et al. 2016). (Reprinted with permission from Li et al. (2016). Copyright (2016) Elsevier)

3.1.1.5 Pesticides and Dyes

The NZVI can completely transform complex chlorinated pesticides, such as γ-hexachlorocyclohexane (known as lindane) in 24 h at a NZVI concentration of 0.1–0.39 g/L (Elliott et al. 2009). Figure 3.7 summarizes the degradation pathway of lindane, yielding γ-3,4,5,6-tetrachlorocyclohexene (TeCCH) as a major intermediate with benzene as a byproduct. The dihalo-elimination of vicinal chlorides from carbons 1 and 2 of lindane is supposed to be the first and rate-limiting step. Initially, lindane must adsorb onto the NZVI surface. Then, an electron from Fe^0 is donated to the surface-associated lindane, forming a neutral radical with the release of a chloride ion. Another electron is then transferred from the transient Fe^+ species to the reacting carbon center of the radical, which then undergoes double bond formation and simultaneous release of chloride from the beta carbon, yielding TeCCH as an intermediate. Noticeably, both chlorine atoms (from carbons 1 and 2 of lindane) occupy axial positions on their respective carbon centers and are oriented antiperiplanar, which maximizes their susceptibility toward reduction. After this initial step, the two subsequent dihalo-elimination steps are believed to occur more rapidly, yielding benzene and chloride as major end products (Elliott et al. 2009).

The NZVI can also decolorize dye-contaminated water. Figure 3.8 is an example of the reductive decolorization of methyl orange (MO). It starts with MO adsorption onto the NZVI surface by the formation of the chelate complex of Fe(II) dye. The radicals of H were generated by the NZVI reduction of water or hydrogen ion. Two H radicals were involved in the cleavage of the azo bond with two more electron transfers from NZVI. An electron is transferred from NZVI to the –N=N– bond, causing the breakage of the bond and the combination of a radical of H. In the meantime, another NZVI integrated another two radicals of H to break the N–N bond under the similar process. Eventually, the azo-double bond –N=N– was disconnected from two different amines. The breakage of the azo-double bond makes the visible absorption peaks at 464 nm vanish.

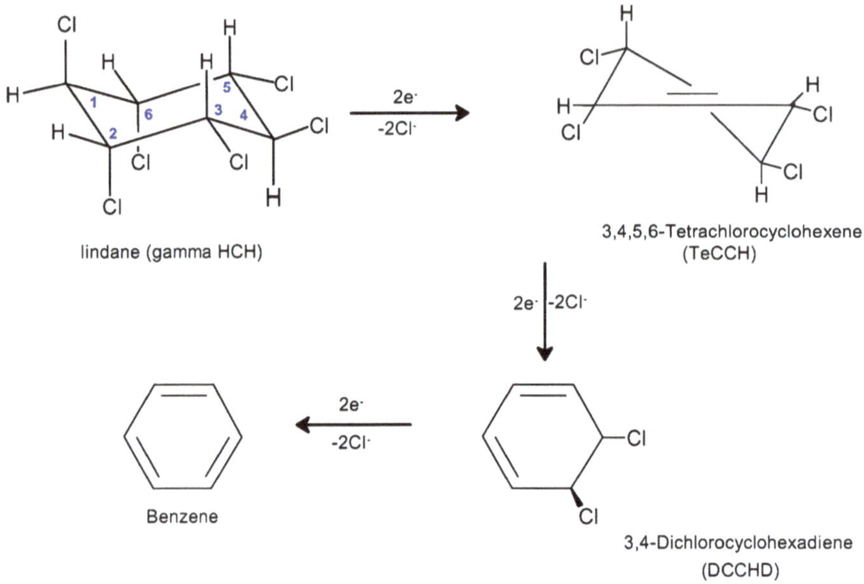

Fig. 3.7 Dihalo-elimination degradation pathway from lindane to benzene (Elliott et al. 2009); with permission from ASCE). (Reprinted with permission from Elliott et al. (2009). Copyright (2016) American Society of Civil Engineers)

3.1.2 Particle Properties Affecting NZVI Reactivity

Table 3.1 summarizes dechlorination rate constants of various contaminants. Noticeably, the rate constants for the same kind of contaminants by different types of NZVI can be substantially different. Although NZVI has two common characteristics, the small size and high specific surface area, NZVI has a vast variety of physicochemical properties, which can enormously affect dechlorination pathways and kinetics. Since the mass of NZVI needed for remediation and thus the cost of the treatment depend on particle reactivity, reactive lifetime, and particle efficiency, all of which are controlled by NZVI physicochemical properties, it is worth elaborating their effects on NZVI performance in this section.

3.1.2.1 Particle Crystallinity and Chemical Composition (Noble Metal)

The previous section discusses the different degradation pathways of various contaminants. Nevertheless, it also notes that different types of bare NZVI, such as Fe^{BH} and Fe^{H2}, have an important contribution to degradation pathways. This is mainly due to the difference in their crystallinity. According to multiple lines of evidence from the transmission electron microscope (TEM), energy-dispersive

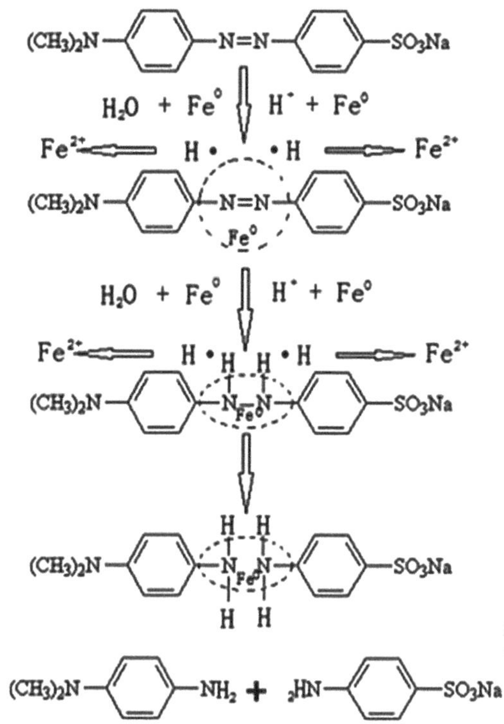

Fig. 3.8 Proposed mechanism for the degradation of MO using Fe^0 nanoparticles (Chen et al. 2011b). (Reprinted with permission from Chen et al. (2011b). Copyright (2011) Elsevier)

spectroscopy (EDS), electron energy-loss spectroscopy (EELS), and X-ray diffraction (XRD) (Liu et al. 2005a, b; Nurmi et al. 2005; see also Chap. 2), the morphology of Fe^{H2} appears to be a crystallographic angular structure rather than a round, amorphous shape-like Fe^{BH}. While no significant oxide shell was observed for Fe ^{BH} (Liu et al. 2005b), the iron oxide phases, such as magnetite and maghemite, were observed as the stabilized shell of Fe^{H2} (Liu et al. 2005b; Nurmi et al. 2005; Baer et al. 2007; Reinsch et al. 2010). This crystallinity and iron oxide shell critically govern particle reactivity, particle lifetime, and particle efficiency (Liu et al. 2005b; Nurmi et al. 2005).

As shown in Table 3.1, using TCE as an example, the k_{SA} value using Fe^{H2} (0.35 g/L) is 4×10^{-4} L/(h m^2), while the k_{SA} value using Fe^{BH} is 2×10^{-3} L/(h m^2) at the same experimental condition. This is a fivefold difference in k_{SA} values (Liu et al. 2005b). The similar trend is also reported for the dechlorination of tetrachloromethane using Fe^{BH} and Fe^{H2} (Nurmi et al. 2005). This is because amorphous Fe^{BH} can utilize H_2 produced from the H^+ reduction for TCE dechlorination through a catalytic hydrodechlorination pathway (Liu et al. 2005a). This increases efficiency of electron utilization for Fe^{BH} because electrons used to produce H_2 can then be used to degrade TCE. This feature is not available in crystalline Fe^{H2}, making its TCE dichlorination rely on β elimination and hydrogenolysis only. Thus, for Fe^{H2}, the electron used to produce H_2 is considered unproductive with respect to dechlorination. Nevertheless, due to their amorphous Fe^0 and the absence of an iron oxide shell, at pH 7, a relatively large fraction of Fe^0

in Fe^{BH} interacted with H^+ and was used to produce H_2, implying that Fe^{BH} has a poor selectivity of electron utilization for dechlorination (Liu et al. 2005a).

In the same conditions, a relatively small fraction of Fe^{H2} was used to produce H_2, implying that Fe^{H2} has a good selectivity of electron utilization for dechlorination (Liu et al. 2005b; Liu and Lowry 2006).

In addition to particle crystallinity, different chemical compositions of Fe^{BH} and Fe^{H2} also affect the electron utilization efficiency. Unlike the case of Fe^{H2}, the formation of iron oxide film around Fe^{BH} was not observed due to the presence of boron in Fe^{BH} particles that facilitated the dissolution of the particles and thus regenerated the reactive sites for TCE dechlorination. Consequently, almost all of the Fe^0 in Fe^{BH} was available for dechlorination (i.e., high particle efficiency ~92%; (Liu et al. 2005a, b)). On the other hand, the iron oxide shell grew as RNIP (commercially available Fe^{H2}) was oxidized, making some fraction of Fe^0 in RNIP unavailable for dechlorination (Liu et al. 2005b). Consequently, particle efficiency went down (only around 52%). However, the disadvantage of Fe^{BH} particles is their relatively short lifetime in comparison to Fe^{H2}. For in situ remediation, a long reactive lifetime is preferable for cost-effective treatment of contaminant plumes and to ensure that NZVI particles do not "burn out" prior to reaching contaminated areas (Liu et al. 2005b).

In addition to the effect of boron in Fe^{BH} on particle dissolution, another example of the influence of chemical composition on NZVI reactivity is the bimetallic NZVI (i.e., NZVI doped with Pd, Pt, Cu, and Ni) to enhance particle reactivity through catalytic pathways (Zhang et al. 1998; Schrick et al. 2002; Zhang 2003). The structure of bimetallic NZVI particles is a reductive Fe^0 core with the shell of inert noble metals (Fig. 3.5b as an example). Because of the presence of noble metals, bimetallic particles can utilize H_2 for hydrodechlorination (Zhang 2003). As shown in Table 3.1, using chlorinated ethane and ethene as examples, k_{SA} or k_{obs} using bimetallic NZVI is around 10–50 times greater than bare NZVI. However, a significant problem with bimetallic NZVI is the observed decrease in reactivity over time due to the deactivation if the thick oxide layer can form and cover the noble metals (Zhu et al. 2006). This deactivation issue is an important obstacle for long-term remediation using bimetallic NZVI.

3.1.2.2 Effect of Polymeric Surface Modification

Polymeric surface modification, as a means to increase NZVI mobility in the subsurface and affinity for specific subsurface contaminants (Chaps. 5 and 6), is essential for in situ remediation using NZVI. However, polymeric surface modification may either enhance or decline NZVI reactivity toward organic contaminants. Since the dechlorination reactions are heterogeneous (the contaminant must contact the particle surface to be degraded), NZVI synthesis in the presence of polymers or polyelectrolytes, such as carboxymethyl cellulose (CMC), guar gum, and polyvinylpyrrolidone (PVP), in a "one pot" manner appears to enhance NZVI reactivity. This surface modification approach yields smaller particles in comparison to the physisorption of polymers or polyelectrolytes onto pre-synthesized NZVI. Conceptually, the Fe^{2+} or $Fe3^{+}$ ions from $FeSO_4$ or $FeCl_3$ may be complexed with polyelectrolytes prior to reductive precipitation using $NaBH_4$. The complexed Fe^{2+} or Fe^{3+} ions behave as nucleation seeding points of NZVI. Consequently, we obtained polymer-modified NZVI with a smaller size that was resistant to aggregation and, thus, more reactive than larger, non-stabilized NZVI. This results in the observed increases in TCE reactivity with polymer-modified Fe-Pd bimetallic nanoparticles at low polyelectrolyte concentrations compared with bare Fe-Pd bimetallic nanoparticles (He and Zhao 2008; Sakulchaicharoen et al. 2010) (Table 3.1).

Nevertheless, the opposite is true for polymeric surface modification of pre-synthesized NZVI, in that polymeric surface modification can decrease the NZVI reactivity via reactive site blocking and mass-transfer resistance. As seen in Fig. 3.9a, the polymeric modified NZVI is covered with a layer of adsorbed macromolecules. According to the Scheutjens and Fleer conceptual model for homopolymer sorption (Scheutjens and Fleer 1979, 1980), charged homopolymers are normally adsorbed onto the surface in the train-loop-tail configuration.

Trains are segments of polymer directly attached to the particle surface, which can block NZVI reactive sites, whereas loops and tails form an extended polyelectrolyte brush away from the surface, which can retard the diffusion of chlorinated volatile organic compounds (CVOCs) from the bulk aqueous solution to the NZVI surface. Similarly, for random copolymer, block copolymer, or grafted polymer, trains or anchoring blocks may block electron transfer sites, while a polymer brush may limit the mass transfer or chlorinated organics to the reactive NZVI surface. Several recent studies support this hypothesis. As summarized in Table 3.1, Phenrat et al. (2009a) revealed that, when the surfaces of pre-synthesized NZVIs were modified by the physisorption of polyelectrolytes, the TCE dechlorination rate constant decreased nonlinearly with the increasing adsorbed mass of the polyelectrolytes, with a maximum 24-fold decrease in reactivity. This is due to reactive site blocking and a decrease in the aqueous TCE concentration at the surfaces of the NZVIs due to the partitioning of TCE to the adsorbed polyelectrolytes. A similar finding was also reported by Wang and Zhou (2010) using solvent-responsive, polymer-coated NZVIs to degrade TCE (Wang and Zhou 2010). Noticeably, for the case of "one pot" polymer-modified NZVI, we believe that polymer still blocks the NZVI reactive site and still resists CVOC mass transfer to the NZVI surface, but the two adverse effects are outweighed by the positive effects from the smaller size of NZVI.

Nevertheless, polymeric surface modification makes the NZVI less sensitive to environmental factors in comparison to bare NZVI. As summarized in the next section, non-reducible ionic species, such as Cl^-, SO_4^{2-}, HCO_3^-, and HPO_4^{2-}, decreased the TCE dechlorination rate constant by bare NZVI up to a factor of seven compared with deionized (DI) water at pH 8.9 (Liu et al. 2007). On the other hand, polymeric surface modification reduces the interaction of NZVIs with nontarget groundwater solutes (organic and ionic species). A possible explanation is the effect of the Donnan potential in the polyelectrolyte layer on ionic solute distributions. The Donnan potential (Fig. 3.9b, c) can decrease the concentration of cationic solutes at

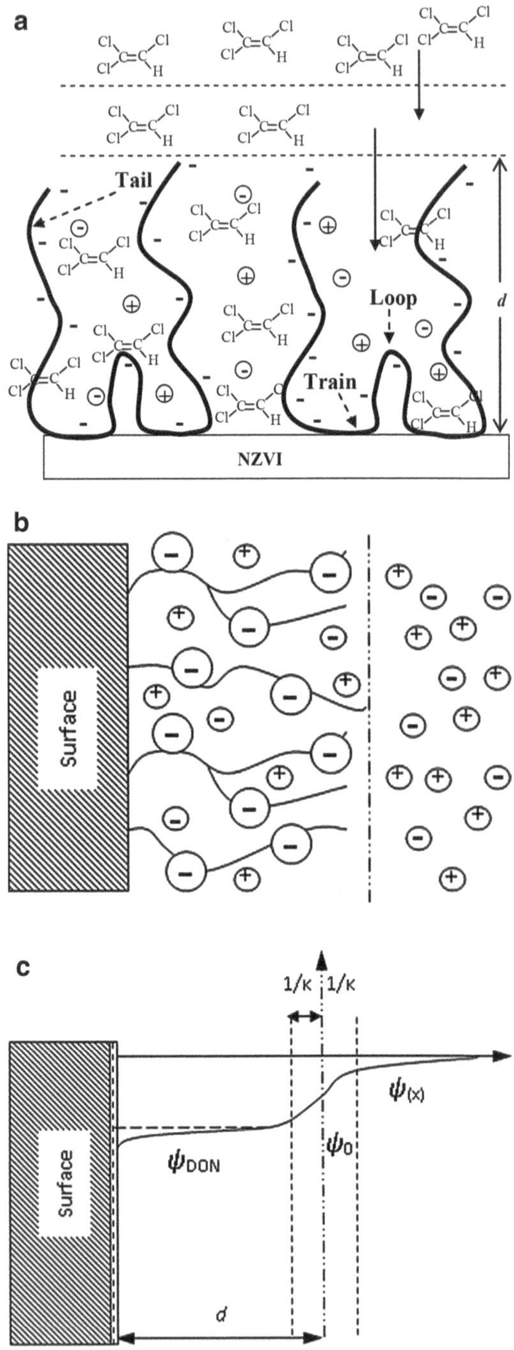

Fig. 3.9 Schematic illustrating site blocking due to adsorbed trains and formation of an extended brush layer (**a**) of loops and tails (Phenrat et al. 2009b) and (**b**, **c**) effect of the Donnan potential in the polyelectrolyte layer on ionic solute distributions (Phenrat et al. 2015). (Reprinted with permission from Phenrat et al. (2009b, 2015). Copyright (2009) American Chemical Society and Copyright (2015) Springer Nature)

the surfaces of NZVIs and possibly reduce their blocking effect compared with that for bare NZVIs. A recent study reported that for the first TCE spike, TCE dechlorination rates using polyaspartate (PAP)-modified NZVIs in the actual groundwater samples were 70–85% of the TCE dechlorination rate using PAP-modified NZVIs in DI water (Phenrat et al. 2015), while the TCE dechlorination rates using bare NZVI in the same groundwater samples for the first TCE spike were around ~22% of the TCE dechlorination rate using bare NZVIs in DI water.

Furthermore, over an intermediate period (30 days), in the presence of groundwater solutes, polyelectrolytes, such as PAP, were desorbed from NZVI and thus restored the reactivity of bare NZVIs with TCE. Evidently, the TCE dechlorination rates using PAP-modified NZVIs in the second and third TCE dechlorination cycles (intermediate-term effect) increased substantially (~100% and 200%, respectively, from the rate of the first spike). The desorption of PAP from the surface of NZVIs over time due to salt-induced desorption is hypothesized to restore NZVI reactivity with TCE (Phenrat et al. 2015). This suggests that the modification of the NZVI surface with small charged macromolecules, such as PAP, helps to deliver NZVIs to the subsurface and restores NZVI reactivity over time due to a gradual PAP desorption in groundwater.

3.1.2.3 Sorptive Support

Another means to increase NZVI mobility in the subsurface and the affinity for specific subsurface contaminants is using NZVI supports, such as submicron- or micron-sized porous silica (Zheng et al. 2008) or activated carbon (Mackenzie et al. 2012). These particles serve as carriers for NZVI transport because incorporation of NZVI into porous particles decreases the extent of magnetic attraction among NZVI particles, leading to less agglomeration and increasing their subsurface mobility. Furthermore, these supports are typically sorptive for organic contaminants; the NZVI composite materials behave as sorptive and reactive remediation agents. Figure 3.10 illustrates the dual-phase TCE removal mechanism by Fe^{BH} entrapped in porous ethyl-functionalized silica (Fe(B)/ethyl-silica system). Noticeably, an immediate sharp decrease of the aqueous TCE concentration to 45% of its original value was observed due to TCE sorption onto the functionalized silica followed by a much slower reaction rate, presumably due to dichlorination. Figure 3.10 also shows the byproduct formation, which supports the sorptive and reactive removal of TCE by the Fe(B)/ethyl-silica system.

Similar behavior is also observed by carbo-iron colloid (CIC), which is Fe^{H2} encapsulated in activated carbon. The sorptive and reactive TCE removal mechanism contributes to the higher reactivity and more NZVI utilization efficiency in comparison to bare NZVI without support. Mackenzie et al. (2012) showed that, for bare Fe^{H2}, dichlorination rate constants (k_{obs}) decreases with the decrease of the NZVI concentration, while the rate constants (k_{obs}) for CIC are insensitive to the CIC concentration in the suspension. This is because almost all TCE is adsorbed to the activated carbon of CIC, and since the dechlorination takes place at the surface, the

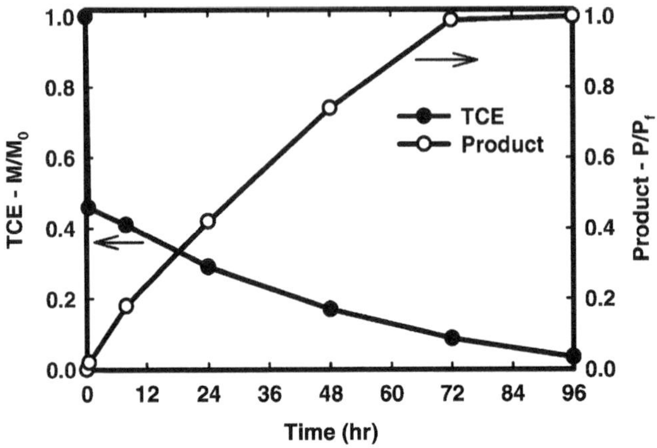

Fig. 3.10 TCE removal from solution and gas product evolution rates for Fe(B)/ethyl–silica, where M/M_0 is the fraction of the original TCE remaining, and P/P_f is the ratio of the gas product peak to the gas product peak at the end of 96 h (Zheng et al. 2008). (Reprinted with permission from Zheng et al. (2008). Copyright (2008) American Chemical Society)

reaction rate is determined by the Fe^0 content of the carbo-iron and not by the total Fe^0 concentration in the suspension (Mackenzie et al. 2012).

3.1.2.4 Aging Effect

Aging or longevity is the change of NZVI during immersion in water mainly by reacting with oxygen, water, target contaminants, or naturally occurring subsurface constituents. It can seriously affect both NZVI morphology and reactivity. Because in situ remediation is a moderate- to long-term operation, NZVI aging should be accounted for in determining the required amount of NZVI for contaminant transformation.

Typically, NZVI has a core (Fe^0) and shell (iron oxide, such as magnetite and maghemite) structure. Aging is a dynamic and complex process conceptually consisting of three processes including (1) breakdown of the existing oxide shell by hydration and auto reduction, (2) oxidation of the freshly exposed underlying Fe^0 coupled with reduction of reactive solutes, such as oxygen or target contaminants, and (3) subsequent cementation by formation of authigenic mixed-valent Fe(II)-Fe (III) phases (Sarathy et al. 2008). The aging phenomenon is substantially affected by the type of NZVI (Fig. 3.11) (Kim et al. 2012).

For Fe^{BH}, the aging model is described by the outward diffusion of the Fe^0 core toward the shell. This results in the formation of hollowed-out iron oxide shells, which are further transformed into sheet- and needle-shaped materials (Liu et al. 2015). Various secondary iron oxide mineral phases were formed at different aging times. For example, at 5 days, the main iron oxide shell is magnetite (Fe_3O_4) and maghemite (γ-Fe_2O_3), accompanied by lepidocrocite (γ-FeOOH). For 10-day

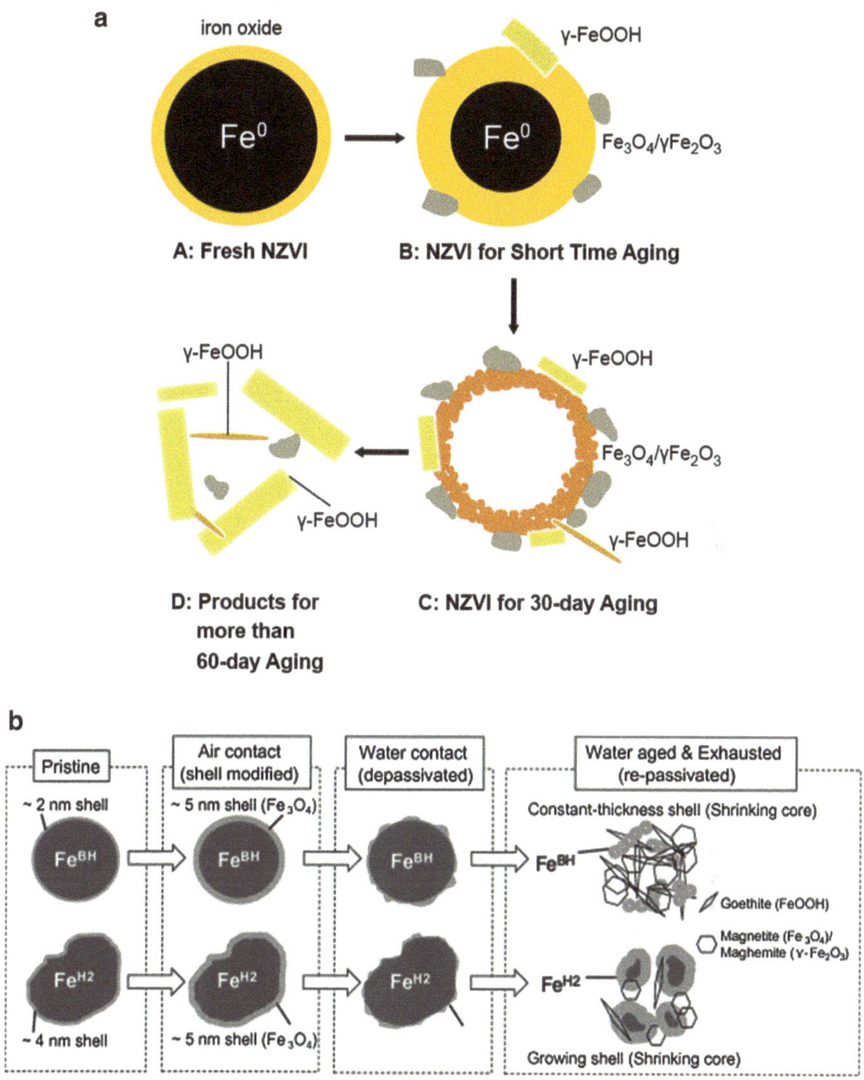

Fig. 3.11 Schematic description of the aging procedures of Fe^{BH} and Fe^{H2} nanoparticles (**a**) from Kim et al. (2012) and (**b**) from Liu et al. (2015). (Reprinted with permission from Kim et al. (2012) and Liu et al. (2015). Copyright (2012 and 2015) Elsevier)

aging, the ferrihydrite and lepidocrocite were dominant. When aged up to 90 days, the products are mainly γ-FeOOH mixed with small amounts of Fe_3O_4 and γ-Fe_2O_3, as also evident by the formation of the sheet- and needle-shaped materials, the typical morphology of lepidocrocite (Liu et al. 2015). Due to their high reactivity and the outward diffusion of Fe^0 core, Fe^{BH} aging can substantially consume Fe^0, resulting in a significant decrease of Fe^{BH} reactivity over time (Wang et al. 2010). A similar decrease in reactivity was also reported in Pd/Fe^{BH} particles in which aging causes both catalyst deactivation (i.e., Pd was completely buried underneath an extensive iron oxide matrix) and Fe^0 depletion (Yan et al. 2010). The CMC-modified Fe^{BH} was affected by aging in a similar manner. Nevertheless, the coating of CMC could slow the aging rate of Fe^{BH} as indicated by the slower drop in Fe^0 intensity in the XRD pattern (Dong et al. 2016). Moreover, CMC was found to influence the transformation of Fe^0 and the formation of iron oxide because greater CMC loading in the suspension results in more the lepidocrocite formation as a corrosion product of CMC-modified Fe^{BH}.

Unlike Fe^{BH}, aging of Fe^{H2} follows a growing shell and shrinking core model with no outward diffusion of Fe^0 core. The periods of increased, declined, and stabilized reactivity were observed during the aging of Fe^{H2}. Liu and Lowry (2006) studied reactivity changes with aging of Fe^{H2} using TCE as a reactivity probe. They observed a significant initial decrease (for the first 10 days) in TCE reaction rate constants ($k_{obs,TCE}$) for Fe^{H2} ($Fe^0 = 48\%$) decreased from 6.2×10^{-3} L/(h·m^2) to 1.0×10^{-3} L/(h·m^2) after 10 days (Liu and Lowry 2006). They interpreted the initial decrease as the healing of defects in the oxide film formed when the particles were removed from the concentrated suspension and dried. This is followed by a constant or slightly increasing rate constant during which the Fe^0 content of the particles decreased by ~40%. Thus, they concluded that the TCE reaction is zero-order with respect to the Fe^0 content of Fe^{H2}.

Eventually, reactivity with TCE ceased after 170 days when the Fe^0 content reached ~4.6%. A similar trend was observed by Sarathy et al. (2008) who used tetrachloromethane as a reactivity probe. They reported the increase of dechlorination rate constants of tetrachloromethane at the initial state (1–2-day exposure of dried Fe^{H2} to DO/DI water) followed by a gradual decrease over 1–6 months of aging (depending on the drying procedure of the Fe^{H2}). The initial increase in reactivity was explained by depassivation of the particles soon after the first immersion of dried Fe^{H2} in water. The depassivation involves breakdown of the iron oxide shell, thereby exposing the underlying Fe^0 core to the target contaminant. The gradual decrease in dechlorination rate constants after 1–2 days suggests the repassivation due to growth of a new oxide shell or transformation of the oxide shell, which gradually stabilizes the electron transfer from the Fe^0 core.

3.1.3 *Environmental Factors Affecting NZVI Reactivity*

Table 3.1 summarizes the batch experiments for reductive treatment of various kinds of organic contaminants using various kinds of NZVI. Most of the studies were conducted in a batch system in DI water at a pH of 8–9, representing an $Fe(OH)_2$/H_2O or Fe_3O_4/H_2O equilibrium. However, NZVI particles are applied in a subsurface environment, which is far more complex than DI water. The presence of dissolved inorganic and organic species, dense nonaqueous phase liquid

(DNAPL), and aquifer materials in the subsurface can physically and chemically affect the performance of NZVI particles. For this reason, this section reviews the effects of these environmental factors on NZVI performance.

3.1.3.1 pH

Groundwater pH typically varies from 6 to 8. The application of NZVI for in situ remediation is normally at only 0.2–0.5 wt %. Thus, because of the high buffer capacity of most soil, the groundwater pH will not change much due to the injection and emplacement of NZVI. Consequently, groundwater pH can substantially affect the electron utilization and lifetime of NZVI (Liu and Lowry 2006). As shown in Eq. (3.3), H^+ can consume Fe^0 to produce H_2. This is controlled by pH and competes with electron utilization for dechlorination (Eq. 3.2). Therefore, at low pH, Fe^0 tends to be utilized for H_2 production rather than dechlorination. For Fe^{H2}, the H_2 evolution rate constant increased by 27-fold (from 0.008 to 0.22 day^{-1}) due to decreasing pH from 8.9 to 6.5 (Liu and Lowry 2006). However, the TCE dechlorination rate constants were only two times higher (Liu and Lowry 2006). The increase of the TCE dichlorination rate constant is 1.44/h with respect to a one unit decrease of pH.

A similar trend was also observed for Fe^{BH} for dichlorination of trichloromethane (Song and Carraway 2006) and 1,1,2,2-tetrachloroethane (Song and Carraway 2005) of which the dichlorination rate constants increase by 0.33 and 0.24/h per one unit decrease of pH, respectively. Rapid H_2 evolution and Fe^0 consumption yield a relatively short lifetime of NZVI. At a pH of 6.5, the reactive lifetime of Fe^{H2} was only around 2 weeks (Liu and Lowry 2006). For this reason, both particle lifetime and efficiency decline due to the decrease of pH. The adverse effect of low pH becomes more severe when NZVI particles are applied to treat groundwater plumes in comparison to the source zone treatment because, in plumes, a low concentration of target contaminant is available for reaction with NZVI, and most Fe^0 will be used for H_2 production. In conclusion, the application of NZVI particles to treat contaminant plumes in aquifers with low pH (~6.5) is not favorable, and additional NZVI injections are expected every few weeks, making it economically unfeasible.

3.1.3.2 Anionic and Cationic Species

Due to geochemical cycles (dissolution and precipitation) of minerals in the subsurface, groundwater normally consists of various anionic species, such as NOO_3^-, Cl^-, SO_4^{2-}, HCO_3^-, and HPO_4^{-2}. Groundwater chemistry can affect NZVI corrosion rate, dechlorination rate, H_2 production, dissolution, and formation of an iron oxide shell. At low concentration (0.2–1 mM), reducible solutes, such as NO_3^-, did not significantly affect the TCE dechlorination rate. However, at high concentration (~5 mM), NO_3^- deactivated Fe^{H2} reactivity toward TCE after 3 days, even though Fe^0 remained in Fe^{H2} (Liu et al. 2007). Similarly, NO_3^-

(7.7 mM) was reported to inhibit dichlorination of hexachlorobenzene using Fe^{BH} (Su et al. 2012b).

Presumably, two possible hypotheses are that, at this high NO_3^- concentration, nitrite may be built up to block reactivity, or nitrate may promote the formation of a passivating and insoluble Fe(III) oxide layer, like maghemite or hematite (Schlicker et al. 2000). However, neither maghemite, hematite, nor goethite was observed using EXAFS (Reinsch et al. 2010), suggesting that the passivating layer may be some other phases or at a too low concentration to be detected using EXAFS. Nevertheless, the passivation by nitrate must happen very rapidly because the Fe^{H2} passivated by NO_3^- contained Fe^0, schwertmannite, and magnetite at the same ratio as fresh Fe^{H2} as if no Fe^0 oxidation took place (Reinsch et al. 2010).

In contrast, non-reducible anions, such as Cl^-, SO_4^{2-}, HCO_3^-, and HPO_4^{-2}, decreased the TCE dechlorination rate up to a factor of seven in comparison to DI water, and the order of their effect follows their affinity of anion complexation to hydrous ferric oxide (i.e., $Cl^- < SO_4^{2-} < HCO_3^- < HPO_4^{2-}$) at pH 8.9 (Liu et al. 2007). This implies that the inhibitory effect of these solutes on TCE degradation may be caused by reactive site blocking due to the formation of Fe-anion complexes on the Fe^{H2} surface. On the other hand, as for dichlorination of HCB using Fe^{BH}, HCO_3^- did not affect the dichlorination rate, while Cl^- and SO_4^{2-} slightly enhanced HCB dechlorination rate constants due to their corrosion promotion (Su et al. 2012b).

Similar to the case of anionic species, cationic species, such as Na^+, Mg^{2+}, Fe^{2+}, Cu^{2+}, Ni^{2+}, Cd^{2+}, and Zn^{2+}, might be released to the groundwater due to geochemical cycles. Furthermore, these pollutants might be found at high concentrations in DNAPL-contaminated areas as co-contaminants. The NZVI has been demonstrated to be effective for the immobilization of various metals (Ponder et al. 2000; Zhang 2003; Dries et al. 2005; Kanel et al. 2005) through reduction (Eq. 3.8), co-precipitation, and surface complexation (Eqs. 3.9 and 3.10) depending on the relative standard potential E^0 between NZVI and metals (Li and Zhang 2007; see Chap. 4).

$$Me^{a+} + be^- \rightarrow Me^{a-b} \tag{3.8}$$

$$\equiv S - OH + Me^{2+} \leftrightarrow \equiv S - OMe^+ + H^+ \tag{3.9}$$

$$2 \equiv S - OH + Me^{2+} \leftrightarrow (\equiv S - O)_2 Me + 2H^+ \tag{3.10}$$

For metal ions such as Zn^{2+} and Cd^{2+}, of which E^0 is very close to or more negative than that of Fe^0, the main removal mechanism is sorption and surface complexation (Li and Zhang 2007). For metal ions such as Ni^{2+} and Pb^{2+}, of which E^0 is slightly more positive than that of Fe^0, the removal mechanism is the combination of reduction and sorption (Li and Zhang 2007). For metal ions such as Cu^{2+}, Ag^+, and Hg^{2+}, of which E^0 is greatly more positive than that of Fe^0, the major removal mechanism is reduction (Li and Zhang 2007). As shown in Eq. (3.8), the reduction of metals consumes electrons (i.e., a competing reaction to dichlorination of chlorinated organics; Eq. 3.2). In addition, the reduction of metals is normally followed by the precipitation of reduced metals on the surface of NZVI. This co-precipitation reaction with surface complexation (Eqs. 3.9, 3.10, and 3.11) can block reactive sites or promote the formation of Fe(III)-metal oxide, a passivating layer, at anodic sites. This decreases NZVI reactivity for dechlorination. For this reason, in general, the presence of metals as co-contaminants adversely affects the dechlorination rate. For example, in the presence of Zn, TCE degradation using ZVI was 2–4 times slower (Dries et al. 2005). Similarly, in the presence of Cr(VI) at a concentration higher than 5 mg/L, the TCE dechlorination rate decreases by a factor of 3–13 (Dries et al. 2005).

Similarly, Fe(II) inhibited the HCB degradation reaction due to passivation layers formed, while Na^+ and Mg^{2+} did not substantially affect the dichlorination (Su et al. 2012b). However, the presence of noble metals including Cu^{2+}, Ni^{2+}, Pd^{2+}, and Pt^{2+} appeared to enhance the dichlorination rate through the catalytic pathway. The presence of Ni (5–100 mg/L) enhanced the TCE dechlorination due to the catalytic hydrodechlorination by bimetallic Fe^0/Ni^0 from the precipitation of Ni^0 on the surface of ZVI (Dries et al. 2005). Similarly, the presence of Cu^{2+} enhanced dichlorination of HCB (Su et al. 2012b). Furthermore, metal ions including Co^{2+}, Cu^{2+}, and Ni^{2+} enhanced the dechlorination of 4-chlorobiphenyl (4-ClBP) by NZVI. The dechlorination percentages of 4-ClBP in the presence of 0.1 mmol/L of Co^{2+}, Cu^{2+}, and Ni^{2+} were 66.1%, 66.0%, and 64.6% in 48 h, and then increased to 67.9%, 71.3%, and 73.5%, after 96 h, respectively (Wang et al. 2011).

3.1.3.3 Natural Organic Matters in Groundwater and Soil

Groundwater naturally contains a significant amount of natural organic matter (NOM) originated from decomposition of animal and plant bodies (Schwarzenbach et al. 2003b). The NOM is a natural charged macromolecule, carrying a net negative charge at a natural pH due to the dissociation of carboxylic groups (Schwarzenbach et al. 2003a). Furthermore, NOM consists of humic and fulvic acids. By operational definition, humic acid is the fraction of NOM that precipitates at pH 2 or lower, while the fulvic acid fraction stays soluble under all pH conditions (Schwarzenbach et al. 2003a). The NOM was found to adsorb various kinds of colloids and nanoparticles (Ramos-Tejada et al. 2003; Hyung et al. 2007). Similarly, carboxylic groups of NOM can specifically adsorb onto the iron oxide surface of Fe^{H2} (Fig. 3.12a). The NOM is an anionic polyelectrolyte, which tends to adsorb onto the substrate in a train-loop-tail configuration (Fleer et al. 1998) similar to the polyelectrolyte discussed above.

There are two different hypotheses of the effects of NOM on ZVI performance. First, NOM can enhance electron transfer and thus ZVI reactivity for pollutant degradation through electron shuttle effects (Tratnyek et al. 2001). Second, adsorbed NOM decreases ZVI reactivity due to reactive site blocking (Tratnyek et al. 2001; Cho and Park 2006; Doong and Lai 2006). Moreover, consisting of the quinone group with standard potential E^0 of 0.23 V, NOM is hypothesized to transfer electrons from ZVI for the dechlorination of chlorinated ethene (Tratnyek et al. 2001). The enhanced

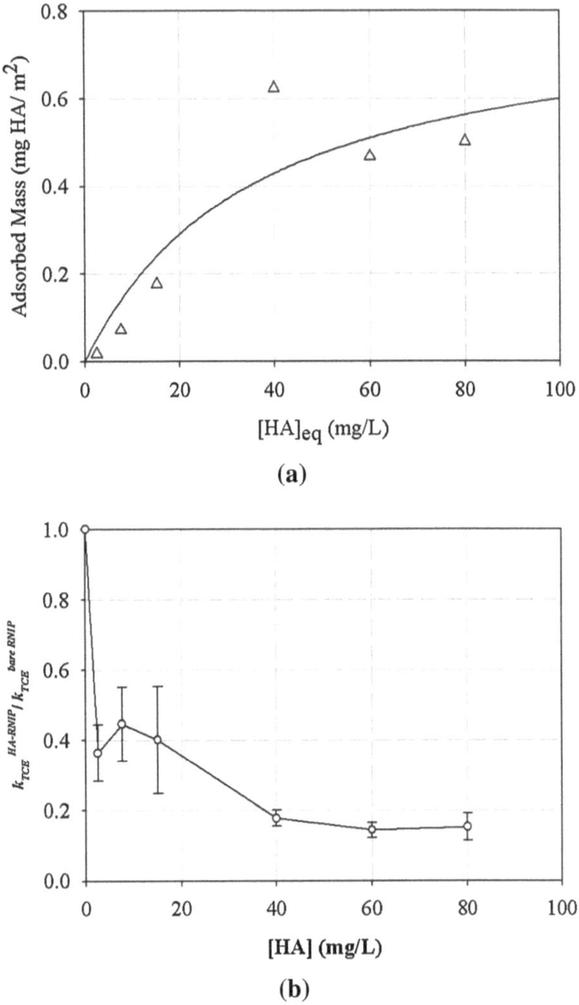

Fig. 3.12 (**a**) Adsorption isotherm of HA on RNIP at pH 8.5; the line illustrates the best fit using the Langmuir isotherm. (RNIP is NZVI formed by reduction using H_2). (b) The relationship between standardized TCE dechlorination rate constants using RNIP and HA concentrations in the aqueous phase

dechlorination due to the presence of humic acid in the ZVI system was observed for PCE dechlorination but not TCE (Cho and Park 2006). In contrast, Tratnyek et al. (2001) reported that TCE degradation kinetics decreased by 21% and 39% in the presence of 20 and 40 mg/L, respectively, of Suwannee River organic matter, presumably due to reactive site blocking.

Figure 3.12b supports and extends the second hypothesis regarding the effect of NOM on TCE dechlorination using NZVI. In the presence of different humic acid concentrations, the TCE dechlorination rates by bare RNIP decreased nonlinearly and exhibited two regions. The TCE dechlorination pathways were not affected, and β elimination remained the dominant dechlorination pathway, yielding acetylene as the reaction intermediate and ethane and ethene as products. Consistent with the

Scheutjens-Fleer theory for homopolymer sorption (Fleer et al. 1998), the nonlinear relationship between the dechlorination rate and the surface excess of adsorbed humic acids suggests that adsorbed humic acids decrease reactivity primarily by blocking reactive surface sites at low surface excess where they adsorb relatively flat onto the Fe^{H2} surface and by site blocking and decreasing TCE availability at high surface excess where humic acids form an extended layer around the particles. This finding is confirmed by a recent study revealing that the presence of Suwannee River humic acids (SRHA) (10 mg/L) decreased TCE (20 mg/L) dechlorination by Fe^{H2} around 23% but did not affect the H_2 production (Chen et al. 2011a). Nevertheless, the mix effect of NOM on NZVI reactivity was also reported. A recent study reported that the presence of humic acid inhibited the reduction of 4-ClBP in the first 4 h, but then significantly accelerated dechlorination by reaching 86.3% in 48 h (Wang et al. 2011).

In addition to the dissolved NOM, the NOM that is adsorbed onto soil or aquifer material can substantially decline dechlorination efficiency. This is an issue of mass transfer of CVOCs from soil to groundwater. The CVOC-sorbed soil may behave as a long-term secondary source, gradually leaching dissolved CVOCs to contaminate the groundwater downgradient. Slow desorption of CVOCs from the soil can result in retardation of reductive detoxification using NZVI. Using TCE as an example, TCE has an arithmetic mean organic carbon partitioning coefficient (k_{oc}) of 86 (ranging from 18.5 to 150). The soil-water partitioning coefficient (k_d) is the k_{oc} x fraction of organic carbon in soil (f_{oc}). This partitioning coefficient can be used to calculate the retardation factor (R) as shown in Eq. 3.11. Subsequently, we can estimate the decrease in TCE dechlorination rate by NZVI in the soil-water system ($k_{\text{TCE-aq-Soil}}$) with R using Eq. 3.12 in comparison to the TCE dechlorination rate constant using NZVI in water (no sorption onto soil; $k_{\text{TCE-aq}}$):

$$R = 1 + \frac{k_d{}^* \rho_b}{n} \tag{3.11}$$

$$k_{\text{TCE-aq-Soil}} = \frac{k_{\text{TCE-aq}}}{R} \tag{3.12}$$

where ρ_b and n are the bulk density of soil and porosity, respectively. Under equilibrium with the partitioning coefficient (k_d) of 1.46 L/kg, R is calculated as 7.28 using Eq. 3.11. Thus, the TCE sorption into soil can decrease the TCE dechlorination rate constant in the soil-water system by 7.28 times in comparison to the system with groundwater alone. This is problematic because, instead of using its reducing power to destroy contaminants, the NZVI reacts with the water to form H_2, which increases the amount of NZVI required for remediation (Liu et al. 2007; Berge and Ramsburg 2010). This is a mass-transfer limitation problem that cannot be solved by modifying NZVI to have greater reactivity, such as by doping with catalysts. Instead, a possible solution is to use NZVI with thermal-enhanced CVOC dissolution or desorption, which will speed up the reaction rate and improve the electron utilization efficiency of the remediation (see Chap. 11).

3.1.3.4 TCE Concentration and the Presence of DNAPL

The NZVI particles are proposed for the remediation of both the DNAPL source zone and groundwater plumes. Therefore, NZVI is subjected to a wide range of CVOC concentration, from low concentration in contaminant plumes to near saturation or even DNAPL in the source zone. The different CVOC concentrations in different zones influence electron utilization, particle efficiency, and the reactive lifetime of NZVI. A recent study (Liu et al. 2007) revealed that changing TCE concentrations from low to medium range (0.03–0.46 mM) insignificantly affected the TCE dechlorination rate using Fe^{H2} at pH 7. However, at higher concentrations (1.3 mM) to TCE water saturation (8.4 mM), the TCE dechlorination rate by Fe^{H2} decreased by a factor of two, presumably due to reactive site blocking by acetylene, an intermediate.

The higher the TCE concentration, the higher the Fe^0 utilization efficiency for TCE dechlorination, which is evident from the decrease of H_2 evolution and the shift of byproduct formation toward an unsaturated byproduct (Liu et al. 2007), e.g., acetylene. At pH 7, 40% of Fe^0 in Fe^{H2} was consumed for H_2 production at a TCE concentration of 0.46 mM, while only 7% of the initial Fe^0 in Fe^{H2} was used for H_2 production at a TCE concentration of 8.4 mM (Liu et al. 2007). Evidently, TCE outcompeted Fe^0 with H^+. Similarly, acetylene accounted for 86% of the products, while ethene and ethane were 10% and 4%, respectively, at the TCE concentration of 8.4 mM (Liu et al. 2007). Acetylene formation requires fewer electrons than ethene and ethane formation, making TCE dechlorination by Fe^{H2} utilize electrons more effectively. The accumulation of acetylene could be the result of TCE saturating reactive sites and blocking acetylene from further transformation.

However, increasing the TCE concentration adversely affects the reactive life time of Fe^{H2}. The reactive lifetime of Fe^{H2} is 10, 40, and 60 days for TCE concentrations of 8.4, 1.3, and 0.46 mM, respectively, at pH 7 buffered by HEPES (Liu et al. 2007). At the DNAPL-water interface without HEPES, TCE dechlorination with Fe^{H2} decreased pH to 4–5. The reactive life time of Fe^{H2} at DNAPL interface was only 5 days, and no Fe^0 remained in the particles. The particle efficiency for TCE dechlorination using Fe^{H2} at the DNAPL-water interface was only ~15% because the local pH decreased and accelerated H_2 production in a much faster rate than increasing TCE dechlorination (see pH effect). This suggests that Fe^{H2} used for source zone treatment will have a relatively short lifetime. Therefore, additional injections for source zone treatment are expected to be more frequent than for plumes.

3.1.3.5 Microorganisms

Various kinds of microorganisms such as sulfate reducers, iron reducers, and methanogens are present under different geochemical conditions in the subsurface.

Some microorganisms such as halorespirers can biologically transform chlorinated organics to notorious byproducts under favorable conditions (Harkness et al. 1999; Cupples et al. 2004). Potential synergic effects of abiotic and biotic remediation via stimulated bioremediation using polymer-modified NZVI, leading to the long-term degradation of chlorinated organics, is an active area of research. Chapters 7 and 10 are devoted to this important combined remedy. The effects of microbes on NZVI reactivity and longevity are also discussed in Chap. 10. A reader should consult the microbiological-related materials in Chaps. 7 and 10 prior to designing an in situ remediation using NZVI since, in a real field implementation, interaction between NZVI and microbes is unavoidable and can substantially affect contaminant treatability.

3.2 Oxidative Transformation of Organic Contaminants Using NZVI-Induced Fenton's Reaction

While being considered a reducing agent, NZVI has also recently gained a lot of interest in the oxidation aspect. In general, NZVI can be used as a reagent in the process of producing oxidative radicals for contaminant degradation. Oxidation approaches of NZVI involve NZVI-induced Fenton's reaction (Xu and Wang 2011; Choi and Lee 2012; Li and Zhu 2014), NZVI under aeration (Taha and Ibrahim 2014a), and NZVI-induced persulfate system (Al-Shamsi and Thomson 2013; Diao et al. 2016). Nevertheless, the NZVI-induced Fenton's reaction is the most studied one. The state-of-the-art progress of the NZVI-induced Fenton's reaction is focused on in this section.

Discovered by Henry J. H. Fenton in 1894, Fenton's reaction is one of the most common chemical oxidation processes in wastewater treatment and site remediation (Ay and Kargi 2010; Petri et al. 2011; Babuponnusami and Muthukumar 2014; Papoutsakis et al. 2016). Typically, a homogeneous Fenton's reaction utilizes ferrous sulfate as a catalytic Fe^{2+} source. An acidic condition is required to maintain Fe^{2+} dissolution. However, acidification and the high dose of mobile ferrous/ferric required to obtain an effective treatment are the major drawbacks of conventional Fenton's reaction (Xu and Wang 2011; Li and Zhu 2014; Yu et al. 2014; Cheng et al. 2015; Usman et al. 2016). For in situ applications, the acidification of the subsurface is probably one of the most challenging problems due to the large buffering capacity at a neutral pH range of the aquifers (Petri et al. 2011). Moreover, acidification might result in the dissolution of metals in the subsurface (Petri et al. 2011). Hence, NZVI-induced Fenton's reaction has emerged as an alternative approach to avoid these problems.

3.2.1 Mechanism

The mechanism of contaminant degradation using NZVI-induced Fenton's reaction is primarily based on highly reactive radicals, which are generated from the reaction between Fe^0 and hydrogen peroxide. The radicals playing the key role in Fenton's reaction include highly reactive hydroxyl radical (•OH) and hydroperoxyl radical (OOH•). Instead of using acidification to maintain dissolved Fe^{2+} concentration, the NZVI Fenton process is self-catalytic, based on oxidative dissolution of NZVI in the presence of H_2O_2. Interfacial H^+ is produced at the NZVI surface to provide appropriate local pH, which continuously releases Fe^{2+} for Fenton's reaction. Babuponnusami and Muthukumar (2012), Chu et al. (2012), and Xu and Wang (2011) indicated the success of NZVI Fenton's reaction at circumneutral pH 6. Moreover, the NZVI Fenton process is more favorable than the conventional approach using dissolved Fe^{2+} because NZVI can be magnetically recovered and reused for several times (Diya'uddeen et al. 2015).

The critical reactions in a heterogeneous NZVI-induced Fenton's system are as follows:

$$Fe^0 + H_2O_2 + 2H^+ \rightarrow Fe^{2+} + 2H_2O \tag{3.13}$$

$$Fe^0 + O_2 + 2H^+ \rightarrow Fe^{2+} + H_2O_2 \tag{3.14}$$

$$Fe^0 + 2H^+ \rightarrow Fe^{2+} + H_2 \tag{3.15}$$

$$Fe^0 + H_2O \rightarrow Fe^{2+} + OH^- \tag{3.16}$$

$$Fe^{2+} + H_2O_2 \rightarrow Fe^{3+} + \cdot OH + OH^- \tag{3.17}$$

$$Fe^{2+} + \cdot OH \rightarrow Fe^{3+} + OH^- \tag{3.18}$$

$$Fe^{3+} + H_2O_2 \rightarrow Fe^{2+} + OOH \cdot + H^+ \tag{3.19}$$

The NZVI-induced Fenton's reaction has been reported to effectively degrade a variety of organic contaminants including textile wastewater, pharmaceuticals, halogenated compounds, and other non-halogenated compounds (Table 3.2). The degradation kinetics was found to fit well with the pseudo-first-order reaction. Degradation rates are mainly in the range of 0.01–0.2/min. However, depending on the reaction conditions, the rates can vary from 0.0064/min to 1.79/min. Zhang et al. (2017) reported that the degradation of norfloxacin was at a rate of 0.0064/min at pH 6, while the degradation rate at pH 3 was 0.1/min. Xu and Wang (2011) presented the 100% removal of 4-chloro-3-methyl phenol within 60 min at the degradation rate of 0.35/min at 0.1 g/L of NZVI and 1.79/min at 0.5 g/L of NZVI. In addition, although the NZVI-induced Fenton's reaction kinetics is pseudo-first-order in most of the cases, some studies suggest that the degradation kinetics is more appropriate for the pseudo-second-order reaction. For instance, Zhou et al. (2015) revealed the pseudo-second-order degradation of 1-alkyl-3-methylimidazolium-bromides at a rate of 0.0415 L/(mM·min).

Table 3.2 Summary of batch experiments for NZVI-induced Fenton's reaction of various organic contaminants

Contaminants	Optimal conditions	Degradation	Notes	References
Organic dyes				
Orange II	Orange II 105 mg/L; NZVI 0.02 g/L; H_2O_2 5.88 mM; pH 3	Orange II >95%; TOC 53%; reaction time 60 min	Pretreatment of NZVI reduction followed by Fenton oxidation improved degradation efficiency by 10%–15%	Moon et al. (2011)
Textile wastewater (reactive black B and polyvinyl alcohol)	COD 750–810 mg/L; NZVI 0.2–0.225 g/L; H_2O_2 30–33 mM; pH 2	COD 76%; reaction time 30 min	Wastewater was pretreated with NZVI reduction, followed by Fenton oxidation, and 10%–15% of COD was removed by NZVI reduction	Yu et al. (2014)
Halogenated compounds				
2,4-Dichloro-phenoxyacetic acid (2,4-D)	2,4-D 235 mg/L; NZVI 0.6 g/L; H_2O_2 16.65 mM; pH 3	TOC 57%; reaction time 60 min	Employment of acoustic cavitation and hydrodynamic cavitation enhanced degradation efficiency by 5%–10%	Bremner et al. (2008)
4-Chlorophenol (4-CP)	4-CP 50–100 mg/L; NZVI 1–5 g/L; H_2O_2 0.016–0.065 mM; pH 3–4	Complete within 30 min; TOC 50%; $k = 0.35–0.58$/min	4-Chlorocatechol (4CC) and ortho-parachlorophenolperoxyl radical (ClPP•) were identified as intermediates and oxidized to aliphatic organic acids (e.g., maleic acid)	Zhou et al. (2008)
4-Chloro-3-methyl phenol (CMP)	CMP 50–150 mg/L; NZVI 0.1–0.5 g/L; H_2O_2 3.0 mM; pH 3.0–6.0	Complete within 60 min; $k = 0.35–1.79$/min	Two-stage reactions: Lag and degradation	Xu and Wang (2011)
Trichloroethylene (TCE)	TCE 70 mg/L; NZVI 8.5 g/L; H_2O_2 188 mM; pH 3.0–6.0; Cu (II) 20 mM	With cu(II): TCE 95% within 10 min $k = 4.90$/min Without cu (II): TCE 25% within 10 min	The presence of cu(II) could significantly enhanced the degradation of TCE by NZVI Fenton's reaction	Choi and lee (2012)

(continued)

Table 3.2 (continued)

Contaminants	Optimal conditions	Degradation	Notes	References
		$k = 0.0194$/min (at pH 3) $k = 0.0318$/min (at pH 6)		
p-chloro-nitroben-zene (p-ClNB)	p-ClNB 60 mg/L; NZVI 0.09–0.36 g/L; H_2O_2 4.9 mM; pH 2–3	Complete within 30 min; $k = 0.18$/min	Degradation was interfered by SO_4^{2-} 1 M, while the effects of Cl^- 1 M and NO_3^- 1M were insignificant. Identified intermediates included azo compounds and aromatic amines. Carboxylic acids were the final products	Li and Zhu (2014)
4-Chlorophenol (4-CP)	4-CP 100 mg/L; NZVI 0.4 g/L; H_2O_2 9.79 mM; pH 3	Complete within 30 min	A presence of Cr(VI) that is greater than 10 mg/L slightly decreased the reaction rate due to Fe (II) consumption	Yin et al. (2014)
2,4-dichloro-phenol (2,4-DCP)	2,4-DCP 50–100 mg/L; NZVI 1.5–2.0 g/L; H_2O_2 10 mM; pH 3	2,4-DCP 88%–92%; TC 25%; TOC 30%; IC 10%; reaction time 180 min; $k = 0.022$–0.029/min	Higher temperatures (10–40 °C) produced higher degradation rates (0.022–0.084/min)	Li et al. (2015)
Chlorpheniramine	Chlorpheniramine 5–15 mg/L; NZVI 22.4 g/L; H_2O_2 0.1 mM; pH 2–3	Complete within 60 min; TOC 21%; $k = 0.093$/min	4-hydroxy-4-methyl-2-pentanone, 2-methylaminopyridine, and NDMA were identified as interme-diates that were resistant to the treatment	Wang et al. (2016b)
Non-halogenated compounds				
Phenol	Phenol 100–400 mg/L; NZVI 0.5 g/L; H_2O_2 14.7 mM; pH 2–6.2 (natural pH); current density 12 mA/cm^2; an 8 W UV lamp ($\lambda = 254$ nm)	Complete within 60 min; $k = 0.0926$/min	Assisted with photo-electro effect	Babuponnusami and Muthukumar (2012)

Palm oil mill effluent	COD 1160 mg/L; NZVI 0.6 g/L; H_2O_2 40 mM; pH 2; Ultrasound 20 kHz	COD 80%; reaction time 120 min	The employment of an ultrasound accelerated the degradation from 24 h (in silence) to 2 h	Taha and Ibrahim (2014a)
Palm oil mill effluent	COD 4568 mg/L; NZVI 3.91 g/L; H_2O_2 54 mM; pH 2; Aeration rate 23.84 L/min	COD up to 71%; reaction time 240 min	The employment of aeration improved the degradation efficiency by 10-20%	Taha and Ibrahim (2014b)
Petroleum refinery wastewater	COD 1259 mg/L; BOD 173 mg/L; Phenol 14.7 mg/L; Oil and grease 233 mg/L; NZVI 1.24 g/L; H_2O_2 222 mM; pH 3	COD 76.5%; BOD 37.6%; TOC 45% phenol 96%; oil and grease 100%; reaction time 120 min	Followed by sequencing batch reactor (SBR). Fenton's process improved the biodegradability of the wastewater for SBR treatment. The BOD_5/COD ratio increased from 0.137 to 0.365, while approx. 130 mg/L of total iron was produced after Fenton's process	Diya'uddeen et al. (2015)
Naphthalene	Naphthalene 200 mg/L; NZVI 1.0 g/L; H_2O_2 10 mM; pH 3	COD 82.7%; reaction time 60 min	The wastewater was pretreated with bacterial strains of Bacillus fusiformis (BFN) for 40 h, followed by Fenton's oxidation	Yu et al. (2015)
1-Alkyl-3-methyl-imidazolium-bromides ([$C_n$mim]Br)	[C_4mim]Br 438 mg/L; NZVI 0.25–1.0 g/L; H_2O_2 40–100 mM; pH 3–4; ultrasound 45 kHz	[$C_n$mim]Br 80%-90%; reaction time 120 min; $k = 0.0415$ L/(mM.min)	Assisted with ultrasound	Zhou et al. (2015)
Phenol	Phenol 25 mg/L; NZVI 1.0 g/L; H_2O_2 30 mM; pH 7; ultrasound 20 kHz	Phenol 75%; reaction time 60 min; $k = 0.0145$/min	Assisted with ultrasonic irradiation	Yehia et al. (2015)
Pharmaceuticals, personal care products, and microbial toxins				
Amoxicillin (AMX)	AMX 50 mg/L; NZVI 0.5 g/L; H_2O_2 6.6 mM; pH 3–4	AMX 86%; reaction time 20 min; $k = 0.15$/min	The degradation was affected by temperature. At 45 °C, H_2O_2 tends to be decomposed	Zha et al. (2014)

(continued)

Table 3.2 (continued)

Contaminants	Optimal conditions	Degradation	Notes	References
Norfloxacin (NOR)	NOR 100 mg/L; NZVI 0.1 g/L; H_2O_2 20 mM; pH 3	NOR 95%; TOC 50%; reaction time 40 min; $k = 0.10$/min (pH 3); $k = 0.063$/min (pH 5); $k = 0.0064$/min (pH 6)	The optimal temperature was 25–35 °C. At either lower or higher temperature, the degradation tends to be inhibited	Zhang et al. (2017)
Microcystin-LR (MC-LR)	MC-LR 5 mg/L; NZVI 0.6 g/L; H_2O_2 6.6 mM; pH 6.86	MC-LR 59.1%; COD 22%; reaction time 250 min; $k = 0.14$/min	The presence of humic acid and oxalate enhanced the degradation efficiency by 10–20%	Wang et al. (2016a)

In general, NZVI-induced Fenton's reaction has high reactivity. However, it is a two-stage reaction, in which the first stage usually shows low degradation rates and is followed by a rapid degradation in the second stage. The lag stage is considered an activation process of the surface iron species where iron dissolution occurs on the NZVI surface and participates in the catalytic reaction. Moreover, despite the high reactivity, the degradation by NZVI-induced Fenton's reaction is usually inhibited quickly ranging from 0.5 to 3 h due to the depletion of H_2O_2. Yu et al. (2014) showed that the textile wastewater removal was inhibited after 40 min gaining 80% efficiency. Cheng et al. (2015) reported the inhibition of pentachlorophenol degradation after an hour with a removal efficiency of 60%. Zha et al. (2014) described that amoxicillin degradation was impeded after 15 min, obtaining approximately 80% removal efficiency. The inhibition even occurs more frequently at a neutral pH (Li and Zhu 2014; Zhou et al. 2015; Wang et al. 2016b). The sequential addition of Fenton's reagent can be a solution to this problem. Munoz et al. (2014) presented the prolongation of homogenous Fenton reaction to treat sawmill wastewater by the sequential addition of H_2O_2.

3.2.2 *Factors Affecting Treatment Efficiency*

3.2.2.1 Reagent Dose

The H_2O_2 acts as a scavenger of hydroxyl radical (OH) (Xu and Wang 2011; Li and Zhu 2014; Li et al. 2015; Wang et al. 2016b) as shown in Eqs. 3.20 and 3.21, while ·OH plays the dominant role in contaminant degradation (Petri et al. 2011). In addition, H_2O_2 can be self-degraded as presented in Eq. 3.22. Therefore, the use of excessive H_2O_2 led to the degradation inefficiency. As a result, sequential H_2O_2 addition, as done in the homogeneous Fenton's process (Martins et al. 2010; Villa et al. 2010; Munoz et al. 2014), is also a possible solution for sustaining NZVI-induced Fenton's reaction as to be discussed next in the treatability of 1,2-DCA.

$$H_2O_2 + \cdot OH \rightarrow HO_2\cdot + H_2O \tag{3.20}$$

$$HO_2\cdot + \cdot OH \rightarrow H_2O + O_2 \tag{3.21}$$

$$H_2O_2 + H_2O_2 \rightarrow 2H_2O + O_2 \tag{3.22}$$

$$Fe^{2+} + H_2O_2 \rightarrow Fe^{3+} + \cdot OH + OH^- \tag{3.24}$$

$$Fe^{2+} + \cdot OH \rightarrow Fe^{3+} + OH^- \tag{3.25}$$

The NZVI was the source to generate ·OH (Li et al. 2015; Wang et al. 2016b) as shown in Eqs. 3.13 and 3.24. Therefore, a higher NZVI concentration provided higher degradation efficiency. However, the excessive iron source has been known as an ·OH scavenger (Xu and Wang 2011; Wang et al. 2016b), which may prohibit the degradation as shown in Eqs. 3.13 and 3.25. Table 3.2 summarizes the optimal reagent dose and conditions for the treatment of various contaminants.

3.2.2.2 Initial pH

Babuponnusami and Muthukumar (2012) and Xu and Wang (2011) also reported that the heterogeneous NZVI Fenton's reaction at a neutral pH succeeded in removing phenol and 4-chloro-3-methyl phenol, respectively. Babuponnusami and Muthukumar (2012) reported 65% of phenol removal using NZVI Fenton's process at a pH 6.2. Chu et al. (2012) showed a decrease of 95% in total phenols and 50% in COD of coking wastewater using iron powder as a Fenton's catalyst at pH 6.5 and 5.4. Xu and Wang (2011) presented the complete degradation of 4-chloro-3-methyl phenol using an NZVI-induced Fenton's system at a pH of 6.1. Some examples of optimal pH values for the treatment of different organic contaminants are summarized in Table 3.2.

3.2.3 Treatability of 1,2–DCA

While reductive dechlorination using NZVI is incapable of detoxifying 1,2-DCA as discussed previously, the Fenton process can. Masten and Butler (1986) suggested the success of 1,2-DCA degradation due to free radicals. Noticeably, Vilve et al. (2010) successfully degraded 1,2-DCA at the laboratory scale using a conventional Fenton process. Recently, Le and Phenrat (2018) evaluated the NZVI-induced Fenton process at a neutral pH to degrade 1,2-DCA at a high concentration (2000 mg/L), representing a dissolved 1,2-DCA concentration close to the DNAPL source zone. Approximately 87% of 1,2-DCA was degraded at a neutral pH, with a pseudo-first-order rate constant of 0.98/h using 10 g/L of NZVI and 200 mM of H_2O_2. However, the reaction was prohibited quickly, within 3 h, presumably due to the rapid depletion of H_2O_2. The application of sequential H_2O_2 addition provided a better approach for preventing rapid inhibition via controlling the H_2O_2 concentration in the system to be sufficient but not in excess, thus resulting in the higher degradation efficiency (the pseudo-first-order rate constant of 0.49/h and 99% degradation in 8 h for 10 g/L NZVI and 200 mM but sequential for 25 mM per 30 min). Using NZVI with sequential H_2O_2 addition (25 mM per 30 min) was also successful in degrading 1,2-DCA sorbed onto soil, yielding 99% removal of 1,2-DCA within 16 h at a rate constant of 0.23/h (Fig. 3.13a), around two times slower than in the system without soil, presumably due to rate-limited 1,2-DCA desorption from the soil.

The NZVI-induced Fenton reaction can be reused for several treatment cycles. Figure 3.13b illustrates the 1,2-DCA degradation kinetics using NZVI-induced Fenton's reaction in three consecutive cycles. In the first cycle of degradation, 99.9% of 1,2-DCA was degraded for 16 h, obtaining the rate constant of 0.49/h.

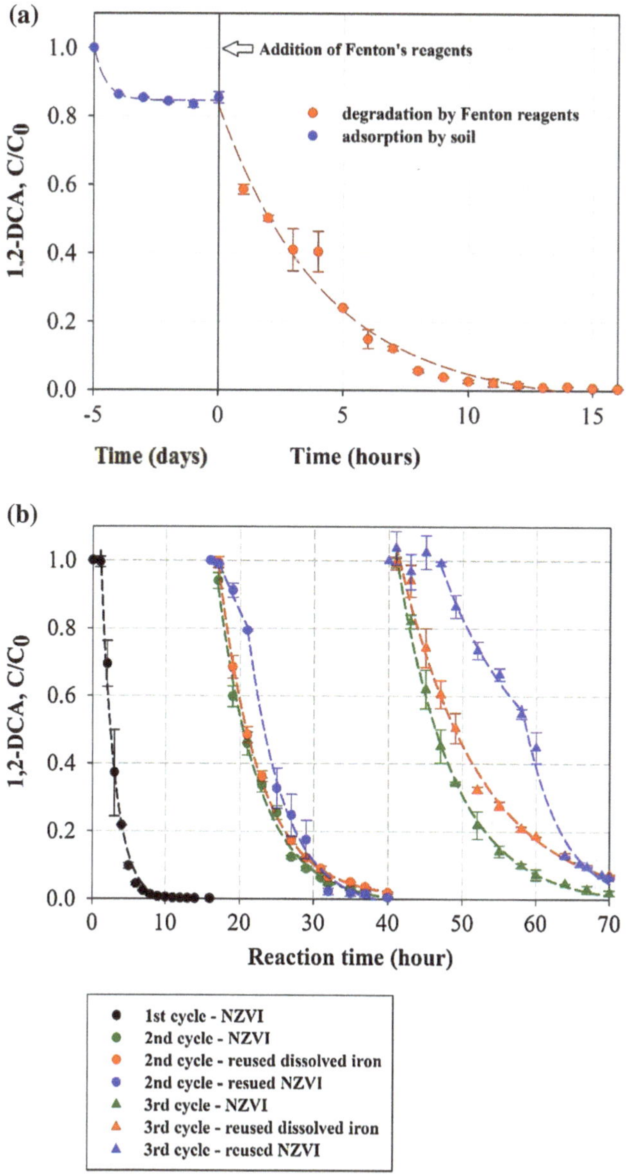

Fig. 3.13 (**a**) 1,2-DCA degradation in soil-groundwater system and (**b**) NZVI reuse for three removal cycles (conditions: natural pH, initial 1,2-DCA 2000 mg/L, NZVI 10 g/L, 25 mM H_2O_2 per 30 min)

When the stock 1,2-DCA solution was added into the reactors to restart the treatment cycles, the degradation rate constant declined almost three times to 0.18/h in Cycle 2 and four times to 0.13/h in Cycle 3. Approximately 99.2% and 98.0% of 1,2-DCA were degraded for 24 h and 30 h in Cycles 2 and 3, respectively. This finding suggests that the heterogeneous NZVI Fenton process is a promising approach for in situ treatment. The NZVI can be emplaced in the subsurface close to the DNAPL source zone, while a small amount of H_2O_2 is recirculated to treat 1,2-DCA in both groundwater and soil.

References

Al-Shamsi, M. A., & Thomson, N. R. (2013). Treatment of organic compounds by activated persulfate using nanoscale zerovalent iron. *Industrial & Engineering Chemistry Research, 52*, 13564–13571.

Amir, A., & Lee, W. (2011). Enhanced reductive dechlorination of tetrachloroethene by nano-sized zerovalent iron with vitamin B12. *Chemical Engineering Journal, 170*, 492–497.

Arnold, W. A., & Roberts, A. L. (2000). Pathway and kinetics of chlorinated ethylene and chlorinated acetylene reaction with Fe(0) particles. *Environmental Science & Technology, 34*, 1794–1805.

Ay, F., & Kargi, F. (2010). Advanced oxidation of amoxicillin by Fenton's reagent treatment. *Journal of Hazardous Materials, 179*, 622–627.

Babuponnusami, A., & Muthukumar, K. (2012). Removal of phenol by heterogenous photo electro Fenton-like process using nano-zero valent iron. *Separation and Purification Technology, 98*, 130–135.

Babuponnusami, A., & Muthukumar, K. (2014). A review on Fenton and improvements to the Fenton process for wastewater treatment. *Journal of Environmental Chemical Engineering, 2*, 557–572.

Baer, D. R., Tratnyek, P. G., Qiang, Y., Amonette, J. E., Linehan, J., Sarathy, V., Nurmi, J. T., Wang, C.-M., & Antony, J. (2007). Synthesis, characterization, and properties of zero-valent iron nanoparticles. In G. E. Fryxell & G. Cao (Eds.), *Environmental applications of nanomaterials*. London: Imperial College Press.

Berge, N. D., & Ramsburg, C. A. (2010). Iron-mediated trichloroethene reduction within nonaqueous phase liquid. *Journal of Contaminant Hydrology, 118*, 105–116.

Bremner, D. H., Carlo, S. D., Chakinala, A. G., & Cravotto, G. (2008). Mineralisation of 2,4-dichlorophenoxyacetic acid by acoustic or hydrodynamic cavitation in conjunction with the advanced Fenton process. *Ultrasonics Sonochemistry, 15*, 416–419.

Cao, J., Xu, R., Tang, H., Tang, S., & Cao, M. (2011). Synthesis of monodispersed CMC-stabilized Fe–Cu bimetal nanoparticles for in situ reductive dechlorination of 1, 2, 4-trichlorobenzene. *The Science of the Total Environment, 409*, 2336–2341.

Chen, J., Xiu, Z., Lowry, G. V., & Alvarez, P. (2011a). Effect of natural organic matter on toxicity and reactivity of nano-scale zero-valent iron. *Water Research, 45*, 1995–2001.

Chen, Z.-X., Jin, X.-Y., Chen, Z., Megharaj, M., & Naidu, R. (2011b). Removal of methyl orange from aqueous solution using bentonite-supported nanoscale zero-valent iron. *Journal of Colloid and Interface Science, 363*, 601–607.

Cheng, R., Cheng, C., Liu, G. H., Zheng, X., Li, G., & Li, J. (2015). Removing pentachlorophenol from water using a nanoscale zero-valent iron/H2O2 system. *Chemosphere, 141*, 138–143.

Cho, H.-H., & Park, J. W. (2006). Sorption and reduction of tetrachloroethylene with zero valent iron and amphiphilic molecules. *Chemosphere, 64*, 1047–1052.

Choi, K., & Lee, W. (2012). Enhanced degradation of trichloroethylene in nano-scale zero-valent iron Fenton system with Cu(II). *Journal of Hazardous Materials, 211-212*, 146–153.

Choi, H., Al-Abed, S. R., Agarwal, S., & Dionysiou, D. (2008). Synthesis of reactive Nano-Fe/Pd bimetallic system-impregnated activated carbon for the simultaneous adsorption and dechlorination of PCBs. *Chemistry of Materials, 20*, 3649–3655.

Chu, L., Wang, J., Dong, J., Liu, H., & Sun, X. (2012). Treatment of coking wastewater by an advanced Fenton oxidation process using iron powder and hydrogen peroxide. *Chemosphere, 86*, 409–414.

Cioslowski, J., Liu, G., & Moncrieff, D. (1997). Thermochemistry of homolytic C-C, C-H, and C-Cl bond dissociations in polychloroethanes: benchmark electronic structure calculations. *Journal of the American Chemical Society, 119*, 11452–114757.

Cupples, A. M., Spormann, A. M., & McCarty, P. L. (2004). Comparative evaluation of chloroethene dechlorination to ethene by dehalococcoides-like microorganisms environ. *Science and Technology, 38*, 4768–4774.

Deng, B., & Hu, S. (2007). Reductive dechlorination of chlorinated solvents on zerovalent iron surfaces. In J. A. Smith & S. E. Burns (Eds.), *Physicochemical Groundwater Remediation* (pp. 139–160). New York: Springer Science & Business Media.

Diao, Z. H., Xu, X. R., Chen, H., Jiang, D., Yang, Y. X., Kong, L. J., Sun, Y. X., Hu, Y. X., Hao, Q. W., & Liu, L. (2016). Simultaneous removal of Cr(VI) and phenol by persulfate activated with bentonite-supported nanoscale zero-valent iron: Reactivity and mechanism. *Journal of Hazardous Materials, 316*, 186–193.

Diya'uddeen, B. H., Rahim Pouran, S., Abdul Aziz, A. R., Nashwan, S. M., Wan Daud, W. M. A., & Shaaban, M. G. (2015). Hybrid of Fenton and sequencing batch reactor for petroleum refinery wastewater treatment. *Journal of Industrial and Engineering Chemistry, 25*, 186–191.

Dong, H., Zhao, F., Zeng, G., Tang, L., Fan, C., Zhang, L., Zeng, Y., He, Q., Xie, Y., & Wu, Y. (2016). Aging study on carboxymethyl cellulose-coated zero-valent iron nanoparticles in water: Chemical transformation and structural evolution. *Journal of Hazardous Materials, 312*, 234–242.

Doong, R.-A., & Lai, Y.-L. (2006). Effect of metal ions and humic acid on the dechlorination of tetrachloroethylene by zerovalent iron. *Chemosphere, 64*, 371–378.

Dries, J., Bastiaens, L., Springael, D., Agathos, S. N., & Diels, L. (2005). Combined removal of chlorinated ethenes and heavy metals by zerovalent iron in batch and continuous flow column systems environ. *Science and Technology, 39*, 8460–8465.

Elliott, D. W., Lien, H.-L., & Zhang, W. X. (2009). Degradation of lindane by zero-valent iron nanoparticles. *Journal of Environmental Engineering, 135*, 317–324.

Fagerlund, F., Illangasekare, T. H., Phenrat, T., Kim, H. J., & Lowry, G. V. (2012). PCE dissolution and simultaneous dechlorination by nanoscale zero-valent iron particles in a DNAPL source zone. *Journal of Contaminant Hydrology, 131*, 9–28.

Fleer, G. J., Cohen Stuart, M. A., Scheutjens, J. M. H. M., Cosgrove, T., & Vincent, B. (1998). *Polymers at interfaces*. New York: Chapman & Hall.

Frost, R. L., Xi, Y., & He, H. (2010). Synthesis, characterization of palygorskite supported zero-valent iron and its application for methylene blue adsorption. *Journal of Colloid and Interface Science, 341*, 153–161.

Garcia, A. N., Boparai, H. K., & O'Carroll, D. M. (2016). Enhanced Dechlorination of 1,2-Dichloroethane by coupled Nano Iron-dithionite treatment. *Environmental Science & Technology, 50*, 5243–5251.

Ghosh, C. K. (2015). Quantum effect on properties of nanomaterials. In A. Sengupta & C. K. Sarkar (Eds.), *Introduction to nano: Basics to nanoscience and nanotechnology* (pp. 73–111). Berlin/Heidelberg: Springer.

Gillham, R. W., & O'Hannesin, S. F. (1994). Enhanced degradation of halogenated aliphatics by zero-valent iron. *Groundwater, 32*, 958–967.

Han, L., Xue, S., Zhao, S., Yan, J., Qian, L., & Chen, M. (2015). Biochar supported nanoscale iron particles for the efficient removal of methyl orange dye in aqueous solutions. *PLoS One, 10*, e0132067.

Harkness, M. R., Bracco, A. A., Brennan, M. J., Jr., DeWeerd, K. A., & Spivack, J. L. (1999). Use of bioaugmentation to stimulate complete reductive dechlorination of trichloroethene in dover soil columns environ. *Science and Technology, 33*, 1100–1109.

He, F., & Zhao, D. (2005). Preparation and characterization of a new class of starch-stabilized bimetallic nanoparticles for degradation of chlorinated hydrocarbons in water. *Environmental Science & Technology, 39*, 3314–3320.

He, F., & Zhao, D. (2008). Hydrodechlorination of trichloroethene using stabilized Fe-Pd nanoparticles: Reaction mechanism and effects of stabilizers, catalysts, and reaction conditions. *Applied Catalysis B: Environmental, 84*, 533–540.

He, F., Zhao, D., Liu, J., & ROberts, C. B. (2007). Stabilization of Fe–Pd nanoparticles with sodium Carboxymethyl cellulose for enhanced transport and Dechlorination of trichloroethylene in soil and groundwater. *Industrial and Engineering Chemistry Research, 46*, 29–34.

Hepure Technology Inc. (2017). *Technical specification sheet: Ferox flow ZVI reactive iron powder*. Flemington: Hepure Technology.

Huang, Q., Liu, W., Peng, P., & Huang, W. (2013). Reductive debromination of tetrabromobisphenol A by Pd/Fe bimetallic catalysts. *Chemosphere, 92*, 1321–1327.

Hyung, H., Fortner, J. D., Hughes, J. B., & Kim, J. H. (2007). Natural organic matter stabilizes carbon nanotubes in the aqueous phase. *Environmental Science & Technology, 41*, 179–184.

Jia, H., & Wang, C. (2015). Dechlorination of chlorinated phenols by subnanoscale Pd0/Fe0intercalated in smectite: Pathway, reactivity, and selectivity. *Journal of Hazardous Materials, 300*, 779–787.

Jiang, Q., Liang, L. H., & Zhao, D. S. (2001). Lattice contraction and surface stress of fcc nanocrystals. *The Journal of Physical Chemistry B, 105*, 6275–6277.

Johnson, T. L., Scherer, M. M., & Tratnyek, P. G. (1996). Kinetics of halogenated organic compound degradation by iron metal. *Environmental Science & Technology, 30*, 2634–2640.

Kanel, S. R., Manning, B., Charlet, L., & Choi, H. (2005). Removal of arsenic(III) from groundwater by nanoscale zero-valent iron. *Environmental Science & Technology, 39*, 1291–1298.

Kim, H.-S., Kim, T., Ahn, J.-Y., Hwang, K.-Y., Park, J.-Y., Lim, T.-T., & Hwang, I. (2012). Aging characteristics and reactivity of two types of nanoscale zero-valent iron particles (FeBH and FeH2) in nitrate reduction. *Chemical Engineering Journal, 197*, 16–23.

Kim, H. J., Leitch, M., Naknakorn, B., Tilton, R. D., & Lowry, G. V. (2017). Effect of emplaced nZVI mass and groundwater velocity on PCE dechlorination and hydrogen evolution in water-saturated sand. *Journal of Hazardous Materials, 322*, 136–144.

Li, X.-Q., & Zhang, W.-X. (2007). Sequestration of metal cations with zerovalent iron nanoparticles-a study with high resolution X-ray photoelectron spectroscopy (HR-XPS) J. *Physical Chemistry C, 111*, 6939–6946.

Li, B., & Zhu, J. (2014). Removal of p-chloronitrobenzene from groundwater: Effectiveness and degradation mechanism of a heterogeneous nanoparticulate zero-valent iron (NZVI)-induced Fenton process. *Chemical Engineering Journal, 255*, 225–232.

Li, R., Gao, Y., Jin, X., Chen, Z., Megharaj, M., & Naidu, R. (2015). Fenton-like oxidation of 2,4-DCP in aqueous solution using iron-based nanoparticles as the heterogeneous catalyst. *Journal of Colloid and Interface Science, 438*, 87–93.

Li, Y., Li, X., Xiao, Y., Wei, C., Han, D., & Huang, W. (2016). Catalytic debromination of tetrabromobisphenol A by Ni/nZVI bimetallic particles. *Chemical Engineering Journal, 284*, 1242–1250.

Li, Y., Li, X., Han, D., Huang, W., & Yang, C. (2017). New insights into the role of Ni loading on the surface structure and the reactivity of nZVI toward tetrabromo- and tetrachlorobisphenol A. *Chemical Engineering Journal, 311*, 173–182.

Lien, H.-L., & Zhang, W. X. (1999). Transformation of chlorinated methanes by nanoscale Iron particles. *Journal of Environmental Engineering, 125*, 1042–1047.

Lien, H.-L., & Zhang, W. X. (2001). Nanoscale iron particles for complete reduction of chlorinated ethenes. *Colloids and Surfaces A, 191*, 97–105.

Lien, H.-L., & Zhang, W. X. (2005). Hydrodechlorination of chlorinated ethanes by nanoscale Pd/Fe bimetallic particles. *Journal of Environmental Engineering, 131*, 4–10.

Liu, Y., & Lowry, G. V. (2006). Effect of particle age (Fe0 content) and solution pH on NZVI reactivity: H2 evolution and TCE dechlorination. *Environmental Science & Technology, 40*, 6085–6090.

Liu, Y., Choi, H., Dionysiou, D., & Lowry, G. V. (2005a). Trichloroethene hydrodechlorination in water by highly disordered monometallic nanoiron. *Chemistry of Materials, 17*, 5315–5322.

Liu, Y., Majetich, S. A., Tilton, R. D., Sholl, D. S., & Lowry, G. V. (2005b). TCE dechlorination rates, pathways, and efficiency of nanoscale iron particles with different properties. *Environmental Science & Technology, 39*, 1338–1345.

Liu, Y., Phenrat, T., & Lowry, G. V. (2007). Effect of TCE concentration and dissolved groundwater solutes on NZVI-promoted TCE dechlorination and H2 evolution. *Environmental Science & Technology, 41*, 7881–7887.

Liu, A., Liu, J., & Zhang, W. X. (2015). Transformation and composition evolution of nanoscale zero valent iron (nZVI) synthesized by borohydride reduction in static water. *Chemosphere, 119*, 1068–1074.

Le, T. S. T., & Phenrat, T. (2018). Sustaining 1,2-Dichloroethane Degradation in Nanoscale Zero-Valent Iron induced Fenton System by Using Sequential H2O2 Addition at Natural pH. Water Research, (Under Review).

Mackenzie, K., Bleyl, S., Georgi, A., & Kopinke, F.-D. (2012). Carbo-Iron – An Fe/AC composite – As alternative to nano-iron for groundwater treatment. *Wat. Res., 46*, 3817–3826.

Maes, A., Reamdinck, H. V., Smith, K., Ossieur, W., Lebbe, L., & Verstraete, W. (2006). Transport and activity of desulfitobacterium dichloroeliminans strain DCA1 during bioaugmentation of 1,2-DCA-contaminated groundwater. *Environmental Science and Technology, 40*, 5544–5552.

Martins, R. C., Rossi, A. F., & Quinta-Ferreira, R. M. (2010). Fenton's oxidation process for phenolic wastewater remediation and biodegradability enhancement. *Journal of Hazardous Materials, 180*, 716–721.

Masten, S. J., & Butler, J. N. (1986). Ultraviolet-enhanced ozonation of organic compounds: 1,2-Dichloroethane and trichloroethylene as model substrates. *Ozone: Science & Engineering, 8*, 339–353.

Mccormick, M. L., & Adriaens, P. (2004). Carbon tetrachloride transformation on the surface of nanoscale biogenic magnetite particles. *Environmental Science & Technology, 38*, 1045–1053.

Moon, B.-H., Park, Y.-B., & Park, K.-H. (2011). Fenton oxidation of Orange II by pre-reduction using nanoscale zero-valent iron. *Desalination, 268*, 249–252.

Munoz, M., Pliego, G., de Pedro, Z. M., Casas, J. A., & Rodriguez, J. J. (2014). Application of intensified Fenton oxidation to the treatment of sawmill wastewater. *Chemosphere, 109*, 34–41.

Naja, G., Halasz, A., Thiboutot, S., Ampleman, G., & Hawari, J. (2008). Degradation of hexahydro-1,3,5-trinitro-1,3,5-triazine (RDX) using zerovalent iron nanoparticles. *Environmental Science & Technology, 42*, 4364–4370.

Nurmi, J. T., Tratnyek, P. G., Sarathy, V., Baer, D. R., Amonette, J. E., Pecher, K., Wang, C., Linehan, J. C., Matson, D. W., Penn, R. L., & Driessen, M. D. (2005). Characterization and properties of metallic iron nanoparticles: Spectroscopy, electrochemistry, and kinetics. *Environmental Science & Technology, 39*, 1221–1230.

Papoutsakis, S., Pulgarin, C., Oller, I., Sánchez-Moreno, R., & Malato, S. (2016). Enhancement of the Fenton and photo-Fenton processes by components found in wastewater from the industrial processing of natural products: The possibilities of cork boiling wastewater reuse. *Chemical Engineering Journal, 304*, 890–896.

Petri, B. G., Watts, R. J., Teel, A. L., & Hugling, S. G. (2011). Fundamentals of ISCO using hydrogen peroxide. In R. L. Siegrist, M. Crimi, & T. L. Simpkin (Eds.), *Situ chemical oxidation for groundwater remediation* (pp. 33–88). New York: Springer Science+Business Media, LLC.

Phenrat, T., Kim, H.-J., Fagerlund, F., Illangasekare, T., Tilton, R. D., & Lowry, G. V. (2009a). Particle size distribution, concentration, and magnetic attraction affect transport of polymer-modified Fe^0 nanoparticles in sand columns. *Environmental Science & Technology, 43*, 5079–5085.

Phenrat, T., Liu, Y., Tilton, R. D., & Lowry, G. V. (2009b). Adsorbed polyelectrolyte coatings decrease Fe^0 nanoparticle reactivity with TCE in water: Conceptual model and mechanisms. *Environmental Science & Technology, 43*, 1507–1514.

Phenrat, T., Schoenfelder, D., Kirschling, T. L., Tilton, R. D., Lowry, G. V. (2015). Adsorbed poly (aspartate) coating limits the adverse effects of dissolved groundwater solutes on Fe^0 nanoparticle reactivity with trichloroethylene. *Environmental Science Pollution Research Vol 25* (8) 7157–7169.

Phenrat, T., Thongboot, T., & Lowry, G. V. (2016). Electromagnetic induction of zerovalent iron (ZVI) powder and nanoscale zerovalent iron (NZVI) particles enhances dechlorination of trichloroethylene in contaminated groundwater and soil: Proof of concept. *Environmental Science & Technology, 50*, 872–880.

Ponder, S. M., Darab, J. G., & Mallouk, T. E. (2000). Remediation of Cr(VI) and Pb(II) aqueous solutions using supported, nanoscale zero-valent Iron. *Environmental Science & Technology, 34*, 2564–2569.

Ramos-Tejada, M. M., Ontiveros, A., Viota, J. L., & Durán, J. D. G. (2003). Interfacial and rheological properties of humic acid/hematite suspensions. *Journal of Colloid and Interface Science, 268*, 85–95.

Reinsch, B. C., Forsberg, B., Penn, R. L., Kim, C. S., & Lowry, G. V. (2010). Chemical transformations during aging of zero-valent iron nanoparticles in the presence of common groundwater dissolved constituents. *Environmental Science & Technology, 44*, 3455–3461.

Reynolds, G. W., Hoff, J. T., & Gillham, R. W. (1990). Sampling bias caused by materials used to monitor halocarbons in groundwater. *Environmental Science & Technology, 24*, 135–142.

Sakulchaicharoen, N., O'Carroll, D. M., & Herrera, J. E. (2010). Enhanced stability and dechlorination activity of pre-synthesis stabilized nanoscale FePd particles. *Journal of Contaminant Hydrology, 118*, 117–127.

Sarathy, V., Tratnyek, P. G., Nurmi, J. T., Baer, D. R., Amonette, J. E., Chun, C. L., Penn, R. L., & Reardon, E. J. (2008). Aging of iron nanoparticles in aqueous solution: Effects on structure and reactivity. *Journal of Physical Chemistry C, 112*, 2286–2293.

Scheutjens, J. M. H. M., & Fleer, G. J. (1979). Statistical theory of the adsorption of interacting chain molecules. 1. Partition function, segment density distribution, and adsorption isotherms. *The Journal of Physical Chemistry, 83*, 1619–1635.

Scheutjens, J. M. H. M., & Fleer, G. J. (1980). Statistical theory of the adsorption of interacting chain molecules. 2. Train, loop, and tail size distribution. *The Journal of Physical Chemistry, 84*, 178–190.

Schlicker, O., Ebert, M., Fruth, M., Weidner, M., Wust, W., & Dahmke, A. (2000). Degradation of TCE with iron: The role of competing chromate and nitrate reduction. *Groundwater, 38*, 403–409.

Schrick, B., Blough, J. L., Jones, A. D., & Mallouk, T. E. (2002). Hydrodechlorination of trichloroethylene to hydrocarbons using bimetallic nickel-iron nanoparticles. *Chemistry of Materials, 14*, 5140–5147.

Schwarzenbach, R. P., Gschwend, P. M., & Imboden, D. M. (2003a). *Environmental organic chemistry* (2nd ed.). Hoboken: Wiley.

Schwarzenbach, R. P., Gschwend, P. M., & Imboden, D. M. (2003b). *Environmental organic chemistry* (2nd ed.). Hoboken: Wiley-Interscience.

Shih, Y.-H., Chen, Y.-C., Chen, M.-Y., Tai, Y.-T., & Tso, C.-P. (2009). Dechlorination of hexachlorobenzene by using nanoscale Fe and nanoscale Pd/Fe bimetallic particles. *Colloids and Surfaces A, 332*, 84–89.

Shu, H.-Y., Chang, M.-C., Chen, C.-C., & Chen, P.-E. (2010). Using resin supported nano zero-valent iron particles for decoloration of acid blue 113 azo dye solution. *Journal of Hazardous Materials, 184*, 499–505.

Song, H., & Carraway, E. R. (2005). Reduction of chlorinated ethanes by nanosized zero-valent iron: Kinetics, pathways, and effects of reaction conditions. *Environmental Science & Technology, 39*, 6237–6245.

Song, H., & Carraway, E. R. (2006). Reduction of chlorinated methanes by nano-sized zero-valent Iron. Kinetics, pathways, and effect of reaction conditions. *Environmental Engineering Science, 23*, 272–284.

Su, C., Puls, R. W., Krug, T. A., Watling, M. T., O'Hara, S. K., Quinn, J. W., & Ruiz, N. W. (2012a). A two and half-year-performance evaluation of a field test on treatment of source zone tetrachloroethene and its chlorinated daughter products using emulsified zero valent iron nanoparticles. *Water Research, 46*, 5071–5084.

Su, Y.-F., Hsu, C.-Y., & Shih, Y.-H. (2012b). Effects of various ions on the dechlorination kinetics of hexachlorobenzene by nanoscale zero-valent iron. *Chemosphere, 88*, 1346–1352.

Sun, Y.-P., Li, X.-Q., Zhang, W. X., & Wang, H. P. (2007). A method for the preparation of stable dispersion of zero-valent iron nanoparticles. *Colloids and Surfaces A, 308*, 60–66.

Scherer, M. M., Balko, B. A., Gallagher, D. A., Tratnyek, P. G. (1998). Correlation Analysis of Rate Constants for Dechlorination by Zero-Valent Iron. Environmental Science Technology, 32 (19), 3026–3033

Taha, M. R., & Ibrahim, A. H. (2014a). Characterization of nano zero-valent iron (nZVI) and its application in sono-Fenton process to remove COD in palm oil mill effluent. *Journal of Environmental Chemical Engineering, 2*, 1–8.

Taha, M. R., & Ibrahim, A. H. (2014b). COD removal from anaerobically treated palm oil mill effluent (AT-POME) via aerated heterogeneous Fenton process: Optimization study. *Journal of Water Process Engineering, 1*, 8–16.

Thompson, J. M., Chrisholm, B. J., & Bezbaruah, A. N. (2010). Reductive dechlorination of chloroacetanilide herbicide (Alachlor) using zero-valent iron nanoparticles. *Environmental Engineering Science, 27*, 227–232.

Tobiszewski, M., & Namieśnik, J. (2012). Abiotic degradation of chlorinated ethanes and ethenes in water. *Environmental Science and Pollution Research, 19*, 1994–2006.

Totten, L. A., & Roberts, A. L. (2010). Calculated one- and two-electron reduction potentials and related molecular descriptors for reduction of alkyl and vinyl halides in water. *Critical Reviews in Environment Science and Technology, 31*, 175–221.

Tratnyek, P. G., & Johnson, R. L. (2006). Nanotechnologies for environmental cleanup. *Nano Today, 1*, 44–48.

Tratnyek, P. G., Scherer, M. M., Deng, B., & Hu, S. (2001). Effects of natural organic matter, anthropogenic surfactants, and model quinones on the reduction of contaminants by zero-valent iron. *Water Research, 35*, 4435–4443.

Usman, M., Hanna, K., & Haderlein, S. (2016). Fenton oxidation to remediate PAHs in contaminated soils: A critical review of major limitations and counter-strategies. *The Science of the Total Environment, 569–570*, 179–190.

Villa, R. D., Trovó, A. G., & Nogueira, R. F. P. (2010). Diesel degradation in soil by Fenton process. *Journal of the Brazilian Chemical Society, 21*, 1088–1095.

Vilve, M., Vilhunen, S., Vepsalainen, M., Kurniawan, T. A., Lehtonen, N., Isomaki, H., & Sillanpaa, M. (2010). Degradation of 1,2-dichloroethane from wash water of ion-exchange resin using Fenton's oxidation. *Environmental Science and Pollution Research International, 17*, 875–884.

Wang, C. B., & Zhang, W. X. (1997). Synthesizing nanoscale iron particles for rapid and complete dechlorination of TCE and PCBs. *Environmental Science & Technology, 31*, 2154–2156.

Wang, W., & Zhou, M. (2010). Degradation of trichloroethylene using solvent-responsive polymer coated Fe nanoparticles. *Colloids and Surfaces A, 369*, 232–239.

Wang, Q., Lee, S., & Choi, H. (2010). Aging study on the structure of Fe0-nanoparticles: Stabilization, characterization, and reactivity. *Journal of Physical Chemistry C, 114*, 2027–2033.

Wang, Y., Zhou, D., Wang, Y., Zhu, X., & Jin, S. (2011). Humic acid and metal ions accelerating the dechlorination of 4-chlorobiphenyl by nanoscale zero-valent iron. *Journal of Environmental Sciences, 23*, 1286–1292.

Wang, F., Wu, Y., Gao, Y., Li, H., & Chen, Z. (2016a). Effect of humic acid, oxalate and phosphate on Fenton-like oxidation of microcystin-LR by nanoscale zero-valent iron. *Separation and Purification Technology, 170*, 337–343.

Wang, L., Yang, J., Li, Y., Lv, J., & Zou, J. (2016b). Removal of chlorpheniramine in a nanoscale zero-valent iron induced heterogeneous Fenton system: Influencing factors and degradation intermediates. *Chemical Engineering Journal, 284*, 1058–1067.

Xu, L., & Wang, J. (2011). A heterogeneous Fenton-like system with nanoparticulate zero-valent iron for removal of 4-chloro-3-methyl phenol. *Journal of Hazardous Materials, 186*, 256–264.

Xu, X., Wo, J., Zhang, J., Wu, Y., & Liu, Y. (2009). Catalytic dechlorination of p-NCB in water by nanoscale Ni/Fe. *Desalination, 242*, 346–354.

Yan, W., Herzing, A. A., Li, X.-Q., Kiely, C. J., & Zhang, W. X. (2010). Structural evolution of Pd-doped nanoscale zero-valent Iron (nZVI) in aqueous media and implications for particle aging and reactivity. *Environmental Science & Technology, 44*, 4288–4294.

Yehia, F. Z., Eshaq, G., Rabie, A. M., Mady, A. H., & ElMetwally, A. E. (2015). Phenol degradation by advanced Fenton process in combination with ultrasonic irradiation. *Egyptian Journal of Petroleum, 24*, 13–18.

Yin, X., Liu, W., & Ni, J. (2014). Removal of coexisting Cr(VI) and 4-chlorophenol through reduction and Fenton reaction in a single system. *Chemical Engineering Journal, 248*, 89–97.

Yu, R. F., Chen, H. W., Cheng, W. P., Lin, Y. J., & Huang, C. L. (2014). Monitoring of ORP, pH and DO in heterogeneous Fenton oxidation using nZVI as a catalyst for the treatment of azo-dye textile wastewater. *Journal of the Taiwan Institute of Chemical Engineers, 45*, 947–954.

Yu, B., Jin, X., Kuang, Y., Megharaj, M., Naidu, R., & Chen, Z. (2015). An integrated biodegradation and nano-oxidation used for the remediation of naphthalene from aqueous solution. *Chemosphere, 141*, 205–211.

Zha, S., Cheng, Y., Gao, Y., Chen, Z., Megharaj, M., & Naidu, R. (2014). Nanoscale zero-valent iron as a catalyst for heterogeneous Fenton oxidation of amoxicillin. *Chemical Engineering Journal, 255*, 141–148.

Zhang, W. (2003). Nanoscale iron particles for environmental remediation: An overview. *Journal of Nanoparticle Research, 5*, 323–332.

Zhang, W.-X., Wang, C.-B., & Lien, H.-L. (1998). Treatment of chlorinated organic comtaminants with nanoscale bimetallic particles. *Catalysis Today, 40*, 387–395.

Zhang, X., Lin, Y.-M., Shan, X.-Q., & Chen, Z.-L. (2010). Degradation of 2,4,6-trinitrotoluene (TNT) from explosive wastewater using nanoscale zero-valent iron. *Chemical Engineering Journal, 158*, 566–570.

Zhang, R., Li, J., Shen, C. L. J., Sun, X., Han, W., & Wang, L. (2013). Reduction of nitrobenzene using nanoscale zero-valent iron confined in channels of ordered mesoporous silica. *Colloids and Surfaces A, 425*, 108–114.

Zhang, W., Gao, H., He, J., Yang, P., Wang, D., Ma, T., Xia, H., & Xu, X. (2017). Removal of norfloxacin using coupled synthesized nanoscale zero-valent iron (nZVI) with H2O2 system: Optimization of operating conditions and degradation pathway. *Separation and Purification Technology, 172*, 158–167.

Zheng, T., Zhan, J., He, J., Day, C., Lu, Y., McPherson, G. L., Piringer, G., & John, V. T. (2008). Reactivity characteristics of nanoscale zerovalent iron – silica composites for trichloroethylene remediation. *Environmental Science & Technology, 42*, 4494–4499.

Zhou, T., Li, Y., Ji, J., Wong, F.-S., & Lu, X. (2008). Oxidation of 4-chlorophenol in a heterogeneous zero valent iron/H2O2 Fenton-like system: Kinetic, pathway and effect factors. *Separation and Purification Technology, 62*, 551–558.

Zhou, H., Han, J., Baig, S. A., & Xu, X. (2011). Dechlorination of 2,4-dichlorophenoxyacetic acid by sodium carboxymethyl cellulose-stabilized Pd/Fe nanoparticles. *Journal of Hazardous Materials, 198*, 7–12.

Zhou, H., Shen, Y., Lv, P., Wang, J., & Li, P. (2015). Degradation pathway and kinetics of 1-alkyl-3-methylimidazolium bromides oxidation in an ultrasonic nanoscale zero-valent iron/hydrogen peroxide system. *Journal of Hazardous Materials, 284*, 241–252.

Zhu, B.-W., Lim, T.-T., & Feng, J. (2006). Reductive dechlorination of 1,2,4-trichlorobenzene with palladized nanoscale Fe0 particles supported on chitosan and silica. *Chemosphere, 65*, 1137–1145.

Zhu, B.-W., Lim, T.-T., & Feng, J. (2008). Influences of amphiphiles on dechlorination of a trichlorobenzene by nanoscale Pd/Fe: Adsorption, reaction kinetics, and interfacial interactions. *Environmental Science & Technology, 42*, 4513–4519.

Zhu, N., Luan, H., Yuan, S., Chen, J., Wu, X., & Wang, L. (2010). Effective dechlorination of HCB by nanoscale Cu/Fe particles. *Journal of Hazardous Materials, 176*, 1101–1105.

Chapter 4
Nanoscale Zerovalent Iron Particles for Treatment of Metalloids

Jan Filip, Jan Kolařík, Eleni Petala, Martin Petr, Ondřej Šráček, and Radek Zbořil

Abstract In the past few decades, the remediation ability of nanoscale zerovalent iron (NZVI) particles has been exploited in both lab-scale and real-world scenarios. These studies and application examples brought about numerous breakthrough results. Therefore, NZVI has proved to be an excellent candidate for the efficient remediation of even challenging and complicated polluted aqueous systems. Herein, we emphasize the treatment of heavy metals (e.g., copper, cobalt, nickel, zinc, uranium, mercury, cadmium, lead, etc., and also hexavalent chromium) and metalloids (e.g., arsenic) as pollutants in water by NZVI. The mechanisms involved in the metal removal by NZVI are described and explained in terms of selectivity and reaction pathways. Analytical aspects, mainly represented by X-ray photoelectron spectroscopy as tool for deep understanding of the mechanism of metal removal, are mentioned, while an extensive report of examples of metal cations that can be removed by NZVI is overviewed. Specifically, the cases of chromium and arsenic removal are analyzed thoroughly, explaining the efficiency of various NZVI-based systems for immobilization and/or reduction of such toxic species. Finally, success stories of pilot and full-scale tests where NZVI was employed for metal removal are presented, describing the conditions, the effects, and the advantages of NZVI in large-scale applications.

Keywords Nanoscale Zerovalent Iron · Metals · Metalloids · Sequestration · Arsenic · Chromium · Pilot scale

J. Filip (✉) · J. Kolařík · E. Petala · M. Petr · O. Šráček · R. Zbořil
Regional Centre of Advanced Technologies and Materials, Departments of Physical Chemistry, Experimental Physics and Geology, Faculty of Science, Palacký University in Olomouc, Olomouc, Czech Republic
e-mail: jan.filip@upol.cz

T. Phenrat, G. V. Lowry (eds.), *Nanoscale Zerovalent Iron Particles for Environmental Restoration*, https://doi.org/10.1007/978-3-319-95340-3_4

4.1 Introduction

Decades of large industrial and urban activities have led to elevated concentrations of a wide range of pollutants in groundwater and wastewater, which affect the health of millions of people worldwide. In this context, heavy metals pose a significant risk because they are very toxic and tend to accumulate in living organisms. Many heavy metals are known to be poisonous, mutagenic, or carcinogenic, such as chromium, copper, nickel, mercury, cadmium, and lead.

Due to their versatility and multimodal action, nanoscale zerovalent iron (NZVI) particles represent one of the most promising materials for water treatment and environmentally friendly remediation processes (Li et al. 2006; Zhao et al. 2016). Besides treatment of organic pollutants (Chap. 3), they have been applied for sequestration of inorganic compounds like heavy metal ions and metalloids, e.g., arsenic (Kanel et al. 2005; Li and Zhang 2007; Karabelli et al. 2008), phosphates (Almeelbi and Bezbaruah 2012), and numerous other inorganic compounds (e.g., perchlorate, nitrate). NZVI has received significant attention, and various promising water treatment approaches have been documented (Fu et al. 2014; Guo et al. 2016).

It has been reported that NZVI has much larger capacity for metal ion removal than conventional technologies, i.e., ion exchange, filtration, electrochemical precipitation, adsorption (e.g., on activated carbon), and bioremediation (Li et al. 2016b). This is explained by the combined effect of reductive properties of metallic iron (i.e., by turning the metals into less toxic and less soluble forms through lowering their oxidation state) with the high sorption capabilities of both NZVI and the reaction products (i.e., the newly formed iron oxyhydroxides (Klimkova et al. 2011)). The applicability of NZVI in environmental remediation and water treatment has been recently reviewed by many authors (Crane and Scott 2012; O'Carroll et al. 2013; Li et al. 2016b; Zou et al. 2016).

The reaction of NZVI with metals is generally fast (compared to reaction with organic pollutants like trichloroethylene (TCE)) and highly efficient as proved by numerous laboratory studies (Han et al. 2016). The NZVI particles evoke an alkaline reaction with water leading to removal of a broad range of toxic elements like heavy or transition metals and metalloids by various mechanisms. The main mechanisms of metal sequestration by NZVI particles include electrochemical reduction, reductive mineral precipitation and coprecipitation (i.e., due to substantial decrease in oxidation–reduction potential, ORP, from oxidative to neutral or anoxic conditions), formation of insoluble hydroxides, and sorption on nanocrystalline iron oxyhydroxides formed as a result of NZVI oxidation (Li and Zhang 2007; Klimkova et al. 2011).

Predominantly, the application of NZVI has been proposed for elimination of water and soil contamination. This occurs by decreasing the concentration of dissolved metals in the treated area through reduction and immobilization or sequestration onto the NZVI. A representative example of NZVI use for such metal removal from complex acid mine water with different extent of dilution is presented in Fig. 4.1. However, in the case of industrial wastewater or large contaminated

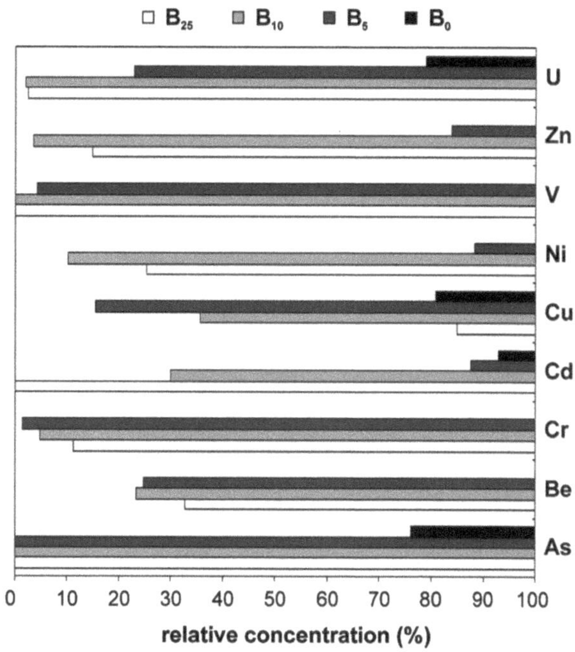

Fig. 4.1 Percentual decrease of metal concentration in raw and diluted acid mine water based on application of NZVI particles (raw water – B_0, diluted 5, 10, and 25 times – B_5, B_{10}, and B_{25}, respectively; note: shown are normalized initial metal concentrations). (Adapted with permission from. Reprinted with permission from Klimkova et al. (2011). Copyright (2011) Elsevier)

areas, such as old mining or industrial sites, where the concentration of dissolved metals is typically high, metals immobilized on the NZVI surface could potentially be concentrated and recycled (Li et al. 2017).

This book chapter summarizes the known mechanisms of metal ion and metalloid removal by NZVI particles as well as their derivatives. Examples of a broad range of laboratory studies on model and real polluted waters are given, with a focus on sequestration of arsenic, copper, and chromium. Moreover, an overview of pilot and full-scale applications of NZVI for metal removal are described. The analytical aspects of metal immobilization by NZVI are also highlighted.

4.2 Mechanisms and Kinetics of Metal Removal by NZVI

The interaction of NZVI with metals follows various physical and chemical reactions. When NZVI particles get into contact with pollutants, they typically possess a core–shell structure. The metallic iron core is surface-oxidized to form a thin iron oxide/oxyhydroxide shell either spontaneously by reaction of the iron surface with water (see (Filip et al. 2014)) or purposefully during NZVI production (see Zboril et al. (2012) and the product NANOFER STAR from NANO IRON Company, Czech Republic). The oxide shell is poorly crystalline to amorphous and could act as an excellent sorbent providing a large number of sorption and coordination sites. On

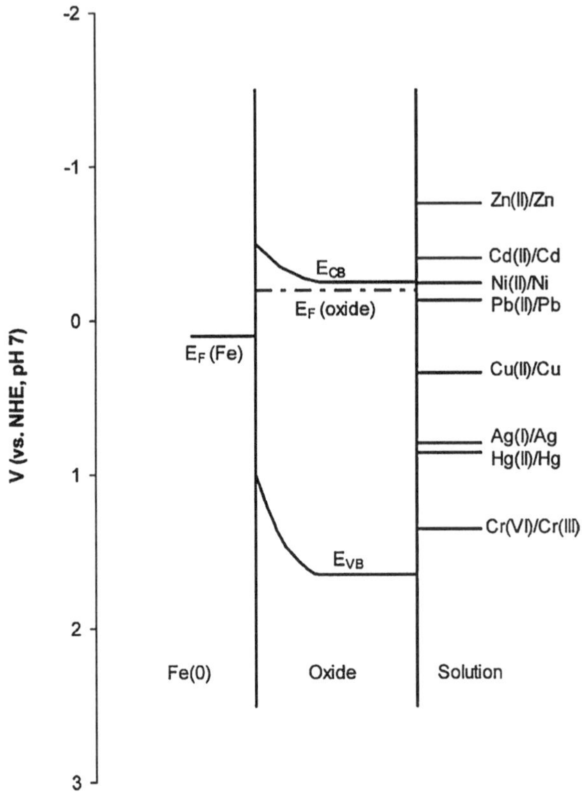

Fig. 4.2 A simplified band diagram for iron/iron oxide/aqueous metal cations (values are for pH = 7). E_F, Fermi energies; E_{CB}, energies of conduction band; and E_{VB}, energies of valence band. (Reprinted with permission from Li and Zhang (2007). Copyright (2007) American Chemical Society)

the other hand, the metallic iron core has strong reducing (i.e., standard potential −440 mV) or electron-donating ability. Moreover, Fe(II) released from NZVI due to its corrosion with water represents an essential component of reduction and coprecipitation (see below). Therefore, such a core–shell structure is unique for concurrent sorption and reductive precipitation of metal ions. Moreover, the oxide shell strongly influences not only the mechanism of NZVI interaction with metal ions, but it also significantly controls the kinetics through electron transfer (ET). The possible ET pathways have been identified as (Li and Zhang 2007) (1) direct ET from NZVI surface through defects in the oxide shell behaving as simple physical barrier; (2) indirect ET through oxide shell via the conduction band, impurity bands, or localized bands; and (3) as the least important, ET from Fe(II) sorbed on lattice surface sites. Figure 4.2 represents a simplified band diagram for core–shell NZVI particles – the details about the ET can be found in Li and Zhang (2007).

Most of the reaction mechanisms between NZVI (i.e., any type of iron, including nano-, micro-, and macroscopic particles) and metals/metalloids can be derived from the basic reaction of iron with water (Filip et al. 2014):

$$Fe(0)_{(s)} + 2H_2O_{(aq)} \rightarrow Fe(II) + H_2 + 2OH^- \quad (4.1)$$

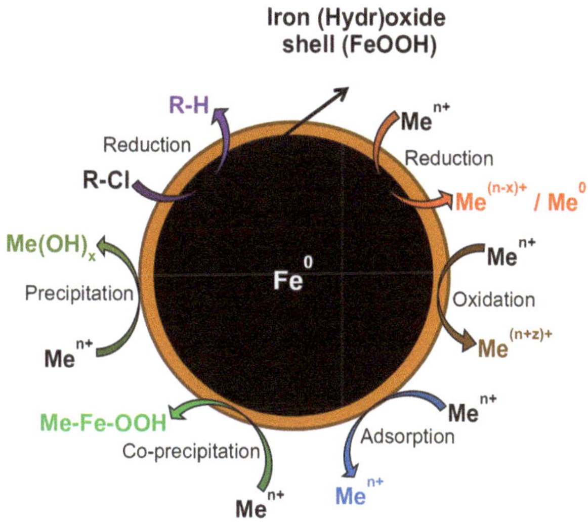

Fig. 4.3 Graphical illustration of possible mechanisms for removal or inorganic and organic pollutants on the surface of core–shell NZVI particles. (Reprinted with permission from O'Carroll et al. (2013). Copyright (2013) Elsevier)

followed by:

$$Fe(II) + 2H_2O \rightarrow Fe(OH)_2 + 2H^+ \tag{4.2}$$

Actually, the main difference between nano- and micro−/macroscale iron lies in the size and specific surface area (and thus, reactivity is dramatically influenced) and different extent of newly formed nanocrystalline reaction products.

The abovementioned reactions clearly demonstrate the possibility of NZVI-mediated acid–base changes that can lead to the increase of pH and decrease of ORP. Moreover, the oxidation of Fe(II) to Fe(III) gives rise to the formation of new nanocrystalline phase(s) (ferric oxides, hydroxides, and oxyhydroxides) with high sorption or coagulation properties. Such reaction products typically precipitate on or close to the surface of NZVI particles. Thus, the main known interaction between NZVI and particular metals can be categorized as (adapted from O'Carroll et al. (2013); the types of reactions are summarized in Fig. 4.3):

1. *Reduction* (typical for Cr, As, Cu, U, Pb, Ni, Se, Co, Pd, Pt, Hg, Ag).
2. *Adsorption* (typical for Cr, As, U, Pb, Ni, Se, Co, Cd, Zn, Ba).
3. *Oxidation/re-oxidation* (typical for As, U, Se, Pb).
4. *Coprecipitation* (typical for Cr, As, Ni, Se).
5. *Precipitation* (typical for Cu, Pb, Cd, Co, Zn).

However, in most cases a combination of more than one of the listed process can occur, simultaneously, during the interaction of NZVI surface with a particular metal (Fig. 4.3). Namely, the process of reduction typically includes two distinct mechanisms: (i) direct reduction of metal ions by NZVI and (ii) primary adsorption of metal ions onto NZVI surface followed by their subsequent reduction (Zou et al. 2016).

Table 4.1 Standard redox potentials (E^0) for selected metals, metalloids, and common organic pollutants (values are for aqueous solution at 25 °C relative to the standard hydrogen electrode)

Aqueous solution	Half reaction	E^0 (V)
Chromium (Cr)	$CrO_4^{2-} + 8H^+ + 3e^- \leftrightarrow Cr^{3+} + 4H_2O$	1.51
Chromium (Cr)	$Cr_2O_7^{2-} + 14H^+ + 6e^- \leftrightarrow 2Cr^{3+} + 7H_2O$	1.36
Platinum (Pt)	$Pt^{2+} + 2e^- \leftrightarrow Pt$	1.19
Palladium (Pd)	$Pd^{2+} + 2e^- \leftrightarrow Pd$	0.92
Mercury (Hg)	$Hg^{2+} + 2e^- \leftrightarrow Hg$	0.86
Silver (Ag)	$Ag^+ + e^- \leftrightarrow Ag$	0.80
Arsenic (As^V)	$H_3AsO_4 + 2H^+ + 2e^- \leftrightarrow HAsO_2 + 4H_2O$	0.56
Copper (Cu)	$Cu^{2+} + 2e^- \leftrightarrow Cu$	0.34
Uranium (U)	$UO_2^{2+} + 4H^+ + 2e^- \leftrightarrow U^{4+} + 2H_2O$	0.27
Arsenic (As^{III})	$H_3AsO_3 + 3H^+ + 3e^- \leftrightarrow As + 3H_2O$	0.24
Copper $(Cu)^+$	$Cu^{2+} + e^- \leftrightarrow Cu^+$	0.16
Lead (Pb)	$Pb^{2+} + 2e^- \leftrightarrow Pb$	−0.13
Nickel (Ni)	$Ni^{2+} + 2e^- \leftrightarrow Ni$	−0.25
Cadmium (Cd)	$Cd^{2+} + 2e^- \leftrightarrow Cd$	−0.40
Iron (Fe)	$Fe^{2+} + 2e^- \leftrightarrow Fe$	−0.44
Zinc (Zn)	$Zn^{2+} + 2e^- \leftrightarrow Zn$	−0.76
Barium (Ba)	$Ba^{2+} + 2e^- \leftrightarrow Ba$	−2.92
1,2-Dichloroethane	$ClH_2C{-}CH_2Cl + 2e^- \leftrightarrow H_2C{=}{=}CH_2 + 2Cl^-$	0.74
Carbon tetrachloride (CT)	$CCl_4 + H^+ + 2e^- \leftrightarrow CHCl_3 + Cl^-$	0.67
Tetrachloroethylene (PCE)	$Cl_2C{=}{=}CCl_2 + H^+ + 2e^- \leftrightarrow Cl_2C{=}{=}CHCl + Cl^-$	0.57
Trichloroethylene (TCE)	$Cl_2C{=}{=}CHCl + H^+ + 2e^- \leftrightarrow Cl_2C{=}{=}CH_2 + Cl^-$	0.53
Vinyl chloride (VC)	$ClHC{=}{=}CH_2 + H^+ + 2e^- \leftrightarrow H_2C{=}{=}CH_2 + Cl^-$	0.45
1,1-Dichloroethene (1,1-DCE)	$Cl_2C{=}{=}CH_2 + H^+ + 2e^- \leftrightarrow ClHC{=}{=}CH_2 + Cl^-$	0.42

Adapted with permission from O'Carroll et al. (2013). Copyright (2013) Elsevier

The mode of interaction largely depends on the standard redox potential (E^0) of the particular metal ion (see Table 4.1). Metals that have E^0 far more positive compared to E^0 of Fe^0 (typically Cr, Hg, Ag, As, Cu, U, and Se) are capable of being electrochemically reduced, and they could also precipitate on or close to the NZVI surface – it is the most common sequestration mechanism of metal ions by means of interaction with NZVI. Metals with E^0 just slightly more positive than Fe^0 (e.g., Pb and Ni) can be both reduced and adsorbed. On the contrary, metal ions with E^0 similar or more negative than Fe^0 are mostly adsorbed on the surface of NZVI (i.e., on the oxide/hydroxide shell) without their reduction; typical examples include Cd and Zn ions (Li and Zhang 2007; O'Carroll et al. 2013).

The mechanism and efficiency of metal sequestration by NZVI are also dependent on the pH of the solution (but less importantly than in the case of common sorbents without multimodal functionality). The fresh NZVI particles typically increase pH of the aqueous solution (see Eq. 4.1) resulting in a negatively charged NZVI surface via deprotonation of oxygen-containing groups (Zou et al. 2016). This promotes the electrostatic adsorption of positively charged metal species. The increase of pH also induces precipitation of some metals in the form of hydroxides simultaneously with the formation of the NZVI passivation shell. Typical examples include mainly Zn and Co and also Be, Ra, Th, Pu, Sr, Mn, and Cs (i.e., those metal ions with E^0 more negative than Fe^0; (O'Carroll et al. 2013)). As an example, the dramatic increase of nickel removal by NZVI in the pH range from 4 to 8 was demonstrated by Li and Zhang (2007). On the other hand, at lower pH the iron corrosion is accelerated; thus an increase in the generation of hydrogen occurs, while the formation of the passivation shell is not favorable. In this case, the reduction is the main mechanism; thus the beneficial degradation of various metals especially in the form of oxyanions, e.g., CrO_4^{2-}, is more favorable at lower pH. In other words, NZVI offers a surface where both reduction and adsorption processes can eliminate a wide range of pollutants.

In oxygen-containing water (i.e., in surface or waste waters), NZVI particles could indirectly act also as oxidant. This unusual mechanism is attributed to production of hydroxyl radicals through Fenton reaction involving hydrogen peroxide and Fe(II) produced during NZVI corrosion – see below (Keenan and Sedlak 2008).

In real polluted waters, the interactions are typically much more complex due to the presence of many coexisting organic and inorganic ions. However, the detailed investigation of reaction mechanism(s) of NZVI with metal-containing acid mine water from in situ chemical leaching of uranium (water from Straz pod Ralskem, Czech Republic) (Klimkova et al. 2011) identified mechanisms similar to the abovementioned examples. Major contaminants in this water included aluminum and sulfates, with As, Be, Cd, Cr, Cu, Ni, U, and Zn as important micropollutants. The following observations were made:

1. Metals reducible by NZVI to a less soluble forms: Cr, Cu, U, and V. Cr(VI) is reduced to Cr(III) precipitating then as insoluble hydroxide; U(VI) in the form of $(UO^{2+}{}_2)$ is reduced to less soluble U(IV) ion (UO^{2+}). Similarly are reduced metals whose standard electrode potential is above the Fe^0/Fe^{2+} pair potential, e.g., Cu^{2+} is reduced to Cu^0 directly at the surface of NZVI particles.
2. Metals, non-reducible by NZVI, which precipitate, adsorb, or are incorporated in the structure of newly formed iron oxyhydroxides due to the increase in pH: Al, Be, Cd, Zn, and Ni. Typical example is precipitation of aluminum (oxo)-hydroxide or adsorption/surface complexation of divalent cations like Cd^{2+} and Zn^{2+} (which are insensitive to reduction by NZVI) on the surface of iron oxyhydroxides.
3. Metal(loid)s with increasing solubility when reduced by NZVI and undergoing subsequent coprecipitation with the iron oxyhydroxides: typical example is reduction of less soluble As(V) to more soluble As(III), when the arsenate(III) tends to be adsorbed and/or coprecipitates with iron oxyhydroxides.

Besides the reaction mechanism, the authors of the abovementioned study observed fast reaction kinetics in the range of minutes or tens of minutes along with an increase in pH from 2.5 to above 4 (depending on the dilution of acid mine water) and a decrease in ORP from +700 mV down to less than +500 mV for nondiluted and close to 0 mV for variously diluted acid mine water samples. The kinetic parameters for the removal of selected metals by NZVI are given in Table 4.2.

Table 4.2 Kinetic parameters for selected metal(loid) removal by NZVI particles. For references see Li et al. (2016b)

Heavy metals	NZVMs	Initial conc.	Model	$q_{e,exp}$ (mg/g)	$q_{e,cal}$ (mg/g)	k_{obs}	R^2	Ref.
Cd(II)	NZVI/CHI	100 mg/L	Pseudo-first-order	124.74	88.32	0.5847 (1/h)	0.9756	Zhou et al. (2012)
			Pseudo-second-order		135.14	0.0105 (g/mg h)	0.9990	
Cr(VI)	NZVI/SDS	140 mg/L	Pseudo-first-order	257.52	41.7317	0.1261 (1/h)	0.8424	Teng et al. (2013)
			Pseudo-second-order		259.7403	0.0152 (g/mg h)	0.9999	
		160 mg/L	Pseudo-first-order	278.14	80.6396	0.4648 (1/h)	0.9672	
			Pseudo-second-order		280.8989	0.0200 (g/mg h)	1.0000	
		180 mg/L	Pseudo-first-order	300.15	14.3397	0.1262 (1/h)	0.4565	
			Pseudo-second-order		300.3003	0.0437 (g/mg h)	1.0000	
		200 mg/L	Pseudo-first-order	336.28	104.3072	0.6653 (1/h)	0.9657	
			Pseudo-second-order		336.7003	0.0256 (g/mg h)	1.0000	
As(V)	NZVI (acid addition) removal capacity of 95.31 (mg/g) 25 min	3000 ppb	Pseudo-first-order	1652 (μg/L)		0.022 (1/min)	0.91	Tanboonchuy et al. (2011)
	NZVI (gas bubbling) removal capacity of 125.3 (mg/g) 25 min	3000 ppb	Pseudo-first-order	153 (μg/L)		0.109 (1/min)	0.97	

Adapted with permission from Li et al. (2016b). Copyright (2016) Elsevier

4.3 Analytical Aspects of Metal Removal by NZVI

As mentioned in the previous section, all types of NZVI–metal interactions involve surface-mediated reactions of NZVI particles. In order to understand such mechanisms, surface-sensitive analytical tools have to be employed. X-ray photoelectron spectroscopy (XPS) also known as electron spectroscopy for chemical analysis (ESCA) is a unique surface-sensitive technique for chemical analysis. XPS is based on irradiating solid samples with soft X-rays and the energy analysis of emitted photoelectrons which comes from the surface of the sample (5–10 nm penetration depth). The obtained spectrum is a plot of the number of detected photoelectrons per energy interval versus their kinetic (or rather binding) energy. The measured kinetic energy is given by the equation:

$$\mathrm{KE} = h\nu - \mathrm{BE} - \Phi_{\mathrm{s}}, \tag{4.3}$$

where hν is the photon energy, BE is the binding energy of the atomic orbital from which the electron originates, and Φ_s is the spectrometer work function. Providing that each element has a unique set of binding energies and that the same atom in the different chemical environment can give rise to discrete components in its core-level spectrum (so-called chemical shift), XPS is an ideal tool for elemental quantification and identification of chemical states of surface atoms (Moulder and Chastain 1992; Briggs 1998).

Although some metallic spectra deconvolutions are straightforward, there are also some examples representing significant difficulties. For example, the 2p spectra of many transition metals contain features like multiplet splitting and are complicated further by shake-up and plasmon loss structures. Good starting points for deeper knowledge of transition metal XPS spectra deconvolutions are published elsewhere (Watts and Wolstenholme 2003; Biesinger et al. 2010, 2011).

Because of the features mentioned above, XPS is a method of choice for determining the valence state of pollutants like Cr, Co, Ni, Cu, As, U, etc. removed by zerovalent iron (for typical XPS spectra of metals on NZVI surface, see Fig. 4.4). From published studies (see, e.g., Li and Zhang (2007)), metal ions on the NZVI surface can be present in different valence states:

1. The same as the metal ion in solution (e.g., Zn^{2+}, Cd^{2+}) as a result of their sorption or surface complexation.
2. Reduced compared to the dissolved form (e.g., Cu^0, Ag^0, Hg^0, and also Cr^{III}) as a result of their electrochemical reduction and precipitation.
3. Both reduced and ionic (e.g., Ni, Pb) suggesting a combination of sorption and partial reduction.

One example for all of the deconvoluted core-level spectrum of a pollutant is shown in Fig. 4.5 representing an experiment of As(V) reduction by NZVI (for experimental details and further results, see section on "Removal of Arsenic by NZVI" and reference Tuček et al. (2017). In this figure, there is deconvoluted 3d

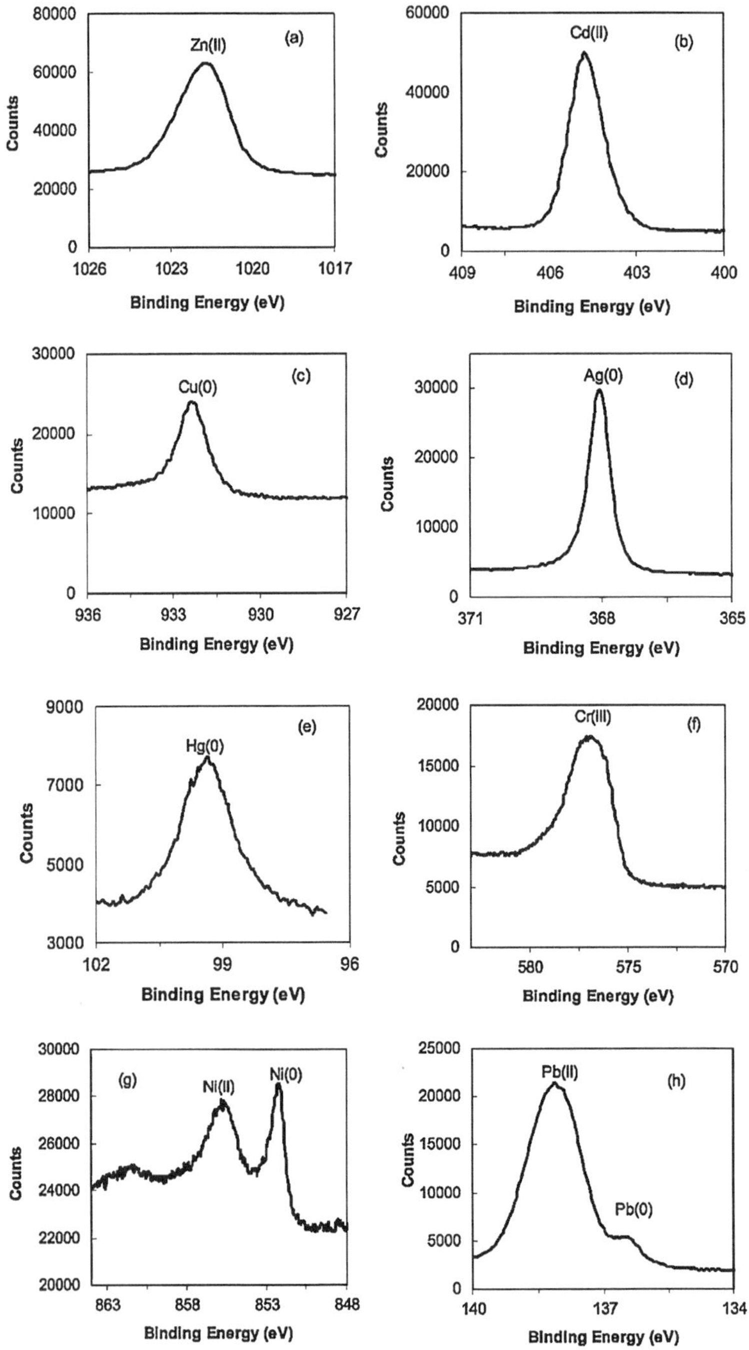

Fig. 4.4 High-resolution XPS spectra for various metals removed by NZVI particles from aqueous solution. (Reprinted with permission from Li and Zhang (2007). Copyright (2007) American Chemical Society)

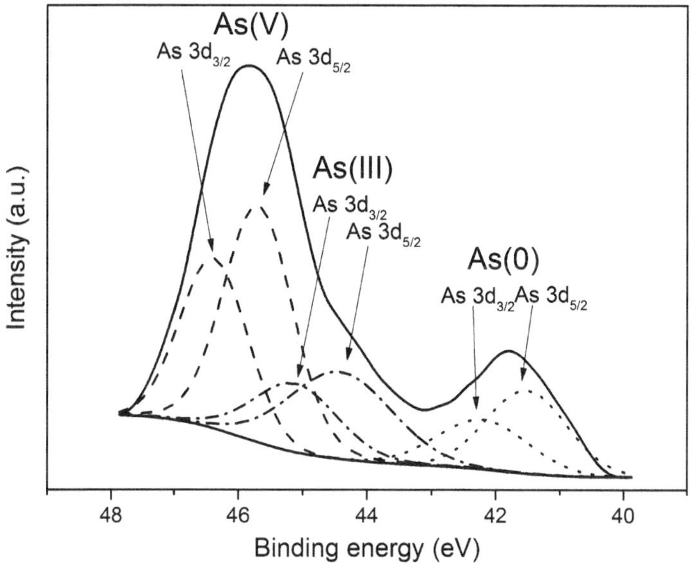

Fig. 4.5 Deconvoluted high-resolution XPS spectrum at the As 3d region

core-level spectrum of arsenic (As 3d) with six spectral components at 41.5 eV, 42.2 eV, 44.4 eV, 45.1 eV, 45.7 eV, and 46.4 eV corresponding to As 3d5/2 (As(0)), As 3d3/2 (As(0)), As 3d5/2 (As(III)), As 3d3/2 (As(III)), As 3d5/2 (As(V)), and As 3d3/2 (As(V)), respectively. Therefore, XPS could be used for detailed monitoring of reaction mechanisms including metal removal by NZVI.

4.4 Examples of Various Metal Cations that Can Be Removed by NZVI

The removal of metal cations like Ni^{2+}, Cu^{2+}, Zn^{2+}, Co^{2+}, Cd^{2+}, and Zn^{2+} by NZVI is highly efficient and characterized by complexation, surface precipitation, high uptake capacities, fast uptake kinetics, and very limited desorption of metal ions from NZVI. The performance of NZVI for removal of such heavy metal cations is typically improved by increasing pH. This is because besides reduction and adsorption, heavy metal ions can be removed by hydroxide precipitation (Eq. 4.4) and coprecipitation at elevated pH values (Dries et al. 2005; Ludwig et al. 2009; Moraci and Calabrò 2010; Bruzzoniti and Fiore 2014; Eglal and Ramamurthy 2014; Statham et al. 2015a, b; Guo et al. 2016; Han et al. 2016):

$$M^{n+} + nOH^{-} \rightarrow M(OH)_n \downarrow \left(M^{n+}\text{represents } Cd^{2+}, Ni^{2+}, Cu^{2+}\text{and } Zn^{2+}\right) \quad (4.4)$$

From the studied ions, Cu^{2+} is much more easily removed than Zn^{2+}, Ni^{2+}, and Cd^{2+}, because the redox potential of copper (Cu^{2+}/Cu) is +0.34 V, being much greater than that of zinc (−0.76 V, Zn^{2+}/Zn), cadmium (−0.40 V, Cd^{2+}/Cd), and nickel (−0.24 V, Ni^{2+}/Ni) (Han et al. 2016):

$$Fe^0 + Cu^{2+} \rightarrow Fe^{2+} + Cu^0 \tag{4.5}$$

In particular, the copper removal mechanism by NZVI, i.e., by both sorption and reduction, led to the emergence of various NZVI-based materials with remarkable properties such as NZVI with various surface morphologies, ZVI composites, or ZVI-based column filters (Xiao et al. 2011; Statham et al. 2015b; Slovák et al. 2016; Tiberg et al. 2016). The removal efficiency of Cu^{2+} (and also other metals) by ZVI can also be enhanced with the application of a weak magnetic field (WMF) (Jiang et al. 2015).

One interesting study, driven by the need for understanding the mechanisms of heavy metals ion removal, has compared the surface reactions on NZVI and bimetallic nanoscale iron/aluminum (NZVI/nZAl) particles for removal of five heavy metal ions (Cu^{2+}, Cr^{6+}, Ni^{2+}, Zn^{2+}, Cd^{2+}) – see Fig. 4.6. The study shows the possibility to tailor the properties of a NZVI-based bimetallic system. Particularly, the reduction of heavy metals ions by ZVI/ZVAl was greater than for ZVI alone. ZVAl can reduce all the five of the heavy metal ions, while ZVI can efficiently reduce only Cr^{6+} and Cu^{2+}. Moreover, the five heavy metal ions can be removed by adsorption onto ZVI/ZVAl or metal hydroxide precipitates, while reduction is occurred by aluminum and iron acting as electron donors. Furthermore, the heavy metal ions can form hydroxide precipitates.

The application of NZVI particles for reduction, precipitation, and removal of ${UO_2}^{2+}$ has been also demonstrated. The Pourbaix diagram indicates that highly soluble and mobile ${UO_2}^{2+}$ could be reduced to insoluble U^{4+} oxides at the potential of the Fe/Fe^{2+} redox couple. Although the solution conditions favor U(VI) as the stable uranium oxidation state, conditions at the zerovalent iron surface theoretically predict that reduction to U(IV) and precipitation as UO_2 is possible. Thermodynamically, reduction to U^{4+} is slightly favorable in strongly acidic media as indicated in Eq. 4.6: (Fiedor et al. 1998; Riba et al. 2008; Li et al. 2015b):

$$Fe^0 + 1.5{UO_2}^{2+} + 6H^+ \rightarrow Fe^{3+} + 1.5U^{4+} + 3H_2O \tag{4.6}$$

Moreover, Sb^{5+} can be also successfully removed by ZVI. Previous studies have demonstrated that NZVI has a strong capability to reduce Sb^{5+} to Sb^{3+}. This efficiency could be significantly enhanced with the application of a weak magnetic field, and it is slightly influenced by coexisting anions. The incorporation of Sb^{5+} into the structure of in situ formed iron (hydr)oxides accounted for the mechanism of Sb^{5+} removal by ZVI, and thus the release of Sb^{5+} from the Sb^{5+}-treated ZVI was negligible (Filella et al. 2002; Dorjee et al. 2015; Li et al. 2015a).

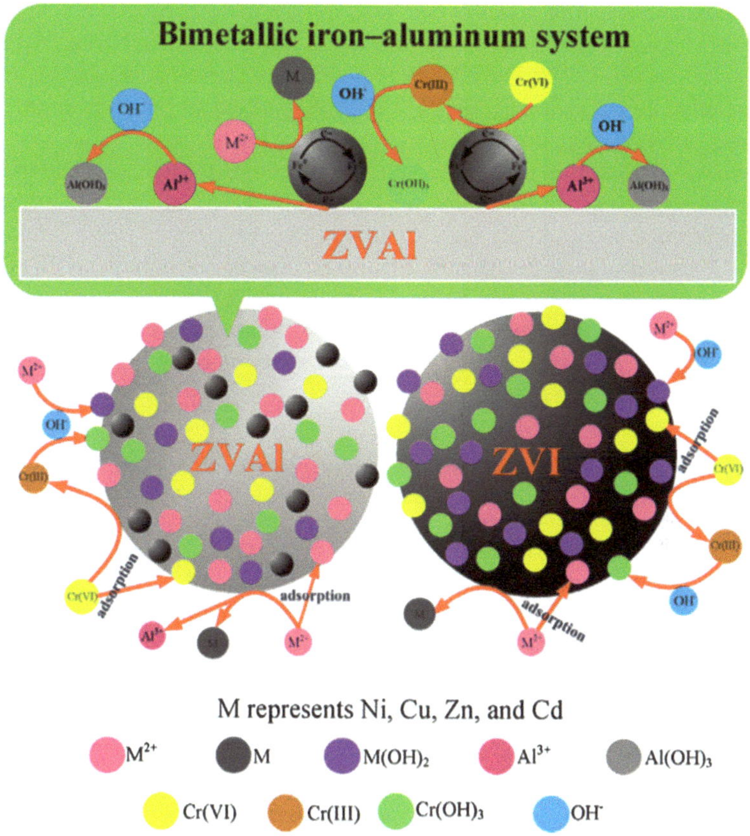

Fig. 4.6 Schematic mechanisms of five heavy metal ions removed by ZVI/ZVAl. (Reprinted with permission from Han et al. (2016). Copyright (2016) Elsevier)

In aquatic ecology, phosphate has been regarded as a nutrient responsible for water eutrophication, which depletes oxygen, affects aquatic life forms, and jeopardizes water quality. The application of NZVI for phosphate removal is highly pH-dependent but only slightly dependent on the ionic strength. The typical coexisting anions like chloride, nitrate, and sulfate did not affect phosphate removal, but removal efficiency was significantly impacted by carbonate anions due to its influence on initial pH of the solution. The main mechanism of phosphate removal includes adsorption and coprecipitation with secondary iron oxyhydroxide phases. The calculated maximum adsorption capacity of NZVI (prepared by reduction of ferric iron by sodium borohydride) can reach 250 mg/g for phosphate in batch experiments (Almeelbi and Bezbaruah 2012; Wen et al. 2014; Sleiman et al. 2016).

Furthermore, the mechanism of Pb^{2+} uptake has been described in several published studies. The adsorption of Pb^{2+} onto the NZVI surface was rapid, with reduction of Pb^{2+} to Pb^0 as the main mechanism (Xi et al. 2010; Arshadi et al. 2014). Besides, NZVI was used for immobilization of other selected specific metal ions, e.g., for reduction and removal of Se^{6+} and Se^{4+} (Zhang et al. 2005; Liang et al. 2013, 2014; Tang et al. 2014a, b) and Mo^{6+} (Huang et al. 2012). Further details on reduction of Se, Ni, U, Cr, and As are reviewed by O'Carroll et al. (2013).

Generally, the capacity of NZVI for metal ion sequestration is higher than 100 mg/g of NZVI, depending on the properties of the particular metal and on the geochemical conditions in the treated solution. Therefore, the total capacity of NZVI is many times higher than common inorganic sorbents including zeolites and also much higher than most of polymeric cation-exchangeable resins (see Li and Zhang (2007)). As an example, comparison of various materials including NZVI used for copper sequestration is listed in Table 4.3. From this comparison it is evident that NZVI particles with a unique 3D hollow-sphere architecture (Slovák et al. 2016) have the highest removal capacity for Cu out of most commonly used sorbents (except graphene-based sorbents).

Table 4.3 Removal capacity values of copper (Cu(II)) by various materials typically used for sequestration of metals (for references see Slovák et al. (2016))

Material	Removal capacity (mg/g)
Functional materials	
Aminated polyacrylonitrile fibers	31.4
Chitosan/cellulose acetate blend hollow fiber membranes	48.2
Aminated polyacrylonitrile nanofibers	116.5
Aniline formaldehyde condensate (AFC) coated on silica gel	76.3
Glutaraldehyde/aminopropyltriethoxysilane/Fe_3O_4 NPs	61.1
Copper ion-imprinted chitosan/*Sargassum* sp. composite	68.6
Poly(hydroxyethyl methacrylate) nanobeads containing imidazole groups	58.0
Pectin-iron oxide magnetic nanocomposite	49.0
Nano-hydroxyapatite/chitosan composite	6.2
Electrosterically stabilized nanocrystalline cellulose	185.0
Magnetic chitosan beads	129.6
Plasma polymer functionalized silica particles	25.0
Carbonaceous sulfur-containing chitosan–Fe(III)	413.2
Polyvinylpyrrolidone-reduced graphene oxide	1689
GO/Fe_3O_4/PEI	157
GO nanosheets cross-linked with poly(allylamine hydrochloride)	349.0
Facial composite adsorbent	176.3
Novel magnetic weak acid resin	267.2
Superabsorbent sodium polyacrylate composite	243.9
Citric acid-modified cellulose	15.1
Multi-amine decorated resin	322.2
Mercapto functionalized palygorskite	30
Gemini surfactant modified rectorite	15.2
Polystyrene supported chitosan thin film	99.8
Esterified orange peel cellulose biosorbent	77.5

(continued)

Table 4.3 (continued)

Material	Removal capacity (mg/g)
Grafted chitosan beads	126.0
Monolithic adsorbents	245.0 (theoretical)
Functional silylating agents containing thiourea	79.4
Chitosan/poly(methacrylic acid) graft copolymer	170
Chitosan/cellulose film combination	122
Amino-functionalized silica	33.5
Biomass	
Fungus *Aspergillus niger*	2.7
Sphaerotilus natans biomass	57.5
Brown marine alga *S. filipendula*	54.0
Sargassum algal biomass	73.1
Brown marine algae (*Ecklonia radiata*)	70.5
Algae *Gelidium*	33.0
Rehydrated biomass of *Spirulina platensis*	96.8
Shorea robusta tree barks	51.4
Pseudomonas pseudoalcaligenes	46.8
Algal biomass	160
Heavy-metal resistant bacterial consortium	450
Cell wall	84
Immobilized microorganisms on polyurethane foam	28.7
Wine-relevant *lactobacilli*	46.2
Posidonia oceanica	85.8
Cyanospira capsulata	240
Pseudomonas syringae	20
Immobilized *Microcystis aeruginosa*	8.9
Aspergillus terreus	15.2
Aspergillus Niger	34.1
Trametes versicolor	60.5
Rhizopus oryzae	43.7
Penicillium chrysogenum	60
Immobilized *T. versicolor* + montmorillonite clay	62.8
Caulerpa lentillifera	0.04
Ulothrix zonata	176.2
Pithophora oedogonia	23.1
Bjerkandera sp.	12.1
Penicillium simplicissimum	112.3
Algal waste gelidium	31.1
Undaria pinnatifida	38.8
Asparagopsis armata	21.3
Fucus vesiculosus	114.9
Enterobacter sp. *J1*	32.5
Pseudomonas putida	6.2

(continued)

Table 4.3 (continued)

Material	Removal capacity (mg/g)
Paenibacillus polymyxa	112
Sphaerotillus natans	60
Sphaerotillus natans gram-negative bacteria	44.5
Baker's yeast	65
Carbon materials/(bio)chars	
Carbon nanotube/calcium alginate composite	67.9
Granular-activated carbon Darco 12–20 mesh	173.9
High-density 3D graphene macroscopic objects	3820
Carbon nanocomposites	256
Mesoporous carbon	117.1
Pinewood char	4.5
Alkali-activated switchgrass biochar	31.0
Thermally treated olive pomace	17.8
Phenolic resin-based carbon	247
Pecan shells	95
Chestnut shell-activated C	100
Grape seed-activated C	48.8
Hazelnut husk	6.7
Peanut shell	50.4
Hazelnut shells impregnated with H_2SO_4	58.3
Hulls of *Ceiba pentandra* (tree)	24.2
Wastes	
Sewage sludge ash	4.1
Partially deacetylated prawn shell	16.9
Pelletized peanut hulls	12.0
Sour orange residue	21.7
Phosphated apple residues	36.2
Spent grain	10.5
Carbonate hydroxyapatite from eggshell waste	142.9
Olive mill residues	13.5
Newspaper pulp	30.0
Pretreated powdered waste sludge	156.0
Sewage sludge	5.7
Pomace ashes	7.0
Activated sludge	35.0
Industrial oil sludge-based adsorbent	85.3
Grape stalk	42.9
Brine sediments (industrial residue)	4.7
Prismatic oyster shell powder	8.9
Sugar beet pulp treated with NaOH and citric acid	119.4
Sugarcane bagasse	9.5
H_3PO_4-activated rubber wood sawdust	5.6

(continued)

Table 4.3 (continued)

Material	Removal capacity (mg/g)
Citric acid functionalizing wheat straw	78.1
Acid-pretreated bivalve mollusk shells	139.0
Tea waste	21.0
Chemically modified corncobs	26.0
Sargassum wightii	115
Garden grass	58.3
Orange peel	86.7
Spent coffee grains	60.4
Natural materials	
Rubber (*Hevea brasiliensis*) leaf powder	8.9
Natural zeolite	1.6
Phosphate rock	10.8
Purolite C100-MB resin	121.5
Mimosa tannin resin	43.7
Valonia tannin resin	44.2
Manganese-coated sand	1.0
Chitosan	85.8
Dolomite–palygorskite clay	225.7
Eucalyptus leaves	1.9
Hydrilla verticillata (plant)	30.8
Palm oil fruit shells	60
Banana peel	8.2
Watermelon rind	5.7
Ash zeolite	27.9
Insoluble wood xanthate	27.8
Other inorganic materials/composites	
Zerovalent iron	96.9
Zerovalent iron/pumice	15.0
Zerovalent iron/sand mixture	13.3
Nanoscale diboron trioxide/titanium dioxide	82
NZVI	250
NZVI + recirculation	343
3D spherical NZVI	623

Adapted with permission from Slovák et al. (2016). Copyright (2016) American Chemical Society

The only identified potential risk of metal ion sequestration by NZVI is the possible re-oxidation and reverse dissolution of some metal ions like Cu^{2+}. However, the dissolved oxidized metals could be adsorbed on the oxide/hydroxide shell of NZVI or on other iron oxide reaction products with high specific surface area (see example of the uranium reduction by NZVI and subsequent re-oxidation (Dickinson and Scott 2010)). This limits the potential release of metal once it is adsorbed.

Below are two detailed examples of metals/metalloids whose removal by NZVI has been successfully studied and tested worldwide. The hazardous effects of arsenic and chromium are described, while the mechanism and examples of their sequestration are also provided.

4.4.1 Arsenic in the Environment

Natural occurrence of arsenic represents a serious environmental problem in many countries (Nordstrom 2002; Smedley and Kinniburgh 2002; Ravenscroft et al. 2009). Arsenic is a well-known carcinogen metalloid. First symptoms of arsenicosis are dermatological manifestations such as hyperkeratosis with characteristic brown spots on palms and soles followed by cancer of the skin. In more advanced stage, there is cancer of the lungs and bladder (Marshall et al. 2007). Also, based on extensive epidemiological study in Antofagasta, north of Chile, consumption of arsenic in drinking water increases risk of acute myocardial mortality and spontaneous abortion (Yuan et al. 2007).

Arsenic in surface water and groundwater occurs predominantly as arsenate, As(V), and arsenite, As(III). Both oxidation states of arsenic form oxyanions (Plant et al. 2005): at about neutral pH, arsenate is present as $H_2AsO_4^-$ and $HAsO_4^{2-}$ and arsenite as $H_3AsO_3^0$. They are typically adsorbed on hydrous ferric oxides (HFO) with higher adsorption affinity for As(V) and decreasing adsorption rate with increasing pH (Smedley and Kinniburgh 2002). Other oxidation states that can exist under reducing conditions are As(-III) and As(0), while methylated species of arsenic such as dimethylarsinate (DMA) are typically found in pond and lake water rich in organic matter (Mandal and Suzuki 2002). In geothermal systems, thioarsines such as $(CH_3)_2AsCl$ have been identified too (Planer-Friedrich et al. 2006).

Typical processes responsible for natural arsenic enrichment are (1) reductive dissolution of ferric minerals with adsorbed and coprecipitated arsenic in regions like Bangladesh and West Bengal (Ahmed et al. 2004); (2) mobilization of arsenic under oxidizing and high pH conditions, e.g., in the Pampean region of Argentina (Bhattacharya et al. 2006); (3) oxidation of sulfidic minerals in bedrock, e.g., in Wisconsin (Schreiber et al. 2000); and (4) geothermal activity, e.g., in the Trans-Mexican Volcanic Belt (Birkle et al. 2010) and Chile (López et al. 2012). Reductive mineral dissolution is by far the most important natural process exposing large populations to arsenic risk.

However, concentrations of dissolved arsenic at mining sites, where sulfidic minerals like arsenopyrite and As-rich pyrite are exposed to oxygen, are several orders of magnitude higher than at natural arsenic occurrence sites. A typical example is the Richmond Mine in California, where mine drainage water is concentrated by evaporation and arsenic concentrations reach up to 850 mg/L (Nordstrom and Alpers 1999). At Snow Lake Pile in Manitoba, Canada, groundwater downgradient of the pile has arsenic concentrations higher than 20 mg/L (Salzsauler et al. 2005). Concentrations of arsenic in pore water at the waste rock pile in Berikul, Siberia, Russia, reach extreme concentrations as high as 22,000 mg/L (Gieré et al. 2003). The primary As(V) mineral precipitating from mine drainage is scorodite, but there is an assemblage of other minerals including kaňkite, bukovskýite, pharmacosiderite, and pitticite (Drahota and Filippi 2009; Kocourková-Víšková et al. 2015). Also, As(V) in mining wastes can be incorporated into beudantite or adsorbed onto the surfaces of jarosite (Gräfe et al. 2008).

The World Health Organization (WHO) and US Environmental Protection Agency (EPA) adapted drinking water guideline of 0.01 mg/L, which replaced the old limit of 0.05 mg/L in 2001 (Agency 2001). Diverse approaches for elimination of arsenic have been explored including adsorption, coagulation/filtration, ion exchange, photooxidation, reverse osmosis, membrane, nanofiltration, and conventional iron/manganese removal processes; here, adsorption is viewed as one of the most affordable, cheap, and effective methods (Mohan and Pittman Jr 2007).

4.4.2 *Removal of Arsenic by NZVI*

The use of Fe(0) to remove arsenic has been actively investigated by many researchers. The surface area of NZVI plays a major role in both the adsorption kinetics and capacities. Recently, nanoscale zerovalent iron was tested for the removal of As(V) and As(III), which is the predominant arsenic species under anoxic groundwater conditions (Kanel et al. 2005, 2006). The mechanisms of the reaction with As species and NZVI depend on whether or not dissolved oxygen is present along with the NZVI as described next.

When using NZVI in oxic conditions for arsenic removal, the oxidation of NZVI occurs, and the reaction mechanisms are controlled mostly by the Fenton reaction:

$$\mathrm{Fe(0)} + \mathrm{O_2} + 2\mathrm{H^+} \rightarrow \mathrm{Fe(II)} + \mathrm{H_2O_2} \tag{4.7}$$

Hydrogen peroxide combines with Fe(II) and Fe(III) corrosion products of NZVI, thereby generating hydroxyl and superoxide radicals via the Fenton reaction:

$$\mathrm{Fe(II)} + \mathrm{H_2O_2} \rightarrow \mathrm{Fe(III)} + \mathrm{HO}^{\bullet} + \mathrm{OH^-} \tag{4.8}$$

$$\mathrm{Fe(II)} + \mathrm{O_2} \rightarrow \mathrm{Fe(III)} + \mathrm{O_2}^{\bullet -} \tag{4.9}$$

$$\mathrm{Fe(II)} + \mathrm{O_2}^{\bullet -} + 2\mathrm{H^+} \rightarrow \mathrm{Fe(III)} + \mathrm{H_2O_2} \tag{4.10}$$

These radicals are viewed as oxidants and are considered to be responsible for oxidation of As(III) to As(V) when arsenite is treated by NZVI. In other words, under oxic conditions, corrosion of NZVI and formation of Fe_3O_4/Fe_2O_3 nanoparticles play a dominant role in the removal of As(III) and As(V) species from aqueous solutions. The removal then involves only adsorption of As(V) species

Table 4.4 Aqueous speciation reactions and surface complexation reactions

Surface complexation reactions	log K
Intrinsic surface complexation reactions	
$FeOH + H^+ = FeOH_2^+$	7.29
$FeOH = FeO^- + H^+$	−8.93
Arsenate adsorption constants	
$HAsO_4^{2-} + FeOH + H^+ = FeHAsO_4^- + H_2O$	30.51
$HAsO_4^{2-} + FeOH + 2H^+ = FeH_2AsO_4 + H_2O$	11.53
Arsenite adsorption constants	
$FeOH + H_3AsO_3 = FeH_2AsO_3 + H_2O$	5.41

Adapted with permission from Tyrovola et al. (2007). Copyright (2007) Elsevier

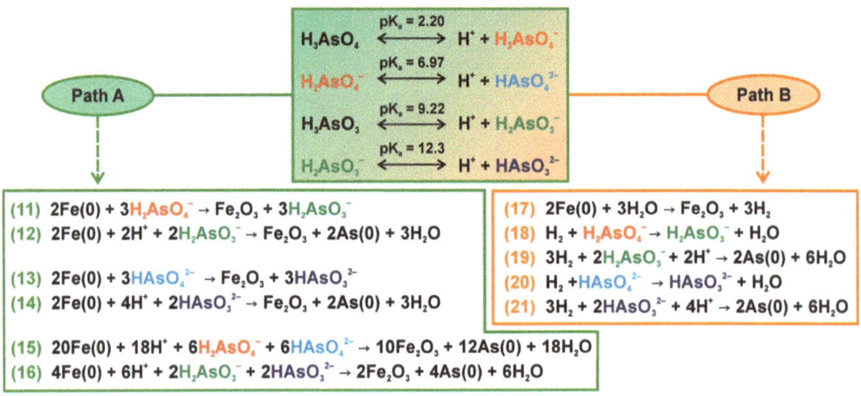

Fig. 4.7 Two possible pathways (Path A and Path B) for As(III) and As(V) reduction to As(0) by NZVI under anoxic conditions. Note: Fe_2O_3 is used to represent an iron oxide phase produced from Fe(0) reacting with arsenite/arsenate, i.e., to specify any possible iron oxide/hydroxide that could be formed under particular pH/redox conditions. (Reprinted with permission from Tuček et al. (2017). Copyright (2017) American Chemical Society)

on the nanoparticles' surfaces (Hug and Leupin 2003; Katsoyiannis et al. 2008; Triszcz et al. 2009; Pang et al. 2011). Aqueous speciation reactions and surface complexation reactions are presented in Table 4.4. Precipitation is assumed to be a kinetically controlled slow process and is modeled as a mass transfer controlled reaction represented in Table 4.4 (Tyrovola et al. 2007).

Under anoxic conditions, the reduction of As(V) to As(0) by Fe(0) is thermodynamically possible. Therefore in anoxic conditions, As(V) and As(III) can be reduced and/or adsorbed onto the nanoparticle surfaces. Figure 4.7 summarizes the two possible pathways by which electron transfer between NZVI and arsenite and/or arsenate pollutants can occur under anoxic conditions (Sasaki et al. 2009; Tuček et al. 2017).

According to a proposed mechanism of As(V) and As(III) removal under anoxic conditions in several published studies, initial breakage of As–O bonds in arsenite

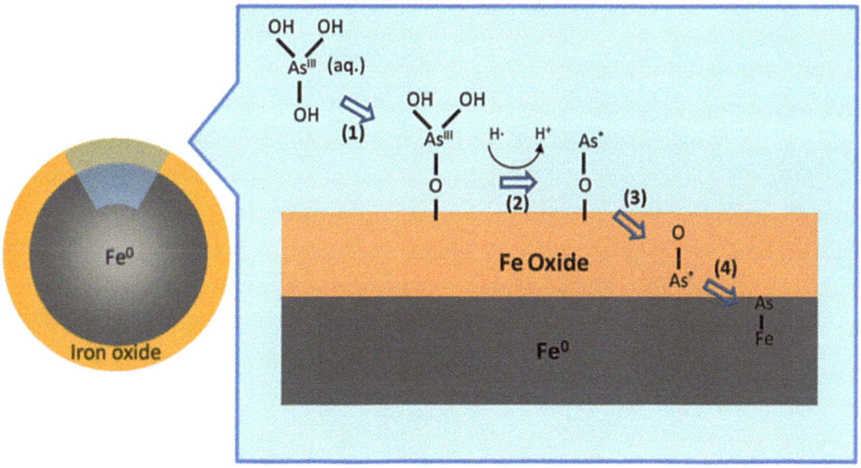

Fig. 4.8 Proposed mechanism of As(III)–NZVI reactions. The process includes (1) rapid adsorption of aqueous As(III) at the oxide surface, (2) reduction of As(III) complexes at the surface, (3) translocation of arsenic across the oxide shell accompanied by further breaking of As–O bonds, and (4) diffusion into the Fe(0) core. Arsenic prefers to accumulate at the surface of Fe(0) forming a thin layer of Fe – As intermetallic; thus the amount of reduced arsenic depends on the amount of NZVI used. As* denotes intermediate states between As(III) and Fe – As intermetallic. (Reprinted with permission from Yan et al. (2012b). Copyright (2012) American Chemical Society)

molecules occurs upon adsorption onto a NZVI surface, followed by further reduction and diffusion of arsenic through the very thin layer of iron oxide formed at the NZVI surface. This process results in the formation of an intermetallic phase of As (0) (i.e., as proved by XPS – see Fig. 4.5), with an NZVI core, i.e., an FeAs alloy. This mechanism was observed on air-stable NZVI particles with a core–shell ZVI architecture (Fig. 4.8) (Farrell et al. 2001; Su and Puls 2001; Bang et al. 2005; Yan et al. 2012a, b; Ling and Zhang 2014).

However, in the case of an oxide-shell-free NZVI (pyrophoric) particles, removal of As(III) and As(V) species under anoxic conditions followed a different pathway. Using XPS, no spectral components corresponding to FeAs alloy or Fe(0) enriched with As were detected in samples incubated under these conditions. In this system, there was a core–shell particle architecture, with a core composed of Fe(0) and a double shell with inner and outer regions composed exclusively of As and iron(III) oxide, respectively (Fig. 4.9) (Tuček et al. 2017).

The effects of Fe/As mass ratio, duration of reaction, and solution pH on arsenic removal efficiency under both oxic and anoxic conditions are demonstrated in Fig. 4.10. Arsenic removal by NZVI is more efficient in an anoxic environment than in an oxic environment. In the oxic and anoxic environments, the removal equilibrium for both As(III) and As(V) species was obtained within 24 h after NZVI addition. Moreover, reaction rate data confirmed that within the first 60 min, the removal rate was slightly faster for As(III) species than for As(V) species. As reported earlier (Bang et al. 2005), this occurs due to the different adsorption

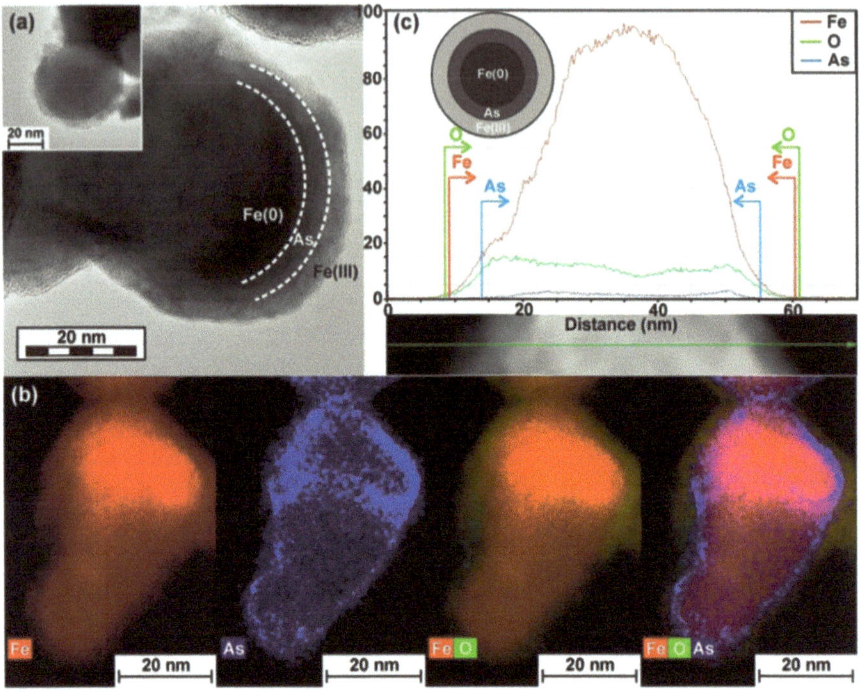

Fig. 4.9 (**a**) Representative HRTEM image of the sample dried after reaction of As(III) and/or As (V) species with oxide-shell-free NZVI under anoxic conditions showing formation of a secondary core–shell architecture. (**b**) EDS chemical mapping of the Fe(0)/As/iron(III) oxide core−shell nanoparticles, showing the Fe(0) core, As attached to the surface of Fe(0) nanoparticles and iron (III) oxide phase covering the adsorbed As. (**c**) Line profile showing the firm locking of As between the Fe(0) core and iron(III) oxide outer shell. (Adapted with permission from Tuček et al. (2017). Copyright (2017) American Chemical Society)

capacity of NZVI for oxidized vs. reduced As species. Under neutral pH, both As (III) and As(V) are adsorbed strongly onto iron oxide surfaces via surface complex formation. The pH-dependent behavior can be explained by ionization of both the adsorbate and the adsorbent causing repulsion at the surface and decreasing the As (III) adsorption (Kanel et al. 2005, 2006; Katsoyiannis et al. 2008; Mamindy-Pajany et al. 2011; Wang et al. 2014; Tuček et al. 2017).

In addition, various ions in natural water could have significant effect on arsenic removal efficiency by NZVI. The presence of borates, organic matter, silicates, and phosphates in water decreases the removal of arsenic. On the other hand, carbonates, sulfates, and nitrate ions at concentrations up to 10 mM have minimal effect on arsenic removal; however higher concentration of anions can enhance the removal rate of arsenic (Kanel et al. 2005, 2006; Biterna et al. 2007). On the other hand, other work (Rodová et al. 2015) has shown a significant decrease of both arsenic (from initial value of 85 mg L^{-1}) and phosphate concentrations from mine water after being amended with NZVI (Kaňk, Czech Republic).

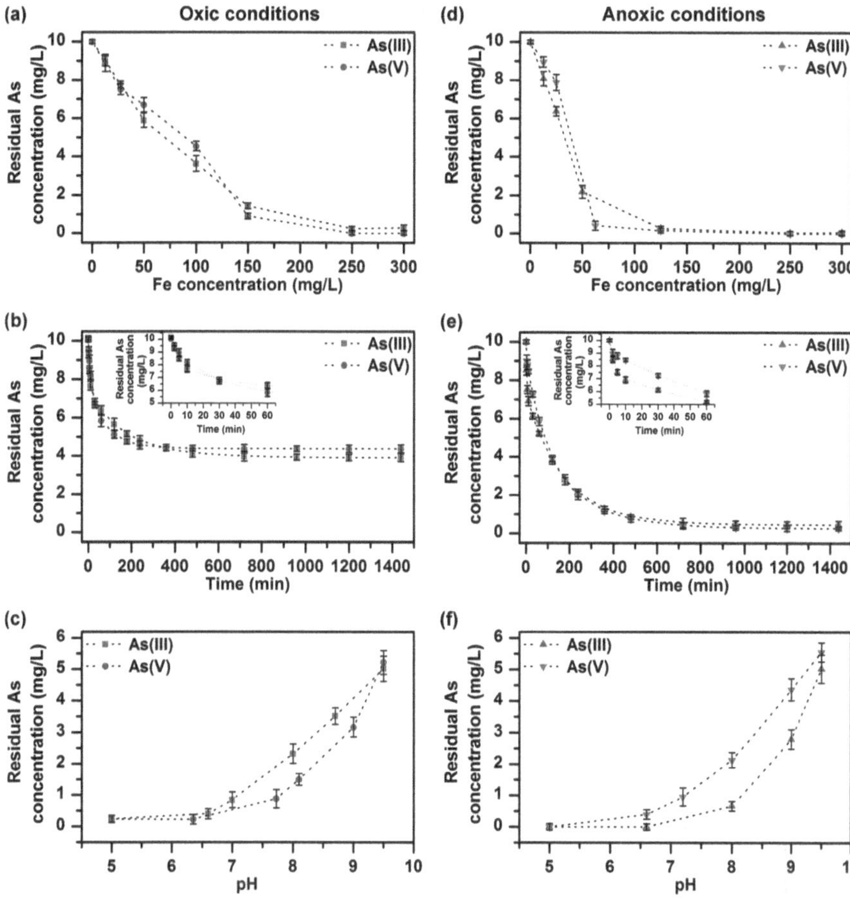

Fig. 4.10 Dependence of residual arsenic concentration on the amount of NZVI added (at pH ≈ 7 and reaction time of 24 h) under (**a**) oxic and (**d**) anoxic conditions. Kinetics of removal of As(III) and As(V) species by NZVI (at pH ≈ 7 and Fe/As = 10:1) under (**b**) oxic and (**e**) anoxic conditions. Dependence of residual arsenic concentration on the pH of the reaction mixture (at Fe/As = 20:1 and reaction time of 24 h) under (**c**) oxic and (**f**) anoxic conditions. (Adapted with permission from Tuček et al. (2017). Copyright (2017) American Chemical Society)

4.5 Removal of Arsenic by NZVI Derivatives

Generally, the capacity of NZVI particles toward removal of metals/metalloids is limited due to their air instability and aggregation. In the past few years, various composites have been tested with the aim to optimize the NZVI-based materials for enhanced performance (see Table 4.5). For example, chitosan-modified zerovalent iron nanoparticles (CIN) were suggested for arsenic removal. This material enhances the stability of Fe(0) nanoparticles. It has been demonstrated that CIN is capable of

Table 4.5 Examples of NZVI-based composites used for As(V)/As(III) removal with capacity values for selected materials

Material	Removal capacity (mg/g) As(V)/As(III)
Chitosan ZVI (CIN) (Gupta et al. 2012)	119/94
ZVI-based SONO filters (Neumann et al. 2013)	–
ZVI-reduced graphite oxide (Wang et al. 2014)	29.0/35.8
Fe-containing mesoporous carbon (FeOMC) (Gu et al. 2007)	5.5–8.0
ZVI on activated carbon (Zhu et al. 2009)	12/18.2
Oxidants with ZVI (Guo et al. 2016)	–

As adsorption in a wide range of pH, and the study revealed that commonly interfering ions like sulfate, phosphate, and silicate marginally affect the adsorption behavior of both arsenite and arsenate (Gupta et al. 2012). Another study revealed a continuously accelerated Fe^0 corrosion driven by common oxidants (i.e., NaClO, $KMnO_4$, or H_2O_2) and thereby the rapid and efficient removal of As(V) by ZVI under certain experimental conditions. ZVI simply coupled with NaClO, $KMnO_4$, or H_2O_2 resulted in almost complete As(V) removal within only 10 min (Guo et al. 2016). Zerovalent iron-activated carbon (NZVI-AC) derivatives increase the specific surface area of NZVI. Activated carbon was used as the supporting material due to its excellent properties and mechanical strength and porous structure (Zhu et al. 2009). For a cost-effective treatment of polluted waters (i.e., mainly in developing countries), sand filters with NZVI parts were designed (Neumann et al. 2013). Other NZVI derivatives improved applicability of ZVI for complicated real waters (Gu et al. 2007; Wang et al. 2014).

Another pathway is to immobilize NZVI particles on fiber filters or membranes (Thekkae Padil et al. 2016). A case study of NZVI-containing electrospun membrane composed of poly[acrylonitrile-co-(methyl acrylate)-co-(itaconic acid)] (Fig. 4.11) proved that this filtration material can be highly promising for repeated (or continuous in the case of flowing media) use in arsenic removal. In this particular case, the sorption capacity was nearly identical about 46 mg/g for both As(III) and As(V) with just a slight decrease in efficiency over five cycles. However, a decrease in efficiency was reported in the presence of some anions (PO_4^{3-}, HCO_3^-, SO_4^{2-}, and NO_3^-) and cations (Fe^{2+}, Ca^{2+}, and Mg^{2+}) (Thekkae Padil et al. 2016).

4.6 Chromium as a Waste

Chromium is a heavy metal that is used in many engineering and chemical industries, such as electroplating facilities, steel and alloy production, pigments, chemical production, and tannery industries (Muthukrishnan and Guha 2008). Thus, chromium is a common groundwater contaminant at many hazardous waste sites. The two most common chromium oxidation states are hexavalent chromium, Cr(VI), and

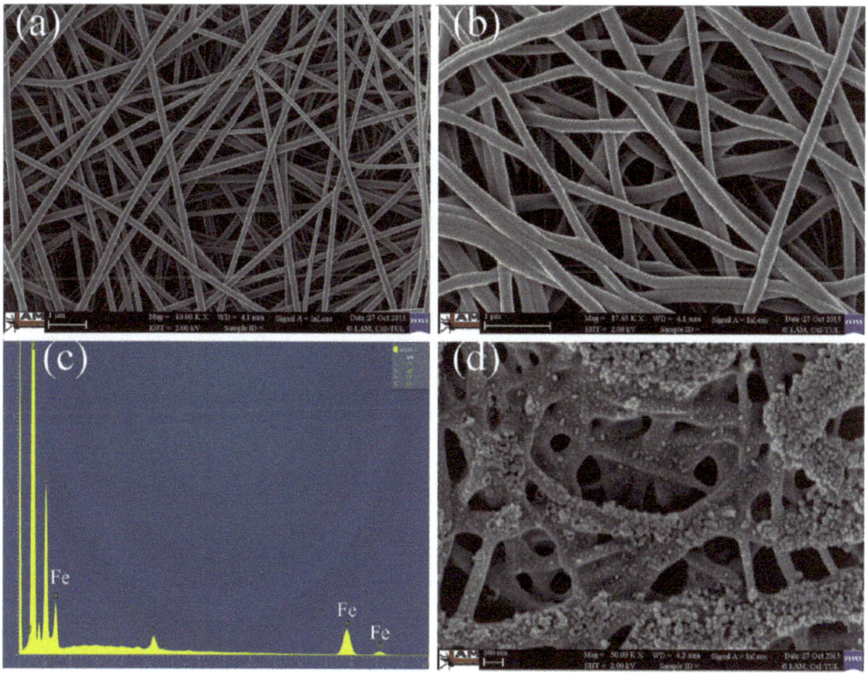

Fig. 4.11 SEM images of electrospun membrane (**a**) and NZVI-decorated electrospun membrane (**b**, **d**) with EDS spectrum for NZVI

the trivalent chromium, Cr(III), with the former being most toxic, carcinogenic, and mutagenic (Weng et al. 2007; Petala et al. 2013). Specifically in humans, Cr(VI) can be very harmful and cause various serious health problems, like diseases to the respiratory tract and to the reproductive and gastroenterological systems. The US Environmental Protection Agency has set the maximum contaminant level of total chromium concentration in drinking water at 0.1 mg/L (2011), whereas the World Health Organization (2003) and European Economic Community recommend a provisional drinking water guideline of 0.05 mg/L (Organization 2011; Agency 2015). Cr(III) occurs naturally in the environment, and it is an essential element in humans and is much less toxic than Cr(VI). Cr(III) is relatively stable and has low solubility ($<10^{-5}$ M) in aqueous solutions (Richard and Bourg 1991). On the other hand, Cr(VI), in the form of chromate anions (e.g., CrO_4^{2-}, $HCrO_4^{-}$, $Cr_2O_7^{2-}$), is highly soluble, especially under acidic conditions, reactive, and more mobile in aquatic systems and a porous medium.

Reduction of Cr(VI) to Cr(III) is considered as an efficient remediation method because the toxicity of Cr(III) is 500–1000 times less than in the case of Cr (VI) (Costa 2003). However, long-term exposure to Cr(III) in large doses can also be considered harmful (Kareus et al. 2001; Stearns et al. 2002). Thus, the ideal Cr (VI) removal strategy should include both reduction and precipitation or adsorption of Cr species, as is provided by NZVI.

4.7 Removal of Chromium by NZVI

Zerovalent iron, especially NZVI, has received significant attention toward Cr(VI) removal due to its remarkable efficiency in reduction of Cr(VI) (Li et al. 2006). Cr(VI) can easily accept the electrons that are provided from ZVI and can be reduced to lower and less toxic oxidation state. So far, many pathways have been suggested in order to explain the reduction of Cr(VI) by NZVI. The reduction of Cr (VI) can be described both as heterogeneous (direct) and homogenous (indirect) (Gheju 2011). The direct route refers to the direct electron transfer from NZVI to Cr(VI). Conversely, indirect electron transfer occurs when iron is first oxidized to the 2+ valent state, which then reacts with Cr(VI). Indirect reduction can also occur when active hydrogen, which is derived from the corrosion of NZVI by water, reacts with Cr(VI). Generally, the overall reaction of Cr(VI) with NZVI can be described by Eq. 4.11:

$$Fe^{0}{}_{(s)} + HCrO_4{}^{-}{}_{(aq)} + 7H^{+}{}_{(aq)} \rightarrow Cr^{3+}{}_{(aq)} + Fe^{3+}{}_{(aq)} + 4H_2O_{(l)} \qquad (4.11)$$

The reaction can also involve species, such as $CrO_4{}^{2-}$ and $Cr_2O_7{}^{2-}$, with the predominant species being dependent on the pH of the solution (Weng et al. 2007).

The overall mechanism of Cr(VI) elimination by NZVI is not only from reduction; a combination of sorption, reduction, and precipitation processes can occur, with the rate and degree depending on parameters, such as pH, concentration of initial Cr(VI), surface area, size and dose of NZVI, and the presence of various dissolved and organic substances. The core of NZVI, formed by metallic iron, acts as an e^- donor and is responsible for the reduction, while the shell serves to conduct electrons and adsorbs (both of Cr(VI) and Cr(III)) where it can precipitate (i.e., $Cr_xFe_{1-x}(OH)_3$ or $Cr_xFe_{1-x}(OOH)$, where $x < 1$) (Li and Zhang 2007). The following reaction is proposed for the formation of the mixed Fe(III)–Cr(III) hydroxide solid precipitates:

$$xCr^{3+} + (1-x)Fe^{3+} + 3OH^{-} \rightarrow Cr_xFe_{1-x}(OH)_3 \qquad (4.12)$$

Hence, the NZVI technology turned out to be an excellent solution to eliminate Cr(VI) not only by reduction and conversion into the less toxic form (i.e., CrIII) but also by promoting its precipitation and limiting its mobility.

The Cr(VI) removal rate by NZVI is often rapid at early times, while for longer times the reaction is decelerated. This could be an effect of the Fe(III)–Cr(III) (oxy) hydroxide formation which can block the surface of the NZVI particles and inhibit or delay the reduction process. A typical example of a Cr(VI) removal kinetics is given in Fig. 4.12. Moreover, among various parameters, the degree of the maximum Cr(VI) removal capacity by NZVI is strongly dependent on the pH of the aqueous solution that is applied. According to many reports, it was found that the Cr(VI) reduction rate is increasing as pH decreases (Gheju 2011). Besides the requirement of hydrogen ions in the reduction of Cr(VI), the following effects

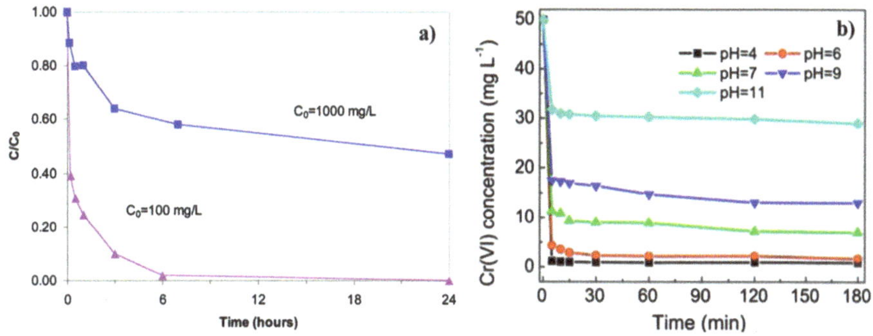

Fig. 4.12 (**a**) Cr(VI) removal kinetics as a function of reaction time when the initial Cr(-VI) concentrations were 100 and 1000 mg/L and NZVI dose was 5.0 g/L (Li et al.). (**b**) Effects of pH on the Cr(VI) removal from aqueous solutions. (Adapted with permission from Li et al. and Dai et al. (2016). Copyright (2008) American Chemical Society and Copyright (2016) Elsevier)

occur when pH values are low: the surface of NZVI is more positively charged, and the dissolution of Fe(III)–Cr(III) (oxy)hydroxides is more favorable; thus, there are more NZVI surface sites available for reaction.

4.8 NZVI Derivatives for Chromium Removal

Advanced materials consisting of NZVI have been developed in the past few years. These materials overcome some of the disadvantages of NZVI, e.g., strong tendency of aggregation, fast oxidation, and rapid sedimentation, and combine removal mechanisms and capacities of different materials. The careful choice and combination of two or more components in a single material, especially at the nanoscale, can result in the development of systems that incorporate the benefits from two different materials and overcome the limitations that each one faces. Thus, approaches involving surface modification of NZVI by electrostatic and/or steric stabilization, such as coating with polyelectrolytes or nonionic surfactants (Alessi and Li 2001), dispersing the particles in oil–water emulsions (Quinn et al. 2005), use of a support material for their synthesis (Ponder et al. 2001; Wu et al. 2012) and bimetallic particles with iron and a second metal (Liu et al. 2014b), are strategies that have been investigated thoroughly. The composites could be a solution to overcoming the magnetic attraction between the iron nanoparticles leading to increases in the surface area or interfacial accessibility, increased mobility and stability, minimized aggregation, and increased sorption capacity and thus significant increase in the reactivity. In Table 4.6 are given some examples of advanced materials based on NZVI that exhibited remarkable Cr(VI) removal ability.

Particularly, the selection of an appropriate surfactant for NZVI modification can dramatically enhance the colloidal stability and mobility. The higher total capacity

Table 4.6 Removal capacity values of Cr(VI) by various NZVI-modified and NZVI-based composites

Material	Removal capacity (mg/g)
NZVI (Li et al. 2008)	50–179.9
Montmorillonite/NZVI (Wu et al. 2015)	410
ZVI/chitosan (Thinh et al. 2013)	55.8
NZVI/Fe_3O_4 (Lv et al. 2012)	100
NZVI/TiO_2 (Petala et al. 2016)	51.6
Sodium dodecyl sulfate (SDS)-modified NZVI (Huang et al. 2015)	253.7
Graphene/NZVI (Jabeen et al. 2011)	162.6
NZVI-immobilized in calcium alginate beads (NZVI–C–A)	320.7
Biofilm-coated NZVI–C–A (Ravikumar et al. 2016)	473.9
Hierarchical lamellar porous carbon (FHLC)/NZVI (Li et al. 2016a)	357.1
Activated carbon (PAC)/NZVI/ag (Kakavandi et al. 2014)	100
Starch-coated NZVI-graphene composite (Kumarathilaka et al. 2016)	143.3
Pumice (P)/NZVI (Liu et al. 2014a; Liu et al. 2015)	106.9/306.6
Microcrystalline cellulose/NZVI (C-NZVI) (Sharma et al. 2015)	66.4
Polyaniline nanofibers/NZVI (Bhaumik et al. 2014)	434.8

for removal of Cr(VI) is attributed not only to the lower agglomeration and higher mobility of the NZVI particles but also to the increased local concentration and sorption of the contaminant on the surface of iron particles (Alessi and Li 2001; Zhang et al. 2002). For instance, Pluronic F-127 copolymer and polyacrylic acid, PAA, as surfactants for Cr(VI) removal were found to be beneficial, while the sedimentation rate of NZVI was significantly decreased (Dong et al. 2016; Wang et al. 2017). On the other hand, bimetallic particles with iron and a second (usually noble) metal, e.g., Pd/Fe, Ni/Fe, or Pt/Fe, have exhibited a significantly higher efficacy toward various pollutants' removal, including Cr(VI) (Chen et al. 2008; Kakavandi et al. 2014). The incorporated metal, even in a small amount, can substantially enhance the overall NZVI efficiency by acting as a catalyst for electron transfer and hydrogenation phenomena (Li et al. 2006). Furthermore, advanced composites, where NZVI is supported on various materials, have shown promise for water treatment (Fig. 4.13). In these composites, NZVI is reported to be less agglomerated and to possess a higher specific surface area, colloidal stability, homogeneous dispersion, and narrower size distribution (Ponder et al. 2000; Petala et al. 2013). The appropriate choice of a supporting material can invoke different removal mechanisms, i.e., absorption, reduction, and/or precipitation or even photocatalysis, as well as combined effects (Baikousi et al. 2012; Zhou et al. 2016). For instance, composites have been made where NZVI is supported on carbon materials that possess high sorption capacity and large surface area. This appears to better disperse the NZVI and prevent agglomeration (Baikousi et al. 2015; Liang et al. 2016). NZVI has also been supported on titanium oxide (Fig. 4.13). This provides synergistic photoreduction and improves the stability of NZVI against aggregation (Petala et al. 2016). These supported materials are important to the development of NZVI because it can make NZVI attractive for a larger number of environmental treatment processes.

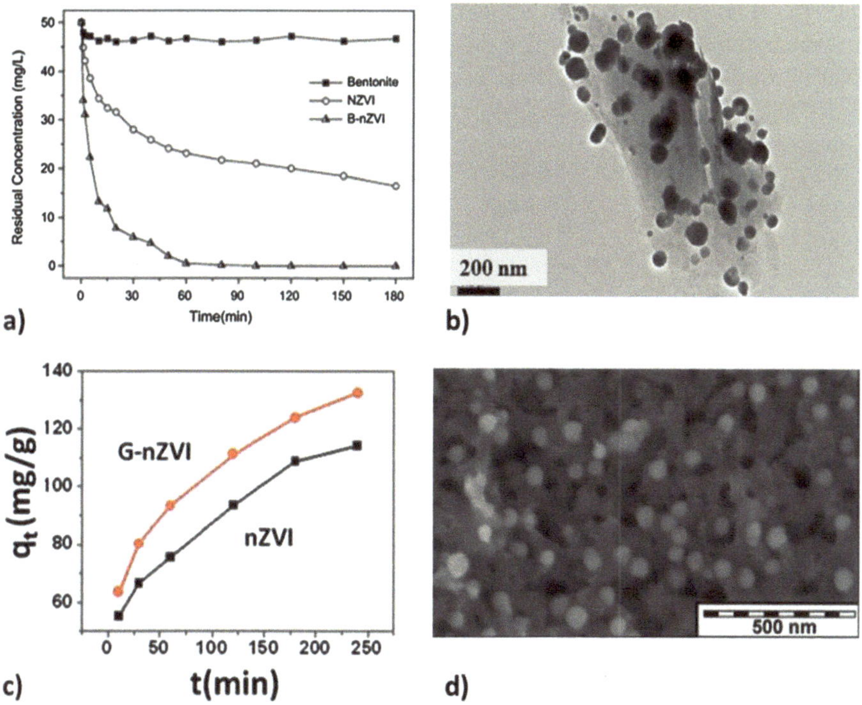

Fig. 4.13 Enhanced removal ability of NZVI composites toward Cr(VI), (**a**) B-NZVI: NZVI supported on bentonite (Shi et al. 2011), (**c**) G-NZVI: NZVI supported on graphene (Jabeen et al. 2011), (**b**) TEM image of NZVI supported on montmorillonite, well dispersed and non-agglomerated (dark spherical particles) (Zhang et al. 2013), (**d**) SEM image of NZVI supported on a porous TiO_2 photocatalyst, uniform size, and well dispersion onto the porous matrix (white particles). (Adapted with permission from Petala et al. (2016). Copyright (2011) Elsevier and Copyright (2013) Royal Society of Chemistry)

4.9 Pilot and Full-Scale Application of NZVI for Metal Removal

The number of the reports of pilot and/or full-scale applications of NZVI for treatment of metal–/metalloid-polluted waters are few in comparison with the hundreds of published laboratory studies demonstrating applicability of NZVI and their composites for elimination of organic pollutants from a broad spectrum of water types. However, the use of NZVI for Cr(VI) reduction in pilot in situ remediation was recently reported by Němeček et al. (2014). The reported contaminated area is located in the northern part of the Czech Republic as a former industrial site where the source of contamination comes from potassium dichromate used for production of Cr(III) in the leather tanning process. Prior to treatment, the Cr (VI) concentration in groundwater was below 3 mg/L, but 2–46 mg/kg in aquifer soil. The contaminated aquifer was 4.5–5.5 m bgs (with a thickness about 5 m) in Quaternary sands and gravels with clayey admixtures. The groundwater had low total mineralization ($Ca–SO_4$ type, <0.3 g/L of total dissolved solids, pH ~5.4, Eh = +450–550 mV, and low TOC <1.5 mg/L) with a flow velocity from 0.2 to 2 m per day and discharging into a river at a distance about 500 m. In August 2012, 120 kg of NZVI (commercial NANOFER 25 from NANO IRON Company, Czech Republic) was dispersed in tap water (the final concentration of 2 g/L) and applied into the saturated zone of the contaminated zone through three injection wells (diameter 80 mm) situated perpendicularly to the groundwater flow with spacing of 2.8 m (i.e., configuration of permeable injected barrier). In addition to injection wells, there were also four monitoring wells: one upgradient and three downgradients.

The evolution of selected physicochemical parameters of groundwater and changes in Cr(VI) concentration, as monitored before and after the NZVI application in all seven wells, is illustrated in Fig. 4.14. The 2D model of the Cr

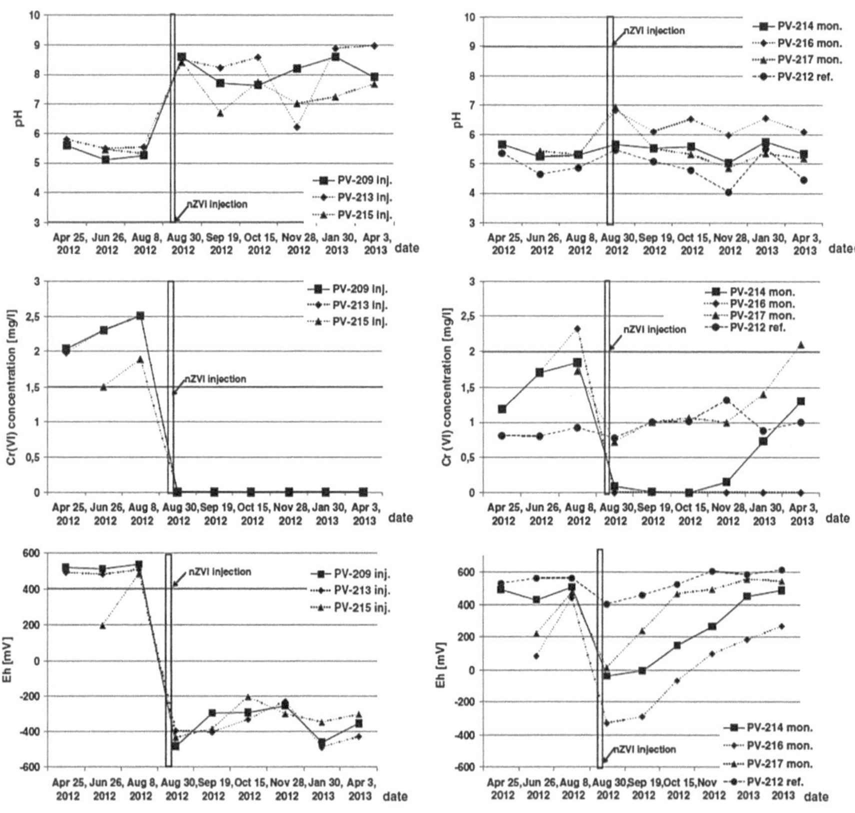

Fig. 4.14 Evolution of pH, ORP, and concentration of Cr(VI) in groundwater during pilot application of NZVI. Left part, injection wells; right part, monitoring wells and reference well. (Reprinted with permission from Němeček et al. (2014). Copyright (2014) Elsevier)

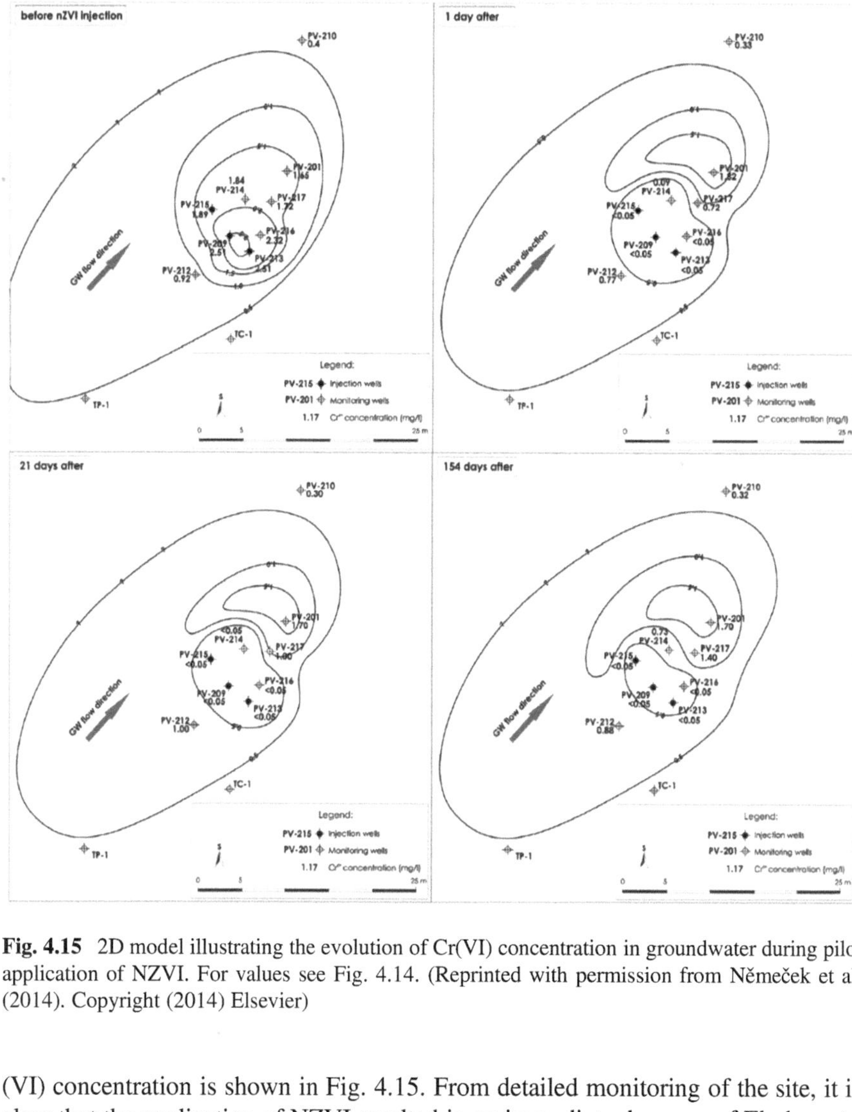

Fig. 4.15 2D model illustrating the evolution of Cr(VI) concentration in groundwater during pilot application of NZVI. For values see Fig. 4.14. (Reprinted with permission from Němeček et al. (2014). Copyright (2014) Elsevier)

(VI) concentration is shown in Fig. 4.15. From detailed monitoring of the site, it is clear that the application of NZVI resulted in an immediate decrease of Eh down to −484 mV (within one day) followed by a rapid decrease in the concentration of both Cr(VI) and also total Cr through geofixation of reduced Cr(III) in the soil matrix in the form of insoluble Fe^{III}–Cr^{III}–oxyhydroxides. The rapid reductive effect of NZVI was observed at a distance of 7 m downgradient. In the injection wells, the evident reductive effect remained over 217 days; however, there was just a temporary effect in the monitoring wells located 4 and 7 m downgradient from the injection point requiring thus another treatment step (see Fig. 4.14).

In the study of Němeček et al. (2014), the authors also provided a detailed assessment of the potential ecotoxicological effects of NZVI on autochthonous

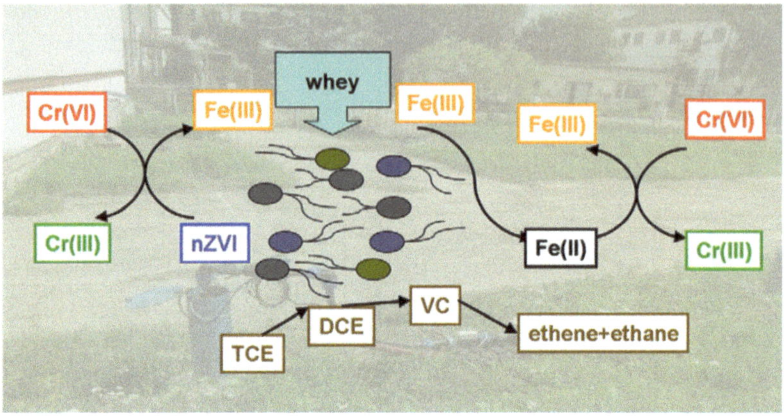

Fig. 4.16 Graphical illustration of possible reactions among NZVI, Cr(VI), TCE, and whey (with highlighted role of microorganisms) leading to decontamination of groundwater. (Reprinted with permission from Němeček et al. (2015). Copyright (2015) Elsevier)

bacterial microflora populations (i.e., microbial cultivation tests, phospholipid fatty acid analysis (PLFA), and ecotoxicological tests with *Vibrio fischeri*). Surprisingly, there was no observable negative impact of NZVI in the treated groundwater, and PLFA of soil samples even indicated stimulation of Gram-positive bacteria growth due to elimination of highly toxic Cr(VI) in aquifer. In a subsequent study of Němeček et al. (2015), microbial growth was stimulated by application of inexpensive organic substrate (cheese whey) 9 months after the application of NZVI (i.e., abiotic phase; see above). The increase in anaerobic bacterial microflora directly or indirectly contributed to biotic reduction of remaining Cr(IV) at the site with broader spatial effects detectable even at a distance of 22 m from the injection wells, with clear long-term effects persisting at least 10 months after injection of whey (note, there is rebound of total Cr in monitoring wells over monitoring period; Fig. 4.17). The whole reaction mechanism (graphically illustrated in Fig. 4.16) included oxidation of Fe^0 to Fe(III) during abiotic phase of Cr(VI) reduction, followed by microbial reduction of Fe(III) to Fe(II) (i.e., mostly by iron-reducing bacteria); the oxidation of Fe(II) then caused further reduction of Cr(VI) even after the consumption of substrate when microbial density decreased – see Fig. 4.17.

The identical approach, which turned out to be highly promising due to partial regenerating of the applied NZVI through successive abiotic and biotic reduction steps, was applied by the same authors at another site in the Czech Republic (Němeček et al. 2016). In this case, the aquifer (4 m thick in Quaternary sandy gravel with silty admixtures and overlaid by clay and clayey loam) was co-contaminated by Cr(VI) (4.4–57 mg/L) and TCE/cis-DCE (400–6526 μg/L of sum of chlorinated ethenes) as a consequence of historical degreasing and chromium coating industrial activities. The groundwater had elevated mineralization (Ca–HCO_3^- type, 0.9–1.2 g/L of total dissolved solids, pH = 6.9–7.0, Eh = +130–490- mV, and low TOC from 1.0 to 5.4 mg/L) with a flow velocity 1.5 m per day

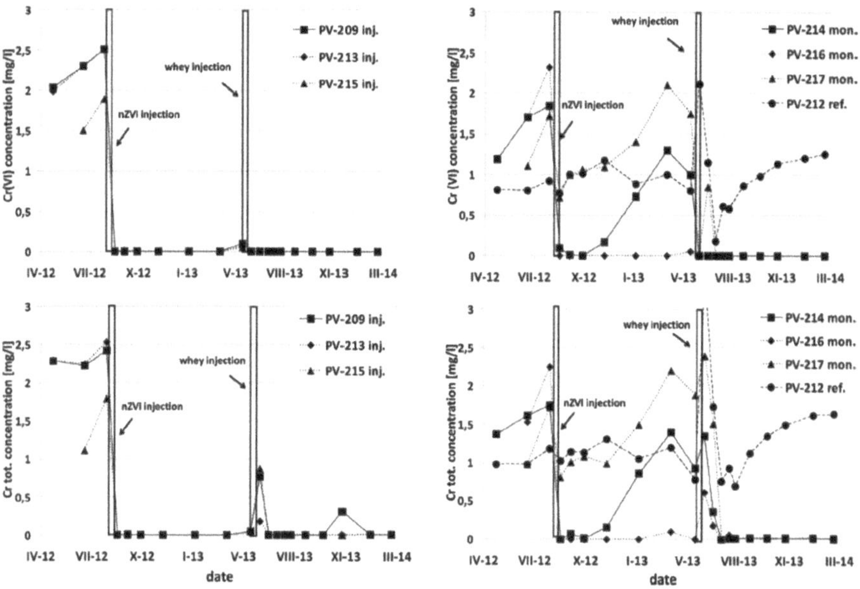

Fig. 4.17 Changes in concentration of Cr(VI) and total Cr in groundwater during pilot application of NZVI (first Roman numeral on the x-axis is the month, and the second number is the year). Left part, injection wells; right part, monitoring wells and reference well. (Reprinted with permission from Němeček et al. (2015). Copyright (2015) Elsevier)

discharging into a river at the distance of 430 m. Initially, two types of NZVI particles (20 kg of NANOFER STAR and then 20 kg of NANOFER 25S, both from NANO IRON Company, Czech Republic) were applied by direct push technology at 4-month intervals at a concentration of 1 g/L and 2 g/L, respectively, into the three boreholes situated perpendicularly to groundwater flow. Subsequently, the whey was applied 2.5 months later using a circulation system resulting in 60 mg/L of TOC in the groundwater (i.e., 8.2 m^3 of whey was injected during 5 weeks). The application of NZVI caused the Eh to decrease below −400 mV, which then stabilized at 50 to 100 mV after injection of whey. The evolution of Cr (VI) concentration during the course of the pilot test is presented in Fig. 4.18, demonstrating the efficiency and sustainability of the combined nano-bio-approach for the treatment of groundwater with mixed contamination. It is worth mentioning that at the end of the pilot test (i.e., after 13 months from the first NZVI injection), the NZVI was almost totally oxidized according to Mössbauer spectroscopy measurement on frozen samples (6% of Fe^0 from Fe_{tot}).

A full-scale application of NZVI for heavy metal removal from waste water of the Jiangxi Copper Company, China, has been reported. The wastewater contained mainly Cu ≈ 103 mg L^{-1}, Ni, Zn, Pb, and also arsenic ≈ 110 mg L^{-1}, Se, and Sb (Li et al. 2017). The wastewater treatment plant consists of several mixed reactors (h 4.8 m, l 4.8 m, and w 3 m with an effective volume about 60 m^3) constructed as separate modules (Fig. 4.19). It was demonstrated that NZVI acted as a highly

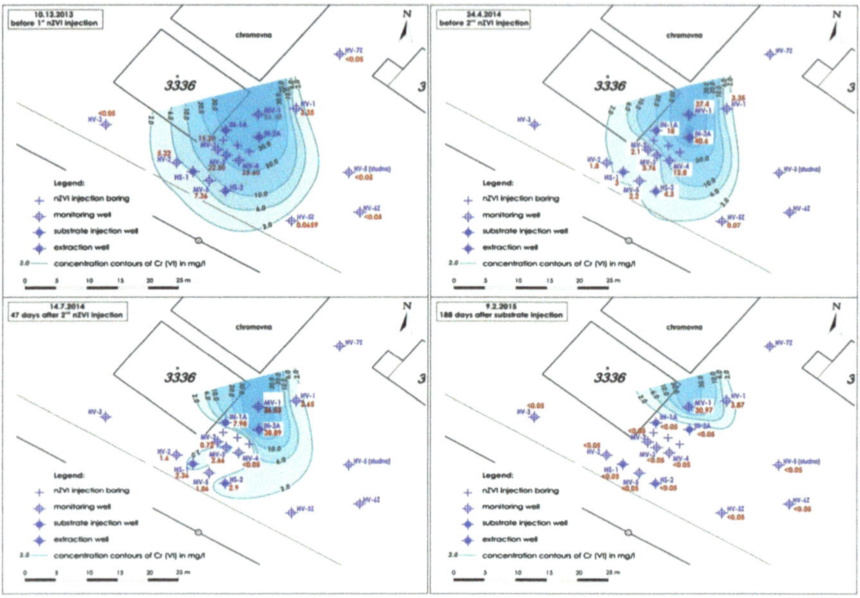

Fig. 4.18 Evolution of Cr(VI) concentration in groundwater based on NZVI injection and application of whey – a 2D model. (Adapted with permission from Němeček et al. (2016). Copyright (2016) Elsevier)

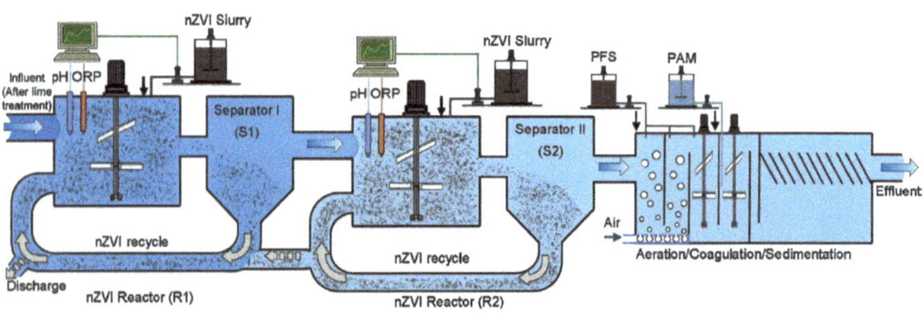

Fig. 4.19 Schematic drawing of pilot plant utilizing NZVI particles for removal of metals from wastewater. (Reprinted with permission from Li et al. (2017). Copyright (2017) Elsevier)

efficient reagent (>99.5% removal of key metals with capacities of 245 mg As and 226 mg of Cu per 1 g of NZVI at the same time) providing a low redox potential and subsequent separation of metals. The performance of the technology over >120 days is shown in Fig. 4.20; the average concentration of Cu(II) was reduced from 103 to 0.16 mg L^{-1} and As(V) from 110 to 0.29 mg L^{-1} (i.e., to a value below the local discharge limits), and other metals and metalloids were reduced successfully below 0.1 mg L^{-1} as well. The advantage of such technology lies in the fact that NZVI keeps a low redox condition in the closed reactors, lowering the required NZVI

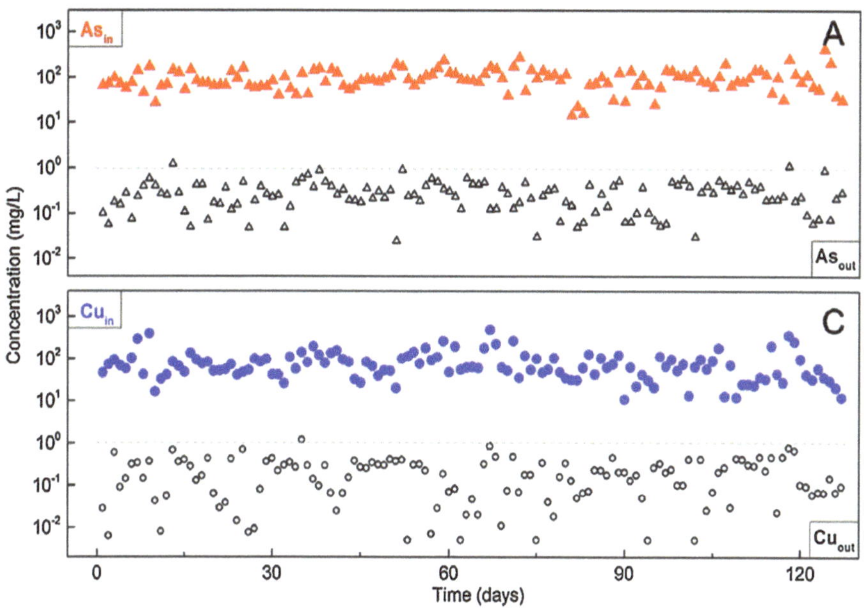

Fig. 4.20 Demonstration of successful long-term performance of wastewater treatment plant (see Fig. 4.18) for simultaneous removal of As and Cu. (Reprinted with permission from Li et al. (2017). Copyright (2017) Elsevier)

dosage over time (>12 months) with partial recycling of NZVI. This example serves as evidence that such technology can be sustainable and cost-effective, utilizing several key aspects of NZVI, including simultaneous removal of different metals by multiple modes of action, fast reactivity, easy separation via gravitational settling, and low NZVI dosing due to high specific surface area of the used NZVI particles. Moreover, the NZVI particles after adsorption of metals were subsequently processed as a valuable metal resource, possessing higher concentration of copper than the original ore processed in the Jiangxi Copper Company (Li et al. 2017).

Similarly, arsenic removal from drinking water was successfully demonstrated within a pilot test in Canada (Gottinger et al. 2010). In this case, a column filter containing 50% sand and 50% of NZVI was integrated into a biological water treatment plant. The arsenic removal reached the value of 99.7% (i.e., below 0.1 mg L^{-1}), accompanied with reduction of ammonia, iron, and manganese concentration.

Acknowledgments This work was supported by grants from the Technology Agency of the Czech Republic "Competence Centers" (project No. TE01020218), Ministry of the Interior of the Czech Republic (project No. VI20162019017), and the Ministry of Education, Youth and Sports of the Czech Republic (project No. LO1305). This work was further supported by Student Project IGA_PrF_2018_015 of Palacký University, Olomouc.

References

Agency U.S.E.P. (2001). *Drinking water arsenic rule history.*

Agency U.S.E.P. (2015). *Chromium in drinking water.*

Ahmed, K. M., Bhattacharya, P., Hasan, M. A., Akhter, S. H., Alam, S. M. M., Bhuyian, M. A. H., Imam, M. B., Khan, A. A., & Sracek, O. (2004). Arsenic enrichment in groundwater of the alluvial aquifers in Bangladesh: An overview. *Applied Geochemistry, 19*, 181–200.

Alessi, D. S., & Li, Z. (2001). Synergistic effect of cationic surfactants on perchloroethylene degradation by zero-valent iron. *Environmental Science & Technology, 35*, 3713–3717.

Almeelbi, T., & Bezbaruah, A. (2012). Aqueous phosphate removal using nanoscale zero-valent iron. *Journal of Nanoparticle Research, 14*, 900.

Arshadi, M., Soleymanzadeh, M., Salvacion, J. W. L., & SalimiVahid, F. (2014). Nanoscale Zero-Valent Iron (NZVI) supported on sineguelas waste for Pb(II) removal from aqueous solution: Kinetics, thermodynamic and mechanism. *Journal of Colloid and Interface Science, 426*, 241–251.

Baikousi, M., Bourlinos, A. B., Douvalis, A., Bakas, T., Anagnostopoulos, D. F., Tuček, J., Šafářová, K., Zboril, R., & Karakassides, M. A. (2012). Synthesis and characterization of γ-Fe2O3/carbon hybrids and their application in removal of hexavalent chromium ions from aqueous solutions. *Langmuir, 28*, 3918–3930.

Baikousi, M., Georgiou, Y., Daikopoulos, C., Bourlinos, A. B., Filip, J., Zbořil, R., Deligiannakis, Y., & Karakassides, M. A. (2015). Synthesis and characterization of robust zero valent iron/mesoporous carbon composites and their applications in arsenic removal. *Carbon, 93*, 636–647.

Bang, S., Johnson, M. D., Korfiatis, G. P., & Meng, X. (2005). Chemical reactions between arsenic and zero-valent iron in water. *Water Research, 39*, 763–770.

Bhattacharya, P., Claesson, M., Bundschuh, J., Sracek, O., Fagerberg, J., Jacks, G., Martin, R. A., Storniolo, A. D. R., & Thir, J. M. (2006). Distribution and mobility of arsenic in the Río Dulce alluvial aquifers in Santiago del Estero Province, Argentina. *Science of the Total Environment, 358*, 97–120.

Bhaumik, M., Choi, H. J., McCrindle, R. I., & Maity, A. (2014). Composite nanofibers prepared from metallic iron nanoparticles and polyaniline: High performance for water treatment applications. *Journal of Colloid and Interface Science, 425*, 75–82.

Biesinger, M. C., Lau, L. W. M., Gerson, A. R., & Smart, R. S. C. (2010). Resolving surface chemical states in XPS analysis of first row transition metals, oxides and hydroxides: Sc, Ti, V, Cu and Zn. *Applied Surface Science, 257*, 887–898.

Biesinger, M. C., Payne, B. P., Grosvenor, A. P., Lau, L. W. M., Gerson, A. R., & Smart, R. S. C. (2011). Resolving surface chemical states in XPS analysis of first row transition metals, oxides and hydroxides: Cr, Mn, Fe, Co and Ni. *Applied Surface Science, 257*, 2717–2730.

Birkle, P., Bundschuh, J., & Sracek, O. (2010). Mechanisms of arsenic enrichment in geothermal and petroleum reservoirs fluids in Mexico. *Water Research, 44*, 5605–5617.

Biterna, M., Arditsoglou, A., Tsikouras, E., & Voutsa, D. (2007). Arsenate removal by zero valent iron: Batch and column tests. *Journal of Hazardous Materials, 149*, 548–552.

Briggs, D. (1998). *Surface analysis of polymers by XPS and static SIMS*. Cambridge University Press.

Bruzzoniti, M. C., & Fiore, S. (2014). Removal of inorganic contaminants from aqueous solutions: Evaluation of the remediation efficiency and of the environmental impact of a zero-valent Iron substrate. *Water, Air, & Soil Pollution, 225*, 2098.

Chen, L.-H., Huang, C.-C., & Lien, H.-L. (2008). Bimetallic iron–aluminum particles for dechlorination of carbon tetrachloride. *Chemosphere, 73*, 692–697.

Costa, M. (2003). Potential hazards of hexavalent chromate in our drinking water. *Toxicology and Applied Pharmacology, 188*, 1–5.

Crane, R. A., & Scott, T. B. (2012). Nanoscale zero-valent iron: Future prospects for an emerging water treatment technology. *Journal of Hazardous Materials, 211–212*, 112–125.

Dai, Y., Hu, Y., Jiang, B., Zou, J., Tian, G., & Fu, H. (2016). Carbothermal synthesis of ordered mesoporous carbon-supported nano zero-valent iron with enhanced stability and activity for hexavalent chromium reduction. *Journal of Hazardous Materials, 309*, 249–258.

Dickinson, M., & Scott, T. B. (2010). The application of zero-valent iron nanoparticles for the remediation of a uranium-contaminated waste effluent. *Journal of Hazardous Materials, 178*, 171–179.

Dong, H., He, Q., Zeng, G., Tang, L., Zhang, C., Xie, Y., Zeng, Y., Zhao, F., & Wu, Y. (2016). Chromate removal by surface-modified nanoscale zero-valent iron: Effect of different surface coatings and water chemistry. *Journal of Colloid and Interface Science, 471*, 7–13.

Dorjee, P., Amarasiriwardena, D., & Xing, B. (2015). Erratum to "Antimony adsorption by zero-valent iron nanoparticles (NZVI): Ion chromatography-inductively coupled plasma mass spectrometry (IC-ICP-MS) study" [Microchem. J. 116 (2014) 15–23]. *Microchemical Journal, 118*, 278.

Drahota, P., & Filippi, M. (2009). Secondary arsenic minerals in the environment: A review. *Environment International, 35*, 1243–1255.

Dries, J., Bastiaens, L., Springael, D., Agathos, S. N., & Diels, L. (2005). Combined removal of chlorinated ethenes and heavy metals by zerovalent iron in batch and continuous flow column systems. *Environmental Science & Technology, 39*, 8460–8465.

Eglal, M. M., & Ramamurthy, A. S. (2014). Nanofer ZVI: Morphology, particle characteristics, kinetics, and applications. *Journal of Nanomaterials, 2014*, 11.

Farrell, J., Wang, J., O'Day, P., & Conklin, M. (2001). Electrochemical and spectroscopic study of arsenate removal from water using zero-valent iron media. *Environmental Science & Technology, 35*, 2026–2032.

Fiedor, J. N., Bostick, W. D., Jarabek, R. J., & Farrell, J. (1998). Understanding the mechanism of uranium removal from groundwater by zero-valent iron using X-ray photoelectron spectroscopy. *Environmental Science & Technology, 32*, 1466–1473.

Filella, M., Belzile, N., & Chen, Y.-W. (2002). Antimony in the environment: A review focused on natural waters: I. Occurrence. *Earth-Science Reviews, 57*, 125–176.

Filip, J., Karlický, F., Marušák, Z., Lazar, P., Černík, M., Otyepka, M., & Zbořil, R. (2014). Anaerobic reaction of nanoscale zerovalent Iron with water: Mechanism and kinetics. *The Journal of Physical Chemistry C, 118*, 13817–13825.

Fu, F., Dionysiou, D. D., & Liu, H. (2014). The use of zero-valent iron for groundwater remediation and wastewater treatment: A review. *Journal of Hazardous Materials, 267*, 194–205.

Gheju, M. (2011). Hexavalent chromium reduction with Zero-Valent Iron (ZVI) in aquatic systems. *Water, Air, & Soil Pollution, 222*, 103–148.

Gieré, R., Sidenko, N. V., & Lazareva, E. V. (2003). The role of secondary minerals in controlling the migration of arsenic and metals from high-sulfide wastes (Berikul gold mine, Siberia). *Applied Geochemistry, 18*, 1347–1359.

Gottinger, A. M., Wild, D. J., McMartin, D., Moldovan, B., & Wang, D. (2010). Development of an Iron-amended biofilter for removal of arsenic from rural Canadian prairie potable water. In C. A. Brebbia & A. M. Marinov (Eds.), *Water pollution X* (pp. 333–344). WIT.

Gräfe, M., Beattie, D. A., Smith, E., Skinner, W. M., & Singh, B. (2008). Copper and arsenate co-sorption at the mineral–water interfaces of goethite and jarosite. *Journal of Colloid and Interface Science, 322*, 399–413.

Gu, Z., Deng, B., & Yang, J. (2007). Synthesis and evaluation of iron-containing ordered mesoporous carbon (FeOMC) for arsenic adsorption. *Microporous and Mesoporous Materials, 102*, 265–273.

Guo, X., Yang, Z., Dong, H., Guan, X., Ren, Q., Lv, X., & Jin, X. (2016). Simple combination of oxidants with zero-valent-iron (ZVI) achieved very rapid and highly efficient removal of heavy metals from water. *Water Research, 88*, 671–680.

Gupta, A., Yunus, M., & Sankararamakrishnan, N. (2012). Zerovalent iron encapsulated chitosan nanospheres – A novel adsorbent for the removal of total inorganic Arsenic from aqueous systems. *Chemosphere, 86*, 150–155.

Han, W., Fu, F., Cheng, Z., Tang, B., & Wu, S. (2016). Studies on the optimum conditions using acid-washed zero-valent iron/aluminum mixtures in permeable reactive barriers for the removal of different heavy metal ions from wastewater. *Journal of Hazardous Materials, 302*, 437–446.

Huang, Y. H., Tang, C., & Zeng, H. (2012). Removing molybdate from water using a hybridized zero-valent iron/magnetite/Fe(II) treatment system. *Chemical Engineering Journal, 200–202*, 257–263.

Huang, D.-L., Chen, G.-M., Zeng, G.-M., Xu, P., Yan, M., Lai, C., Zhang, C., Li, N.-J., Cheng, M., He, X.-X., & He, Y. (2015). Synthesis and application of modified zero-valent iron nanoparticles for removal of hexavalent chromium from wastewater. *Water, Air, & Soil Pollution, 226*, 375.

Hug, S. J., & Leupin, O. (2003). Iron-catalyzed oxidation of arsenic(III) by oxygen and by hydrogen peroxide: pH-dependent formation of oxidants in the Fenton reaction. *Environmental Science & Technology, 37*, 2734–2742.

Jabeen, H., Chandra, V., Jung, S., Lee, J. W., Kim, K. S., & Kim, S. B. (2011). Enhanced Cr (vi) removal using iron nanoparticle decorated graphene. *Nanoscale, 3*, 3583–3585.

Jiang, X., Qiao, J., Lo, I. M. C., Wang, L., Guan, X., Lu, Z., Zhou, G., & Xu, C. (2015). Enhanced paramagnetic Cu2+ ions removal by coupling a weak magnetic field with zero valent iron. *Journal of Hazardous Materials, 283*, 880–887.

Kakavandi, B., Kalantary, R. R., Farzadkia, M., Mahvi, A. H., Esrafili, A., Azari, A., Yari, A. R., & Javid, A. B. (2014). Enhanced chromium (VI) removal using activated carbon modified by zero valent iron and silver bimetallic nanoparticles. *Journal of Environmental Health Science and Engineering, 12*, 115.

Kanel, S. R., Manning, B., Charlet, L., & Choi, H. (2005). Removal of arsenic(III) from groundwater by nanoscale zero-valent iron. *Environmental Science & Technology, 39*, 1291–1298.

Kanel, S. R., Grenèche, J.-M., & Choi, H. (2006). Arsenic(V) removal from groundwater using nano scale zero-valent Iron as a colloidal reactive barrier material. *Environmental Science & Technology, 40*, 2045–2050.

Karabelli, D., Üzüm, Ç., Shahwan, T., Eroğlu, A. E., Scott, T. B., Hallam, K. R., & Lieberwirth, I. (2008). Batch removal of aqueous Cu2+ ions using nanoparticles of zero-valent iron: A study of the capacity and mechanism of uptake. *Industrial & Engineering Chemistry Research, 47*, 4758–4764.

Kareus, S. A., Kelley, C., Walton, H. S., & Sinclair, P. R. (2001). Release of Cr(III) from Cr(III) picolinate upon metabolic activation. *Journal of Hazardous Materials, 84*, 163–174.

Katsoyiannis, I. A., Ruettimann, T., & Hug, S. J. (2008). pH dependence of Fenton reagent generation and as(III) oxidation and removal by corrosion of zero valent iron in aerated water. *Environmental Science & Technology, 42*, 7424–7430.

Keenan, C. R., & Sedlak, D. L. (2008). Factors affecting the yield of oxidants from the reaction of nanoparticulate zero-valent iron and oxygen. *Environmental Science & Technology, 42*, 1262–1267.

Klimkova, S., Cernik, M., Lacinova, L., Filip, J., Jancik, D., & Zboril, R. (2011). Zero-valent iron nanoparticles in treatment of acid mine water from *in situ* uranium leaching. *Chemosphere, 82*, 1178–1184.

Kocourková-Víšková, E., Loun, J., Sracek, O., Houzar, S., & Filip, J. (2015). Secondary arsenic minerals and arsenic mobility in a historical waste rock pile at Kaňk near Kutná Hora, Czech Republic. *Mineralogy and Petrology, 109*, 17–33.

Kumarathilaka, P., Jayaweera, V., Wijesekara, H., Kottegoda, I. R. M., Rosa, S. R. D., & Vithanage, M. (2016). Insights into starch coated nanozero valent iron-graphene composite for Cr(VI) removal from aqueous medium. *Journal of Nanomaterials, 2016*, 10.

Li, X.-q., & Zhang, W.-x. (2007). Sequestration of metal cations with Zerovalent Iron Nanoparticles A study with high resolution X-ray Photoelectron Spectroscopy (HR-XPS). *The Journal of Physical Chemistry C, 111*, 6939–6946.

Li, X.-q., Elliott, D. W., & Zhang, W.-x. (2006). Zero-valent Iron nanoparticles for abatement of environmental pollutants: Materials and engineering aspects. *Critical Reviews in Solid State and Materials Sciences, 31*, 111–122.
Li, X.-q., Cao, J., & Zhang, W.-x. (2008). Stoichiometry of Cr(VI) immobilization using nanoscale Zerovalent Iron (NZVI): A study with high-resolution X-ray photoelectron spectroscopy (HR-XPS). *Industrial Engineering Chemistry Research, 47*, 2131–2139.
Li, J., Bao, H., Xiong, X., Sun, Y., & Guan, X. (2015a). Effective Sb(V) immobilization from water by zero-valent iron with weak magnetic field. *Separation and Purification Technology, 151*, 276–283.
Li, Z.-J., Wang, L., Yuan, L.-Y., Xiao, C.-L., Mei, L., Zheng, L.-R., Zhang, J., Yang, J.-H., Zhao, Y.-L., Zhu, Z.-T., Chai, Z.-F., & Shi, W.-Q. (2015b). Efficient removal of uranium from aqueous solution by zero-valent iron nanoparticle and its graphene composite. *Journal of Hazardous Materials, 290*, 26–33.
Li, C., Huang, B., Li, C., Chen, X., & Huang, Y. (2016a). In situ formation of nanoscale zero-value iron on fish-scale-based porous carbon for Cr(VI) adsorption. *Water Science and Technology, 73*, 2237–2243.
Li, L., Hu, J., Shi, X., Fan, M., Luo, J., & Wei, X. (2016b). Nanoscale zero-valent metals: A review of synthesis, characterization, and applications to environmental remediation. *Environmental Science and Pollution Research, 23*, 17880–17900.
Li, S., Wang, W., Liang, F., & Zhang, W.-x. (2017). Heavy metal removal using nanoscale zero-valent iron (NZVI): Theory and application. *Journal of Hazardous Materials, 322*(Part A), 163–171.
Liang, L., Yang, W., Guan, X., Li, J., Xu, Z., Wu, J., Huang, Y., & Zhang, X. (2013). Kinetics and mechanisms of pH-dependent selenite removal by zero valent iron. *Water Research, 47*, 5846–5855.
Liang, L., Sun, W., Guan, X., Huang, Y., Choi, W., Bao, H., Li, L., & Jiang, Z. (2014). Weak magnetic field significantly enhances selenite removal kinetics by zero valent iron. *Water Research, 49*, 371–380.
Liang, Z., Wen, Q., Wang, X., Zhang, F., & Yu, Y. (2016). Chemically stable and reusable nano zero-valent iron/graphite-like carbon nitride nanohybrid for efficient photocatalytic treatment of Cr(VI) and rhodamine B under visible light. *Applied Surface Science, 386*, 451–459.
Ling, L., & Zhang, W.-X. (2014). Sequestration of arsenate in zero-valent iron nanoparticles: Visualization of Intraparticle reactions at angstrom resolution. *Environmental Science & Technology Letters, 1*, 305–309.
Liu, T., Wang, Z.-L., Yan, X., & Zhang, B. (2014a). Removal of mercury (II) and chromium (VI) from wastewater using a new and effective composite: Pumice-supported nanoscale zero-valent iron. *Chemical Engineering Journal, 245*, 34–40.
Liu, W.-J., Qian, T.-T., & Jiang, H. (2014b). Bimetallic Fe nanoparticles: Recent advances in synthesis and application in catalytic elimination of environmental pollutants. *Chemical Engineering Journal, 236*, 448–463.
Liu, T., Wang, Z.-L., & Sun, Y. (2015). Manipulating the morphology of nanoscale zero-valent iron on pumice for removal of heavy metals from wastewater. *Chemical Engineering Journal, 263*, 55–61.
López, D. L., Bundschuh, J., Birkle, P., Armienta, M. A., Cumbal, L., Sracek, O., Cornejo, L., & Ormachea, M. (2012). Arsenic in volcanic geothermal fluids of Latin America. *Science of the Total Environment, 429*, 57–75.
Ludwig, R. D., Smyth, D. J. A., Blowes, D. W., Spink, L. E., Wilkin, R. T., Jewett, D. G., & Weisener, C. J. (2009). Treatment of arsenic, heavy metals, and acidity using a mixed ZVI-compost PRB. *Environmental Science & Technology, 43*, 1970–1976.
Lv, X., Xu, J., Jiang, G., Tang, J., & Xu, X. (2012). Highly active nanoscale zero-valent iron (NZVI)–Fe3O4 nanocomposites for the removal of chromium(VI) from aqueous solutions. *Journal of Colloid and Interface Science, 369*, 460–469.

Mamindy-Pajany, Y., Hurel, C., Marmier, N., & Roméo, M. (2011). Arsenic (V) adsorption from aqueous solution onto goethite, hematite, magnetite and zero-valent iron: Effects of pH, concentration and reversibility. *Desalination, 281*, 93–99.

Mandal, B. K., & Suzuki, K. T. (2002). Arsenic round the world: A review. *Talanta, 58*, 201–235.

Marshall, G., Ferreccio, C., Yuan, Y., Bates, M. N., Steinmaus, C., Selvin, S., Liaw, J., & Smith, A. H. (2007). Fifty-year study of lung and bladder cancer mortality in Chile related to arsenic in drinking water. *JNCI: Journal of the National Cancer Institute, 99*, 920–928.

Mohan, D., & Pittman, C. U., Jr. (2007). Arsenic removal from water/wastewater using adsorbents—A critical review. *Journal of Hazardous Materials, 142*, 1–53.

Moraci, N., & Calabrò, P. S. (2010). Heavy metals removal and hydraulic performance in zero-valent iron/pumice permeable reactive barriers. *Journal of Environmental Management, 91*, 2336–2341.

Moulder, J. F., & Chastain, J. (1992). *Handbook of X-ray photoelectron spectroscopy: A reference book of standard spectra for identification and interpretation of XPS data*. Physical Electronics Division, Perkin-Elmer Corporation.

Muthukrishnan, M., & Guha, B. K. (2008). Effect of pH on rejection of hexavalent chromium by nanofiltration. *Desalination, 219*, 171–178.

Němeček, J., Lhotský, O., & Cajthaml, T. (2014). Nanoscale zero-valent iron application for *in situ* reduction of hexavalent chromium and its effects on indigenous microorganism populations. *Science of the Total Environment, 485–486*, 739–747.

Němeček, J., Pokorný, P., Lacinová, L., Černík, M., Masopustová, Z., Lhotský, O., Filipová, A., & Cajthaml, T. (2015). Combined abiotic and biotic *in situ* reduction of hexavalent chromium in groundwater using NZVI and whey: A remedial pilot test. *Journal of Hazardous Materials, 300*, 670–679.

Němeček, J., Pokorný, P., Lhotský, O., Knytl, V., Najmanová, P., Steinová, J., Černík, M., Filipová, A., Filip, J., & Cajthaml, T. (2016). Combined nano-biotechnology for *in situ* remediation of mixed contamination of groundwater by hexavalent chromium and chlorinated solvents. *Science of the Total Environment, 563–564*, 822–834.

Neumann, A., Kaegi, R., Voegelin, A., Hussam, A., Munir, A. K. M., & Hug, S. J. (2013). Arsenic removal with composite iron matrix filters in Bangladesh: A field and laboratory study. *Environmental Science & Technology, 47*, 4544–4554.

Nordstrom, D. K. (2002). Worldwide occurrences of arsenic in ground water. *Science, 296*, 2143–2145.

Nordstrom, D. K., & Alpers, C. N. (1999). Negative pH, efflorescent mineralogy, and consequences for environmental restoration at the Iron Mountain Superfund site, California. *Proceedings of the National Academy of Sciences, 96*, 3455–3462.

O'Carroll, D., Sleep, B., Krol, M., Boparai, H., & Kocur, C. (2013). Nanoscale zero valent iron and bimetallic particles for contaminated site remediation. *Advances in Water Resources, 51*, 104–122.

Organization W.H. (2011). *Guidelines for drinking-water quality* (4th ed.).

Pang, S.-Y., Jiang, J., & Ma, J. (2011). Oxidation of sulfoxides and arsenic(III) in corrosion of nanoscale zero valent iron by oxygen: Evidence against Ferryl ions (Fe(IV)) as active intermediates in Fenton reaction. *Environmental Science & Technology, 45*, 307–312.

Petala, E., Dimos, K., Douvalis, A., Bakas, T., Tucek, J., Zbořil, R., & Karakassides, M. A. (2013). Nanoscale zero-valent iron supported on mesoporous silica: Characterization and reactivity for Cr(VI) removal from aqueous solution. *Journal of Hazardous Materials, 261*, 295–306.

Petala, E., Baikousi, M., Karakassides, M. A., Zoppellaro, G., Filip, J., Tucek, J., Vasilopoulos, K. C., Pechousek, J., & Zboril, R. (2016). Synthesis, physical properties and application of the zero-valent iron/titanium dioxide heterocomposite having high activity for the sustainable photocatalytic removal of hexavalent chromium in water. *Physical Chemistry Chemical Physics, 18*, 10637–10646.

Planer-Friedrich, B., Lehr, C., Matschullat, J., Merkel, B. J., Nordstrom, D. K., & Sandstrom, M. W. (2006). Speciation of volatile arsenic at geothermal features in Yellowstone National Park. *Geochimica et Cosmochimica Acta, 70*, 2480–2491.

Plant, J. A., Kinniburgh, D. G., Smedley, P. L., Fordyce, F. M., & Klinck, B. A. (2005). Environmental geochemistry. *Elsevier, 9*, 17–66.
Ponder, S. M., Darab, J. G., & Mallouk, T. E. (2000). Remediation of Cr(VI) and Pb(II) aqueous solutions using supported, nanoscale zero-valent iron. *Environmental Science & Technology, 34*, 2564–2569.
Ponder, S. M., Darab, J. G., Bucher, J., Caulder, D., Craig, I., Davis, L., Edelstein, N., Lukens, W., Nitsche, H., Rao, L., Shuh, D. K., & Mallouk, T. E. (2001). Surface chemistry and electrochemistry of supported zerovalent iron nanoparticles in the remediation of aqueous metal contaminants. *Chemistry of Materials, 13*, 479–486.
Quinn, J., Geiger, C., Clausen, C., Brooks, K., Coon, C., O'Hara, S., Krug, T., Major, D., Yoon, W.-S., Gavaskar, A., & Holdsworth, T. (2005). Field demonstration of DNAPL dehalogenation using emulsified zero-valent iron. *Environmental Science & Technology, 39*, 1309–1318.
Ravenscroft, P., Brammer, H., & Richards, K. (2009). *Arsenic pollution: A global synthesis*. Chichester: Wiley-Blackwell.
Ravikumar, K. V. G., Kumar, D., Kumar, G., Mrudula, P., Natarajan, C., & Mukherjee, A. (2016). Enhanced Cr(VI) removal by nanozerovalent iron-immobilized alginate beads in the presence of a biofilm in a continuous-flow reactor. *Industrial & Engineering Chemistry Research, 55*, 5973–5982.
Riba, O., Scott, T. B., Vala Ragnarsdottir, K., & Allen, G. C. (2008). Reaction mechanism of uranyl in the presence of zero-valent iron nanoparticles. *Geochimica et Cosmochimica Acta, 72*, 4047–4057.
Richard, F. C., & Bourg, A. C. M. (1991). Aqueous geochemistry of chromium: A review. *Water Research, 25*, 807–816.
Rodová, A., Filip, J., & Černík, M. (2015). Arsenic immobilization by nanoscale zero-valent iron/ Immobilizacja Arsenu Przez Nanożelazo Na Zerowym Stopniu Utlenienia. *Ecological Chemistry and Engineering S, 22*(1), 45–59.
Salzsauler, K. A., Sidenko, N. V., & Sherriff, B. L. (2005). Arsenic mobility in alteration products of sulfide-rich, arsenopyrite-bearing mine wastes, Snow Lake, Manitoba, Canada. *Applied Geochemistry, 20*, 2303–2314.
Sasaki, K., Nakano, H., Wilopo, W., Miura, Y., & Hirajima, T. (2009). Sorption and speciation of arsenic by zero-valent iron. *Colloids and Surfaces A: Physicochemical and Engineering Aspects, 347*, 8–17.
Schreiber, M. E., Simo, J. A., & Freiberg, P. G. (2000). Stratigraphic and geochemical controls on naturally occurring arsenic in groundwater, eastern Wisconsin, USA. *Hydrogeology Journal, 8*, 161–176.
Sharma, A. K., Kumar, R., Mittal, S., Hussain, S., Arora, M., Sharma, R. C., & Babu, J. N. (2015). In situ reductive regeneration of zerovalent iron nanoparticles immobilized on cellulose for atom efficient Cr(vi) adsorption. *RSC Advances, 5*, 89441–89446.
Shi, L.-N., Lin, Y.-M., Zhang, X., & Chen, Z.-l. (2011). Synthesis, characterization and kinetics of bentonite supported NZVI for the removal of Cr(VI) from aqueous solution. *Chemical Engineering Journal, 171*, 612–617.
Sleiman, N., Deluchat, V., Wazne, M., Mallet, M., Courtin-Nomade, A., Kazpard, V., & Baudu, M. (2016). Phosphate removal from aqueous solution using ZVI/sand bed reactor: Behavior and mechanism. *Water Research, 99*, 56–65.
Slovák, P., Malina, O., Kašlík, J., Tomanec, O., Tuček, J., Petr, M., Filip, J., Zoppellaro, G., & Zbořil, R. (2016). Zero-valent iron nanoparticles with unique spherical 3D architectures encode superior efficiency in copper entrapment. *ACS Sustainable Chemistry & Engineering, 4*, 2748–2753.
Smedley, P. L., & Kinniburgh, D. G. (2002). A review of the source, behaviour and distribution of arsenic in natural waters. *Applied Geochemistry, 17*, 517–568.
Statham, T. M., Mason, L. R., Mumford, K. A., & Stevens, G. W. (2015a). The specific reactive surface area of granular zero-valent iron in metal contaminant removal: Column experiments and modelling. *Water Research, 77*, 24–34.

Statham, T. M., Mumford, K. A., Rayner, J. L., & Stevens, G. W. (2015b). Removal of copper and zinc from ground water by granular zero-valent iron: A dynamic freeze–thaw permeable reactive barrier laboratory experiment. *Cold Regions Science and Technology, 110*, 120–128.

Stearns, D. M., Silveira, S. M., Wolf, K. K., & Luke, A. M. (2002). Chromium(III) tris(picolinate) is mutagenic at the hypoxanthine (guanine) phosphoribosyltransferase locus in Chinese hamster ovary cells. *Mutation Research/Genetic Toxicology and Environmental Mutagenesis, 513*, 135–142.

Su, C., & Puls, R. W. (2001). Arsenate and arsenite removal by zerovalent iron: Kinetics, redox transformation, and implications for in situ groundwater remediation. *Environmental Science & Technology, 35*, 1487–1492.

Tanboonchuy, V., Hsu, J. C., Grisdanurak, N., & Liao, C. H. (2011). Gas-bubbled nano zero-valent iron process for high concentration arsenate removal. *Journal of Hazardous Matererials, 186*, 2123–2128.

Tang, C., Huang, Y. H., Zeng, H., & Zhang, Z. (2014a). Promotion effect of Mn2+ and Co2+ on selenate reduction by zero-valent iron. *Chemical Engineering Journal, 244*, 97–104.

Tang, C., Huang, Y. H., Zeng, H., & Zhang, Z. (2014b). Reductive removal of selenate by zero-valent iron: The roles of aqueous Fe2+ and corrosion products, and selenate removal mechanisms. *Water Research, 67*, 166–174.

Teng, H., Xu, S., Zhao, C., Lv, F., & Liu, H. (2013). Removal of hexavalent chromium from aqueous solutions by sodium dodecyl sulfate stabilized nano zero-valent iron: a kinetics, equilibrium, thermodynamics study. *Seperation Science and Technology, 48*, 1729–1737.

Thekkae Padil, V. V., Filip, J., Suresh, K. I., Waclawek, S., & Cernik, M. (2016). Electrospun membrane composed of poly[acrylonitrile-co-(methyl acrylate)-co-(itaconic acid)] terpolymer and ZVI nanoparticles and its application for the removal of arsenic from water. *RSC Advances, 6*, 110288–110300.

Thinh, N. N., Hanh, P. T. B., Ha, L. T. T., Anh, L. N., Hoang, T. V., Hoang, V. D., Dang, L. H., Khoi, N. V., & Lam, T. D. (2013). Magnetic chitosan nanoparticles for removal of Cr(VI) from aqueous solution. *Materials Science and Engineering: C, 33*, 1214–1218.

Tiberg, C., Kumpiene, J., Gustafsson, J. P., Marsz, A., Persson, I., Mench, M., & Kleja, D. B. (2016). Immobilization of Cu and As in two contaminated soils with zero-valent iron – Long-term performance and mechanisms. *Applied Geochemistry, 67*, 144–152.

Triszcz, J. M., Porta, A., & Einschlag, F. S. G. (2009). Effect of operating conditions on iron corrosion rates in zero-valent iron systems for arsenic removal. *Chemical Engineering Journal, 150*, 431–439.

Tuček, J., Prucek, R., Kolařík, J., Zoppellaro, G., Petr, M., Filip, J., Sharma, V. K., & Zbořil, R. (2017). Zero-valent iron nanoparticles reduce arsenites and arsenates to As(0) firmly embedded in Core–Shell superstructure: Challenging strategy of arsenic treatment under anoxic conditions. *ACS Sustainable Chemistry & Engineering, 5*, 3027–3038.

Tyrovola, K., Peroulaki, E., & Nikolaidis, N. P. (2007). Modeling of arsenic immobilization by zero valent iron. *European Journal of Soil Biology, 43*, 356–367.

Wang, C., Luo, H., Zhang, Z., Wu, Y., Zhang, J., & Chen, S. (2014). Removal of As(III) and As (V) from aqueous solutions using nanoscale zero valent iron-reduced graphite oxide modified composites. *Journal of Hazardous Materials, 268*, 124–131.

Wang, X., Cong, S., Wang, P., Ma, J., Liu, H., & Ning, P. (2017). Novel green micelles Pluronic F-127 coating performance on nano zero-valent iron: Enhanced reactivity and innovative kinetics. *Separation and Purification Technology, 174*, 174–182.

Watts, J. F., & Wolstenholme, J. (2003). *An introduction to surface analysis by XPS and AES*. Wiley.

Wen, Z., Zhang, Y., & Dai, C. (2014). Removal of phosphate from aqueous solution using nanoscale zerovalent iron (NZVI). *Colloids and Surfaces A: Physicochemical and Engineering Aspects, 457*, 433–440.

Weng, C.-H., Lin, Y.-T., Lin, T. Y., & Kao, C. M. (2007). Enhancement of electrokinetic remediation of hyper-Cr(VI) contaminated clay by zero-valent iron. *Journal of Hazardous Materials, 149*, 292–302.

Wu, P., Li, S., Ju, L., Zhu, N., Wu, J., Li, P., & Dang, Z. (2012). Mechanism of the reduction of hexavalent chromium by organo-montmorillonite supported iron nanoparticles. *Journal of Hazardous Materials, 219*(220), 283–288.

Wu, L., Liao, L., Lv, G., & Qin, F. (2015). Stability and pH-independence of nano-zero-valent iron intercalated montmorillonite and its application on Cr(VI) removal. *Journal of Contaminant Hydrology, 179*, 1–9.

Xi, Y., Mallavarapu, M., & Naidu, R. (2010). Reduction and adsorption of Pb2+ in aqueous solution by nano-zero-valent iron—A SEM, TEM and XPS study. *Materials Research Bulletin, 45*, 1361–1367.

Xiao, S., Ma, H., Shen, M., Wang, S., Huang, Q., & Shi, X. (2011). Excellent copper(II) removal using zero-valent iron nanoparticle-immobilized hybrid electrospun polymer nanofibrous mats. *Colloids and Surfaces A: Physicochemical and Engineering Aspects, 381*, 48–54.

Yan, W., Ramos, M. A. V., Koel, B. E., & Zhang, W.-X. (2012a). As(III) sequestration by iron nanoparticles: Study of solid-phase redox transformations with X-ray photoelectron spectroscopy. *The Journal of Physical Chemistry C, 116*, 5303–5311.

Yan, W., Vasic, R., Frenkel, A. I., & Koel, B. E. (2012b). Intraparticle reduction of Arsenite (As (III)) by nanoscale Zerovalent Iron (NZVI) investigated with in situ X-ray absorption spectroscopy. *Environmental Science & Technology, 46*, 7018–7026.

Yuan, Y., Marshall, G., Ferreccio, C., Steinmaus, C., Selvin, S., Liaw, J., Bates, M. N., & Smith, A. H. (2007). Acute myocardial infarction mortality in comparison with lung and bladder cancer mortality in arsenic-exposed region II of Chile from 1950 to 2000. *American Journal of Epidemiology, 166*, 1381–1391.

Zboril, R., Andrle, M., Oplustil, F., Machala, L., Tucek, J., Filip, J., Marusak, Z., & Sharma, V. K. (2012). Treatment of chemical warfare agents by zero-valent iron nanoparticles and ferrate(VI)/(III) composite. *Journal of Hazardous Materials, 211–212*, 126–130.

Zhang, P., Tao, X., Li, Z., & Bowman, R. S. (2002). Enhanced perchloroethylene reduction in column systems using surfactant-modified zeolite/zero-valent iron pellets. *Environmental Science & Technology, 36*, 3597–3603.

Zhang, Y., Amrhein, C., & Frankenberger, W. T., Jr. (2005). Effect of arsenate and molybdate on removal of selenate from an aqueous solution by zero-valent iron. *Science of the Total Environment, 350*, 1–11.

Zhang, Y.-Y., Jiang, H., Zhang, Y., & Xie, J.-F. (2013). The dispersity-dependent interaction between montmorillonite supported NZVI and Cr(VI) in aqueous solution. *Chemical Engineering Journal, 229*, 412–419.

Zhao, X., Liu, W., Cai, Z., Han, B., Qian, T., & Zhao, D. (2016). An overview of preparation and applications of stabilized zero-valent iron nanoparticles for soil and groundwater remediation. *Water Research, 100*, 245–266.

Zhou, J., Ren, F., Wu, W., Zhang, S., Xiao, X., Xu, J., & Jiang, C. (2012). Controllable synthesis and catalysis application of hierarchical PS/Au core-shell nanocomposites. *Journal of Colloid and Interface Science, 387*, 47–55.

Zhou, Q., Li, J., Wang, M., & Zhao, D. (2016). Iron-based magnetic nanomaterials and their environmental applications. *Critical Reviews in Environmental Science and Technology, 46*, 783–826.

Zhu, H., Jia, Y., Wu, X., & Wang, H. (2009). Removal of arsenic from water by supported nano zero-valent iron on activated carbon. *Journal of Hazardous Materials, 172*, 1591–1596.

Zou, Y., Wang, X., Khan, A., Wang, P., Liu, Y., Alsaedi, A., Hayat, T., & Wang, X. (2016). Environmental remediation and application of nanoscale zero-valent iron and its composites for the removal of heavy metal ions: A review. *Environmental Science & Technology, 50*, 7290–7304.

Chapter 5
Colloidal and Surface Science and Engineering for Bare and Polymer-Modified NZVI Applications: Dispersion Stability, Mobility in Porous Media, and Contaminant Specificity

Tanapon Phenrat, Gregory V. Lowry, and Peyman Babakhani

Abstract This chapter summarizes the fundamentals of colloidal and surface science for understanding poor deliverability and mobility of bare NZVI in the subsurface for in situ remediation. The role of three factors, namely, intrinsic magnetic attraction of NZVI, high particle concentration for remedial application, and unfavorable environmental conditions, on the poor NZVI mobility is fundamentally explained. Moreover, this chapter also describes the state of the art in using polymeric surface modification to provide NZVI characteristics needed for effective in situ remediation (i.e., enhancing NZVI dispersion stability and mobility in porous media and maximizing their affinity to target dense nonaqueous phase liquid source zones). It explains the physicochemical reasons for how polymeric surface modification can overcome dispersion instability, extensive deposition, and contaminant non-targetability.

T. Phenrat (✉)
Department of Civil Engineering, Environmental Engineering Program, Naresuan University, Phitsanulok, Thailand

Center of Excellence for Sustainability of Health, Environment and Industry (SHEI), Faculty of Engineering, Naresuan University, Phitsanulok, Thailand

G. V. Lowry
Center for Environmental Implications of Nanotechnology (CEINT), Durham, NC, USA

Department of Civil & Environmental Engineering, Carnegie Mellon University, Pittsburgh, PA, USA

P. Babakhani
Center for Engineering Sustainability, School of Engineering, University of Liverpool, Liverpool, Merseyside, UK

Institute of Environmental Engineering, National Chiao Tung University, Hsinchu City, Taiwan

T. Phenrat, G. V. Lowry (eds.), *Nanoscale Zerovalent Iron Particles for Environmental Restoration*, https://doi.org/10.1007/978-3-319-95340-3_5

Keywords Nanoscale zerovalent iron · Polymer surface modification · Aggregation · Contaminant targeting · Colloidal science · Dispersion stability · DLVO

5.1 Why Does Colloidal and Surface Science and Engineering Matter for NZVI Applications?

After first introduced in 1996 (Wang and Zhang 1997), nanoscale zerovalent iron (NZVI) has been a promising remediation agent for a broad range of researchers and practitioners who want to explore a source-zone remediation route. Similar to bulk zerovalent iron (ZVI) powder, NZVI is capable of reductive dechlorination of chlorinated organics (Chap. 3) and immobilization of metals via reductive precipitation and surface complexation (Chap. 4). Nevertheless, compared to the ZVI that is typically used for contaminated plume interception via a permeable reactive barrier (PRB), the small size of NZVI gives plenty of opportunity for in situ active treatment. First, being "nano" results in a high surface area, leading to higher contaminant degradation rates or increased adsorption capacity per mass of the remediation agents compared to the bulk material. The small size of nanomaterials also offers the *potential* for their injection into the subsurface for in situ remediation (Tratnyek and Johnson 2006; Lowry 2007). This feature is particularly attractive for treating deep contaminant source areas (e.g., dense nonaqueous phase liquid or DNAPL), which are costly to remediate and technically challenging, and it is difficult (if possible) to meet cleanup targets in a reasonable amount of time (Miller et al. 1990; Leeson et al. 2003; US.EPA 2003; Grant and Gerhard 2007).

Originally, bare NZVI was proposed for in situ remediation. The first pilot remediation using bare NZVI was conducted in 2001 (Elliott and Zhang 2001). Nevertheless, high reactivity and small size alone are insufficient to make NZVI a good in situ remediation agent. The NZVI must also be readily dispersible in water such that it can migrate through water-saturated porous media to the contaminated area and must be able to target specific contaminants, such as DNAPL (see Fig. 5.1). Not long after laboratory and field trials, researchers realized the limited mobility of bare NZVI in saturated porous media (i.e., practical transport distances of only a few centimeters or less for bare, unsupported particles; Schrick et al. 2004; Saleh et al. 2007).

There are two explanations for the poor transport of NZVI in water-saturated porous media. First, NZVI can substantially be filtered from the aqueous phase by attachment to aquifer materials (Schrick et al. 2004; Saleh et al. 2007; Phenrat et al. 2009a). Second, agglomeration and subsequent gelation (formation of a particle network) could cause pore plugging, which limits transport (Phenrat et al. 2007; 2009a; Saleh et al. 2007); see some modeling approaches in Chap. 6. In addition, bare NZVI has no affinity to DNAPL, making it incapable of NAPL targeting, which limits the source-zone remediation potential since NZVI may just flow past the source zone, even when being injected close to the source-zone vicinity (Baumann et al. 2005).

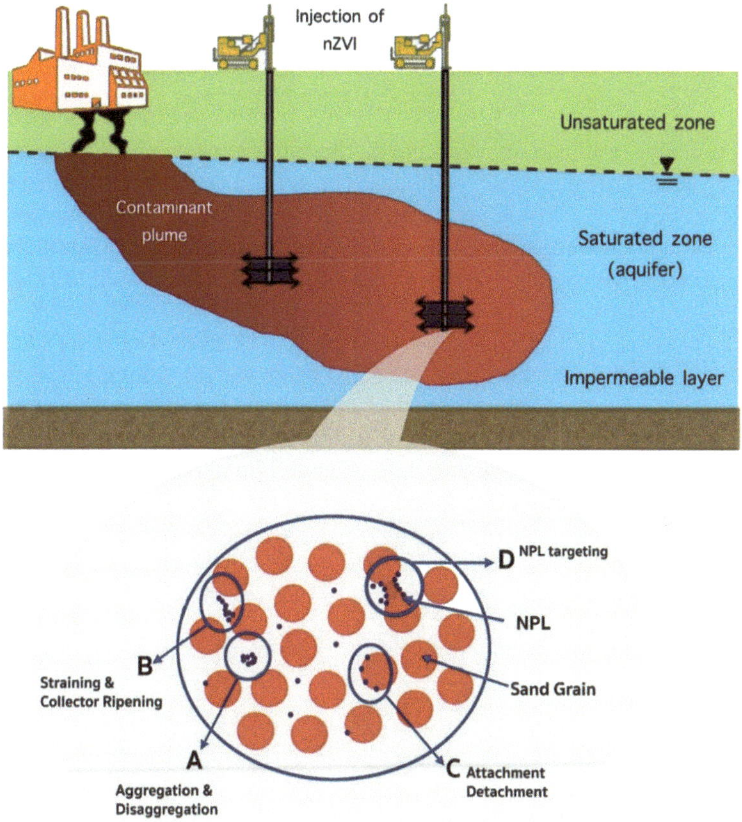

Fig. 5.1 Schematic of the colloidal and surface phenomena control of subsurface remediation using NZVI injection. (Modified from Lowry 2007)

This chapter will utilize colloidal and surface science to illustrate these technical difficulties. Obviously, enhancing NZVI dispersion stability and mobility in porous media and maximizing their affinity to target NAPL are necessary to ensure the success of in situ groundwater remediation using NZVI. Polymeric surface modification (Saleh et al. 2005, 2007; Phenrat et al. 2008; 2009b) is one of the most attractive alternatives to improve performance of NZVI regarding these aspects. Thus, this chapter describes the state of the art in using colloidal and surface engineering via polymeric surface modification of NZVI to provide NZVI characteristics needed for effective in situ remediation. It explains the physicochemical reasons for how polymeric surface modification can overcome the following challenges: dispersion instability, extensive deposition, and contaminant non-targetability.

5.2 Why Does Bare NZVI Extensively Agglomerate and Immobilize in the Subsurface Environment?

Excessive agglomeration of bare NZVI is due to the combined effects of three factors, namely, magnetic attraction, high concentration for remedial application, and unfavorable environmental conditions. The synergetic roles of the three factors might be explored by considering the rate of change of particle number concentration during agglomeration. Such a model is known as the population balance equation (PBE), which was originally developed by Smoluchowski (1917) in the context of coagulation (Smoluchowski 1917; Chandrasekhar 1943).

Other discretization approaches that are deemed to be more suitable for sectioning the particle size dimension grid rather than considering the full ranges of particle sizes were developed by Hounslow et al. (1988), Lister et al. (1995), and Kumar and Ramkrishna (1996a, b). One of the most widely accepted models of this family is known as the fixed pivot (FP) approach (Kumar and Ramkrishna 1996a), which can preserve two characteristics of the formulation integral, such as mass and population, and is claimed to be capable of being used on any arbitrary particle size grid. This model is given as:

$$\frac{dN_k}{dt} = \sum_{\substack{j,i \\ x_{k-1} \le (x_j + x_i) \le x_{k+1}}}^{j \ge k} \left[1 - \frac{1}{2}\delta_{j,i}\right] \eta_k \alpha_{x_j,x_i} \beta_{x_j,x_i} N_j N_i - N_k \sum_i \alpha_{x_k,x_i} \beta_{x_k,x_i} N_i \tag{5.1}$$

where N_k is the particle number concentration of agglomerates consisting of k primary particles [L^{-3}], β is the collision frequency, α is collision efficiency factor, x is representative volume of each size class in the grid, δ is Kronecker delta, and η_k is:

$$\eta_k = \begin{cases} \dfrac{x_{k+1} - (x_j + x_i)}{x_{k+1} - x_k}, & x_k \le (x_j + x_i) \le x_{k+1} \\ \dfrac{(x_j + x_i) - x_{k-1}}{x_k - x_{k-1}}, & x_{k-1} \le (x_j + x_i) \le x_k \end{cases} \tag{5.2}$$

Moreover, NZVI's magnetic attraction and unfavorable environmental conditions promote NZVI agglomeration via increasing the collision efficiency factor (α), while high NZVI particle concentration for remedial application facilitates the agglomeration via increasing the collision frequency factor (β) and decreasing the number of small size N_j while increasing the number of larger size N_i. A detailed discussion of each factor follows.

5.2.1 Magnetic Attraction

First, NZVI's magnetic attraction directly accelerates NZVI agglomeration, which subsequently promotes NZVI deposition in saturated porous media. When two surfaces (i.e., NZVI particle-particle surfaces or NZVI particle surface of aquifer material) approach one another, if attractive forces outweigh repulsive forces, the two surfaces will stick once collided. For natural colloids and various nanomaterials, the probability of agglomeration and deposition is typically modeled by classical DLVO theory (Evans and Wennerstrom 1999).

In this theory, the van der Waals forces (V_{vdW}) are the primary attractive force, while repulsive forces are derived from the electrostatic double layer (V_{ES}) (Fig. 5.2). The V_{vdW} originates from the sum of interactions between atomic and molecular constitutes of the two approaching particles or surfaces (including permanent dipoles, induced dipoles, and fluctuating dipoles), while V_{ES} originates from the excess osmotic pressure of ionic constitutes surrounding the charged surface when they are overlapping (Fig. 5.3). The V_{vdW} attractive force between two spherical particles of radius R_1 and R_2 or between a spherical particle of radius R and a flat surface (a representative of a large collector grain (aquifer material) relative to a nanoparticle) is given in Eqs. (5.3) and (5.4), respectively (Evans and Wennerstrom 1999):

$$V_{\substack{\text{vdw} \\ \text{sphere-} \\ \text{sphere}}}(h) = -\frac{A}{6h}\frac{R_1 R_2}{R_1 + R_2} \tag{5.3}$$

$$V_{\substack{\text{vdw} \\ \textit{sphere-} \\ \textit{wall}}}(h) = -\frac{AR}{6h}, \tag{5.4}$$

where A is the Hamaker constant, which is 1×10^{-19} N·m for NZVI (Phenrat et al. 2007), and h (m) is the separation distance between two interacting surfaces. The V_{vdW} decays with increasing h and increases in magnitude and extent as the particle size increases (Fig. 5.4). This attractive energy promotes agglomeration and deposition.

Electrostatic repulsion between two identical particles of radius R and between a particle and a flat surface is given in Eqs. (5.5) and (5.6), respectively (Evans and Wennerstrom 1999):

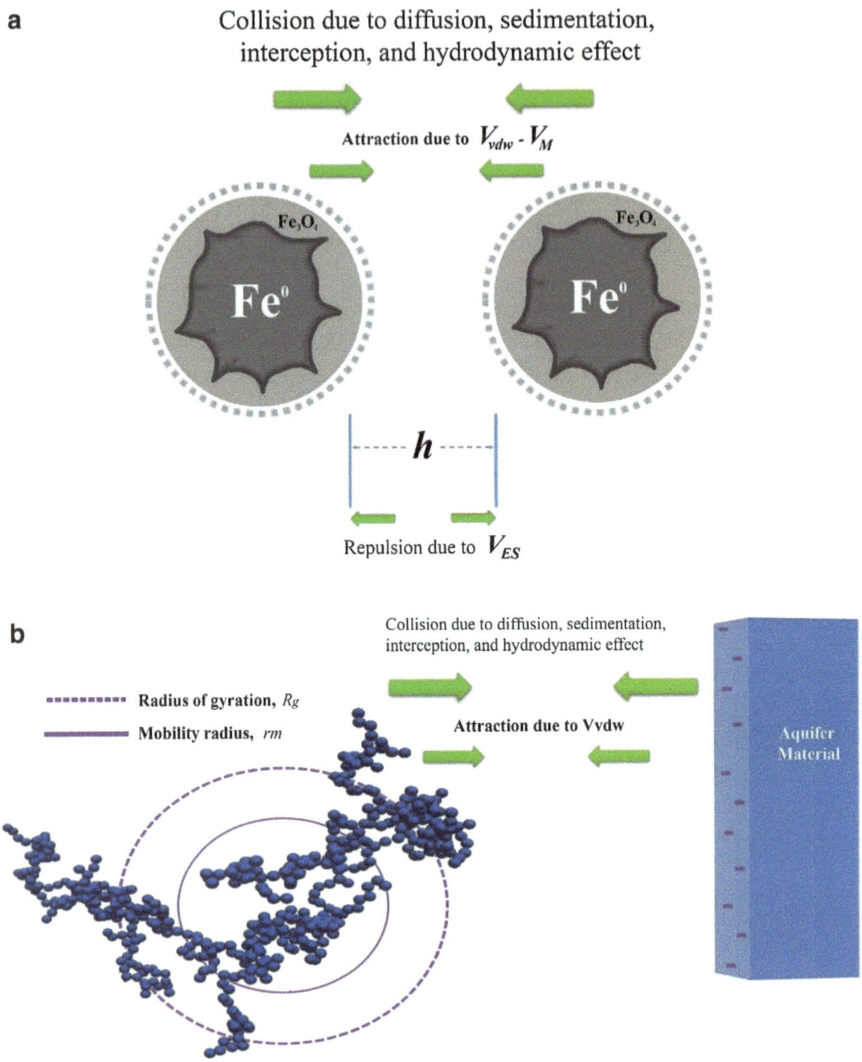

Fig. 5.2 Components of classical and extended DLVO theory (with V_M) between (**a**) bare particle-particle and (**b**) bare cluster-aquifer material

$$V_{\substack{\text{ES}\\\text{sphere-}\\\text{sphere}}} = 2\pi\varepsilon_r\varepsilon_0 R\zeta_1^2 \ln\left[1+e^{-\kappa h}\right] \tag{5.5}$$

$$V_{\substack{\text{ES}\\\text{sphere-}\\\text{wall}}} = \pi\varepsilon_r\varepsilon_0 R\left[2\zeta_1\zeta_2 \ln\left(\frac{1+e^{-\kappa h}}{1-e^{-\kappa h}}\right) + \left(\zeta_1^2+\zeta_2^2\right)\ln\left(1-e^{-2\kappa h}\right)\right], \tag{5.6}$$

Fig. 5.3 Origins of V_{vdW} and V_{ES}

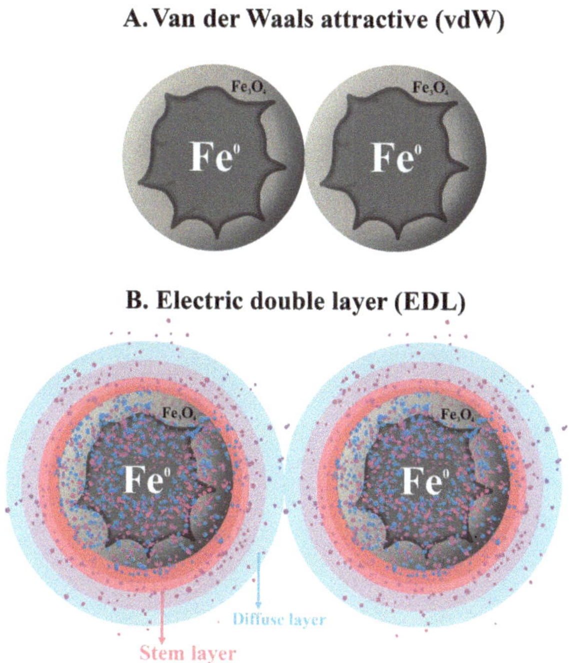

where ε_r is the relative dielectric constant of the liquid, ε_0 is the permittivity of the vacuum, and ζ_1 and ζ_2 are the zeta potentials of a particle and a collector, respectively. The zeta potential of bare NZVI is typically −32 mV at 1 mM $NaHCO_3$, while κ is the inverse Debye length, which, for symmetrical (*z*-*z*) electrolytes, can be expressed as follows (Evans and Wennerstrom 1999):

$$\kappa = \sqrt{\frac{e^2 \Sigma n_i z_i^2}{\varepsilon_r \varepsilon_0 k_B T}}, \tag{5.7}$$

where e is the electron charge, n_i is the number concentration of ion i in the bulk solution, and z_i is the valence of ion i. The V_{ES} decays exponentially with separation distance (h) and decreases in magnitude as the Debye length ($1/\kappa$) decreases. In addition, V_{ES} decreases as the particle size decreases (Fig. 5.4). This repulsive energy inhibits aggregation and deposition. Because of their small size, the energy barrier for NZVI to resist aggregation and deposition may be less than that of micron-sized particles of the same surface charge.

The total force (V_T) according to the DLVO theory is the sum of V_{ES} and V_{vdW}, where V_{ES} is a short-ranged force in comparison to V_{vdW}. Thus, at conditions favorable for agglomeration (high ionic strength and large particle size and low surface charge with high Hamaker constant), V_{vdW} outweighs V_{ES} at any distance, leading to agglomeration at the primary minimum with no energy barrier ($V_{T,\max} \leq 0$)

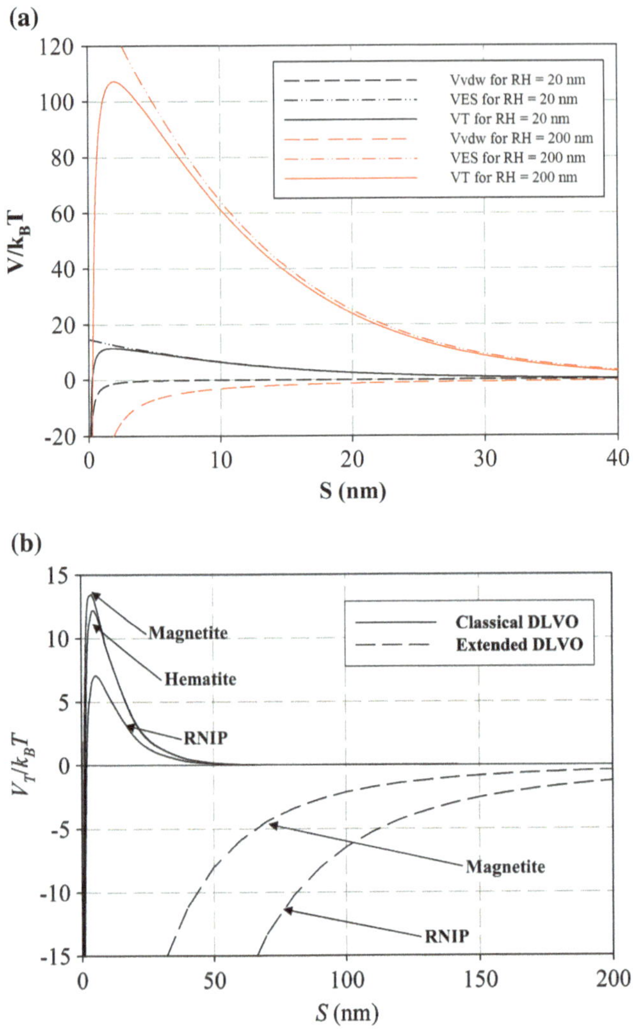

Fig. 5.4 (**a**) Classical DLVO theory for NZVI at two particle sizes ($R_H = 20$ and 200 nm). (**b**) Classical and extended DLVO theories for NZVI (RNIP), magnetite, and hematite at $R_H = 20$ nm (Phenrat et al. 2007). (Reprinted with permission from Phenrat et al. 2007. Copyright (2007) American Chemical Society)

to resist agglomeration. This condition is called mass transfer-limited agglomeration (or diffusion-limited agglomeration (DLA)), meaning that any collision of the two particles will result in agglomeration (see Fig. 5.5) (Elimelech et al. 1998a; Zhu et al. 2014). On the other hand, at conditions unfavorable for agglomeration (high surface charge of nanoparticles, low ionic strength, and small particles with low Hamaker constant), V_{ES} at some closer distance outweighs V_{vdW}, creating an energy barrier to encounter agglomeration (Fig. 5.4a). This condition is called reaction-limited

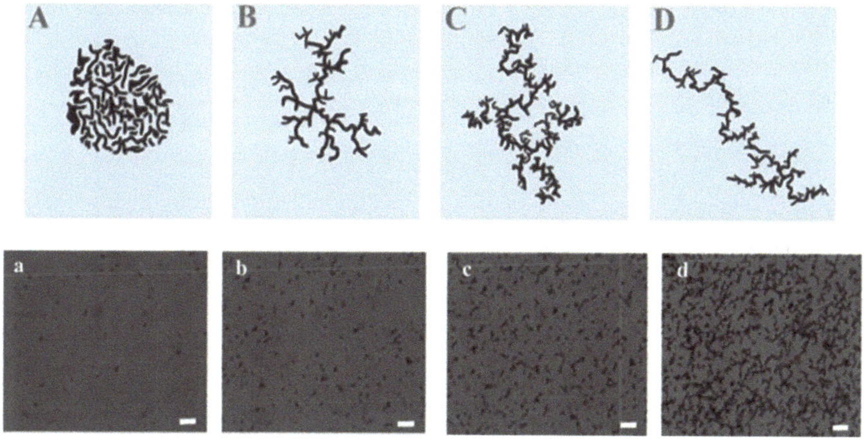

Fig. 5.5 Diffusion-limited agglomeration (DLA) and diffusion-limited cluster-cluster aggregation (DLCCA) regimes of bare NZVI (Phenrat et al. 2007). Reprinted with permission from Phenrat et al. 2007. Copyright (2007) American Chemical Society

agglomeration (Fig. 5.5), meaning that the agglomeration will happen only when the collision results in a collision energy greater than the energy barrier ($V_{T,\,max}$) to yield the attachment of the two surfaces. Using kinetics theory, the flow rate of particles aggregating on a single particle F can be estimated as in Eq. (5.8):

$$F = -\frac{8\pi D N_0}{\int_{2R}^{\infty} \exp\left(\frac{V_T}{k_B T}\right)\frac{dh}{h^2}} \tag{5.8}$$

where D is diffusion coefficient of particles, N_0 is the initial particle concentration in the dispersion, and h is the distance between the particles aggregating on a single particle. For $V_{T,\,max} = 0$ (i.e., DLA), $F = 8\pi D N_0(2R)$ where R is the radius of the particles.

On the other hand, for RLA, one can calculate the stability ratio (W) as defined by Eq. (5.9) (Baalousha 2017):

$$W \sim \frac{1}{2\kappa R}\exp\left(\frac{V_{T,\,max}}{k_B T}\right) \tag{5.9}$$

Applying classical DLVO theory to bare NZVI particle-particle interaction and assuming spherical particles of the radius = 20 nm, an energy barrier of ~$7.0 k_B T$ is predicted (Fig. 5.4a), where k_B is the Boltzmann constant (1.38×10^{-23} m^2 kg s^{-2} K^{-1}). This energy barrier corresponds to a stability ratio of ~90 (i.e., agglomeration of particles with this energy barrier is predicted to be 90 times slower than for particles aggregating under DLA, which has no energy barrier), suggesting that the dispersions of bare NZVI should be stable.

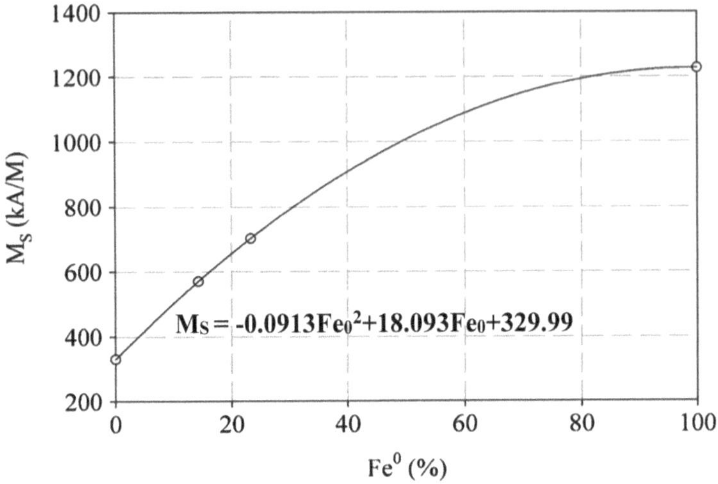

Fig. 5.6 Empirical relationship between the saturation magnetization *Ms* and the Fe^0 content of the particles. The curve was determined from *Ms* measurements on magnetite, on two RNIP particles having different Fe^0 content, and on elemental iron (100% Fe^0) (Phenrat et al. 2008). (Reprinted with permission from Phenrat et al. 2008. Copyright (2008) Springer Nature)

Nevertheless, this theoretical prediction contrasts with the actual observation of rapid agglomeration of bare NZVI. This is because, besides V_{vdW} and V_{ES}, NZVI appears to have an intrinsic permanent magnetic dipole moment, even in the absence of an applied magnetic field (Phenrat et al. 2007), which plays a substantial role in agglomeration and deposition. When particle dipoles are oriented in a head-to-tail configuration, the maximum magnetic attraction energy (V_{M}) can be expressed as (Phenrat et al. 2007):

$$V_{\mathrm{M}} = \frac{-8\pi\mu_0 M_{\mathrm{s}}^2 R^3}{9\left(\frac{h}{R}+2\right)^3}, \tag{5.10}$$

where μ_0 is the permeability of the vacuum and M_{s} is the saturation magnetization of the particles. In addition, M_{s} is a function of the Fe^0 content of the particles (Fig. 5.6). Thus, V_{M} is a function of NZVI age (freshly prepared or aged with interaction with water and air). Considering the magnetic attraction, the extended DLVO theory can successfully predict the instability of bare NZVI as experimentally observed (Fig. 5.4b). Considering Eq. (5.10), the size of magnetic nanoparticles, such as NZVI, significantly affects aggregation and deposition (Phenrat et al. 2007) because V_{M} increases in magnitude with a particle radius to the sixth power. The NZVI is polydisperse, and particle size (hydrodynamic radius (R_{H})) ranges from 5 to 40 nm (Phenrat et al. 2009a). Large NZVI particles in the dispersion may behave as a magnetic seeding, catalyzing agglomeration and leading to the formation of large NZVI clusters. This agrees with the two-phase pattern of NZVI agglomeration. In

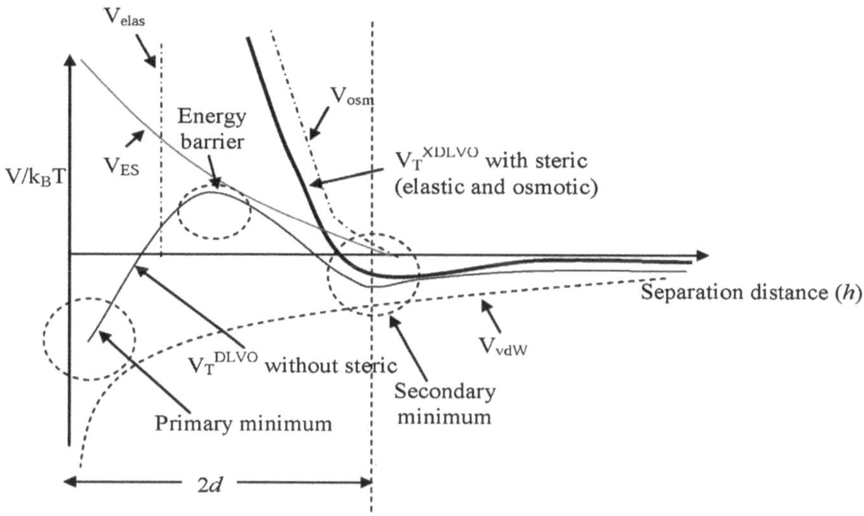

Fig. 5.7 Hypothetical interaction energy profiles showing various components of the particle-collector/particle-particle interaction energy profile including total interaction energy with and without the electrosteric repulsive forces afforded by adsorbed or grafted polyelectrolytes (V_{osm} and V_{elas}) (Phenrat and Lowry 2009)

the first state, the individual NZVI (radius of 5–40 nm with an average radius = 20 nm) rapidly agglomerates to form discrete, micron-sized aggregates (~1.2 μm). This is followed by the linking of each of these aggregates into chains or gelation (15–50 μm) based on the diffusion-limited cluster-to-cluster aggregation (DLCCA) behavior (Fig. 5.5). This leads to bare NZVI dispersion instability, as the gravitational flux of NZVI cluster settling outweighs their diffusional flux due to Brownian motion.

Magnetic-promoting NZVI agglomeration also subsequently accelerates NZVI attachment to aquifer material. Although the interaction between NZVI cluster and surface of aquifer materials (modeled by silica sand) involves no magnetic attraction, the much larger size of NZVI clusters in comparison to individual NZVI particles results in favorable particle-surface attachment via the primary and secondary minimum depending on the increased sizes as a result of increased V_{vdW}. Figure 5.7 shows examples of total energy of interaction between bare NZVI particles and NZVI clusters and a flat surface. The sum of the interaction energy results in an attractive primary minimum where NZVI clusters are deposited (Phenrat et al. 2007). Typically, the attachment under this attractive primary minimum is irreversible and could promote pore clogging (see Chap. 6).

5.2.2 High NZVI Particle Concentration

The subsurface restoration requires the delivery of a high concentration of NZVI favorable for NZVI agglomeration. While the past studies of colloidal and nanoparticle mobility in porous media were typically conducted at low (natural-relevant) particle concentration (<30 mg/L) to assess the risk of contaminant exposure via contaminant-bound colloids, the studies of NZVI mobility in porous media must be based on engineering design to deliver relatively high NZVI particle concentration into the subsurface to degrade or immobilize contaminants of concern. This is because several recent studies estimated that the amount of NZVI to completely degrade TCE is 1:1 by mass (Liu et al. 2005). Similarly, to completely immobilize metals, an NZVI-to-metal ratio of 67 (He et al. 2010) or 33 (Busch et al. 2015) has been used in field trials. Thus, NZVI is typically delivered into the subsurface at the concentration of 1–20 g/L, and an average injection concentration of ~6 g/L has been found, considering several field studies using NZVI (Elliott and Zhang 2001; Henn and Waddill 2006; Bennett et al. 2010; He et al. 2010; Wei et al. 2012; Johnson et al. 2013; Jordan et al. 2013; Köber et al. 2014; Kocur et al. 2014). This corresponds to roughly 5×10^{20} to 1×10^{22} individual NZVI particles (radius = 40 nm) per liter. Consequently, NZVI can collide with one another in the dispersion or in the pore water of porous media much easier than in the case of natural colloidal particles or incidental nanoparticles at a low particle concentration (<30 mg/L or $<1.6 \times 10^{19}$ particles/L). This can be logically explained by the concept of mean free path (λ), which is defined as the average distance between collisions for two adjacent nanoparticles (Eq. 5.11):

$$\lambda = \frac{1}{4\sqrt{2}\pi n_{\mathrm{v}} R^2}, \tag{5.11}$$

where n_{v} is the number of particles per unit volume and R is the radius of the individual or primary nanoparticles. Thus, the more concentrated the NZVI dispersion, the greater the n_{v}, the smaller the λ, and the greater the chance of collision. This high change of particle collision superimposes the attractive sum of forces dominated by V_{M}, which will lead to extensive NZVI agglomeration that is unfavorable for subsurface delivery.

The frequency of collisions between NZVI particles and/or agglomerates depends on the structure and porosity of the agglomerates since the hydrodynamic drag on porous agglomerates is significantly different from that of impermeable, individual particles. The particles may collide with one another via three main mechanisms: perikinetic aggregation, orthokinetic aggregation, and differential sedimentation. The perikinetic mechanism is caused by the Brownian motion of particles. The collision frequency factor for perikinetic aggregation (Brownian motion, $\beta_{i,j}{}^{\mathrm{Br}}$) is given in Eq. (5.12):

$$\beta_{i,j}^{\mathrm{Br}} = \frac{2k_bT}{3\mu}\left(v_i^{1/D_f} + v_j^{1/D_f}\right)\left(\frac{1}{\Omega_i}v_i^{-1/D_f} + \frac{1}{\Omega_j}v_j^{-1/D_f}\right), \tag{5.12}$$

where μ is the dynamic viscosity of the suspending fluid and v_i and v_j are the collision volume of NZVI agglomerates belonging to sections i and j, respectively. In addition, D_{f} is the fractal dimension of agglomerates, and Ω_i and Ω_j are the drag force correction factors for agglomerates belonging to sections i and j, defined as the ratio of the force exerted by the fluid on a permeable agglomerate to that on an impermeable particle or agglomerate (Neale et al. 1973; Thill et al. 2001; Aziz et al. 2003; Vikesland et al. 2016):

$$\Omega = \frac{2\xi^2\left(1 - \frac{\tanh(\xi)}{\xi}\right)}{2\xi^2 + 3\left(1 - \frac{\tanh(\xi)}{\xi}\right)}, \tag{5.13}$$

where ξ is the nondimensional permeability of the porous agglomerate given as follows:

$$\xi = \frac{a}{\omega^{1/2}}, \tag{5.14}$$

where a and ω are the radius and the intrinsic permeability of agglomerates, respectively. The collision volume (v) of an agglomerate is defined by Eq. (5.15):

$$v = \frac{\pi}{6}(d_0)^{3-D_{\mathrm{f}}}(d)^{D_{\mathrm{f}}}, \tag{5.15}$$

where d_0 is the diameter of the primary particles and d is the diameter of each agglomerate or the agglomerate class diameter. In aggregation modeling, typically, the particle volume dimension is discretized as either an arithmetic series of agglomerates volumes (v_0, $2v_0$, $3v_0$, $4v_0$, …) or a geometric series of agglomerate volumes ($v_{i+1}/v_i = 2^{1/q}$, where q is the geometric grid discretization factor; Babakhani et al. 2018). The geometric series is more practical in aggregation modeling of NZVI since less computation effort is required for this discretization than that of an arithmetic series. After taking such an assumption for the volume of aggregates, the diameter of each agglomerate can be calculated from Eq. (5.15). Having the number fraction, volume, or mass concentration of each agglomerate class, the particle size distribution (PSD) can be determined for the dispersion. Given the mass concentration of each class (C_i), the average diameter (D_{geom}) of PSD can be then obtained via a geometric mean formulation (Merkus 2009; Babakhani et al. 2018):

$$D_{\text{geom}} = \exp\left(\frac{\sum_{i=1}^{k_{\max}} C_i d_i^3 \ \ln \ d_i}{\sum_{i=1}^{k_{\max}} C_i d_i^3}\right), \tag{5.16}$$

This mean diameter can be interpreted as a diameter of gyration. A relationship is available for determining the hydrodynamic diameter (D_H) based on this mean diameter (Lattuada et al. 2004).

The orthokinetic aggregation results from any motion or flow in a fluid that can cause shear stress (Eq. 5.17). The collision frequency factor for orthokinetic aggregation ($\beta_{i,j}{}^{Sh}$) is given by Babakhani et al. (2018):

$$\beta_{i,j}^{\text{Sh}} = \frac{G}{\pi} v_0^{1-(3/D_{\text{f}})} \left(\eta_{\text{c}_i}^{1/2} \ v_i^{1/D_{\text{f}}} + \eta_{\text{c}_j}^{1/2} \ v_j^{1/D_{\text{f}}}\right)^3, \tag{5.17}$$

where G is the volume-averaged fluid velocity gradient or shear rate, v_0 is the volume of individual primary particles, and η_{cfi} is the fluid collection efficiency of an agglomerate, defined as the ratio of the fluid flow moving through the agglomerate to that approaching it. This can be determined from the Brinkman equation (Thill et al. 2001; Jeldres et al. 2015):

$$\eta_{\text{c}} = 1 - \frac{d_{\text{c}}}{\xi} - \frac{c_{\text{c}}}{\xi^3}, \tag{5.18}$$

where

$$d_{\text{c}} = \frac{3}{J}\xi^3\left(1 - \frac{\tanh(\xi)}{\xi}\right), \tag{5.19}$$

$$c_{\text{c}} = -\frac{1}{J}\left(\xi^5 + 6\xi^3 - \frac{\tanh(\xi)}{\xi}\left(3\xi^5 + 6\xi^3\right)\right), \tag{5.20}$$

$$J = 2\xi^2 + 3 - 3\frac{\tanh(\xi)}{\xi}, \tag{5.21}$$

Lastly, differential sedimentation is another mechanism of aggregation that occurs when the particle sizes are different. The collision frequency factor for differential sedimentation ($\beta_{i,j}{}^{\text{DS}}$) is given by Babakhani et al. (2018):

$$\beta_{i,j}^{\text{DS}} = \frac{3}{2}\left(\frac{\pi}{6}\right)^{1/3} v_0^{2/3-2/D_{\text{f}}} \left(\eta_{\text{c}_i}^{1/2} \ v_i^{1/D_{\text{f}}} + \eta_{\text{c}_j}^{1/2} \ v_j^{1/D_{\text{f}}}\right)^2 \left|U_i - U_j\right|, \tag{5.22}$$

where g is the acceleration due to gravity and U_i and U_j are the settling velocities of agglomerates. Recently, Babakhani et al. (2018) examined four different approaches to the settling velocity of nanoparticle agglomerates and found that a power-law formulation is the best model for the settling velocity of nanoparticles:

$$\frac{U}{U_0} = \left(\frac{d}{d_0}\right)^{D_f}. \tag{5.23}$$

Both the drag force correction factor and the fluid collection of an agglomerate depend on agglomerate permeability, which in turn is a function of the solid volume fraction of the agglomerate. There are several empirical expressions for estimating agglomerate permeability (Logan and Hunt 1987; Johnson et al. 1996; Veerapaneni and Wiesner 1996; Li and Logan 2001; Thill et al. 2001). Babakhani et al. (2018) also investigated two types of these expressions in addition to the case with the assumption of no permeability for colliding agglomerates. They found that the Brinkman permeability model (Veerapaneni and Wiesner 1996; Li and Logan 2001; Thill et al. 2001) performs best for calculating the collision rates of agglomerates. This model is given as follows (Jeldres et al. 2015):

$$\omega = \frac{(d_0)^2}{72}\left(3 + \frac{3}{(1-\varphi)} - \left(\frac{8}{(1-\varphi)} - 3\right)^{\frac{1}{3}}\right), \tag{5.24}$$

where φ is porosity, which is determined from the following (Lee et al. 2000; Sterling et al. 2005; Jeldres et al. 2015):

$$\varphi = 1 - \left(\frac{d}{d_0}\right)^{D_f - 3}. \tag{5.25}$$

Using a model developed for the solution of Eq. (5.1) with the β expressions from Eqs. (5.12), (5.17), and (5.22) (Babakhani et al. 2018) and considering the following equation for sedimentation of particles, the effects of different factors on mean size, aqueous total mass concentration, and PSD can be examined:

$$\frac{dN_k}{dt} = -\frac{U_k}{Z_s} N_k, \tag{5.26}$$

where Z_s is the sedimentation depth and U_k is the settling velocity of the agglomerates in class k.

The outcomes of this investigation are shown in Fig. 5.8 for the effects of the initial concentration. According to Fig. 5.8, with the increase of the initial mass concentration of NZVI in the dispersion by three orders of magnitude, the average size of the particles, which is progressively growing, reaches approximately two orders of magnitude larger than that at low concentration. This causes a much higher sedimentation rate of agglomerates in the case of high concentration. Overall, the most important factor in triggering bare NZVI agglomeration is the amount of suspended concentration, which is dynamically changing with time.

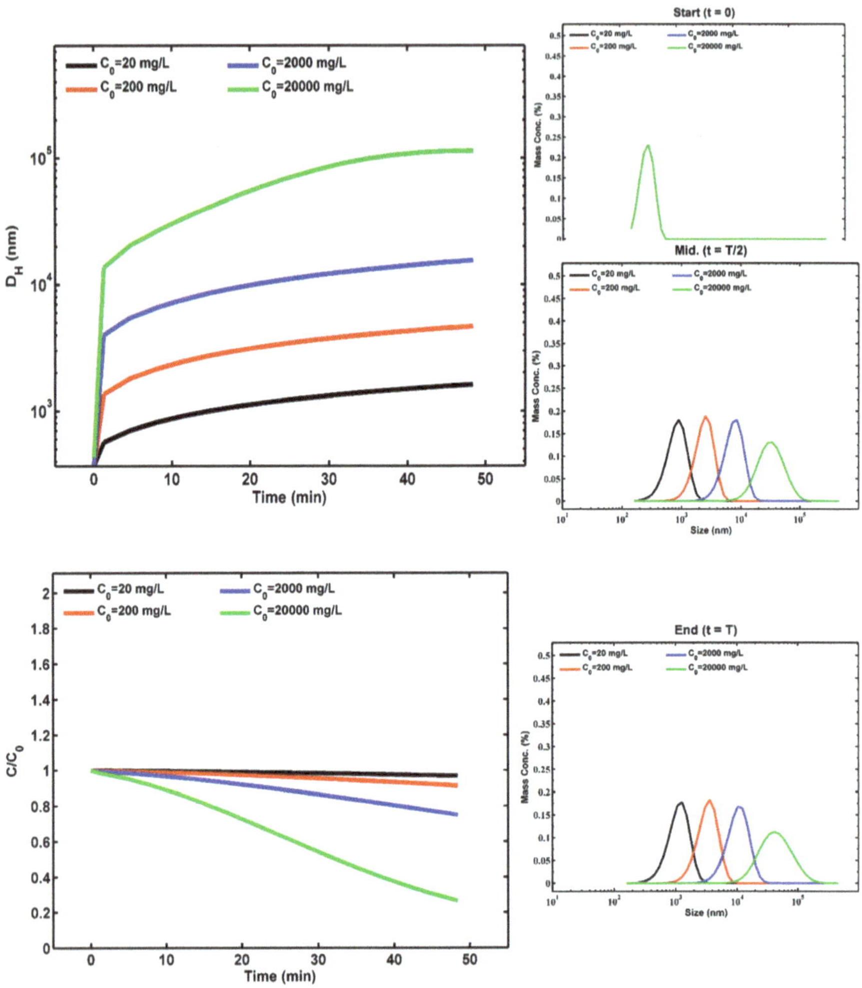

Fig. 5.8 Effect of initial mass concentration (C_0) on mean particle size growth, nondimensionalized suspended concentration, and PSD over time. The PSD graphs are shown at the beginning, middle, and end of simulation period (50 min). Other input parameters in the model are attachment efficiency $\alpha = 1$, fractal dimension $D_f = 2.0$, the geometric grid discretization factor $q = 2$, primary particle radius $a_0 = 80$ nm, particle density $\rho_p = 7874$ kg/m^3, sedimentation depth 2.3 cm, and shear rate $G = 0$

5.2.3 *Unfavorable Environmental Conditions*

While attractive forces that destabilize NZVI dispersion, such as V_{vdW} and V_M, are insensitive to environmental parameters, as the only force to battle bare NZVI agglomeration, V_{ES} is very sensitive to environmental parameters, especially pore water solution chemistry. As shown in Eqs. (5.5) and (5.6), the magnitude of V_{ES} is a function of the zeta potentials of a particle (for agglomeration) and of the zeta potential of a collector (for particle deposition in porous media). On the other hand, the extent of V_{ES} is a function of the Debye length ($1/\kappa$). The greater the zeta potential, the greater the magnitude of V_{ES}. Similarly, the greater the Debye length, the greater the extent of V_{ES}. However, in natural geochemical conditions, pore water typically contains electrolytes, such as NO^{3-}, PO_4^{-3}, Ca^{2+}, and Na^+. These ionic species can seriously decrease both magnitude and the extent of V_{ES}. Although indifferent electrolytes including Ca^{2+} and Na^+ will not substantially sorb on to the surface of NZVI, their presence at high concentrations in pore water can suppress the Debye length by the charge screening process. As shown in Eq. (5.7), multivalent ions, such as Ca^{2+} and Mg^{2+}, in groundwater are four times more effective in the screening charge of NZVI than monovalent ions, such as Na^+. This charge screening phenomenon will seriously decrease the extent of $V_{ES,}$ rendering lower resistance against agglomeration and more irreversible agglomeration and deposition. Furthermore, divalent cations, such as Ca^{2+}, in the presence of any polymer, such as natural organic matter (NOM), can bring about substantial agglomeration and deposition due to bridging (Chen and Elimelech 2006; Wang et al. 2011; Torkzaban et al. 2012). Despite that, these polymers can also have a stabilizing effect, as will be discussed later (Phenrat et al. 2010b).

Although the existence of magnetic force-driven agglomeration and subsequent deposition is deemed to play a key role in poor mobility of bare NZVI in porous media, there is a range of other factors in natural environmental conditions of soils that might contribute to this behavior. These features include various heterogeneities resulting from surface roughness (Yao et al. 1971; Ryan and Elimelech 1996; Elimelech et al. 1998b; Tufenkji and Elimelech 2004), NOM (Hotze et al. 2010; Phenrat et al. 2010b), iron oxyhydroxide coating (Tian et al. 2012; Wang et al. 2012; Liu et al. 2013), silylation (Elimelech et al. 2000), extracellular polymeric substances, biofilm (Jiang et al. 2013; Li et al. 2013; Mitzel and Tufenkji 2014; He et al. 2015; Basnet et al. 2016), and other minerals, such as clays, which have a high surface area and a highly asymmetric platelet with amphoteric behavior of the edges (having both positive and negative charges for the same geochemical conditions, leading to ambivalent interactions with NZVI at the same time; Tombacz and Szekeres 2004; Holmboe et al. 2009; Amorós et al. 2010; Chowdhury et al. 2012; Kim et al. 2012; Tsujimoto et al. 2013).

Moreover, in field conditions, porous media may involve mobile and immobile domains (Krupp et al. 1972; Van Genuchten 1981; Bradford et al. 2003; Bradford et al. 2011). The NZVI, due to its higher diffusion than larger colloids, is prone to diffuse into the immobile domains more easily and be trapped permanently or temporarily. Recent studies (Torkzaban et al. 2008; Liang et al. 2013; Molnar et al. 2015) indicate that the effect of heterogeneities, such as micro- and nano-surface roughness, is considerable for nanoparticle transport in porous media. However, a more recent study (Babakhani et al. 2017) suggested that the role of these heterogeneities is overcome by the influence of the secondary minimum energy well.

Although the research on the effect of biofilm-coated porous media is in an infancy level, current reports (Tripathi et al. 2011; Basnet et al. 2016) suggest that NZVI and other nanoparticles showed elevated retention (e.g., up to 26-fold in the presence of $CaCl_2$) in biofilm-coated porous media. In the presence of $CaCl_2$, one of the reasons for enhanced deposition of NZVI was attributed to bridging with functional groups in biofilm matrices (Basnet et al. 2016). Therefore, greater retention of bare NZVI, disregarding the agglomeration and its subsequent mechanisms, can be such surface heterogeneities. Babakhani et al. (2017) tried to lump such heterogeneities into a single parameter in a meta-analysis based on a dataset of 493 published column experiment outcomes. They found that the average sensitivity of the proposed heterogeneity parameter was even greater than parameters like dispersivity. In their investigation, Babakhani et al. (2017) considered 20 environmental factors affecting nano-sized particle transport in porous media. Their results showed that the most crucial factor in controlling effluent concentration (C/C_0) was the grain zeta potential and the most important factor in controlling the attachment rate was the nanoparticle coating concentration followed by the grain zeta potential. Therefore, surface interaction forces are extremely important in marked retention of bare NZVI in field porous media and need to be engineered for successful emplacement of this technology in contaminated subsurface environments.

5.3 What Is Polymeric Surface Modification?

Polymeric surface modification has long been receiving a great deal of attention as a means to enhance the dispersion stability of colloidal particles in various industrial applications from food processing to drilling fluid (Evans and Wennerstrom 1999). Thus, unsurprisingly, it is the very first approach ever tried for improving NZVI dispersion stability (Schrick et al. 2004; Saleh et al. 2005). A polymer is a macromolecule consisting of a repetition of smaller chemical units, monomers (Fig. 5.9). The number of units in a given chain is called the degree of polymerization, D_p. A polymer in which some monomer units are ionized and charged is referred to as a polyelectrolyte. Polyelectrolytes often achieve a net negative charge from carboxylate, sulfate, or sulfonate groups ($-COO^-$, $-SO_4^-$, and $-SO_3^-$, respectively) or a net positive charge from ammonium or protonated amines. Chapter 2 provides the detailed synthesis and properties of polymer-modified NZVI. Nevertheless, Table 5.1 summarizes various polyelectrolytes used for surface modification/functionalization of NZVI for groundwater remediation.

Both simple homopolymer and advanced block copolymer have been used for NZVI modification (Fig. 5.9c). Homopolymers consist of the same monomer units in the macromolecule, while block copolymers consist of one repeating monomer unit followed by one or more blocks of different repeating units. Block copolymers have

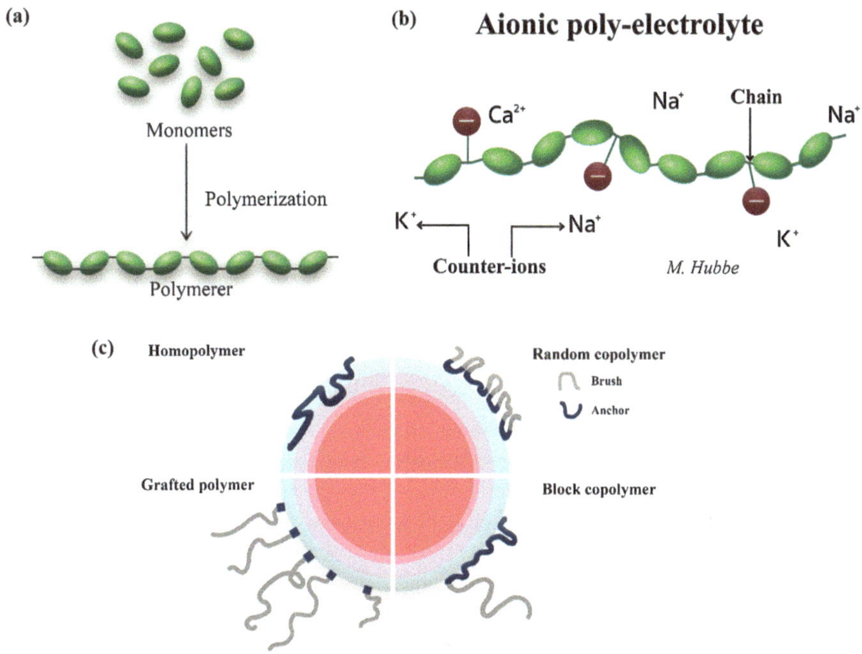

Fig. 5.9 (**a**) Polymerization to form polymer, (**b**) charged polymer is polyelectrolyte, and (**c**) sorption configurations of various types of polymer

advantages over homopolymers in that different functionalities for different tasks can be built into different blocks. Homopolymers are normally sorbed onto the surface in the train-loop-tail orientation (Fig. 5.9c). Trains are sequences in contact with the sorbed surface. Loops and tails are attached but extend from the surface. Block copolymers adsorb similarly, but block copolymers can be designed to anchor to the surface and theoretically can control the adsorbed layer configuration (Saleh et al. 2005). For example, poly(methacrylic acid) (PMAA; Table 5.1), which has a specific sorption affinity on the oxide surface, is used as an anchoring block of the trick block copolymer for NZVI surface modification. The chemical and structural complexity of different polymers influences their ability to stabilize nanoparticle dispersions, enhance transport, and afford contaminant targeting (Saleh et al. 2005). It should be noted that only negatively charged polyelectrolytes or uncharged polymers are used for this purpose because aquifer materials at a neutral pH are normally negatively charged. The mobility of positively charged polyelectrolyte-modified nanoparticles would be limited due to the electrostatic attraction between nanoparticles and aquifer materials.

Table 5.1 Polymeric modifiers used to modify NZVI for subsurface remediation and their laboratory reported ability to enhance colloidal stability and transport in porous media (Phenrat et al. 2011a)

	Performance			
Surface modifier	**Stabilization**	**Transport**	**Reactivity**	**Targetability**
Polymers				
Polyethylene glycol (PEG)	Good[a]	Poor[a,b]	–	–
Polyvinyl alcohol (PVA)	Poor[a]	Suspected to be poor	–	–
Guar gum (Tiraferri et al. 2008; Tiraferri and Sethi 2009)	Good	Moderate	–	–
Xanthan gum (shear thinning) (Comba and Sethi 2009; Vecchia et al. 2009)	Excellent	Excellent	–	–
	Performance			
Polyelectrolytes	**Stabilization**	**Transport**	**Reactivity**	**Targetability**
Triblock copolymers[c]				
$PMAA_{48}$-$PMMA_{17}$-PSS_{650} (Saleh et al. 2005, 2007, 2008)	Excellent	Excellent	Good	Good
$PMAA_{42}$-$PMMA_{26}$-PSS_{462} (Saleh et al. 2005, 2007, 2008)	Good	Excellent	Good	Moderate
Polystyrene (PSS) (Phenrat et al. 2008; Phenrat and Kumloet 2016)	Excellent	Good	Good	Good
Polyaspartate (PAP) (Phenrat et al. 2008, 2015)	Excellent	Good	Moderate	Good
Carboxymethyl cellulose (CMC) (He et al. 2007; Phenrat et al. 2008)	Fair to excellent	Poor to excellent	Moderate	Good
Polyacrylic acid (PAA) (Schrick et al. 2004; Yang et al. 2007)	Excellent	Good	Good	–
Mixture of PAA-PSS-bentonite[d] (Hydutsky et al. 2007)	–	Excellent	–	–

[a]Unpublished data from our laboratory
[b]Poor transportability of PEG-modified nanoparticles is attributed to the strong specific interaction between PEG and silica sand
[c]Triblock copolymer-functionalized NZVI can form stable picking emulsions of DNAPL (TCE)-water
[d]The difference in the performance of CMC between refs. (Phenrat et al. 2008) and (Hydutsky et al. 2007) is presumably due to different modification approaches

5.3.1 How Can Polymeric/Polyelectrolyte Surface Modification Enhance NZVI Dispersion Stability and Mobility in the Subsurface?

Polymeric/polyelectrolyte surface modification can improve NZVI dispersion stability and mobility in porous media by introducing an additional (electro)steric repulsion to counter V_{vdW} and V_M forces that promote agglomeration and deposition.

Steric repulsion has two origins. The first component is called osmotic repulsion (V_{osm}), which is from the overlap of the polymer layers on two approaching surfaces. This overlap increases the local polymer segment concentration and thus increases the local osmotic pressure in the overlap region. The second component is called elastic repulsion (V_{elas}), which is from further compression of the adsorbed polyelectrolyte layers below the thickness of the unperturbed layer (d), leading to a loss of entropy. The two components are mathematically illustrated in Eqs. (5.27) and (5.28), respectively (Phenrat et al. 2008):

$$\begin{aligned}
&\frac{V_{\mathrm{osm}}}{k_{\mathrm{B}}T} = 0, \quad 2d \leq h \\
&\frac{V_{\mathrm{osm}}}{k_{\mathrm{B}}T} = \frac{a4\pi}{\upsilon_1}{\varphi_{\mathrm{p}}}^2\left(\frac{1}{2}-\chi\right)\left(d-\frac{h}{2}\right)^2, \quad d \leq h < 2d \\
&\frac{V_{\mathrm{osm}}}{k_{\mathrm{B}}T} = \frac{a4\pi}{\upsilon_1}{\varphi_{\mathrm{p}}}^2\left(\frac{1}{2}-\chi\right)d^2\left(\frac{h}{2d}-\frac{1}{4}-\ln\left(\frac{h}{d}\right)\right), \quad h < d,
\end{aligned} \tag{5.27}$$

where χ is the Flory-Huggins solvency parameter, ϕ_{p} is the calculated volume fraction of polymer within the brush layer, d is the thickness of the brush, and υ_1 is the volume of a solvent molecule. Moreover, Eq. (5.28) is given as:

$$\begin{aligned}
\frac{V_{\mathrm{elas}}}{k_{\mathrm{B}}T} = 0,\ d \leq h \frac{V_{\mathrm{elas}}}{k_{\mathrm{B}}T} &= \left(\frac{2\pi a}{M_{\mathrm{w}}}\varphi_{\mathrm{p}}d^2\rho_{\mathrm{p}}\right)\left(\frac{h}{d}\ln\left(\frac{h}{d}\left(\frac{3-h/d}{2}\right)^2\right)\right. \\
&\left.-6\ln\left(\frac{3-h/d}{2}\right)+3\left(1+\frac{h}{d}\right)^2\right), d > h,
\end{aligned} \tag{5.28}$$

where M_{W} is the molecular weight of the polyelectrolyte and ρ_{p} is its density. For polyelectrolyte, the charged portion of the macromolecule also causes an electrostatic component (Eq. 5.5), which is superimposed with the steric repulsions to become electrosteric repulsion.

The range and magnitude of the electrosteric repulsion between a pair of particles or a particle and a collector depend on the surface concentration of adsorbed polyelectrolyte and extension and charge density of the adsorbed polyelectrolyte layer. The mass and configuration of polyelectrolytes adsorbed onto the NZVI surface are governed by the molecular weight, the ionization and charge density of the macromolecule, the charge density and polarity of the NZVI surface, the solvent quality, and the ionic strength. The conformation of the adsorbed polymer can also contribute to the magnitude of the electrosteric repulsion. However, the range of electrosteric repulsion is solely controlled by the adsorbed layer thickness (d) of polyelectrolyte on the NZVI. As shown in Fig. 5.7 (${V_{\mathrm{T}}}^{\mathrm{XDLVO}}$ with steric), for separation distances beyond $2d$, V_{vdW} attractive forces are dominant and result in the net attractive potential. However, when the polyelectrolyte-modified particles

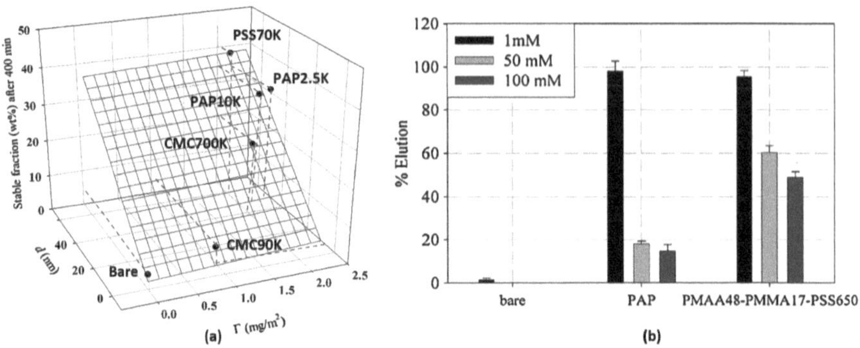

Fig. 5.10 (**a**) Correlation between the colloidally stable fraction (wt%) of nanoparticles and the measured surface excess (Γ, mg/m^2) and layer thickness (*d*, nm) of each adsorbed polyelectrolyte (Phenrat et al. 2008). Reprinted with permission from Phenrat et al. 2008. Copyright (2008) Springer Nature. (**b**) Effect of ionic strength on the elution of bare and modified RNIP through a 12.5 cm column with water-saturated silica sand having a porosity of 0.33. Particle mass concentration is 3 g/L. All samples also contained 1 mM $NaHCO_3$ to control pH at 7.4. The approach velocity was 9.5 m/d (Saleh et al. 2007)

approach one another at a distance of less than 2*d*, there is a large energy barrier due to the adsorbed layer. In fact, the polymer coatings make it very difficult for these particles to aggregate in a primary minimum, so all agglomeration or deposition must occur in under a shallower secondary minimum. This has very important consequences regarding the dispersion stability and mobility of nanoparticles in the subsurface because agglomeration and attachment under a secondary minimum are subjected to disaggregation and detachment due to shear force and Brownian energy (see Chap. 6). For this reason, aggregation and mobility of surface-modified nanoparticles should be largely controlled by the adsorbed polyelectrolyte layer properties. A recent study (Phenrat et al. 2008) correlated the adsorbed polyelectrolyte layer properties (*d* estimated using Ohshima's soft particle analysis) and the adsorbed polyelectrolyte mass (Γ) on NZVI to their dispersion stabilities (Fig. 5.10a). The higher the adsorbed polyelectrolyte mass and the more extended the layer thickness, the larger the magnitude and extent of electrosteric repulsion and the smaller the secondary minimum well for agglomeration. For a similar reason, considering the particle-collector interaction under the influence of adsorbed polymer layers, collector ripening and pore plugging are not expected for the transport of colloidally stable polyelectrolyte-modified nanoparticles in saturated porous media. Figure 5.10b illustrates the enhanced mobility of NZVI particles modified by various surface modifiers compared to bare NZVI particles (see more discussion on Chap. 6). The influence of electrosteric stabilization on the enhancement of dispersion stability and nanoparticle transport is also generally observed as summarized in Table 5.1.

5.3.2 How Can Polymeric Surface Modification Help NZVI Dispersion Resist Unfavorable Environmental Conditions?

Another unique feature of electrosteric stabilization with implications on subsurface restoration using NZVI is the ability to maintain the strong repulsive forces at a high ionic strength and in the presence of multivalent ionic species. Aggregation and deposition of electrostatically stabilized nanoparticles are sensitive to ionic strength and divalent cations due to charge neutralization and shrinking of the Debye length. Electrosterically stabilized NZVI are much less sensitive to ionic strengths and types of ionic species as can be seen from Eqs. (5.27) and (5.28) in which V_{osm} and V_{elas} are not a function of ionic strengths. The sensitivity of mobility of polyelectrolyte-modified NZVI on the change of ionic strength is also seen in Fig. 5.10b. The transportability (reported as percent elution) of NZVI modified by large polyelectrolyte (PMAA48-PMMA17-PSS650) is much less sensitive to the increase of ionic strength in comparison to bare NZVI and NZVI modified by small polyelectrolytes (PAP) (Saleh et al. 2007). A review of the applicable colloidal forces and influences of these forces on the fate of nanoparticles in the environment is available (Louie et al. 2016).

In addition to the ionic species in the subsurface, polyelectrolytes, such as CMC, can enhance the transportability of NZVI even within biofilm-coated porous media, although the mobility significantly decreases compared to transport in clean sand (Basnet et al. 2016). The higher retention in biofilm-coated sand was attributed to the agglomeration of NZVI and subsequent straining. Thus, CMC could, to some extent, help stabilize NZVI, thereby improving its mobility.

Golmohamadi et al. (2013) suggested that the self-diffusion of nanoparticles in biofilm is mainly controlled by the particle size rather than electrostatic interactions. These authors mentioned that particles >66 nm can be excluded from the polymer network, whereas the self-diffusion coefficient of nanoparticles <10 nm significantly decreased in biofilm compared to that in water. The presence of polymer in NZVI dispersion may alter NZVI diffusion and therefore its mobility, especially in dual-domain porous media.

Mun et al. (2013) found that nanoparticle diffusion is retarded within water-soluble polymer solutions (PEO, PV, PAA, and HEC within a concentration range of 0.5–1 g/L) predominantly due to changes in medium viscosity, particle size, and existence of specific interactions between nanoparticles and components of the medium. However, some coating, such as polyethylene glycol (PEG) on the surface of particles, may enhance their diffusion because of the lubricating effect of the coating (Mun et al. 2013). Therefore, the addition of polymer to NZVI dispersion, depending on the amount of adsorbed polymer and free polymer in the solution, may lead to lower or higher diffusion of NZVI into the immobile zones of porous media compared to pure NZVI dispersion, thereby enhancing or reducing the mobility of polymer-modified NZVI dispersions.

Phenrat et al. (2010a) investigated the effect of excess polymer in NZVI dispersion on the transport of NZVI through a heterogeneously packed 2D media. They found that excess polymer in a concentrated NZVI dispersion decreased its transport. The excess polymer concentration (~0.13 g/L) in a system with high NZVI concentration (6 g/L) promoted agglomeration via depletion flocculation, which decreased transport. Depletion flocculation is driven by the size exclusion of unadsorbed, large macromolecule coils from the space between two particles dispersed in the solution. This results in osmotic attraction between two particles sharing the space as a result of the flow of solution (polymer-free) out of the interparticle space driven by the osmotic gradient. The occurrence and magnitude of depletion flocculation will depend on the polymer concentration, the polymer MW, and the particle concentration (Phenrat et al. 2010a). In the light of that study, it appears that the reduced diffusivity of NZVI in the presence of excess polymer and the consequent expected decreased diffusion of NZVI into immobile zones of the flow cell, which should lead to enhanced transport, are not considerable effects compared to the enhanced agglomeration and consequent reduced mobility.

There are, of course, other effects of free polymer in dispersion. Higher concentration of free polymer can elevate the viscosity of the fluid (Becker et al. 2015). It can block the attachment surfaces in competition with the nanoparticles (Wang et al. 2014; Becker et al. 2015). In the presence of divalent cations, polymer bridging can bring about a substantial level of agglomeration and deposition (Chen and Elimelech 2006; Wang et al. 2011; Torkzaban et al. 2012). The meta-analysis by Babakhani et al. (2017) revealed that the effects of free polymer on nanoparticle transport are dominated by enhanced agglomeration and subsequent reduced mobility. Overall, the free polymer concentration appeared to have the least contribution to the transport of nanoparticles among 20 environmental factors analyzed in that study because of several potentially conflicting ways that excess polymer influences mobility.

5.4 Targeting the DNAPL Source Zone by Polymer-Modified NZVI

Chlorinated DNAPLs, entrapped in the source-zone subsurface, slowly dissolve and generate a toxic plume of contaminated groundwater. A novel conceptual model of active and selective DNAPL source-zone removal was made possible by polymeric surface modification. This involves the delivery of polymer-modified NZVI to target

DNAPL sources and subsequently perform reductive dechlorination. This original idea was proposed by Saleh and coworkers (Saleh et al. 2005).

The original concept requires three different polymer blocks designed to provide three different functionalities for three different tasks. The PMAA block serves to anchor the triblock copolymers to the NZVI particles. Hydrophobic attractions arising from the poly(methyl methacrylate) (PMMA) block provide the strong affinity to the NAPL and create a low polarity region that hinders water access to the NZVI particles with the goal of minimizing ZVI oxidation during transport in the soil before it reaches the NAPL. The strongly charged PSS block serves to provide strong electrosteric interparticle repulsions that promote colloidal stability and electrosteric repulsion from the negatively charged surfaces that are predominant in the subsurface at near neutral pH. In water, the PMMA block is collapsed, while the extended PSS block provides electrosteric protection. At the DNAPL/water interface, the PMMA block swells in the organic solvent and anchors the particle at the interface (Fig. 5.11a). The PSS solubility in the organic phase is too weak to allow full passage of the NZVI through the interface into the bulk NAPL phase. The NZVI nanoparticles modified by triblock copolymers with this architecture and chemical composition successfully anchored to DNAPL/water interfaces in the laboratory as evaluated from the Pickering emulsion formation (Fig. 5.11b).

Recently, Phenrat et al. (2011a) and Phenrat and Kumloet (2016) demonstrated that NZVI modified by olefin-maleic acid copolymer and PSS (homopolyelectrolyte) can also form a Pickering emulsion. Phenrat et al. (2011a) also demonstrated that this concept can be successfully applied in an upscaled, more realistic situation. They delivered NZVI modified by olefin-maleic acid copolymer into a 2D tank with an NAPL source zone (Fig. 5.12). They illustrated that NZVI modified by olefin-maleic acid copolymer could target entrapped NAPL source zones at various NAPL saturations in 2D flow-through porous media, which is more relevant to the field conditions (Figs. 5.12 and 5.11c).

However, delivering NZVI to the chlorinated DNAPL/water interface appears to be insufficient for effective DNAPL source-zone removal. Although NZVI increased the efficient use of Fe^0 for reductive dechlorination at the DNAPL/water interface (Liu et al. 2007), under groundwater flow conditions, DNAPL dissolution to the aqueous phase is a rate-limited step of dechlorination (Miller et al. 1990; Saba and Illangasekare 2000). This is because the reductive dechlorination reaction is surface mediated; therefore, contaminants must be dissolved from the DNAPL to transport and adsorb onto the NZVI surface for electron transference. For this reason, the rate of DNAPL degradation is limited by the rate of mass transfer to the aqueous phase (Berge and Ramsburg 2010; Taghavy et al. 2010; Fagerlund et al. 2012; Phenrat et al. 2016). Rather, the NZVI reacts with water to form H_2, which increases the amount of NZVI required for remediation. Therefore, Phenrat's research group proposed a novel use of the electromagnetic field with NZVI to enhance the DNAPL dissolution and to speed up the reaction rate and improve the electron efficiency of the remediation. This novel concept of using polymer-modified NZVI that is capable of targeting NAPL with electromagnetic field is presented in Chap. 11.

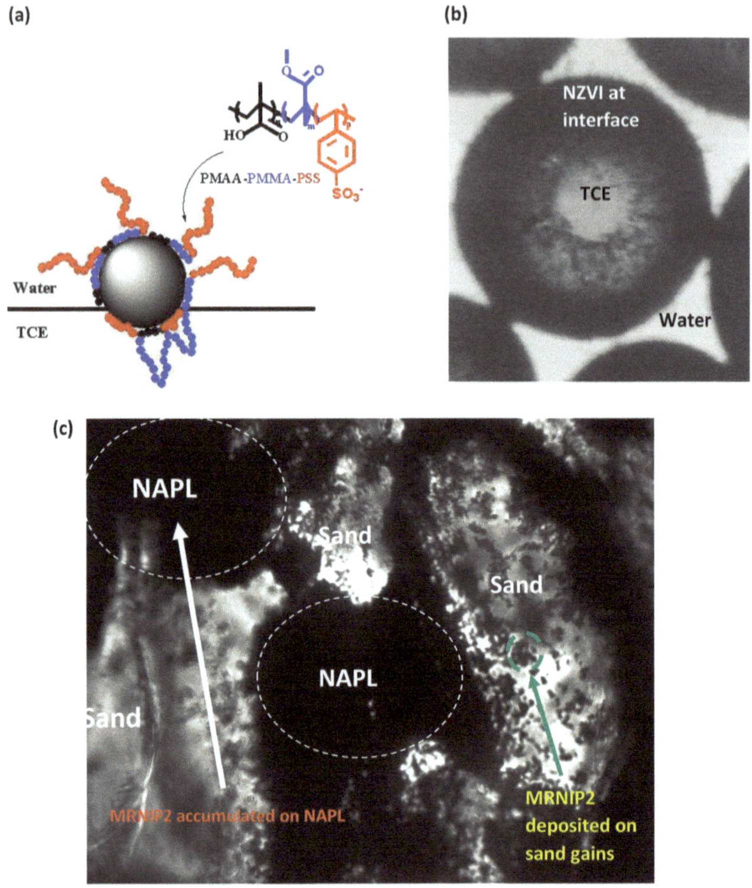

Fig. 5.11 (**a**) A schematic diagram illustrating the proposed polyelectrolyte response used to anchor particles at the DNAPL/water interface (Saleh et al. 2005). (**b**) Micrographs of emulsified TCE droplets in water stabilized by $PMAA_{42}$-$PMMA_{26}$-PSS_{462} triblock copolymer-modified NZVI nanoparticles at the TCE-water interface. (Reprinted with permission from Saleh et al. 2005. Copyright (2005) American Chemical Society). (**c**) A micrograph of a dodecane source taken from 2D cell experiment after in situ targeting experiment (Fig. 5.12) with MRNIP2 (NZVI modified by olefin-maleic acid copolymer). (**c**) shows the accumulation of MRNIP2 at dodecane interface rather than the deposition on sand grain which is in good agreement with the macroscopic observation. The scale bars are 10 μm

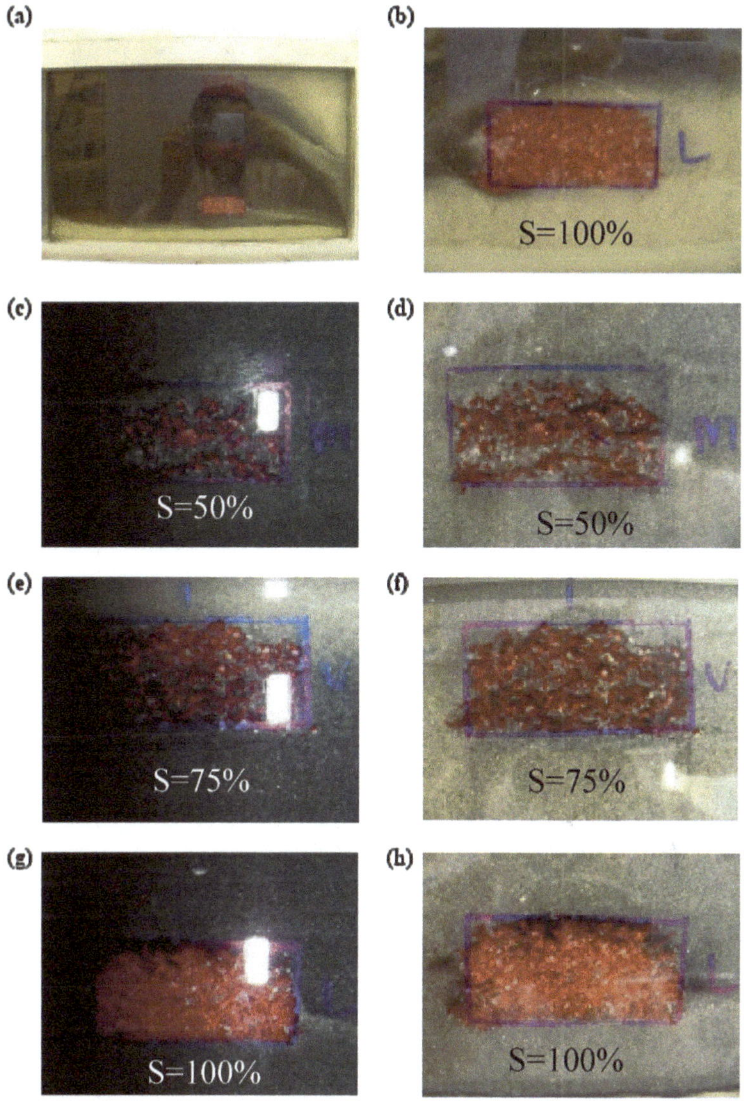

Fig. 5.12 (**a**) MRNIP2 (NZVI modified by olefin-maleic acid copolymer) (15 g/L) transported to the NAPL sources (allowing 24 h without flow prior to flushing). (**b**) MRNIP2 bypassing the source of S = 100%. (**c**) MRNIP2 in the source zone of S = 50%, (**e**) $S = 75\%$, and (**g**) $S = 100\%$. MRNIP2 remaining in the source zone of (**d**) $S = 50\%$, (**f**) $S = 75\%$, and (**h**) $S = 100\%$ after flushing (Phenrat et al. 2011b). (Reprinted with permission from Phenrat et al. (2011b). Copyright (2011) American Chemical Society)

References

Amorós, J. L., Beltrán, V., Sanz, V., & Jarque, J. C. (2010). Electrokinetic and rheological properties of highly concentrated kaolin dispersions: Influence of particle volume fraction and dispersant concentration. *Applied Clay Science, 49*, 33–43.

Aziz, J. J., Serra, C. A., & Wiesner, M. R. (2003). Hydrodynamics of permeable aggregates in differential sedimentation. *Environmental Engineering Science, 20*, 21–31.

Baalousha, M. (2017). Effect of nanomaterial and media physicochemical properties on nanomaterial aggregation kinetics. *NanoImpact, 6*, 55–68.

Babakhani, P., Bridge, J., Doong, R.-A., & Phenrat, T. (2017). Parameterization and prediction of nanoparticle transport in porous media: A reanalysis using artificial neural network. *Water Resources Research*. https://doi.org/10.1002/2016WR020358.

Babakhani, P., Doong, R.-A., & Bridge, J. (2018). Significance of early and late stages of coupled aggregation and sedimentation in the fate of nanoparticles: Measurement and modeling. *Environmental Science Technology, 52*(15), 8419–8428.

Basnet, M., Gershanov, A., Wilkinson, K. J., Ghoshal, S., & Tufenkji, N. (2016). Interaction between palladium-doped zerovalent iron nanoparticles and biofilm in granular porous media: Characterization, transport and viability. *Environmental Science: Nano, 3*, 127.

Baumann, T., Keller, A. A., Auset-Vallejo, M., & Lowry, G. V. (2005). Micromodel study of transport issues during TCE dechlorination by ZVI colloids. American Geophysical Union Fall Meeting, San Francisco, CA.

Becker, M. D., Wang, Y., Pennell, K. D., & Abriola, L. M. (2015). A multi-constituent site blocking model for nanoparticle and stabilizing agent transport in porous media. *Environmental Science: Nano, 2*, 155–166.

Bennett, P., He, F., Zhao, D., Aiken, B., & Feldman, L. (2010). In situ testing of metallic iron nanoparticle mobility and reactivity in a shallow granular aquifer. *Journal of Contaminant Hydrology, 116*, 35–46.

Berge, N. D., & Ramsburg, C. A. (2010). Iron-mediated trichloroethene reduction within nonaqueous phase liquid. *Journal of Contaminant Hydrology, 118*, 105–116.

Bradford, S. A., Simunek, J., Bettahar, M., van Genuchten, M. T., & Yates, S. R. (2003). Modeling colloid attachment, straining, and exclusion in saturated porous media. *Environmental Science & Technology, 37*, 2242–2250.

Bradford, S. A., Torkzaban, S., & Simunek, J. (2011). Modeling colloid transport and retention in saturated porous media under unfavorable attachment conditions. *Water Resources Research, 47*, W10503.

Busch, J., Meißner, T., Potthoff, A., Bleyl, S., Georgi, A., Mackenzie, K., Trabitzsch, R., Werban, U., & Oswald, S. E. (2015). A field investigation on transport of carbon-supported nanoscale zero-valent iron (nZVI) in groundwater. *Journal of Contaminant Hydrology, 181*, 59–68.

Chandrasekhar, S. (1943). Stochastic problems in physics and astronomy. *Reviews of Modern Physics, 15*, 1.

Chen, K. L., & Elimelech, M. (2006). Aggregation and deposition kinetics of fullerene (C60) nanoparticles. *Langmuir, 22*, 10994–11001.

Chowdhury, A. I. A., O'Carroll, D. M., Xu, Y., & Sleep, B. E. (2012). Electrophoresis enhanced transport of nano-scale zero valent iron. *Advances in Water Resources, 40*, 71–82.

Comba, S., & Sethi, R. (2009). Stabilization of highly concentrated suspensions of iron nanoparticles using shear-thinning gels of xanthan gum. *Water Research, 43*, 3717–3726.

Elimelech, M., Gregory, J., & Jia, X. (1998a). *Particle deposition and aggregation: Measurement, modelling and simulation*. Woburn: Butterworth-Heinemann.

Elimelech, M., Jia, X., Gregory, J., & Williams, R. (1998b). *Particle deposition and aggregation: Measurement, modelling and simulation*. Amsterdam: Elsevier.

Elimelech, M., Nagai, M., Ko, C.-H., & Ryan, J. N. (2000). Relative insignificance of mineral grain zeta potential to colloid transport in geochemically heterogeneous porous media. *Environmental Science & Technology, 34*, 2143–2148.

Elliott, D. W., & Zhang, W.-X. (2001). Field assessment of nanoscale bimetallic particles for groundwater treatment. *Environmental Science & Technology, 35*, 4922–4926.

Evans, D. F., & Wennerstrom, H. (1999). *The colloidal domain; where physics, chemistry, biology, and technology meet.* New York: Wiley-VCH.

Fagerlund, F., Illangasekare, T. H., Phenrat, T., Kim, H.-J., & Lowry, G. V. (2012). PCE dissolution and simultaneous dechlorination by nanoscale zero-valent iron particles in a DNAPL source zone. *Journal of Contaminant Hydrology, 131*, 9–28.

Golmohamadi, M., Clark, R. J., Veinot, J. G. C., & Wilkinson, K. J. (2013). The role of charge on the diffusion of solutes and nanoparticles (silicon nanocrystals, nTiO2, nAu) in a biofilm. *Environmental Chemistry, 10*, 34–41.

Grant, G. P., & Gerhard, J. I. (2007). Simulating the dissolution of a complex dense nonaqueous phase liquid source zone: 1. Model to predict interfacial area. *Water Resources Research, 43*, W12410.

He, F., Zhao, D., Liu, J., & Roberts, C. B. (2007). Stabilization of Fe-Pd nanoparticles with sodium carboxymethyl cellulose for enhanced transport and dechlorination of trichloroethylene in soil and groundwater. *Industrial and Engineering Chemistry Research, 46*, 29–34.

He, F., Zhao, D., & Paul, C. (2010). Field assessment of carboxymethyl cellulose stabilized iron nanoparticles for in situ destruction of chlorinated solvents in source zones. *Water Research, 44*, 2360–2370.

He, J.-Z., Li, C.-C., Wang, D.-J., & Zhou, D.-M. (2015). Biofilms and extracellular polymeric substances mediate the transport of graphene oxide nanoparticles in saturated porous media. *Journal of Hazardous Materials, 300*, 467–474.

Henn, K. W., & Waddill, D. W. (2006). Utilization of nanoscale zero-valent iron for source remediation—A case study. *Remediation Journal, 16*, 57–77.

Holmboe, M., Wold, S., Jonsson, M., & Garcia-Garcia, S. (2009). Effects of γ-irradiation on the stability of colloidal Na+-Montmorillonite dispersions. *Applied Clay Science, 43*, 86–90.

Hotze, E. M., Phenrat, T., & Lowry, G. V. (2010). Nanoparticle aggregation: Challenges to understanding transport and reactivity in the environment. *Journal of Environmental Quality, 39*, 1909–1924.

Hounslow, M. J., Ryall, R. L., & Marshall, V. R. (1988). A discretized population balance for nucleation, growth, and aggregation. *AICHE Journal, 34*, 1821–1832.

Hydutsky, B. W., Mack, E. J., Beckerman, B. B., Skluzacek, J. M., & Mallouk, T. E. (2007). Optimization of nano- and microiron transport through sand columns using polyelectrolyte mixtures environ. *Science and Technology, 41*, 6418–6424.

Jeldres, R. I., Concha, F., & Toledo, P. G. (2015). Population balance modelling of particle flocculation with attention to aggregate restructuring and permeability. *Advances in Colloid and Interface Science, 224*, 62–71.

Jiang, X., Wang, X., Tong, M., & Kim, H. (2013). Initial transport and retention behaviors of ZnO nanoparticles in quartz sand porous media coated with Escherichia coli biofilm. *Environmental Pollution, 174*, 38–49.

Johnson, C. P., Li, X., & Logan, B. E. (1996). Settling velocities of fractal aggregates. *Environmental Science & Technology, 30*, 1911–1918.

Johnson, R. L., Nurmi, J. T., O'Brien Johnson, G. S., Fan, D., O'Brien Johnson, R. L., Shi, Z., Salter-Blanc, A. J., Tratnyek, P. G., & Lowry, G. V. (2013). Field-scale transport and transformation of carboxymethylcellulose-stabilized nano zero-valent iron. *Environmental Science & Technology, 47*, 1573–1580.

Jordan, M., Shetty, N., Zenker, M. J., & Brownfield, C. (2013). Remediation of a former dry cleaner using nanoscale zero valent iron. *Remediation Journal, 24*, 31–48.

Kim, H.-J., Phenrat, T., Tilton, R. D., & Lowry, G. V. (2012). Effect of kaolinite, silica fines and pH on transport of polymer-modified zero valent iron nano-particles in heterogeneous porous media. *Journal of Colloid and Interface Science, 370*, 1–10.

Köber, R., Hollert, H., Hornbruch, G., Jekel, M., Kamptner, A., Klaas, N., Maes, H., Mangold, K. M., Martac, E., & Matheis, A. (2014). Nanoscale zero-valent iron flakes for groundwater treatment. *Environmental Earth Sciences, 72*, 3339–3352.

Kocur, C. M., Chowdhury, A. I., Sakulchaicharoen, N., Boparai, H. K., Weber, K. P., Sharma, P., Krol, M. M., Austrins, L., Peace, C., & Sleep, B. E. (2014). Characterization of nZVI mobility in a field scale test. *Environmental Science & Technology, 48*, 2862–2869.

Krupp, H. K., Biggar, J. W., & Nielsen, D. R. (1972). Relative flow rates of salt and water in soil. *Soil Science Society of America Journal, 36*, 412–417.

Kumar, S., & Ramkrishna, D. (1996a). On the solution of population balance equations by discretization—I. A fixed pivot technique. *Chemical Engineering Science, 51*, 1311–1332.

Kumar, S., & Ramkrishna, D. (1996b). On the solution of population balance equations by discretization—II. A moving pivot technique. *Chemical Engineering Science, 51*, 1333–1342.

Lattuada, M., Wu, H., Sandkühler, P., Sefcik, J., & Morbidelli, M. (2004). Modelling of aggregation kinetics of colloidal systems and its validation by light scattering measurements. *Chemical Engineering Science, 59*, 1783–1798.

Lee, D. G., Bonner, J. S., Garton, L. S., Ernest, A. N. S., & Autenrieth, R. L. (2000). Modeling coagulation kinetics incorporating fractal theories: A fractal rectilinear approach. *Water Research, 34*, 1987–2000.

Leeson, A., Kavanaugh, M. C., Marqusee, J. A., Smith, B., Stroo, H., Unger, M., Vogel, C., & Ward, C. H. (2003). Remediating chlorinated source zones. *Environmental Science & Technology, 37*, 224A–230A.

Li, X.-Y., & Logan, B. E. (2001). Permeability of fractal aggregates. *Water Research, 35*, 3373–3380.

Li, Z., Hassan, A. A., Sahle-Demessie, E., & Sorial, G. A. (2013). Transport of nanoparticles with dispersant through biofilm coated drinking water sand filters. *Water Research, 47*, 6457–6466.

Liang, Y., Bradford, S. A., Simunek, J., Vereecken, H., & Klumpp, E. (2013). Sensitivity of the transport and retention of stabilized silver nanoparticles to physicochemical factors. *Water Research, 47*, 2572–2582.

Lister, J., Smit, D., & Hounslow, M. (1995). Adjustable discretized population balance for growth and aggregation. *AICHE Journal, 41*, 591–603.

Liu, Y., Choi, H., Dionysiou, D., & Lowry, G. V. (2005). TCE hydrodechlorination in water by highly disordered monometallic nanoiron. *Chemistry of Materials, 17*, 5315–5322.

Liu, Y., Phenrat, T., Lowry, G. V. (2007). Effect of TCE concentration and dissolved groundwater solutes on NZVI-promoted TCE dechlorination and H2 evolution. *Environmental Science and Technology, 41*, 7881–7887.

Liu, L., Gao, B., Wu, L., Yang, L., Zhou, Z., & Wang, H. (2013). Effects of pH and surface metal oxyhydroxides on deposition and transport of carboxyl-functionalized graphene in saturated porous media. *Journal of Nanoparticle Research, 15*, 1–8.

Logan, B. E., & Hunt, J. R. (1987). Advantages to microbes of growth in permeable aggregates in marine systems. *Limnology and Oceanography, 32*, 1034–1048.

Lowry, G. V. (2007). Nanomaterials for groundwater remediation. In M. R. Wiesner & J.-Y. Bottero (Eds.), *Environmental nanotechnology: Applications and impacts of nanomaterials*. New York: McGraw-Hill.

Merkus, H. G. (2009). *Particle size measurements: Fundamentals, practice, quality*. New York: Springer Science & Business Media.

Louie, S., Tilton, R., Lowry, G. V. (2016). Nanoparticle Macromolecular Coatings and Their Impact on Critical Physicochemical Processes Controlling Environmental Fate and Effects of Nanomaterials. ES: Nano. *3*, 283–310

Miller, C. T., Poirer-McNeill, M. M., & Mayer, A. S. (1990). Dissolution of trapped nonaqueous phase liquids: Mass transfer characteristics. *Water Resources Research, 26*, 2783–2796.

Mitzel, M. R., & Tufenkji, N. (2014). Transport of industrial PVP-stabilized silver nanoparticles in saturated quartz sand coated with Pseudomonas aeruginosa PAO1 biofilm of variable age. *Environmental Science & Technology, 48*, 2715–2723.

Molnar, I. L., Johnson, W. P., Gerhard, J. I., Willson, C. S., & O'Carroll, D. M. (2015). Predicting colloid transport through saturated porous media: A critical review. *Water Resources Research, 51*, 6804.
Mun, E. A., Hannell, C., Rogers, S. E., Hole, P., Williams, A. C., & Khutoryanskiy, V. V. (2013). On the role of specific interactions in the diffusion of nanoparticles in aqueous polymer solutions. *Langmuir, 30*, 308–317.
Neale, G., Epstein, N., & Nader, W. (1973). Creeping flow relative to permeable spheres. *Chemical Engineering Science, 28*, 1865–1874.
Phenrat, T., & Kumloet, I. (2016). Electromagnetic induction of nanoscale zerovalent iron particles accelerates the degradation of chlorinated dense non-aqueous phase liquid: Proof of concept. *Water Research, 107*, 19–28.
Phenrat, T., & Lowry, G. V. (2009). Physicochemistry of polyelectrolyte coatings that increase stability, mobility, and contaminant specificity of reactive nanoparticles used for groundwater remediatio. In N. Savage, M. Diallo, J. Duncan, A. Street, & R. Sustich (Eds.), *Nanotechnology applications: Solutions for improving water quality* (pp. 249–267). New York: William Andrew Publisher.
Phenrat, T., Saleh, N., Sirk, K., Tilton, R. D., & Lowry, G. V. (2007). Aggregation and sedimentation of aqueous nanoscale zerovalent iron dispersions. *Environmental Science & Technology, 41*, 284–290.
Phenrat, T., Saleh, N., Sirk, K., Kim, H.-J., Tilton, R. D., & Lowry, G. V. (2008). Stabilization of aqueous nanoscale zerovalent iron dispersions by anionic polyelectrolytes: Adsorbed anionic polyelectrolyte layer properties and their effect on aggregation and sedimentation. *Journal of Nanoparticle Research, 10*, 795–814.
Phenrat, T., Kim, H.-J., Fagerlund, F., Illangasekare, T., Tilton, R. D., & Lowry, G. V. (2009a). Particle size distribution, concentration, and magnetic attraction affect transport of polymer-modified Fe^0 nanoparticles in sand columns. *Environmental Science & Technology, 43*, 5079–5085.
Phenrat, T., Liu, Y., Tilton, R. D., & Lowry, G. V. (2009b). Adsorbed polyelectrolyte coatings decrease Fe^0 nanoparticle reactivity with TCE in water: Conceptual model and mechanisms. *Environmental Science & Technology, 43*, 1507–1514.
Phenrat, T., Cihan, A., Kim, H.-J., Mital, M., Illangasekare, T., & Lowry, G. V. (2010a). Transport and deposition of polymer-modified FeO nanoparticles in 2-D heterogeneous porous media: Effects of particle concentration, FeO content, and coatings. *Environmental Science & Technology, 44*, 9086–9093.
Phenrat, T., Song, J. E., Cisneros, C. M., Schoenfelder, D. P., Tilton, R. D., & Lowry, G. V. (2010b). Estimating attachment of nano- and submicrometer-particles coated with organic macromolecules in porous media: Development of an empirical model. *Environmental Science & Technology, 44*, 4531–4538.
Phenrat, T., Crimi, M., Illanagasekare, T., & Lowry, G. V. (2011a). Reactive nanoparticles for the treatment of chlorinated dense nonaqueous phase liquids in soil and groundwater. In M. Ram, E. S. Andreescu, & D. Hanming (Eds.), *Nanotechnology for environmental decontamination* (pp. 271–322). New York: McGraw-Hill Publisher.
Phenrat, T., Fagerlund, F., Illanagasekare, T., Lowry, G. V., & Tilton, R. D. (2011b). Polymer-modified FeO nanoparticles target entrapped NAPL in two dimensional porous media: Effect of particle concentration, NAPL saturation, and injection strategy. *Environmental Science & Technology, 45*, 6102–6109.
Phenrat, T., Schoenfelder, D., Kirschling, T. L., Tilton, R. D., & Lowry, G. V. (2015). Adsorbed poly(aspartate) coating limits the adverse effects of dissolved groundwater solutes on FeO nanoparticle reactivity with trichloroethylene. *Environmental Science and Pollution Research, 25*, 7157–7169.
Phenrat, T., Thongboot, T., & Lowry, G. V. (2016). Electromagnetic induction of zerovalent iron (ZVI) powder and nanoscale zerovalent iron (NZVI) particles enhances dechlorination of

trichloroethylene in contaminated groundwater and soil: Proof of concept. *Environmental Science & Technology, 50*, 872–880.
Ryan, J. N., & Elimelech, M. (1996). Colloid mobilization and transport in groundwater. *Colloids and Surfaces A, 107*, 1–56.
Saba, T., & Illangasekare, T. H. (2000). Effect of groundwater flow dimensionality on mass transfer from entrapped nonaqueous phase liquid contaminants. *Water Resources Research, 36*, 971–980.
Saleh, N., Phenrat, T., Sirk, K., Dufour, B., Ok, J., Sarbu, T., Matyjaszewski, K., Tilton, R. D., & Lowry, G. V. (2005). Adsorbed triblock copolymers deliver reactive iron nanoparticles to the oil/water interface. *Nano Letters, 5*, 2489–2494.
Saleh, N., Sirk, K., Liu, Y., Phenrat, T., Dufour, B., Matyjaszewski, K., Tilton, R. D., & Lowry, G. V. (2007). Surface modifications enhance nanoiron transport and NAPL targeting in saturated porous media. *Environmental Engineering Science, 24*, 45–57.
Saleh, N., Kim, H.-J., Phenrat, T., Matyjaszewski, K., Tilton, R. D., & Lowry, G. V. (2008). Ionic strength and composition affect the mobility of surface-modified Fe0 nanoparticles in water-saturated sand columns environ. *Science and Technology, 42*, 3349–3355.
Schrick, B., Hydutsky, B. W., Blough, J. L., & Mallouk, T. E. (2004). Delivery vehicles for zerovalent metal nanoparticles in soil and groundwater. *Chemistry of Materials, 16*, 2187–2193.
Smoluchowski, M. (1917). Versuch einer mathematischen Theorie der Koagulationskinetik kolloider Lösungen. *Zeitschrift fuer Physikalische Chemie, 92*, 129–168.
Sterling, M. C., Bonner, J. S., Ernest, A. N. S., Page, C. A., & Autenrieth, R. L. (2005). Application of fractal flocculation and vertical transport model to aquatic sol–sediment systems. *Water Research, 39*, 1818–1830.
Taghavy, A., Costanza, J., Pennell, K. D., & Abrio, L. M. (2010). Effectiveness of nanoscale zero-valent iron for treatment of a PCE–DNAPL source zone. *Journal of Contaminant Hydrology, 118*, 128–142.
Thill, A., Moustier, S., Aziz, J., Wiesner, M. R., & Bottero, J. Y. (2001). Flocs restructuring during aggregation: Experimental evidence and numerical simulation. *Journal of Colloid and Interface Science, 243*, 171–182.
Tian, Y., Gao, B., Wang, Y., Morales, V. L., Carpena, R. M., Huang, Q., & Yang, L. (2012). Deposition and transport of functionalized carbon nanotubes in water-saturated sand columns. *Journal of Hazardous Materials, 213*, 265–272.
Tiraferri, A., & Sethi, R. (2009). Enhanced transport of zerovalent iron nanoparticles in saturated porous media by guar gum. *Journal of Nanoparticle Research, 11*, 635–645.
Tiraferri, A., Chen, K.-L., Sethi, R., & Elimelech, M. (2008). Reduced aggregation and sedimentation of zero-valent iron nanoparticles in the presence of guar gum. *Journal of Colloid and Interface Science, 324*, 71–79.
Tombacz, E., & Szekeres, M. (2004). Colloidal behavior of aqueous montmorillonite suspensions: The specific role of pH in the presence of indifferent electrolytes. *Applied Clay Science, 27*, 75–94.
Torkzaban, S., Tazehkand, S. S., Walker, S. L., & Bradford, S. A. (2008). Transport and fate of bacteria in porous media: Coupled effects of chemical conditions and pore space geometry. *Water Resources Research, 44*, W04403.
Torkzaban, S., Wan, J., Tokunaga, T. K., & Bradford, S. A. (2012). Impacts of bridging complexation on the transport of surface-modified nanoparticles in saturated sand. *Journal of Contaminant Hydrology, 136*, 86–95.
Tratnyek, P. G., & Johnson, R. L. (2006). Nanotechnologies for environmental cleanup. *Nano Today, 1*, 44–48.
Tripathi, S., Champagne, D., & Tufenkji, N. (2011). Transport behavior of selected nanoparticles with different surface coatings in granular porous media coated with Pseudomonas aeruginosa biofilm. *Environmental Science & Technology, 46*, 6942–6949.

Tsujimoto, Y., Chassagne, C., & Adachi, Y. (2013). Dielectric and electrophoretic response of montmorillonite particles as function of ionic strength. *Journal of Colloid and Interface Science, 404*, 72–79.

Tufenkji, N., & Elimelech, M. (2004). Correlation equation for predicting single-collector efficiency in physicochemical filtration in saturated porous media. *Environmental Science & Technology, 38*, 529–536.

US.EPA. (2003). *The DNAPL remediation challenge: Is there a case for source depletion?* (p. 107). Cincinnati: National Risk Management Research Laboratory, Office of Research and Development, US.EPA.

Van Genuchten, M. T. (1981). *Non-equilibrium transport parameters from miscible displacement experiments*. Riverside: U.S. Salinity Laboratory.

Vecchia, E. D., Luna, M., & Sethi, R. (2009). Transport in porous media of highly concentrated iron micro- and nanoparticles in the presence of xanthan gum. *Environmental Science & Technology, 43*, 8942–8947.

Veerapaneni, S., & Wiesner, M. R. (1996). Hydrodynamics of fractal aggregates with radially varying permeability. *Journal of Colloid and Interface Science, 177*, 45–57.

Vikesland, P. J., Rebodos, R. L., Bottero, J. Y., Rose, J., & Masion, A. (2016). Aggregation and sedimentation of magnetite nanoparticle clusters. *Environmental Science: Nano, 3*, 567–577.

Wang, C. B., & Zhang, W. X. (1997). Synthesizing nanoscale iron particles for rapid and complete dechlorination of TCE and PCBs. *Environmental Science & Technology, 31*, 2154–2156.

Wang, D., Paradelo, M., Bradford, S. A., Peijnenburg, W. J. G. M., Chu, L., & Zhou, D. (2011). Facilitated transport of Cu with hydroxyapatite nanoparticles in saturated sand: Effects of solution ionic strength and composition. *Water Research, 45*, 5905–5915.

Wang, D., Bradford, S. A., Harvey, R. W., Gao, B., Cang, L., & Zhou, D. (2012). Humic acid facilitates the transport of ARS-labeled hydroxyapatite nanoparticles in iron oxyhydroxide-coated sand. *Environmental Science & Technology, 46*, 2738–2745.

Wang, D., Ge, L., He, J., Zhang, W., Jaisi, D. P., & Zhou, D. (2014). Hyperexponential and nonmonotonic retention of polyvinylpyrrolidone-coated silver nanoparticles in an Ultisol. *Journal of Contaminant Hydrology, 164*, 35–48.

Wei, Y.-T., Wu, S.-C., Yang, S.-W., Che, C.-H., Lien, H.-L., & Huang, D.-H. (2012). Biodegradable surfactant stabilized nanoscale zero-valent iron for in situ treatment of vinyl chloride and 1, 2-dichloroethane. *Journal of Hazardous Materials, 211*, 373–380.

Yang, G. C. C., Tu, H.-C., & Hung, C.-H. (2007). Stability of nanoiron slurries and their transport in the subsurface environment. *Separation and Purification Technology, 58*, 166–172.

Yao, K.-M., Habibian, M. T., & O'Melia, C. R. (1971). Water and waste water filtration. Concepts and applications. *Environmental Science & Technology, 5*, 1105–1112.

Zhu, X., Chen, H., Li, W., He, Y., Brookes, P. C., & Xu, J. (2014). Aggregation kinetics of natural soil nanoparticles in different electrolytes. *European Journal of Soil Science, 65*, 206–217.

Chapter 6
Mechanistic, Mechanistic-Based Empirical, and Continuum-Based Concepts and Models for the Transport of Polyelectrolyte-Modified Nanoscale Zerovalent Iron (NZVI) in Saturated Porous Media

Tanapon Phenrat, Peyman Babakhani, Jonathan Bridge, Ruey-an Doong, and Gregory V. Lowry

Abstract Controlled emplacement of polyelectrolyte-modified NZVI at a high particle concentration (1–10 g/L) is needed for effective in situ subsurface remediation. For this reason, a modeling tool capable of predicting polyelectrolyte-modified NZVI transport is imperative. However, the deep bed filtration theory is invalid for this purpose because several phenomena governing the transport of polyelectrolyte-modified NZVI in saturated porous media, including detachment, particle

T. Phenrat (✉)
Department of Civil Engineering, Environmental Engineering Program, Naresuan University, Phitsanulok, Thailand

Center of Excellence for Sustainability of Health, Environment and Industry (SHEI), Faculty of Engineering, Naresuan University, Phitsanulok, Thailand

P. Babakhani
Department of Biomedical Engineering and Environmental Sciences, National Tsing Hua University, Hsinchu City, Taiwan

Department of the Natural and Built Environment, Sheffield Hallam University, Sheffield, UK

J. Bridge
Department of the Natural and Built Environment, Sheffield Hallam University, Sheffield, UK

R.-a. Doong
Department of Biomedical Engineering and Environmental Sciences, National Tsing Hua University, Hsinchu City, Taiwan

Institute of Environmental Engineering, National Chiao Tung University, Hsinchu City, Taiwan

G. V. Lowry
Center for Environmental Implications of Nanotechnology (CEINT), Durham, NC, USA

Department of Civil & Environmental Engineering, Carnegie Mellon University, Pittsburgh, PA, USA

T. Phenrat, G. V. Lowry (eds.), *Nanoscale Zerovalent Iron Particles for Environmental Restoration*, https://doi.org/10.1007/978-3-319-95340-3_6

agglomeration, straining, and porous media ripening, violate the fundamental assumption of such a classical theory. Thus, this chapter critically reviews the literature of each phenomenon with various kinds of nanoparticles with a special focus on polyelectrolyte-modified NZVI. Then, each phenomenon is elaborated using three kinds of mathematical models, including mechanistic (such as extended DLVO theory), mechanistic-based empirical (correlations to predict NZVI agglomeration and deposition), and continuum-based (Eulerian continuum-based models). These proposed modeling tools can be applied at various scales from column experiments (1-D) to field-scaled operations (3-D) for designing NZVI injection and emplacement in the subsurface.

Keywords Nanoscale zerovalent iron · Transport in porous media · Groundwater · Mechanism · Empirical · Numerical · Modeling tools

6.1 Introduction

Controlled emplacement of polyelectrolyte-modified nanoscale zerovalent iron (polyelectrolyte-modified NZVI) at high particle concentration (1–10 g/L) is needed for effective in situ subsurface remediation. Understanding the fate and transport of polymer-modified NZVI at low particle concentration is essential for assessing potential off-site migration of NZVI for risk management purposes. While it is feasible to model the fate and transport of various nanoparticles (NPs), deep bed filtration theory alone is not capable of estimating the transport and deposition of either concentrated or diluted polyelectrolyte-modified NZVI. This is because several other important phenomena including detachment, particle agglomeration, straining, and porous media ripening take place during transport, in violation of a fundamental assumption of the theory (Phenrat et al. 2009, 2010a, b, c). This chapter reviews all the underlying phenomena that can possibly affect the behavior of polyelectrolyte-modified NZVI in subsurface porous media (Tratnyek and Johnson 2006; Phenrat et al. 2009, 2010a, b, c; Praetorius et al. 2012; Ehtesabi et al. 2013; Hashemi et al. 2013; Bardos et al. 2014; Babakhani et al. 2018; Yu et al. 2015b; Baalousha et al. 2016; Peijnenburg et al. 2016). The mechanics of all relevant phenomena will be explained, and empirical and continuum-based approaches will be used to provide computational tools. This knowledge will facilitate more effective and precise descriptions and predictions about the fate and transport of polyelectrolyte-modified NZVI in porous media, thereby facilitating the development of NZVI delivery tools for subsurface restoration and regulations to halt conversion of NZVI from an emerging solution (Bottero et al. 2015) to an emerging threat (Tian et al. 2010; Baalousha et al. 2016) (Chap. 15).

6.2 Mechanistic vs Mechanistic-Based Empirical vs Continuum-Scale Models

Modeling tools for assessing NZVI delivery and exposure in porous media may be categorized into three major groups: (1) mechanistic models, (2) empirical models, and (3) continuum-scale models. Selection of the most appropriate model for the situation and modeling purpose is a trade-off between simplicity and accuracy (Comba and Braun 2012; Dale et al. 2015a) and will depend on the scale of the assessment (particle scale, pore scale, laboratory scale, or field scale) and the required resolution/precision of the outcomes.

Mechanistic models focus on individual particles and consider the force, torque, and energy of the particle and interacting media (McDowell-Boyer et al. 1986; Adamczyk and Weroński 1999; Schijven and Hassanizadeh 2000; Grasso et al. 2002; Bradford and Torkzaban 2008; Keir et al. 2009; Phenrat et al. 2009; Hotze et al. 2010; Petosa et al. 2010; Adamczyk et al. 2013; Molnar et al. 2015). Derjaguin-Landau-Verwey-Overbeek (DLVO) theory (Chap. 5) is perhaps the most frequently used mechanistic model for transport of NZVI in porous media. Although this is a lively area of ongoing research about NZVI and other NPs, mechanistic models are typically used to explain phenomena affecting NZVI transport rather than to design NZVI delivery, presumably due to the technical challenges posed when several phenomena occur simultaneously during transport of NZVI in porous media. Moreover, the intrinsic complexities and heterogeneities prevalent in real environmental conditions, such as surface roughness (Yao et al. 1971; Ryan and Elimelech 1996; Elimelech et al. 1998; Tufenkji and Elimelech 2004a), natural organic matter (NOM) (Hotze et al. 2010; Phenrat et al. 2010c), iron oxyhydroxide coating (Tian et al. 2012; Wang et al. 2012b; Liu et al. 2013), silylation (Elimelech et al. 2000), extracellular polymeric substances, and biofilm (Jiang et al. 2013; Li et al. 2013; He et al. 2015), make it difficult to use mechanistic models that impose assumptions of uniformity. Obviously, the next challenge in terms of these models is the development of mechanistic descriptions of particle interactions with heterogeneities at a range of scales.

Mechanistic-based empirical models are an alternative to pure mechanistic models. They are empirical in nature but are also based on a mechanical understanding of relevant parameters and phenomena. Elimelech (Eq. 6.6) (Elimelech 1992) was the first researcher who proposed a mechanistic-based empirical model for predicting deposition of electrostatically stabilized (bare) colloids. Subsequently, Bia and Tien (Eq. 6.7) (Bai and Tien 1996, 1999) proposed semiempirical approaches using dimensionless parameters based on DLVO forces to model the collision efficiency of colloids in porous media. The correlations successfully modeled sticking coefficients of several sets of deep bed filtration experimental data for bare colloidal particles. Phenrat et al. (2010c) further extended this approach to model transport of polyelectrolyte-modified NZVI at both low and high particle concentrations. These models can be applied at various scales from column experiment to field-scale operations, as discussed in this chapter.

Continuum-based models, on the other hand, are defined as partial differential equations based on the continuity principle (either the mass conservation law (mass balance) or particle number/volume balance) over the bulk spatial and temporal domains of the system which are either continuous or numerically discretized continuous domains (Van Genuchten and Wierenga 1976; Jacobson 2005; Adamczyk et al. 2013; Kelly et al. 2013; Mehmani and Balhoff 2015b; Molnar et al. 2015). This contrasts with pore-scale models, in which the domain has a discrete configuration of solid and fluid (Mehmani and Balhoff 2015b). This chapter focuses on Eulerian continuum-based models and thus excludes particle tracking analysis (generally Lagrangian methods) (Taghavy et al. 2013, 2015), as it considers particles a separate constituent from fluid. Continuum models have been widely used for simulating miscible contaminants and classical colloid transport problems over the past decades, and more recently they have been increasingly popular for modeling NP mobility behaviors in the environment (Van Genuchten and Wierenga 1976; Saiers and Hornberger 1996; Zheng and Wang 1999b; Dale et al. 2015a, b). These models can be applied at various scales, such as pore scale (Seetha et al. 2015), column experiments (e.g., Bradford et al. 2003; Babakhani et al. 2018; Bradford and Torkzaban 2015; Chrysikopoulos and Katzourakis 2015), bench-scale experiments (Babakhani et al. 2018), mesocosm experiments (Therezien et al. 2014; Baalousha et al. 2016), and field-scale operations (Cullen et al. 2010; Krol et al. 2013). This allows continuum-based models to be applied not only for designing NZVI emplacement but also for delineating the risk of NZVI release to the environment (Kelly et al. 2013; Nowack et al. 2015).

In contrast to mechanistic approaches which deal with forces or energy, continuum models deal with rates. These models depend on the capability of receiving different types of data based on particle number concentration (e.g., Song and Elimelech 1993; Kretzschmar et al. 1999; Bradford et al. 2003, 2006a; Babakhani et al. 2018) or particle mass concentration (e.g., Saiers et al. 1994; Kretzschmar et al. 1999; Sun et al. 2001) or the ability to transform data variables internally (Babakhani et al. 2018; Taghavy et al. 2015) and can be calibrated and validated against a wide variety of available data types obtained from simplified single-process or multi-process laboratory experiments or realistic environmentally relevant field measurements (Cullen et al. 2010; Krol et al. 2013; Babakhani et al. 2018; Chowdhury et al. 2015). More noteworthy is their ability to describe several transport phenomena simultaneously (e.g., Saiers et al. 1994; Bradford et al. 2003, 2006b; Scheibe and Wood 2003; Bradford and Bettahar 2006; Shen et al. 2008; Tosco and Sethi 2010; Raychoudhury et al. 2012; Hosseini and Tosco 2013; Wang et al. 2014b; Babakhani et al. 2018). This chapter will focus on discussing transport phenomena on a continuum scale and incorporating them into a model for NZVI applications.

6.3 Transport Mechanisms and Continuum Modeling Approaches

According to a literature survey, Fig. 6.1 illustrates NZVI transport phenomena evaluated over the past decade in comparison to those for other types of NPs. Table 6.1 summarizes continuum-based modeling types applied for the transport of NZVI in porous media. Table 6.2 summarizes literature pertaining to NZVI transport that is reviewed in this chapter. This section describes each phenomenon separately by first mechanistically explaining their physics using mechanical models, then providing mechanistic-based empirical models if available, and finally summarizing their respective continuum modeling approaches.

6.3.1 Advection and Dispersion

Advection and dispersion are core phenomena in every solute and NP transport in subsurface and porous media. Advection is the mechanism of solute or NP transport with the average velocity of flowing groundwater. Mechanical dispersion (Zheng and Wang 1999b) acts to spread the solute and NP suspension parallel with and normal to the direction of flow because of the difference between the real velocity of the water inside the pores and the mean groundwater velocity, which arises from microscale velocity variation both within and between pores of different sizes

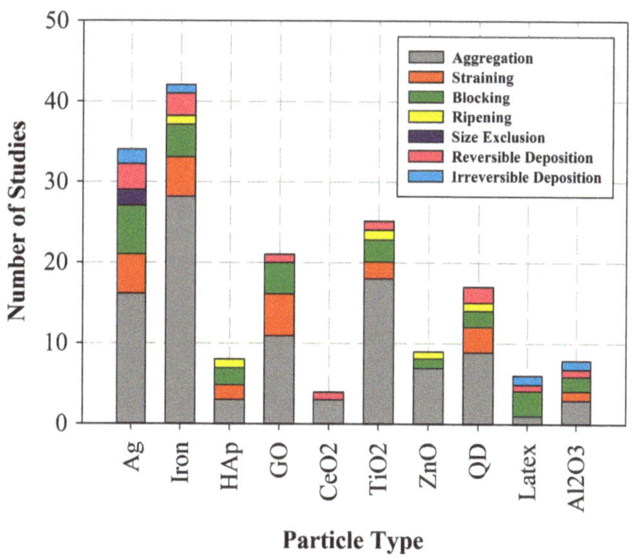

Fig. 6.1 NZVI (iron) transport phenomena evaluated over the past decade in comparison to those for other types of NPs

Table 6.1 List of continuum-based modeling types so far applied for the transport of nanoparticles in porous media[a]

Model type no.	Colloid transport model equations	Modeled mechanisms other than advection and dispersion	Fitting parameters other than dispersion	Reference
i	(b) $\frac{\rho_b}{\varepsilon}\frac{\partial S}{\partial t} = K_{att}C$	Attachment	K_{att}	Iwasaki et al. (1937), Herzig et al. (1970), Tufenkji and Elimelech (2004b)
ii	(b) $\frac{\rho_b}{\varepsilon}\frac{\partial S}{\partial t} = K_{att}C - \frac{\rho_b}{\varepsilon}K_{det}S$	Attachment/detachment	K_{att}, K_{det}	de Marsily (1986), Kretzschmar et al. (1999)
iii	(b) $\frac{\rho_b}{\varepsilon}\frac{\partial S}{\partial t} = K_{att}C - \frac{\rho_b}{\varepsilon}f_r K_{det}S$	Attachment and dual-site detachment	K_{att}, K_{det}, f_r	Jiang et al. (2012a)
iv	(b) $\frac{\partial S}{\partial t} = \frac{\partial S_1}{\partial t} + \frac{\partial S_2}{\partial t}$ (c) $\frac{\rho_b}{\varepsilon}\frac{\partial S_1}{\partial t} = K_{att,1}C - \frac{\rho_b}{\varepsilon}K_{det,1}S_1$ (d) $\frac{\rho_b}{\varepsilon}\frac{\partial S_1}{\partial t} = K_{att,2}C - \frac{\rho_b}{\varepsilon}K_{det,2}S_2$	Dual-site attachment/detachment	$K_{att,1}$, $K_{det,1}$, $K_{att,2}$, $K_{det,2}$	Schijven et al. (2002)
v	(b) $\frac{\partial S}{\partial t} = \frac{\partial S_1}{\partial t} + \frac{\partial S_2}{\partial t}$ (c) $\frac{\rho_b}{\varepsilon}\frac{\partial S_1}{\partial t} + \frac{\rho_b}{\varepsilon}\frac{\partial S_2}{\partial t} = K_{att,1}\psi_{b,1}C + K_{att,2}\psi_{b,2}C$ (d) $\psi_{b,1} = \left(1 - \frac{S_1}{S_{m,1}}\right)$ (e) $\psi_{b,2} = \left(1 - \frac{S_2}{S_{m,2}}\right)$	Dual-site with site-blocking kinetic attachment for available and not available favorable sites	$K_{att,1}$, $K_{att,2}$, $S_{m,1}$, $S_{m,2}$	Sasidharan et al. (2014)
vi	(a) $\frac{\rho_b}{\varepsilon}\frac{\partial S}{\partial t} = K_{rel}\frac{\rho_b}{\varepsilon}(S - S_{eq})H_0(S - S_{eq})$ (b) $S_{eq} = f_{nr}S_i$	Release after perturbation	K_{rel}, f_{nr}	Torkzaban et al. (2013), Bradford et al. (2014), Torkzaban et al. (2015)
vii	(c) $\frac{\partial C}{\partial t}\left(1 + \frac{\rho_b}{\varepsilon}K_d\right) = D\frac{\partial^2 C}{\partial x^2} - V\frac{\partial C}{\partial x} - K_{att}C$	Attachment, adsorption (retardation)	K_{att}, K_d	Harvey and Garabedian (1991), He et al. (2009)

viii	(a) $\frac{\partial C}{\partial t}+\frac{\rho_b}{\varepsilon}\frac{\partial S}{\partial t}=D\frac{\partial^2 C}{\partial x^2}-V\frac{\partial C}{\partial x}-\lambda_1 C-\lambda_2\frac{\rho_b}{\varepsilon}S$ (b) $\frac{\rho_b}{\varepsilon}\frac{\partial S}{\partial t}=K_{att}C-\frac{\rho_b}{\varepsilon}K_{det}S$ (c) $C=\frac{\widehat{C}}{\frac{4}{3}\pi r^3\rho_p}$ (d) $S=\frac{\widehat{S}}{\frac{4}{3}\pi r^3\rho_p}$ (e) $r=r_0\,e^{\frac{\lambda_1 t}{3}}$	Agglomeration, attachment/detachment, and irreversible deposition	K_{att}, K_{det}, λ_1, λ_2	Babakhani et al. (2015)
ix	(b) $\frac{\rho_b}{\varepsilon}\frac{\partial S}{\partial t}=K_{att}\psi_x C-\frac{\rho_b}{\varepsilon}K_{det}S$ (c) $\psi_x=\left(\frac{d_c+x}{d_c}\right)^{-\beta}$	Depth-dependent (straining) attachment together with detachment	K_{att}, K_{det}, β	Bradford et al. (2003)
x	(b) $\frac{\partial S}{\partial t}=\frac{\partial S_1}{\partial t}+\frac{\partial S_2}{\partial t}$ (c) $\frac{\rho_b}{\varepsilon}\frac{\partial S_1}{\partial t}=K_{att,1}\psi_b C-\frac{\rho_b}{\varepsilon}K_{det,1}S_1$ (d) $\frac{\rho_b}{\varepsilon}\frac{\partial S_2}{\partial t}=K_{att,2}\psi_x C-\frac{\rho_b}{\varepsilon}K_{det,2}S_2$ (e) $\psi_b=\left(1-\frac{S_1}{S_m}\right)$ (f) $\psi_x=\left(\frac{d_c+x}{d_c}\right)^{-\beta}$	Dual-site with site-blocking attachment and depth-dependent (straining), detachment	$K_{att,1}$, $K_{det,1}$, $K_{att,2}$, $K_{det,2}$, S_m, β	Wang et al. (2012d)
xi	(b) $\frac{\rho_b}{\varepsilon}\frac{\partial S}{\partial t}=K_{att}\psi_b\psi_x C-\frac{\rho_b}{\varepsilon}K_{det}S$ (c) $\psi_b=\left(1-\frac{S}{S_m}\right)$ (d) $\psi_x=\left(\frac{d_c+x}{d_c}\right)^{-\beta}$	Site-blocking, depth-dependent (straining) attachment together with detachment	K_{att}, K_{det}, S_m, β	Bradford and Bettahar (2006)
xii	(b) $\frac{\rho_b}{\varepsilon}\frac{\partial S}{\partial t}=K_{att}\psi_x C-\frac{\rho_b}{\varepsilon}K_{det}S+\rho_b K_{rip}\,S\,C$ (c) $\psi_x=\left(\frac{d_c+x}{d_c}\right)^{-\beta}$	Ripening, depth-dependent (straining) attachment together with detachment	K_{att}, K_{det}, K_{rip}, β	Bradford and Bettahar (2006), Wang et al. (2014b)

(continued)

Table 6.1 (continued)

Model type no.	Colloid transport model equations	Modeled mechanisms other than advection and dispersion	Fitting parameters other than dispersion	Reference
xiii	(a) $\frac{\partial C_1}{\partial t} + \frac{\rho_b}{\varepsilon}\frac{\partial S_1}{\partial t} = D\frac{\partial^2 C_1}{\partial x^2} - V\frac{\partial C_1}{\partial x} - \frac{\rho_b}{\varepsilon}K_{12}F_p$ (b) $\frac{\partial C_2}{\partial t} + \frac{\rho_b}{\varepsilon}\frac{\partial S_2}{\partial t} = D\frac{\partial^2 C_2}{\partial x^2} - V\frac{\partial C_2}{\partial x} - \frac{\rho_b}{\varepsilon}K_{12}F_p$ (c) $\frac{\rho_b}{\varepsilon}\frac{\partial S_1}{\partial t} = K_{att,1}\psi_b\psi_x C_1 - \frac{\rho_b}{\varepsilon}K_{det,1}S_1 - \frac{\rho_b}{\varepsilon}K_{12}F_p$ (d) $\frac{\rho_b}{\varepsilon}\frac{\partial S_2}{\partial t} = K_{att,2}C_2 - \frac{\rho_b}{\varepsilon}K_{det,2}S_2$ (e) $\psi_b = \left(1 - \frac{S_1}{S_m}\right)$ (f) $\psi_x = \left(\frac{d_c + x}{d_c}\right)^{-\beta}$ (g) $F_p = \max\left(S_1 - S_{crit1}, 0\right)$	Dual-species for transport of aggregated and non-aggregated species with site-blocking attachment and depth-dependent (straining), detachment	$K_{att,1}$, $K_{det,1}$, $K_{att,2}$, $K_{det,2}$, S_m, β, K_{12}, S_{crit1}	Bradford et al. (2006b)
xiv	(b) $\frac{\partial S}{\partial t} = \frac{\partial S_1}{\partial t} + \frac{\partial S_2}{\partial t}$ (c) $\frac{\rho_b}{\varepsilon}\frac{\partial S_1}{\partial t} = K_{att,1}\psi_1 C - \frac{\rho_b}{\varepsilon}K_{det,1}S_1$ (d) $\frac{\rho_b}{\varepsilon}\frac{\partial S_2}{\partial t} = K_{att,2}\psi_2 C - \frac{\rho_b}{\varepsilon}K_{det,2}S_2$ (e) $\psi_1 = \left(1 + A_1 S_1^{\beta_1}\right)$ (f) $\psi_2 = \left(\frac{d_c + x}{d_c}\right)^{\beta_2}$	Dual-site with site-blocking/ripening attachment and depth-dependent (straining), detachment	$K_{att,1}$, $K_{det,1}$, $K_{att,2}$, $K_{det,2}$, A_1, β_1, β_2	Tosco and Sethi (2010), Hosseini and Tosco (2013)
xv	(b) $\frac{\rho_b}{\varepsilon}\frac{\partial S}{\partial t} = K_{att}\psi_b C - \frac{\rho_b}{\varepsilon}K_{det}S$ (c) $\psi_b = \left(1 - \frac{S}{S_m}\right)$	Site-blocking attachment together with detachment	K_{att}, K_{det}, S_m	Saiers et al. (1994)
xvi	(b) $\frac{\partial S}{\partial t} = \frac{\partial S_1}{\partial t} + \frac{\partial S_2}{\partial t}$ (c) $\frac{\rho_b}{\varepsilon}\frac{\partial S_1}{\partial t} = K_{att,1}\psi_b C$ (d) $\frac{\rho_b}{\varepsilon}\frac{\partial S_2}{\partial t} = K_{att,2}\psi_x C$ (e) $\psi_b = \left(1 - \frac{S_1}{S_m}\right)$ (f) $\psi_x = \max\left(1, S_2^{S_m'}\right)$	Dual-site with site-blocking attachment and straining at very high ionic strength	$K_{att,1}$, $K_{det,1}$, $K_{att,2}$, $K_{det,2}$, S_m, β	Qi et al. (2014b)

Note: Model xii also includes modifications in the flow condition for changes in porosity and permeability due to clogging, which are not presented here.

[a]Wherever Eq. (a) is not presented, it must be considered the same as Eq. (6.1) given in Sect. 6.3.1. x is distance in the porous media [with the dimension of L]; t is time elapsed [T]; C with dimension of [NL^{-3}; N represents the particle number, not a dimension]; and S with dimension of [NM^{-1}] are the particle number concentrations of fluid-phase particles and deposited-phase particles, respectively, according to a number of references (e.g., Song and Elimelech 1993; Kretzschmar et al. 1999; Bradford et al. 2003; Bradford et al. 2006a; Babakhani et al. 2015), or C with the dimension of [ML^{-3}] and S with a dimension of [MM^{-1}] are the particle mass concentrations of fluid-phase particles and deposited-phase particles according to other references (e.g., Saiers et al. 1994; Kretzschmar et al. 1999; Sun et al. 2001); V is the interstitial particle velocity or pore water velocity [LT^{-1}]; D is the hydrodynamic dispersion coefficient [L^2T^{-1}]; ε is the bed porosity [—]; ρ_b is the porous medium bulk density [ML^{-3}]; K_{att} and K_{det} are the attachment and detachment rate constants, respectively [T^{-1}]; f_r is the fraction of reversibly retained particles [—]; ψ_b is the Langmuirian blocking function related to the fraction of porous medium available for deposition [—]; S_m is the maximum retained-particle phase concentration, i.e., colloidal retention capacity [NM^{-1} or MM^{-1}]; ψ_x is the depth-dependent retention function commonly used for hyper-exponential retention profile modeling [—]; d_c is the granular media average diameter [L]; β is the empirical depth-dependent retention parameter [—]; K_{rip} is the particle-particle interaction rate coefficient between fluid-phase particles and deposited-phase particles [L^3T^{-1}]; F_p is the aggregated species production function [NM^{-1}]; S_{crit1} is the critical concentration of S_1 when production starts [NM^{-1}]; K_{12} is the first-order production coefficient to account for the release rate of aggregated species [T^{-1}]; A_1 and β_1 are the multiplier and exponent coefficients, respectively [—], which define the interaction dynamics, i.e., ripening for $A_1 > 0$ and $\beta_1 > 0$ or blocking for $\beta_1 = 1$ and $A_1 = -1/S_m$; ψ_1 is the expression for the effect of ripening or blocking which is equal to ψ_b when $\beta_1 = 1$ and $A_1 = -1/S_m$; ψ_2 is the expression for the effect of straining, equal to ψ_x if assuming $\beta_2 = -\beta$; $\hat{C}$ is the mass concentration of fluid-phase particles [ML^{-3}]; $\hat{S}$ is the mass concentration of deposited-phase particles [MM^{-1}]; r is the average radii of particles in the fluid phase and deposited phase [L]; r_0 is the average radius of particles (or agglomerates) at $t = 0$ [L]; ρ_p is the average density of the particles or aggregates [ML^{-3}]; λ_1 is the pseudo-first-order reaction rate [T^{-1}], which stands for the decay in population of particles due to agglomeration; λ_2 is the pseudo-first-order reaction rate [T^{-1}], which stands for the decay in the population of deposited, detachable particles and represents irreversible deposition; K_d is the well-known distribution coefficient of the nanoparticles between the aqueous and solid phases [L^3M^{-1}]; K_{rel} is the release rate constant; S_{eq} [NM^{-1} or MM^{-1}] is the equilibrium value of S in the new steady state following perturbation; $H_o(S - S_{eq})$ is the Heaviside function (acting as a canceling function) which is equal to one when $S > S_{eq}$ and zero when $S \leq S_{eq}$; S_i [NM^{-1} or MM^{-1}] is the value of S before the perturbation; f_{nr} [–] is the fraction of deposited particles that are not released with the perturbation in the system conditions; and indexes 1 and 2 in dual-site models (types (iv), (x), (xiv), (xvi), and (v)) stand for sites 1 and 2, and in dual-species model (type (xiii)) stand for species 1 and 2.

Table 6.2 Summary of the NZVI transport studies, experimental conditions, and outcomes together with the major observed phenomena, the shapes of the breakthrough curve and residual concentration profile, and types of the applied continuum model

No.	NP type	Coating and free polymer	IS species	pH	IS (mM)	C_0 (mg/L)	d_p/d_g [a]	C/C_0 (%)	BTC plateau shape[b]	RP shape	Major underlying phenomena	Model type (s) used	Ref.
1	NZVI	PAA	$NaClO_4$	8.6	1	200	3.1E-04–3.5E-04	18–70	Continuous, flat	NM	Aggregation, irreversible deposition	i	Laumann et al. (2013)
2	Iron oxide NPs	PAA	NC	NC	NC	20–500	2.7E-05–4.7E-05	88–97	Continuous, flat	NM	Irreversible deposition	i	Golzar et al. (2014)
3	NZVI	CMC	NC	8.5	30–160	100–2500	5.4E-05–1.3E-04	80	Continuous, descending at low velocity	NM	Aggregation and settling in the feeding stock dispersion	i With modified boundary condition	Kocur et al. (2013)
4	NZVI	Pure PAA-NZVI; NOM; Lignin sulfonate; CMC; HA	$NaHCO_3$	7.9–8.3	1	200	1.6E-03–1.9E-03	22–53	Continuous, flat	NM	Irreversible deposition	i	Laumann et al. (2014)
5	NZVI	CMC	KCl	8.1	15	200	1.6E-05–5.8E-05	69–99	Symmetric, flat	NM	Filtration due to Brownian diffusion and gravitational sedimentation	vii	He et al. (2009)
6	NZVI	CMC	$NaHCO_3$	7.4	0.1	85–1700	2.2E-04–1.2E-03	55–97	Predominately ascending	Hyper-exponential	Straining, detachment	x without ψ_b and K_{det2}	Raychoudhury et al. (2014)
7	NZVI	Dispersed in xanthan solution	NC	NC	0–13	20000	1.0E-04	86–90	Flat	Hyper-exponential	Attachment, detachment (tailing), blocking, ripening, and clogging	xiv	Tosco et al. (2010), (Tosco and Sethi 2010)

8	NZVI	Cu	Natural groundwater	7.2	40	8000	8.4E-05–8.7E-05	61–90	Ascending at low concentration (2 g/L), descending at middle concentration (5 g/L), and multiple peaks and tailing at very high concentrations (8 and 12 g/L)	NC	Clogging after ripening	xiv	Hosseini and Tosco (2013)
9	NZVI	Poly(styrene sulfonate) (PSS)	5 mM NaCl 5 – mM $NaHCO_3$	8	10	30–6000	1.6E-04–2.1E-3	61.5–97	Flat	NM	Agglomeration and subsequent irreversible deposition of agglomerates; disaggregation Identified by the model	ii; viii	Babakhani et al. (2015)
10	NZVI	CMC	$NaHCO_3$	7.4	0.1	70–725	5.1E-04	55–72	Continuous, ascending	NM	Aggregation, detachment	viii	Babakhani et al. (2015) and Raychoudhury et al. (2012)

NM not measured, *NA* not applicable, *NC* not clear, *PVP* polyvinylpyrrolidone, *CMC* carboxy-methyl-cellulose, *PAA* Polyacrylic acid, *SDBS* anionic sodium dodecyl benzene sulfonate, *CTAB* cationic cetyltrimethylammonium bromide, *COOHP* carboxyl derivatized polymer, *IS* ionic strength

[a]d_p/d_g is calculated based on the ratio of the average hydrodynamic diameter to the average grain size which might not be according to the what the authors reported, i.e., they might have calculated d_p/d_g based on other types of size measurement techniques than hydrodynamic diameter used here

[b]BTC BTC is breakthrough curve plateau shape mostly considered at intermediate or favorable retention condition

(Prieve and Hoysan 1978; de Marsily 1986; Howington et al. 1997; Zheng and Wang 1999b; James and Chrysikopoulos 2003; Fallah et al. 2012; Mehmani and Balhoff 2015a, b). Adding the molecular diffusion to the mechanical dispersion will result in hydrodynamic dispersion (Zheng and Wang 1999b).

In case of slow groundwater velocity, such as flow through compacted porous media, dispersion is of low significance but can nevertheless result from the molecular diffusion due to concentration gradients (Zheng and Wang 1999a; Huang et al. 2011). Accordingly, the transport of solutes and NPs through porous media in the most simplified form might be described by the partial differential "advection-dispersion" equation (ADE) (Iwasaki et al. 1937; Herzig et al. 1970; Van Genuchten and Wierenga 1976; Harvey and Garabedian 1991; Zheng and Wang 1999a; Zhang et al. 2010; Huang et al. 2011):

$$\frac{\partial C}{\partial t} + \frac{\rho_b}{\varepsilon}\frac{\partial S}{\partial t} = D\frac{\partial^2 C}{\partial x^2} - V\frac{\partial C}{\partial x} \tag{6.1}$$

where x is distance in the porous media [with the dimension of L], t is time elapsed [T], C [N L^{-3}; N represents the particle number], and S [NM^{-1}] are the particle number concentrations of fluid-phase particles and deposited-phase particles, respectively (e.g., Song and Elimelech 1993; Kretzschmar et al. 1999; Bradford et al. 2003, 2006a; Babakhani et al. 2018), or alternatively C [ML^{-3}] and S [MM^{-1}] are the particle mass concentrations of fluid-phase particles and deposited-phase particles, respectively (e.g., Saiers et al. 1994; Kretzschmar et al. 1999; Sun et al. 2001); V [LT^{-1}] is the pore water velocity (also known as real water velocity, linear groundwater flow velocity, seepage water velocity, interstitial velocity, or advection velocity). This contrasts with the Darcy velocity (also known as the superficial velocity or approach velocity) which is equal to the porosity times the pore water velocity [LT^{-1}]. D is the hydrodynamic dispersion or simply dispersion coefficient [L^2T^{-1}]; ε is the bed porosity [—]; and ρ_b is the porous medium bulk density [ML^{-3}].

In a one-dimensional uniform flow field such as column experiments, transverse dispersion is not considered, and thus the dispersion coefficient has one longitudinal component related to pore water velocity (Wang and Anderson 1995; Zheng and Wang 1999b):

$$D = \alpha_L V + D^* \tag{6.2}$$

where α_L is longitudinal dispersivity [L] and D^* is the effective molecular diffusion coefficient [L^2T^{-1}]. α_L is typically an intrinsic property of the porous medium as well as a function of scale (Wang and Anderson 1995; Keller et al. 2004; Illangasekare et al. 2010; Chrysikopoulos and Katzourakis 2015). Thus, the common practice for both NPs and reactive solute transport is to determine this coefficient based on the breakthrough data of nonreactive solute transport through any given porous media (Parker et al. 1984; Harvey and Garabedian 1991; Wang and Anderson 1995; Chrysikopoulos and Katzourakis 2015). In the case of NP transport, strictly speaking, the dispersion coefficient should be obtained from the NP breakthrough data rather than from nonreactive solute transport due to the potential for particulate-specific dispersion mechanisms such as size exclusion and preferential flow (discussed in later sections) (Bradford et al. 2003; James and Chrysikopoulos 2003; Keller et al. 2004; Chrysikopoulos and Katzourakis 2015). For example, recently Chrysikopoulos and Katzourakis (2015) in a meta-study of 48 different breakthrough curve (BTC) datasets found that dispersivity can exhibit positive correlation with particle size and pore water velocity.

6.3.2 Attachment and Detachment

6.3.2.1 Irreversible Deposition

Solute mass transfer from liquid to solid phases is generally dominated by sorption which is modeled as a reversible kinetic process (Limousin et al. 2007). However, NZVI interacts with solid surfaces of porous media by a number of mechanisms which may yield attachment that is practically irreversible. In other words, solute retention which appears irreversible at short timescales can be considered reversible at larger time scales under the same conditions (Limousin et al. 2007), whereas irreversibly attached NZVI may not be detached unless a significant physical or chemical perturbation occurs in the conditions of the system (Ryan and Elimelech 1996). Irreversible deposition or physicochemical filtration was well described by colloid filtration theory (CFT) originally by drawing an analogy with the transport in flocculation process (Yao et al. 1971) by trajectory analysis (Rajagopalan and Tien 1976) and generally by using filtration mass balance equations in a sphere around a single collector (Logan et al. 1995a). According to this theory, the extent of migration of colloids in saturated porous media can be estimated from the collision efficiency (or sticking coefficient) (α) which represents the probability that a particle successfully attaches to the surface of a collector (e.g., sand grain) once they come into contact. α is defined as the ratio between the actual dimensionless deposition rate of particles (η) and the dimensionless deposition rate under barrierless interaction (η_0), where all collisions result in attachment:

$$\alpha = \frac{\eta}{\eta_0} \tag{6.3}$$

Filtration theory relies upon estimation of two parameters, the single-collector attachment efficiency, α, and the single-collector collision or contact efficiency, η_0. The single-collector collision frequency, η_0, can be calculated using a mechanistic-based empirical model such as Tufenkji and Elimelech's correlation (Tufenkji and Elimelech 2004a):

$$\eta_0 = 2.4A_S^{1/3}N_R^{-0.081}N_{Pe}^{-0.715}N_{vdW}^{0.052} + 0.55A_SN_R^{1.675}N_A^{0.125} + 0.22N_R^{-0.24}N_G^{1.11}N_{vdW}^{0.053} \quad (6.4)$$

A_S is a porosity-dependent parameter, N_R is an aspect ratio, N_{Pe} is the Peclet number, N_{vdw} is the van der Waals number, N_A is the attraction number, and N_G is the gravity number. The first, second, and third terms represent the transport of particles to a collector via diffusion, interception, and gravity, respectively (Tufenkji and Elimelech 2004a). This single-collector collision or contact efficiency describes the probability of that collision occurring in the first place.

On the other hand, the second parameter (α) describes the probability of any collision between a particle and a collector resulting in attachment. Practically, α is obtained from particle breakthrough at steady state, C, in a packed column (a clean bed media of length L packed with a collector with an average radius a_c, porosity ε, and initial particle concentration C_0) (Eq. 6.5):

$$\alpha_{exp} = -\ln\left(\frac{C}{C_0}\right)\frac{4a_c}{3(1-\varepsilon)\eta_0 L} \quad (6.5)$$

Similar to the DLVO theory elaborated in Chap. 5, for bare (uncoated) colloidal particles, α is theoretically a function of surface forces that act between a particle and a collector, namely, electrostatic double-layer (EDL) repulsion and van der Waals attraction (Elimelech et al. 1995), when considering a general, simplified case (i.e., neglecting chemical and physical heterogeneity of a collector and hydrodynamic and diffusion forces). Irreversible attachment is supposed to be driven by the primary minimum (see Fig. 5.7 in Chap. 5). Using dimensional analysis, Elimelech (1992) and Bai and Tien (1996, 1999) proposed dimensionless parameters (Table 6.3) and mechanistic-based empirical correlations (Eqs. 6.6 and 6.7, respectively) to estimate the collision efficiency of electrostatically stabilized nanoparticles (without surface coatings) in such simplified porous media:

$$\alpha = 2.57 \times 10^{-2}N_{col}^{1.19} \quad (6.6)$$

$$\alpha = 2.527 \times 10^{-3}N_{Lo}^{0.7031}N_{E1}^{-0.3121}N_{E2}^{3.5111}N_{DL}^{1.352} \quad (6.7)$$

The Elimelech correlation accounts for screened electrostatic double-layer (EDL) repulsion and van der Waals attractions. These are captured by the particle zeta potential and solution Debye screening length and the Hamaker constant, respectively. N_{col} is essentially the ratio of the Hamaker constant (H) that controls van der Waals attraction over zeta potentials of particles and collectors (ζ_p and ζ_g) that determine EDL repulsions. Bai and Tien empirically account for similar interactions in a more detailed way.

The set of dimensionless parameters (N_{Lo}, N_{E1}, N_{E2}, N_{DL}) proposed by Bai and Tien (1996, 1999) in Eq. (6.7) is the deconvolution of N_{col} with each dimensionless

Table 6.3 Dimensionless parameters (Elimelech 1992; Bai and Tien 1996, 1999; Tufenkji and Elimelech 2004a) used in Eqs. (6.4)–(6.7)

Notation	Definition	Formula
N_{col}	Stability parameter	$\frac{\kappa H}{\varepsilon_0 \varepsilon_r \zeta_p \zeta_g}$
N_{Lo}	London number	$\frac{4H}{9\pi\mu d_p^2 u_s}$
N_{E1}	First electrokinetic parameter	$\frac{\varepsilon_0 \varepsilon_r \left(\zeta_p^2 + \zeta_g^2\right)}{3\pi\mu u_s d_p}$
N_{E2}	Second electrokinetic parameter	$\frac{2\zeta_p \zeta_g}{\left(\zeta_p^2 + \zeta_g^2\right)}$
N_{DL}	Double-layer force parameter	κd_p
N_{R}	Aspect ratio	$\frac{d_p}{d_c}$
N_{Pe}	Peclet number	$\frac{u_s d_c}{D_\infty}$
N_{vdW}	van der Waals number	$\frac{H}{k_b T}$
N_{A}	Attraction number	$\frac{H}{3\pi\mu d_p^2 u_s}$
N_{G}	Gravity number	$\frac{d_p^2(\rho_p - \rho_f)g}{18\mu u_s}$
As	Porosity-dependent parameter	$\frac{2(1-\gamma^5)}{2-3\gamma+3\gamma^5-2\gamma^6}$

Note: d_p is particle diameter; d_c is collector diameter; D_∞ is the bulk diffusion coefficient (described by Stokes-Einstein equation); ε_0 and ε_r are permittivity of vacuum and dielectric constant, respectively; g is the gravitational acceleration; H is the Hamaker constant; k_b is the Boltzmann constant; κ is the Debye parameter; ρ_p and ρ_f are the density of particle and fluid, respectively; T is fluid absolute temperature; u_s is fluid superficial velocity; μ is viscosity of fluid; ζ_p and ζ_g are electrical surface potential of particle and collector, respectively; and γ is $(1 - f)^{1/3}$, where f is porosity

parameter representing a specific physical phenomenon. N_{Lo} and the product of N_{E1}, N_{E2}, and N_{DL} represent the van der Waals attraction and electrostatic repulsion, respectively, over the influence of fluid velocity on particle deposition due to interception (Bai and Tien 1996; Tufenkji and Elimelech 2004a). N_{DL} accounts for electrostatic screening to scale N_{E1} and N_{E2} for a particular solution chemistry. These correlations reasonably predict the collision efficiencies of electrostatically stabilized colloids in porous media under environmentally relevant conditions (Elimelech 1992; Bai and Tien 1996, 1999). Nevertheless, Phenrat et al. (2010c) reported that neither Elimelech's nor Bai and Tien's correlation adequately predicts the collision efficiencies of polymer-coated NZVI (even at low concentration when NZVI agglomeration is not significant). Presumably, the correlations neglect electrosteric repulsions and the decreased friction afforded by such polymeric coatings that can inhibit attachment to surfaces as well as allow reversible attachment under a secondary minimum. As a result, these correlations systematically significantly overestimate the attachment efficiency of NZVI coated with NOM or polymer

(Phenrat et al. 2010c). This would tend to underestimate the transport distance and overestimate the filtration efficiency of these particles.

In continuum-based models, the definition of the attachment term depends on the conceptualization of deposition phenomena. The simplified form of the governing ADE equation for colloid transport in porous media can be written with a first-order irreversible attachment term as model type (i), Table 6.1 (Iwasaki et al. 1937; Herzig et al. 1970). In this model, K_{att} is the attachment rate coefficient $[T^{-1}]$ which can be related to the parameters of CFT via the following equation (Harvey and Garabedian 1991; Elimelech et al. 1998; O'Carroll et al. 2004; Tufenkji and Elimelech 2004a; Johnson et al. 2007; O'Carroll et al. 2012):

$$K_{att} = \frac{3(1-\varepsilon)}{2d_{50}} \alpha \eta_0 V \tag{6.8}$$

where d_{50} is the median porous media grain size [L] and ε is the porosity of porous media [—]. Although the parameter K_{att} can be expressed via the theoretical relationship of Eq. (6.8), determination of α still requires experimental data. On this basis, throughout this chapter we directly deal with the parameter K_{att} instead of α (Bradford and Torkzaban 2015; Seetha et al. 2015). From the continuum modeling viewpoint, type (i) models only describe irreversible deposition and do not consider subsequent detachment. In a particular experiment where other mechanisms of transport are not operative, increases in the value of K_{att} lead to abatement of breakthrough curve (BTC), which appears as a total drop of the plateau as shown in Fig. 6.2 with the curves generated by numerical solution of type (i) models using the MT3DMS model code (Babakhani et al. 2018). Table 6.2 shows that 44% of NZVI continuum modeling studies used type (i) models. It is evident that most used very simplified experimental conditions, highlighting that this model and thereby CFT are strictly limited in scope of application to steady flow through idealized, uniform media. There have already been several excellent review papers (Ives 1970; McDowell-Boyer et al. 1986; Schijven and Hassanizadeh 2000; Jegatheesan and Vigneswaran 2005; Bradford and Torkzaban 2008; Keir et al. 2009; Petosa et al. 2010; Bradford et al. 2014; Cornelis 2015; Molnar et al. 2015) discussing the deposition mechanisms of NZVI and other NPs in porous media. From the perspective of continuum modeling, however, it is apparent that deposition cannot be considered in isolation except in the most artificial circumstances. For example, using classical filtration theory to calculate α and η_0 by neglecting agglomeration and secondary minimum caused by adsorbed polymer layer on NZVI is inappropriate for describing the transport of polyelectrolyte-modified NZVI at high concentration where agglomeration does occur (Phenrat et al. 2009, 2010a, b) as to be discussed next.

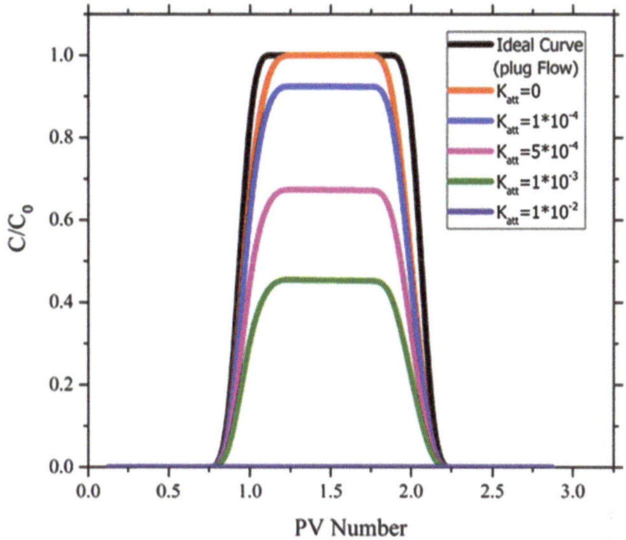

Fig. 6.2 Breakthrough curves generated by numerical solution of model type (i) with MT3DMS model code following Babakhani et al. (2015) for various K_{att} values (S^{-1}). Other parameters of flow and porous media conditions were selected according to Babakhani et al. (2015). (Reprinted with permission from Babakhani et al. (2017). Copyright (2017) Elsevier)

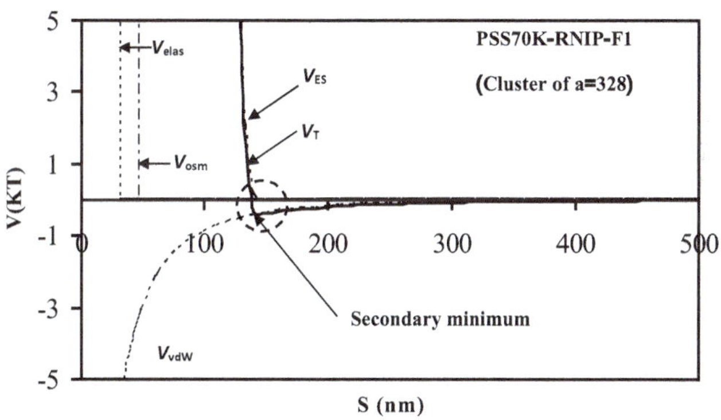

Fig. 6.3 Representative potential energy of particle-collector interaction calculated by extended DLVO for a cluster with $a = 328$ representing PSS70K-RNIP-F1 dispersion (Phenrat et al. 2009)

6.3.2.2 Detachment

When the NZVI particle association with the solid surface is not strong enough to be permanently retained, particles can be subjected to reversible deposition and thereby detachment. Particles trapped at a shallow secondary minimum (Fig. 6.3) due to energy barriers imposed by electrosteric repulsions or, as recently found, even at a shallow primary minimum resulting from nanoscale surface heterogeneities (Torkzaban and Bradford 2016; Wang et al. 2016) will be maintained just temporarily. Phenrat et al. (2009) revealed that detachment of polystyrene sulfonate (PSS)-modified NZVI transporting through a sand-packed column contributed to lower collision efficiency than what is theoretically predicted by classical DLVO theory. They hypothesized that electrosteric repulsions due to an adsorbed PSS layer on NZVI caused NZVI attachment on sand grains under a shallow secondary minimum according to extended DLVO calculation (Fig. 6.3). In a flow field, the deposited NZVI under a shallow secondary minimum can be re-entrained when the applied torque (T_{applied}) overcomes the adhesive torque (T_{adhesive}) due to the particle-collector interaction. The T_{applied} induced by hydrodynamic drag force on retained particles at the collector surface can be represented as (Bergendahl and Grasso 2000; Li et al. 2005; Torkzaban et al. 2007; Phenrat et al. 2010a):

$$T_{\text{applied}} = 1.4aF_{\text{D}} \tag{6.9}$$

where a is the radius of retained colloid particles/aggregates and F_{D} is the drag force acting on attached NZVI which can be expressed as Eq. (6.10) (Torkzaban et al. 2007):

$$F_{\text{D}} = 10.205\pi\mu\left(\frac{\partial v}{\partial r}\right)a^2 \tag{6.10}$$

where μ is the fluid viscosity and $\partial v/\partial r$ is the shear experienced by a retained particle at the pore wall. Assuming a constricted tube model represents the void space between collectors in the porous media, and $\partial v/\partial r$ can be expressed as Eq. (6.11) (Bergendahl and Grasso 2000):

$$\frac{\partial v}{\partial r} = \frac{Q/N_{\text{pore}}}{(\pi/4)d_z^2}\frac{4(d_z/2 - a)}{(d_z/2)^2} \tag{6.11}$$

d_z is the diameter of pore in the constricted tube model used to calculate ν_{colloid}, the fluid velocity at colloid (Bergendahl and Grasso 2000). Figure 6.4 illustrates d_z as a function of z from the center of a collector:

$$d_z = 2\left\{\frac{d_{\text{max}}}{2} + \left[4\left(\frac{d_{\text{c}}}{2} - \frac{d_{\text{max}}}{2}\right)\left(0.5 - \frac{z}{h}\right)^2\right]\right\} \tag{6.12}$$

$$\nu_{\text{colloid}} = \frac{Q/N_{\text{pore}}}{(\pi/4)d_z^2}\frac{4(d_z/2 - a_{\text{colloid}})}{(d_z/2)^2} \tag{6.13}$$

d_{max} (Eq. 6.12) is the maximum pore diameter.

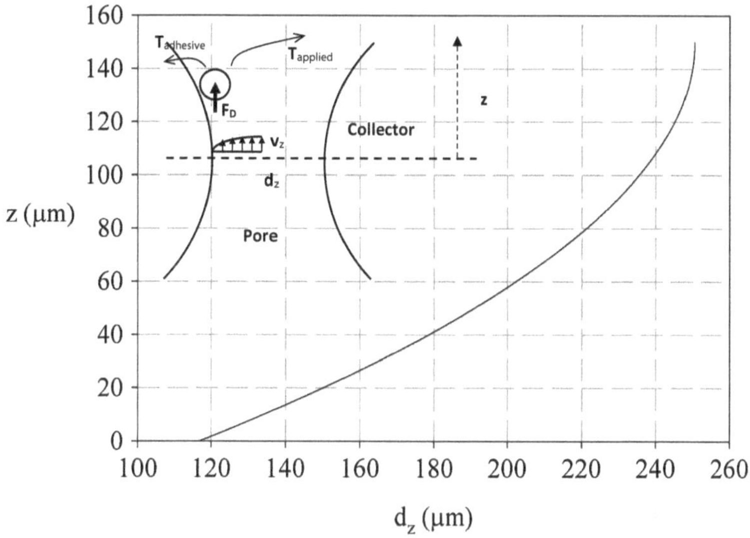

Fig. 6.4 Diameter of pore (d_z) as a function of distance from the center of a collector (z) used in this study assuming the constricted tube model as a representation of the void space between collectors in the porous media

$$d_{max} = 2.141 d_c \tag{6.14}$$

On the other hand, $T_{adhesive}$ is expressed as the adhesive force (F_A) acting on a lever arm (l_x) (Torkzaban et al. 2007). In the case of PSS-modified NZVI, the adhesive force (F_A) that acts on the attached particles is calculated from the absolute value of a secondary minimum (Θ_{min}) predicted by extended DLVO potential profiles (Fig. 6.3) using Eq. (6.15):

$$F_A = \frac{\Theta_{min}}{h_{min}} \tag{6.15}$$

where h_{min} is the separation distance at the secondary minimum between the colloid and the collector surface.

l_x represents the radius of the colloid-surface contact area and is estimated using the theory of Johnson, Kendall, and Roberts (Eq. 6.16) (Torkzaban et al. 2007). K is the composite Young's modulus. For deposition of PSS70K-modified nanoparticles on silica sand in this study, a value of $K = 4.014 \times 10^9$ N m^{-2} for glass bead collectors and polystyrene is used (Bergendahl and Grasso 2000):

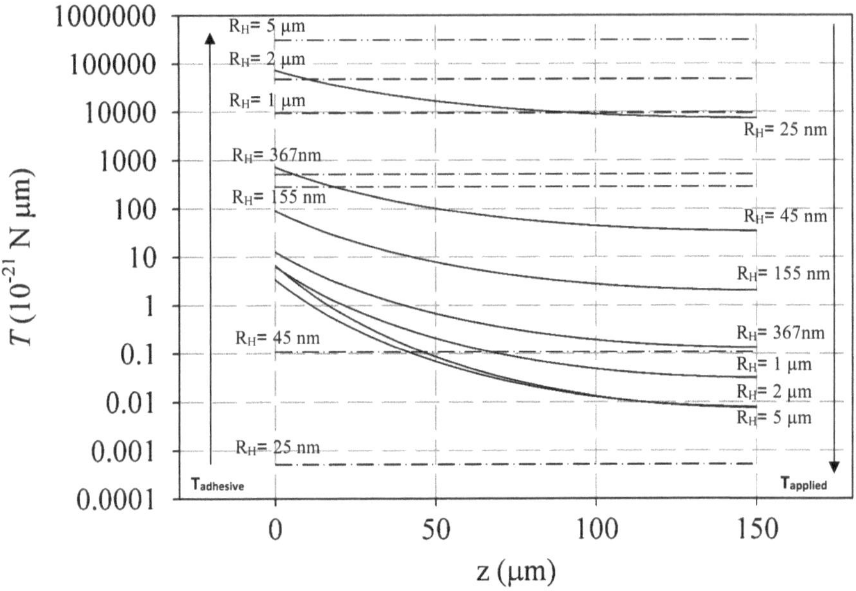

Fig. 6.5 The calculated T_{applied} (solid lines) and T_{adhesive} (dashed lines) for different sizes of PSS-modified NZVI (as individual particles and aggregates) for the transport conditions used in this study assuming the constricted tube model (Fig. 6.4) as a representation of the void space between collectors in the porous media (Phenrat et al. 2009)

$$l_x = \left(\frac{F_{\text{A}}a}{4K}\right)^{1/3} \tag{6.16}$$

Figure 6.5 illustrates the magnitude of T_{adhesive} and T_{applied} as a function of particle/aggregate sizes for PSS-modified NZVI with the measured adsorbed layer properties under the hydrodynamic and geochemical conditions used in Phenrat et al. (2009). This simplified theoretical analysis reveals that T_{applied} for the individual PSS70K-NZVI of hydrodynamic radius (R_{H}) = 25 and 45 nm is seven and three orders of magnitude greater than its T_{adhesive}. Therefore, detachment takes place and promotes NZVI mobility in porous media. This agrees with the low sticking coefficient of PSS70K-F1 (R_{H} = 25 nm) and –F2 (R_{H} = 45 nm) at low particle concentration (30 mg/L) where agglomeration does not occur. Nevertheless, for delivery of PSS-modified NZVI at high particle concentration, agglomeration plays a substantial role on T_{adhesive} and T_{applied} and NZVI detachment as will be discussed next.

As for mechanistic-based empirical models, Phenrat et al. (2010c) extended Bai and Tien's correlation to include the decrease of the friction force afforded by the adsorbed macromolecules which subsequently results in greater potential of particle rolling or detachment promoted by fluid drag force. They developed a dimensionless number, layer-electrokinetic parameter (N_{LEK}) (6.17 and 6.18), and corresponding empirical correlation for predicting collusion efficiency developed by the Buckingham-Π approach (Eq. 6.19). The new correlations (Eq. 6.19) can estimate experimental collision efficiency (α_{exp}) well. They illustrated that including N_{LEK} into the semiempirical model significantly improves estimation of α_{exp} over the Bai and Tien or Elimelech models for the 80 combinations of particle and coating types used:

$$N_{LEK} = \frac{d_p d_M^2 u_s \Gamma N_a \rho_p}{\mu M_W} \tag{6.17}$$

$$d_M = d_{M0} \left[\frac{I}{I_{ave}} \right]^{-2/3} \tag{6.18}$$

$$\alpha_{pre} = 10^{-1.35} N_{LO}^{0.39} N_{E1}^{-1.17} N_{LEK}^{-0.10} \tag{6.19}$$

where N_a is Avogadro's number.

Ohshima's approach assumes that the adsorbed layer thickness, d_{M0}, is constant over the range of salt concentrations used in the analysis. Thus, d_{M0} from Ohshima's analysis is the *average* layer thickness at the *average* ionic strength (I_{ave}) used in the analysis. Monte Carlo simulations, mean field theory, and self-consistent field theory (Hariharan et al. 1998) indicate that the layer thickness of adsorbed polyelectrolyte (d_M) decreases with increasing ionic strength (I) (Argiller and Tirrell 1992; Hariharan et al. 1998; Abu-Lail and Camesano 2003) to the power 2/3, i.e., $d_M \sim I^{-2/3}$. The layer thickness (d_M) at a particular salt concentration (I) used in Eq. (6.17) was therefore scaled using Eq. (6.18).

As shown in Eqs. (6.17)–(6.19), steric repulsions are a function of particle and adsorbed layer properties including d_p, d_M, Γ, ρ_p (polymer density), and M_W (molecular weight of polymer). The rolling and detachment promoted by the reduction of the friction force is a function of fluid viscosity (μ) and fluid velocity (u_s) in addition to the particle and adsorbed layer properties mentioned above. The dimensionless layer-electrokinetic parameter (N_{LEK}), representing both electrosteric repulsions and the decrease in friction force, can be expressed as a function of d_p, d_M, Γ, ρ_p, and M_W, μ, and u_s.

In addition to detachment resulting in low collision efficiency, studies commonly identify three forms of NP detachments which affect the shape of the breakthrough curve (BTC): (1) detachment shown by tailing in the BTC, (2) detachment in the form of a retarded or delayed BTC, and (3) detachment or release in the form of separated peaks following the BTC. The first type of detachment emerges in the form of tailing in the BTC where the feeding solution into the column inlet is switched from the injecting particle dispersion to the particle-free solution with the same ionic strength and pH (two-phase injection experiments). This is the most common reference to "detachment" and has been observed for various NPs such as polyelectrolyte-modified NZVI (Tosco and Sethi 2010; Raychoudhury et al. 2014), carbon nanotubes (Kasel et al. 2013), silica (Saiers et al. 1994), nanoporous silicate (Shang et al. 2013), and Ag NPs (Flory et al. 2013; Kanel et al. 2013; Neukum et al. 2014a). This is also known as re-entrainment and has been observed as dramatically extended BTC tailing which has been observed for a wide variety of colloids (Zhang et al. 2001a; b; Johnson et al. 2007; Landkamer et al. 2013). The underlying mechanism for this kind of detachment is commonly sought in the hydrodynamic forces.

Based on continuum-scale models, the governing set of transport equations considering kinetic detachment of this type is given as type (ii) models (Table 6.1) (de Marsily 1986; Kretzschmar et al. 1999), as well as types (iii) and (iv) models accounting for dual-site attachment/detachment. A solution of type (ii) sets of equations holding K_{att} constant while varying K_{det} was conducted with the aid of MT3DMS code (Zheng and Wang 1999b) and is presented in Fig. 6.6a. The results show that rising K_{det} not only causes the emergence of tailing in BTC but also leads to an asymmetric rise in the plateau of BTC higher at the side of the descending limb, until finally the plateau becomes flat and BTC resembles the conservative transport at the largest value of K_{det}.

In the second type of detachment, retardation appears as a horizontal shift in the colloid/NP BTC compared to conservative transport BTC (Fig. 6.6b). This has been observed for iron NPs (Harendra and Vipulanandan 2010), Ag NPs (Flory et al. 2013; Kanel et al. 2013; Liang et al. 2013a), carboxyl-modified latex NPs (Sasidharan et al. 2014), HAP NPs (Wang et al. 2012d), and QDs (Torkzaban et al. 2010b; Wang et al. 2013; Kini et al. 2014). The most pronounced occurrence of the retardation has been observed for NPs with infinitesimally small size (<10 nm) such as QDs, particularly at elevated IS, decreased pH (Torkzaban et al. 2010b, 2012; Wang et al. 2013), smaller sizes of porous media grains (Torkzaban et al. 2010b; Wang et al. 2014d), or increased temperature (Kini et al. 2014). It should be noted that although here we categorize retardation as a kind of detachment, in several studies which have observed retardation, "detachment" has been ruled out because the tailing in the BTC was not observed (He et al. 2009; Torkzaban et al. 2010a, b; Torkzaban et al. 2012, 2013; Wang et al. 2013, 2014d; Sasidharan et al. 2014; Treumann et al. 2014).

Nevertheless our view, based on the sum of evidence, is that retardation results from reversible interaction of particles with the porous media surfaces, i.e., attachment and detachment, the net result of which will control the final transport of particles (Babakhani et al. 2018). Retardation in its linear form is mathematically related to the most common parameter in the literature of reactive transport modeling in groundwater known as distribution coefficient or partition coefficient K_{d}. The use of K_{d}, however, requires applying an assumption of equilibrium between the dissolved and sorbed phases of solute contaminant (or attached and mobile phases of colloids/NPs). The respective model of this parameter is given in Table 6.1, as type (v). The solution to this model using MT3DMS code is shown in Fig. 6.6b. This

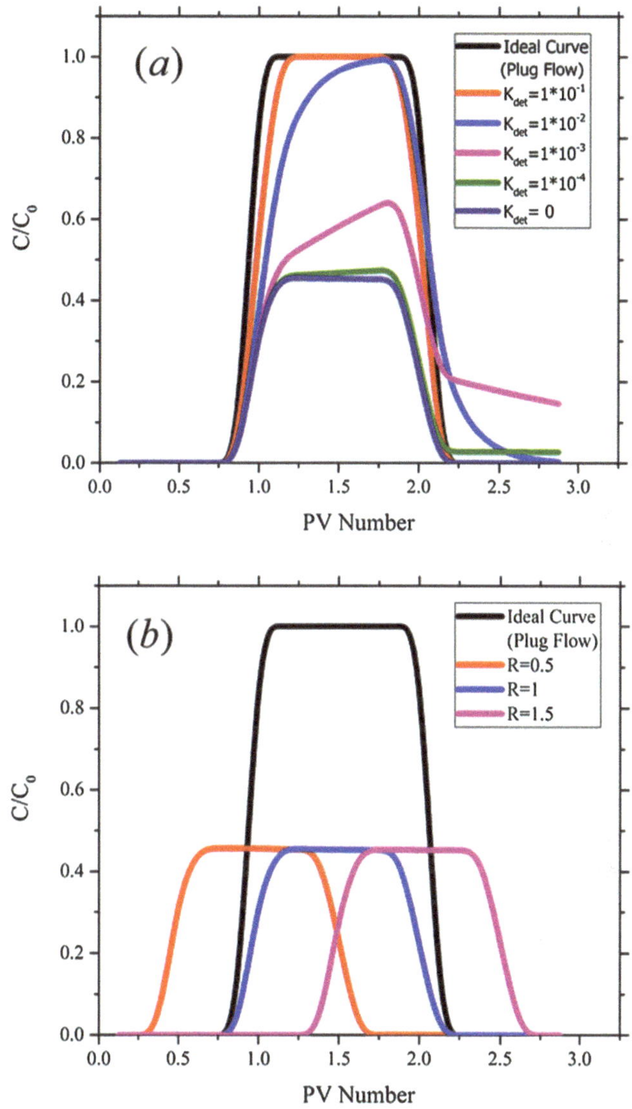

Fig. 6.6 Breakthrough curves generated by numerical solution of (**a**) model type (iv) from Table 6.1, for various K_{det} values (s^{-1}) and a fixed $K_{\mathrm{att}} = 0.001\ s^{-1}$, and (**b**) model type (v) from Table 6.1, for various retardation factors and a fixed $K_{\mathrm{att}} = 0.001\ s^{-1}$ with MT3DMS model code following Babakhani et al. (2015). Other parameters of flow and porous media conditions were selected according to Babakhani et al. (2015) (or refer to Table 6.1). (Reprinted with permission from Babakhani et al. (2017). Copyright (2017) Elsevier)

figure elucidates how retardation of BTC occurs in combination with a significant amount of irreversible attachment ($K_{att} = 1 \times 10^{-3}\ s^{-1}$).

The third type of detachment, which has been referred to as release or remobilization, is related to the cases where after injection of the particle dispersion (first phase) and particle-free background solution (second phase), a solution with a lower ionic strength (IS) or a different pH is flushed through the porous media container. The released mass can be seen in the BTC in the form of separated peak (s) following the primary breakthrough (Torkzaban et al. 2010a, 2013, 2015; Godinez and Darnault 2011; Li et al. 2011; Jones and Su 2012; Lanphere et al. 2013; Bradford et al. 2015). The model for simulating the release is given as type (vi) model in Table 6.1 (Torkzaban et al. 2013, 2015; Bradford et al. 2014). This equation assumes that remobilized particles do not reattach again, with the rationale that conditions following a perturbation that causes release are less favorable for reattachment than the original conditions (Torkzaban et al. 2015). This type of model has not yet been investigated for polyelectrolyte-modified NZVI.

6.3.3 Straining

In addition to the filtration mechanisms which are driven by the interfacial forces between particles and the porous media surfaces, retention of particles in porous media can be driven by physical straining (Bradford et al. 2002; Xu et al. 2006). Straining is defined as the process of physical trapping of particles in pore throats which are narrower than the size of particles to be passed through or in the case of high concentrations of particles, straining is deemed to occur as a result of concurrent arrival of many particles at a pore opening and consequent jamming and arching which can eventually lead to clogging of the filtration surface (Ives 1970; Jegatheesan and Vigneswaran 2005). In engineered filtration systems, such physical straining results in a continuous compressible cake (Cleasby and Baumann 1962; Ives 1963) or a mat with holes (Ives 1963).

Observed in experimental micromodels by the aid of pore visualization (Auset and Keller 2006), straining can happen in cases where polyelectrolyte-modified NZVI is injected into subsurface media in the near vicinity of the injection well (e.g., Elliott and Zhang 2001; Edmiston et al. 2011; Rosansky et al. 2013; Köber et al. 2014). Occasionally, straining has been further subdivided into two mechanisms of wedging and bridging. Wedging is the trapping of particles at two bounding surfaces without interference of particles, while bridging is the simultaneous arriving and accumulation of particles at the pore constriction (Bradford and Torkzaban 2008; Zhang et al. 2012; Dong et al. 2015). In reality, even for ideal spherical NPs and uniform grains, the pores are not uniform because of the existence of narrow spaces at the contact angle of the spheres which brings about a differential capacity of straining for different sizes of NPs. Herzig et al. (1970) illustrated different retention sites for straining as (Fig. 6.7) (1) surface sites, (2) crevice sites, (3) constriction sites, and (4) cavern sites. This initiated the development and the use of a

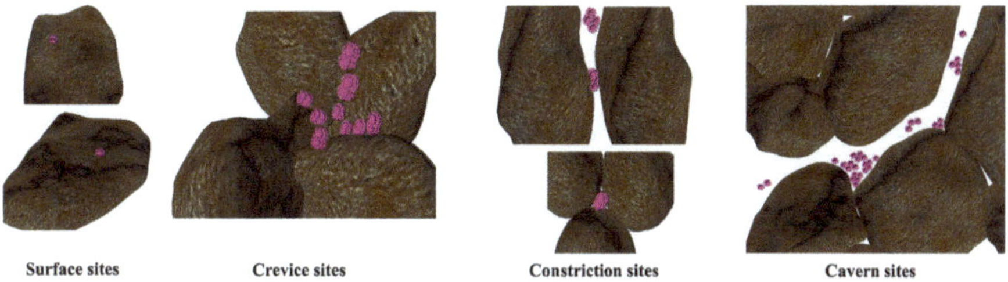

Fig. 6.7 Possible retention sites for straining from Herzig et al. (1970). (Reprinted with permission from Babakhani et al. (2017). Copyright (2017) Elsevier)

criterion for straining model selection by the ratio of particle diameter to grain diameter (d_p/d_g) (Xu et al. 2006; Shen et al. 2008; Xu et al. 2008). This critical ratio for straining as obtained from experimental results simultaneously under the influence of physicochemical filtration and straining retention was found to be 0.0017 (Tan et al. 1994; Bradford et al. 2002, 2003; Li et al. 2004; Foppen et al. 2005). The idea of a critical straining threshold is important in developing the conceptual model of NP transport with regard to straining, as many papers have identified (Choy et al. 2008; Fang et al. 2009, 2013, 2015a, b; Li et al. 2011; Jiang et al. 2012a, b; Cornelis et al. 2013; El Badawy et al. 2013; Jiang et al. 2013; Neukum et al. 2014b; Qi et al. 2014a, b; He et al. 2015; Sun et al. 2015a; Wang et al. 2015b, c) or rejected the influence of straining based on this criteria (Wang et al. 2011; Jones and Su 2014; Kurlanda-Witek et al. 2014; Toloni et al. 2014).

In addition to this criterion, another sign for identifying the straining is the shape of the retained colloid mass profile as a function of distance from the inlet (RCP), particularly the observation of hyper-exponential behavior of the RCP (a marked decreasing rate of deposition with distance from inlet according to an inverse power law) by a number of authors (Bradford and Bettahar 2006; Bradford et al. 2006b; Bradford and Toride 2007; Gargiulo et al. 2007, 2008; Ben-Moshe et al. 2010; Wang et al. 2011, 2012c; Kasel et al. 2013), even though this behavior of the RCP can also be attributed to other factors and mechanisms such as surface roughness (Bradford et al. 2002; Shellenberger and Logan 2002; Bradford et al. 2003; Yoon et al. 2006), concurrent aggregation (Chen and Elimelech 2006, 2007; Chatterjee and Gupta 2009; Chatterjee et al. 2010), colloid population heterogeneity (Tong and Johnson 2007; Jones and Su 2012), variations in the pore-scale velocity (Bradford et al. 2011a, b; Liang et al. 2013b), and chemical heterogeneity (Bolster et al. 1999; Li et al. 2004; Tufenkji and Elimelech 2005). Besides, monotonic and/or non-monotonic RCPs were also observed when the straining was the dominant mechanism of retention (Bradford et al. 2006b; Xu et al. 2006; Xu and Saiers 2009; Porubcan and Xu 2011). Non-monotonic RCP demonstrates a peak of concentration somewhere between the inlet and the outlet of the porous media domain (Bradford et al. 2006b). Despite extensive allusions to the reflection of transport and deposition mechanisms in the RCP shape among literature studies (e.g., Bradford and Bettahar 2006; Bradford et al. 2006b; Bradford and Toride 2007; Gargiulo et al.

2007, 2008; Wang et al. 2011, 2012c; Kasel et al. 2013), there have been fewer attempts to evaluate the slope of the BTC plateau in respect of the underlying mechanisms of transport. More frequently, however, straining has evidently come with an ascending BTC plateau including that of polyelectrolyte-modified NZVI (Raychoudhury et al. 2014).

From the perspective of continuum-scale modeling, straining has been described either with a depth-dependent decaying exponential function (Bradford et al. 2003) or with a concentration-dependent decaying exponential function (Xu et al. 2006, 2008; Xu and Saiers 2009). The former includes one empirical parameter, β, standing for the shape of the colloid spatial distribution and one extra variable of x standing for the distance from the inlet. This function, given below, is in fact able to predict the depth-dependent RCP (Bradford et al. 2003):

$$\psi_x = \left(\frac{d_c + x}{d_c}\right)^{-\beta} \tag{6.20}$$

where d_c is size of grains (collector) representing the pore length which is oftentimes considered as d_{50}, the median size of the porous media grains. The implementation of this function in the advection-dispersion model is its multiplication in the term for attachment (in K_{att}) as given in model types (ix), (x), (xi), (xii), (xiii), and (xiv) in Table 6.1. When β is equal to zero ($\psi_x = 1$), the decrease of retained concentration with distance is exponential which is the case for clean bed filtration theory (Bradford and Bettahar 2006).

On the other hand, the concentration-dependent decaying exponential expression includes two parameters, a rate coefficient parameter, k_0, standing for straining kinetic rate coefficient [T^{-1}] and the coefficient for the exponential decline in straining rates, λ, with the same unit as retained concentration variable, S, [NM^{-1} or MM^{-1}] (Xu et al. 2006):

$$\frac{\rho_b}{\varepsilon}\frac{\partial S}{\partial t} = k_0\, C\, e^{-S/\lambda} \tag{6.21}$$

This expression is solved within the advection-dispersion equation. When $S \approx 0$, the equation becomes the same as that for clean bed colloid filtration theory.

The depth-dependent expression used by Bradford et al. (2003) has become more popular for simulation of colloid and NP straining to the extent that most of the recent studies on NP transport have utilized this expression to fit the BTC/RCP of NP transport through column experiments, e.g., for hydroxyapatite NPs (Wang et al. 2012b, c, d, 2014a, b, 2015b, c) and silver NPs (Liang et al. 2013b; Wang et al. 2014a). More frequent applications of the expression developed by Bradford et al. (2003) might be because of two reasons. The first reason is that the depth-dependent expression of Bradford et al. (2003) uses only one parameter (β), which can be estimated based on the observed shape of the RCP and even in most of the studies as either a fixed value of 0.432 (Bradford et al. 2003; Bradford and Bettahar 2006;

Wang et al. 2011, 2012b; c, d, 2014a; b, 2015a; Fang et al. 2013; Liang et al. 2013b; Qi et al. 2014a, b; Fan et al. 2015a, b) or 1.532 (e.g., Liang et al. 2013a), which has been used without adding an extra free parameter to the parameter estimation procedure. For polyelectrolyte-modified NZVI, however, this parameter has been considered as a free parameter to be estimated in terms of inverse modeling, and the values obtained comprise of 0.609 for CMC-modified NZVI (Raychoudhury et al. 2014). The second reason is that the depth-dependent model considers a decaying attachment rate with a distance which is favorable for capturing the hyper-exponential behavior of RCP, frequently witnessed in the transport of colloids and NPs subjected to concurrent physicochemical filtration and straining.

On the other hand, in studies that have used the concentration-dependent model, the effect of physicochemical filtration was turned off, by thorough cleaning of the sand and the use of deionized water (DI) as the dispersant solution, in order to maximize electrostatic repulsion between the colloids and sand surface thereby to expect the mere influence of straining unambiguously (Xu et al. 2006, 2008; Xu and Saiers 2009; Porubcan and Xu 2011; Torkzaban et al. 2012). This brought about a nearly monotonic-shaped RCP in those studies. It is still not clear whether this model can fit the BTCs and RCPs for conditions where the concurrent influences of physicochemical filtration and straining exist. It should be noted that the straining model used by Bradford et al. (2003) has been criticized for not having enough power to describe the real straining phenomena (Johnson et al. 2011). In other words, whenever the depth-dependent model has been fitted to the experimental data with hyper-exponential RCP, it is obscured, recalling the ambiguities regarding the relevancy of straining to the RCP behavior, whether in reality the model describes the straining phenomena or merely the depth-dependent behavior (Johnson et al. 2011). Therefore, there is a crucial need to develop more rigorous conceptual models that can consider the concurrent effect of various phenomena occurring for NP during transport in porous media in order to decrease the uncertainty regarding the underlying mechanisms. This problem becomes even more convoluted when other phenomena such as blocking/ripening are involved, as addressed in the next section.

6.3.4 *Site-Blocking and Ripening*

6.3.4.1 Site-Blocking

The surface of porous media may have a limited capacity for the adsorption/attachment of solute/particles, and once this capacity is filled, the adsorption/attachment of further solute/particles is hindered by the presence of previously sorbed/attached materials. The most popular approach for modeling this quality, called the site-blocking effect, is the Langmuir approach for solutes/gases (Langmuir 1918) and for colloids (Adamczyk et al. 1994) adsorption onto solid surfaces. This is used to model the site-blocking effect of colloidal particle attachment during transport in porous media via the following relationship (Saiers et al. 1994):

$$\psi_{\mathrm{b}} = \left(1 - \frac{S}{S_{\mathrm{m}}}\right) \tag{6.22}$$

where ψ_b is the Langmuirian blocking function related to the fraction of porous medium available for deposition [—] and S_m is the maximum retained-particle phase concentration, i.e., colloidal retention capacity [NM^{-1} or MM^{-1}]. When used in the advection-dispersion model, this expression will result in second-order colloid deposition kinetics limited by aqueous phase as well as solid phase concentrations (Skopp 1986; Saiers et al. 1994). The BTC produced by this model does not maintain a plateau but instead gradually climbs toward the peak (ascending plateau). This gradual rising plateau is a sign of decline in the deposition rate with increases in the amount of attached phase particles, i.e., the blocking effect (Saiers et al. 1994).

Accordingly, many papers, by observing the gradual increasing plateau of the BTC, applied the Langmuirian approach to the continuum model (model types (xv), (x), (xi), (xiii), (xvi), and (v), Table 6.1) in order to simulate the transport of, e.g., Ag NPs (Braun et al. 2014; Neukum et al. 2014b), GO (Feriancikova and Xu 2012; Liu et al. 2015; Sun et al. 2015b), CeO_2 NPs (Li et al. 2011), TiO_2 NPs (Saiers et al. 1994; Toloni et al. 2014), QD (Torkzaban et al. 2012, 2013; Wang et al. 2013, d), and CML (Torkzaban et al. 2012). Based on the satisfactory model fitting results, these studies proposed the blocking of the physicochemical attachment sites as underlying phenomena for transport of NPs in porous media. However, the application of this model has not been limited to blocking due to physicochemical deposition but also been widely applied for describing the debilitating capacity of the straining sites over time via model types (xi) and (x), Table 6.1. In addition, in a few studies (Wang et al. 2012b, d), the Langmuirian type site-blocking function was combined with the depth-dependent model of straining in the form of dual-deposition sites (model types (*x*), Table 6.1) – one for time-dependent retention and another for depth-dependent deposition, without any mention to specific phenomena of straining or blocking.

It is very important to note that an arising plateau of the BTC should not be always taken as a sign of site-blocking phenomena. Recalling the generated BTC by using the simple nonequilibrium attachment/detachment model (model type (ii), Table 6.1) in Fig. 6.6a, ascending BTC plateaus can also emerge with increasing K_{det} values as described earlier. Nevertheless, these shapes of BTC, resulting from simple attachment/detachment models, come with tailing which may not be the case when the underlying transport process is site blocking (e.g., Feriancikova and Xu 2012; Torkzaban et al. 2013; Neukum et al. 2014b; Qi et al. 2014a, b; Sasidharan et al. 2014; Toloni et al. 2014; Wang et al. 2014d; Sun et al. 2015b) or may have occurred simultaneously, as modeled via both K_{det} and Langmuirian function at the same time (Tosco and Sethi 2010; Fan et al. 2015a; He et al. 2015).

From a mechanistic standpoint, the Langmuir function can be criticized for being a linear function of the surface coverage (Johnson and Elimelech 1995). Therefore, a dynamic blocking function based on the nonequilibrium model of random sequential adsorption (RSA) mechanics as a nonlinear power law function of surface coverage was developed in order to account for the real surface exclusion effects introduced by larger colloids (Schaaf and Talbot 1989; Johnson and Elimelech 1995). Being a sort of mechanistic approach, this model obviated the need for parameter estimation. This model was even employed to simulate the transport of colloids in heterogeneous porous media (Sun et al. 2001). Nevertheless, this type of modeling approach generally comes with the disadvantages of ignoring the specific interactions of attaching particles with interfaces and with previously deposited particles, as well as inefficiency of their application for non-spherical particles (Adamczyk et al. 2013). Furthermore, they have the limitation of being strictly valid for just the irreversible deposition (Adamczyk et al. 2013). A detailed discussion of this model can be found in Adamczyk et al. (2013).

Noticeably, no NZVI transport studies report the site-blocking effect. Presumably, NZVI is intrinsic to magnetic particles, and particle-particle interaction involving magnetic attraction is much stronger than particle-collector interaction which involves only van der Waals attraction. Thus, deposited NZVI on a collector surface will favor further deposition of suspended NZVI on the NZVI-occupied surface rather than blocking of the deposition site. Thus, for cases of NZVI (especially bare NZVI), ripening rather than blocking has been reported.

6.3.4.2 Ripening

Ripening is the opposite of the blocking phenomena, i.e., it is expected to occur when the particle-particle interactions/associations on the surface of porous media are stronger than the particle-surface interactions/associations. Also, in contrast to the blocking mechanism, in which the deposition rate decreases with time, ripening causes an increase in the deposition rate with time. This leads to a dropping plateau of the BTC in the case of ripening (Deshpande and Shonnard 1999; Kretzschmar et al. 1999; Bradford et al. 2003, 2006a; Nascimento et al. 2006; Bradford and Torkzaban 2008; Shen et al. 2008; Chatterjee et al. 2010; Chen et al. 2011; Hosseini and Tosco 2013; Bradford et al. 2014; Gastone et al. 2014; Tosco et al. 2014; Wang et al. 2014b, 2015b; Basnet et al. 2015).

Ripening has been observed for NPs such as NZVI (Tiraferri and Sethi 2009; Tosco and Sethi 2010; Basnet et al. 2013, 2015; Hosseini and Tosco 2013; Tosco et al. 2014), HAP NPs (Wang et al. 2012b, 2014b, 2015b, c), TiO_2 (Chen et al. 2011, 2012; Wang et al. 2014c), nano-C_{60} (Cheng et al. 2005), CeO_2 NP (Lv et al. 2014), and ZnO NP (Jiang et al. 2012a, b) typically at ISs in the range 10–20 mM NaCl or 1 mM $CaCl_2$ (e.g., Chen et al. 2011, 2012; Jiang et al. 2012b). The increase in the inflow concentration of bare NZVI strongly affected ripening and can result in clogging of porous media (Fig. 6.8).

In the context of continuum-based modeling of NPs, two modeling approaches have generally been applied to ripening. First, a more robust form of Eq. (6.22), already capable of modeling the blocking effect, was introduced to model the ripening phenomena alternatively, given as Eq. (e) in model type (xiv), Table 6.1

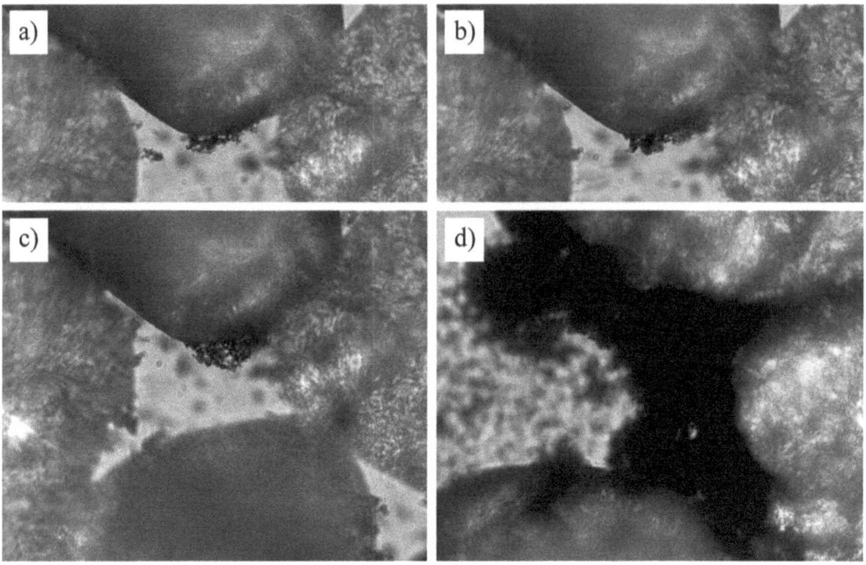

Fig. 6.8 Micro-fluidic cell images of bare NZVI ripening followed by clogging (**a**) after 1 min, (**b**) after 5 min, (**c**) after 10 min, and (**d**) after 15 min (Saleh et al. 2007)

(Shen et al. 2008; Tosco and Sethi 2010; Hosseini and Tosco 2013). This equation represents ripening when $\beta_1 > 0$ and $A_1 > 0$ or blocking when $\beta_1 = 1$ and $A_1 = 1/S_m$. Second, ripening has been modeled by adding a second-order term between the aqueous phase and the attached phase to the mass balance equation of the attached phase as given in Eq. (b) in model type (xii), Table 6.1 (Bradford and Bettahar 2006; Wang et al. 2014b).

Since ripening takes place in physicochemical conditions that are also favorable for aggregation, these terms sometimes have been used interchangeably, or in other words, aggregation in porous media has sometimes been alluded to as ripening (Shen et al. 2008; Tong et al. 2008; Wang et al. 2014c). Yet, it should be clarified that aggregation, by itself, can occur both in the aqueous phase and the attached or immobilized phase separately. In the aqueous phase, aggregation results from the common mechanisms of aggregation, namely, perikinetic aggregation, differential sedimentation, and orthokinetic aggregation (Elimelech et al. 1998; Babakhani et al. 2018), while in the attached phase, aggregation can result from the movement and/or rolling of the particles weakly attached to the secondary minimum of the collector surface and translating along the surface due to hydrodynamic drag forces and eventually their accumulation near the rear of the collector or at grain-to-grain contact areas (Bradford et al. 2006b; Kuznar and Elimelech 2007; Phenrat et al. 2009, 2010b; Bradford et al. 2011b; Sun et al. 2015b).

On the other hand, ripening arises from the interactions of depositing particles in the aqueous phase, with the surface of the previously deposited particles in the retained phase, leading to multilayer accumulations of particles on the surface of the collector (Kretzschmar et al. 1999; Nascimento et al. 2006; Chen et al. 2011; Basnet et al. 2013; Wang et al. 2014b, 2015b). This also gives physical meaning to the second type of ripening model outlined above in which ripening is described by a second-order term standing for the interactions between the fluid phase and the attached phase.

6.3.5 Aggregation/Agglomeration

Homoaggregation or agglomeration of polyelectrolyte-modified NZVI during transport in porous media plays a major role in the attachment, detachment, straining, and ripening discussed above since these phenomena are partly governed by particle or agglomerate sizes. Due to the intrinsic magnetic attraction of NZVI, agglomeration is substantial even with polymeric surface modification (see Chap. 5) as it can be visually observed with bare eyes. Thus, NZVI literature has repeatedly studied agglomeration effect on NZVI transport in porous media (Table 6.2).

The terms "agglomeration" and "dispersion stability" have been used interchangeably, but they are two distinct processes; agglomeration only considers the attachment of particles to each other, while dispersion stability considers both their attachment and their subsequent sedimentation and removal from the aqueous media (Friedlander 1960a, b; Jeffrey 1981; Abel et al. 1994; Risovic and Martinis 1994; Tang and Raper 2002; Phenrat et al. 2007). This difference is more pronounced in the context of porous media. In particular, the restricted length scales for sedimentation to take place in pores and the fact that they are advecting through the pores may distinguish agglomeration in porous media from that in bulk aqueous scale (Phenrat et al. 2009).

Thus far, the most common model for simulating the aggregation of colloids or NPs is the Smoluchowski model (Smoluchowski 1917; Hunt 1982; Elimelech et al. 1998; Raychoudhury et al. 2012; Quik et al. 2014a, b; Taghavy et al. 2015). This model is based on the superposition of three mechanisms: perikinetic, orthokinetic, and differential sedimentation. The perikinetic mechanism of aggregation involves particle-particle collisions caused by the Brownian motion of particles. Orthokinetic aggregation involves collisions arising from any motion or flow in a fluid that can cause shear stress. Differential sedimentation occurs when the particle sizes are so different that larger particles settle faster than smaller ones and collide with those in their paths (Elimelech et al. 1998; Babakhani et al. 2018). Detailed accounts of aggregation mechanisms, e.g., consideration of the fractal nature of aggregates or of gravitational sedimentation in aqueous media, are available elsewhere (Lee et al. 2000, 2002; Liu et al. 2011).

Agglomeration of NPs, in spite of being the focus of ongoing studies in aqueous media (Liu et al. 2011; Quik et al. 2014a, b; Therezien et al. 2014; Goudeli et al. 2015; Yu et al. 2015a), has been widely ignored by models of transport in porous media. Many modeling studies on NP transport reported to date have overlooked agglomeration by assuming that in the selected experimental conditions, aggregation might have not been operative. This rationale may be justified by several observations including (1) high experimental values of zeta potential or stability analysis based on critical coagulation concentration (CCC) or DLVO theory (El Badawy et al. 2013; Kini et al. 2014; Toloni et al. 2014), (2) the lack of significant size change observed under equivalent conditions in batch experiments (Li et al. 2011, 2013; Rahman et al. 2013; Torkzaban et al. 2013; Braun et al. 2014; Neukum et al. 2014b; Rahman et al. 2014; Sasidharan et al. 2014), or (3) the lack of any significant change in the concentration of suspended NP in such batch experiments, indicating the absence of sedimentation (e.g., He et al. 2015). Nevertheless, agglomeration cannot be ignored in the case of NZVI applications for in situ remediation using high particle concentration as elaborated in Chap. 5.

Agglomeration of polyelectrolyte-modified NZVI can affect the deposition of NZVI during their transport through porous media in a number of ways. First, the increase in size due to agglomeration can affect particle transport to a collector via diffusion, sedimentation, and interception as represented by a single-collector collision frequency (η_0) (Eq. 6.4). NZVI agglomeration reduces diffusivity, thereby decreasing the collision frequency with a collector surface arising from diffusive transport, although the effect can deviate from Stokes-Einstein predictions where agglomeration results in high-porosity clusters with lower drag coefficients than solid particles (Lin and Wiesner 2012).

Increasing the size of NZVI agglomerates can enhance NZVI transport to collector surface via sedimentation and interception. Additionally, the formation of porous agglomerates can also bring about a two to three times higher sedimentation velocity than that predicted by Stokes' law for impermeable spheres of equivalent mass and size (Li and Logan 1997). The porosity and thus settling velocity of fractal agglomerates can rise as their size grows because of the change in the fractal dimension that indicates their density (Johnson et al. 1996; Li and Logan 2001; Chakraborti et al. 2003; Meng et al. 2013). The observed single-collector collision frequency of the agglomerate (η_0') is a function of agglomerate size (d_{agg}) and can be expressed as in Eq. (6.23). Please note that the "prime" notation on η0' is used to distinguish between filtration theory that assumes no agglomeration and the modified filtration theory presented in this study which assumes rapid steady state agglomeration.

$$\eta_0{}' = \left[\frac{67.39 H^{0.05} A_s{}^{1/3}}{(k_B T)^{-0.66} d_c^{0.63} u_s^{0.72}}\right] d_{agg}^{-0.796} + \left[\frac{0.985 H^{0.12} A_s}{u_s^{0.12} d_c^{1.68}}\right] d_{agg}^{1.425} + \left[\frac{3.16 \times 10^{-5} H^{0.05} d_c^{0.24}}{(k_B T)^{0.05} u_s^{1.11}}\right] d_{agg}^{1.98} \tag{6.23}$$

Additionally, for NZVI particles which collide with collector surfaces, increases in the size of agglomerates can increase the magnitude of the secondary minimum according to DLVO interaction energy, thereby increasing attachment rate and decreasing attachment reversibility. Phenrat et al. (2009) revealed that attachment of polystyrene sulfonate (PSS)-modified NZVI with the polydisperse R_H from 15 to 260 nm (PSS-RNIP-F1) transporting through a sand-packed column was much greater of PSS-modified NZVI with the monodisperse R_H of 24 nm (PSS-RNIP-F3) contrary to extended DLVO calculation predicting similar transportability of individual PSS-modified NZVI for both sizes.

While they theoretically had similar electrosteric repulsions due to similar properties of adsorbed polyelectrolyte layers, PSS-modified NZVI with polydisperse R_H from 15 to 260 nm shared much greater magnetic attraction in comparison to PSS-modified NZVI with the monodisperse R_H of 24 nm. During the transport, van der Waals and magnetic attractions of polydisperse PSS-modified NZVI with R_H from 15 to 260 nm induced agglomeration under the much deeper secondary minimum. The shear flow in the porous media causes both agglomeration and disagglomeration until the agglomerates reach a stable size. PSS70K-RNIP-F1 with greater fraction of large particles/agglomerates initially forms stable agglomerates with a larger average size than PSS70K-RNIP-F2 and F3, respectively.

As discussed earlier, in a flow field, the deposited NZVI under a shallow secondary minimum can be re-entrained when the applied torque (T_{applied}) overcomes the adhesive torque (T_{adhesive}) due to the particle-collector interaction. As shown in Fig. 6.5, the greater the particle or agglomerate sizes, the larger the T_{adhesive} and the lower the T_{applied}. Similarly, the T_{adhesive} for PSS70K-RNIP aggregates of $R_H = 155$ and 367 nm (the representative of the first peak of PSS70K-RNIP-F1 and PSS70K-RNIP–F2, respectively) is two and three orders of magnitude greater than its T_{applied}.

Thus, irreversible deposition is predicted in contrast to those of non-aggregating PSS-modified NZVI. This model predicts that the larger the size of aggregates transported through porous media for a given flow condition, geochemical condition, and surface properties, the higher the probability of irreversible deposition, i.e., $T_{\text{adhesive}} \gg T_{\text{applied}}$, as shown in the case of R_H from 1 to 5 μm.

Consequently, Phenrat et al. (2009) proposed that an alternative conceptual model of particle agglomeration and subsequent deposition (Fig. 6.9) is more appropriate for explaining the deposition behavior of polyelectrolyte-modified NZVI. This conceptual model assumes that stable agglomerates form quickly under a secondary minimum attachment promoted by intrinsic magnetic attraction among NZVI. The agglomerate size is a result of the balance between van der Waals and magnetic attractions between polyelectrolyte-modified NZVI, electrosteric repulsive forces from the adsorbed polyelectrolytes, and the induced fluid shear in pore spaces. The agglomerates transport and deposit onto a collector (e.g., a sand grain). Larger agglomerate sizes yield higher attachment efficiency and lower the deposition reversibility as explained above. This conceptual model is consistent with the observed deposition and attachment reversibility of PSS-RNIP-F1, PSS-RNIP-F2, and PSS-RNIP-F3 where PSS-RNIP with larger particles and higher Fe^0 content (-F1 and -F2) formed larger stable agglomerates due to stronger magnetic attraction between particles.

Additionally, using regression analysis of the set of column data, Phenrat et al. (2010b) propose a mechanistic-based empirical model to predict a dimensionless

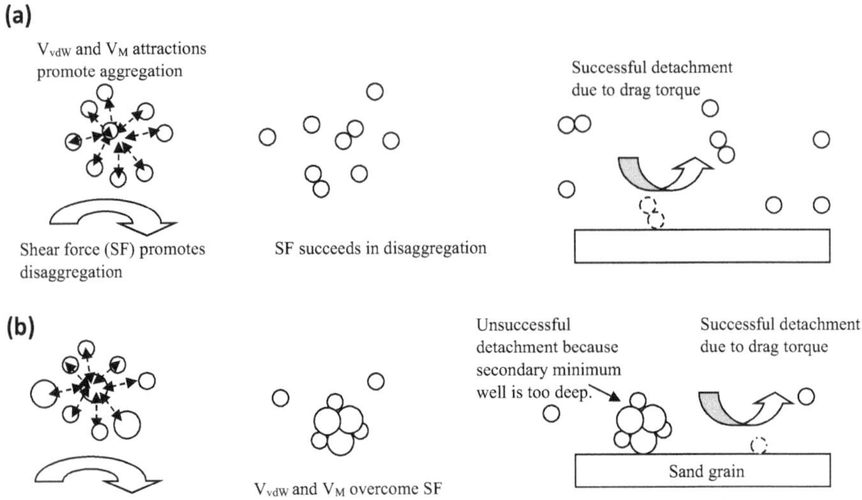

Fig. 6.9 Conceptual model of aggregation of polydisperse NZVI followed by deposition of clusters under a secondary minimum. This model conceptually considers the balance of shear (which is a function of fluid velocity) and total attraction energy between particle-particle or particle-aggregate (which is a function of ionic strength, van der Waals attraction, magnetic attraction, and electrosteric repulsions) on aggregation and disaggregation together with the balance between secondary minimum between clusters and collectors and drag torque on attachment and detachment; (**a**) is for small size, less polydisperse surface-modified NZVI and (**b**) is for large size, more polydisperse surface-modified NZVI (Phenrat et al. 2009). (Reprinted with permission from Phenrat et al. (2009). Copyright (2009) American Chemical Society)

agglomerate size (d_{agg}/d_p) as a function of the transport conditions, NaCl concentration, shear rate in porous media (s^{-1}), and properties of the polymeric surface modifiers (Eq. 6.24).

For NaCl:

$$\frac{d_{agg}}{d_p} = 4.24 \times 10^{-15} \frac{\mu_0^{1.28} M_s^{2.55} M_w^{0.10}}{\eta_v^{2.26} N_{avo}^{0.3} d_{M0}^{1.2} \gamma_s^{0.3} [\mathrm{NaCl}]^{*0.2}} \tag{6.24}$$

where μ_0 is magnetic permeability in vacuum (1.25×10^{-6} N/A^2) and N_{avo} is Avogadro's number (6.03×10^{23} mole^{-1}). [NaCl]* in this equation is in mole/m^3. M_S (in A/m) is NZVI saturation magnetization. d_{M0} (in m) is adsorbed polymer layer thickness determined from Ohshima's soft particle analysis. M_W is molecular weight of polymeric surface modifier and η_v is fluid viscosity. The empirical unit of the constant in Eq. (6.24) is kg$^{-0.88}$ m$^{4.13}$ mole$^{-0.3}$.

Once we can predict agglomerate size, we can also predict the attachment efficiency of the agglomerate (α') from Eq. (6.19) by substituting d_p with d_{agg} to yield:

$$\alpha' = \left[\frac{6.29 \times 10^7 H^{0.39} u_{\mathrm{s}}^{0.68} \eta_{\mathrm{v}}^{0.88} M_{\mathrm{W}}^{0.10}}{\rho_{\mathrm{poly}}^{0.10} \Gamma^{0.10} d_{\mathrm{M}}^{0.20} \left(\zeta_{\mathrm{p}}^2 + \zeta_{\mathrm{g}}^2 \right)^{1.17}} \right] d_{\mathrm{agg}}^{0.29} \tag{6.25}$$

The set of mechanistic-based correlations (Eqs. 6.23–6.25) can serve as a tool for preliminary design of emplacement strategies for polyelectrolyte-modified NZVI at high particle concentration in porous media using estimates of approach velocity, grain size, ionic strength, and properties of NZVI particles (particle size and Fe^0 content) and adsorbed polyelectrolyte layers (layer thickness and surface excess). The application of this set of equations is done by using Eq. (6.24) to estimate the *apparent average* size of stable aggregates formed in the system, and then using Eqs. (6.23) and (6.25) to determine the collision frequency (η_0') and collision efficiency (α') for NZVI agglomerates, respectively. Then, the transport distance (L) for a desired C/C_0 can be calculated from Eq. (6.5) by substituting α_{CF} and η_0 with α' and η_0', respectively. In addition, this set of correlations can be used in the design of controlled emplacement strategies of polyelectrolyte-modified NZVI for in situ remediation (see Sect. 6.4).

For continuum-scale modeling, Babakhani et al. (2018) modified widely used solute transport models, MODFLOW and MT3DMS, in order to incorporate the influence of agglomeration with other transport phenomena including attachment, detachment, and subsequent irreversible deposition of agglomerates for polyelectrolyte-modified NZVI at low and high particle concentrations. Accordingly, they proposed modified advection-dispersion-reaction (ADR) and first-order, reversible, kinetic-reaction (FRK) equations to capture these processes:

$$\frac{\partial N}{\partial t} + \frac{\rho_{\mathrm{b}}}{\varepsilon} \frac{\partial \bar{N}}{\partial t} = D \frac{\partial^2 N}{\partial x^2} - V \frac{\partial N}{\partial x} - \lambda_1 N - \lambda_2 \frac{\rho_{\mathrm{b}}}{\varepsilon} \bar{N} \tag{6.26}$$

$$\frac{\rho_{\mathrm{b}}}{\varepsilon} \frac{\partial \bar{N}}{\partial t} = K_{\mathrm{att}} N - \frac{\rho_{\mathrm{b}}}{\varepsilon} K_{\mathrm{det}} \bar{N} \tag{6.27}$$

where N [with a dimension of L^{-3}] and $\bar{N}$ [with a dimension of M^{-1}] are the particle number concentrations of fluid-phase particles and deposited particles, respectively; V is the interstitial particle velocity; D is the hydrodynamic dispersion coefficient; ε is the bed porosity; ρ_{b} is the porous medium bulk density; λ_1 is the pseudo first-order reaction rate [T^{-1}], which stands for the decay in population of particles due to agglomeration (mechanism #2); λ_2 is the pseudo first-order reaction rate [T^{-1}] which stands for the decay in the population of deposited, detachable particles and represents irreversible deposition (mechanism #3); and K_{att} and K_{det} are the attachment and detachment rate constants [T^{-1}], respectively (mechanism #1). It should be noted that, throughout this section, the word *particles* refers to both primary particles and agglomerates (clusters), which are differentiated by size. Over the use of equilibrium adsorption assumption, instead of first-order, reversible, kinetic-reaction equation (nonequilibrium), a linear reversible deposition model (equilibrium) was also investigated in this study that uses one parameter of K_d, the distribution coefficient [$L^3\ M^{-1}$] (He et al. 2009).

While the concentration variable in Eqs. (6.26) and (6.27) is expressed in terms of particle number concentration, it may be more convenient in many situations to work with mass concentration. Mass concentration data from laboratory and field measurements is typically more readily available than particle number concentration data. In addition, widely used flow and transport models such as MODFLOW/MT3D are also based on mass concentration. Furthermore, a recent study on the transport of NPs (Wang et al. 2012a) implies that different interpretations can occur when either of mass or particle number concentrations are used. Hence, it is useful to be able to convert particle number concentration to mass concentration. To accomplish this, the following two equations were used:

$$N = \frac{C}{\frac{4}{3}\pi\, r^3 \rho_{\mathrm{fe}}} \tag{6.28}$$

$$\bar{N} = \frac{\bar{C}}{\frac{4}{3}\pi \bar{r}^3 \rho_{\mathrm{fe}}} \tag{6.29}$$

where C is the mass concentration of fluid-phase particles (ML^{-3}); $\bar{C}$ is the mass concentration of deposited-phase particles (MM^{-1}); r and $\bar{r}$ are the average radii of particles in the fluid phase and deposited phase, respectively; and ρ_{fe} is the average density of the particles or aggregates.

It should be noted that assuming the shape of the particles or agglomerates in Eqs. (6.28) and (6.29) to be spherical is a common assumption for polymer-coated NZVI particles (Raychoudhury et al. 2012; Taghavy et al. 2014). In order to reduce the computational burden in this study, the average particle size is considered instead of particle size distribution. On the other hand, in the previous paper, it was revealed that dispersions with different degrees of polydispersivity have different agglomeration and deposition behaviors compared to monodisperse dispersions. Therefore, the impact of polydispersivity is investigated by fitting the model against different breakthrough data sets obtained for dispersions with various degrees of initial polydispersivity. After inserting Eqs. 6.28 and (6.29) into Eqs. (6.26) and (6.27), there were four variables in the resulting equations: C, $\bar{C}$, r, and $\bar{r}$. In order to solve the equations for these unknowns, two additional equations were required, one of which was obtained by considering that the change in the particles' radii is governed by agglomeration. Since Eq. (6.26) assumes that the population of particles decays at a rate of λ_1 due to agglomeration, Babakhani et al. (2018) considered a pseudo first-order reaction equation to represent the agglomeration process:

$$\frac{\partial N}{\partial t} = -\lambda_1 t \tag{6.30}$$

Integrating Eq. (6.30) with the initial condition that $N = N_0$ at $t = 0$ yielded:

$$N = N_0 e^{-\lambda_1 t} \tag{6.31}$$

Equation (6.31) relies on a simple pseudo first-order reaction model to represent agglomeration (model type (viii), Table 6.1). The basis of common agglomeration models such as the Smoluchowski model is a second-order rate equation (Holthoff et al. 1996, 1997; Elimelech et al. 1998; Szilagyi et al. 2014) although there is a limited number of studies which used a first-order formulation for agglomeration (Swift and Friedlander 1964; Birkner and Morgan 1968; Logan et al. 1995b; Baalousha 2009; Kocur et al. 2013). Babakhani et al. (2018) showed that the use of a pseudo first-order model could successfully describe the agglomeration of polyelectrolyte-modified NZVI in porous media conditions of homogenous one-dimensional domains. Then, by inserting Eq. (6.28) into Eq. (6.31), the following equation was obtained to account for the change in the average particle radius due to agglomeration:

$$r = r_0 \sqrt[3]{e^{\lambda_1 t} \frac{C}{C_0}} \tag{6.32}$$

where r_0 is the average radius of particles (or agglomerates) at $t = 0$ and C and C_0 are the mass concentrations of fluid-phase particles at a given t and $t = 0$, respectively. It should be mentioned that Eq. (6.32) was developed solely based on agglomeration and that the parameter C/C_0, which emerged in this equation, must assumed unity, because the change in the mass concentration cannot reflect any agglomeration effect. As for this second additional equation, it was assumed that the radius of the suspended particles was equal to the radius of the deposited, detachable particles. This simplified assumption is used because particles are continually being exchanged between the suspended phase and the deposited, detachable phase, considering the fact that those particles that are irreversibly attached behave as if they were eliminated from the system (a typical assumption of perfect sink models) (Ryan and Elimelech 1996).

Ultimately, three equations remaining to be solved simultaneously include Eqs. (6.26), (6.27), (6.32). In laboratory particle transport experiments, the transport parameters related to the attachment and agglomeration (K_{att}, K_{det}, λ_1, and λ_2) are typically unknown. Therefore, an iterative procedure was developed to optimize these parameters with respect to observed concentration data.

The steps of the iterative algorithm for calibrating the parameters of the model are as follows:

(i) First, assume that there is neither NZVI agglomeration nor irreversible deposition in the model's porous media; thus, the value of r would be equal to the value of r_0, which is equal to the average radius of the particles at time zero. Then, fit the standard numerical code to the observations, which are in terms of mass concentration, and obtain the parameters of the model (K_{att}, K_{det}, λ_1, and λ_2).

(ii) Use Eq. (6.32) to calculate r with the λ_1 value obtained in the previous step; assume that t is equal to the average retention time of the particles in the model domain and C/C_0 is equal to unity.
(iii) Convert the inflow concentration and the concentration data observed at the outlet(s) from mass concentration to number concentration using Eq. (6.28) and the radius of particles (aggregates) obtained from the previous step.
(iv) Fit the model output data (in number concentration) to the observed data, and optimize the parameters of the model using the WinPEST model.
(v) Beginning at step ii, iterate with the new values of λ_1 through step iv, until the difference in the values of λ_1 for two successive iterations becomes negligible.
(vi) As a convergence criteria for λ_1, a value of 1% was considered to be sufficient, because it induced a similar amount of error in the estimated size of agglomerates which is relatively minor in comparison to other possible uncertainties of typical experimental and modeling procedures.

Figure 6.10 shows that the MODFLOW together with M3TD modified by Eq. (6.26) to Eq. (6.32) successfully modeled experimental results of PSS-modified NZVI transport in one-dimensional porous media both at low particle concentration (0.03 g/L) where agglomeration is not significant and at high particle concentration (1–6 g/L) where agglomeration plays an important role.

6.4 Upscaling Transport Models of Polyelectrolyte-Modified NZVI

Although experiments to develop mechanistic-based models and continuum models (in Table 6.1) were one-dimensional, it is clear that for field application, we need two-dimensional (2-D) or three-dimensional (3-D) transport modeling platforms for NZVI delivery and emplacement design. Thus, in the final section of this chapter, we aim to present examples of integrating various mechanistic-based models and continuum models representing multiple contemporaneous transport phenomena together to develop a modeling tool for simulation and prediction of polymer-modified NZVI delivery in subsurface. This can be done by using a multi-physics modeling platform such as COMSOL or using a well-known MODFLOW/MT3D as follows:

6.4.1 COMSOL-Based Transport Model for Predicting the Emplacement and Transport of Polymer-Modified NZVI in 2-D Heterogeneous Porous Media

In Sect. 6.3.5, we described the development of an empirical model (Eq. 6.23–6.25) that allows for the prediction of polymer-modified NZVI transport in porous media

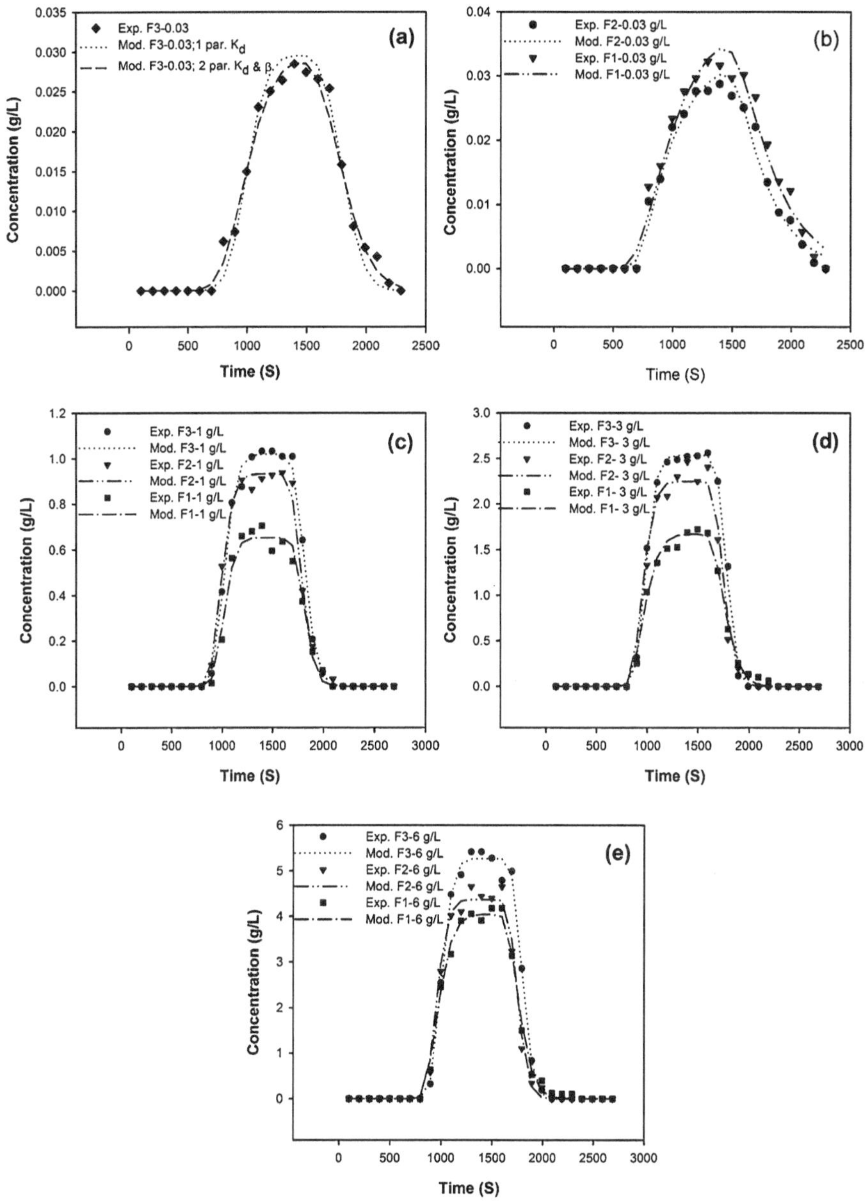

Fig. 6.10 Comparison of the experimental and modeled breakthrough curves of the 1-D model: (a) F3, F2, and F1 at the low concentration of 0.03 g/L, (b) F3, F2, and F1 at the high concentration of 1 g/L, (c) F3, F2, and F1 at the high concentration of 3 g/L, and (d) F3, F2, and F1 at the high concentration of 6 g/L. The symbols represent the experimental (Exp.) data, and the lines represent the modeled (Mod.) data. The experimental data were taken from Phenrat et al. (2009), which had been conducted in columns (25.5 cm long with internal diameters of 1.02 cm) packed with spherical silica sand ($d_{50} = 300$ μm) at a pore water velocity of 3.2×10^{-4} m/s (Babakhani et al. 2015)

at high particle concentrations. However, it was validated only for homogeneous sand and one-dimensional flow. To enable the model to predict the transport and deposition of NZVI in heterogeneous porous media with varying flow velocities, we coupled the mechanistic-based empirical correlations for estimating stable aggregate sizes and their deposition in porous media (Eq. 6.23–6.25) with a two-dimensional flow field simulation using COMSOL. To demonstrate the performance of the model, we compared the measured transport and deposition of polymer-modified NZVI transport in 2-D heterogeneous porous media with modeled results using the same hydrogeochemical conditions as in the experiment. This experiment is described in detail by Phenrat et al. (2010a).

Figure 6.11 shows the experimental and modeling results of polyelectrolyte-modified NZVI (MRNIP2) transport at different pore volumes (PV). The experimental and modeling results of MRNIP2 transport are similar. This suggests that the COMSOL-based model using the mechanistic-based empirical correlations for estimating stable aggregate sizes and of those aggregates in porous media is a promising tool for designing delivery schemes for polymer-modified NZVI in porous media when the nanoparticle properties and subsurface properties are known. We also tried to couple other available colloidal transport models (such as standard and modified filtration models that account for steric effects on deposition but ignore agglomeration) with the COMSOL model. However, these models could not simulate the experimental results. We assumed that a filtration model alone (ignoring steric effect and agglomeration) underestimates MRNIP2 transport, while a modified filtration model with the effects of steric repulsions on deposition but ignoring agglomeration overestimates transport.

6.4.2 *MODFLOW/MT3D-Based Transport Model for Predicting the Emplacement and Transport of Polymer-Modified NZVI in 2-D Heterogeneous Porous Media and in 3-D for Design of NZVI Injection*

The MODFLOW-based model can be modified to account for all the processes affecting the transport of polymer-modified NZVI, including advection, dispersion, attachment, detachment, and agglomeration as discussed previously. As with using MODFLOW together with M3TD modified by Eq. (6.26) to Eq. (6.32) to model transport of PSS-modified NZVI in 1-D porous media, the modeling platform can successfully model experimental results of PSS-modified NZVI transport in 2-D heterogeneous porous media.

Figure 6.12 compares the resulting concentration contours based on the λ_1 values obtained in the initial and final iterations with photographs of the washed MRNIP2 plume (experimental results). These results show an overall improvement in the agreement between the experimental and simulated shapes of the NZVI plume after estimating λ_1 values through the iterative procedure. However, about a twofold

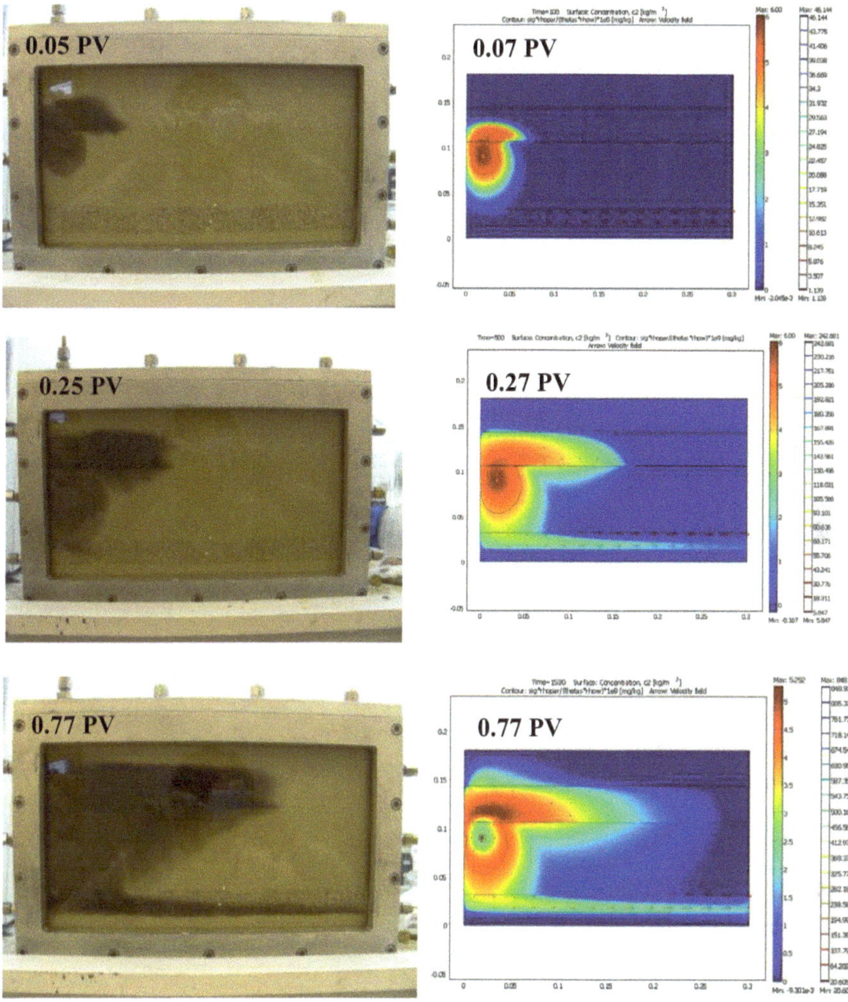

Fig. 6.11 Comparison between experimental and modeling results using COMSOL together with mechanistic-based models for MRNIP2 transport in 2-D heterogeneous porous media at similar pore volumes (PV)

overestimation is obvious in the right side of the medium layer, which may have been due to the uncertainties found in this layer.

Even with some uncertainty, this approach, owing to the use of standard MODFLOW modules, provides substantially simplified NZVI modeling to remediation practitioners and researchers who are already familiar with MODFLOW. Figure 6.13 illustrates an example of using MODFLOW and M3TD with the

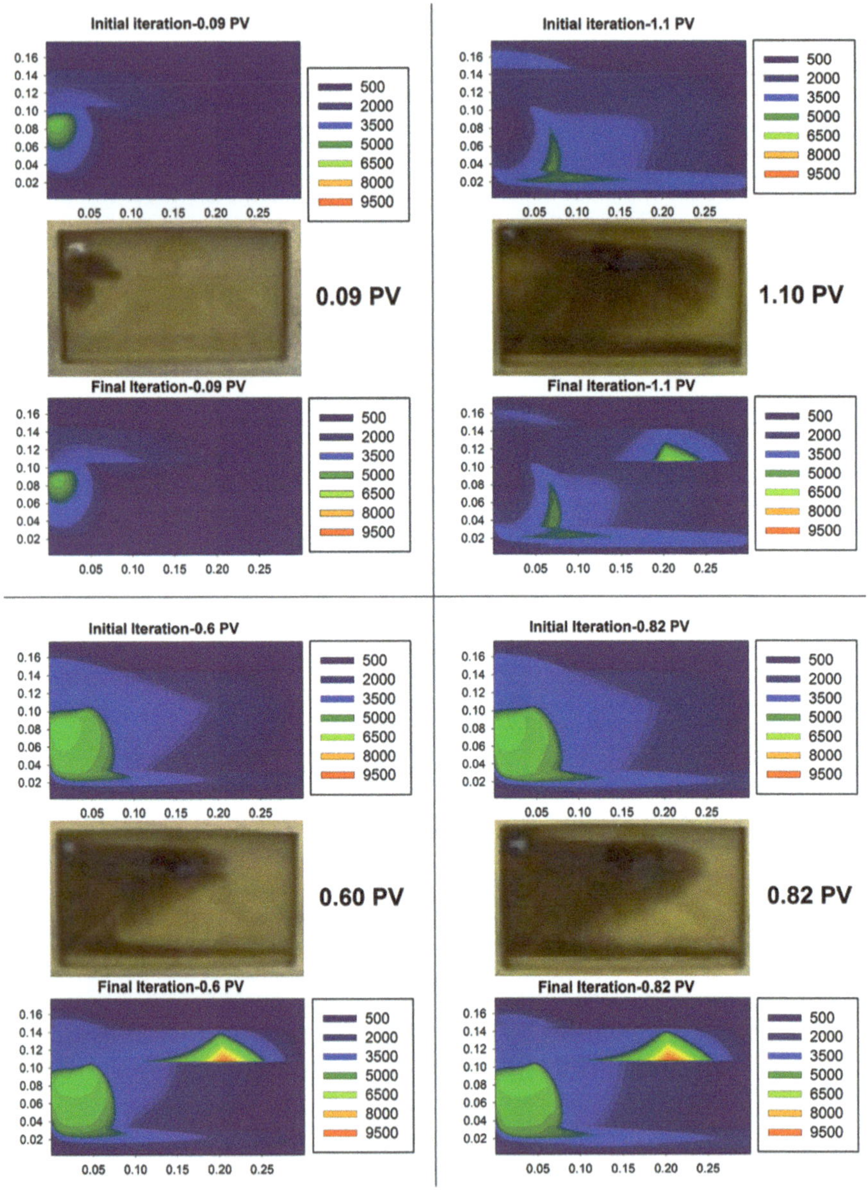

Fig. 6.12 Comparison of the experimental photographs with the modeled, color-shaded plots for washed MRNIP2 at the high concentration of 6 g/L. The experimental photographs were taken from Phenrat et al. (2010a). The upper, color-shaded plots of each photo were obtained from the initial iteration of the iterative procedure, while the lower plots were obtained from the final iteration (Babakhani et al. 2015)

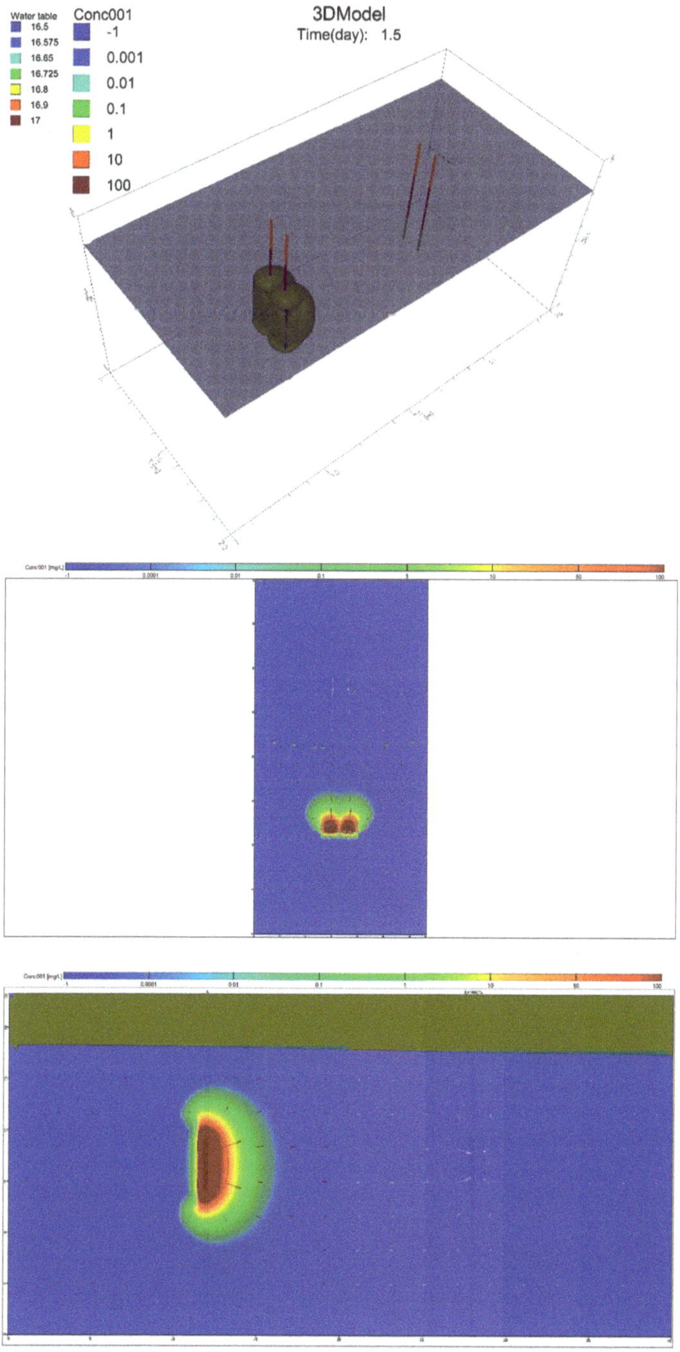

Fig. 6.13 An example of using MODFLOW and M3TD with the modified protocol proposed by Babakhani et al. (2015) to design an injection and extraction of polyelectrolyte-modified NZVI in subsurface for in situ remediation. In the investigated scenario, an impermeable cutoff wall has been added to the up-gradient of the injection wells to assess the enhancement in migration distance of NZVI

modified protocol proposed by Babakhani et al. (2018) to design an injection and extraction of polyelectrolyte-modified NZVI in subsurface for emplacing a reactive NZVI treatment zone for in situ groundwater remediation.

Acknowledgments This work was supported in part by (1) the Thailand Research Fund (TRF) MRG5680129; (2) the National Nanotechnology Center (Thailand), a member of the National Science and Technology Development Agency, through grant number P-11-00989; (3) the National Research Council (R2556B070); and (4) Taiwan's Ministry of Science and Technology (MOST) under grant no. 104-2221-E-009-020-MY3.

References

Abel, J. S., Stangle, G. C., Schilling, C. H., & Aksay, I. A. (1994). Sedimentation in flocculating colloidal suspensions. *Journal of Materials Research, 9*, 451–461.

Abu-Lail, N. I., & Camesano, T. A. (2003). Role of ionic strength on the relationship of biopolymer conformation, dlvo contributions, and steric interactions to bioadhesion of Pseudomonas putidaKT2442. *Biomacromolecules, 4*, 1000–1012.

Adamczyk, Z., Nattich-Rak, M., Sadowska, M., Michna, A., & Szczepaniak, K. (2013). Mechanisms of nanoparticle and bioparticle deposition–Kinetic aspects. *Colloids and Surfaces A: Physicochemical and Engineering Aspects, 439*, 3–22.

Adamczyk, Z., Siwek, B., Zembala, M., & Belouschek, P. (1994). Kinetics of localized adsorption of colloid particles. *Advances in Colloid and Interface Science, 48*, 151–280.

Adamczyk, Z., & Weroński, P. (1999). Application of the DLVO theory for particle deposition problems. *Advances in Colloid and Interface Science, 83*, 137–226.

Argiller, J. F., & Tirrell, M. (1992). Adsorption of water soluble ionic/hydrophobic diblock copolymer on a hydrophobic surface. *Theoretica Chimica Acta, 82*, 343–350.

Auset, M., & Keller, A. A. (2006). Pore-scale visualization of colloid straining and filtration in saturated porous media using micromodels. *Water Resources Research, 42*, W12S02.

Baalousha, M. (2009). Aggregation and disaggregation of iron oxide nanoparticles: Influence of particle concentration, pH and natural organic matter. *Science of the Total Environment, 407*, 2093–2101.

Baalousha, M., Cornelis, G., Kuhlbusch, T., Lynch, I., Nickel, C., Peijnenburg, W., & van den Brink, N. (2016). Modeling nanomaterials fate and uptake in the environment: Current knowledge and future trends. *Environmental Science: Nano, 3*, 323–345.

Babakhani, P., Bridge, J., Doong, R.-A., & Phenrat, T. (2017). Continuum-based models and concepts for the transport of nanoparticles in saturated porous media: A state-of-the-science review. *Advances in Colloid and Interface Science, 246*, 75–104.

Babakhani, P., Fagerlund, F., Shamsai, A., Lowry, G. V., & Phenrat, T. (2018). Modified MODFLOW-based model for simulating the agglomeration and transport of polymer-modified Fe^0 nanoparticles in saturated porous media. *Environmental Science and Pollution Research, 25*(8), 7180–7199.

Bai, R., & Tien, C. (1996). A new correlation for the initial filter coefficient under unfavorable surface interactions. *Journal of Colloid and Interface Science, 179*, 631–634.

Bai, R., & Tien, C. (1999). Particle deposition under unfavorable surface interactions. *Journal of Colloid and Interface Science, 218*, 488–499.

Bardos, P., Bone, B., Daly, P., Elliott, D., Jones, S., Lowry, G., & Merly, C. (2014). *A risk/benefit appraisal for the application of nano-scale zero valent iron (nZVI) for the remediation of contaminated sites, "Taking Nanotechnological Remediation Processes from Lab Scale to End User Applications for the Restoration of a Clean Environment"*. NANOREM, Supporting MS3, EU, 7th FP, NMP.2012.1.2.

Basnet, M., Di Tommaso, C., Ghoshal, S., & Tufenkji, N. (2015). Reduced transport potential of a palladium-doped zero valent iron nanoparticle in a water saturated loamy sand. *Water Research, 68*, 354–363.

Basnet, M., Ghoshal, S., & Tufenkji, N. (2013). Rhamnolipid biosurfactant and soy protein act as effective stabilizers in the aggregation and transport of palladium-doped zerovalent iron nanoparticles in saturated porous media. *Environmental Science & Technology, 47*, 13355–13364.

Ben-Moshe, T., Dror, I., & Berkowitz, B. (2010). Transport of metal oxide nanoparticles in saturated porous media. *Chemosphere, 81*, 387–393.

Bergendahl, J., & Grasso, D. (2000). Prediction of colloid detachment in a model porous media: hydrodynamics. *Chemical Engineering Science, 55*, 1523–1532.

Birkner, F. B., & Morgan, J. J. (1968). Polymer flocculation kinetics of dilute colloidal suspensions. *Journal (American Water Works Association), 60*, 175–191.

Bolster, C. H., Mills, A. L., Hornberger, G. M., & Herman, J. S. (1999). Spatial distribution of deposited bacteria following miscible displacement experiments in intact cores. *Water Resources Research, 35*, 1797–1807.

Bottero, J.-Y., Auffan, M., Borschnek, D., Chaurand, P., Labille, J., Levard, C., Masion, A., Tella, M., Rose, J., & Wiesner, M. R. (2015). Nanotechnology, global development in the frame of environmental risk forecasting. A necessity of interdisciplinary researches. *Comptes Rendus Geoscience, 347*, 35–42.

Bradford, S. A., & Bettahar, M. (2006). Concentration dependent transport of colloids in saturated porous media. *Journal of Contaminant Hydrology, 82*, 99–117.

Bradford, S. A., Simunek, J., Bettahar, M., van Genuchten, M. T., & Yates, S. R. (2003). Modeling colloid attachment, straining, and exclusion in saturated porous media. *Environmental Science & Technology, 37*, 2242–2250.

Bradford, S. A., Simunek, J., Bettahar, M., van Genuchten, M. T., & Yates, S. R. (2006a). Significance of straining in colloid deposition: Evidence and implications. *Water Resources Research, 42*, W12S15.

Bradford, S. A., Simunek, J., & Walker, S. L. (2006b). Transport and straining of E. coli O157: H7 in saturated porous media. *Water Resources Research, 42*, W12S12.

Bradford, S. A., & Toride, N. (2007). A stochastic model for colloid transport and deposition. *Journal of Environmental Quality, 36*, 1346–1356.

Bradford, S. A., & Torkzaban, S. (2008). Colloid transport and retention in unsaturated porous media: A review of interface-, collector-, and pore-scale processes and models. *Vadose Zone Journal, 7*, 667–681.

Bradford, S. A., & Torkzaban, S. (2015). Determining parameters and mechanisms of colloid retention and release in porous media. *Langmuir, 31*, 12096–12105.

Bradford, S. A., Torkzaban, S., Leij, F., & Simunek, J. (2015). Equilibrium and kinetic models for colloid release under transient solution chemistry conditions. *Journal of Contaminant Hydrology, 181*, 141–152.

Bradford, S. A., Torkzaban, S., & Simunek, J. (2011a). Modeling colloid transport and retention in saturated porous media under unfavorable attachment conditions. *Water Resources Research, 47*, W10503.

Bradford, S. A., Torkzaban, S., & Wiegmann, A. (2011b). Pore-scale simulations to determine the applied hydrodynamic torque and colloid immobilization. *Vadose Zone Journal, 10*, 252–261.

Bradford, S. A., Wang, Y., Kim, H., Torkzaban, S., & Šimůnek, J. (2014). Modeling microorganism transport and survival in the subsurface. *Journal of Environmental Quality, 43*, 421–440.

Bradford, S. A., Yates, S. R., Bettahar, M., & Simunek, J. (2002). Physical factors affecting the transport and fate of colloids in saturated porous media. *Water Resources Research, 38*, 63–61.

Braun, A., Klumpp, E., Azzam, R., & Neukum, C. (2014). Transport and deposition of stabilized engineered silver nanoparticles in water saturated loamy sand and silty loam. *Science of the Total Environment, 535*, 102–112.

Chakraborti, R. K., Gardner, K. H., Atkinson, J. F., & Van Benschoten, J. E. (2003). Changes in fractal dimension during aggregation. *Water Research, 37*, 873–883.
Chatterjee, J., Abdulkareem, S., & Gupta, S. K. (2010). Estimation of colloidal deposition from heterogeneous populations. *Water Research, 44*, 3365–3374.
Chatterjee, J., & Gupta, S. K. (2009). An agglomeration-based model for colloid filtration. *Environmental Science & Technology, 43*, 3694–3699.
Chen, G., Liu, X., & Su, C. (2011). Transport and retention of TiO2 rutile nanoparticles in saturated porous media under low-ionic-strength conditions: measurements and mechanisms. *Langmuir, 27*, 5393–5402.
Chen, G., Liu, X., & Su, C. (2012). Distinct effects of humic acid on transport and retention of tio2 rutile nanoparticles in saturated sand columns. *Environmental Science & Technology, 46*, 7142–7150.
Chen, K. L., & Elimelech, M. (2006). Aggregation and deposition kinetics of fullerene (C60) nanoparticles. *Langmuir, 22*, 10994–11001.
Chen, K. L., & Elimelech, M. (2007). Influence of humic acid on the aggregation kinetics of fullerene (C 60) nanoparticles in monovalent and divalent electrolyte solutions. *Journal of Colloid and Interface Science, 309*, 126–134.
Cheng, X., Kan, A. T., & Tomson, M. B. (2005). Study of C 60 transport in porous media and the effect of sorbed C 60 on naphthalene transport. *Journal of Materials Research, 20*, 3244–3254.
Chowdhury, A. I. A., Krol, M. M., Kocur, C. M., Boparai, H. K., Weber, K. P., Sleep, B. E., & O'Carroll, D. M. (2015). NZVI injection into variably saturated soils: Field and modeling study. *Journal of Contaminant Hydrology, 183*, 16–28.
Choy, C. C., Wazne, M., & Meng, X. (2008). Application of an empirical transport model to simulate retention of nanocrystalline titanium dioxide in sand columns. *Chemosphere, 71*, 1794–1801.
Chrysikopoulos, C. V., & Katzourakis, V. E. (2015). Colloid particle size-dependent dispersivity. *Water Resources Research, 51*, 4668.
Cleasby, J. L., & Baumann, E. R. (1962). Selection of sand filtration rates. *Journal (American Water Works Association), 54*, 579–602.
Comba, S., & Braun, J. (2012). A new physical model based on cascading column experiments to reproduce the radial flow and transport of micro-iron particles. *Journal of Contaminant Hydrology, 140*, 1–11.
Cornelis, G. (2015). Fate descriptors for engineered nanoparticles: the good, the bad, and the ugly. *Environmental Science: Nano, 2*, 19–26.
Cornelis, G., Pang, L., Doolette, C., Kirby, J. K., & McLaughlin, M. J. (2013). Transport of silver nanoparticles in saturated columns of natural soils. *Science of the Total Environment, 463*, 120–130.
Cullen, E., O'Carroll, D. M., Yanful, E. K., & Sleep, B. (2010). Simulation of the subsurface mobility of carbon nanoparticles at the field scale. *Advances in Water Resources, 33*, 361–371.
Dale, A., Casman, E. A., Lowry, G. V., Lead, J. R., Viparelli, E., & Baalousha, M. A. (2015a). Modeling nanomaterial environmental fate in aquatic systems. *Environmental Science & Technology, 49*, 2587–2593.
Dale, A. L., Lowry, G. V., & Casman, E. A. (2015b). Much ado about α: reframing the debate over appropriate fate descriptors in nanoparticle environmental risk modeling. *Environmental Science: Nano, 2*, 27–32.
de Marsily, G. (1986). *Quantitative hydrogeology; groundwater hydrology for engineers*. New York: Academic Press.
Deshpande, P. A., & Shonnard, D. R. (1999). Modeling the effects of systematic variation in ionic strength on the attachment kinetics of Pseudomonas fluorescens UPER-1 in saturated sand columns. *Water Resources Research, 35*, 1619–1627.
Dong, H., Zeng, G., Zhang, C., Liang, J., Ahmad, K., Xu, P., He, X., & Lai, M. (2015). Interaction between Cu 2+ and different types of surface-modified nanoscale zero-valent iron during their transport in porous media. *Journal of Environmental Sciences, 32*, 180–188.

Edmiston, P. L., Osborne, C., Reinbold, K. P., Pickett, D. C., & Underwood, L. A. (2011). Pilot scale testing composite swellable organosilica nanoscale zero-valent iron—Iron-Osorb®—for in situ remediation of trichloroethylene. *Remediation Journal, 22*, 105–123.

Ehtesabi, H., Ahadian, M. M., Taghikhani, V., & Ghazanfari, M. H. (2013). Enhanced heavy oil recovery in sandstone cores using tio2 nanofluids. *Energy & Fuels, 28*, 423–430.

El Badawy, A. M., Aly Hassan, A., Scheckel, K. G., Suidan, M. T., & Tolaymat, T. M. (2013). Key factors controlling the transport of silver nanoparticles in porous media. *Environmental Science & Technology, 47*, 4039–4045.

Elimelech, M. (1992). Predicting collision efficiencies of colloidal particles in porous media. *Water Research, 26*, 1–8.

Elimelech, M., Gregory, J., Jia, X., & Williams, R. (1995). *Particle deposition and aggregation: Measurement, modeling, and simulation*. Boston: Butterworth-Heinemann.

Elimelech, M., Jia, X., Gregory, J., & Williams, R. (1998). *Particle deposition and aggregation: Measurement, modelling and simulation*. Amsterdam: Elsevier.

Elimelech, M., Nagai, M., Ko, C.-H., & Ryan, J. N. (2000). Relative insignificance of mineral grain zeta potential to colloid transport in geochemically heterogeneous porous media. *Environmental Science & Technology, 34*, 2143–2148.

Elliott, D. W., & Zhang, W.-X. (2001). Field assessment of nanoscale bimetallic particles for groundwater treatment. *Environmental Science & Technology, 35*, 4922–4926.

Fallah, H., Fallah, A., Rahmani, A., Afkhami, M., & Ahmadi, A. (2012). Size exclusion mechanism, suspension flow through porous medium. *International Journal of Modern Nonlinear Theory and Application, 1*, 113.

Fan, W., Jiang, X., Lu, Y., Huo, M., Lin, S., & Geng, Z. (2015a). Effects of surfactants on graphene oxide nanoparticles transport in saturated porous media. *Journal of Environmental Sciences, 35*, 12–19.

Fan, W., Jiang, X. H., Yang, W., Geng, Z., Huo, M. X., Liu, Z. M., & Zhou, H. (2015b). Transport of graphene oxide in saturated porous media: Effect of cation composition in mixed Na–Ca electrolyte systems. *Science of the Total Environment, 511*, 509–515.

Fang, J., Shan, X.-Q., Wen, B., Lin, J.-M., & Owens, G. (2009). Stability of titania nanoparticles in soil suspensions and transport in saturated homogeneous soil columns. *Environmental Pollution, 157*, 1101–1109.

Fang, J., Xu, M.-J., Wang, D.-J., Wen, B., & Han, J.-Y. (2013). Modeling the transport of TiO2 nanoparticle aggregates in saturated and unsaturated granular media: effects of ionic strength and pH. *Water Research, 47*, 1399–1408.

Feriancikova, L., & Xu, S. (2012). Deposition and remobilization of graphene oxide within saturated sand packs. *Journal of Hazardous Materials, 235*, 194–200.

Flory, J., Kanel, S. R., Racz, L., Impellitteri, C. A., Silva, R. G., & Goltz, M. N. (2013). Influence of pH on the transport of silver nanoparticles in saturated porous media: laboratory experiments and modeling. *Journal of Nanoparticle Research, 15*, 1–11.

Foppen, J. W. A., Mporokoso, A., & Schijven, J. F. (2005). Determining straining of Escherichia coli from breakthrough curves. *Journal of Contaminant Hydrology, 76*, 191–210.

Friedlander, S. K. (1960a). On the particle-size spectrum of atmospheric aerosols. *Journal of Meteorology, 17*, 373–374.

Friedlander, S. K. (1960b). Similarity considerations for the particle-size spectrum of a coagulating, sedimenting aerosol. *Journal of Meteorology, 17*, 479–483.

Gargiulo, G., Bradford, S., Šimůnek, J., Ustohal, P., Vereecken, H., & Klumpp, E. (2007). Bacteria transport and deposition under unsaturated conditions: The role of the matrix grain size and the bacteria surface protein. *Journal of Contaminant Hydrology, 92*, 255–273.

Gargiulo, G., Bradford, S. A., Simunek, J., Ustohal, P., Vereecken, H., & Klumpp, E. (2008). Bacteria transport and deposition under unsaturated flow conditions: The role of water content and bacteria surface hydrophobicity. *Vadose Zone Journal, 7*, 406–419.

Gastone, F., Tosco, T., & Sethi, R. (2014). Guar gum solutions for improved delivery of iron particles in porous media (Part 1): Porous medium rheology and guar gum-induced clogging. *Journal of contaminant hydrology, 166*, 23–33.

Godinez, I. G., & Darnault, C. J. G. (2011). Aggregation and transport of nano-TiO2 in saturated porous media: effects of pH, surfactants and flow velocity. *Water Research, 45*, 839–851.

Golzar, M., Saghravani, S. F., & Azhdari Moghaddam, M. (2014). Experimental study and numerical solution of poly acrylic acid supported magnetite nanoparticles transport in a one-dimensional porous media. *Advances in Materials Science and Engineering, 2014*, 8.

Goudeli, E., Eggersdorfer, M. L., & Pratsinis, S. E. (2015). Coagulation–Agglomeration of fractal-like particles: Structure and self-preserving size distribution. *Langmuir, 31*, 1320–1327.

Grasso, D., Subramaniam, K., Butkus, M., Strevett, K., & Bergendahl, J. (2002). A review of non-DLVO interactions in environmental colloidal systems. *Reviews in Environmental Science and Biotechnology, 1*, 17–38.

Harendra, S., & Vipulanandan, C. (2010). Fe/Ni bimetallic particles transport in columns packed with sandy clay soil. *Industrial & Engineering Chemistry Research, 50*, 404–411.

Hariharan, R., Biver, C., Mays, J., & Russel, W. B. (1998). Ionic strength and curvature effects in flat and highly curved polyelectrolyte brushes. *Macromolecules, 31*, 7506–7513.

Harvey, R. W., & Garabedian, S. P. (1991). Use of colloid filtration theory in modeling movement of bacteria through a contaminated sandy aquifer. *Environmental Science & Technology, 25*, 178–185.

Hashemi, R., Nassar, N. N., & Pereira Almao, P. (2013). Enhanced heavy oil recovery by in situ prepared ultradispersed multimetallic nanoparticles: A study of hot fluid flooding for athabasca bitumen recovery. *Energy & Fuels, 27*, 2194–2201.

He, F., Zhang, M., Qian, T., & Zhao, D. (2009). Transport of carboxymethyl cellulose stabilized iron nanoparticles in porous media: Column experiments and modeling. *Journal of Colloid and Interface Science, 334*, 96–102.

He, J.-Z., Li, C.-C., Wang, D.-J., & Zhou, D.-M. (2015). Biofilms and extracellular polymeric substances mediate the transport of graphene oxide nanoparticles in saturated porous media. *Journal of Hazardous Materials, 300*, 467–474.

Herzig, J. P., Leclerc, D. M., & Goff, P. L. (1970). Flow of suspensions through porous media—Application to deep filtration. *Industrial & Engineering Chemistry, 62*, 8–35.

Holthoff, H., Egelhaaf, S. U., Borkovec, M., Schurtenberger, P., & Sticher, H. (1996). Coagulation rate measurements of colloidal particles by simultaneous static and dynamic light scattering. *Langmuir, 12*, 5541–5549.

Holthoff, H., Schmitt, A., Fernández-Barbero, A., Borkovec, M., ángel Cabrerizo-Vılchez, M., Schurtenberger, P., & Hidalgo-Alvarez, R. (1997). Measurement of absolute coagulation rate constants for colloidal particles: comparison of single and multiparticle light scattering techniques. *Journal of Colloid and Interface Science, 192*, 463–470.

Hosseini, S. M., & Tosco, T. (2013). Transport and retention of high concentrated nano-Fe/Cu particles through highly flow-rated packed sand column. *Water Research, 47*, 326–338.

Hotze, E. M., Phenrat, T., & Lowry, G. V. (2010). Nanoparticle aggregation: Challenges to understanding transport and reactivity in the environment. *Journal of Environmental Quality, 39*, 1909–1924.

Howington, S. E., Peters, J. F., & Illangasekare, T. H. (1997). *Discrete network modeling for field-scale flow and transport through porous media*. DTIC Document.

Huang, P. M., Li, Y., & Sumner, M. E. (2011). *Handbook of soil sciences: Properties and processes*. New York: CRC Press Taylor and Francis Group an informa business.

Hunt, J. R. (1982). Self-similar particle-size distributions during coagulation: Theory and experimental verification. *Journal of Fluid Mechanics, 122*, 169–185.

Illangasekare, T. H., Frippiat, C. C., & Fuˇcík, R. (2010). *Dispersion and mass transfer coefficients in groundwater of near-surface geologic formations*. Boca Raton: CRC Press/Taylor and Francis Group.

Ives, K. J. (1963). Simplified rational analysis of filter behaviour. In *ICE proceedings* (pp. 345–364). Thomas Telford.
Ives, K. J. (1970). Rapid filtration. *Water Research, 4*, 201–223.
Iwasaki, T., Slade Jr., J. J., & Stanley, W. E. (1937). Some notes on sand filtration [with Discussion]. *Journal (American Water Works Association), 29*, 1591–1602.
Jacobson, M. Z. (2005). *Fundamentals of atmospheric modeling*. New York: Cambridge University Press.
James, S. C., & Chrysikopoulos, C. V. (2003). Effective velocity and effective dispersion coefficient for finite-sized particles flowing in a uniform fracture. *Journal of Colloid and Interface Science, 263*, 288–295.
Jeffrey, D. J. (1981). Quasi-stationary approximations for the size distribution of aerosols. *Journal of the Atmospheric Sciences, 38*, 2440–2443.
Jegatheesan, V., & Vigneswaran, S. (2005). Deep bed filtration: Mathematical models and observations. *Critical Reviews in Environmental Science and Technology, 35*, 515–569.
Jiang, X., Tong, M., & Kim, H. (2012a). Influence of natural organic matter on the transport and deposition of zinc oxide nanoparticles in saturated porous media. *Journal of Colloid and Interface Science, 386*, 34–43.
Jiang, X., Tong, M., Lu, R., & Kim, H. (2012b). Transport and deposition of ZnO nanoparticles in saturated porous media. *Colloids and Surfaces A: Physicochemical and Engineering Aspects, 401*, 29–37.
Jiang, X., Wang, X., Tong, M., & Kim, H. (2013). Initial transport and retention behaviors of ZnO nanoparticles in quartz sand porous media coated with Escherichia coli biofilm. *Environmental Pollution, 174*, 38–49.
Johnson, C. P., Li, X., & Logan, B. E. (1996). Settling velocities of fractal aggregates. *Environmental Science & Technology, 30*, 1911–1918.
Johnson, P. R., & Elimelech, M. (1995). Dynamics of colloid deposition in porous media: Blocking based on random sequential adsorption. *Langmuir, 11*, 801–812.
Johnson, W. P., Li, X., & Assemi, S. (2007). Deposition and re-entrainment dynamics of microbes and non-biological colloids during non-perturbed transport in porous media in the presence of an energy barrier to deposition. *Advances in Water Resources, 30*, 1432–1454.
Johnson, W. P., Ma, H., & Pazmino, E. (2011). Straining credibility: A general comment regarding common arguments used to infer straining as the mechanism of colloid retention in porous media. *Environmental Science & Technology, 45*, 3831–3832.
Jones, E. H., & Su, C. (2012). Fate and transport of elemental copper (Cu 0) nanoparticles through saturated porous media in the presence of organic materials. *Water Research, 46*, 2445–2456.
Jones, E. H., & Su, C. (2014). Transport and retention of zinc oxide nanoparticles in porous media: Effects of natural organic matter versus natural organic ligands at circumneutral pH. *Journal of Hazardous Materials, 275*, 79–88.
Kanel, S. R., Flory, J., Meyerhoefer, A., Fraley, J. L., Sizemore, I. E., & Goltz, M. N. (2013). Influence of natural organic matter on fate and transport of silver nanoparticles in saturated porous media: laboratory experiments and modeling. *Journal of Nanoparticle Research, 17*, 1–13.
Kasel, D., Bradford, S. A., Šimůnek, J., Heggen, M., Vereecken, H., & Klumpp, E. (2013). Transport and retention of multi-walled carbon nanotubes in saturated porous media: Effects of input concentration and grain size. *Water Research, 47*, 933–944.
Keir, G., Jegatheesan, V., & Vigneswaran, S. (2009). Deep bed filtration: modeling theory and practice. In V. Saravanamuthu (Ed.), *Water and wastewater treatment technologies* (pp. 263–307). Oxford, UK: Eolss Publishers.
Keller, A. A., Sirivithayapakorn, S., & Chrysikopoulos, C. V. (2004). Early breakthrough of colloids and bacteriophage MS2 in a water-saturated sand column. *Water Resources Research, 40*, W08304.
Kelly, R. A., Jakeman, A. J., Barreteau, O., Borsuk, M. E., ElSawah, S., Hamilton, S. H., Henriksen, H. J., Kuikka, S., Maier, H. R., & Rizzoli, A. E. (2013). Selecting among five common modelling approaches for integrated environmental assessment and management. *Environmental Modelling & Software, 47*, 159–181.

Kini, G. C., Yu, J., Wang, L., Kan, A. T., Biswal, S. L., Tour, J. M., Tomson, M. B., & Wong, M. S. (2014). Salt-and temperature-stable quantum dot nanoparticles for porous media flow. *Colloids and Surfaces A: Physicochemical and Engineering Aspects, 443*, 492–500.

Köber, R., Hollert, H., Hornbruch, G., Jekel, M., Kamptner, A., Klaas, N., Maes, H., Mangold, K. M., Martac, E., & Matheis, A. (2014). Nanoscale zero-valent iron flakes for groundwater treatment. *Environmental Earth Sciences, 72*, 3339–3352.

Kocur, C. M., O'Carroll, D. M., & Sleep, B. E. (2013). Impact of nZVI stability on mobility in porous media. *Journal of Contaminant Hydrology, 145*, 17–25.

Kretzschmar, R., Borkovec, M., Grolimund, D., & Elimelech, M. (1999). Mobile subsurface colloids and their role in contaminant transport. *Advances in Agronomy, 66*, 121–193.

Krol, M. M., Oleniuk, A. J., Kocur, C. M., Sleep, B. E., Bennett, P., Xiong, Z., & O'Carroll, D. M. (2013). A field-validated model for in situ transport of polymer-stabilized nZVI and implications for subsurface injection. *Environmental Science & Technology, 47*, 7332–7340.

Kurlanda-Witek, H., Ngwenya, B. T., & Butler, I. B. (2014). Transport of bare and capped zinc oxide nanoparticles is dependent on porous medium composition. *Journal of Contaminant Hydrology, 162*, 17–26.

Kuznar, Z. A., & Elimelech, M. (2007). Direct microscopic observation of particle deposition in porous media: Role of the secondary energy minimum. *Colloids and Surfaces A: Physicochemical and Engineering Aspects, 294*, 156–162.

Landkamer, L. L., Harvey, R. W., Scheibe, T. D., & Ryan, J. N. (2013). Colloid transport in saturated porous media: Elimination of attachment efficiency in a new colloid transport model. *Water Resources Research, 49*, 2952–2965.

Langmuir, I. (1918). The adsorption of gases on plane surfaces of glass, mica and platinum. *Journal of the American Chemical Society, 40*, 1361–1403.

Lanphere, J. D., Luth, C. J., & Walker, S. L. (2013). Effects of solution chemistry on the transport of graphene oxide in saturated porous media. *Environmental Science & Technology, 47*, 4255–4261.

Laumann, S., Micić, V., & Hofmann, T. (2014). Mobility enhancement of nanoscale zero-valent iron in carbonate porous media through co-injection of polyelectrolytes. *Water Research, 50*, 70–79.

Laumann, S., Micić, V., Lowry, G. V., & Hofmann, T. (2013). Carbonate minerals in porous media decrease mobility of polyacrylic acid modified zero-valent iron nanoparticles used for groundwater remediation. *Environmental Pollution, 179*, 53–60.

Lee, D. G., Bonner, J. S., Garton, L. S., Ernest, A. N. S., & Autenrieth, R. L. (2000). Modeling coagulation kinetics incorporating fractal theories: A fractal rectilinear approach. *Water Research, 34*, 1987–2000.

Lee, D. G., Bonner, J. S., Garton, L. S., Ernest, A. N. S., & Autenrieth, R. L. (2002). Modeling coagulation kinetics incorporating fractal theories: Comparison with observed data. *Water Research, 36*, 1056–1066.

Li, X.-Y., & Logan, B. E. (2001). Permeability of fractal aggregates. *Water Research, 35*, 3373–3380.

Li, X., & Logan, B. E. (1997). Collision frequencies of fractal aggregates with small particles by differential sedimentation. *Environmental Science & Technology, 31*, 1229–1236.

Li, X., Scheibe, T. D., & Johnson, W. P. (2004). Apparent decreases in colloid deposition rate coefficients with distance of transport under unfavorable deposition conditions: A general phenomenon. *Environmental Science & Technology, 38*, 5616–5625.

Li, X., Zhang, P., Lin, C. L., & Johnson, W. P. (2005). Role of hydrodynamic drag on microsphere deposition and re-entrainment in porous media under unfavorable conditions. *Environmental Science & Technology, 39*, 4012–4020.

Li, Z., Hassan, A. A., Sahle-Demessie, E., & Sorial, G. A. (2013). Transport of nanoparticles with dispersant through biofilm coated drinking water sand filters. *Water Research, 47*, 6457–6466.

Li, Z., Sahle-Demessie, E., Hassan, A. A., & Sorial, G. A. (2011). Transport and deposition of CeO2 nanoparticles in water-saturated porous media. *Water Research, 45*, 4409–4418.

Liang, Y., Bradford, S. A., Simunek, J., Heggen, M., Vereecken, H., & Klumpp, E. (2013a). Retention and remobilization of stabilized silver nanoparticles in an undisturbed loamy sand soil. *Environmental Science & Technology, 47*, 12229–12237.

Liang, Y., Bradford, S. A., Simunek, J., Vereecken, H., & Klumpp, E. (2013b). Sensitivity of the transport and retention of stabilized silver nanoparticles to physicochemical factors. *Water Research, 47*, 2572–2582.

Limousin, G., Gaudet, J. P., Charlet, L., Szenknect, S., Barthès, V., & Krimissa, M. (2007). Sorption isotherms: A review on physical bases, modeling and measurement. *Applied Geochemistry, 22*, 249–275.

Lin, S., & Wiesner, M. R. (2012). Deposition of aggregated nanoparticles – A theoretical and experimental study on the effect of aggregation state on the affinity between nanoparticles and a collector surface. *Environmental Science & Technology, 46*, 13270–13277.

Liu, H. H., Surawanvijit, S., Rallo, R., Orkoulas, G., & Cohen, Y. (2011). Analysis of nanoparticle agglomeration in aqueous suspensions via constant-number Monte Carlo simulation. *Environmental Science & Technology, 45*, 9284–9292.

Liu, L., Gao, B., Wu, L., Sun, Y., & Zhou, Z. (2015). Effects of surfactant type and concentration on graphene retention and transport in saturated porous media. *Chemical Engineering Journal, 262*, 1187–1191.

Liu, L., Gao, B., Wu, L., Yang, L., Zhou, Z., & Wang, H. (2013). Effects of pH and surface metal oxyhydroxides on deposition and transport of carboxyl-functionalized graphene in saturated porous media. *Journal of Nanoparticle Research, 15*, 1–8.

Logan, B. E., Jewett, D. G., Arnold, R. G., Bouwer, E. J., & O'Melia, C. R. (1995a). Clarification of clean-bed filtration models. *Journal of Environmental Engineering, 121*, 869–873.

Logan, B. E., Passow, U., Alldredge, A. L., Grossartt, H.-P., & Simont, M. (1995b). Rapid formation and sedimentation of large aggregates is predictable from coagulation rates (half-lives) of transparent exopolymer particles (TEP). *Deep Sea Research Part II: Topical Studies in Oceanography, 42*, 203–214.

Lv, X., Gao, B., Sun, Y., Shi, X., Xu, H., & Wu, J. (2014). Effects of humic acid and solution chemistry on the retention and transport of cerium dioxide nanoparticles in saturated porous media. *Water, Air, & Soil Pollution, 225*, 1–9.

McDowell-Boyer, L. M., Hunt, J. R., & Sitar, N. (1986). Particle transport through porous media. *Water Resources Research, 22*, 1901–1921.

Mehmani, Y., & Balhoff, M. T. (2015a). Eulerian network modeling of longitudinal dispersion. *Water Resources Research, 51*, 8586–8606.

Mehmani, Y., & Balhoff, M. T. (2015b). Mesoscale and hybrid models of fluid flow and solute transport. *Reviews in Mineralogy and Geochemistry, 80*, 433–459.

Meng, Z., Hashmi, S. M., & Elimelech, M. (2013). Aggregation rate and fractal dimension of fullerene nanoparticles via simultaneous multiangle static and dynamic light scattering measurement. *Journal of Colloid and Interface Science, 392*, 27–33.

Molnar, I. L., Johnson, W. P., Gerhard, J. I., Willson, C. S., & O'Carroll, D. M. (2015). Predicting colloid transport through saturated porous media: A critical review. *Water Resources Research, 51*, 6804–6845.

Nascimento, A. G., Totola, M. R., Souza, C. S., Borges, M. T., & Borges, A. C. (2006). Temporal and spatial dynamics of blocking and ripening effects on bacterial transport through a porous system: A possible explanation for CFT deviation. *Colloids and Surfaces B: Biointerfaces, 53*, 241–244.

Neukum, C., Braun, A., & Azzam, R. (2014a). Transport of engineered silver (Ag) nanoparticles through partially fractured sandstones. *Journal of Contaminant Hydrology, 164*, 181–192.

Neukum, C., Braun, A., & Azzam, R. (2014b). Transport of stabilized engineered silver (Ag) nanoparticles through porous sandstones. *Journal of Contaminant Hydrology, 158*, 1–13.

Nowack, B., Baalousha, M., Bornhöft, N., Chaudhry, Q., Cornelis, G., Cotterill, J., Gondikas, A., Hassellöv, M., Lead, J., & Mitrano, D. M. (2015). Progress towards the validation of modeled

environmental concentrations of engineered nanomaterials by analytical measurements. *Environmental Science: Nano, 2*, 421.
O'Carroll, D. M., Bradford, S. A., & Abriola, L. M. (2004). Infiltration of PCE in a system containing spatial wettability variations. *Journal of Contaminant Hydrology, 73*, 39–63.
O'Carroll, D., Sleep, B., Krol, M., Boparai, H., & Kocur, C. (2012). Nanoscale zero valent iron and bimetallic particles for contaminated site remediation. *Advances in Water Resources, 51*, 104–122.
Parker, J. C., Van Genuchten, T., & Virginia Agricultural Experiment, S. (1984). *Determining transport parameters from laboratory and field tracer experiments*. Virginia: Virginia Agricultural Experiment Station.
Peijnenburg, W., Praetorius, A., Scott-Fordsmand, J., & Cornelis, G. (2016). Fate assessment of engineered nanoparticles in solids dominated media–Current insights and the way forward. *Environmental Pollution, 218*, 1365–1369.
Petosa, A. R., Jaisi, D. P., Quevedo, I. R., Elimelech, M., & Tufenkji, N. (2010). Aggregation and deposition of engineered nanomaterials in aquatic environments: Role of physicochemical interactions. *Environmental Science & Technology, 44*, 6532–6549.
Phenrat, T., Cihan, A., Kim, H.-J., Mital, M., Illangasekare, T., & Lowry, G. V. (2010a). Transport and deposition of polymer-modified Fe0 Nanoparticles in 2-D heterogeneous porous media: Effects of particle concentration, Fe0 content, and coatings. *Environmental Science & Technology, 44*, 9086–9093.
Phenrat, T., Kim, H.-J., Fagerlund, F., Illangasekare, T., & Lowry, G. V. (2010b). Empirical correlations to estimate agglomerate size and deposition during injection of a polyelectrolyte-modified Fe0 nanoparticle at high particle concentration in saturated sand. *Journal of Contaminant Hydrology, 118*, 152–164.
Phenrat, T., Kim, H.-J., Fagerlund, F., Illangasekare, T., Tilton, R. D., & Lowry, G. V. (2009). Particle size distribution, concentration, and magnetic attraction affect transport of polymer-modified Fe^0 nanoparticles in sand columns. *Environmental Science & Technology, 43*, 5079–5085.
Phenrat, T., Saleh, N., Sirk, K., Tilton, R. D., & Lowry, G. V. (2007). Aggregation and sedimentation of aqueous nanoscale zerovalent iron dispersions. *Environmental Science & Technology, 41*, 284–290.
Phenrat, T., Song, J. E., Cisneros, C. M., Schoenfelder, D. P., Tilton, R. D., & Lowry, G. V. (2010c). Estimating attachment of nano- and submicrometer-particles coated with organic macromolecules in porous media: development of an empirical model. *Environmental Science & Technology, 44*, 4531–4538.
Porubcan, A. A., & Xu, S. (2011). Colloid straining within saturated heterogeneous porous media. *Water Research, 45*, 1796–1806.
Praetorius, A., Scheringer, M., & Hungerbühler, K. (2012). Development of environmental fate models for engineered nanoparticles – A case study of tio2 nanoparticles in the rhine river. *Environmental Science & Technology, 46*, 6705–6713.
Prieve, D. C., & Hoysan, P. M. (1978). Role of colloidal forces in hydrodynamic chromatography. *Journal of Colloid and Interface Science, 64*, 201–213.
Qi, Z., Zhang, L., & Chen, W. (2014a). Transport of graphene oxide nanoparticles in saturated sandy soil. *Environmental Science: Processes & Impacts, 16*, 2268–2277.
Qi, Z., Zhang, L., Wang, F., Hou, L., & Chen, W. (2014b). Factors controlling transport of graphene oxide nanoparticles in saturated sand columns. *Environmental Toxicology and Chemistry, 33*, 998–1004.
Quik, J. T. K., van De Meent, D., & Koelmans, A. A. (2014a). Simplifying modeling of nanoparticle aggregation–sedimentation behavior in environmental systems: A theoretical analysis. *Water Research, 62*, 193–201.
Quik, J. T. K., Velzeboer, I., Wouterse, M., Koelmans, A. A., & Van de Meent, D. (2014b). Heteroaggregation and sedimentation rates for nanomaterials in natural waters. *Water Research, 48*, 269–279.

Rahman, T., George, J., & Shipley, H. J. (2013). Transport of aluminum oxide nanoparticles in saturated sand: Effects of ionic strength, flow rate, and nanoparticle concentration. *Science of the Total Environment, 463–464*, 565–571.
Rahman, T., Millwater, H., & Shipley, H. J. (2014). Modeling and sensitivity analysis on the transport of aluminum oxide nanoparticles in saturated sand: Effects of ionic strength, flow rate, and nanoparticle concentration. *Science of the Total Environment, 499*, 402–412.
Rajagopalan, R., & Tien, C. (1976). Trajectory analysis of deep-bed filtration with the sphere-in-cell porous media model. *AIChE Journal, 22*, 523–533.
Raychoudhury, T., Tufenkji, N., & Ghoshal, S. (2012). Aggregation and deposition kinetics of carboxymethyl cellulose-modified zero-valent iron nanoparticles in porous media. *Water Research, 46*, 1735–1744.
Raychoudhury, T., Tufenkji, N., & Ghoshal, S. (2014). Straining of polyelectrolyte-stabilized nanoscale zero valent iron particles during transport through granular porous media. *Water Research, 50*, 80–90.
Risovic, D., & Martinis, M. (1994). The role of coagulation and sedimentation mechanisms in the two-component model of sea-particle size distribution. *Fizika, 3*, 103–118.
Rosansky, S., Condit, W., Sirabian, R., 2013. Best practices for injection and distribution of amendments. Technical Report TR-NAVFAC-EXWC-EV.
Ryan, J. N., & Elimelech, M. (1996). Colloid mobilization and transport in groundwater. *Colloids and Surfaces A: Physicochemical and Engineering Aspects, 107*, 1–56.
Saiers, J. E., & Hornberger, G. M. (1996). Migration of 137Cs through quartz sand: Experimental results and modeling approaches. *Journal of Contaminant Hydrology, 22*, 255–270.
Saiers, J. E., Hornberger, G. M., & Liang, L. (1994). First-and second-order kinetics approaches for modeling the transport of colloidal particles in porous media. *Water Resources Research, 30*, 2499–2506.
Saleh, N., Sirk, K., Liu, Y., Phenrat, T., Dufour, B., Matyjaszewski, K., Tilton, R. D., & Lowry, G. V. (2007). Surface modifications enhance nanoiron transport and NAPL targeting in saturated porous media. *Environmental Engineering Science, 24*, 45–57.
Sasidharan, S., Torkzaban, S., Bradford, S. A., Dillon, P. J., & Cook, P. G. (2014). Coupled effects of hydrodynamic and solution chemistry on long-term nanoparticle transport and deposition in saturated porous media. *Colloids and Surfaces A: Physicochemical and Engineering Aspects, 457*, 169–179.
Schaaf, P., & Talbot, J. (1989). Surface exclusion effects in adsorption processes. *The Journal of Chemical Physics, 91*, 4401–4409.
Scheibe, T. D., & Wood, B. D. (2003). A particle-based model of size or anion exclusion with application to microbial transport in porous media. *Water Resources Research, 39*, 1080.
Schijven, J. F., Hassanizadeh, S. M., & de Bruin, R. H. A. M. (2002). Two-site kinetic modeling of bacteriophages transport through columns of saturated dune sand. *Journal of Contaminant Hydrology, 57*, 259–279.
Schijven, J. K., & Hassanizadeh, S. M. (2000). Removal of viruses by soil passage: Overview of modeling, processes, and parameters. *Critical Reviews in Environmental Science and Technology, 30*, 49–127.
Seetha, N., Majid Hassanizadeh, S., Kumar, M., & Raoof, A. (2015). Correlation equations for average deposition rate coefficients of nanoparticles in a cylindrical pore. *Water Resources Research, 51*, 8034–8059.
Shang, J., Liu, C., & Wang, Z. (2013). Transport and retention of engineered nanoporous particles in porous media: effects of concentration and flow dynamics. *Colloids and Surfaces A: Physicochemical and Engineering Aspects, 417*, 89–98.
Shellenberger, K., & Logan, B. E. (2002). Effect of molecular scale roughness of glass beads on colloidal and bacterial deposition. *Environmental Science & Technology, 36*, 184–189.
Shen, C., Huang, Y., Li, B., & Jin, Y. (2008). Effects of solution chemistry on straining of colloids in porous media under unfavorable conditions. *Water Resources Research, 44*, W05419.

Skopp, J. (1986). Analysis of time-dependent chemical processes in soils. *Journal of Environmental Quality, 15*, 205–213.
Smoluchowski, M. (1917). Versuch einer mathematischen Theorie der Koagulationskinetik kolloider Lösungen. *Zeitschrift fuer Physikalische Chemie, 92*, 129–168.
Song, L., & Elimelech, M. (1993). Dynamics of colloid deposition in porous media: modeling the role of retained particles. *Colloids and Surfaces A, 73*, 49–63.
Sun, N., Elimelech, M., Sun, N.-Z., & Ryan, J. N. (2001). A novel two-dimensional model for colloid transport in physically and geochemically heterogeneous porous media. *Journal of Contaminant Hydrology, 49*, 173–199.
Sun, P., Shijirbaatar, A., Fang, J., Owens, G., Lin, D., & Zhang, K. (2015a). Distinguishable transport behavior of zinc oxide nanoparticles in silica sand and soil columns. *Science of the Total Environment, 505*, 189–198.
Sun, Y., Gao, B., Bradford, S. A., Wu, L., Chen, H., Shi, X., & Wu, J. (2015b). Transport, retention, and size perturbation of graphene oxide in saturated porous media: Effects of input concentration and grain size. *Water Research, 68*, 24–33.
Swift, D. L., & Friedlander, S. K. (1964). The coagulation of hydrosols by Brownian motion and laminar shear flow. *Journal of Colloid Science, 19*, 621–647.
Szilagyi, I., Szabo, T., Desert, A., Trefalt, G., Oncsik, T., & Borkovec, M. (2014). Particle aggregation mechanisms in ionic liquids. *Physical Chemistry Chemical Physics, 16*, 9515–9524.
Taghavy, A., Mittelman, A., Wang, Y., Pennell, K. D., & Abriola, L. M. (2013). Mathematical modeling of the transport and dissolution of citrate-stabilized silver nanoparticles in porous media. *Environmental Science & Technology, 47*, 8499–8507.
Taghavy, A., Pennell, K. D., & Abriola, L. M. (2014). Modeling coupled nanoparticle aggregation and transport in porous media: A Lagrangian approach. *Journal of Contaminant Hydrology, 172*, 48–60.
Taghavy, A., Pennell, K. D., & Abriola, L. M. (2015). Modeling coupled nanoparticle aggregation and transport in porous media: A Lagrangian approach. *Journal of Contaminant Hydrology, 172*, 48–60.
Tan, Y., Gannon, J. T., Baveye, P., & Alexander, M. (1994). Transport of bacteria in an aquifer sand: Experiments and model simulations. *Water Resources Research, 30*, 3243–3252.
Tang, P., & Raper, J. A. (2002). Modelling the settling behaviour of fractal aggregates–A review. *Powder Technology, 123*, 114–125.
Therezien, M., Thill, A., & Wiesner, M. R. (2014). Importance of heterogeneous aggregation for NP fate in natural and engineered systems. *Science of the Total Environment, 485*, 309–318.
Tian, Y., Gao, B., Silvera-Batista, C., & Ziegler, K. J. (2010). Transport of engineered nanoparticles in saturated porous media. *Journal of Nanoparticle Research, 12*, 2371–2380.
Tian, Y., Gao, B., Wang, Y., Morales, V. L., Carpena, R. M., Huang, Q., & Yang, L. (2012). Deposition and transport of functionalized carbon nanotubes in water-saturated sand columns. *Journal of Hazardous Materials, 213*, 265–272.
Tiraferri, A., & Sethi, R. (2009). Enhanced transport of zerovalent iron nanoparticles in saturated porous media by guar gum. *Journal of Nanoparticle Research, 11*, 635–645.
Toloni, I., Lehmann, F., & Ackerer, P. (2014). Modeling the effects of water velocity on TiO2 nanoparticles transport in saturated porous media. *Journal of Contaminant Hydrology, 171*, 42–48.
Tong, M., & Johnson, W. P. (2007). Colloid population heterogeneity drives hyperexponential deviation from classic filtration theory. *Environmental Science & Technology, 41*, 493–499.
Tong, M., Ma, H., & Johnson, W. P. (2008). Funneling of flow into grain-to-grain contacts drives colloid–colloid aggregation in the presence of an energy barrier. *Environmental Science & Technology, 42*, 2826–2832.
Torkzaban, S., & Bradford, S. A. (2016). Critical role of surface roughness on colloid retention and release in porous media. *Water Research, 88*, 274–284.
Torkzaban, S., Bradford, S. A., Vanderzalm, J. L., Patterson, B. M., Harris, B., & Prommer, H. (2015). Colloid release and clogging in porous media: Effects of solution ionic strength and flow velocity. *Journal of Contaminant Hydrology, 181*, 161–171.

Torkzaban, S., Bradford, S. A., & Walker, S. L. (2007). Resolving the coupled effects of hydrodynamics and DLVO forces on colloid attachment in porous media. *Langmuir, 23*, 9652–9660.
Torkzaban, S., Bradford, S. A., Wan, J., Tokunaga, T., & Masoudih, A. (2013). Release of quantum dot nanoparticles in porous media: Role of cation exchange and aging time. *Environmental Science & Technology, 47*, 11528–11536.
Torkzaban, S., Kim, H. N., Simunek, J., & Bradford, S. A. (2010a). Hysteresis of colloid retention and release in saturated porous media during transients in solution chemistry. *Environmental Science & Technology, 44*, 1662–1669.
Torkzaban, S., Kim, Y., Mulvihill, M., Wan, J., & Tokunaga, T. K. (2010b). Transport and deposition of functionalized CdTe nanoparticles in saturated porous media. *Journal of Contaminant Hydrology, 118*, 208–217.
Torkzaban, S., Wan, J., Tokunaga, T. K., & Bradford, S. A. (2012). Impacts of bridging complexation on the transport of surface-modified nanoparticles in saturated sand. *Journal of Contaminant Hydrology, 136*, 86–95.
Tosco, T., Gastone, F., & Sethi, R. (2014). Guar gum solutions for improved delivery of iron particles in porous media (Part 2): Iron transport tests and modeling in radial geometry. *Journal of Contaminant Hydrology, 166*, 34–51.
Tosco, T., & Sethi, R. (2010). Transport of non-Newtonian suspensions of highly concentrated micro-and nanoscale iron particles in porous media: A modeling approach. *Environmental Science & Technology, 44*, 9062–9068.
Tratnyek, P. G., & Johnson, R. L. (2006). Nanotechnologies for environmental cleanup. *Nano Today, 1*, 44–48.
Treumann, S., Torkzaban, S., Bradford, S. A., Visalakshan, R. M., & Page, D. (2014). An explanation for differences in the process of colloid adsorption in batch and column studies. *Journal of Contaminant Hydrology, 164*, 219–229.
Tufenkji, N., & Elimelech, M. (2004a). Correlation equation for predicting single-collector efficiency in physicochemical filtration in saturated porous media. *Environmental Science & Technology, 38*, 529–536.
Tufenkji, N., & Elimelech, M. (2004b). Deviation from the classical colloid filtration theory in the presence of repulsive DLVO interactions. *Langmuir, 20*, 10818–10828.
Tufenkji, N., & Elimelech, M. (2005). Spatial distributions of Cryptosporidium oocysts in porous media: Evidence for dual mode deposition. *Environmental Science & Technology, 39*, 3620–3629.
Van Genuchten, M. T., & Wierenga, P. J. (1976). Mass transfer studies in sorbing porous media I. Analytical solutions. *Soil Science Society of America Journal, 40*, 473–480.
Wang, C., Bobba, A. D., Attinti, R., Shen, C., Lazouskaya, V., Wang, L.-P., & Jin, Y. (2012a). Retention and transport of silica nanoparticles in saturated porous media: Effect of concentration and particle size. *Environmental Science & Technology, 46*, 7151–7158.
Wang, D., Bradford, S. A., Harvey, R. W., Gao, B., Cang, L., & Zhou, D. (2012b). Humic acid facilitates the transport of ARS-labeled hydroxyapatite nanoparticles in iron oxyhydroxide-coated sand. *Environmental Science & Technology, 46*, 2738–2745.
Wang, D., Bradford, S. A., Harvey, R. W., Hao, X., & Zhou, D. (2012c). Transport of ARS-labeled hydroxyapatite nanoparticles in saturated granular media is influenced by surface charge variability even in the presence of humic acid. *Journal of Hazardous Materials, 229*, 170–176.
Wang, D., Bradford, S. A., Paradelo, M., Peijnenburg, W. J. G. M., & Zhou, D. (2012d). Facilitated transport of copper with hydroxyapatite nanoparticles in saturated sand. *Soil Science Society of America Journal, 76*, 375–388.
Wang, D., Ge, L., He, J., Zhang, W., Jaisi, D. P., & Zhou, D. (2014a). Hyperexponential and nonmonotonic retention of polyvinylpyrrolidone-coated silver nanoparticles in an Ultisol. *Journal of Contaminant Hydrology, 164*, 35–48.
Wang, D., Jaisi, D. P., Yan, J., Jin, Y., & Zhou, D. (2015a). Transport and retention of polyvinylpyrrolidone-coated silver nanoparticles in natural soils. *Vadose Zone Journal, 14*, 2–13.

Wang, D., Jin, Y., & Jaisi, D. (2015b). Cotransport of hydroxyapatite nanoparticles and hematite colloids in saturated porous media: Mechanistic insights from mathematical modeling and phosphate oxygen isotope fractionation. *Journal of Contaminant Hydrology, 182*, 194–209.
Wang, D., Jin, Y., & Jaisi, D. P. (2015c). Effect of size-selective retention on the cotransport of hydroxyapatite and goethite nanoparticles in saturated porous media. *Environmental Science & Technology, 49*, 8461–8470.
Wang, D., Paradelo, M., Bradford, S. A., Peijnenburg, W. J. G. M., Chu, L., & Zhou, D. (2011). Facilitated transport of Cu with hydroxyapatite nanoparticles in saturated sand: Effects of solution ionic strength and composition. *Water Research, 45*, 5905–5915.
Wang, D., Su, C., Liu, C., & Zhou, D. (2014b). Transport of fluorescently labeled hydroxyapatite nanoparticles in saturated granular media at environmentally relevant concentrations of surfactants. *Colloids and Surfaces A: Physicochemical and Engineering Aspects, 457*, 58–66.
Wang, H. F., & Anderson, M. P. (1995). *Introduction to groundwater modeling: Finite difference and finite element methods*. New York: Academic Press.
Wang, X., Cai, L., Han, P., Lin, D., Kim, H., & Tong, M. (2014c). Cotransport of multi-walled carbon nanotubes and titanium dioxide nanoparticles in saturated porous media. *Environmental Pollution, 195*, 31–38.
Wang, Y., Becker, M. D., Colvin, V. L., Abriola, L. M., & Pennell, K. D. (2014d). Influence of residual polymer on nanoparticle deposition in porous media. *Environmental Science & Technology, 48*, 10664–10671.
Wang, Y., Zhu, H., Becker, M. D., Englehart, J., Abriola, L. M., Colvin, V. L., & Pennell, K. D. (2013). Effect of surface coating composition on quantum dot mobility in porous media. *Journal of Nanoparticle Research, 15*, 1–16.
Wang, Z., Jin, Y., Shen, C., Li, T., Huang, Y., & Li, B. (2016). Spontaneous detachment of colloids from primary energy minima by brownian diffusion. *PloS one, 11*, e0147368.
Xu, S., Gao, B., & Saiers, J. E. (2006). Straining of colloidal particles in saturated porous media. *Water Resources Research, 42*, W12S16.
Xu, S., Liao, Q., & Saiers, J. E. (2008). Straining of nonspherical colloids in saturated porous media. *Environmental Science & Technology, 42*, 771–778.
Xu, S., & Saiers, J. E. (2009). Colloid straining within water-saturated porous media: Effects of colloid size nonuniformity. *Water Resources Research, 45*, W05501.
Yao, K.-M., Habibian, M. T., & O'Melia, C. R. (1971). Water and waste water filtration. Concepts and applications. *Environmental Science & Technology, 5*, 1105–1112.
Yoon, J. S., Germaine, J. T., & Culligan, P. J. (2006). Visualization of particle behavior within a porous medium: Mechanisms for particle filtration and retardation during downward transport. *Water Resources Research, 42*, 1–16.
Yu, H., Fu, J., Dang, L., Cheong, Y., Tan, H., & Wei, H. (2015a). Prediction of the particle size distribution parameters in a high shear granulation process using a key parameter definition combined artificial neural network model. *Industrial & Engineering Chemistry Research, 54*, 10825–10834.
Yu, H., He, Y., Li, P., Li, S., Zhang, T., Rodriguez-Pin, E., Du, S., Wang, C., Cheng, S., & Bielawski, C. W. (2015b). Flow enhancement of water-based nanoparticle dispersion through microscale sedimentary rocks. *Scientific Reports, 5*, 8702.
Zhang, L., Hou, L., Wang, L., Kan, A. T., Chen, W., & Tomson, M. B. (2012). Transport of fullerene nanoparticles (n C60) in saturated sand and sandy soil: Controlling factors and modeling. *Environmental Science & Technology, 46*, 7230–7238.
Zhang, P., Johnson, W. P., Piana, M. J., Fuller, C. C., & Naftz, D. L. (2001a). Potential artifacts in interpretation of differential breakthrough of colloids and dissolved tracers in the context of transport in a zero-valent iron permeable reactive barrier. *Groundwater, 39*, 831–840.
Zhang, P., Johnson, W. P., Scheibe, T. D., Choi, K. H., Dobbs, F. C., & Mailloux, B. J. (2001b). Extended tailing of bacteria following breakthrough at the Narrow Channel Focus Area, Oyster, Virginia. *Water Resources Research, 37*, 2687–2698.

Zhang, W., Jianzhi, N., Morales, V. L., Chen, X., Hay, A. G., Lehmann, J., & Steenhuis, T. S. (2010). Transport and retention of biochar particles in porous media: Effect of pH, ionic strength, and particle size. *Ecohydrology, 3*, 497–508.

Zheng, C., & Wang, P. (1999a). *MT3DMS, A modular three-dimensional multi-species transport model for simulation of advection, dispersion and chemical reactions of contaminants in groundwater systems; documentation and user's guide*. U.S. Army Corpes Engineers, Engineer Research and Development Center, Contract Report SERDP-99-1, Vicksburg, MS, 202.

Zheng, C., & Wang, P. P. (1999b). *A modular three-dimensional multi-species transport model for simulation of advection, dispersion and chemical reactions of contaminants in groundwater systems; documentation and user's guide*. US Army Engineer Research and Development Center Contract Report SERDP-99-1, Vicksburg, Mississippi, USA.

Chapter 7
Moving into the Third Decade of Nanoscale Zero-Valent Iron (NZVI) Development: Best Practices for Field Implementation

Chris M. Kocur, Brent E. Sleep, and Denis M. O'Carroll

Abstract This chapter provides an overview of environmental restoration efforts involving the application of NZVI. The chapter focuses on the novel application techniques aimed at improving the delivery, characterization, and effectiveness of NZVI, drawing on over two decades of peer-reviewed literature. Stressing a base of knowledge through detailed site characterization toward a site conceptual model, this chapter discusses delivery techniques, options for NZVI formulation, and challenges associated with different site conditions. NZVI particle types and injection characteristics are covered along with field-ready analytical capabilities for NZVI detection and characterization. The chapter also highlights cases where remote sensing and modeling have been used to better understand NZVI delivery. Lessons learned from past field studies are discussed and will become increasingly relevant as the industry gears up for a renaissance of NZVI use. Growing confidence in the use cases for stabilized NZVI, the synergistic application of ZVI + bioremediation and technological advances such as sulfidation will catch the eye of practitioners and site managers into the future and lead to more innovation.

Keywords Nanoscale zerovalent iron · Field implementation · Best practice · Combined remedies with NZVI

C. M. Kocur (✉)
Department of Civil & Environmental Engineering, Western University, London, ON, Canada

School of Public Health, Oregon Health & Science University, Portland, OR, USA
e-mail: ckocur@uwo.ca

B. E. Sleep
Department of Civil Engineering, University of Toronto, Toronto, ON, Canada

D. M. O'Carroll
Water Research Laboratory, Connected Waters Initiative, Manly Vale, NSW, Australia

School of Civil and Environmental Engineering, University of New South Wales, Manly Vale, NSW, Australia

T. Phenrat, G. V. Lowry (eds.), *Nanoscale Zerovalent Iron Particles for Environmental Restoration*, https://doi.org/10.1007/978-3-319-95340-3_7

7.1 Field-Scale Application of NZVI

The use of NZVI in the field has many complicating factors that must be considered in the design of an appropriate treatment technology including, e.g., dilution, oxidation, and interaction with ions. This is compounded by the fact that adequate evaluation of remediation performance is often hampered by poor site characterization (Preslo et al. 2005; Kueper 2014). In this chapter, we use examples from previous NZVI field studies to illustrate important considerations involved in assessing delivery and performance monitoring following NZVI application. The different metrics that have been used to evaluate NZVI remediation performance will be compared and discussed. Examples of site characterization techniques are used to stress the importance of developing a conceptual model prior to field activities. Strategies will be discussed for best assessing the site prior to amendment injection, during injection, and during long-term monitoring.

7.2 Conceptual Site Model

A site conceptual model is important to develop prior to testing and field-scale implementation of NZVI. A robust conceptual model can be developed after rigorous site characterization and is necessary to achieve the desired remediation outcomes.

Characterization must include identification of the sources of contamination, the potential avenues of migration, and transfer between phases, and pre-remediation conditions on site are imperative in selecting a remediation alternative and remediation monitoring plan. Knowledge of the distribution of contaminants within high and low permeability zones allows for targeted remediation, and an understanding of the current state of contaminant fluxes can allow for better prediction of the fate of contaminants on site. The age of the site and the mass of contaminants that were released are also important factors to consider in site characterization.

The 14-compartment model in Fig. 7.1 is an example of a contaminated site conceptual model for a DNAPL-impacted site (ITRC 2011), which is the most prevalent target compound for NZVI remediation operations. The model accounts for the contaminant mass distributed in difference phases (Table 7.1) with different subsurface formations simplified as high and low hydraulic conductivity (K) zones. Initially, contaminants enter the system through high-K zones, perhaps as a DNAPL, and then sorb and diffuse into lower K media. The model can therefore allow for the understanding of contaminant fluxes that occur as the contamination persists, migrates, and degrades. An accurate conceptual model, including knowledge of how the contamination impacted the site in the source zone, how long the contamination has persisted, and to what length the aqueous plume extends, as well as an accurate profile of the lithology and media, is the first step in effective remediation. This allows for strategic selection of a remediation approach and allows for the

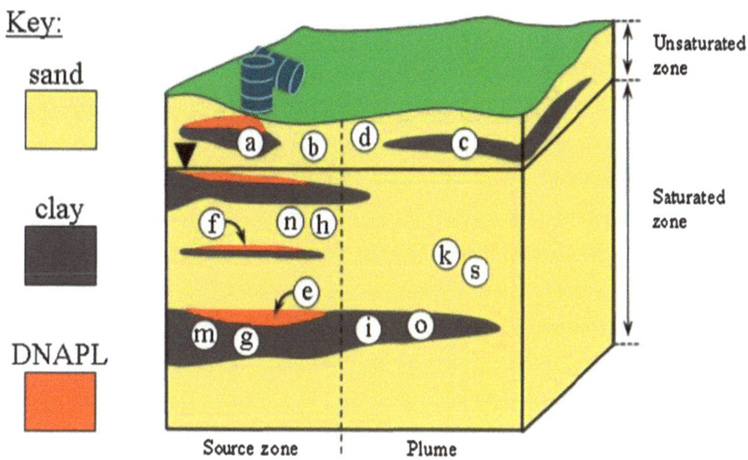

Fig. 7.1 Site conceptual model of subsurface contamination showing all possible phases of contamination within the 14-compartment model: DNAPL, sorbed, aqueous, and vapor. (Adapted from ITRC 2011)

Table 7.1 Phases of contamination in the site conceptual model laid out in the 14-compartment model. This 14-compartment model is an example tool for evaluating chlorinated solvent sites with NZVI. The shaded boxes indicate contamination that can be targeted with NZVI

Zone	Source zone		Plume	
	Lower K	Higher K	Lower K	Higher K
Vapor	a	b	c	d
DNAPL	e	f		
Aqueous	g	h	i	k
Sorbed	m	n	o	s

implementation of technologies that will most effectively remove the highest priority contaminants. This will also facilitate an assessment of remediation performance and the identification of remediation stop points or assist in the transition to another remediation technology in the treatment train.

7.2.1 ZVI Alternatives for Remediation

ZVI-based technologies can treat a broad range of contaminated site problems. Table 7.2 shows a comparative summary of the properties, attributes, and limitations

Table 7.2 Comparative summary of ZVI remediation alternatives in wide use and in development

Type of ZVI	Granular	MZVI	NZVI			
	ZVI		NZVI (bare)	With stabilizers	With carbon substrate	Bimetallic
Size range	Silt to medium sand	1–1000 μm	5–100 nm	Tunable depending on stabilizer (5–100 nm)		
Specific surface area	Low	Moderate	Variable	High, depending on stabilizer		
Mobility	Not mobile	Very limited	Very limited	Moderate to high depending on stabilizer		
Longevity of reaction	Years of use in PRBs, depending on conditions	Months to years, susceptible to passivation	Weeks to months, depending on passivation and method of synthesis	Days to weeks, depending on contaminate loading	Days to weeks, longer if coupled with biodegradable stabilizer	Weeks to months due to more efficient use of electrons
Reaction rate	Slow compared to alternatives	Moderate, depending on surface area	High with many contaminants	High, depending on stabilizer		Rapid reaction with most chlorinated contaminants
Aggregation	NA	NA	Rapid, especially at high concentrations	Low, forms stable aggregates depending on stabilizer		
Use cases (targeted zone)	Soil mixing (i,k,o,s) PRBs (i,k), fracture and injection(g)	Soil mixing (i,k), fracture and injection (g), PRBs (i, k), slurry walls (i,k), high viscosity injection (f,h,n)	High viscosity injection(f,h,n), EZVI (f,h,n)	Hot spot injection/ source zone (e,f,g,h,m, n), high contaminant flux zones (h,n,k,s)	Hot spot injection/ biostimulation (e, f,g,h,m,n)	High contaminant concentrations, hotspots troubled contaminants, tight timeline (g,h,m,n)

of ZVI-based technologies. The conditions of use will be site-specific and depend on many factors; however, general criteria for the effective use of each ZVI variant have been briefly outlined in Table 7.2 as referred to Fig. 7.1. NZVI is particularly attractive due to its ability to treat higher concentrations of contaminants close to the source zone (see also Chaps. 3 and 5) and also its ability to create ideal conditions for in situ bioremediation (see also Chap. 10).

7.2.2 *Role of NZVI*

Advances in delivery techniques and particle mobility have allowed for the successful application of NZVI at dozens of sites in North America, Asia, and Europe. At many of these sites, NZVI was not applied for full-scale remediation but as pilot trials to develop a knowledge set for future applications. What is becoming apparent from the numerous examples of NZVI application is the necessity for a scalable and reliable deployment in order to treat problem areas of contamination. These "hot spots" may need to be treated due to accidental spills or remobilization of NAPL, or there may be areas not remediated using other technologies, under characterized source zones, or secondary sources of contamination. In all these scenarios, NZVI provides a viable alternative when remediation goals require a truncated timeline. Figure 7.2 presents several examples of contamination scenarios for which there are limited viable remediation options. In addition, NZVI application requires minimal aboveground infrastructure and can be effective for remediation around active infrastructure and utilities.

7.3 Site Characterization

Site investigation and use of historical activity records can identify the potential sources of contamination. Site topology and lithology information can be gathered on regional, local, and site levels from borings and can be used to build a conceptual site model. Through further investigation, more detailed elements of the conceptual site model can be identified and decisions about remediation made. Site characteristics including the type of subsurface media, hydraulic gradient, property boundaries, and infrastructure help inform selection and design of remediation alternatives. The decision on whether an amendment injection remediation approach is appropriate requires information related to whether the permeable media is appropriate for delivery. This section provides a summary of some of the site characterization methods that should be considered prior to the application of NZVI. The lithology, depth of contamination, target contaminants, and possible co-contaminants, as well as site-specific considerations (e.g., below- and aboveground infrastructure), will strongly influence design.

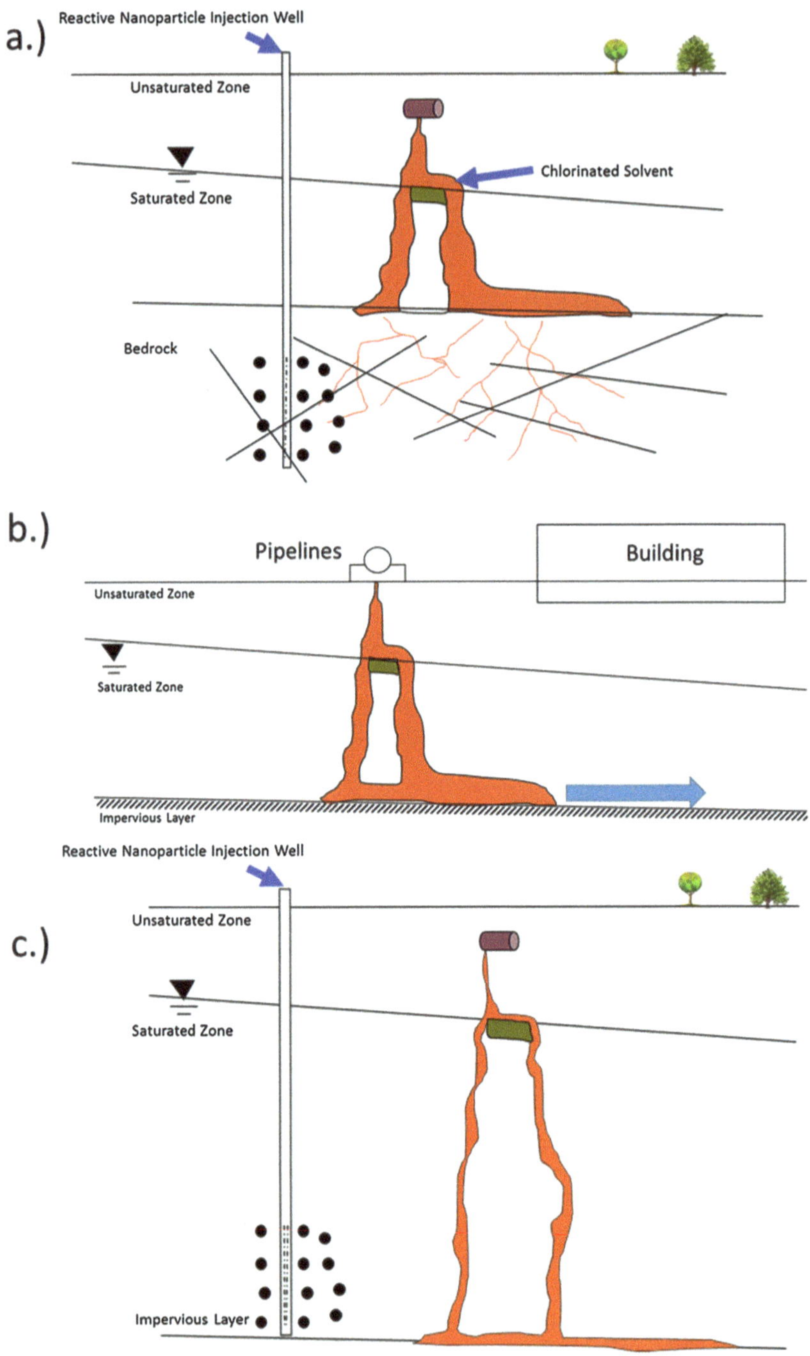

Fig. 7.2 Three cases where NZVI may be a suitable remediation alternative. (**a**) contamination in fractured bedrock. (**b**) existing critical infrastructure is too expensive to dismantle prior to remediation. (**c**) contamination reaches depths unattainable by other remediation methods

7.3.1 Conceptual Model Development

It is essential to delineate site contamination. Site lithology is often logged in detail during borehole investigations. In the past, this information has been used to guide engineers and geologists about where to install water wells or petroleum hydrocarbon extraction networks. For environmental remediation, knowledge of the geological, lithological, and hydrological conditions is not enough to inform decisions. Delineation of contamination is also required at the same level of detail.

Monitoring wells for environmental contamination investigation require a modified approach than that for drinking water extraction wells. Decades of research and development of site investigation on superfund sites have found that traditional monitoring wells are incapable of adequately characterizing contamination. Figure 7.3 provides an example of typical long-screen monitoring wells and some of the shortcomings when applied to environmental remediation (Einarson 2005). Table 7.3 describes the implications of different screen intervals on site characterization. The example contaminant contours can represent any soluble compound sampled in an aqueous matrix. For example, for NZVI, the solute could be chlorinated volatile organic compounds (CVOCs) (e.g., TCE). Well screen lengths can be as long as 20 feet on some sites which can lead to large discrepancies between the conceptual site model and actual site conditions.

Use of a transect of multilevel wells can provide a detailed conceptual understanding of the extent of contamination. Figure 7.4 is an example transect detailing a CVOC plume traveling through Section A-A'. This level of detail helps delineate more precisely plume extent. This type of plume delineation can be combined with

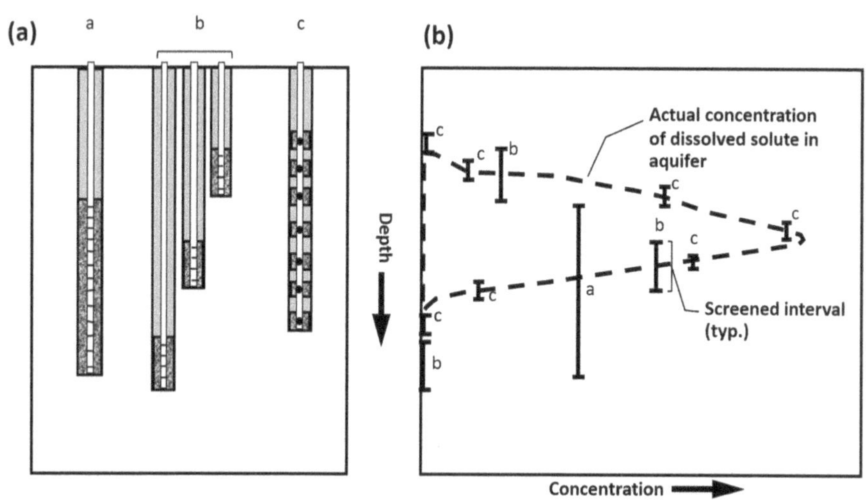

Fig. 7.3 (**a**) Illustration of three different well screen lengths (long-screened interval (L), medium (M), and multilevel (N)) and (**b**) an example of possible concentrations of solute measured with these different screened intervals. (Adapted from Einarson 2005)

Table 7.3 Impact of monitoring well design on conceptual model of the CVOC contamination on site

Bias in monitoring well	Implications on site conceptual model	Resulting consequences on NZVI remediation
Long well screen dilutes CVOCs in samples	Maximum concentration is underestimated	NZVI dose may not treat most of the contaminated zone
Well screen only partially captures contaminated interval	Position of maximum CVOC concentration is inaccurate	Targeted NZVI injection depth may miss contamination
Complex layering of media presents a long well screen	No information about contaminant distribution	Injection may only enter the most conductive media but miss the contaminated zone
Long well installed and a vertical hydraulic gradient is present	Contamination may migrate upward or downward within well screen	NZVI injection (often viscous) may mobilize contamination downward

Modified from Einarson (2005)

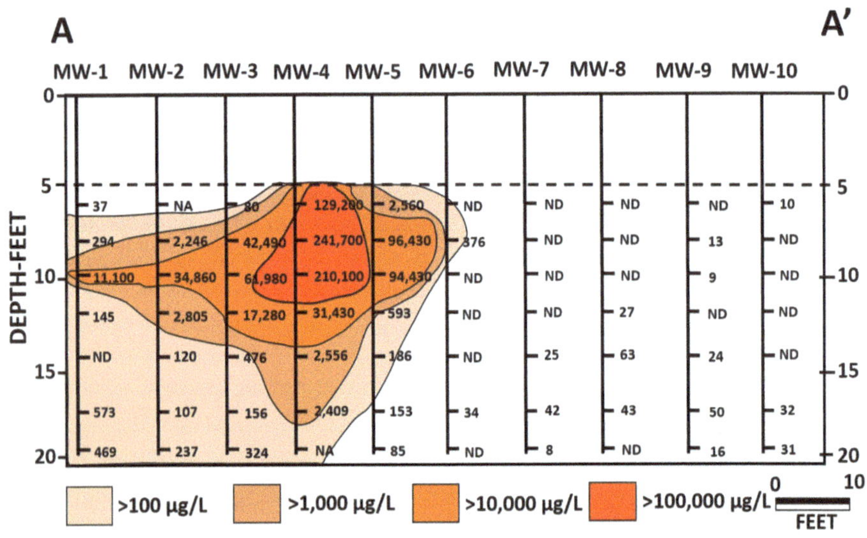

Fig. 7.4 Transect of a groundwater plume with total CVOC concentrations shaded at different contour intervals. The transect was installed transverse to the direction of travel of a groundwater plume. (Adapted from Einarson 2005)

known hydraulic gradient and estimates of hydraulic conductivity to estimate contaminant flux across the transect. The transect can also be oriented longitudinally to the direction of groundwater flow to delineate plume length, potentially providing information related to plume attenuation. The same transect approach can be used to delineate contamination within a source zone. NZVI and other amendments can be targeted to zones where elevated concentrations are detected.

An EZVI field application has taken advantage of the multilevel well approach to estimate mass flux using integral pump tests (Krug et al. 2010). Two monitoring wells were installed with a small transect of three multilevel wells in between. Over time, the multilevel concentrations were monitored, while the groundwater velocity was controlled across the transect using the two surrounding monitoring wells. Over a course of several years, the mass flux was periodically measured in this manner, providing a method for NAPL mass loss determination (SERDP report). Generally, more informed decisions regarding remediation success can be made using contamination profiles along transects in comparison with traditional long-screen wells.

7.3.2 Well Methods

Quantification of subsurface aquifer properties (e.g., storage, hydraulic conductivity, and groundwater direction) is vital in conceptual model development. There is a range of established techniques to rigorously evaluate subsurface hydrogeological parameters of interest (Fetter 2001; Pinder and Celia 2006). Pumping tests can provide information about hydraulic conductivity or transmissivity, specific yield (for unconfined aquifers) or storage (for confined aquifers), and connectivity of wells or layers. Slug tests can be performed for a screened well, within an open borehole, or through the use of packers along multiple discrete intervals to provide estimates of transmissivity and ultimately hydraulic conductivity (Butler 1997). Other considerations related to development of a site hydraulic conceptual model include the position and seasonal fluctuation in the water table. The conceptual model may have to be modified on some sites as seasonal changes in the water table can change the groundwater flow direction and magnitude.

Tracer testing can be used to identify horizontal hydraulic conductivity using arrival times between wells. This information can subsequently be used to benchmark solute transport against a conservative tracer (Bennett et al. 2010; He et al. 2010). This helps quantify the extent of adsorption and desorption of a given media. Tracers are used in column tests to compare water flow to NZVI transport and have been used as the basis for comparison in pilot-scale tests for NZVI injection as well (He et al. 2010; Johnson et al. 2013; Kocur et al. 2014).

Constituents in the injection fluid can also be used as conservative tracers. The electrical signal of the amendment injected can be used as a tracer to monitor extent of migration (Wei et al. 2010; Johnson et al. 2013). The extent of polymer amendment transport has also been demonstrated to correlate well with NZVI travel (Kocur et al. 2014). Reactive tracers employing tracers that possess known sorption properties to the oil/water interface can be used in combination with a conservative tracer to quantify the NAPL/water surface area (Annable et al. 1998). The effects of any tracer on NZVI reactivity and mobility should be considered prior to field application.

7.3.3 Emerging Methods of Investigation

Alternative methods for site investigation may greatly aid or replace conventional methods, depending on the situation. When permanent well infrastructure is not feasible (e.g., creeks or riverbeds, other low-lying areas prone to flooding) or permitted by regulators, alternative techniques can be the only methods available for site characterization. These alternative methods include a range of probe techniques:

- Waterloo profiler – Geoprobe tip with machined holes and screen allowing for relative hydraulic conductivity testing at discrete intervals as the probe is driven into the subsurface. The peristaltic pump at the surface can reverse direction allowing sample collection. Geochemical analysis can indicate lithology (e.g., using electrical conductivity, EC).
- Membrane interface probe (MIP) – Membrane sleeve that allows for volatile compounds sampling as the probe is driven into the ground. When the tip encounters high aqueous concentrations or NAPL, the inert gas pumped on the inside of the membrane will carry the volatile compounds to the surface for analysis on a portable gas chromatograph with the detector of choice.
- Hydrosparge – Probe allows water to enter an internal chamber at a discrete depth where gas is bubbled through the sample and passed to the surface where a gas analysis system determines the volatile compounds in the sample.
- In situ solvent injection and extraction – Similar to the hydrosparge technique but a solvent is released into the formation through the drive tip, then gathered by the probe, and analyzed at the surface.
- Downhole microscope/camera – Visual evidence of NAPL in the subsurface at depth based on the color (usually weathered and black).
- Laser-induced florescence (LIF) – Emission of a specific wavelength that is compound specific and florescent detection can be used to identify polychlorinated biphenyls and polycyclic aromatic hydrocarbons in the subsurface.

A relatively recent example of the application of a probe at an NZVI field site is the study of Bennett et al. (2010) who successfully used a MIP for aqueous phase sampling prior to NZVI injection into a layered sand aquifer that was contaminated with PCE and TCE (Fig. 7.5). More detailed descriptions of probes are available (Kram 2005). Another suite of site characterization techniques are geophysical methods that can be implemented aboveground or in boreholes. Information from geophysical methods can be used to provide a continuum of data on the site when properly calibrated to borehole data (Benson 2005). Probe techniques have the added benefit of reducing waste as only samples are brought up to the surface. This can reduce exposure at highly contaminated sites.

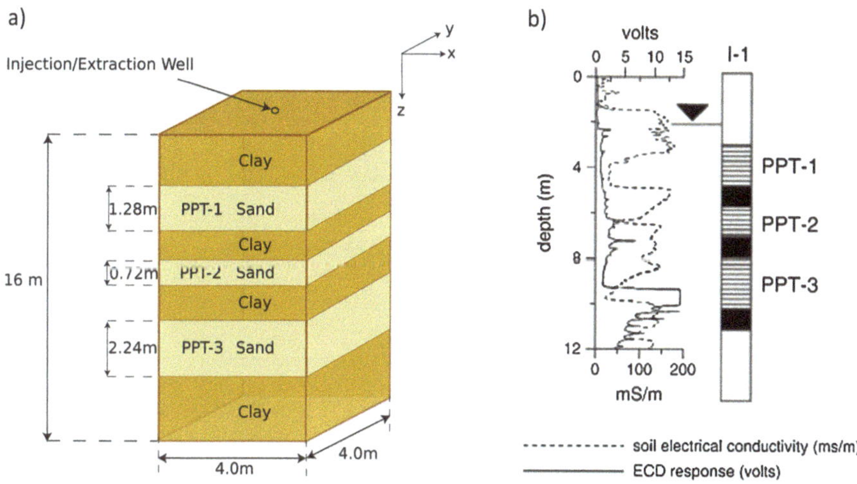

Fig. 7.5 Schematic showing the lithology of the site (**a**) and the results of the electrical conductivity and membrane interface probe (**b**) (Bennett et al. 2010; Krol et al. 2013). (Reprinted with permission from Bennett et al. (2010) and Krol et al. (2013). Copyright (2010) Elsevier and Copyright (2013) American Chemical Society)

7.3.4 Prescreening Mobility of NZVI in Porous Media

Transport through idealized porous media has been tested for a number of NZVI formulations, and a growing understanding of NZVI/media interaction exists. However, laboratory testing using permeable media recovered from the site of interest under the range of conditions that are planned at the site is the best way to predetermine how mobile NZVI in the porous media in the field. Kocur et al. (2013) discussed the importance of replicating field conditions in laboratory tests to avoid overestimating NZVI transport in the field.

A number of studies have used unrealistically high pore water velocities in laboratory experiments when attempting to determine the extent of field transport (Zhan et al. 2008). During site investigation, core logs, and testing, can be used to estimate the range of hydraulic conductivity expected on site and select appropriate conditions for column experiments. Additionally, groundwater and soil constituents that could impact NZVI mobility should be identified (e.g., high clay content and mineralogical anomalies). For example, NZVI had limited mobility in soils rich in calcium carbonates (Laumann et al. 2013). Previous studies have also noted a decline in mobility in natural soils (Schrick et al. 2004; He et al. 2009).

The design and upscaling of NZVI remediation systems from batch, to laboratory, to pilot scale has been done for a range of NZVI formulations and site conditions (see also Chap. 6). Bennett et al. (2010) provide a model for pilot testing using a single well pilot testing method. Coupling NZVI injection and extraction generated important information related to NZVI sorption, interaction, and reaction with porous media. This is particularly important on sites that differ from previous studies.

7.4 Evaluating Injection and Mobility of NZVI

The injection of NZVI at an injection well is governed by Darcy's law:

$$q = \frac{k\rho g}{\mu}\frac{dh}{dr} \tag{7.1}$$

where q is the Darcy velocity, k is the permeability, ρ is the fluid density, μ is the fluid viscosity, h is hydraulic head, and r is the radius from the well.

NZVI can be injected using the following techniques:

- Constant head injection with a constant head at the well or from a reservoir.
- Gravity feed injection is a constant head injection that is limited by the height of the injection well, allowing only head from the standpipe (Fig. 7.6).
- Constant flux injection ensures that the Darcy flux is constant throughout the injection by maintaining the appropriate head.

In column experiments, linear head loss across a column is determined from Darcy's law, resulting in a constant velocity throughout the column. Radial flow from an injection well is often idealized as cylindrical flow emanating from the well. In radial flow, as the radius from the well increases, the Darcy velocity decreases due to a head drop throughout the system. Thus the radius of travel from a well can be predicted using:

Fig. 7.6 Photos of a constant head injection showing the reservoir (left), pump (center), and flush-mount wellhead (right)

$$ROI_{\text{Fluid}} = \sqrt{\frac{V}{\pi n L} + r_{\text{well}}^2} \tag{7.2}$$

where ROI_{Fluid} is the maximum extent or the injection due to injection, V is the volume injected, L is the screen length, and n is the porosity (Bennett et al. 2010).

7.4.1 *Radius of Influence*

The first NZVI field trials employed gravity injection and reported travel distances of NZVI only a few feet due to poor NZVI suspension stability (Elliott and Zhang 2001). Unfortunately, in these studies, the methods of evaluating NZVI transport were not well established, and it is likely that the radius of influence was likely closer to a few inches rather than to a few feet (Sun et al. 2007). Advances in particle stability resulted in NZVI suspensions that were easier to handle and inject (He et al. 2009; He et al. 2010). The most successful method to inject a large volume of NZVI at a high rate is to increase the injection head resulting in a greater flux. This can be achieved using constant head or constant flux injection techniques yielding a radius of influence on the order of 5–7 feet, (Quinn et al. 2005, Krug et al. 2010) as compared to the 2–3 feet with gravity injection (He et al. 2010; Krug et al. 2010). Shear thinning polymers are now being used for MZVI injection under pressure (Köber et al. 2014; Velimirovic et al. 2014; Luna et al. 2015). Krug et al. (2010) tested injection alternatives prior to selecting full-scale delivery method for EZVI. Luna et al. (2015) tested pressurized injection methods and investigated which pressure discharge profile was able to inject the most stabilized MZVI on their pilot site.

7.4.2 *Effect of Viscosity*

Increasing NZVI suspension viscosity is advantageous as it improves NZVI suspension stability. This has been noted in numerous laboratory studies that have used polymers (Tiraferri and Sethi 2009; Kocur et al. 2013), shear thinning polymers (Tosco and Sethi 2010), and an oil-in-water emulsion (Berge and Ramsburg 2009). Although a viscosity increase potentially increases the radius of influence due to improved mobility, it has been noted that even a moderate increase in viscosity can significantly increase injection pressure (Krol et al. 2013). Injection of a viscous fluid also improves the stability of the injection front; however, increases in viscosity could also result in mobilization of NAPL (Pennell et al. 1994; Johnson et al. 2009; Abriola and Pennell 2011).

7.4.3 Delivery Options in Less Permeable Media

Low permeability zones, when contaminated directly or through long-term diffusive transport from higher permeability zones, can act as long-term sources of contamination. Back diffusion has long been established as a cause of long asymptotic concentration decline. This causes particular problems in site closure, and a reliance on concentration-based criteria can result in extended monitoring because of the diffusive flux from low permeability zones. Advanced techniques for accessing and remediating less permeable zones are being explored using several different approaches.

Delivery of electron donor using shear thinning polymers has been investigated for many remediation technologies, including NZVI (Cantrell et al. 1997a, 1997b; Comba and Sethi 2009; Vecchia et al. 2009; Truex et al. 2011b) and MZVI (Cantrell et al. 1997b; Vecchia et al. 2009; Truex et al. 2011a). Shear thinning polymers have been investigated for the injection of NZVI in column experiments (Zhong et al. 2008; Comba et al. 2012), in sandbox experiments (Oostrom et al. 2007), and in the field (Truex et al. 2011a; Truex et al. 2011b). The shear rate is related to velocity (v), permeability (k), and porosity (n) (Martel et al. 1998).

$$\text{Shear Rate }(\Upsilon) = \alpha \frac{4v}{\sqrt{8\frac{k}{n}}} \tag{7.3}$$

Although more complex shear rate relationships have been described for porous media (Phenrat et al. 2009a), Eq. 7.3 demonstrates that shear rate increases with velocity. Thus, a shear thinning agent in the injection will result in a lower viscosity when shear rate is high (i.e., while passing through low permeability media), making it more likely for the fluid to pass through less permeable media. Injection with the aid of shear thinning fluids will cause more uniformly distributed amendment along the injection front in heterogeneous media because higher shear forces will thin the fluid in lower permeability media (Zhong et al. 2008). This generally causes an equalizing sweeping effect on the injection front and increases penetration into low permeability lenses, areas that would otherwise be bypassed.

Amendments can also be delivered through low permeability media with the aid of electrokinetics. In this case, electrodes are used to apply a direct current and transmit charged particles through the subsurface. This has been utilized for heavy metals and waste rock remediation (Acar and Alshawabkeh 1993; Mulligan et al. 2001). Electrokinetic-enhanced NZVI transport through low permeability media has been demonstrated in laboratory studies (Chowdhury et al. 2012). This technology has not been demonstrated for field-scale transport of NZVI; however, it has been used for the delivery of bioremediation amendments in the field (Mao et al. 2012).

Pneumatic and hydraulic fracturing techniques have been adopted from the oil and gas extraction industry to enhance subsurface ZVI distribution. Fracturing of media increases the permeability by intentionally exceeding the overburden effective stress, fracturing the media (Murdoch and Slack 2002). Microscale and granular ZVI can be delivered in this manner. The technique can also introduce sand or other permeable media into the fracture creating a conduit that remains permeable. Such techniques are effective for creating delivery pathways for electron donors in tight media; however, this technique is subject to frequent daylighting of fractures.

7.5 NZVI Detection and Evaluation Methods in Field Samples to Assess Mobility

The most reliable methods of quantifying NZVI mobility have, to date, involved analysis of well-preserved samples or analysis of samples very soon after sampling. Characterization and analytical techniques can also take advantage of the highly reactive nature of NZVI to indirectly observe NZVI transport. In a field setting, this approach of indirect observation (i.e., monitoring geochemical changes due to NZVI oxidation and contaminant degradation at monitoring wells) is often the only available means of evaluation. Observational methods are particularly useful for providing screening level analysis (Fig. 7.7) but should be complemented with additional analysis and characterization to gain a better conceptual understanding of governing NZVI processes on site.

Fig. 7.7 Sampling on a NZVI test site (left) and on-site screening and sample preservation for laboratory analysis (right)

7.5.1 Challenges of Field Data Collection

Use of monitoring wells inherently assumes that this monitoring point is representative of a larger domain and that an analyte is intercepted by the well. As such, it can be challenging to use monitoring wells to definitively determine transport of NZVI or any amendment. Non-detection of NZVI can be due to the injection solution bypassing a well that is installed in a less permeable zone. Heterogeneity or flow along preferential pathways may also allow for NZVI to travel further than detected. This can be a particular problem when budgetary constraints limit spatial sample resolution. Detection of dissolved Fe at a well can be misinterpreted as NZVI travel; however, it could represent the transport of NZVI oxidation products. In order to definitively determine the extent to which NZVI has traveled and deposited in the subsurface, additional analysis in the laboratory should be undertaken to validate field measurements. The effects of sample dilution should be incorporated into analysis. Some particle analysis methods require a minimum number of particles (i.e., particle counts), and dilution of any analyte only serves to make detection more difficult. Well dilution can be mitigated by identifying the highly transmissive zones, prior to injection, and discrete interval sampling through the use of multilevel wells.

Wells are used for several purposes during site characterization, remediation, and in post-remediation monitoring. Total Fe has been used in many column studies to delineate NZVI distribution (Saleh et al. 2008; He et al. 2009; Phenrat et al. 2009a; Raychoudhury et al. 2012; Kocur et al. 2013), and thus many field studies extend total Fe analysis to delineate NZVI distribution following injection (Henn and Waddill 2006; He et al. 2010). This method has been criticized as NZVI is rapidly oxidized, and quantification of total Fe is not necessarily an indicator of NZVI particle transport because geochemical transformations in the subsurface render oxidized forms of iron more mobile than zero-valent forms (Shi et al. 2015).

Kocur et al. (2014) measured total Fe and the Fe^0 content following injection of NZVI at a contaminated site. After 24 h of injection, samples were collected at a monitoring well where NZVI had traveled, and it was found that NZVI content decreased from 55% to only 17% in the particles. Column studies evaluating NZVI transport are commonly conducted for relatively short durations and under anoxic conditions as such it is assumed that NZVI oxidation over the course of an experiment can be neglected. Field studies, however, involving longer duration injections and inherently less controlled conditions require analysis to assess NZVI particle transport under injection conditions and background conditions on site.

Geochemical indicators have also been widely used as evidence of NZVI mobility in the field (Henn and Waddill 2006; He et al. 2010; Wei et al. 2010; Johnson et al. 2013; Kocur et al. 2014). Fe^0 oxidation increases pH and decreases ORP as it rapidly consumes of dissolved oxygen. Geochemical indicators of NZVI oxidation are not definitive evidence that NZVI is present, only that the vicinity of the well has been impacted by oxidation products. Geochemical data is best suited as complementary data, or for screening purposes, to identify the best time to collect additional samples for further analysis. Early field studies, and many applications of PRBs, rely on ORP threshold values in order to ensure that reductive dechlorination is occurring. Thermodynamically, low ORP conditions are necessary for reductive elimination of chlorinated compounds; however, as Bennett et al. (2010) discussed, low ORP does not necessarily mean that reductive dechlorination is occurring, only that it is possible. Reductive dehalogenation and beta-elimination both require electron and hydrogen; thus, evidence of their presence should also be explored. Recent work by Shi et al. (2011, 2015) discussed the importance of additional analysis to support ORP measurements.

Conservative tracers (e.g., conservative ions) are another way to gather supporting information to improve the understanding of NZVI mobility. Many studies have demonstrated a strong correlation between tracer migration and total Fe distribution in aqueous samples during or immediately after injection. Bromide has been used as a tracer during injection of NZVI at a field test (He et al. 2010) and in push-pull tests (Bennett et al. 2010). Specific conductance has been used as a tracer when SO_4^{2-} and Na^+ were the predominant ions in the system (Johnson et al. 2013). Bennett et al. (2010) also showed a strong correlation between Fe and TOC immediately following injection due to the signal produced by the stabilizing polymer, a relationship that was shown to be useful in calibrating viscosity (Krol et al. 2013). Wei et al. (2010) suggested total Fe correlated with suspended and total solids as an analog of mobility.

7.5.2 *Intrusive Methods of Evaluating and Characterizing Particles*

NZVI particles have been characterized using numerous approaches and analytical techniques in studies ranging from batch to field scale. The addition of polymeric coatings will alter NZVI suspension properties and will therefore have an effect on the interpretation of results. Below is a summary and critical analysis of the most widely used methods for NZVI characterization (see more details of particle characterization in Chap. 2).

7.5.2.1 Optical Methods

A jet black color of an aqueous solution is often an indication of NZVI presence (Wang and Zhang 1997; Zhang and Manthiram 1997; Elliott and Zhang 2001). NZVI is visible, depending on the Fe^0 content, at concentrations as low as 20 mg/L (Shi et al. 2015). Visual indication of NZVI has been used in interpretation of 2-D sandbox experiments (Kanel et al. 2008; Phenrat et al. 2010a) and provides evidence of NZVI transport in column experiments as shown in Fig. 7.8. Visual techniques have also been used as evidence of NZVI transport in field studies (Henn and Waddill 2006; He et al. 2010; Kocur et al. 2014). Figure 7.9 shows a jet black NZVI slurry prior to

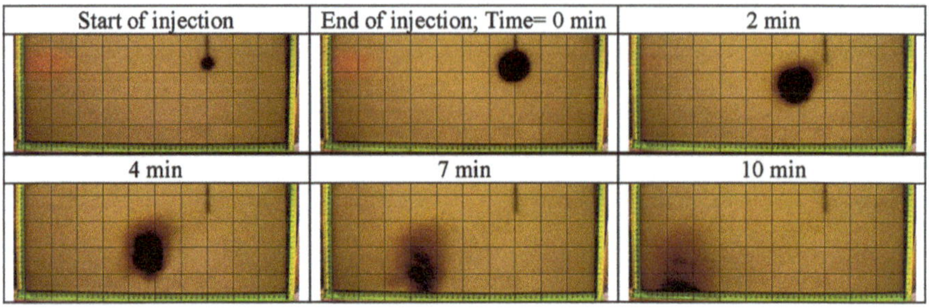

Fig. 7.8 Demonstration of 4 g/L NZVI stabilized with PAA moving through a heterogeneous sandbox (Kanel 2008). (Reprinted with permission from Kanel et al. (2008). Copyright (2008) American Chemical Society)

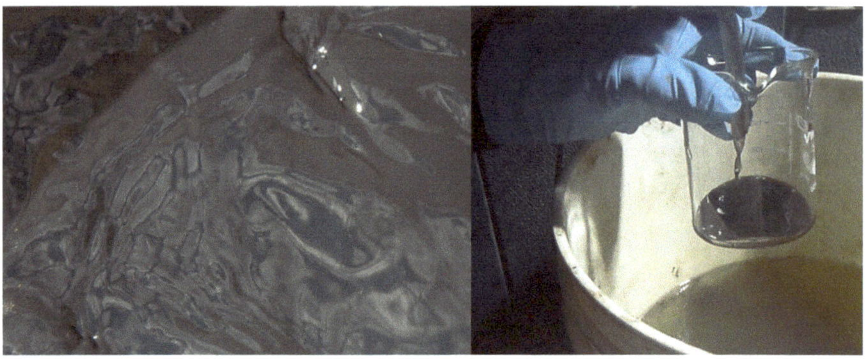

Fig. 7.9 NZVI slurry synthesized at 1 g/L in 0.8% wt CMC prior to injection (left). NZVI sample collection at a monitoring well downstream of the injection (right)

injection and a black solution sampled at a downstream monitoring well. Field samples can, however, contain silt and naturally occurring matter that can interfere with visual techniques. Fe sulfides, other metals (e.g., Mn), and other black or dark substances may be interpreted as NZVI; thus, caution is advisable in relying on only visual evidence. Geochemical evidence of NZVI transport will typically accompany or precede visual observations of NZVI in the field.

More advanced optical methods rely on absorbance quantification. UV/vis techniques specific to NZVI have been used to determine Fe concentrations in stability studies (Saleh et al. 2005; He et al. 2007; Phenrat et al. 2007; Phenrat et al. 2008; Tiraferri et al. 2008; Comba and Sethi 2009; Johnson et al. 2009; Raychoudhury et al. 2010; Sakulchaicharoen et al. 2010; Kocur et al. 2013) and have been extended to laboratory column studies by calibrating absorbance of light at 508 nm (Schrick et al. 2004; Saleh et al. 2007; Saleh et al. 2008; He et al. 2009; Johnson et al. 2009; Phenrat et al. 2009a; Tiraferri and Sethi 2009; Raychoudhury et al. 2010; Kocur et al. 2013). In well-controlled studies in the laboratory, the use of total Fe to quantify

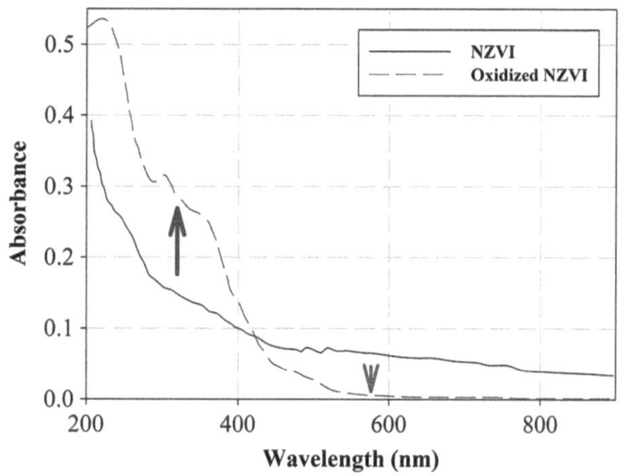

Fig. 7.10 UV/vis spectra of NZVI (black) and oxidized ZVI (brown), showing the peaks used in two wavelength calibration. (Johnson et al. 2013)

NZVI transport is considered to be acceptable. This assumption should be used with caution in field studies or when there is sufficient time and conditions favorable to NZVI oxidation. When a natural suspended matter matrix is involved, as would be the case in field investigation, additional spectrographic considerations must be employed to isolate the NZVI particle absorbance from the matrix.

When possible, a standard method (e.g., ferrozine method (Stookey 1970; Viollier et al. 2000)) should be used to distinguish Fe phases to determine the extent of oxidation and the remaining NZVI content. Field sampling techniques have prescribed collection and Fe preservation techniques. The first sample will be filtered using a 0.45 μm or 0.2 μm filter, and the second will be preserved unfiltered. The dissolved species, predominantly ferrous iron, will be preserved in the filtered sample, while the unfiltered sample will constitute the total Fe. These methods for quantification of total Fe and dissolved Fe rely on the operational definition of dissolved and solid Fe based on the filter size; however, NZVI can pass through such filter sizes, complicating analysis.

Oxidized iron and NZVI can be quantitatively distinguished using UV/vis spectroscopy (Fig. 7.10) (Johnson et al. 2013). NZVI is quantified at 800 nm and oxidized Fe at 325 nm. Use of UV/vis methods can require dilution, complicating analysis. It is important to isolate the absorbance of engineered particles from the matrix that may include natural colloids. The most reliable approach would be to employ multiple lines of evidence to quantify NZVI.

7.5.2.2 Scattering Methods

Dynamic light scattering (DLS) and measurement of zeta potential are light scattering methods that have been used as bulk particle characterization analyses. Zeta potential is calculated based on the Schmolukowski equation and measurement of the electrophoretic mobility. This technique is a well-established tool in characterizing colloids (Nurmi et al. 2005) and has been extended to NZVI characterization in laboratory experiments (Saleh et al. 2005; Saleh et al. 2007; Phenrat et al. 2008; Tiraferri et al. 2008; Chowdhury et al. 2012; Kocur et al. 2013). DLS has been used to characterize NZVI and other nanoparticles (Nurmi et al. 2005; Saleh et al. 2008; Tiraferri et al. 2008; Tiraferri and Sethi 2009; Chowdhury et al. 2012; Kocur et al. 2013). When there is a limited amount of larger background particles present, these techniques are transferable to field analysis (Kocur et al. 2014). However, the presence of larger native colloids will greatly skew the scattered light, and the smaller NZVI particles will not be measured accurately. For polymer-coated particles, converting the measured electrophoretic mobility to a zeta potential is not as straightforward as for uncoated particles, so the measured zeta potential should be reported as apparent zeta potential (Saleh et al. 2008; Lowry et al. 2016).

Other novel methods for particle charge and size measurement have been employed. Acoustic methods have been used to determine NZVI size distribution and estimation of apparent zeta potential in a custom-made apparatus (Dukhin et al. 2001; Sun et al. 2006). Refractive laser analysis (e.g., nanoparticle tracking analysis, NTA) is a microscopic method capable of calculating zeta potential and particle size for low concentrations of nanoparticles, including NZVI (Raychoudhury et al. 2012; Adeleye et al. 2013; Raychoudhury et al. 2014). NTA tracks single particles using image analysis to calculate the parameters. NTA analysis is less affected by the presence of naturally occurring colloids than is DLS and may provide an alternative to DLS for field samples.

Limitations of scattering methods are similar to UV/vis for field methods, often requiring dilution, and the methods are not able to distinguish between the type of colloids in the system. The particle concentrations must be low to allow light transmission, but the suspension must contain sufficient particle counts for measuring particle scattering. DLS analysis yields erroneously small hydrodynamic diameters when concentration is too high due to scattering of light off of multiple particles. The limitations of NTA have not been tested in a field setting. For all optical techniques, application to field samples requires a careful sampling procedure to minimize the presence of sediment, silt, or clay in samples. A sampling program that facilitates low-flow sampling techniques should produce samples suitable for UV/vis and light scattering methods.

7.5.2.3 Microscopic and Energy Techniques

Classical light microscopy cannot resolve nanoscale dimensions; thus, the techniques in this section focus on electron emission and X-ray microscopy. The characterization tools summarized in this section, provided in Table 7.4, have been used for characterization of NZVI in laboratory experiments (see also Chap. 2). The methods also lend themselves to field studies to a limited degree as care must be taken in sampling, and interpretation can become more difficult as the field matrix is

Table 7.4 Analytical electron spectroscopic and X-ray tools for characterizing NZVI

Technique		Analysis	Reference
TEM	Transmission electron microscopy	Size and morphology detection by phase-contrast focusing	(Schrick et al. 2004; Sakulchaicharoen et al. 2010; Kocur et al. 2014)
SEM	Scanning electron microscopy	Topography of larger sample area	(Wei et al. 2010)
STEM	Scanning transmission electron microscopy	Size, morphology, topology narrow beam transmission	(Chang and Kang 2009; Yan et al. 2010)
EDS	Energy dispersive X-ray spectroscopy	Elemental composition using scanning X-ray	(Sun et al. 2006; Sun et al. 2008; Chang and Kang 2009; Yan et al. 2010; Kocur et al. 2014)
XRD	X-ray diffraction	Identification of crystalline structure in nanoparticle	(Nurmi et al. 2005; Sun et al. 2006; Sun et al. 2008)
XPS	X-ray photospectroscopy	Surficial elemental composition	(Nurmi et al. 2005; Li and Zhang 2007)
EXAFS/ XANES	Extended X-ray absorption spectroscopy/X-ray absorption near-edge structure spectroscopy	Determination of oxidation state of the surface	(Nurmi et al. 2005; Sun et al. 2006; Li and Zhang 2007; Reinsch et al. 2010)

more complex than that present in lab samples. These measurements also subject to sample bias given that only a small number of particles are analyzed.

7.5.2.4 Geochemical Parameters

There is no standard method for reporting geochemical results, and although successful application has been reported in several cases, the metrics of success are unclear. ORP has commonly been reported in field studies investigating NZVI transport (Elliott and Zhang 2001; Henn and Waddill 2006). The determination of low ORP may not immediately point to effective transport in the subsurface; rather it provides evidence that NZVI is in the vicinity and has been oxidized by water following Eq. 7.4.

$$Fe^0 + 2H_2O \rightarrow Fe^{2+} + H_2 \uparrow + 2\,OH^- \quad (7.4)$$

A pH change in the system can occur due to Eq. 7.4 along with the reaction of NZVI with dissolved oxygen.

$$Fe^0 + \frac{1}{2}O_2 + H_2O \rightarrow Fe^{2+} + 2OH^- \quad (7.5)$$

pH and ORP have been used as an indicator of NZVI mobility (Henn and Waddill 2006; He et al. 2010; Wei et al. 2010; Kocur et al. 2014). This should be used with caution as reducing conditions (less than −50 mV) may not necessarily indicate the presence of NZVI or the potential for reductive dechlorination (Bennett et al. 2010), especially given that ORP readings are affected by all species in the sample and not only NZVI. Shi et al., in several studies, have investigated uses of ORP probes that lead to poor measurements of redox potential. For example, NZVI particles can attach to the probe yielding abnormally low ORP (Shi et al. 2011; Shi et al. 2015). Given this, rotating disk electrodes have been proposed in place of stationary electrodes for measurement of redox couples that are associated with NZVI formulations.

Screening methods for the determination of the natural reductant demand on site are limited. ORP provides a measurement of the geochemical conditions on site during site investigation; however, the measurement does not estimate the capacity of reducing species. For example, the use of NZVI has been called into question where high levels of nitrates in the groundwater persist as the (hydr)oxide surface of NZVI can become passivated (Reinsch et al. 2010); thus, nitrates should be prescreened prior to NZVI application. Geochemical characterization during site investigation has been proposed using assays for chemical reductant demand (Tratnyek et al. 2014). Such a technique is available for in situ chemical oxidant demand (Haselow et al. 2003) and provides a tool for technology selection. Similar reductant demand tools could be built for screening sites for the feasibility of NZVI application. Such standard methodology would systematically evaluate compatibility issues.

7.5.2.5 Determining Fe^0 Content

Determining and distinguishing all phases of Fe can be difficult. In an environmentally relevant system, Fe^0 is rapidly oxidized to Fe^{2+}, Fe^{3+}, and Fe (hydr)oxides. Qualitative measurements, like TEM, can quantify the core/shell structure of NZVI particles as well as size and shell thickness (Matheson and Tratnyek 1994; Wang and Zhang 1997; Zhang and Manthiram 1997; Schrick et al. 2004; Nurmi et al. 2005). However, other analytical tools summarized above are needed to determine elemental compositions and structure of these surface layers. Groundwater composition impacts the formation of different oxide phases in the shell (e.g., magnetite and maghemite) (Reinsch et al. 2010). Reinsch et al. (2010) also suggest that the entire ZVI core will become oxidized when dissolved oxygen is present except in cases where high nitrates passivate the surface. Laboratory evaluation allows for investigation of surface layer formation under controlled conditions; however, field applications must rely on bulk measurements to determine NZVI reaction potential.

ZVI content can be determined through digestion with acid to completely transform Fe^0 to Fe^{2+} following Eq. 7.4. Analysis of the H_2 gas evolved from the complete oxidation of iron particles along with stoichiometric calculations, based on knowledge of the total Fe, will yield the Fe^0 content in the sample. Several methods have been proposed for analysis of the evolved hydrogen including gas chromatography (Liu et al. 2005; Liu and Lowry 2006) as well as gas displacement for large sample volumes.

7.5.3 Measurement of Immobile Fe^0 in Aquifer Media

NZVI interactions with porous media need to be considered when assessing the Fe^0 content of emplaced NZVI. Bench-scale tests have provided black-box observations of interaction with clay and fine soils, showing that there is more interaction than with homogeneous glass beads or sands (Schrick et al. 2004; He et al. 2009). Other studies have reported increased interaction with silica surfaces at high ionic strengths (Saleh et al. 2008; He et al. 2009; Phenrat et al. 2010b; Laumann et al. 2013). Laboratory transport experiments with natural soils and NZVI retention in field studies suggest a significant portion of the NZVI will become associated with the solid phase in the subsurface. Bennett et al. (2010) and Kocur et al. (2014) have suggested that NZVI is not mobile over the long term and under natural flow conditions. This stationary NZVI is not accounted for in aqueous sampling methods or in many analysis techniques; however, this NZVI could still be reactive toward the target contaminant. This suggests that soil surface analysis techniques should be considered to assess the reactive potential of injected NZVI. In all cases, multiple lines of evidence should be used to validate the mobility of NZVI.

7.5.4 Nonintrusive Methods of NZVI Determination

Alongside the development of particle characterization and identification techniques discussed in Sect. 7.5.2, there is a need to link these results to continuum scale evaluation techniques (Shi et al. 2015). Remote sensing techniques, sensors, and probes are being explored for their utility in linking bulk fluid properties to NZVI delivery and in some cases reaction and oxidation.

Geophysical methods can take advantage of the differential attenuation of electrochemical and electromagnetic properties between the solute or constituent in the plume (e.g., NZVI, NaBr) and the surrounding groundwater. Different media and fluids will have different properties including electrical conductivity and dielectric constants. Complications can arise in implementing geophysical methods as fluid or solid properties can limit use. For example, ground-penetrating radar (GPR) and electrical resistance tomography (ERT) are best suited to shallow subsurface cases. Shi et al. (2015) review several types of geophysical techniques specifically for ZVI applications. Large-scale use for ZVI has been limited to verification of PRB placement (Slater and Binley 2003) although laboratory studies have been extended to track emplacement and geochemical changes in iron column experiments (Slater et al. 2005; Wu et al. 2005; Wu et al. 2006; Wu et al. 2008; Wu et al. 2009). Geophysical methods have been used to identify nanoparticles in laboratory studies with induced polarization (Joyce et al. 2012). Induced polarization has been previously established as a tool in PRB column experiments (Slater et al. 2005).

The limitations of resistivity methods and subsurface tomography are associated with the proximity of measurements to the targeted ZVI (Shi et al. 2015). Thus, chemical threshold indicators for isolating specific redox couples have been proposed as a complementary method for sensing NZVI distribution and oxidation (Jones and Ingle Jr 2001; Tratnyek et al. 2001; Jones and Ingle Jr 2005). The use of specific redox couples as indicators in reactive chemical probes can differentiate target reducing species by color (e.g., resazurin indicator in batch tests with iron filings) (Tratnyek et al. 2001). These reactive redox couples have potential to determine the specific redox state of the injected amendment. This provides an advantage over some redox electrodes, as static electrodes can result in erroneously low ORP measurements. The effective use of redox electrodes has been explored in laboratory studies as a mean for interpreting ORP measurements used to characterize the results of injecting NZVI into groundwater (Shi et al. 2011; Su et al. 2014).

Magnetic susceptibility has also recently been studied as a means of remote geophysical detection of ZVI (Vecchia et al. 2009; Buchau et al. 2010; Köber et al. 2014). Vecchia et al. (2009) quantified ZVI mass using magnetic resonance in column experiments. A field study by Köber et al. (2014) characterized soil cores following the injection of nanoscale/microscale iron filings with a magnetic susceptibility coil developed to delineate total Fe^0 deposited in the subsurface (Fig. 7.11). The study was able to determine the distribution of iron filings at a resolution of <10 cm along soil cores. Similar technique has also been demonstrated for ZVI fracture injection profiling (Arnason et al. 2014). Buchau et al. (2010) developed inductive coils as remote sensors for the in situ detection of ZVI. The planned emplacement of the coils prior to subsurface injection provided quantitative evaluation of ZVI migration. Single particle inductively coupled plasma mass spectroscopy techniques have also been developed with the potential of evaluating nanoparticles (e.g., NZVI in environmental samples) (Lee et al. 2014). A combination of techniques may provide a clearer conceptual model of NZVI transport, deposition, and reaction in the subsurface.

7.5.5 *Modeling Field-Scale NZVI Transport*

Accurate prediction of NZVI transport in porous media relies on many important parameters and a process level understanding of complex interconnected physical and chemical mechanisms governing transport (see Chap. 6). To date, there has not been a comprehensive NZVI model that incorporates every aspect of delivery (see

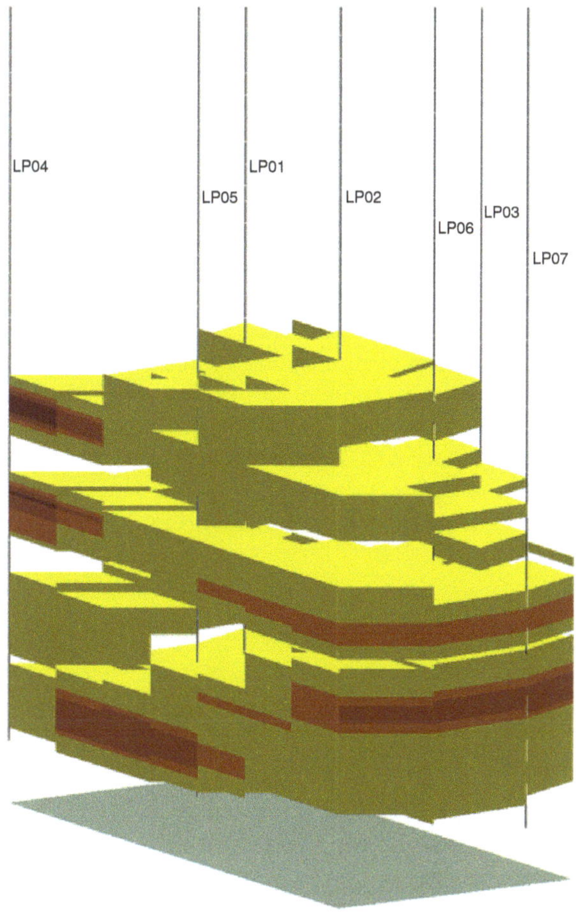

Fig. 7.11 Rendered N/MZVI deposition profile of the subsurface using magnetic susceptibility on soil cores (Köber et al. 2014). Yellow >1 g/kg, red >5 g/kg. (Reprinted with permission from Köber et al. (2014). Copyright (2014) Springer Nature)

Chap. 6), contaminant interaction (see Chap. 5), and degradation (see Chap. 3); however, models are currently available that capture many of the important physical and chemical processes that occur.

Laboratory studies have investigated the transport of NZVI in column studies and in 2-D sandboxes. Many of these studies used colloid filtration theory to predict retention of particles in porous media. Colloid filtration theory (CFT) was originally developed to predict colloid retention in water systems (Yao et al. 1971) but has been extended to include a broad range of colloid-collector conditions (Bai and Tien 1996, 1999; Tufenkji and Elimelech 2004) including NZVI transport experiments (Saleh et al. 2008; He et al. 2009; Johnson et al. 2009; Phenrat et al. 2009a; Tiraferri and Sethi 2009). Care must be taken applying these models to field conditions as the conditions used for model verification can be different than those in the field.

For example, CFT assumes particle stability against aggregation; however, NZVI particles are prone to aggregation. CF also assumes "clean bed" conditions, i.e., that the collector particles are not altered by deposition of particles. This is not likely to be the case for injection of a concentrated slurry with high deposition rates. Some have included aggregation and aggregate settling through an empirical correlation (Phenrat et al. 2010b) or through the manipulation of the already widely adopted predictive physiochemical estimation of colloid interaction forces to include interparticle interaction (Petosa et al. 2010). Stabilizers that inhibit particle-particle interaction, however, allow for extension of CFT to polymer/NZVI systems because stable aggregates are formed (Baalousha 2009; Kocur et al. 2013). Kocur et al. (2013) showed that inclusion of aggregation and sedimentation along with CFT yields good prediction of NZVI transport behavior. Other mathematical formulations have been suggested including general nanoparticle aggregation (Chatterjee and Gupta 2009) and other retention mechanisms (e.g., straining and blocking) (Tosco and Sethi 2010; Köber et al. 2014). These colloid retention mechanisms have been proposed to be operable in NZVI systems (Raychoudhury et al. 2014). Krol et al. (2013) presented a CFT model capable of field-scale predictive modeling of NZVI application during a push-pull test. The model was successful in predicting the transport of NZVI in a sandy aquifer (Fig. 7.12).

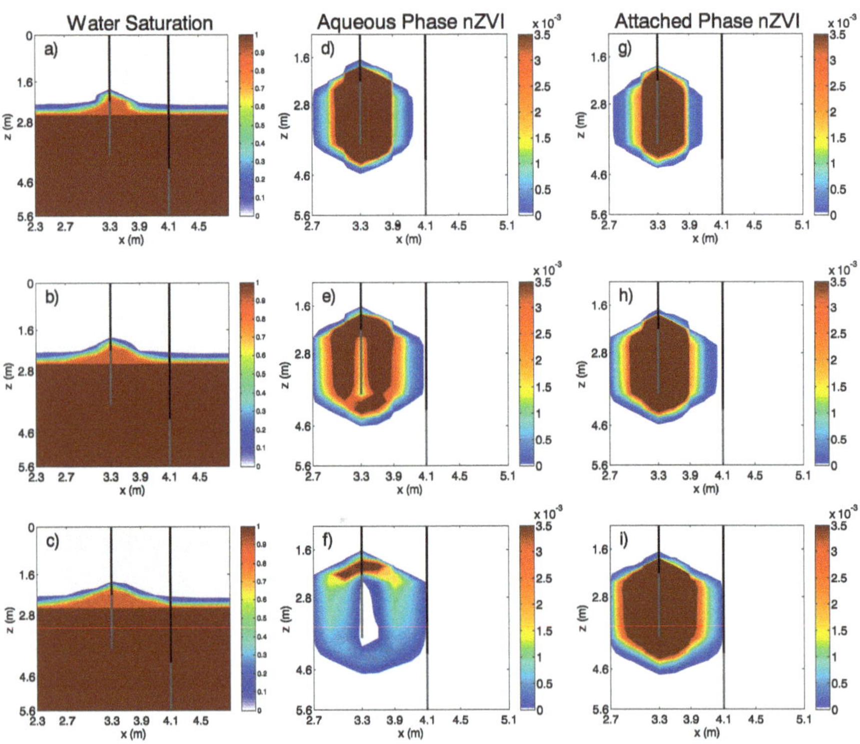

Fig. 7.12 Simulated changes in water saturation (a) $t = 43$ min (end of P1); (b) $t = 76$ min (end of P2); (c) $t = 126$ min (end of P3); aqueous NZVI mass (in moles) at: (d) $t = 43$ min; e) $t = 76$ min; (f) $t = 126$ min; attached NZVI mass (in moles) at: (g) $t = 43$ min; h) $t = 76$ min; (i) $t = 126$ min for field case simulation (Chowdhury et al. 2015). (Reprinted with permission from Chowdhury et al. (2015). Copyright (2015) Elsevier)

Other modeling approaches include simulation of the bulk movement of the fluid containing the nanoparticles. This is particularly useful in the design of NZVI delivery systems. For example, consideration of the density of an NZVI suspension in SEAWAT yielded good predictions of the plugging behavior observed in 2-D sandbox experiments in homogeneous porous media (Kanel et al. 2008). Analogs of heterogeneous porous media have been investigated in small-scale 2-D experiments (Phenrat et al. 2010a; Phenrat et al. 2011) and at the pilot scale in a packed aquifer model (Johnson et al. 2013). Phenrat et al. (2010a, 2010b) showed that a predictive model could capture NZVI transport in a 2-D sandbox. Modeling the interaction and reaction of NZVI with contaminants has been limited as few studies consider mass transfer or reaction. Phenrat et al. (2011) modeled the interaction of NZVI and PCE in laboratory experiments (Fagerlund et al. 2012) showing the improved interaction between PCE and NZVI provided by functionalized polymer coatings. NZVI transport experiments through 1-D columns at residual PCE saturation were used to validate a 1-D reactive transport model (Taghavy et al. 2010). Using a sensitivity analysis, Taghavy et al. (2010) investigated the contact time required to achieve PCE degradation to ethene. These types of simulations aid in the design and testing of remediation strategies prior to implementation.

7.6 Lessons Learned from NZVI Field Applications

7.6.1 Occupational Health and Safety Considerations

There are precautions that should be taken when handling NZVI or other reactive particles. These precautions and the risk associated with the work stem from the novelty of the technology and the unknown repercussions of exposure to these substances. Toxicologists have not yet had sufficient time to fully characterize the effects of NZVI on human and ecosystem health, although there is some indication that NZVI can have toxic effects (Keenan et al. 2009). Dermal and respiratory protection should be employed on site when there is potential for exposure to NZVI.

NZVI may come to the field site in one of several forms: raw materials may be delivered for on-site synthesis (Fig. 7.13); slurry may arrive in containers ready for injection; or dry powders may be supplied for mixing of the slurry on site. Regardless of the form in which NZVI is supplied, appropriate protection should be taken.

Dry ZVI can oxidize rapidly when exposed to the atmosphere. The increased surface area associated with small particle sizes allows this process to occur spontaneously, creating flames and sparks. This can be of particular concern for dry NZVI that has been shipped in an inert atmosphere, e.g., N_2. Vendors can provide air-stabilized NZVI that has a Fe-oxide shell, making them less reactive in air (NanoIron, S. R. O.). Large quantities of slurry that have dried may also be subject to rapid oxidation (Fig. 7.14). Slurry that is preserved in ethanol or another solvent should be handled according to the Material Safety Data Sheets. In all cases, the supplier or vendor should provide information on handling in order to maintain quality and also instruction on safe handling of the material.

Fig. 7.13 Authors undertaking NZVI synthesis in London, Ontario, Canada, prior to a test injection

Fig. 7.14 Photo of dry non-stabilized NZVI rapidly oxidizing as it is poured out of a beaker. (www.nanoiron.cz)

The on-site synthesis of NZVI using chemical precipitation requires a strong reductant. The reaction below involves the reduction of dissolved ferrous iron with a reducing agent, sodium borohydride ($NaBH_4$), forming solid ZVI particles in aqueous solution:

$$2Fe^{2+}{}_{(aq)} + BH_4^-(aq) + 3H_2O_{(l)} \rightarrow 2Fe^0{}_{(s)} + H_2BO_3^-(aq) + 4H^+{}_{(aq)} + 2H_{2(g)} \quad (7.8)$$

The hydrogen that is off-gassed from this reaction should be managed and monitored during the synthesis process. A gas meter can be used to monitor the hydrogen and oxygen in the working zones. Hydrogen will be generated during the preparation of $NaBH_4$ in the chemical handling area and during the reduction of Fe^{2+} to Fe^0 in the reaction vessel. Air monitoring in these areas, as well as at the perimeter of an out-gas emission area, during chemical handling and reaction will ensure that hydrogen levels satisfy safe working conditions and is well under the Lower Explosive Limit (LEL) (4% of total gas for hydrogen). All sources of ignition (e.g., lighters, cell phones, smoking, electrical equipment that sparks, heat, generators, vehicle engines, etc.) and flammable materials (e.g., flammable solvents, flammable clothes, paper) should be removed from the working zones and kept at a safe distance.

7.6.2 *Considerations for Field Applications*

Interpretation of field data requires a validated conceptual model. This includes knowledge of the contribution of natural reductant demand and additional metals or contaminants that may interfere with target contaminant degradation. By-products may build up due to competition for reactive sites when high CVOC concentrations are present on site or in multicomponent systems (Arnold and Roberts 2000). By-products may also already be present on site due to microbially mediated degradation. Appropriate bench- scale testing with soil and groundwater from the site may indicate the potential for impacts from by-products, but reproducing the exact field-scale conditions (inflowing water and microbiology) is difficult.

7.6.2.1 Dose

The appropriate NZVI dose for a given level of contamination on a per soil volume unit basis has not been systematically studied. And therefore quantitative metrics for dose are not available. Vendors and practitioners may have rules of thumb for specific cases; however, an appropriate dose is likely site-specific, as no two scenarios are identical. Fe/soil ratio of 0.004 or 4 g/ kg of soil has been used as a qualitative threshold to achieve low ORP (e.g., −400 mV) throughout a treatment area (Gavaskar et al. 2005). This estimated loading of Fe^0 has not significantly changed since initial PRB installations (Cantrell et al. 1995) although some granular PRB loadings can be as high as 10–22% wt of soil (Gillham et al. 2010). PRB operation at a loading of 0.4% Fe^0 has been estimated to be sufficient to last for several decades if this loading has been delivered in columns (Kaplan et al. 1996; Cantrell et al. 1997a). In situ soil mixing applications target Fe loadings between 0.5% and 2% (Tratnyek et al. 2014).

7.6.2.2 Delivery

There is concern that NZVI and MZVI cannot be delivered at the required dose to treat in situ source zones. Gavaskar et al. (2005) commented that Fe/soil ratios associated with soil mixing and PRB applications in the literature were not attainable via injection methods. This is based on the limited success in early field studies prior to the development of polymer coatings. Considering that many pilot injections involve NZVI at a dose of 0.3–1 g/L, there may be a need for multiple injections. There have been significant advances in the delivery of MZVI and NZVI (Quinn et al. 2005; O'Hara et al. 2006; Truex et al. 2011a; Truex et al. 2011b; Su et al. 2012; Köber et al. 2014; Luna et al. 2015). The addition of a stabilizer has the added advantage of allowing more time for delivery than with an unstable slurry (Kocur et al. 2013) and the ability to stabilize high concentrations of NZVI (Comba and Sethi 2009). Improved stability leads to better mobility as discussed previously (Saleh et al. 2008; He et al. 2009; Phenrat et al. 2009a; Kocur et al. 2013). This provides the opportunity to inject more NZVI mass initially, either through multiple injections or with a more concentrated slurry. The potential for plugging due to multiple injections must be considered. Methods of injecting amendments into the ground have been developed; however, methods of evaluating and accurately analyzing the delivery and performance are not as developed.

7.6.2.3 Longevity of NZVI Reactivity

Depending on the mass of Fe^0 that is injected, loss of electron donors prior to contaminant mass destruction will lead to a rebound in contaminant concentrations. This is commonly reported in early field studies with NZVI where between 80% and 99% reduction in VOC concentrations within the vicinity of injection wells was observed but was followed by rebound several months later (Mace 2006). This is caused by the return of contaminated pore water that is pushed out of the test area by the injection process (Bennett et al. 2010). In these cases, it is likely that dilution of the injection mixture will cause the oxidation of NZVI with dissolved oxygen and water following Eqs 7.4 and 7.5, causing Fe^0 mass consumption. It has been suggested that NZVI injection is best utilized for stationary phase contamination (Bennett 2010). Estimates of reaction longevity vary and depend on subsurface conditions and the amount of NZVI injected. Batch studies have determined that

NZVI is reactive for months but only reactive in sediment aqueous mixtures for about 1 month (Adeleye et al. 2013). In situ reaction longevity, however, has yet to be fully explored. Contaminant interaction and degradation with deposited NZVI is not well studied and remains an open question. This limitation is an important knowledge gap that must be addressed to identify application strategies and conditions that provide optimal remediation strategies using NZVI.

7.6.2.4 Site Screening Using Laboratory Tests

Prior to the implementation of NZVI at a field site, subsurface conditions can be evaluated for NZVI compatibility. Batch tests can be designed to evaluate the feasibility of NZVI with site-specific contaminants. Geochemical parameters are a good starting point as NZVI will work best in a reducing environment, with a neutral or slightly basic pH. Quantifying the natural reductant demand and simple estimates of contaminant mass distribution in the subsurface can be used for technology selection. Batch tests can evaluate the need for pretreatment or the addition of a catalyst. These batch tests can also guide monitoring protocol and methods as well as NZVI dose. The NZVI formulation can be tested in laboratory column tests in model media or site media to determine mobility. The slurry can be injected using several different methods to determine the best method of delivery or to screen contractors with different equipment. Considering the inevitable need for some polishing treatment prior to site closure, the site groundwater can be screened to determine the impact to microbial species.

7.6.3 NZVI Performance Evaluation

A growing number of field tests have been performed and summaries of many of these injections are available (Chap. 1). In Europe alone, more than 15 pilot- and full-scale injections have taken place (Mueller et al. 2012). As discussed in this chapter, a number of methods have been developed for delivering different formulations and types of NZVI. Unfortunately, the rigorous methods that are essential for the evaluation of NZVI technology have not evolved at the same pace. Often field injections assume that successful injection and delivery to the subsurface are enough to remediate the site; however, several decades ago, performance of in situ chemical oxidants was evaluated in a similar manner. At many sites where oxidants were used, concentrations rebounded to pretreatment concentrations in relatively short periods of time. Methods of site characterization have improved since this time. Performance evaluation of the mobility and in situ degradation of NZVI are just as important as optimizing the injection and delivery of NZVI, necessitating the development of more rigorous monitoring methods.

7.6.4 NZVI and Combined Remedies for Field-Scale Remediation

According to the environmental security technology certification program (ESTCP), the most widely used remediation technologies are thermal, enhanced bioremediation, monitored natural attenuation, and in situ chemical oxidation. These technologies can be reliably employed for the cleanup of contaminated sites along with many other technological alternatives that have been around for longer than chemical reduction using NZVI. However, NZVI has many important benefits and can be used in the treatment train coupled with many of these technologies.

Source removal approaches that treat highly NAPL-saturated media (e.g., thermal treatment or flushing technologies) would precede NZVI injection in a treatment train. NAPL source removal by flushing, solvent extraction, and hot water flushing may represent opportunities for coupled use with NZVI injection. Flushing techniques work well for NAPL pools and higher residual NAPL but less well with lower-residual and sorbed contaminants. The infrastructure that would be used for solvent flushing would be similar to NZVI injection wells leading to cost savings as installed wells can be used for both purposes. Another example of sequential treatment was investigated by coupling NZVI and electrical resistive heating (ERH). In order to increase contaminant interaction with NZVI by increasing the mass flux from NAPL residual, in situ ERH has been pilot tested as a coupled technology with NZVI (Truex et al. 2011a). NZVI application would degrade contaminants that were not removed during flooding or heating.

Sequential treatment using monitored natural attenuation may also couple with NZVI application in a treatment train. NZVI that is deposited in the subsurface will oxidize to Fe^{2+} and furthermore to Fe^{3+} species serving as an electron donor. This approach has been demonstrated to degrade low levels of aqueous contaminants as they flow through the reduced media (i.e., in situ redox manipulation) (Szecsody et al. 2004). Such a technique can also immobilize contaminant metals (Cantrell et al. 1995; Li and Zhang 2007; O'Carroll et al. 2013).

Enhanced bioremediation by organohalide-respiring microorganisms has the most potential when coupled with NZVI application. The reducing conditions created as a result of the oxidation of NZVI and the hydrogen that is produced eliminate any significant acclimatization period required for bioremediation. Unintended microbially mediated degradation has been observed following NZVI/CMC injection (He et al. 2010). ZVI emulsified in vegetable oil has also shown to abiotically degrade PCE and TCE, while biodegradation products (e.g., methane and volatile fatty acids) increased over a 2-year period, suggesting microbially mediated degradation was stimulated (Su et al. 2012).

Laboratory studies have shown that naturally occurring microorganisms will utilize the cathodically produced hydrogen in anaerobic systems. Increased microbial degradation has been attributed to abiotic redox and electron donor processes derived from NZVI oxidation in water (Lampron et al. 1998, 2001; Fernandez-Sanchez et al. 2004; Jeon et al. 2013). This has led to number of studies investigating the effects of NZVI on microorganisms. Stimulation of bioactivity by ZVI was noted in microcosms evaluating PRB performance (Lampron et al. 1998, 2001; Fernandez-Sanchez et al. 2004) and in enriched cultures isolating *Dehalobacter* in the degradation of carbon tetrachloride (Lee et al. 2015), *Dehalococcoides* in the degradation of chlorinated ethenes (Rosenthal et al. 2004) and the degradation of chlorobenzenes (Rosenthal et al. 2004; Zhu et al. 2012; Zhu et al. 2013). General dehydrogenase activity was enhanced in soil activity tests with NZVI (Cullen et al. 2011) in addition to upregulation of reductive dehydrogenases genes, *tceA* and *vcrA* (Xiu et al. 2010a) in microcosms.

Many of these studies are limited as they evaluate single organisms or evaluate single organism functions. However, they are important in the development of remediation biomarkers. Specific studies have noted that enhanced microbial activity is aided by the polymer coating that stabilizes the nanoparticles (Zhou et al. 2014), while others have suggested that the polymer screens the organism from harmful oxidative stress (Li et al. 2010; Xiu et al. 2010a). It has also been reported that this biostimulation occurs only after a lag period (Xiu et al. 2010b). These remediation-focused papers are accompanied by numerous toxicity studies focused on evaluating the oxidative stress and negative impacts of NZVI on bacteria (Jiang et al. 2013), aquatic organisms (Li et al. 2009), soils organisms (Pawlett et al. 2013; Saccà et al. 2014), and mammalian cells (Phenrat et al. 2009b) and human cells (Sun et al. 2011).

Fewer studies have taken a focus on the effects of NZVI on microbial community. Kirschling et al. (2010) observed changes in Eubacteria diversity in aquifer media microcosms, but no change in abundance was observed due to exposure to NZVI. Barnes et al. (2010) studied aquifer-derived microbial microcosms, finding that NZVI inhibited chlorinated ethene degradation. Zaa et al. (2010) comprehensively characterized the organohalide-respiring microorganisms in a chlorinated ethene plume undergoing natural attenuation. The most rigorous community-focused analysis was reported in soil mixing mesocosm experiments evaluating pesticide degradation resulting in shifting microbial communities in real soil (Tilston et al. 2013). These studies provide a step forward, evaluating field relevant soils and groundwater in their laboratory studies. However, it is difficult to replicate field relevant exposure conditions on microbial communities in microcosms.

References

Abriola, L. R. A., & Pennell, K. (2011). *Final report: Development and optimization of targeted nanoscale Iron delivery methods for treatment of NAPL source zones*. Strategic Environmental Research and Development Program. Tufts University.

Acar, Y. B., & Alshawabkeh, A. N. (1993). Principles of electrokinetic remediation. *Environmental Science & Technology, 27*, 2638–2647.

Adeleye, A., Keller, A., Miller, R., & Lenihan, H. (2013). Persistence of commercial nanoscaled zero-valent iron (nZVI) and by-products. *Journal of Nanoparticle Research, 15*, 1–18.

Annable, M. D., Jawitz, J. W., Rao, P. S. C., Dai, D. P., Kim, H., & Wood, A. L. (1998). Field evaluation of interfacial and partitioning tracers for characterization of effective NAPL-water contact areas. *Ground Water, 36*, 495–502.

Arnason, J. G., Harkness, M., & Butler-Veytia, B. (2014). Evaluating the subsurface distribution of zero-valent Iron using magnetic susceptibility. *Groundwater Monitoring & Remediation, 34*, 96–106.

Arnold, W. A., & Roberts, A. L. (2000). Pathways and kinetics of chlorinated ethylene and chlorinated acetylene reaction with Fe(0) particles. *Environmental Science & Technology, 34*, 1794–1805.

Baalousha, M. (2009). Aggregation and disaggregation of iron oxide nanoparticles: Influence of particle concentration, pH and natural organic matter. *Science of the Total Environment, 407*, 2093–2101.

Bai, R., & Tien, C. (1996). A new correlation for the initial filter coefficient under unfavorable surface interactions. *Journal of Colloid and Interface Science, 179*, 631–634.

Bai, R., & Tien, C. (1999). Particle deposition under unfavorable surface interactions. *Journal of Colloid and Interface Science, 218*, 488–499.

Barnes, R. J., Riba, O., Gardner, M. N., Singer, A. C., Jackman, S. A., & Thompson, I. P. (2010). Inhibition of biological TCE and sulphate reduction in the presence of iron nanoparticles. *Chemosphere, 80*(5), 554–562.

Bennett, P., He, F., Zhao, D., Aiken, B., & Feldman, L. (2010). In situ testing of metallic iron nanoparticle mobility and reactivity in a shallow granular aquifer. *Journal of Contaminant Hydrology, 116*, 35.

Benson, R. C. (2005). *Practical handbook of environmental site characterization and groundwater monitoring* (2nd ed.pp. 249–295). CRC Press: Boca Raton, FL.

Berge, N. D., & Ramsburg, C. A. (2009). Oil-in-water emulsions for encapsulated delivery of reactive iron particles. *Environmental Science & Technology, 43*, 5060–5066.

Buchau, A., Rucker, W. M., De Boer, C. V., & Klaas, N. (2010). Inductive detection and concentration measurement of nano sized zero valent iron in the subsurface. *IET Science, Measurement and Technology, 4*, 289–297.

Butler, J. J. (1997). *The design, performance, and analysis of slug tests*. Boca Raton: Taylor & Francis.

Cantrell, K. J., Kaplan, D. I., & Gilmore, T. J. (1997a). Injection of colloidal Fe-0 particles in sand with shear-thinning fluids. *Journal of Environmental Engineering-ASCE, 123*, 786–791.

Cantrell, K. J., Kaplan, D. I., & Gilmore, T. J. (1997b). Injection of colloidal size particles of Fe0 in porous media with shear thinning fluids as a method to emplace a permeable reactive zone. *Land Contamination and Reclamation, 5*, 253–257.

Cantrell, K. J., Kaplan, D. I., & Wietsma, T. W. (1995). Zero-valent iron for the in situ remediation of selected metals in groundwater. *Journal of Hazardous Materials, 42*, 201–212.

Chang, M. C., & Kang, H. Y. (2009). Remediation of pyrene-contaminated soil by synthesized nanoscale zero-valent iron particles. *Journal of Environmental Science and Health, Part A, 44*, 576–582.

Chatterjee, J., & Gupta, S. K. (2009). An agglomeration-based model for colloid filtration. *Environmental Science & Technology, 43*, 3694.

Chowdhury, A. I. A., Krol, M. M., Kocur, C. M., Boparai, H. K., Weber, K. P., Sleep, B. E., & O'Carroll, D. M. (2015). nZVI injection into variably saturated soils: Field and modeling study. *Journal of Contaminant Hydrology, 183*, 16–28.

Chowdhury, A. I. A., O'Carroll, D. M., Xu, Y., & Sleep, B. E. (2012). Electrophoresis enhanced transport of nano-scale zero valent iron. *Advances in Water Resources, 40*, 71–82.

Comba, S., Dalmazzo, D., Santagata, E., & Sethi, R. (2012). Rheological characterization of xanthan suspensions of nanoscale iron for injection in porous media. *Journal of Hazardous Materials, 185*, 598–605.

Comba, S., & Sethi, R. (2009). Stabilization of highly concentrated suspensions of iron nanoparticles using shear-thinning gels of xanthan gum. *Water Research, 43*, 3717–3726.

Cullen, L. G., Tilston, E. L., Mitchell, G. R., Collins, C. D., & Shaw, L. J. (2011). Assessing the impact of nano- and micro-scale zerovalent iron particles on soil microbial activities: Particle reactivity interferes with assay conditions and interpretation of genuine microbial effects. *Chemosphere, 82*, 1675–1682.

Dukhin, A. S., Goetz, P. J., & Truesdail, S. (2001). Titration of concentrated dispersions using electroacoustic ζ-potential probe. *Langmuir, 17*, 964–968.

Einarson, M. (2005). *Multilevel ground-water monitoring. Practical handbook of environmental site characterization and ground-water monitoring* (2nd ed.pp. 807–848). CRC Press: Boca Raton, FL.

Elliott, D. W., & Zhang, W.-X. (2001). Field assessment of nanoscale bimetallic particles for groundwater treatment. *Environmental Science & Technology, 35*, 4922–4926.

Fagerlund, F., Illangasekare, T. H., Phenrat, T., Kim, H. J., & Lowry, G. V. (2012). PCE dissolution and simultaneous dechlorination by nanoscale zero-valent iron particles in a DNAPL source zone. *Journal of Contaminant Hydrology, 131*, 9–28.

Fernandez-Sanchez, J. M., Sawvel, E. J., & Alvarez, P. J. J. (2004). Effect of Fe0 quantity on the efficiency of integrated microbial-Fe0 treatment processes. *Chemosphere, 54*, 823–829.

Fetter, C. W. (2001). *Applied hydrogeology* (4th ed.). Inc, Upper Saddle River: Prentice-Hall.

Gavaskar, A., Tatar, L., & Condit, W. (2005). *Contract report: Cost and performance report: Nanoscale zerovalent iron technologies for source remediation*. Port Huenema: NAVFAC: Naval Facilities Engineering Command.

Gillham, R., Vogan, J., Gui, L., Duchene, M., & Son, J. (2010). Iron barrier walls for chlorinated solvent remediantion. In H. F. Stroo & C. H. Ward (Eds.), *In situ remediation of chlorinated solvent plumes* (pp. 537–571). New York: Springer Science Media.

Haselow, J. S., Siegrist, R. L., Crimi, M., & Jarosch, T. (2003). Estimating the total oxidant demand for in situ chemical oxidation design. *Remediation Journal, 13*, 5–16.

He, F., Zhang, M., Qian, T., & Zhao, D. (2009). Transport of carboxymethyl cellulose stabilized iron nanoparticles in porous media: Column experiments and modeling. *Journal of Colloid and Interface Science, 334*, 96–102.

He, F., Zhao, D., Liu, J., & Roberts, C. B. (2007). Stabilization of Fe-Pd nanoparticles with sodium carboxymethyl cellulose for enhanced transport and dechlorination of trichloroethylene in soil and groundwater. *Industrial & Engineering Chemical Research, 46*, 29–34.

He, F., Zhao, D., & Paul, C. (2010). Field assessment of carboxymethyl cellulose stabilized iron nanoparticles for in situ destruction of chlorinated solvents in source zones. *Water Research, 44*, 2360–2370.

Henn, K. W., & Waddill, D. W. (2006). Utilization of nanoscale zero-valent iron for source remediation - a case study. *Remediation Journal, 16*, 57–77.

ITRC. (2011). *Integrated DNAPL site strategy*. Washington, DC: Interstate Technology & Regulatory Council, Integrated DNAPL Site Strategy Team.

Jeon, J.-R., Murugesan, K., Nam, I.-H., & Chang, Y.-S. (2013). Coupling microbial catabolic actions with abiotic redox processes: A new recipe for persistent organic pollutant (POP) removal. *Biotechnology Advances, 31*, 246–256.

Jiang, C., Liu, Y., Chen, Z., Megharaj, M., & Naidu, R. (2013). Impact of iron-based nanoparticles on microbial denitrification by Paracoccus sp. strain YF1. *Aquatic Toxicology, 142–143*, 329–335.

Johnson, R. L., Johnson, G. O. B., Nurmi, J. T., & Tratnyek, P. G. (2009). Natural organic matter enhanced mobility of nano zerovalent iron. *Environmental Science & Technology, 43*, 5455–5460.

Johnson, R. L., Nurmi, J. T., O'Brien Johnson, G. S., Fan, D., O'Brien Johnson, R. L., Shi, Z., Salter-Blanc, A. J., Tratnyek, P. G., & Lowry, G. V. (2013). Field-scale transport and transformation of carboxymethylcellulose-stabilized nano zero-valent iron. *Environmental Science & Technology, 47*, 1573–1580.

Jones, B. D., & Ingle, J. D., Jr. (2001). Evaluation of immobilized redox indicators as reversible, in situ redox sensors for determining Fe(III)-reducing conditions in environmental samples. *Talanta, 55*, 699–714.

Jones, B. D., & Ingle, J. D., Jr. (2005). Evaluation of redox indicators for determining sulfate-reducing and dechlorinating conditions. *Water Research, 39*, 4343–4354.

Joyce, R. A., Glaser, D. R., Werkema, D. D., & Atekwana, E. A. (2012). Spectral induced polarization response to nanoparticles in a saturated sand matrix. *Journal of Applied Geophysics, 77*, 63–71.

Kanel, S. R., Goswami, R. R., Clement, T. P., Barnett, M. O., & Zhao, D. (2008). Two dimensional transport characteristics of surface stabilized zero-valent iron nanoparticles in porous media. *Environmental Science & Technology, 42*, 896–900.

Kaplan, D. I., Cantrell, K. J., Wietsma, T. W., & Potter, M. A. (1996). Retention of zero-valent iron colloids by sand columns: Application to chemical barrier formation. *Journal of Environmental Quality, 25*, 1086–1094.

Keenan, C. R., Goth-Goldstein, R., Lucas, D., & Sedlak, D. L. (2009). Oxidative stress induced by zero-valent Iron nanoparticles and Fe(II) in human bronchial epithelial cells. *Environmental Science & Technology, 43*, 4555–4560.

Kirschling, T., Gregory, K., Minkley, N., Lowry, G., & Tilton, R. (2010). Impact of nanoscale zerovalent iron on geochemistry and microbial populations. *Environmental Science & Technology, 44*, 3474–3480.

Köber, R., Hollert, H., Hornbruch, G., Jekel, M., Kamptner, A., Klaas, N., Maes, H., Mangold, K. M., Martac, E., Matheis, A., Paar, H., Schäffer, A., Schell, H., Schiwy, A., Schmidt, K. R., Strutz, T. J., Thümmler, S., Tiehm, A., & Braun, J. (2014). Nanoscale zero-valent iron flakes for groundwater treatment. *Environment and Earth Science, 72*, 3339–3352.

Kocur, C. M., Chowdhury, A. I., Sakulchaicharoen, N., Boparai, H. K., Weber, K. P., Sharma, P., Krol, M. M., Austrins, L., Peace, C., Sleep, B. E., & O'Carroll, D. M. (2014). Characterization of nZVI mobility in a field scale test. *Environmental Science & Technology, 48*, 2862–2869.

Kocur, C. M., O'Carroll, D. M., & Sleep, B. E. (2013). Impact of nZVI stability on mobility in porous media. *Journal of Contaminant Hydrology, 145*, 17–25.

Kram, M. L. (2005). *Dnapl characterization methods and approaches. Practical handbook of environmental site characterization and ground-water monitoring* (2nd ed.pp. 473–515). CRC Press: Boca Raton, FL.

Krol, M. M., Oleniuk, A. J., Kocur, C. M., Sleep, B. E., Bennett, P., Zhong, X., & O'Carroll, D. M. (2013). A field-validated model for in situ transport of polymer-stabilized nZVI and implications for subsurface injection. *Environmental Science & Technology, 47*, 7332–7340.

Krug, T., O'Hara, S., Watling, M., & Quinn, J. (2010). *Final report: Emulsified zero-valent nano-scale iron treatment of chlorinated solvent DNAPL source areas ESTCP* (763 pp.). Washington DC.

Kueper, B. H., Stroo, H. F., Vogel, C. M., Ward, C. H. (2014). *Chlorinated solvent source zone remediation*. Springer, New York, 713 pp.

Lampron, K. J., Chiu, P. C., & Cha, D. K. (1998). Biological reduction of trichloroethene supported by Fe(0). *Bioremediation Journal, 2*, 175–181.

Lampron, K. J., Chiu, P. C., & Cha, D. K. (2001). Reductive dehalogenation of chlorinated ethenes with elemental iron: The role of microorganisms. *Water Research, 35*, 3077–3084.

Laumann, S., Micić, V., Lowry, G. V., & Hofmann, T. (2013). Carbonate minerals in porous media decrease mobility of polyacrylic acid modified zero-valent iron nanoparticles used for groundwater remediation. *Environmental Pollution, 179*, 53–60.

Lee, M., Wells, E., Wong, Y. K., Koenig, J., Adrian, L., Richnow, H. H., & Manefield, M. (2015). Relative contributions of dehalobacter and zerovalent iron in the degradation of chlorinated methanes. *Environmental Science & Technology, 49*, 4481–4489.

Lee, S., Bi, X., Reed, R. B., Ranville, J. F., Herckes, P., & Westerhoff, P. (2014). Nanoparticle size detection limits by single particle ICP-MS for 40 elements. *Environmental Science & Technology, 48*, 10291–10300.

Li, H., Zhou, Q., Wu, Y., Fu, J., Wang, T., & Jiang, G. (2009). Effects of waterborne nano-iron on medaka (Oryzias latipes): Antioxidant enzymatic activity, lipid peroxidation and histopathology. *Ecotoxicology and Environmental Safety, 72*, 684–692.

Li, X.-Q., & Zhang, W.-X. (2007). Sequestration of metal cations with zerovalent iron nanoparticles - A study with high resolution X-ray photoelectron spectroscopy (HR-XPS). *Journal of Physical Chemistry C, 111*, 6939–6946.

Li, Z. Q., Greden, K., Alvarez, P. J. J., Gregory, K. B., & Lowry, G. V. (2010). Adsorbed polymer and NOM limits adhesion and toxicity of nano scale zerovalent iron to E. coli. *Environmental Science & Technology, 44*, 3462–3467.

Liu, Y., & Lowry, G. V. (2006). Effect of particle age (Fe0 content) and solution pH on NZVI reactivity: H2 evolution and TCE dechlorination. *Environmental Science & Technology, 40*, 6085–6090.

Liu, Y., Majetich, S. A., Tilton, R. D., Sholl, D. S., & Lowry, G. V. (2005). TCE Dechlorination rates, pathways, and efficiency of nanoscale Iron particles with different properties. *Environmental Science & Technology, 39*, 1338–1345.

Lowry, G. V., Hill, R. J., Harper, S., Rawle, A. F., Hendren, C. O., Klaessig, F., Nobbmann, U., Syare, P., & Rumble, J. (2016). Guidance for measuring, interpreting, and reporting zeta potential measurements for environmental nanotechnology and nanotoxicology. *Environmental Science. Nano, 3*, 953–965.

Luna, M., Gastone, F., Tosco, T., Sethi, R., Velimirovic, M., Gemoets, J., Muyshondt, R., Sapion, H., Klaas, N., & Bastiaens, L. (2015). Pressure-controlled injection of guar gum stabilized microscale zerovalent iron for groundwater remediation. *Journal of Contaminant Hydrology, 181*, 46.

Mace, C. (2006). Controlling groundwater VOCs: Do nanoscale ZVI particles have any advantages over microscale ZVI of BNP. *Pollution Engineering, 38*, 24–28.

Mao, X., Wang, J., Ciblak, A., Cox, E. E., Riis, C., Terkelsen, M., Gent, D. B., & Alshawabkeh, A. N. (2012). Electrokinetic-enhanced bioaugmentation for remediation of chlorinated solvents contaminated clay. *Journal of Hazardous Materials, 213–214*, 311–317.

Martel, K. E., Martel, R., Lefebvre, R., & Gélinas, P. J. (1998). Laboratory study of polymer solutions used for mobility control during in situ NAPL recovery. *Ground Water Monitoring & Remediation, 18*, 103–113.

Matheson, L. J., & Tratnyek, P. G. (1994). Reductive dehalogenation of chlorinated methanes by iron metal. *Environmental Science & Technology, 28*, 2045–2053.

Mueller, N. C., Braun, J., Bruns, J., Cernik, M., Rissing, P., Rickerby, D., & Nowack, B. (2012). Application of nanoscale zero valent iron (NZVI) for groundwater remediation in Europe. *Environmental Science and Pollution Research, 19*, 550–558.

Mulligan, C. N., Yong, R. N., & Gibbs, B. F. (2001). Remediation technologies for metal-contaminated soils and groundwater: An evaluation. *Engineering Geology, 60*, 193–207.

Murdoch, L., & Slack, W. (2002). Forms of hydraulic fractures in shallow fine-grained formations. *Journal of Geotechnical and Geoenvironmental Engineering, 128*, 479–487.

Nurmi, J. T., Tratnyek, P. G., Sarathy, V., Baer, D. R., Amonette, J. E., Pecher, K., Wang, C., Linehan, J. C., Matson, D. W., Penn, R. L., & Driessen, M. D. (2005). Characterization and properties of metallic Iron nanoparticles: Spectroscopy, electrochemistry, and kinetics. *Environmental Science & Technology, 39*, 1221–1230.

O'Carroll, D., Sleep, B., Krol, M., Boparai, H., & Kocur, C. (2013). Nanoscale zero valent iron and bimetallic particles for contaminated site remediation. *Advances in Water Resources, 51*, 104–122.

O'Hara, S., Krug, T., Quinn, J., Clausen, C., & Geiger, C. (2006). Field and laboratory evaluation of the treatment of DNAPL source zones using emulsified zero-valent iron. *Remediation Journal, 16*, 35–56.

Oostrom, M., Wietsma, T. W., Covert, M. A., & Vermeul, V. R. (2007). Zero-valent iron emplacement in permeable porous media using polymer additions. *Ground Water Monitoring and Remediation, 27*, 122–130.

Pawlett, M., Ritz, K., Dorey, R. A., Rocks, S., Ramsden, J., & Harris, J. A. (2013). The impact of zero-valent iron nanoparticles upon soil microbial communities is context dependent. *Environmental Science and Pollution Research International, 20*, 1041–1049.

Pennell, K. D., Jin, M., Abriola, L. M., & Pope, G. A. (1994). Surfactant enhanced remediation of soil columns contaminated by residual tetrachloroethylene. *Journal of Contaminant Hydrology, 16*, 35–53.

Petosa, A. R., Jaisi, D. P., Quevedo, I. R., Elimelech, M., & Tufenkji, N. (2010). Aggregation and deposition of engineered nanomaterials in aquatic environments: Role of physicochemical interactions. *Environmental Science & Technology, 44*, 6532–6549.

Phenrat, T., Cihan, A., Kim, H.-J., Mital, M., Illangasekare, T., & Lowry, G. V. (2010a). Transport and deposition of polymer-modified Fe0 nanoparticles in 2-D heterogeneous porous media: Effects of particle concentration, Fe0 content, and coatings. *Environmental Science & Technology, 44*, 9086–9093.

Phenrat, T., Fagerlund, F., Illangasekare, T., Lowry, G. V., & Tilton, R. D. (2011). Polymer-modified Fe0 nanoparticles target entrapped NAPL in two dimensional porous media: Effect of particle concentration, NAPL saturation, and injection strategy. *Environmental Science & Technology, 45*, 6102–6109.

Phenrat, T., Kim, H.-J., Fagerlund, F., Illangasekare, T., Tilton, R. D., & Lowry, G. V. (2009a). Particle size distribution, concentration, and magnetic attraction affect transport of polymer-modified Fe0 nanoparticles in sand columns. *Environmental Science & Technology, 43*, 5079–5085.

Phenrat, T., Kim, H. J., Fagerlund, F., Illangasekare, T., & Lowry, G. V. (2010b). Empirical correlations to estimate agglomerate size and deposition during injection of a polyelectrolyte-modified Fe-0 nanoparticle at high particle concentration in saturated sand. *Journal of Contaminant Hydrology, 118*, 152–164.

Phenrat, T., Long, T. C., Lowry, G. V., & Veronesi, B. (2009b). Partial oxidation ("Aging") and surface modification decrease the toxicity of nanosized zerovalent iron. *Environmental Science & Technology, 43*, 195–200.

Phenrat, T., Saleh, N., Sirk, K., Kim, H.-J., Tilton, R., & Lowry, G. (2008). Stabilization of aqueous nanoscale zerovalent iron dispersions by anionic polyelectrolytes: Adsorbed anionic polyelectrolyte layer properties and their effect on aggregation and sedimentation. *Journal of Nanoparticle Research, 10*, 795–814.

Phenrat, T., Saleh, N., Sirk, K., Tilton, R. D., & Lowry, G. V. (2007). Aggregation and sedimentation of aqueous nanoscale zerovalent iron dispersions. *Environmental Science & Technology, 41*, 284–290.

Pinder, G. F., & Celia, M. A. (2006). *Subsurface hydrology*. Hoboken: Wiley.

Preslo, L. M., Nielsen, G. L., & Nielsen, D. M. (2005). *Environmental site characterization. Practical handbook of environmental site characterization and ground-water monitoring* (2nd ed.pp. 35–205). CRC Press: Boca Raton, FL.

Quinn, J., Geiger, C., Clausen, C., Brooks, K., Coon, C., O'Hara, S., Krug, T., Major, D., Yoon, W.-S., Gavaskar, A., & Holdsworth, T. (2005). Field demonstration of DNAPL dehalogenation using emulsified zero-valent iron. *Environmental Science & Technology, 39*, 1309–1318.

Raychoudhury, T., Naja, G., & Ghoshal, S. (2010). Assessment of transport of two polyelectrolyte-stabilized zero-valent iron nanoparticles in porous media. *Journal of Contaminant Hydrology, 118*, 143–151.

Raychoudhury, T., Tufenkji, N., & Ghoshal, S. (2012). Aggregation and deposition kinetics of carboxymethyl cellulose-modified zero-valent iron nanoparticles in porous media. *Water Research, 46*, 1735–1744.

Raychoudhury, T., Tufenkji, N., & Ghoshal, S. (2014). Straining of polyelectrolyte-stabilized nanoscale zero valent iron particles during transport through granular porous media. *Water Research, 50*, 80–89.

Reinsch, B. C., Forsberg, B., Penn, R. L., Kim, C. S., & Lowry, G. V. (2010). Chemical transformations during aging of zerovalent iron nanoparticles in the presence of common groundwater dissolved constituents. *Environmental Science & Technology, 44*, 3455–3461.

Rosenthal, H., Adrian, L., & Steiof, M. (2004). Dechlorination of PCE in the presence of Fe0 enhanced by a mixed culture containing two Dehalococcoides strains. *Chemosphere, 55*, 661–669.
Saccà, M. L., Fajardo, C., Costa, G., Lobo, C., Nande, M., & Martin, M. (2014). Integrating classical and molecular approaches to evaluate the impact of nanosized zero-valent iron (nZVI) on soil organisms. *Chemosphere, 104*, 184–189.
Sakulchaicharoen, N., O'Carroll, D. M., & Herrera, J. E. (2010). Enhanced stability and dechlorination activity of pre-synthesis stabilized nanoscale FePd particles. *Journal of Contaminant Hydrology, 118*, 117–127.
Saleh, N., Kim, H.-J., Phenrat, T., Matyjaszewski, K., Tilton, R. D., & Lowry, G. V. (2008). Ionic strength and composition affect the mobility of surface-modified Fe0 nanoparticles in water-saturated sand columns. *Environmental Science & Technology, 42*, 3349–3355.
Saleh, N., Phenrat, T., Sirk, K., Dufour, B., Ok, J., Sarbu, T., Matyjaszewski, K., Tilton, R. D., & Lowry, G. V. (2005). Adsorbed triblock copolymers deliver reactive iron nanoparticles to the oil/water interface. *Nano Letters, 5*, 2489–2494.
Saleh, N., Sirk, K., Liu, Y., Phenrat, T., Dufour, B., Matyjaszewski, K., Tilton, R. D., & Lowry, G. V. (2007). Surface modifications enhance nanoiron transport and NAPL targeting in saturated porous media. *Environmental Engineering Science, 24*, 45–57.
Schrick, B., Hydutsky, B. W., Blough, J. L., & Mallouk, T. E. (2004). Delivery vehicles for zerovalent metal nanoparticles in soil and groundwater. *Chemistry of Materials, 16*, 2187–2193.
Shi, Z., Fan, D., Johnson, R. L., Tratnyek, P. G., Nurmi, J. T., Wu, Y., & Williams, K. H. (2015). Methods for characterizing the fate and effects of nano zerovalent iron during groundwater remediation. *Journal of Contaminant Hydrology, 181*, 17–35.
Shi, Z., Nurmi, J. T., & Tratnyek, P. G. (2011). Effects of nano zero-valent iron on oxidation reduction potential. *Environmental Science & Technology, 45*, 1586–1592.
Slater, L., & Binley, A. (2003). Evaluation of permeable reactive barrier (PRB) integrity using electrical imaging methods. *Geophysics, 68*, 911–921.
Slater, L. D., Choi, J., & Wu, Y. (2005). Electrical properties of iron-sand columns: Implications for induced polarization investigation and performance monitoring of iron-wall barriers. *Geophysics, 70*, G87–G94.
Stookey, L. L. (1970). Ferrozine - a new spectrophotometric reagent for iron. *Analytical Chemistry, 42*, 779.
Su, C., Puls, R. W., Krug, T. A., Watling, M. T., O'Hara, S. K., Quinn, J. W., & Ruiz, N. E. (2012). A two and half-year-performance evaluation of a field test on treatment of source zone tetrachloroethene and its chlorinated daughter products using emulsified zero valent iron nanoparticles. *Water Research, 46*, 5071–5084.
Su, Y., Adeleye, A. S., Zhou, X., Dai, C., Zhang, W., Keller, A. A., & Zhang, Y. (2014). Effects of nitrate on the treatment of lead contaminated groundwater by nanoscale zerovalent iron. *Journal of Hazardous Materials, 280*, 504–513.
Sun, J., Wang, S., Zhao, D., Hun, F. H., Weng, L., & Liu, H. (2011). Cytotoxicity, permeability, and inflammation of metal oxide nanoparticles in human cardiac microvascular endothelial cells: Cytotoxicity, permeability, and inflammation of metal oxide nanoparticles. *Cell Biology and Toxicology, 27*, 333–342.
Sun, Q., Feitz, A. J., Guan, J., & Waite, T. D. (2008). Comparison of the reactivity of nanosized zero-valent iron (nZVI) particles produced by borohydride and dithionite reduction of iron salts. *Nano, 3*, 341–349.
Sun, Y.-P., Li, X.-Q., Cao, J., Zhang, W.-X., & Wang, H. P. (2006). Characterization of zero-valent iron nanoparticles. *Advances in Colloid and Interface Science, 120*, 47–56.
Sun, Y.-P., Li, X.-Q., Zhang, W.-X., & Wang, H. P. (2007). A method for the preparation of stable dispersion of zero-valent iron nanoparticles. *Colloids and Surfaces A: Physicochemical and Engineering Aspects, 308*, 60–66.
Szecsody, J. E., Fruchter, J. S., Williams, M. D., Vermeul, V. R., & Sklarew, D. (2004). In situ chemical reduction of aquifer sediments: Enhancement of reactive iron phases and TCE dechlorination. *Environmental Science and Technology, 38*, 4656–4663.

Taghavy, A., Costanza, J., Pennell, K. D., & Abriola, L. M. (2010). Effectiveness of nanoscale zero-valent iron for treatment of a PCE-DNAPL source zone. *Journal of Contaminant Hydrology, 118*, 128–142.

Tilston, E. L., Collins, C. D., Mitchell, G. R., Princivalle, J., & Shaw, L. J. (2013). Nanoscale zerovalent iron alters soil bacterial community structure and inhibits chloroaromatic biodegradation potential in Aroclor 1242-contaminated soil. *Environmental Pollution, 173*, 38–46.

Tiraferri, A., Chen, K. L., Sethi, R., & Elimelech, M. (2008). Reduced aggregation and sedimentation of zero-valent iron nanoparticles in the presence of guar gum. *Journal of Colloid and Interface Science, 324*, 71–79.

Tiraferri, A., & Sethi, R. (2009). Enhanced transport of zerovalent iron nanoparticles in saturated porous media by guar gum. *Journal of Nanoparticle Research, 11*, 635–645.

Tosco, T., & Sethi, R. (2010). Transport of non-Newtonian suspensions of highly concentrated micro- and nanoscale Iron particles in porous media: A modeling approach. *Environmental Science & Technology, 44*, 9062–9068.

Tratnyek, P. G., Johnson, R. L., Lowry, G. V., & Brown, R. A. (2014). *In situ chemical reduction for source remediation*. Springer.

Tratnyek, P. G., Reilkoff, T., Lemon, A., Scherer, M., Balko, B., Feik, L., & Henegar, B. (2001). Visualizing redox chemistry: Probing environmental oxidation–reduction reactions with indicator dyes. *The Chemical Educator, 6*, 172–179.

Truex, M. J., Macbeth, T. W., Vermeul, V. R., Fritz, B. G., Mendoza, D. P., Mackley, R. D., Wietsma, T. W., Sandberg, G., Powell, T., Powers, J., Pitre, E., Michalsen, M., Ballock-Dixon, S. J., Zhong, L., & Oostrom, M. (2011a). Demonstration of combined zero-valent Iron and electrical resistance heating for in situ trichloroethene remediation. *Environmental Science & Technology, 45*, 5346–5351.

Truex, M. J., Vermeul, V. R., Mendoza, D. P., Fritz, B. G., Mackley, R. D., Oostrom, M., Wietsma, T. W., & Macbeth, T. W. (2011b). Injection of zero-valent iron into an unconfined aquifer using shear-thinning fluids. *Ground Water Monitoring and Remediation, 31*, 50–58.

Tufenkji, N., & Elimelech, M. (2004). Correlation equation for predicting single-collector efficiency in physicochemical filtration in saturated porous media. *Environmental Science & Technology, 38*, 529–536.

Vecchia, E. D., Luna, M., & Sethi, R. (2009). Transport in porous media of highly concentrated iron micro- and nanoparticles in the presence of xanthan gum. *Environmental Science & Technology, 43*, 8942.

Velimirovic, M., Tosco, T., Uyttebroek, M., Luna, M., Gastone, F., De Boer, C., Klaas, N., Sapion, H., Eisenmann, H., Larsson, P.-O., Braun, J., Sethi, R., & Bastiaens, L. (2014). Field assessment of guar gum stabilized microscale zerovalent iron particles for in-situ remediation of 1,1,1-trichloroethane. *Journal of Contaminant Hydrology, 164*, 88–99.

Viollier, E., Inglett, P. W., Hunter, K., Roychoudhury, A. N., & Van Cappellen, P. (2000). The ferrozine method revisited: Fe(II)/Fe(III) determination in natural waters. *Applied Geochemistry, 15*, 785–790.

Wang, C.-B., & Zhang, W.-X. (1997). Synthesizing nanoscale Iron particles for rapid and complete dechlorination of TCE and PCBs. *Environmental Science & Technology, 31*, 2154–2156.

Wei, Y. T., Wu, S. C., Chou, C. M., Che, C. H., Tsai, S. M., & Lien, H. L. (2010). Influence of nanoscale zero-valent iron on geochemical properties of groundwater and vinyl chloride degradation: A field case study. *Water Research, 44*, 131–140.

Wu, Y., Slater, L., Versteeg, R., & LaBrecque, D. (2008). A comparison of the low frequency electrical signatures of iron oxide versus calcite precipitation in granular zero valent iron columns. *Journal of Contaminant Hydrology, 95*, 154–167.

Wu, Y., Slater, L. D., & Korte, N. (2005). Effect of precipitation on low frequency electrical properties of zerovalent iron columns. *Environmental Science and Technology, 39*, 9197–9204.

Wu, Y., Slaters, L. D., & Korte, N. (2006). Low frequency electrical properties of corroded iron barrier cores. *Environmental Science and Technology, 40*, 2254–2261.

Wu, Y., Versteeg, R., Slater, L., & LaBrecque, D. (2009). Calcite precipitation dominates the electrical signatures of zero valent iron columns under simulated field conditions. *Journal of Contaminant Hydrology, 106*, 131–143.

Xiu, Z.-M., Gregory, K. B., Lowry, G. V., & Alvarez, P. J. J. (2010a). Effect of bare and coated nanoscale zerovalent iron on tceA and vcrA gene expression in Dehalococcoides spp. *Environmental Science & Technology, 44*, 7647–7651.

Xiu, Z.-M., Jin, Z.-H., Li, T.-L., Mahendra, S., Lowry, G. V., & Alvarez, P. J. J. (2010b). Effects of nano-scale zero-valent iron particles on a mixed culture dechlorinating trichloroethylene. *Bioresource Technology, 101*, 1141–1146.

Yan, W., Herzing, A. A., Kiely, C. J., & Zhang, W.-X. (2010). Nanoscale zero-valent iron (nZVI): Aspects of the core-shell structure and reactions with inorganic species in water. *Journal of Contaminant Hydrology, 118*, 96–104.

Yao, K.-M., Habibian, M. T., & O'Melia, C. R. (1971). Water and waste water filtration. Concepts and applications. *Environmental Science & Technology, 5*, 1105–1112.

Zaa, C. L. Y., McLean, J. E., Dupont, R. R., Norton, J. M., & Sorensen, D. L. (2010). Dechlorinating and iron reducing bacteria distribution in a TCE-contaminated aquifer. *Ground Water Monit. Remediat., 30*, 46–57. https://doi.org/10.1111/j.1745-6592.2009.01268.x.

Zhan, J., Zheng, T., Piringer, G., Day, C., McPherson, G. L., Lu, Y., Papadopoulos, K., & John, V. T. (2008). Transport characteristics of nanoscale functional zerovalent iron/silica composites for in situ remediation of trichloroethylene. *Environmental Science & Technology, 42*, 8871–8876.

Zhang, L., & Manthiram, A. (1997). Chains composed of nanosize metal particles and identifying the factors driving their formation. *Applied Physics Letters, 70*, 2469–2471.

Zhong, L., Oostrom, M., Wietsma, T. W., & Covert, M. A. (2008). Enhanced remedial amendment delivery through fluid viscosity modifications: Experiments and numerical simulations. *Journal of Contaminant Hydrology, 101*, 29–41.

Zhou, L., Thanh, T. L., Gong, J., Kim, J.-H., Kim, E.-J., & Chang, Y.-S. (2014). Carboxymethyl cellulose coating decreases toxicity and oxidizing capacity of nanoscale zerovalent iron. *Chemosphere, 104*, 155–161.

Zhu, L., Lin, H.-Z., Qi, J.-Q., Xu, X.-Y., & Qi, H.-Y. (2012). Effect of H2 on reductive transformation of p-ClNB in a combined ZVI–anaerobic sludge system. *Water Research, 46*, 6291–6299.

Zhu, L., Lin, H., Qi, J., & Xu, X. (2013). Enhanced transformation and dechlorination of p-chloronitrobenzene in the combined ZVI–anaerobic sludge system. *Environmental Science and Pollution Research, 20*, 6119–6127.

Chapter 8
Experiences from Pilot- and Large-Scale Demonstration Sites from Across the Globe Including Combined Remedies with NZVI

Johannes Bruns, Florin Gheorghiu, Michael Borda, and Julian Bosch

Abstract This chapter provides an overview of environmental restoration efforts involving the application of NZVI and NZVI composite materials. The chapter focuses on the novel application techniques aimed at improving the delivery, characterization and effectiveness of particles, drawing on direct field experience of the authors as well as review of over two decades of peer-reviewed literature. Stressing a base of knowledge through detailed site characterization towards a conceptual site model, this chapter's main emphasis is on delivery techniques, options for particle formulation and challenges associated with different site conditions that would influence particle transport to the target contaminated areas, as well as monitoring requirements to evaluate treatment effectiveness. Particle types and injection characteristics are covered along with an evaluation of geological, hydrogeological, geochemical and microbiological conditions favourable for NZVI injection. The synergy between NZVI and bioremediation treatments is discussed along with observed changes in microbiological activity following particle treatments. This chapter also provides highlights of laboratory data needs, general permitting requirements and health and safety issues related to NZVI. The chapter concludes with a discussion of potential technology improvements that will likely improve the effectiveness and implementability of NZVI treatments.

Keywords Nanoscale Zerovalent Iron · Field Implementation · Combined Remedies with NZVI

J. Bruns (✉) · J. Bosch
Intrapore GmbH, Essen, Germany
e-mail: johannes.bruns@intrapore.com; julian.bosch@intrapore.com

F. Gheorghiu
Applied Testing and Geosciences, LLC, Bridgeport, PA, USA
e-mail: florin@appliedtesting.com

M. Borda
University of Pennsylvania, Philadelphia, PA, USA

T. Phenrat, G. V. Lowry (eds.), *Nanoscale Zerovalent Iron Particles for Environmental Restoration*, https://doi.org/10.1007/978-3-319-95340-3_8

8.1 Introduction

The use of zero-valent iron (Fe^0) to treat sites impacted by chlorinated solvents is an established and accepted technology in the environmental remediation industry. It is effective for complete reductive dechlorination of chlorinated organic compounds, e.g. chlorinated ethenes and chlorinated ethanes in contaminated soils, sediments, groundwater and wastewater (Chap. 3). Nanoparticle field applications for groundwater decontamination purposes were largely first performed in North America in the early 2000s, shortly after applications in Europe followed. To date three particle groups are used for remediation purposes in natural groundwater systems.

The first nanomaterial applied was zero-valent iron (NZVI), used for in situ reduction of volatile chlorinated hydrocarbons (VCHC) and heavy metal fixing. Fast chemical reaction and high performance in contamination reduction, complex handling in the field and short travel distances in the aquifer characterize this material available under several trade names. In North America additional catalysts (e.g. palladium) are used to accelerate contaminant reduction or to reduce recalcitrant constituents using NZVI alone. In Europe these kinds of additives are not currently licensable for in situ applications.

R&D activities in Europe and North America lead to improvement of available particles and forced the development of new, advanced nanomaterials like Carbo-Iron® and intraBlue®. Carbo-Iron® is a composite material of colloidal activated carbon and embedded nano-iron structures for VCHC reduction (Chap. 2). For Carbo-Iron®, supporting the NZVI on a larger carbon colloid maintained the high chemical reactivity of iron in water, but enabling the reagent is both more mobile in the aquifer and easier to handle during application. The colloidal, nanoscale *iron oxide* particle intraBlue® serves as electron acceptor for microbial hydrocarbon degradation, and therefore indirect oxidative contaminant removal. After injection, the material adsorbs to the soil matrix and acts as a non-toxic, sustained and rebound-free stimulator for microbial activities.

The three particle groups, NZVI, supported NZVI, and nanoscale iron oxides, have different characteristics, address various fields of application and are primary materials currently available for commercial use. The following explanations summarize the experiences with these materials made during large-scale field demonstrations and site remediation over the last 20 years. Other particles show promising degradation effects on various contaminants in laboratory tests, but they are not yet suitable for commercial applications: there is need for further R&D, and the materials have to successfully complete field pilot tests before they are placed on the market. A successful pilot test is one in which the ability of the material to be emplaced and to degrade contaminants in situ is verified.

To warrant a cost-effective and successful field use, the application of nanoparticles for site remediation should be based on a stepwise approach. Usually the following steps are taken:

- Site suitability study
- Detailed site investigation

- Contaminant degradation tests
- Authority licensing
- Application conception
- Particle design and production
- Particle injection
- Performance monitoring

Economic and ecologic aspects, resources conservation and requirements of sustainable remediation procedures have led to an increasing number of in situ applications of nanoparticles worldwide. This trend is estimated to continue, especially since the fears of negative consequences of engineered nanomaterials are allayed, and more regulators are now supporting this technology.

This chapter presents the experience and lessons learned from field implementations of nanoparticle-based remedies carried out worldwide. The structure of the chapter is organized to include an overview of the in situ treatment and the details needed for the treatment design such as slurry composition, slurry injection methods specific to unconsolidated sediments and fractured bedrock, estimation of particle quantities and observations related to treatment longevity. Finally, a section on the needs of future technology and implementation improvements viewed from the perspective of field practitioners is added.

8.2 Site Investigation and Data Needs for Successful Injection of NZVI

In situ contaminant treatment using nanoparticles consists of the injection of a water-based suspension into the subsurface. Normally stabilizers are added to keep the slurry stable, and sometimes catalysts are added to accelerate contaminant reduction or to degrade recalcitrant constituents.

Typically, dissemination of nanoparticles in the subsurface can be achieved by multiple methods including hydraulic soil fracturing and injection of NZVI/ZVI sand mixtures; however, this section includes references to particle slurry gravity feed, pressurized injections and pumping and reinjection of closed loops using wells or possibly recharge galleries (i.e. trenches). Pressurized injections can be implemented under slight injection pressures, or using higher pressures designed to reopen existing or induced bedrock fractures.

The goal of nanoparticle injections is to obtain a well-dispersed reactant slurry in the subsurface either throughout the affected aquifer or at least in the area with most contaminant mass. As a result, the main or initial target of NZVI injections should be the source zone or hotspots and highly permeable features, if present.

The design of nanoparticle injection is highly dependent on local geologic and hydrogeologic heterogeneities. Well screen location and particle mass injection should target the permeable sections of the stratigraphy. It is important that a sufficient number of boreholes/wells in a proposed remedial area are installed and that the boreholes be logged in detail and stratigraphic correlations are made. Detailed stratigraphic correlations allow for placement of monitoring wells downgradient of the injection area with screens set at the appropriate level. Water level monitoring and detailed understanding of groundwater flow, groundwater fluctuations and variability in groundwater gradients and flow direction are also important for the same reasons discussed above. Tracer tests and detailed monitoring with multi-parameter data loggers are recommended prior and during nanoparticle injection. Continuous data logger monitoring allows for refinements of the hydrogeological conditions in the remediation area and help the interpretation of the geochemical changes following nanoparticle application (Fig. 8.1). Not understanding the complexities of local stratigraphy and groundwater flow migration may result in monitoring of zones not affected by nanoparticle injections, or potentially to miss delivering NZVI to the areas of interest.

Injection well drilling, installation and development are critically important for the successful implementation of a nanoparticle remedy. The mobility of the particles can be obstructed by the presence of clay lenses that could generate a smearing zone, improper well screen design, incomplete well development and other factors. A nanoparticle remedy should also take into account the need for injection well rehabilitation and well replacements.

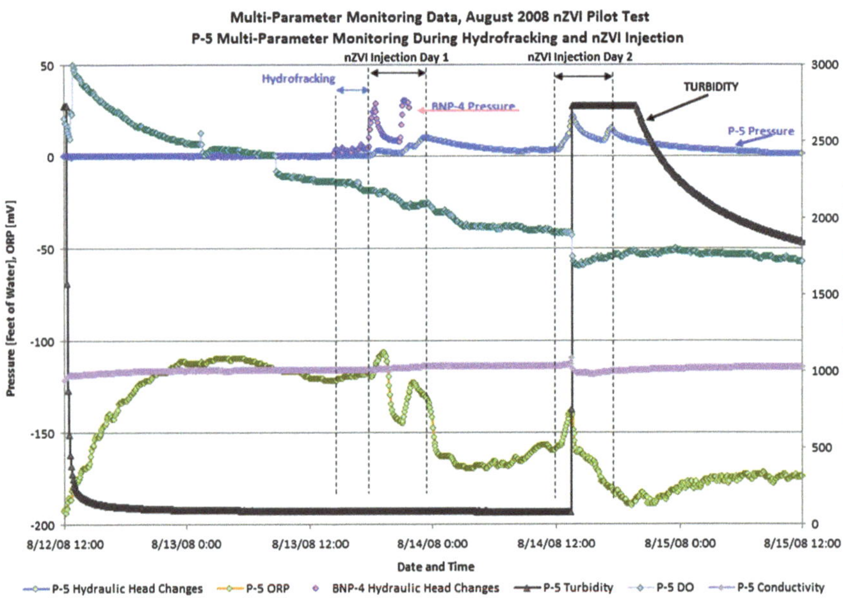

Fig. 8.1 Example of multi-parameter monitoring conducted in monitoring well P-5 located approximately 10 m downgradient of injection well BNP-4 (Gheorghiu and Plant 2010). ORP and DO are the main indicators of NZVI slurry migration showing lower DO and lower ORP. Turbidity provided in this case is an indication of fracture propagation reaching monitoring well P-5 corroborated by a sudden increase in water level in P-5. Conductivity does not provide sufficient change to indicate the NZVI slurry presence

The development of a conceptual site model (CSM) is an essential step in the successful implementation of nanomaterial-based remedies. All information about the underground conditions are basis for the development of a CSM that supports the understanding of those subsurface structures and conditions that influence the contamination distribution and that favour or hinder the particle dissemination. Geological systems are complex, and the site investigation has to be performed in a way that provides all of the necessary data for a successful remediation action including monitoring all contaminants present and detailed geological, geochemical, hydraulic and microbiological conditions. In addition to site utilization, infrastructure and technical site conditions have to be known and used as a basis for designing an effective and sustainable remediation strategy (Fig. 8.2).

The first step of a successful remediation strategy is to determine whether any nano-remediation strategy would work at a specific site, including identifying which particle species could be effective for the contamination degradation. Nanomaterials have different activity spectrums, and they are often suitable only for particular contaminants, contaminant groups or site conditions. In detail the study should geologically characterize the local formations, tectonics, hydraulic conductivities, groundwater flow direction and velocity and hydrochemical conditions of the regional groundwater regime and local aquifer(s), and it should describe the contaminant spectrum in all relevant underground compartments (groundwater, soil, soil vapour) as well as pathways of its mobility. An environmental site assessment should evaluate potential contaminant impacts on protected properties like human health, drinking water, soil, air, etc. This initial evaluation should lead to a go/no-go decision regarding the applicability of nano-remediation. If the site appears suitable for nanomaterials, a treatability study may be warranted (see Sect. 8.3).

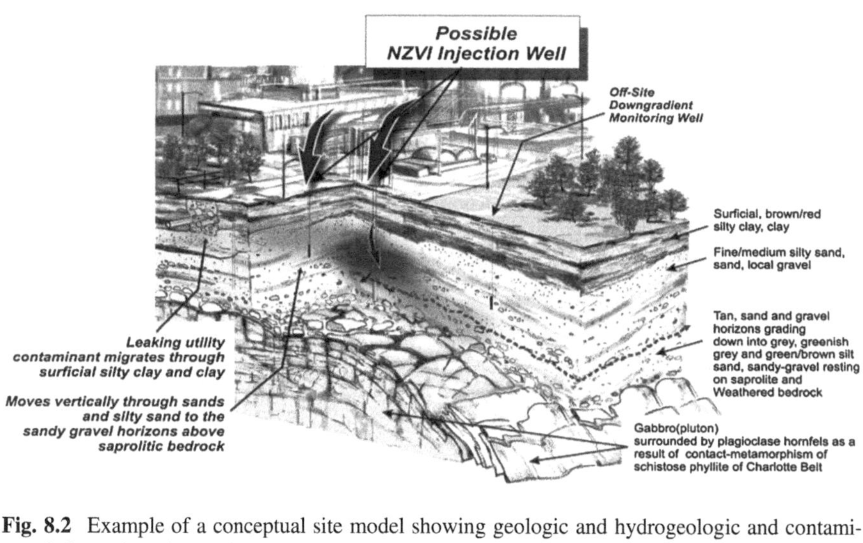

Fig. 8.2 Example of a conceptual site model showing geologic and hydrogeologic and contaminant information along with conceptual design information of an NZVI injection remedy (Macé et al. 2006)

With respect to the process of particle injection in the subsurface, the main significant underground characteristics are related to the geological, hydrogeological and contaminant conditions. Geologically the underground can be grouped as unconsolidated materials (gravel, sand, silt and clay) and consolidated materials (sedimentary, igneous or metamorphic bedrock). Corresponding to these types of unconsolidated and consolidated materials, the groundwater flow can be porous media flow or fracture flow. Favourable subsurface conditions for NZVI injection include presence of higher permeability materials such as sands and gravels or highly fractured consolidated materials. Injection in lower permeability materials requires closely spaced injection points or soil/bedrock fracturing.

If nano-remediation on site is determined to be feasible (e.g. nanoparticles can degrade the site-specific contaminants, most contaminant mass is present in permeable materials, or if induced fractures are cost-effective), a site investigation programme has to be performed to fill information gaps prior to the injection conception and design. The investigations should provide information on the spatial contaminant distribution and physico-chemical parameters (pH, redox potential, dissolved oxygen, conductivity, temperature, etc.) within the underground formations and show the spatiotemporal hydraulic and hydrodynamic pattern on site. In addition the groundwater microbiology in the contaminated source and plume area has to be known to describe biological activities in the aquifer and evaluate their potential interactions with the particle application. In addition to the site investigations, laboratory treatability tests with groundwater samples should be performed to determine contamination degradation patterns and cross-sensitivity reactions.

All information gathered is used for the remediation planning process and particle injection design. Degradation takes place only if the nanomaterial comes in contact with the contaminants in the source or plume area. Technically this is induced by injecting or infiltrating slurry into the contaminated areas of the aquifer; here ideally the nanomaterial travels into/through the pore space or fractures where the contaminant sits, settles and reacts with the pollution and/or stimulates microbial contaminant removal. The travel behaviour is the key to an effective degradation mechanism: only a large radius of influence ensures a sufficient contact area between nanoparticle and contaminant. The ability to deliver NZVI over distances more than a few decimetres (no surface modifiers) or a few metres (surfactants) is currently a limitation of the method. However, some modifications of NZVI, e.g. Carbo-Iron® and intraBlue®, can travel distances up to 20 m in natural aquifer systems and in lab tests. Advances in materials design and injection methods will continue to improve the ability to deliver NZVI.

Knowledge of hydraulically effective layers in the underground, especially the differentiation between gravel, sand (fine, medium, coarse), silt and clay in unconsolidated materials and the fracture flow characteristics in consolidated materials and determination of aquifer parameters, is needed to bring the nanomaterial in contact with the contaminants. This requires that the spatial distribution of heterogeneities be understood. Further, a contaminant mass balance and a groundwater flow and contaminant transport model for the remediation area is important for injection planning, stoichiometric calculations, contaminant transport estimation and monitoring design.

Data needs in unconsolidated sediments include discharge velocity and groundwater flow velocity, hydraulic conductivity of the target injection zone as well as overall and hydraulic conductivity distribution in the impacted groundwater zone. These data are used to estimate particle slurry injection rates and dissemination around injection wells, considering the particle-specific travel characteristics. These data can be collected using standard hydrogeologic tests. Short-term to extended time constant rate injection/withdrawal tests are preferred because they provide quantification of the skin effect (i.e. skin factor, a dimensionless pressure drop caused by a flow restriction in the near-wellbore region) of the injection well. A large skin effect may prevent effective injection of particle slurry. In addition to the needs and for unconsolidated sediments, there are specific additional data needed for the implementation of particle injection in fractured bedrock. Data needs include fracture orientation, fracture aperture, fracture transmissivity/hydraulic conductivity, fracture reopening pressure and orientation and magnitude of minimum principal stress. In addition, basic water level data from wells and screened in known fracture systems or hydrostratigraphic units are needed.

Fracture orientation, aperture and transmissivity are needed to identify and quantify discrete fracture networks (DFNs) which allow for understanding groundwater and contaminant migration pathways. Fracture reopening pressures and orientation and magnitude of the minimum principal stress are data needed to design the injection pressures using existing fractures as the particle slurry dispersion mechanism in the subsurface. Water level data is needed to develop an understanding of the hydraulic pressure distribution in the subsurface and groundwater flow direction.

Fracture orientation and fracture aperture can be obtained from surface mapping of outcrops, downhole televiewer logging (optical or acoustic televiewer) or measurements performed on oriented cores. Fracture transmissivity/hydraulic conductivity can be obtained from drill stem tests/packer tests from downhole heat-pulse flow metre logging. Fracture reopening pressure and orientation and magnitude of minimum principal stress can be obtained from in situ stress tests combined with pre- and postfracturing downhole televiewer logging.

8.3 Laboratory Investigations and Field Verification

The successful field application of nanoparticle remediation materials requires a prescribed set of laboratory investigations. The test results are used for all aspects of the remediation process, from environmental impact assessments and risk evaluations over mass calculations and stoichiometric calculations to conceptual site model development, technical injection design and contamination reduction process evaluations during field monitoring.

These analyses cover:

- General groundwater characterization
- Contamination spectrum determination

- Contaminant degradation tests with different particle concentrations and suspension stabilizers (carboxymethyl cellulose (CMC), humic substances, organic polymers, etc.)
- Particle reactivity tests (contaminant degradation vs. time per particle mass)
- Particle selection (optimal size, coatings and additives for reaction control and slurry stabilization)
- Suspension additive tests for suspension stabilization and particle travel optimization
- Particle transport behaviour in column or pilot test tank experiments
- SEM/TEM investigations on selected particles and sediments to assess the impacts of particle injection on geomorphology, especially for monitoring purposes
- Granulometry and grain size distribution, pore size/pore space distribution and ecotoxicological tests

An increasingly important part of tests comprises investigations of microbiological communities and microbial activities in soil and groundwater to evaluate their role in natural attenuation processes on site and investigations of microbiological balance modifications over time as a consequence of nanoparticle application. For NZVI the influence on microbiology is partly understood (see Chap. 10); for the new composite material Carbo-Iron®, latest investigations point to increases in microbial degradation (Mackenzie et al. 2014); and the iron oxide material intraBlue® was developed specifically to enhance oxidative microbiological degradation processes.

For NZVI, its coupling and interaction with microbiological activities are known, and long-term experience exists within the scope of large-scale remediation. Theoretical recognition of the synergy between NZVI and the microbiome was recognized in early Records of Decision (ROD) approved by the US EPA and Underground Injection Control Application and permits in which a transition from an NZVI remedy to a bio-remedy was envisioned. These synergies are described in detail in Chap. 10.

During the largest pilot-scale NZVI injection performed to date (~4500 kg NZVI injected), a thorough microbiological assessment was performed illustrating the transition of abiotic VCHC degradation to enhanced anaerobic bioremediation of VCHCs (Hains 2008). Microbial counts, polymerase chain reaction (PCR) and denatured gradient gel electrophoresis (DGGE) analyses were conducted before and after NZVI injections. Before treatment, bacterial analysis indicated predominance of aerobic bacteria and a low bacterial population density. Traces of *Dehalococcoides* and *Dehalobacter* were detected. PCR and DGGE analysis confirmed that, immediately following NZVI injections, population density became much less diverse. With time, population density increased significantly but presented a different set of dominant bacterial species than prior to NZVI injection. This behaviour is consistent with laboratory-based mesocosms evaluating the effect of NZVI on microbial community structure in sediments collected from three different contaminated sites (Kirschling et al. 2010). Microbial counts confirmed that aerobic heterotrophic bacterial population remained relatively stable, but iron-reducing bacteria, sulphate-reducing bacteria and nitrifying bacteria increased significantly in the treatment zone. In general, bacterial population increased by five orders of magnitude compared to the upgradient bacterial population. *Dehalococcoides* was detected in the treatment zone, but its distribution was non-homogeneous over the treatment zone.

This transition in microbial communities is driven by the redox condition of the aquifer (Fig. 8.3) due to NZVI reactivity (ORP values changed from +200 mV to −500 mV upon injection), coupled with the addition of a complex source of soluble and sparingly soluble carbon (soy protein) which acts as an electron donor. This transition and degradation of VCHCs continued for over 1 year and achieved a groundwater concentration target of five parts per billion (ppb) for trichloroethene (TCE) without build-up of intermediate degradation products (Hains 2008). This suggests that a combined remedy of NZVI injection and long-term enhanced bioremediation may be a strong candidate technology for a number of VCHC-impacted sites (Figs. 8.3 and 8.4).

The abiotic treatment of VCHCs using NZVI is driven by the direct contact of NZVI particles with the VCHC and consequent electron transfer. With particle mobility being low in many cases, and the high reactivity of NZVI particles with water potentially placing an upper limit of the functional treatment time of this technology, direct-contact abiotic NZVI treatment may not be a sustainable remedial option, without periodic successive injections. This alternative can become costly. In addition, aquifer clogging and limited electron transfer efficiency can begin to render multiple injections unsuccessful. In contrast to abiotic NZVI treatment, the presence and reactivity of NZVI in the subsurface cause broad changes in the geochemistry which is not direct-contact driven. Changes in ORP to more reducing conditions, reduction of dissolved oxygen (DO) and removal of other terminal electron acceptors (TEAs) can occur over a significantly larger area than direct-contact abiotic treatment.

Typically under these scenarios, the capacity of the system for intrinsic bioremediation is high; however, the systems are typically carbon limited. The addition of an electron donor (e.g. soy protein) can stimulate the indigenous microorganisms and result in successful enhanced bioremediation over a significantly larger treatment area than the initial NZVI impact (Fig. 8.4). A phased approach may be applicable and sustainable as NZVI can be injected in source areas to treat high dissolved

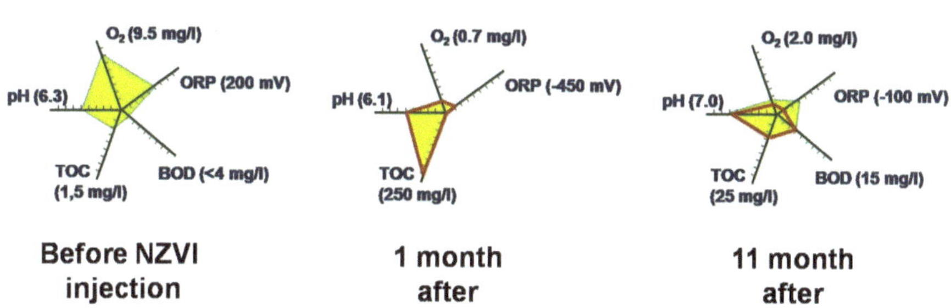

Fig. 8.3 Graphical representation of the geochemical changes that occur after the injection of NZVI into a groundwater system (Borda et al. 2009)

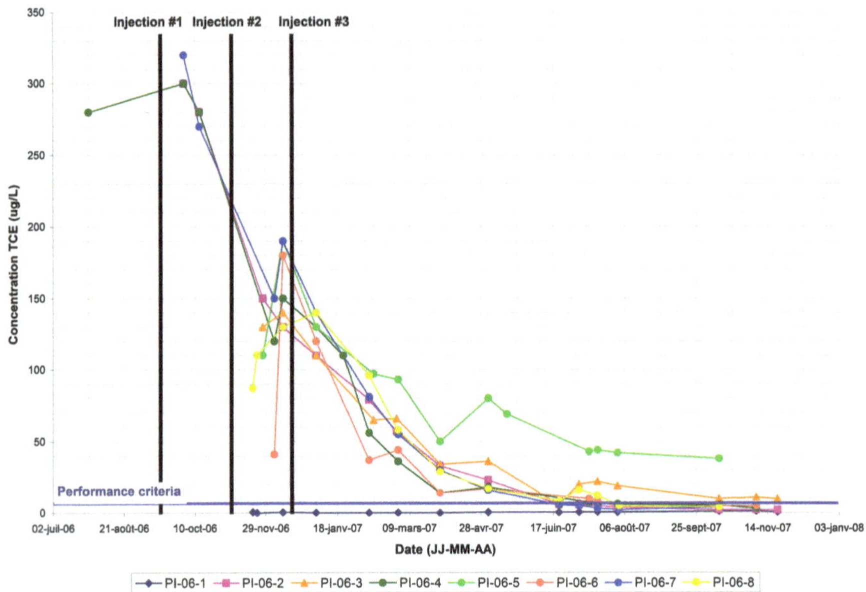

Fig. 8.4 Pilot-scale study showing prolonged treatment (over 18 months) of TCE showing short-term treatment with abiotic NZVI reactions, then long-term treatment with abiotic and enhanced bioremediation (Hains 2008)

concentration of VCHCs abiotically and to condition the aquifer for enhance bioremediation. Further treatments of electron donor may continue for significantly longer time frames than NZVI alone allowing for less frequent NZVI injections to condition the aquifer.

8.4 Permitting and Health and Safety

All nanoparticle field applications have to be permitted by federal, state or local authorities. The permitting procedure for nano-remediation in general requires a formal remediation plan according to the local, state or national requirements. Typically required documents include:

- Detailed descriptions of the site:
  - Utilization
  - Contamination
  - Environmental impact and risk assessment
- Remediation measures:
  - Schedule
  - Particle characterization:

  - Design
  - Production
  - Effect on pollution including by-product generation
  - Toxicology data
  - Travel behaviour within the aquifer
  - Effective reactive lifetime

- Injection mass calculations
- Site injection schema
- Ecotoxicological impact evaluations
- Remediation success estimates
- Monitoring plan descriptions:

  – Maps
  – Schedule
  – Parameters

To avoid holding time of the NZVI, the remediation planning should allow time for the permitting procedure. Toxicology, particle travel behaviour within the aquifer, by-products of contaminant degradation and reactive lifetime considerations of the particles used are important factors for the regulator's internal decision-making process and permit stipulations. Early involvement of the regulatory authority in the remediation planning could benefit (speed up) the process.

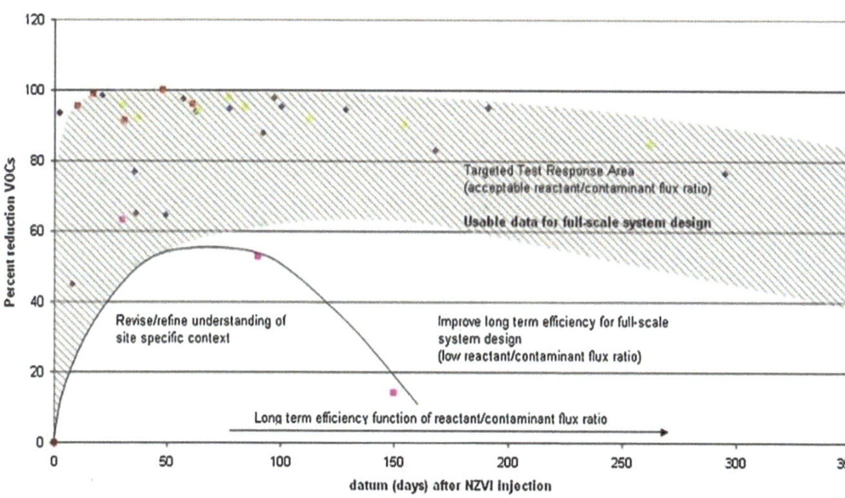

Fig. 8.5 Graph drawn from data obtained in the injection wells and monitoring wells used during pilot-scale tests undertaken by Golder Associates at locations around the world. Target test response area illustrates the optimum treatment conditions that can be achieved using NZVI technology with 60% to 100% treatment efficiency and long-term (<1 yr) treatment times. Rapid rebound of contaminants after treatment (illustrated by the black curve) suggests that the CSM be refined. (Modified from Macé et al. (2006))

Rough estimates for the in situ reactive lifetime of zero-valent iron in Carbo-Iron® particles resulted in a time span of >100 days so far. The nanosized, colloidal iron oxides for bioremediation, intraBlue®, have shown a reactive lifetime of more than 1 year. The effective reactive lifetime for NZVI particles is illustrated by the graph on Fig. 8.5 (Macé et al. 2006).

In terms of ecotoxicology, Carbo-Iron® and intraBlue® have been tested thoroughly. Investigations of ecotoxicological impacts (acute and chronic toxicity) of Carbo-Iron® particles on fish embryos, daphnia and other organisms have shown no toxic influence, although the organisms partly incorporated the particles or were tainted with it. Tests with adsorptive enriched PCE on aged Carbo-Iron® lead to an increased lethal concentration of PCE (Bleyl 2014), i.e. lower toxicity. Ecotoxicological tests of intraBlue® particles on plant roots, bacteria and nematodes (Höss et al. 2015) have been used to estimate critical exposure levels. So far, no relevant toxicity has been observed for these materials.

Based on the ecotoxicological tests, permitting for the particles has been non-problematic in recent field applications.

Finally it should be pointed out that the permitting process has also to consider basic interests of the party obliged to undertake remediation: although this is public information, sometimes companies affected are not happy to have their sites featured in publications. Thus, much of the practical experience with NZVI is unfortunately not well-documented in the scientific literature.

8.5 Particle Transport

As described above, the injection and movement of NZVI is important for its success as a remediation strategy. Particle travel in natural aquifer systems primarily is induced by flow and gravity forces within pores or fractures. The mobility is hindered by magnetic properties (and resulting aggregation) and relatively low surface charge of the particle; these characteristics lead to an interaction between the particle and encircling mineralogical materials with its own geophysical features, so that particles adhere to the surrounding area. In addition the nanoparticle shape affects the travel properties: spherical particles travel farther than other forms. Particle density and agglomeration is another travel obstacle: porous particles like the composite material Carbo-Iron® have a lower density and a more negative surface charge compared to other particles like NZVI and subsequently have lower rates of agglomeration. Lab tests and field findings confirm higher travel distances of Carbo-Iron® compared to NZVI. Recent research suggests that agglomeration may be a significant cause to the limited mobility observed in NZVI applications. This suggests that methods to prevent agglomeration could further improve NZVI delivery.

Within the aquifer particle travel happens primarily with the injected water along areas of high hydraulic permeability. In unconsolidated sediments, gravel and sand are preferred pathways compared to silt and clay. Coarse sand layers are preferred compared to fine grain areas. In unconsolidated material, grain size distribution and pore volume influence the travel behaviour; in consolidated material the fracture networks and its aperture affect the travel. In some respect the particle travel routes follow the preferred contamination migration pathways. Thus it happens, e.g. in complex structured aquifers, that particles circulate around thick silt and clay lenses and are detected below this layers. To evaluate main particle injection areas and migration pathways, it is important to consider the subsurface in all three dimensions. Besides the geological structures, site hydraulics affects the particle spreading. Especially spatial and temporal changes of groundwater flow directions and artesian conditions have to be understood for the detailed injection planning. By first approximation, the overall particle transport direction can be derived from the main groundwater flow direction.

To reach an optimal remediation result, particle features like surface charge and magnetic properties have to be tailored for best performance under the conditions of the natural system, particularly to optimize the particle spreading. The travel behaviour of the three commercially used particle groups NZVI, Carbo-Iron® and intraBlue® in natural subsurface systems is different, according to their different shape, size and surface charge.

For NZVI the authors draw from their experience with travel behaviour for the last 20 years. During the processing of crushed NZVI materials, it has been observed that the primary particle size range of NZVI produced is approximately 50–100 nm. Particles then begin to agglomerate, based on surface charge and on magnetic properties of ZVI, forming secondary particle sizes that may approach the micron-scale (Fig. 8.6).

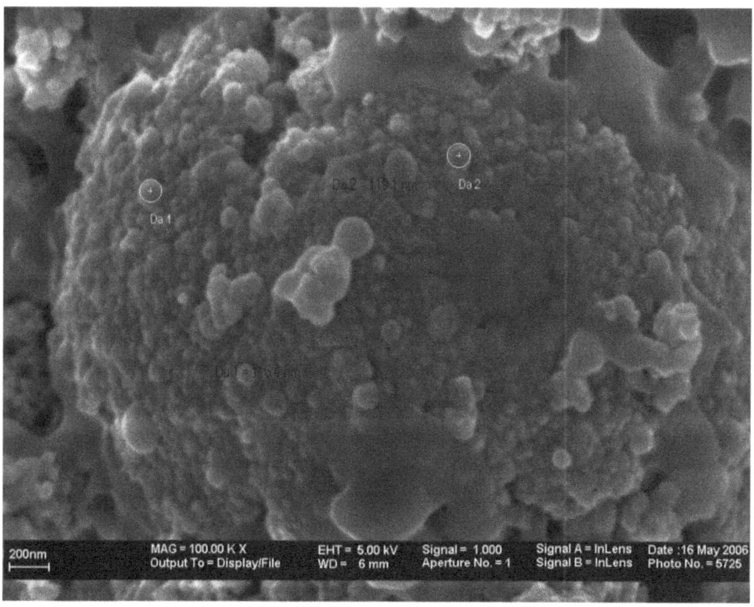

Fig. 8.6 Agglomerated nanoscale NZVI particles (average diameter of population 1, 115 nm; average diameter of population 2, 120 nm) forming a secondary spherical micron-scale structure (Bruns 2013)

Much of the focus of field implementation has been on using food-grade additives to limit the degree of NZVI agglomeration to promote their distribution in the subsurface. Along with using surface modifiers, a critical factor that has been observed in enhancing the mobility of some products is the use of freshly produced NZVI particles. Because the processes that cause NZVI to become immobile (e.g. agglomeration) are kinetically controlled, using material that is less than 1 week old has been observed to significantly impact the quality of the product. Recent advancements in identification and use of surface modifiers as well as successful field demonstration using mechanical dispersion of the NZVI slurry as part of the pressurized injection processes have reduced concerns related to NZVI agglomeration, low subsurface dispersion and particle settling. Using appropriate – in Europe currently not licensable – surfactants, the current technology allows for gravity-fed and pressurized injection in unconsolidated sediments with maximum radii of influence (ROI) on the order of 3–6 m. Without surface modifiers, travel distances of only a few decimetres are reached on average. In fractured bedrock, hydraulic fracturing and injection have shown particle travel distances of about 15 m upgradient and 35 m downgradient from the injection point.

For the composite material Carbo-Iron®, travel behaviour has been determined in a diverse set of testing conditions and scale in situ in unconsolidated sediments. The particle size can be adapted to the requirements given by the natural aquifer system by milling the activated charcoal (ACC) carrier to the optimal grain size. In addition the iron structures are placed within the carbon grain, so that the properties of the external ACC surface are maintained and the carbon framework inhibits aggregation.

To stabilize the Carbo-Iron® slurry, carboxy methyl cellulose (CMC) is added. CMC is a macromolecule that sorbs with a maximum load of 7% by mass to the Carbo-Iron® particle surface, coats the particle and keeps it in suspension and thus mobile in the aquifer. Decreasing CMC concentration leads to increasing particle sedimentation, and for $c_{CMC} < 5\%$ by mass, significant sedimentation is observed. In natural systems, the loss of CMC after injection either by desorption or biodegradation results in destabilization and retardation of particle spreading. By appropriate design of the CMC particle wrapping to control hydrophobicity and steric stabilization, the travel behaviour can be controlled and accustomed to the remediation needs for contaminated source or plume areas. Based on lab investigations, the CMC loss and thus particle fixing happens within about 10 days. Furthermore travel distances of Carbo-Iron® depend on the content of carbonate in groundwater: travel distances are lower in hard water.

Generally depending on the local permeability and hydrodynamics of the aquifer, horizontal particle travel distances on the order of 10 m and vertical spreading of 25 m for mixed-grained sands with embedded fine sand and silt layers (average permeability coefficient 10^{-4} m/s) had been determined by field measurements. The travel behaviour additionally can be improved by (repeated) rinsing the slurry injection points with CMC-containing water. In summary, Carbo-Iron® is significantly more mobile compared to conventional NZVI particles.

The travel behaviour of the newly applied iron oxide particle intraBlue® is also known from a diverse set of testing in unconsolidated sediments. Based on cascading column tests, particle travel distances on the order of 5 m horizontal and 20 m vertical spreading depending on the aquifer conditions can be expected. However, the first field findings indicated 3 m spreading in mixed sands. After injection intraBlue® stayed within the injection zone. Hydraulic remobilization of deposited particles is quite unlikely under natural conditions. However, sampling of groundwater from multilevel system wells (CMT) showed that particles located near the sampling port could be mobilized due to the higher flow velocity at the sampling port.

8.6 Particle Injection: Theoretical NZVI Mass Requirements and Injection Techniques

Particle mass calculation is a basic requirement to reach a given remediation target. Along with developing a concise conceptual site model and consideration of the important geological and hydrogeological conditions at the site, a critical step in evaluating the technological feasibility of using nanoparticles as a remedial technology is the determination of the required particle dosage. This determination relies on knowledge and evaluation of the site geochemistry and has broad implications for the effectiveness of the remedy and the overall remediation cost.

The mass of NZVI necessary for treatment can be estimated by calculating the stoichiometric natural reductant demand (NRD), i.e. the concentration of electrons (e^-) necessary (donated from NZVI) to overcome the electron demand from naturally occurring redox species in the system. These naturally occurring species include, but are not limited to, iron (Fe), manganese (Mn), nitrate (NO^{3-}) and dissolved oxygen (DO).

These types of calculations require standard water quality data, knowledge of the approximate treatment zone volume, hydraulic parameters including conductivity (k), hydraulic gradient (i) and total porosity (n_t) and contaminant data. The stoichiometric electron demand provided by NZVI is assumed to be two electron equivalents per mole (e-equiv/mol), although a treatment efficiency factor is typically used to address the fact that the transfer of electrons from NZVI is inefficient.

It is critical to determine the total NRD compared to the electron demand from contaminants to evaluate the cost impact of side reactions. However, there may be occasions where the express purpose of the remedial action is to radically change the system's geochemistry and this can be achieved using NZVI.

An example of an injection condition where NZVI can be effectively placed in the subsurface is a condition where the aquifer is predominantly sand and gravel with limited silt and no clay. Under these conditions, large quantities of particle material can be injected and potentially overcome NRD concerns. With modifications considering the individual physico-chemical properties and travel behaviour of other materials, the method is applicable for any particle to be injected.

8.6.1 Suspension Preparation

The particle slurry as an important part of the remediation chain has to be prepared and handled carefully, especially stabilization against agglomeration is critical.

For practical reasons the preparation should be done on site directly before injection using a mobile injection platform to perform the injection fast and reach the particular injection point of the array smoothly. Based on the calculated particle mass and slurry particle concentration, the total volume of slurry water for injection is calculated. Additional water requirements may be needed if the injection is followed with chase water. Normally the slurry is prepared by dispersion of particles into deoxygenized, drinking water containing the stabilizer using a high-speed industry disperser forming a 10–20 g particles/L suspension. The stabilizer concentration (CMC, humins, poly(acrylic acid) (PAA), others) is determined experimentally in laboratory columns and has to be carefully controlled. Particularly for Carbo-Iron®, slight differences in the stabilizer/particle ratio may result in significant differences of the particle migration. Depending on the physico-chemical properties of the particle used (size, density, surface charge), mechanical mixers within the tanks may be needed to prevent agglomeration and settling of the slurry during the pre-injection phase.

As a final step directly before injection, the slurry should be homogenized again (Fig. 8.7). To prevent consumption of the zero-valent iron in the particles, it is crucial to keep the injection equipment during the entire preparation and injection process under deoxygenized conditions. Normally nitrogen is used which can easily applied by aerators within the slurry tanks and disperser unit.

To check the slurry quality, physico-chemical parameters (particle specific: oxygen content, pH, temperature, redox potential, conductivity) should be measured. Oxygen consumes zero-valent iron; consequently it should be as low as possible in the slurry. Under field conditions, an oxygen content of 0.1 mg/L is acceptable. To control the technical injection process, flow velocity and injection pressure should be regulated considering the local geohydraulic conditions as described next. To prevent sediment wash out, hydraulic shear failure occurrence and tunnel erosion in soil and loose rock aquifers and a hydraulic displacement of the contamination, the injection pressure should not strongly exceed the local hydrostatic pressure, and the flow velocity should be adjusted to the capacity given by the natural permeability. Also the frame conditions of the injection method (e.g. direct-push injection, infiltration via groundwater wells with or without packer sealing, drainage infiltration) have to be considered to achieve an even particle distribution over the contaminated area.

8.6.2 Injection Methods

The injection methods of slurry vary depending on the geological and hydrogeological site conditions and the contamination distribution. The methods described below have been developed for NZVI slurry injection. However, with

Fig. 8.7 Example of slurry field mixing and injection equipment (Doose and Bruns 2014). The NZVI slurry is a mixture of water with NZVI particles typically up to 10–20 g/L of NZVI powder mixed with surface modifiers to maintain NZVI particles in suspension and prevent NZVI agglomeration, and additional catalysts (e.g. palladium acetate) to accelerate contaminant reduction or to reduce recalcitrant constituents

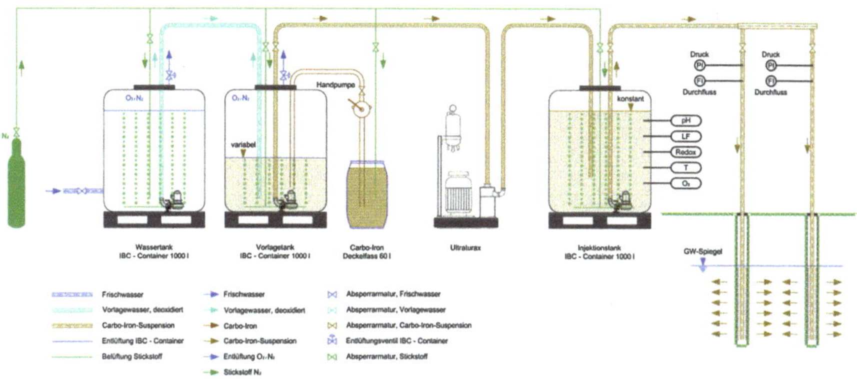

Fig. 8.8 Flow chart of an injection equipment recently used to slurry Carbo-Iron® under anoxic conditions and inject it into an aquifer modified according to (Doose and Bruns 2014)

slight modifications which consider the specifications of other particle types, the methods are generally applicable. Aquifers in different types of subsurface as unconsolidated sediments and consolidated fractured bedrock require different injection methods.

Figure 8.8 shows a flow chart of a typical injection equipment as recently used for on-site preparation of Carbo-Iron® particle slurry and injection.

Depending on the technical specifications on site, contamination distribution, hydraulics and hydrodynamics, the goal of the remediation (e.g. source or plume remediation) and any other requirement of CSM or remediation concept, there are diverse methods used to bring the particles into the subsurface. A few of them are:

- Infiltration without pressure via full-screened wells
- Injection with variable pressure in wells with/without packer use

- Partly screened wells
- Surface infiltration via drainage pipes
- Direct push and fracking systems
- Injection starting upgradient/downgradient
- Array injection in source areas
- Time interval injection with ring injection to fence source and prevent contamination spreading during central source injections

An inappropriate injection technique may result in hydraulic pollution displacement and may cause shear failure and tunnel/soil piping erosion within unconsolidated sediments. In any case, methods and techniques have to be customized to the site requirements. If necessary, test injections should be performed to fine-tune injection techniques and equipment operation.

Dissemination of NZVI in the subsurface can be achieved by multiple methods including mixing NZVI or coarser ZVI powders in an excavation backfill as a polishing step following hotspot source excavation. However, the following includes references to NZVI slurry gravity feed, pressurized injections and pumping and reinjection of closed loops using wells or possibly recharge galleries (i.e. trenches). Other slurry injection techniques include hydraulic soil fracturing and injection of NZVI/ZVI sand mixtures. Experiments have also been conducted using electrokinetic approaches to disseminate NZVI particles in low permeable materials (i.e. silt and clays).

Injection Methods for Unconsolidated Sediments This section presents the typical slurry injection methods for various geologic materials including gravels and coarse sands and finer grain sediments. In addition, slurry injections in loess are also discussed.

Injections in Gravels and Coarse Sands The slurry can be readily injected in well-sorted gravels and coarse sands using injection wells. The injection technique is typically gravity flow. Pressurized injections are not needed because the injection flow rates are sufficiently high to maintain a reduced time needed to deliver in the subsurface the designed particle quantities in a cost-effective manner.

Injections in Fine Grain Sediments (i.e. Medium to Fine Sands, Silts and Clays) Injections in finer grain sediments become progressively more difficult with the decrease in particle size. Moreover, the hydraulic conductivity following NZVI injections in medium to fine sands is generally lower by approximately one order of magnitude. As a result, pressurized injections using direct-push drilling and slurry delivery techniques are recommended in this instance. Pressurized injection should account for depth of the screened interval of the well and should not exceed pressures calculated as depth to the top of the well screen measured in feet multiplied by 1 pound per square inch (PSI). One PSI per foot is the approximate lithostatic pressure or vertical stress of a soil column of 1 foot. To better focus the particle emplacement, membrane interface probe (MIP) investigation prior to injections is recommended to focus the treatment to intervals with high concentrations of contaminants or slight increase in hydraulic conductivity. In this way the main contaminant mass or the slightly higher transmissivity of contaminant migration pathways is addressed.

Injections in Loess A particular case of unconsolidated sediments is the loess which is a clastic, predominantly silt-sized sediment that is formed by the accumulation of wind-blown dust. It is loosely cemented by calcium carbonate. Important for groundwater and contaminant migration pathways are properties typical to loess materials including:

- The loess is highly porous and is traversed by vertical capillaries that permit the sediment to fracture and readily transmit water/contaminants vertically downward.
- The loess includes palaeosol layers, typically impermeable allowing the development of perched water saturation.
- Within loess there are layers of calcium carbonate concretions (i.e. loess kindchen) with typical higher horizontal hydraulic conductivity allowing lateral water/contaminant movement.

The presence of these discontinuities typically results in a stair-step-like distribution of contaminants that migrate downward from a release, and then the lateral migration is enhanced by the presence of calcium carbonate concretions (loess kindchen), while the vertical downward migration is interrupted by the palaeosols that may not be continuous and the cycle of vertical lateral migration is repeated.

MIP investigations of loess layers provide a good three-dimensional distribution of contaminants and vertical heterogeneities. MIP investigation is a recommended first step prior to NZVI or any other in situ injection treatment of groundwater present within loess materials. Slurry injections targeted for the permeable concretion layers may be conducted using gravity flow or low level pressure injections using dedicated injection wells. For the remainder of the loess mass, dedicated injection well may also be used if sufficient vertical fractures are intercepted by the injection well; however, targeted closer spaced direct-push pressurized injections are more feasible in this instance.

Injection Methods for Fractured Bedrock Slurry injection methods in fractured bedrock include gravity or pressurized injections. Open bedrock wells are preferred to allow well rehabilitation especially in context with NZVI injection over time because they can be aggressively redeveloped or hydraulically fractured. Gravity flow via injection wells is possible in solution-enhanced fractures or karst conduits. Dissemination of slurries in fracture bedrock requires in most cases pressurized injections.

Enhancing Delivery with Hydraulic Fracturing Hydraulic fracturing can be used to initiate new fractures in a bedrock aquifer to increase the transmissivity and enhance the ability to inject reagents into the subsurface. A typical hydraulic fracturing programme consists of sealing off a short segment (1–10 ft) of a borehole at a desired depth (using inflatable packers), injecting fluid (typically water) into the isolated zone at a sufficient rate to raise the hydraulic pressure rapidly and bring

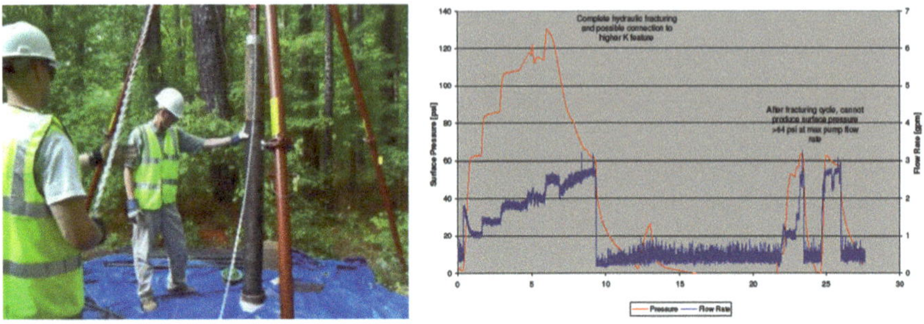

Fig. 8.9 Field implementation and monitoring of hydraulic fracturing of bedrock prior to NZVI injection (Gheorghiu et al. 2010)

about hydraulic fracturing of the borehole wall. Hydraulic fracturing occurs when the fluid pressure in the isolated portion of the borehole reaches a critical level, called the breakdown pressure (Pc). At this breakdown pressure, the rock fractures causing hydraulic fluid loss and a drop in pressure (Fig. 8.9).

8.7 Monitoring

The monitoring of geohydraulic, physico-chemical, chemical and microbiological data before, during and after particle injection provides information on subsurface conditions and its changes induced by particle effects. The variations in groundwater, soil vapour and soil and changes of microorganism communities have to be traced and recorded. In the course of evaluating spatial and temporal effects of the particle injection, seasonal changes of the natural system (e.g. groundwater hydraulics and chemistry) have to be considered.

The number of monitoring points, distribution network and construction as well as sampling procedures has an important influence on obtaining quality monitoring data. The most important compartment for monitoring measures is the groundwater; thus all groundwater wells have to be located in locations and layers that are affected by the contamination of NZVI injection and installed and equipped considering geological and hydrogeological conditions and requirements of the conceptual site model. Also soil vapour and soil has to be part of the monitoring chain. Coupled groundwater/soil vapour wells could provide balanced state evaluations for these two compartments. A complete data set of hydraulic and chemical conditions which show a detailed picture of the initial state and changes during the whole remediation procedure is an essential basis for a comprehensive evaluation of contamination degradation. In addition it serves as input data for calibration and validation of the groundwater flow and contaminant transport model created during the site investigation phase. The modelling results enable a detailed forecast of the environmental development on site.

The monitoring key dates should be scheduled flexible, and subsequent monitoring measures should be adjusted accordingly: observation periods and sampling events have to cover the needs given by injection deadlines and corresponding lead time periods. Furthermore acquisition of data fluctuation ranges (background) has to be ensured before particle injection. Considering the extent of the remediation area, groundwater and soil vapour sampling campaigns with increasing time intervals after baseline and time zero sampling may range from a few days up to some month, based on aquifer conditions and hydrodynamics on site.

The spatial acquisition of chemical and hydraulic data needs groundwater wells with sampling ports covering all relevant layers down to the aquifer basis. Standard multilevel wells and CMT wells allow these measurements. To provide complete documentation, physico-chemical data like pH, conductivity, oxygen concentration, redox potential, temperature and water table level should be recorded using multiparameter probes with adjustable scanning frequency, so that the data density can be increased when needed, e.g. in hydraulically interesting phases or during the injection process. In complex aquifers, in situ measurements of groundwater flow directions over depth could help with measurement planning or interpretation of findings. These must be taken at the correct times as significant variations in groundwater flow within an aquifer are often based on geologic structures, as well as local precipitation influences, river management, groundwater extraction, etc., in the experience of the authors.

In general monitoring must cover all relevant aspects to evaluate the state and results of the remediation, but considering cost aspects, the extent should not exceed the actual needs. Specifics of the analyses spectrum have to be derived and fixed based on the results of the detailed site investigation as described earlier.

8.8 Technology Improvements

From the point of view of field practitioners, some technology improvements are needed to improve the technical performance of NZVI and nanoparticle-based remediation technologies. Although the newly developed particles like the composite material Carbo-Iron® and iron oxide particles like intraBlue® are easier to manage than NZVI, i.e. they show reasonable travel behaviour and contain no questionable additives, the reactivity and injection technologies have to be customized to specific field site requirements. For Carbo-Iron® knowledge about the effect on and support of on-site bioactivity have to be deepened.

For NZVI several improvements are needed. There is a need for more effective particle and slurry additives, especially ones that can further exploit the synergies between particles and biological treatments. Development of procedures to mechanically enhance NZVI mobility, potential use of electrokinetics and measures that will lower overall remedy cost is needed.

The current use of additives to enhance the mobility of NZVI particles in the subsurface represents a facet of the technology where significant advancements may still exist. Although, poly(acrylic acid) (PAA) and soy-based products have been approved and used in the United States and in Canada, progress should be made to develop better surface modifiers with even less impact to the natural system. Future development of "green" polymers may advance this ability to improve the mobility of NZVI and achieve better contaminant targeting in the subsurface with more sustainable technology. This also would support regulator acceptance and licensability in Europe and other parts of the world. These advancements will also be strongly tied to the further development of coupling NZVI with enhanced bioremediation as the additives used to enhance mobility also represent soluble forms of carbon that can enhance bioremediation.

Along with the use of additives, mechanically enhancing mobility shows a great deal of promise for the further development of this technology. Golder has used hydraulic fracturing to enhance mobility in bedrock; however, the use of fracturing technologies in unconsolidated sediments may represent a potential opportunity to enhance the dispersion of NZVI in a number of systems.

As mentioned previously, further research on larger-sized ZVI may change the paradigm that smaller particles are necessary for environmental applications. Similar mobility and reactivity may be achieved with larger particle size materials (~500 nm to 1 μm) with the continued use of surface modifiers and catalysts resulting in significantly lowered costs and potentially eliminating perceived risk.

Although NZVI has shown significant progress on sites with unconsolidated sediments, it has performed well in formations that trend towards larger particle size (i.e. gravel and sand). Typically, in situ treatments have suffered from limited applicability in complex geology and also where nonaqueous phase liquid (NAPL) is present. A potential remediation option that has shown promise is the use of electrokinetics to deliver NZVI into DNAPL and dissolved phase contaminant zones with low transmissivity. The electrokinetic movement of particles can overcome the limitations of conventional injections and physical flow to achieve rapid and uniform contact of NZVI with targeted compounds.

Finally, it is important to note that during the early development of the nano-remediation technology, a number of myths became widely prevalent and continue to influence the opinion of regulatory agencies and policy-makers. These include the concept that nanoscale materials travel indefinitely in groundwater and can treat elevated concentrations of contaminants over highly dispersed areas. This provoked many people to demand more information on the fate and transport of particles and created a great deal of interest in the human health and environmental risk of using this technology. Research and field observations have demonstrated the relatively low mobility of these particles in the subsurface, and therefore the relatively low potential for unwanted migration off site.

Over all the years of applying NZVI, one of the most critical design issues of this technology is particle mobility. In fact, as stated above, a number of additives have been used to enhance the mobility of particles to achieve even modest dispersion in the subsurface. The future of this technology is strongly tied to the ability of researchers and professionals applying this technology, to continue to better understand and report information observed during NZVI work. It is also important that this technology finds its place as a remediation tool which under the appropriate site conditions can be highly successful as a treatment technology.

However, this is not the remediation panacea that it was first thought to be and is not applicable at all VCHC-impacted sites. For all nanoparticle-based approaches, a thorough conceptual site model must be developed, and all geologic, hydrogeologic, geochemical and microbiological information must be considered to properly evaluate the feasibility of this technology. Finally, rigorous field-scale pilot testing must be performed to determine the applicability of the technology on a small scale under actual site conditions. Only then can a successful full-scale application be designed and implemented.

References

Bleyl, S. (2014). *Einsatz von Nanomaterialien im Grundwasser – Chancen & Risiken, FachDialog 2 Nanotechnologie und aquatische Umwelt*. Berlin.

Borda, M. J., Venkatakrishnan, R., & Gheorghiu, F. (2009). Status of NZVI technology lessons learned from North America and international implementations. In C. L. Geiger & K. M. Carvalho-Knighton (Eds.), *Environmental applications of nanoscale and microscale reactive metal particles* (pp. 219–232). New York: American Chemical Society.

Bruns, J. (2013). NanoIron for in situ remediation. In: *International workshop: NanoEcotoxEthics, responsible nanoremediation?*. Oslo.

Doose, H., & Bruns, J. (2014). *Fe-NANOSIT, Nanopartikel und Nanokomposit¬strukturen zur Schadstoffentfernung aus Grund- und Abwässern, Teilvorhaben*. Abschlussbericht: Technologische Lösungen zur Grundwassersanierung.

Gheorghiu, F., & Plant, S. (2010). *Nano scale zero-valent iron for site remediation*. Brownfield Briefing, Environmental Analyst, Shrewsbury, United Kingdom.

Gheorghiu, F., Borda, M.J., & Plant, S. (2010). *Remediating at nano scale*. Brownfield Briefing, Remediation Solutions, Environmental Analyst, Shrewsbury, United Kingdom.

Hains, S. (2008). *Implementation of nZVI reactive zone for the treatment of TCE in a deep aquifer*.

Höss, S., Fritzsche, A., Meyer, C., Bosch, J., Meckenstock, R. U., & Totsche, K. U. (2015). Size- and composition-dependent toxicity of synthetic and soil-derived Fe oxide colloids for the nematode Caenorhabditis elegans. *Environmental Science & Technology, 49*, 544–552.

Kirschling, T., Gregory, K., Minkley, N., Lowry, G., & Tilton, R. (2010). Impact of nanoscale zero valent iron on geochemistry and microbial populations. *Environmental Science & Technology, 44*, 3474–3480.

Macé, C., Gheorghiu, F., Desrocher, S., Kane, A., Pupeza, M., Cernik, M., Kvapil, P., Venkatakrishnan, R., & Zhang, W.-X. (2006). Nanotechnology and groundwater remediation, a step forward in technology understanding. *Remediation Journal, 16*, 23–33.

Mackenzie, K., Bleyl, S., Kopinke, F. -D., Doose, H., & Bruns, J. (2014). *Proceedings 9th international conference on chlorinated and recalcitrant compounds*. Battelle, Monterey.

Chapter 9
Sulfide-Modified NZVI (S-NZVI): Synthesis, Characterization, and Reactivity

Yiming Su, Gregory V. Lowry, David Jassby, and Yalei Zhang

Abstract Sulfide-modified nanoscale zerovalent iron (S-NZVI) is attracting more and more attention due to its ease of production, improved reactivity with various pollutants (e.g., trichloroethene, diclofenac, cadmium, chromate), and most importantly, the selectivity to pollutants over water. Although there are some microstructural differences between nanoparticles derived from one-pot and two-step synthesis methods, with optimal S/Fe molar ratio (during preparation), both types of S-NZVI can achieve much higher pollutant removal efficacy than unmodified NZVI. For dechlorination, sulfidation not only inhibits the reaction between Fe (0) and H_2O but creates a nucleophilic zone on the particle surface which is favorable for β-elimination. The latter change endows S-NZVI with capacity to degrade particular pollutants which previously cannot be removed by NZVI. For metal ion removal, besides the increased metal removal capacity, the chemical stability of metal-NZVI is also enhanced through sulfidation. Further, sulfidation is beneficial to heterogeneous Fenton-like reactions. With the presence of dissolved oxygen, S-NZVI generates much more hydroxyl radicals for pollutant degradation through a one-electron transfer pathway than NZVI. Although results from lab-scale studies are very encouraging, there is still lack of pilot-scale field test demonstrating the efficacy of S-NZVI in the field. Fate and transport of S-NZVI in subsurface and how S-NZVI (with pollutants) affects microbial community are still largely missing.

Keywords Nanoscale zerovalent iron · Sulfidation · Selectivity · Synthesis · Reactivity · Particle life-time

Y. Su (✉) · D. Jassby
Department of Civil and Environmental Engineering, University of California, Los Angeles, CA, USA
e-mail: yimingsu88@ucla.edu

G. V. Lowry
Civil & Environmental Engineering, Carnegie Mellon University, Pittsburgh, PA, USA

Y. Zhang
State Key Laboratory of Pollution Control and Resource Reuse, Tongji University, Shanghai, China

T. Phenrat, G. V. Lowry (eds.), *Nanoscale Zerovalent Iron Particles for Environmental Restoration*, https://doi.org/10.1007/978-3-319-95340-3_9

9.1 Introduction

9.1.1 History of FeS as a Reductant

Iron sulfides comprise a varied group of minerals, e.g., iron monosulfide (mackinawite, $Fe_{(0.91-1.15)}S$), greigite (Fe_3S_4), pyrite (FeS_2), smythite (rhombohedral Fe_9S_{11}), marcasite (orthorhombic dimorph of pyrite), and cubic FeS_c (Rickard and Luther 2007). This group of iron sulfides are ubiquitous in the environment, especially under sulfate-reducing conditions. Among this group, mackinawite (amorphous FeS) and greigite are generally considered as the main components of acid volatile sulfide minerals in sediments and as the precursors to pyrite formation (Morse and Arakaki 1993). They have been shown to be important in metal immobilization (e.g., mercury, chromium, arsenic, uranium, and neptunium) (Bostick and Fendorf 2003; Jeong et al. 2007b; Moyes et al. 2002; Özverdi and Erdem 2006; Shao et al. 2016) and organic contaminant degradation (e.g., chlorinated organic compounds, nitroaromatic compounds, polychlorinated biphenyls) (Butler and Hayes 1998, 1999; Jeong et al. 2007a, b; Kriegman-King and Reinhard 1994; Lee and Batchelor 2002). In addition, they control the bioavailability of these metals in many environments (Chapman et al. 1998; Morse 1994; Suzuki 2001).

The coexistence of heavy metals (other than iron) in FeS minerals directed Morese and Arakaki to study the reactivity of FeS with other metals. They assessed the ability of mackinawite to remove divalent metals from water in 1993 (Morse and Arakaki 1993). Since then, the interaction between arsenite, mercury, and other metals with iron sulfides, particularly mackinawite and pyrite, has been widely investigated (Bostick and Fendorf 2003; Jeong et al. 2007a, b; Özverdi and Erdem 2006), and the broad spectrum reactivity of FeS minerals shows great promise for iron sulfides in remediating heavy metal pollution.

The other important application of iron sulfides in pollutant removal is the dechlorination of chlorinated organic contaminants. Both mackinawite and amorphous FeS are highly reactive with chlorinated organic compounds. Tetrachloroethene (PCE) is mainly reduced by mackinawite to acetylene via β-elimination (dominating reaction), to trichloroethene (TCE) via hydrogenolysis, and to 1,1-dichloroethene (DCE) via α-elimination; TCE transforms to acetylene also via β-elimination (dominating reaction) and to DCE via hydrogenolysis (Jeong et al. 2007a, b). However, while acetylene still could undergo further transformation, transformation of DCE was not observed (Jeong et al. 2007a, b). Similarly, 1,1-dichloroethane and 1,2-dichloroethane (1,1-DCA and 1,2-DCA) and 1,1,2-trichloroethane (1,1,2-TCA) showed no appreciable transformation by FeS, while pentachloroethane (PCA), tetrachloroethane (TeCA), 1,1,1-TCA, and carbon tetrachloride were readily transformed by FeS (Butler and Hayes 2000). The dechlorination rate of these reactions was demonstrated to be strongly pH dependent (Butler and Hayes 1998, 2001); over the pH range of 7.1–9.5, the dechlorination rate increased with increasing pH (Butler and Hayes 1998). This enhancement has been attributed to the acceleration of the deprotonation process at high pH

(Prince and Dutton 1976). Above pH 8, it is very likely that FeS has to release one additional proton upon oxidation (Eq. 9.1); below pH 8, oxidation is associated with an electron release (Eq. 9.2) (Prince and Dutton 1976).

$$FeS(H)_{reduced} \rightarrow FeS_{oxidized} + e^- + H^+ \quad pH > 8 \tag{9.1}$$

$$FeS(H)_{reduced} \rightarrow FeS(H^+)_{oxidized} + e^- \quad pH < 8 \tag{9.2}$$

While the dechlorination pathways during reaction with FeS are relatively well delineated, the underlying mechanism of the reaction is still under debate. It is widely accepted that the reducing capability is related to the separation of iron *d* orbitals and sulfur orbitals. However, how the electron transfers from FeS (Fe (II) or S(-II)) to the adsorbed contaminant is not clear. Both surface-bound Fe (II) (Lee and Batchelor 2002) and S(-II) (Kriegman-King and Reinhard 1992) were found to be able to catalyze the dechlorination reaction. Further, there is a weak electron transfer from the metal to the sulfur atom (Oudar 1980). This electron transfer is related to the atomic density of the plane, which means comparing dechlorination performance after adding S(-II) or Fe(II) into the system cannot truly reflect the source of electrons (Oudar 1980).

Compared to the reducing capacity of nanoscale zerovalent iron (NZVI) and Fe (II) hydroxide, FeS suffers from a lack of available electrons at the same metal dosage. However, while NZVI is highly reactive, it lacks the selectivity for pollutants, with the majority of NZVI consumed by reactions with water and oxygen, which greatly decreases the reaction efficiency (Eq. 9.3) (Liu et al. 2013) of NZVI. Despite the relatively lower dechlorination rates afforded by FeS compared to NZVI, the low reactivity of FeS with numerous nontarget ionic and organic species in groundwater (Butler and Hayes 1998) suggests that FeS can have a higher *E* value than NZVI. FeS-mediated reductive dechlorination could also be an important attenuation process in natural systems. Thus, combining the advantages of FeS and NZVI into a single particle could produce a better material for environmental remediation.

$$E = \frac{N}{M} \times 100\% \tag{9.3}$$

N: the number of the electrons used for pollutants reduction
M: the number of entire electrons consumed

9.1.2 Development of Sulfide-Modified NZVI

The ability of sulfide treatment to enhance the reactivity of zerovalent iron toward chlorinated compounds was first described by Butler and Hayes (2001). The dechlorination rate of TCE by micron-sized Fe(0) was enhanced by pre-treatment of the

Table 9.1 Rate constants for transformation of TCE by 100 g/L iron metal (Butler and Hayes 2001)

Condition	K_{obs} (h^{-1})
Fisher iron	3.49 (±0.50) × 10^{-3}
Fisher iron (1 mM NaHS)	6.50 (±0.59) × 10^{-3}
Peerless iron	1.77 (±0.51) × 10^{-3}
Peerless iron (1 mM NaHS)	3.78 (±0.89) × 10^{-3}

iron with a 1 mM NaHS solution. The reactivity of Fe(0) reported by Peerless and Fischer was increased by about a factor of two after treatment with sulfide (Table 9.1). In 2011, Kim et al. reported the synthesis of FeS/Fe nanoparticles through a facile one-pot method, with the TCE reduction rate significantly improved after sulfidation compared to the NZVI alone (Kim et al. 2011). However, the authors did not provide a detailed explanation for the observed enhancement. The underlying mechanism for the increased dechlorination rate was reported by Fan et al. (2016); through sulfidation, the reaction between S-NZVI and H_2O was inhibited relative to NZVI, while the reactivity of S-NZVI with TCE remained high. Thus, sulfidation appears to have increased the selectivity of the reaction toward TCE over water, relative to NZVI, which leads to the observed increase in *E*.

9.1.3 Summary of Recent Findings on Reactivity of S-NZVI

Since the first report of S-NZVI in 2001, there have been additional studies, including nine studies on 2016 and four studies in 2017 (as of April), suggesting a significant interest in this material, largely due to the promise of enhanced selectivity toward contaminants relative to NZVI (Fig. 9.1). Besides chlorinated carbon compounds (typically TCE), the reactivity of S-NZVI toward different types of contaminants has been assessed, e.g., pertechnetate (Fan et al. 2013), chromate (Du et al. 2016), cadmium (Su et al. 2015), *p*-nitrophenol (PNP) (Tang et al. 2016), tetrabromobisphenol A (TBBPA) (Li et al. 2016), azo dye (Xu et al. 2016), rhodamine B (Zhang et al. 2016), and diclofenac (DCF) (Song et al. 2017). In general, the reactivity of S-NZVI with organic pollutants increased with increasing sulfidation. However, a maximum is most often observed, with increased sulfidation leading to a decline in reactivity as the particles become more FeS and less NZVI. The optimal Fe/S ratio depends to some degree on the identity of the target contaminant. Furthermore, compared to NZVI, the reactivity of S-NZVI is more robust against effects from common ions/anions in groundwater. However, doping different trace metals can result in different reactivity of S-NZVI, e.g., Mn^{2+} and Cu^{2+} decrease TCE reduction rates, while Pd^{2+}, Co^{2+}, and Ni^{2+} increase them, suggesting both hydrogen activation and electron transfer play important roles in TCE reduction (Kim et al. 2014).

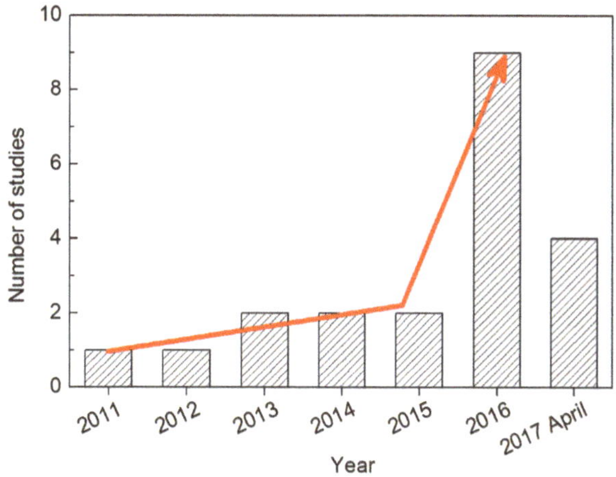

Fig. 9.1 Number of published papers on S-NZVI since 2011

9.2 Synthesis Methods

As previously described, introducing iron sulfide (typically FeS) into NZVI has proven to have many advantages, especially increased reaction selectivity over water and reactivity enhancement compared to that of NZVI. Two primary synthesis methods have emerged for sulfidation of NZVI: a one-pot method and a two-step method. However, these two methods result in slightly different S-NZVI. Ideally, a core-shell structure, comprising of Fe(0) core covered by a FeS shell, is preferred. This type of structure protects the NZVI core from reaction with water, with the FeS shell being reactive toward contaminants. Unfortunately, neither of these methods appear to afford this ideal structure, and so far, deduced from the results of different studies (Han and Yan 2016; Su et al. 2015; Tang et al. 2016), a mixture shell composed of iron oxide and iron sulfide is often observed. The one-pot method results in the S-NZVI containing other types of iron sulfides such as Fe_3S_4 or iron polysulfide (Su et al. 2015), while the two-step method may form FeS out of the iron oxide shell (originally emerging from the surface oxidation of NZVI) and not directly out of the Fe(0) core (Fan et al. 2013). Although there are structural (and even compositional) differences between the S-NZVI made from these two methods, the reactivity is similar (Han and Yan 2016). Each of the methods is described in more detail below.

9.2.1 One-Pot Method

In this synthesis method, dissolved iron is reduced in the presence of a reduced sulfur species. Generally, a mixture of sodium borohydride and sodium dithionite (or thiosulfate) is used to reduce ferric iron or ferrous iron through titration (Kim et al. 2011). By varying the amount of sodium dithionite/thiosulfate to dissolved iron (S/Fe ratio), different degrees of sulfidation of NZVI can be obtained. The S/Fe molar ratio correlates with the degree of sulfidation. However, the actual S/Fe ratio in the synthesized nanoparticles is different because of side reactions (e.g., SO_3^{2-} and SO_4^{2-} formation) (Su et al. 2015). However, when S/Fe ratio is above ~0.2–0.3 (Gong et al. 2017; Su et al. 2015), this process becomes unreliable, and the synthesized nanomaterials are nonmagnetic. This change might be related to the molar ratio of sodium dithionite to sodium borohydride in the synthesis. Sodium borohydride directly reduces ferric/ferrous iron to Fe(0) (Su et al. 2016), while dithionite cannot (Ma et al. 2016). With the increasing addition of sodium dithionite, the iron reduction pathway is modified, and iron hydroxide, rather than Fe(0), forms first in the synthesis process (Su et al. 2016). Using a pure sodium dithionite solution (no borohydride) results in a mixture of iron oxide and iron sulfides (Ma et al. 2016). To obtain magnetic S-NZVI at high S/Fe ratio (above ~0.2–0.3), nano-seeding (i.e., dosing a very small amount of nano-SiO_2, nano-TiO_2, or nano-Al_2O_3) into the reductant solution has been demonstrated to be an effective strategy (Su et al. 2016). In addition, while not common, it is possible to synthesize S-NZVI by ball milling iron in the presence of added sulfide (Li et al. 2009).

9.2.2 Two-Step Method

The two-step method involves two separate processes: NZVI synthesis and subsequent sulfidation. Generally, NZVI is produced via borohydride reduction, followed by sulfidation through mixing the NZVI suspension with a Na_2S solution under sonication (Fan et al. 2013). S-NZVI with different S/Fe molar ratios can be synthesized by altering the NZVI to S ratio. In addition, there are two slightly different sulfidation methods that are used; one uses sodium dithionite or sodium thiosulfate instead of Na_2S (Han and Yan 2016); the other is adding both ferrous salt and Na_2S into a NZVI suspension (Du et al. 2016). Although this two-step method transforms some NZVI to FeS, it is difficult to generate a homogeneous shell of FeS around the NZVI core, especially at low S/Fe ratios. Instead of forming an FeS shell, it is likely that FeS islands grow on the surface. Unlike the one-pot method, the two-step method should (in theory) produce highly sulfidated NZVI. However, when only Na_2S was added as the sulfidation species, and the S/Fe ratio increased from 0.112 to 1.12, the FeS content in S-NZVI, as shown by Mössbauer spectroscopy analysis, remained relatively stable (8%, FeS; 92% Fe(0)) (Fan et al. 2013). This might be due to the lack of surface Fe(II) species, and the formation of FeS prevents the further reaction between S^{2-} and iron.

9.3 Physical and Chemical Properties of S-NZVI

9.3.1 Morphology of S-NZVI

Typically, NZVI has an Fe(0) core covered by a shell of iron oxide, and its structure has been studied at the atomic scale (Ling and Zhang 2014a, b; Liu and Zhang 2014) (Fig. 9.2). However, for S-NZVI, to the best of our knowledge, there is no study focusing on the microstructure at the atomic scale. Detailed information, especially the 3-D distribution of S and O on nanoparticles, still remains unknown, although it is very important for explaining the different reaction mechanisms. Studies indicate that the two different synthesis methods could result in different microstructures. During the one-pot synthesis process, there are some side reactions (Su et al. 2015), which not only affect the formation of the Fe(0) core but also affect shell development (e.g., inhibiting FeOOH formation) and formation of flake-like structures

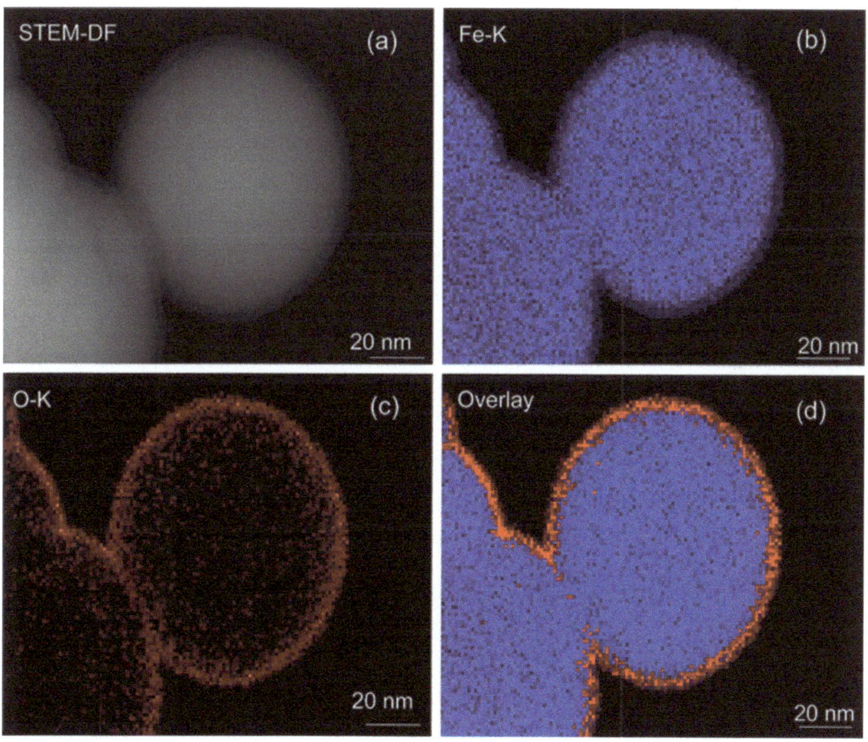

Fig. 9.2 NZVI images obtained by scanning transmission electron microscopy at high-angle annular dark-field model (**a**) and elemental mappings of Fe K_α (**b**) and O K_α (**c**) and their overlay (**d**) with energy-dispersive X-ray spectroscopy. (Reprinted with permission from Liu and Zhang (2014). Copyright (2014) Royal Society of Chemistry)

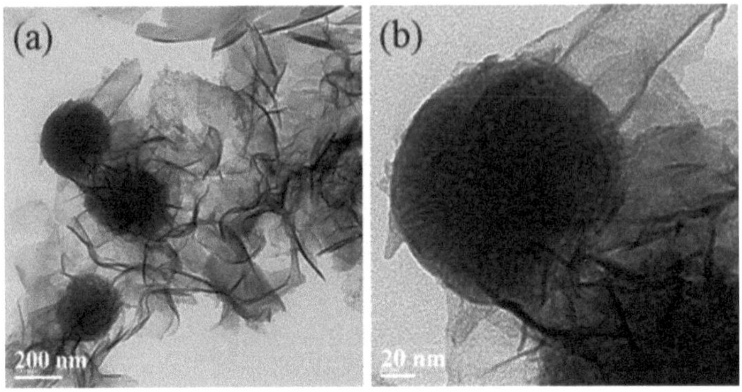

Fig. 9.3 S-NZVI obtained from a one-pot method (**a**) particle and flake-like structure (**b**) particle structure. (Reprinted with permission from Su et al. (2015). Copyright (2015) Elsevier)

(Fig. 9.3) (Song et al. 2017). Even though the same method was employed for S-NZVI synthesis, different S/Fe and $NaBH_4$/dithionite ratios can lead to different microstructures, e.g., change of shell structure (Fig. 9.4) (Su et al. 2015). This might be the reason why different microstructure of S-NZVI was observed in different studies.

With regard to the two-step method, it was reported that FeS was randomly located on the surface of NZVI instead of forming a homogeneous shell (Li et al. 2009). Similar to nanomaterials obtained through a one-pot method, different S/Fe ratios during synthesis also affect the original core-shell structure of S-NZVI: at a S/Fe ratio of 0.04, the shell was still identified, but at 0.08, it was not (Fig. 9.4). Notably, the flake-like structure was not observed in S-NZVI obtained from a two-step method. Another important issue remaining to be addressed is the variation of S distribution during the aging process. Given that newly adsorbed elements (such as U(VI) (Ling and Zhang 2015), Cr(VI) (Ling and Zhang 2014a), As(III, VI) (Ling and Zhang 2014b), Se(VI) (Xia et al. 2017)) on NZVI can diffuse from the surface to the interface between the core and shell, and even to the inner core, it is likely that S also undergoes intra-diffusion. Thus, the reaction time between NZVI and S(-II) during synthesis (two-step method, varying from 10 min (Rajajayavel and Ghoshal 2015) to 1 day (Fan et al. 2013)) needs to be considered for explaining the morphology and further reactivity difference of S-NZVI from different studies. In addition, S(-II) has a higher affinity to Fe than O(-II) (Song et al. 2017), which might affect the location of FeS even though iron oxide formed before sulfidation. It has been proven that the structural evolution of Pd-NZVI has a great impact on its reactivity. However, to date, there is still a lack of studies that systematically investigate the relationship between structural evolution and reactivity of S-NZVI.

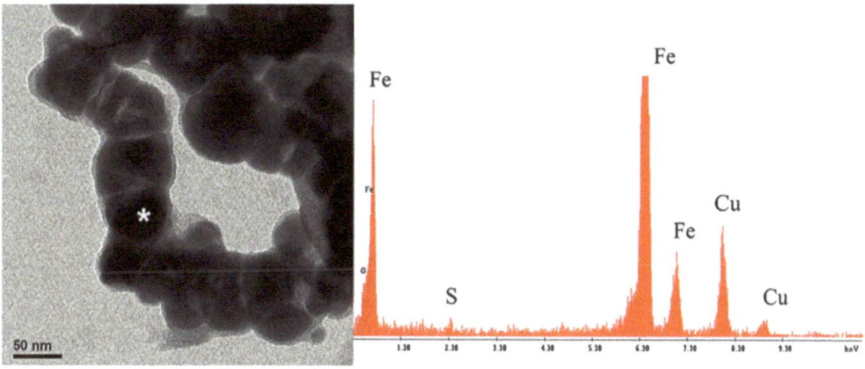

(a)

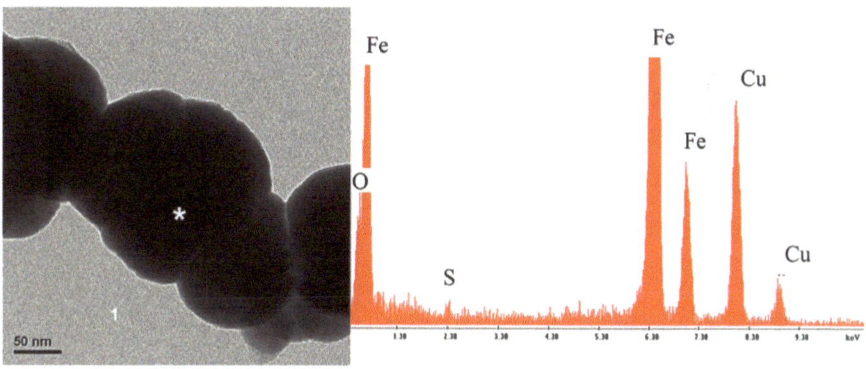

(b)

Fig. 9.4 S-NZVI obtained from a two-step method: (**a**) S/Fe = 0.04; (**b**) S/Fe = 0.08 (EDS image on right represents the selected area marked with * on left image). (Reprinted with permission from Rajajayavel and Ghoshal (2015). Copyright (2015) Elsevier)

9.3.2 Physical Properties

Sulfidation has been found to impact the size of nanoparticles. S-NZVI derived from both methods have larger sizes than pristine NZVI (Song et al. 2017), although the reason for this growth might be different. In the one-pot method, dithionite may inhibit the nucleation of Fe(0), which in turn contributes to the growth of fewer, but larger, Fe(0) nanoparticles (Su et al. 2016). In the two-step synthesis method, adding sulfide salt may stimulate further growth of NZVI. Although the size of the particles is larger than standard NZVI, due to the abundance of flake-like structures, S-NZVI obtained from the one-pot method have much higher surface area than NZVI (70.4 vs. 37.4 m^2/g (Su et al. 2015); 42.2 vs. 24.9 m^2/g (Kim et al. 2011)).

Since sulfidation imparts NZVI with FeS surface groups, the isoelectric point (IEP) of S-NZVI is different from NZVI. It was often reported that the IEP of S-NZVI was lower than NZVI (pH, ~8.0) as iron sulfides had low IEP, but the IEP values obtained from different studies vary, e.g., 4.2 (Kim et al. 2013), ~5 (Su et al. 2015), 5.8 (Tang et al. 2016), and 7.0 (Gong et al. 2017), which may result from the different extent of sulfidation and/or the different working conditions (such as different solution composition, different ionic strength) during measurement. In addition, sulfidation also lowers the saturation magnetization of NZVI (120 emu/g) to 103.1 emu/g (Kim et al. 2011), which might reduce the particle aggregation (Song et al. 2017).

9.3.3 Chemical Composition

The reactivity of S-NZVI has been shown to be dependent of the S/Fe ratio (Rajajayavel and Ghoshal 2015; Song et al. 2017; Su et al. 2015). Therefore, elemental analysis is needed to determine the S/Fe ratio and demonstrate that FeS forms on NZVI, especially for the S-NZVI harvested from the one-pot method. Normally, XRD, XPS, Mössbauer spectroscopy, and XANES are employed to study the composition of S-NZVI. Due to the poor crystalline structure (and potentially low content of FeS), XRD does not provide insightful information on FeS formation, but TEM with X-ray diffraction (at selected areas) are sometimes helpful (Fan et al. 2013). XPS can provide information on surface composition and is a useful method to identify FeS species on the surface. Based on XPS measurements, S-NZVI synthesized using the one-pot method had S_2^{2-}, S_n^{2-}, S^0, and even SO_3^{2-}/SO_4^{2-} surface groups (Su et al. 2015; Tang et al. 2016); for those from the two-step method, FeS is the only form of iron sulfide identified on surface, but it was reported that FeS_2 will form under high degrees of sulfidation (Rajajayavel and Ghoshal 2015). However, studies showed that FeS was not evenly distributed in the two-step method (Fan et al. 2013) while the one-pot method resulted in a more uniform distribution of sulfide groups (Su et al. 2015; Tang et al. 2016). In general, with respect to the composition of the shell, results from studies (Fan et al. 2014; Kim et al. 2011; Song et al. 2017; Su et al. 2015) indicate that a heterogeneous shell of iron oxide and iron sulfide is formed.

To investigate the composition of S-NZVI on a whole, XANES or Mössbauer spectroscopy have been successfully employed. While Mössbauer analysis showed that FeS content remained at approximately 8% even though S/Fe ratio increased from 0.112 to 1.12 in the two-step synthesis process (without adding Fe(II) during sulfidation) (Fan et al. 2013), the XANES analysis suggested that FeS increased from 0% to 11.3% when S/Fe ratio increased from 0 to 0.255 in one-pot process (Song et al. 2017). Under optimal conditions, iron oxide on S-NZVI could be greatly restricted (Song et al. 2017). In addition, in the one-pot method, a low extent of sulfidation is able to increase the content of Fe(0) (Song et al. 2017).

9.4 Reactivity of S-NZVI with Selected Environmental Contaminants

Although the reactivity of S-NZVI is largely dependent on S/Fe molar ratio, the optimal extent of sulfidation greatly enhances the reactivity with a range of environmental pollutants, such as halogenated carbon compounds (Table 9.2) and other recalcitrant organic pollutants (Table 9.3) and metals (Table 9.4). More importantly, sulfidation of NZVI appears to improve the selectivity of NZVI toward contaminants relative to water (Fan et al. 2016). This improved selectivity will likely make NZVI even more attractive as an in situ remediation agent if it indeed lengthens its reactive lifetime (Fan et al. 2016). It may also enhance the chemical stability of the adsorbed or reduced heavy metal species (as metal sulfides), making it more attractive for in situ stabilization of heavy metal contamination (Su et al. 2015).

9.4.1 Reactivity with Chlorinated Organic Contaminants

TCE and PCE are often detected in groundwater due to their widespread application in industrial and commercial processes and their recalcitrance to natural degradation processes. Compared to biological process (Field and Sierra-Alvarez 2004), in situ iron-based techniques, especially FeS (Jeong et al. 2007a, b) and NZVI (Liu et al. 2007), provide faster reaction rates. Although both FeS and NZVI can efficiently reduce chlorinated organic contaminants, the degradation mechanism is different. Taking TCE as an example, it is reduced to acetylene through β-elimination by FeS, with detailed transformation pathways available in the study of Jeong et al. (2007a). When using the unsulfidized NZVI, no acetylene was detected, and the detected reaction products were ethane, ethene, and C3–C6 hydrocarbons (with mass recovery of approximately 36%) (Han and Yan 2016). It has been suggested that the active H atom density on surface of FeS is extremely low (Benziger and Madix 1980) (which restricts the hydrogenolysis process), whereas the active H atom density on NZVI is high (Wang and Farrell 2003).

In the case of S-NZVI, the results so far suggest that it integrates the advantages of both FeS and NZVI. This may indicate that the S-NZVI formed is a not strictly Fe(0)/FeS core-shell product but rather is a heterogeneous mixture of both NZVI/Fe-oxide and FeS surfaces. For NZVI (Fig. 9.5a), having high H_{ad} atom density, more than 95% of the reducing equivalents are consumed by water-induced corrosion, although complete TCE reduction was also observed (Fan et al. 2016). This high reactivity of NZVI with water and its impact on the reactive lifetime of NZVI is a well-known limitation of the technology (Liu et al. 2005; Su et al. 2014a, b; Zhang et al. 2013). After sulfidation (with an initial S/Fe molar ratio above 0.3), water-induced corrosion was significantly inhibited, while TCE reduction was not, which greatly improves the electron efficiency. However, the dechlorination pathway was altered (Fig. 9.5b) (Fan et al. 2016). H_2 generation from the Fe(0) surface could be

Table 9.2 Summary of studies using S-NZVI with halogenated organic contaminants

Method	S/Fe molar ratio	Contaminant	Reaction condition	Key results	Explanation for enhanced reactivity over NZVI	References
One-pot	0–0.29	TCE	Anoxic	S-NZVI has better TCE degradation performance than NZVI. S-NZVI with a S/Fe molar ratio of 0.12 has the highest reaction rate	High electron conductivity of FeS; high affinity between TCE and FeS	Kim et al. (2011)
One-pot	0.102	TCE	Anoxic pH 6.3–9.0	For S-NZVI, TCE removal rate increased with increasing pH; Reactivity was unaffected by ionic strength; Ca^{2+} and Mg^{2+} increased the dechlorination rate, while humic acid inhibited the reaction	Higher electron transfer at high pH levels	Kim et al. (2013)
Two-step	0.02–0.091	TCE	Anoxic pH 7.0–11.0	S/Fe ratios in the range of 0.04–0.08 provided the highest TCE dechlorination rates, and rates decreased at both higher and lower S/Fe	Greater electron transfer through FeS layer than Fe-oxide layer. Greater binding of TCE on the reactive sites of the iron sulfide outer layer	Rajajayavel and Ghoshal (2015)
One-pot / two-step	0.33; 1	TCE	Anoxic pH 7.2	Sulfidation imparts NZVI with TCE selectivity over water	Sulfidation at S/Fe molar ratios of $\geq$0.3 effectively eliminates reaction with water but retains significant reactivity with TCE	Fan et al. (2016)
One-pot / two-step	0.00125–0.75	TCE	Anoxic	Up to 60-fold increase in the TCE reaction rate was observed upon sulfidation. TCE reaction rates were found to depend strongly on sulfur to iron ratio	Sulfidation facilitates electron transfer. Sulfur in NZVI poisons hydrogen recombination	Han and Yan (2016)
One-pot	0.51, 0.12, 0.24	TBBPA	Anoxic	The reactivity of s-NZVI was 1.7 times higher than NZVI for TBBPA	Greater electron transfer capacity of S-NZVI	Li et al. (2016)

Table 9.3 Summary of studies using S-NZVI with some recalcitrant organic pollutants

Method	S/Fe molar ratio	Contaminant	Reaction condition	Key results	Explanation for enhanced reactivity over NZVI	References
One-pot	0.14	PNP	Anoxic/oxic; pH 6.7–9.11	S-NZVI enhanced PNP degradation compared to NZVI; in aerobic S-NZVI, and in anoxic or aerobic NZVI systems, reactivity decreased as pH increased. Oxygen improved the degradation of PNP by excessive amounts of hydroxyl radicals in slightly acidic conditions	Enhanced electron transfer; increased free radical generation	Tang et al. (2016)
One-pot	0–0.34	DCF	Oxic; pH 3.5–8.5	Dithionite sulfidizes NZVI and stimulates the crystal growth of Fe(0) and restrains FeOOH formation. S-NZVI showed greater reactivity with DCF compared to NZVI	Heterogeneous layer of iron sulfide and iron oxide restricts the direct reaction between oxygen and Fe(0) but facilitates electron transfer from the Fe(0) core to Fe(III), producing a considerable amount of surface-bound Fe(II)	Song et al. (2017)
One-pot	0.04	BA	pH 2.0–9.0	Complete oxidation of benzoic acid, and its transformed product, hydroxybenzoic acid, was observed for S-NZVI/persulfate. Alkaline pH lowered the reactivity of NZVI/persulfate, but not s-NZVI/persulfate	Both sulfate and hydroxyl radicals were involved in the NZVI/persulfate system, but hydroxyl radical was the primary oxidant in the S-NZVI/persulfate system	Rayaroth et al. (2017)

Table 9.4 Summary of studies using S-NZVI with radioactive/heavy metals

Method	S/Fe	Contaminant	Reaction condition	Key results	Explanation for enhanced reactivity over NZVI	References
Two-step	0–1.12	TcO_4^-	Anoxic; pH 7.9	Mössbauer spectra showed that S content did not increase when the S/Fe ratio increased from 0.112 to 1.12. Tc removal rate increased up to a S/Fe ratio of 0.224 and then decreased	High affinity of TcO_4^- for FeS than iron oxide; formed Tc sulfide instead of Tc oxide on S-NZVI with high extent of sulfidation	Fan et al. (2013)
Two-step	0–1.12	TcO_4^-	Oxic	Tc(VII) reduced by s-NZVI has a substantially slower reoxidation rate than Tc(VII) reduced by NZVI only	Initial inhibition of Tc(IV) dissolution at S/Fe = 0.112 is due to the redox buffer capacity of FeS. TcS_2 is more stable	Fan et al. (2014)
One-pot	0–0.34	Cd^{2+}	Anoxic/oxic; pH 4.0–9.0	S-NZVI had greater capacity for Cd removal than NZVI. Oxygen affected the structure of S-NZVI but enhanced Cd removal capacity to about 120 mg/g. Particle aging had no negative effect on the removal capacity of S-NZVI, and Cd-containing mixtures remained stable over 2 months	S^{2-} has a high affinity for Cd^{2+}, and the chemical stability of CdS is high	Su et al. (2015)
Two-step	0.067–0.33	$Cr_2O_7^{2-}$	Anoxic/oxic; pH 3.0–11.0	The reactivity of Cr(VI) with s-NZVI increased with an increasing FeS-to-Fe^0 ratio from 0/1 to 1/9 and then decreased for the FeS-to-Fe^0 ratio of 1/5 or 1/3. Increasing pH and DO results in a decrease of removal ratio	Electrostatic adsorption of negatively charged dichromate to the positively charged S-NZVI; enhanced electron transfer	Du et al. (2016)
One-pot	0–0.207	$Cr_2O_7^{2-}$	pH 3.5–9.0	Increasing the S/Fe molar ratio from 0 to 0.2 decreased Cr(VI) removal compared to NZVI. Low pH and initial Cr (VI) concentration favored Cr(VI) removal	Removal mechanism mainly involves adsorption, reduction of Cr(VI) to Cr(III), and subsequent immobilization in the solid phase of $(Cr_xFe_{1-x})(OH)_3$ and $FeCr_2S_4$	Gong et al. (2017)
One-pot (with nano-seeding)	0.5	Cd^{2+}	Anoxic	Increased nanoparticle seeding results in formation of more Fe^0 crystals for high sulfidation extent. High extent of sulfidation increases metal removal capacity and metal-nanoparticle stability	Sulfide has high affinity to metal ions	Su et al. (2016)

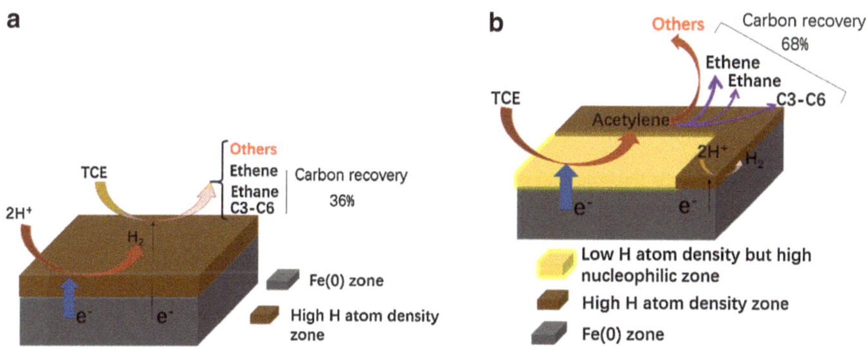

Fig. 9.5 Schematic of reactions of TCE on NZVI (**a**) and S-NZVI (**b**). The ability of the S-NZVI to promote reaction with TCE and limit reaction with water suggests that the S-NZVI formed is not a strictly Fe(0)/FeS core-shell product but rather is a heterogeneous mixture of both NZVI/Fe-oxide and FeS surfaces

described as a recombination of H atoms, modeled by a simple second-order reaction on the Fe(100) plane (Benziger and Madix 1980). The most important change induced by the addition of S atoms on the Fe surface is the decrease of H atom density, since 1 S atom can block 10–20 H atom adsorption sites (Bartholomew and Bowman 1985). In addition, S atoms are able to decrease H atom mobility through steric and long-range electronic effects (Benziger and Madix 1980). Seen from these two aspects, it is more reasonable to use the actual S/Fe molar ratio on the Fe surface instead of the initial S/Fe molar ratio during synthesis of the materials (there was S left as unreacted) to explain the degradation performance. Above a certain surface coverage of S atom, the decreased H atom density and H mobility effectively restrains H_2 generation from water-induced corrosion.

During TCE dechlorination, S atoms on the Fe surface appear to induce the formation of acetylene through β-elimination. Using FeS alone, electron quantity is limited, but for S-NZVI, the inner core of Fe(0) serves as an electron reservoir. Considering the good conductivity of FeS (Kim et al. 2011), the dechlorination rate through β-elimination is accelerated. Nevertheless, S results in low H atom density, which is not favorable for further hydrogenolysis of acetylene. It is very likely that formation of ethene and ethane from acetylene occurs at high H atom density zone. Thus, it is reasonable to conclude that sulfidation accelerates β-elimination but not hydrogenolysis (Han and Yan 2016). The formation of acetylene by NZVI has been previously observed (Liu et al. 2005). This pathway may be a result of S in the matrix of that NZVI material.

So far studies have only focused on the influence of H atom density on Fe (0) surface after sulfidation. The presence of S atoms on the Fe surface also likely affects the dissociation process of carbon compounds. Given that S^{2-} functions as nucleophile (Kriegman-King and Reinhard 1992), the different degrees of positive charge of C atom in different chlorinated hydrocarbon will likely lead to different affinity between the hydrocarbon and surface of S-NZVI. Subsequently, outer-sphere binding will be affected, and different reaction rates with different hydrocarbons should be observed. Pollutants that readily bind onto S^{2-} sites and can undergo reduction without the help of active H atom should have higher degradation rates on S-NZVI than those that do not. However, this effect has not been thoroughly investigated.

9.4.2 *Organic Compound Degradation Through Fenton-Like Reaction*

S-NZVI also alters the Fenton-like reactivity of NZVI. NZVI-based heterogeneous Fenton-like systems have been reported since 2004 (Joo et al. 2004) and have been employed to treat a wide range of organic pollutants, such as 4-chlorophenol, *p*-chloronitrobenzene, pentachlorophenol, 4-chloro-3-methyl phenol, molinate, EDTA, and bisphenol A (Ai et al. 2013; Feitz et al. 2005; Joo et al. 2004; Li and Zhu 2014; Li et al. 2015; Noradoun et al. 2003; Noradoun and Cheng 2005; Xu and Wang 2011). Both hydroxyl radical (acidic condition) and ferryl radicals (alkaline condition) were identified in a NZVI/O_2 system (Keenan and Sedlak 2008), but in general, for degradation of persistent organic contaminants, hydroxyl radicals are more effective (Pang et al. 2011). In the presence of oxygen, H_2O_2 is generated through a two-electron transfer process (Eqs. 9.4 and 9.5), and hydroxyl radicals are produced from the subsequent reaction between H_2O_2 and Fe(II) (Eq. 9.6) (Ai et al. 2007, 2013). The most important feature of NZVI/O_2 systems is hydroxyl radical production via surface-bound Fe(II), which accounts for the pollutant degradation at near-neutral pH. However, in a NZVI/O_2 system, a four-electron transfer process (Eq. 9.7) is usually the dominant reaction (Feitz et al. 2005), and H_2 generation (Eq. 9.8) also cannot be ignored especially at low pH (Liu and Lowry 2006), both of which result in the nonproductive loss of electrons. Therefore, the NZVI/O_2 system has low yield of free radicals.

$$Fe^0 + O_2 \rightarrow {O_2}^{2-} + Fe^{2+} \tag{9.4}$$

$${O_2}^{2-} + 2H^+ \rightarrow H_2O_2 \tag{9.5}$$

$$Fe^{2+} + H_2O_2 \rightarrow Fe^{3+} + {}^{\bullet}OH + OH^- \tag{9.6}$$

$$2Fe^0 + O_2 + H_2O \rightarrow 2Fe^{2+} + 4OH^- \tag{9.7}$$

$$Fe^0 + 2H_2O \rightarrow Fe^{2+} + H_2 + 2OH^- \tag{9.8}$$

To enhance free radical generation, modifying electron transfer on the core-shell surface of NZVI is a good strategy. Ai et al. synthesized core-shell Fe@Fe_2O_3 nanowires, using the Fe_2O_3 shell to prevent the direct reaction between Fe(0) and O_2 (Ai et al. 2007, 2013; Wang et al. 2014). In their study, they proposed that O_2 was catalyzed by surface-bound ferrous atoms through a one-electron transfer pathway (Eqs. 9.9–9.11).

$$\equiv \mathrm{Fe} - \mathrm{O} - \mathrm{Fe}^{2+} + \mathrm{O}_{2,\mathrm{ad}} \rightarrow \equiv \mathrm{Fe} - \mathrm{O} - \mathrm{Fe}^{3+} + {}^{\bullet}\mathrm{O2}^{-} \quad (9.9)$$

$$\equiv \mathrm{Fe}^{2+} + {}^{\bullet}\mathrm{O2}^{-} \rightarrow \equiv \mathrm{Fe}^{3+} + \mathrm{H}_2\mathrm{O}_2 \quad (9.10)$$

$$\mathrm{Fe}^{2+} + \mathrm{H}_2\mathrm{O}_2 \rightarrow \mathrm{Fe}^{3+} + {}^{\bullet}\mathrm{OH} + \mathrm{OH}^{-} \quad (9.11)$$

After sulfidation, the PNP and DCF removal efficiency (Table 9.3) was improved (Song et al. 2017; Tang et al. 2016), which was attributed to the higher yield of hydroxyl radicals. However, Fe(II) concentration in the supernatant S-NZVI/O_2 system was equal to that in NZVI/O_2 system, indicating that Fe(II) in solution was not the reason for the enhanced free radical yield (Song et al. 2017). Thus, the greater amount of surface-bound Fe(II) on S-NZVI was mainly responsible for the greater generation of hydroxyl radical compared to NZVI alone. Song et al. (2017) found that a one-electron transfer pathway via surface-bound Fe(II) was the main process for hydroxyl radical generation. According to their study, the reaction between Fe (0) and O_2 in the S-NZVI/O_2 system was much lower than that in NZVI/O_2 system (even though S-NZVI does not have the fine core-shell structure like Fe@Fe_2O_3 nanowires), indicating that the four-electron transfer reaction is effectively inhibited. Moreover, sulfidation is able to depress H_2 generation effectively (Eq. 9.8), contributing to the preservation of Fe(0) content. In addition, FeS has good conductivity, which means that it can conduct electrons from the Fe(0) core to surface Fe(III), producing Fe(II) (Eq. 9.12). The preserved Fe(0) from the inhibition of Eqs. (9.7) and (9.8) could provide abundant electrons for Eq. (9.12), and the produced Fe(II) is beneficial to Fenton-like reaction. The modified electron transfer pathway is illustrated in Fig. 9.6.

$$\mathrm{Fe}^{0} + 2 \equiv \mathrm{Fe} - \mathrm{S} - \mathrm{Fe}^{3+} \rightarrow 2 \equiv \mathrm{Fe} - \mathrm{S} - \mathrm{Fe}^{2+} + \mathrm{Fe}^{2+} \quad (9.12)$$

S-NZVI has been used to activate persulfate (PS) for benzoic acid (BA) degradation (Rayaroth et al. 2017). Complete degradation of BA was observed in both NZVI/PS and S-NZVI/PS system, and over 75% of TOC removal was achieved. While both sulfate and hydroxyl radicals were responsible for BA degradation for NZVI/PS, hydroxyl radicals were the main species for S-NZVI/PS. For NZVI/PS, sulfate radicals could be activated through surface-bound Fe(II) after $S_2O_8{}^{2-}$ dissociation as shown in Eq. (9.13). However, in the S-NZVI/PS system, it seems S(-II) atoms block the adsorption sites for $S_2O_8{}^{2-}$ due to steric and long-range electronic effects. The advantage of the S-NZVI/PS system over the NZVI/PS system is that under alkaline conditions S-NZVI/PS is still able to degrade BA while NZVI/PS not. The proposed reaction for hydroxyl radical generation in S-NZVI/PS system is shown in Eq. (9.14). It is also possible that S-NZVI simply prevents the formation of Fe-oxyhydroxides, keeping the material more reactive. The mechanisms behind the increased reactivity of S-NZVI are not well described.

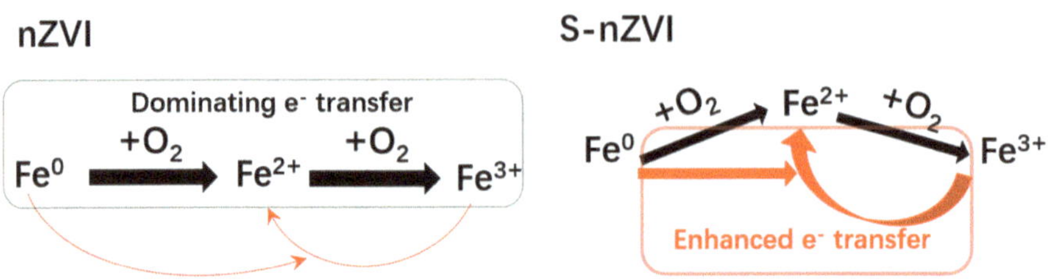

Fig. 9.6 Schematic showing electron transfer modification in the Fenton-like reaction through sulfidation

$$\equiv Fe(II) + S_2O_8{}^{2-}{}_{(aq)} \rightarrow \equiv Fe(II)O_3SO_2SO_3{}^{-} \rightarrow \equiv Fe(III) + SO_4{}^{\cdot -} + SO_4{}^{2-}{}_{(aq)} \quad (9.13)$$

$$Fe^0 + S_2O_8{}^{2-}{}_{(aq)} + 2H_2O \rightarrow Fe^{2+} + 2{}^{\bullet}OH + 2SO_4{}^{2-} + 2H^+ \quad (9.14)$$

9.4.3 Radioactive/Heavy Metal (Me) Immobilization

Heavy metal pollution is a severe problem around the world and is a great threat to public health. Both FeS and NZVI have long been successfully employed to sequester Me ions or Me-oxygen anions from aqueous media (Bostick and Fendorf 2003; Miller et al. 1998; Morse and Arakaki 1993; Moyes et al. 2002; Özverdi and Erdem 2006; Su et al. 2014a, b, 2015; Zhang et al. 2014). In general, adsorption, precipitation, complexation/substitution, and reduction are the main removal mechanisms for Me ions/anions. However, the existing form of immobilized Me on NZVI or S-NZVI is very different. On NZVI, some metals are reduced to zerovalent metals, but more commonly Me atoms bind with O atoms to form metal oxides or mixed Fe-Me-oxides (Li and Zhang 2007; Li et al. 2008). In contrast, on S-NZVI, Me atom may prefer S atoms to O atom, and with the increase of S/Fe ratio, the majority of the immobilized Me will shift from metal oxide to metal sulfides (Fan et al. 2013, 2014). The increase of the S/Fe ratio leads to different S-NZVI with different removal capacity (Su et al. 2015). For S-NZVI obtained from the one-pot synthesis method, increasing sulfidation leads to first a decline and then an increase in Cd(II) and Cr(VI) removal (Gong et al. 2017; Su et al. 2015). However, for S-NZVI obtained from the two-step synthesis method (adding Fe(II) and S(-II) to form FeS), Cr(VI) removal capacity increases first and then decreases (Du et al. 2016). It seems that different synthesis methods result in different morphologies of S-NZVI, which subsequently affect its reactivity. Nevertheless, metal removal capacity could potentially be much higher than NZVI if the S-NZVI is synthesized with the optimal S/Fe ratio.

The presence of FeS on the NZVI can influence the metal reduction process. The standard oxidation-reduction potential of Fe/Fe(II) is −0.447 V. Similar to NZVI, S-NZVI can reduce those metals with a standard oxidation-reduction potential above iron (e.g., Ni(II), Cu(II) Ag(I)), but not those below it (e.g., Zn(II), Al(III)) (Li and Zhang 2007). The presence of FeS can influence the reduction process in following two possible ways:

1. FeS contributes to Me reduction driven by Fe(0). Here, FeS inhibits the reaction between H_2O and Fe(0), which means that more electrons are available for metal reduction. Additionally, FeS is able to conduct electrons from Fe(0) to Me on the S-NZVI surface.
2. FeS participates in the reduction. Here, not only can Fe(II) be oxidized to Fe(III) but S(-II) can be oxidized to many different states under different conditions (such as different pH). It is reported that Cr(VI), As(III), Se(IV), Tc(VII), and U (VI) can be efficiently reduced by FeS (Bostick and Fendorf 2003; Du et al. 2016; Gong et al. 2016; Moyes et al. 2002).

Another important advantage of S-NZVI in metal immobilization over NZVI is the enhanced chemical stability of sequestered metal. The main concern about the application of NZVI in metal-contaminated groundwater remediation is the re-release of metal ions from NZVI upon oxidation. The oxidation of NZVI by nitrate, dissolved oxygen, and any other oxidants could change the composition of the particles and oxidize and release adsorbed metals (Su et al. 2014a, 2015; Zhang et al. 2014). Results so far on metal removal by S-NZVI show that sulfidation enhanced the chemical stability of the sequestered metals. Fan et al. (2014) reported that FeS on S-NZVI can inhibit Tc(IV) reoxidation. Their results show persistence of FeS after 24 h oxidation but complete oxidation after 120 h oxidation. While in their experiment TcS_2 previously formed under anoxic conditions transformed into TcO_2 after 120 h of oxidation, Tc dissolution was not recorded. The authors proposed that speciation transformation (TcS_2 to TcO_2) retarded the dissolution process of Tc. Another study by Su et al. (2015) reported that 2 h oxidation did not cause the re-release of Cd(II) ions from S-NZVI, whereas it did in a NZVI reaction system. Further, they found that the 2 h oxidation destroyed Fe-S bonds and stimulated additional formation of CdS, increasing Cd(II) removal efficiency by 50%. Hence, S-NZVI might be better than NZVI in heavy metal-polluted groundwater remediation.

However, metal removal mechanism by S-NZVI, especially for reduction (e.g., Cr(VI)), has not been clearly elucidated. There was an effort made by Du et al. (2016) in which they used 1,10-phenanthroline to quench Fe(II) and subsequently block Cr(VI) reduction through Fe(II). They found that abundant surface-bound Fe (II) in S-NZVI system contributed significantly to Cr(VI) reduction. However, the authors did not provide data on the NZVI reaction system for comparison. The role of FeS on metal adsorption and reduction processes and how FeS affects the inner diffusion of metals on NZVI remains largely unexplored.

9.5 Implications for Field Application of S-NZVI

Lab-based studies on pollutant removal indicate that S-NZVI is more effective than NZVI for groundwater remediation. However, studies on the mobility, reactivity, and interaction between microorganisms and S-NZVI in the subsurface are crucial for field applications.

9.5.1 *Mobility*

Pristine NZVI in field applications suffers from low mobility in porous media. This is due to a combination of effects including deposition onto aquifer solids, aggregation and settling due to the van der Waals forces, electrical double layer interactions, and magnetic attractive forces. Based on previously published studies (Kim et al. 2013; Song et al. 2017), sulfidation can improve electrostatic repulsion (between pH 7 and 9) and lower the saturation magnetization to a certain degree, both of which help to alleviate the aggregation of nanoparticles. Although sulfidation does improve the stability of the nanoparticle suspensions in the presence of Cl^-, CO_3^{2-}, SO_4^{2-}, and NO_3^-, 5 mM Ca^{2+} still caused rapid aggregation and sedimentation. Unfortunately, groundwater often contains appreciable concentrations of Ca^{2+}, which will affect the mobility of S-NZVI. On the basis of previous studies on NZVI, the most efficient method for enhancing mobility is polymer modification (Kocur et al. 2013). Compared with the "few inches to a few feet" traveling distance of NZVI, carboxymethylcellulose (CMC)-modified NZVI traveled at least 1 m to the contaminated source zone at significant Fe^0 concentrations (Kocur et al. 2014; Krol et al. 2013). Although no study has been carried out to study the mobility of S-NZVI in the subsurface, it can be deduced that for practical applications, surface modification will be needed.

9.5.2 *Reactivity*

Unlike Pd-doped NZVI, the reactivity of S-NZVI toward pollutants in situ is largely unaffected by groundwater constituents, which is one of the main advantages of this material. In the study of Kim et al. (2014), it was reported (seen Fig. 9.7) that most groundwater solutes had limited effect on the reduction rate of TCE by S-NZVI compared with NZVI, and both surface-area-normalized and mass-normalized rate constants of S-NZVI were approximately 1 order of magnitude higher than those of NZVI. In 2015, Su et al. (2015) also reported that S-NZVI was able to efficiently sequester Cd(II) ions from groundwater, river water, and even seawater, without re-release of Cd(II) after 2 months. As to heterogeneous Fenton-like reaction driven by S-NZVI, while common ions/anions had a relatively small effect on this process,

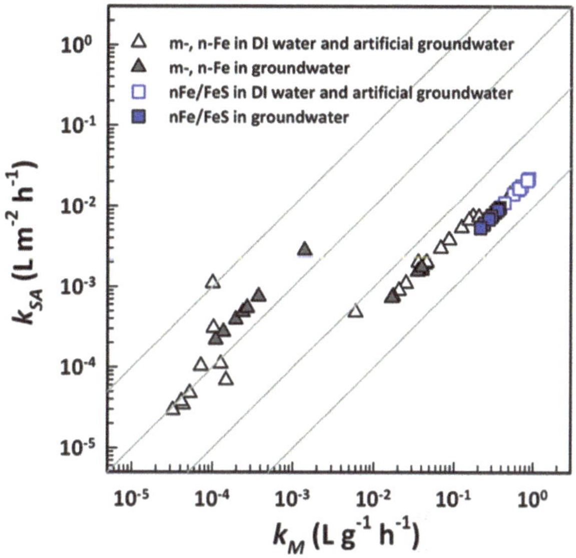

Fig. 9.7 A logarithmic–logarithmic plot of surface-area-normalized rate constants (k_{SA}, L m^{-2} h^{-1}) versus mass-normalized rate constants (k_M, L g^{-1} h^{-1}) showing TCE reduction rates by nFe/FeS (squares). (Reprinted with permission from Kim et al. (2014). Copyright (2014) American Chemical Society)

the most influential factor is pH. As long as the pH was below 7, high removal of DCF could be achieved (Song et al. 2017).

The reactivity of S-NZVI with 1,2-dichloroethane (1,2-DCA) was demonstrated in the field in Canada (Nunez Garcia et al. 2016). While NZVI was not able to degrade 1,2-DCA, S-NZVI degraded >90% 1,2-DCA in a year with observed pseudo-first-order rate constants ranging from 3.8×10^{-3} to 7.8×10^{-3} d^{-1}. Fe (0) in S-NZVI was still present even after more than 1 year, which may be due to the inhibition of water-induced corrosion. Both nucleophilic substitution and reductive dechlorination were responsible for the 1,2-DCA dechlorination.

9.5.3 Interactions Between S-NZVI and Subsurface Microorganism

NZVI can create conditions for anaerobic microorganisms (Kirschling et al. 2010). In a recent paper describing the field application of CMC-NZVI, its injection created conditions favoring the growth of native populations of organohalide-respiring microorganisms, which resulted in a 1 order of magnitude increase of the abundance of *Dehalococcoides* spp. (Kocur et al. 2015). The authors successfully proved that NZVI injection not only gave rise to the abiotic degradation of volatile chlorinated compounds but stimulated a considerable amount of biotic degradation.

Although there is no study on the influence of S-NZVI to microorganism in the subsurface, the toxicity of S-NZVI on algae has been evaluated (Adeleye et al. 2016). It was shown that even a high concentration of 180 ppm of S-NZVI (modified with nanosilica during synthesis) did not cause an obvious negative impact on the growth of algae. The dissolved organic matter produced by algae not only alleviate the toxicity of S-NZVI but stabilize the nanoparticles and lower the oxidation of S-NZVI (which may prove advantageous for pollutant degradation in practical remediation). It is reasonable to assume that S-NZVI might have a similar effect on the microorganism as NZVI, but studies on this topic have yet to be conducted.

9.6 Conclusions and Future Research Needs

9.6.1 Main Conclusion

S-NZVI was shown to be highly reactive toward a wide range of pollutants, including halogenated carbon compounds, some persistent pharmaceuticals, and metal ions/metal-containing anions. In this chapter, the following conclusions can be summarized:

1. Both the one-pot synthesis method and the two-step synthesis methods can produce highly reactive S-NZVI. However, different synthesis procedures result in different morphologies of S-NZVI. S-NZVI generated from one-pot method doesn't have a neat structure of Fe(0) core and iron oxide shell. S is very much likely homogeneously distributed on the whole particle (inside and outside), and flake-like structure was often observed. For those obtained from the two-step method, although the initial core-shell structure could be affected by the S/Fe ratio, S is believed to be distributed on the outside layer of the particle, and no flake-like structures were observed.
2. Different degrees of sulfidation result in different surface S atom density, which subsequently leads to different reactivity of S-NZVI toward contaminants. S atoms can preferentially adsorb on certain sites, which can ① inhibit adsorption (which normally occurs on these sites and nearby 10–20 sites) but also ② create new specific adsorption sites due to its nucleophilic character. Influence ① lowers the related reaction kinetics (not necessarily only means the pollutant reaction but also the H_2 generation from H_2O); influence ② could stimulate another immobilization or reaction pathway. Different S/Fe molar ratios mean different density of surface S atoms, which results in the different extent of influence ① and ②. With an optimal extent of sulfidation, the reactivity of NZVI can be significantly enhanced, and unlike NZVI, it is highly resistant to many environmental interferences, such as common ions/anions.
3. Sulfidation creates a nucleophilic zone on the particle surface with lower density of H atoms compared to NZVI, which lowers the recombination of H atoms to form H_2 (water-induced corrosion). However, it enhances the β-elimination pathway for halogenated carbon compounds by facilitating the adsorption between positively charged carbons (α position) and the nucleophile (S^{2-}). After β-elimination, hydrogenolysis is then readily achieved on areas of the surfaces with higher H atom density. This modified pathway not only accelerates

the dechlorination rate but also provides a chance to degrade pollutants that previously cannot be degraded by NZVI, such as 1,2-DCA.

4. In heterogeneous Fenton-like reactions, S-NZVI generates much more hydroxyl radicals for pollutant degradation through a one-electron transfer. Sulfidation effectively inhibits the direct reaction between oxygen and Fe(0), preserving a considerable amount of Fe(0) for ferric iron reduction to surface-bound ferrous iron (Fe(II)), which plays an important role in hydroxyl radical generation. While common ions/anions have a small effect on pollutant removal performance, pH above 7 has a clear inhibition effect.
5. S-NZVI has a higher removal capacity for metal ions compared to NZVI, and the chemical stability of the formed metal sulfides is high. With an increase of sulfidation, metal oxides will transform into metal sulfides, indicating that S(-II) is highly involved in the metal removal pathway. The abundant surface-bound Fe (II) on S-NZVI may play an important role in metal immobilization process, and the high conductivity of FeS may facilitate electron transfer during reduction. Common ions/anions have limited effect on metal removal capacity of S-NZVI, and S-NZVI can sequester metal ions over a wide range of solution pH.
6. S-NZVI might stimulate the growth of certain species of microorganism, which can further improve pollutant removal. The toxicity of S-NZVI toward algae is low. Organic matter derived from algae can stabilize S-NZVI and lower the oxidation rate, which might be beneficial for pollutant reduction.

9.7 Future Research

S-NZVI shows great promise in the application of in situ remediation. Future research may include:

1. A better assessment of the microstructure of S-NZVI under different degrees of sulfidation and structure evolution during reaction and aging
2. Elucidating the role of sulfur in the dechlorination process, including its influence on carbon atom dissociation and H atom dissociation/recombination
3. Understanding the immobilization pathway of metals (especially the metals which can be reduced) via S-NZVI
4. Assessing the mobility of S-NZVI in subsurface and its reactivity over time
5. Determining the ability of S-NZVI to biostimulate in situ reactions of interest

References

Adeleye, A. S., Stevenson, L. M., Su, Y., Nisbet, R. M., Zhang, Y., & Keller, A. A. (2016). Influence of phytoplankton on fate and effects of modified zerovalent iron nanoparticles. *Environmental Science & Technology, 50*, 5597–5605. https://doi.org/10.1021/acs.est.5b06251.

Ai, Z., Lu, L., Li, J., Zhang, L., Qiu, J., & Wu, M. (2007). Fe@Fe_2O_3 Core-shell nanowires as iron reagent. 1. Efficient degradation of rhodamine by a novel Sono-Fenton process. *Journal of Physical Chemistry C, 111*, 4087–4093. https://doi.org/10.1021/jp065559l.

Ai, Z., Gao, Z., Zhang, L., He, W., & Yin, J. J. (2013). Core-shell structure dependent reactivity of Fe@Fe_2O_3 nanowires on aerobic degradation of 4-chlorophenol. *Environmental Science & Technology, 47*, 5344–5352. https://doi.org/10.1021/es4005202.

Bartholomew, C. H., & Bowman, R. M. (1985). Sulfur poisoning of cobalt and iron fischer-tropsch catalysts. *Applied Catalysis, 15*, 59–67. https://doi.org/10.1016/S0166-9834(00)81487-6.

Benziger, J., & Madix, R. J. (1980). The effects of carbon, oxygen, sulfur and potassium adlayers on CO and H2 adsorption on Fe(100). *Surface Science, 94*, 119–153. https://doi.org/10.1016/0039-6028(80)90160-0.

Bostick, B. C., & Fendorf, S. (2003). Arsenite sorption on troilite (FeS) and pyrite (FeS_2). *Geochimica et Cosmochimica Acta, 67*, 909–921. https://doi.org/10.1016/S0016-7037(02)01170-5.

Butler, E. C., & Hayes, K. F. (1998). Effects of solution composition on the reductive dechlorination of hexachloroethane by iron sulfide, preprint extended abstracts. *Environmental Science & Technology, 32*, 1276–1284.

Butler, E. C., & Hayes, K. F. (1999). Kinetics of the transformation of trichloroethylene and tetrachloroethylene by iron sulfide. *Environmental Science & Technology, 33*, 2021–2027. https://doi.org/10.1021/es9809455.

Butler, E. C., & Hayes, K. F. (2000). Kinetics of the transformation of halogenated aliphatic compounds by iron sulfide. *Environmental Science & Technology, 34*, 422–429.

Butler, E. C., & Hayes, K. F. (2001). Factors influencing rates and products in the transformation of trichloroethylene by iron sulfide and iron metal. *Environmental Science & Technology, 35*, 3884–3891. https://doi.org/10.1021/es010620f.

Chapman, P. M., Wang, F., Janssen, C., Persoone, G., & Allen, H. E. (1998). Ecotoxicology of metals in aquatic sediments: Binding and release, bioavailability, risk assessment, and remediation. *Canadian Journal of Fisheries and Aquatic Sciences, 55*, 2221–2243. https://doi.org/10.1139/f98-145.

Du, J., Bao, J., Lu, C., & Werner, D. (2016). Reductive sequestration of chromate by hierarchical FeS@Fe0 particles. *Water Research, 102*, 73–81. https://doi.org/10.1016/j.watres.2016.06.009.

Fan, D., Anitori, R. P., Tebo, B. M., Tratnyek, P. G., Pacheco, J. S. L., Kukkadapu, R. K., Engelhard, M. H., Bowden, M. E., Kovarik, L., & Arey, B. W. (2013). Reductive sequestration of pertechnetate (99TcO4-) by nano zerovalent iron (NZVI) transformed by abiotic sulfide. *Environmental Science & Technology, 47*, 5302–5310. https://doi.org/10.1021/es304829z.

Fan, D., Anitori, R. P., Tebo, B. M., Tratnyek, P. G., Lezama Pacheco, J. S., Kukkadapu, R. K., Kovarik, L., Engelhard, M. H., & Bowden, M. E. (2014). Oxidative remobilization of technetium sequestered by sulfide-transformed nano zerovalent iron. *Environmental Science & Technology, 48*, 7409–7417. https://doi.org/10.1021/es501607s.

Fan, D., O'Brien Johnson, G., Tratnyek, P. G., & Johnson, R. L. (2016). Sulfidation of nano zerovalent iron (NZVI) for improved selectivity during *in situ* chemical reduction (ISCR). *Environmental Science & Technology, 50*, 9558–9565. https://doi.org/10.1021/acs.est.6b02170.

Feitz, A. J., Joo, S. H., Guan, J., Sun, Q., Sedlak, D. L., & Waite, T. D. (2005). Oxidative transformation of contaminants using colloidal zero-valent iron. *Colloids and Surfaces A: Physicochemical and Engineering Aspects, 265*, 88–94. https://doi.org/10.1016/j.colsurfa.2005.01.038.

Field, J. A., & Sierra-Alvarez, R. (2004). Biodegradability of chlorinated solvents and related chlorinated aliphatic compounds. *Reviews in Environmental Science and Biotechnology, 3*, 185–254. https://doi.org/10.1007/s11157-004-4733-8.

Gong, Y., Tang, J., & Zhao, D. (2016). Application of iron sulfide particles for groundwater and soil remediation: A review. *Water Research, 89*, 309–320. https://doi.org/10.1016/j.watres.2015.11.063.

Gong, Y., Gai, L., Tang, J., Fu, J., Wang, Q., & Zeng, E. Y. (2017). Reduction of Cr(VI) in simulated groundwater by FeS-coated iron magnetic nanoparticles. *Science of the Total Environment, 595*, 743–751. https://doi.org/10.1016/j.scitotenv.2017.03.282.

Han, Y., & Yan, W. (2016). Reductive dechlorination of trichloroethene by zero-valent iron nanoparticles: Reactivity enhancement through sulfidation treatment. *Environmental Science & Technology, 50*, 12992–13001. https://doi.org/10.1021/acs.est.6b03997.

Jeong, H. Y., Kim, H., & Hayes, K. F. (2007a). Reductive dechlorination pathways of tetrachloroethylene and subsequent transformation of their dechlorination products by mackinawite (FeS) in the presence of metals. *Environmental Science & Technology, 41*, 7736–7743.

Jeong, H. Y., Klaue, B., Blum, J. D., & Hayes, K. F. (2007b). Sorption of mercuric ion by synthetic nanocrystalline mackinawite (FeS). *Environmental Science & Technology, 41*, 7699–7705. https://doi.org/10.1021/es070289l.

Joo, S. H., Feitz, A. J., & Waite, T. D. (2004). Oxidative degradation of the carbothioate herbicide, molinate, using nanoscale zero-valent iron. *Environmental Science & Technology, 38*, 2242–2247. https://doi.org/10.1021/es035157g.

Keenan, C. R., & Sedlak, D. L. (2008). Factors affecting the yield of oxidants from the reaction of nanoparticulate zero-valent iron and oxygen. *Environmental Science & Technology, 42*, 1262–1267. https://doi.org/10.1021/es7025664.

Kim, E. J., Kim, J. H., Azad, A. M., & Chang, Y. S. (2011). Facile synthesis and characterization of Fe/FeS nanoparticles for environmental applications. *ACS Applied Materials & Interfaces, 3*, 1457–1462. https://doi.org/10.1021/am200016v.

Kim, E. J., Murugesan, K., Kim, J. H., Tratnyek, P. G., & Chang, Y. S. (2013). Remediation of trichloroethylene by FeS-coated iron nanoparticles in simulated and real groundwater: Effects of water chemistry. *Industrial and Engineering Chemistry Research, 52*, 9343–9350. https://doi.org/10.1021/ie400165a.

Kim, E., Kim, J., & Chang, Y. (2014). Effects of metal ions on the reactivity and corrosion electrochemistry of Fe/FeS nanoparticles. *Environmental Science & Technology, 48*, 4002–4011.

Kirschling, T. L., Gregory, K. B., Minkley, E. G., Lowry, G. V., & Tilton, R. D. (2010). Impact of nanoscale zero valent iron on geochemistry and microbial populations in trichloroethylene contaminated aquifer materials. *Environmental Science & Technology, 44*, 3474–3480. https://doi.org/10.1021/es903744f.

Kocur, C. M., O'Carroll, D. M., & Sleep, B. E. (2013). Impact of NZVI stability on mobility in porous media. *Journal of Contaminant Hydrology, 145*, 17–25. https://doi.org/10.1016/j.jconhyd.2012.11.001.

Kocur, C. M., Chowdhury, A. I., Sakulchaicharoen, N., Boparai, H. K., Weber, K. P., Sharma, P., Krol, M. M., Austrins, L., Peace, C., Sleep, B. E., & O'Carroll, D. M. (2014). Characterization of NZVI mobility in a field scale test. *Environmental Science & Technology, 48*, 2862–2869. https://doi.org/10.1021/es4044209.

Kocur, C. M. D., Lomheim, L., Boparai, H. K., Chowdhury, A. I. A., Weber, K. P., Austrins, L. M., Edwards, E. A., Sleep, B. E., & O'Carroll, D. M. (2015). Contributions of abiotic and biotic dechlorination following carboxymethyl cellulose stabilized nanoscale zero valent iron injection. *Environmental Science & Technology, 49*, 8648–8656. https://doi.org/10.1021/acs.est.5b00719.

Kriegman-King, M. R., & Reinhard, M. (1992). Transformation of carbon tetrachloride in the presence of sulfide, biotite, and vermiculite. *Environmental Science & Technology, 2206*, 2198–2206. https://doi.org/10.1021/es00035a019.

Kriegman-King, M. R., & Reinhard, M. (1994). Transformation of carbon tetrachloride by pyrite in aqueous solution. *Environmental Science & Technology, 28*, 692–700. https://doi.org/10.1021/es00053a025.

Krol, M. M., Oleniuk, A. J., Kocur, C. M., Sleep, B. E., Bennett, P., Xiong, Z., & O'Carroll, D. M. (2013). A field-validated model for *in situ* transport of polymer-stabilized NZVI and

implications for subsurface injection. *Environmental Science & Technology, 47*, 7332–7340. https://doi.org/10.1021/es3041412.
Lee, W., & Batchelor, B. (2002). Abiotic reductive dechlorination of chlorinated ethylenes by iron-bearing soil minerals. 1. Pyrite and magnetite. *Environmental Science & Technology, 36*, 5147–5154. https://doi.org/10.1021/es025836b.
Li, X., & Zhang, W. (2007). Sequestration of metal cations with zerovalent iron nanoparticles: A study with high resolution X-ray photoelectron spectroscopy (HR-XPS). *The Journal of Physical Chemistry, 111*(19), 6939–6946. https://doi.org/10.1021/jp0702189.
Li, B., & Zhu, J. (2014). Removal of p-chloronitrobenzene from groundwater: Effectiveness and degradation mechanism of a heterogeneous nanoparticulate zero-valent iron (NZVI)-induced Fenton process. *Chemical Engineering Journal, 255*, 225–232. https://doi.org/10.1016/j.cej.2014.06.013.
Li, X. Q., Cao, J., & Zhang, W. X. (2008). Stoichiometry of Cr(VI) immobilization using nanoscale zero valent iron (NZVI): A study with high-resolution X-ray photoelectron spectroscopy (HR-XPS). *Industrial and Engineering Chemistry Research, 47*, 2131–2139. https://doi.org/10.1021/ie061655x.
Li, S., Yan, W., & Zhang, W. (2009). Solvent-free production of nanoscale zero-valent iron (NZVI) with precision milling. *Green Chemistry, 11*, 1618. https://doi.org/10.1039/b913056j.
Li, R., Jin, X., Megharaj, M., Naidu, R., & Chen, Z. (2015). Heterogeneous Fenton oxidation of 2,4-dichlorophenol using iron-based nanoparticles and persulfate system. *Chemical Engineering Journal, 264*, 587–594. https://doi.org/10.1016/j.cej.2014.11.128.
Li, D., Mao, Z., Zhong, Y., Huang, W., Wu, Y., & Peng, P. (2016). Reductive transformation of tetrabromobisphenol a by sulfidated nano zerovalent iron. *Water Research, 103*, 1–9. https://doi.org/10.1016/j.watres.2016.07.003.
Ling, L., & Zhang, W. (2014a). Mapping the reactions of hexavalent chromium [Cr(VI)] in iron nanoparticles using spherical aberration corrected scanning transmission electron microscopy (Cs-STEM). *Analytical Methods, 6*, 3211. https://doi.org/10.1039/C4AY00075G.
Ling, L., & Zhang, W. X. (2014b). Sequestration of arsenate in zero-valent iron nanoparticles: Visualization of intraparticle reactions at angstrom resolution. *Environmental Science & Technology Letters, 1*, 305–309. https://doi.org/10.1021/ez5001512.
Ling, L., & Zhang, W. X. (2015). Enrichment and encapsulation of uranium with iron nanoparticle. *Journal of the American Chemical Society, 137*, 2788–2791. https://doi.org/10.1021/ja510488r.
Liu, Y., & Lowry, G. V. (2006). Effect of particle age (Fe0 content) and solution pH on NZVI reactivity: H2 evolution and TCE dechlorination. *Environmental Science & Technology, 40*, 6085–6090. https://doi.org/10.1021/es060685o.
Liu, A., & Zhang, W. (2014). Fine structural features of nanoscale zero-valent iron characterized by spherical aberration corrected scanning transmission electron microscopy (Cs-STEM). *Analyst, 139*, 4512. https://doi.org/10.1039/C4AN00679H.
Liu, Y., Majetich, S. A., Tilton, R. D., Sholl, D. S., & Lowry, G. V. (2005). TCE dechlorination rates, pathways, and efficiency of nanoscale iron particles with different properties. *Environmental Science & Technology, 39*, 1338–1345. https://doi.org/10.1021/es049195r.
Liu, Y., Phenrat, T., & Lowry, G. V. (2007). Effect of TCE concentration and dissolved groundwater solutes on NZVI-promoted TCE dechlorination and H2 evolution. *Environmental Science & Technology, 41*, 7881–7887. https://doi.org/10.1021/es0711967.
Liu, H., Wang, Q., Wang, C., & Li, X. Z. (2013). Electron efficiency of zero-valent iron for groundwater remediation and wastewater treatment. *Chemical Engineering Journal, 215–216*, 90–95. https://doi.org/10.1016/j.cej.2012.11.010.
Ma, X., He, D., Jones, A. M., Collins, R. N., & Waite, T. D. (2016). Reductive reactivity of borohydride- and dithionite-synthesized iron-based nanoparticles: A comparative study. *Journal of Hazardous Materials, 303*, 101–110. https://doi.org/10.1016/j.jhazmat.2015.10.009.
Miller, P. L., Vasudevan, D., Gschwend, P. M., & Roberts, A. L. (1998). Transformation of hexachloroethane in a sulfidic natural water. *Environmental Science & Technology, 32*, 1269–1275. https://doi.org/10.1021/es970687w.

Morse, J. W. (1994). Interactions of trace metals with authigenic sulfide minerals: Implications for their bioavailability. *Marine Chemistry, 46*, 1–6. https://doi.org/10.1016/0304-4203(94)90040-X.

Morse, J. W., & Arakaki, T. (1993). Adsorption and coprecipitation of divalent metals with mackinawite (FeS). *Geochimica et Cosmochimica Acta, 57*, 3635–3640. https://doi.org/10.1016/0016-7037(93)90145-M.

Moyes, L. N., Jones, M. J., Reed, W. A., Livens, F. R., Charnock, J. M., Mosselmans, J. F. W., Hennig, C., Vaughan, D. J., & Pattrick, R. A. D. (2002). An X-ray absorption spectroscopy, study of neptunium(V) reactions with mackinawite (FeS). *Environmental Science & Technology, 36*, 179–183. https://doi.org/10.1021/es0100928.

Noradoun, C. E., & Cheng, I. F. (2005). EDTA degradation induced by oxygen activation in a zerovalent iron/air/water system. *Environmental Science & Technology, 39*, 7158–7163. https://doi.org/10.1021/es050137v.

Noradoun, C., Engelmann, M. D., McLaughlin, M., Hutcheson, R., Breen, K., Paszczynski, A., & Cheng, I. F. (2003). Destruction of chlorinated phenols by dioxygen activation under aqueous room temperature and pressure conditions. *Industrial and Engineering Chemistry Research, 42*, 5024–5030. https://doi.org/10.1021/ie030076e.

Nunez Garcia, A., Boparai, H. K., & Ocarroll, D. M. (2016). Enhanced dechlorination of 1,2-dichloroethane by coupled nano iron-dithionite treatment. *Environmental Science & Technology, 50*, 5243–5251. https://doi.org/10.1021/acs.est.6b00734.

Oudar, J. (1980). Sulfur adsorption and poisoning of metallic catalysts. *Catalysis Reviews-Science and Engineering, 22*, 171–195. https://doi.org/10.1080/03602458008066533.

Özverdi, A., & Erdem, M. (2006). Cu2+, Cd2+ and Pb2+ adsorption from aqueous solutions by pyrite and synthetic iron sulphide. *Journal of Hazardous Materials, 137*, 626–632. https://doi.org/10.1016/j.jhazmat.2006.02.051.

Pang, S. Y., Jiang, J., & Ma, J. (2011). Oxidation of sulfoxides and arsenic(III) in corrosion of nanoscale zero valent iron by oxygen: Evidence against ferryl ions (Fe(IV)) as active intermediates in Fenton reaction. *Environmental Science & Technology, 45*, 307–312. https://doi.org/10.1021/es102401d.

Prince, R. C., & Dutton, P. L. (1976). Further studies on the rieske iron-sulfur center in mitochondrial and photosynthetic systems: A pK on the oxidized form. *FEBS Letters, 65*, 117–119.

Rajajayavel, S. R. C., & Ghoshal, S. (2015). Enhanced reductive dechlorination of trichloroethylene by sulfidated nanoscale zerovalent iron. *Water Research, 78*, 144–153. https://doi.org/10.1016/j.watres.2015.04.009.

Rayaroth, M. P., Lee, C.-S., Aravind, U. K., Aravindakumar, C. T., & Chang, Y.-S. (2017). Oxidative degradation of benzoic acid using Fe0- and sulfidized Fe0-activated persulfate: A comparative study. *Chemical Engineering Journal, 315*, 426–436. https://doi.org/10.1016/j.cej.2017.01.031.

Rickard, D., & Luther, G. W. (2007). Chemistry of iron sulfides. *Chemical Reviews, 107*, 514–562. https://doi.org/10.1021/cr0503658.

Shao, D., Ren, X., Wen, J., Hu, S., Xiong, J., Jiang, T., Wang, X., & Wang, X. (2016). Immobilization of uranium by biomaterial stabilized FeS nanoparticles: Effects of stabilizer and enrichment mechanism. *Journal of Hazardous Materials, 302*, 1–9. https://doi.org/10.1016/j.jhazmat.2015.09.043.

Song, S., Su, Y., Adeleye, A. S., Zhang, Y., & Zhou, X. (2017). Optimal design and characterization of sulfide-modified nanoscale zerovalent iron for diclofenac removal. *Applied Catalysis B: Environmental, 201*, 211–220. https://doi.org/10.1016/j.apcatb.2016.07.055.

Su, Y., Adeleye, A. S., Huang, Y., Sun, X., Dai, C., Zhou, X., Zhang, Y., & Keller, A. A. (2014a). Simultaneous removal of cadmium and nitrate in aqueous media by nanoscale zerovalent iron (NZVI) and au doped NZVI particles. *Water Research, 63*, 102–111. https://doi.org/10.1016/j.watres.2014.06.008.

Su, Y., Adeleye, A. S., Zhou, X., Dai, C., Zhang, W., Keller, A. A., & Zhang, Y. (2014b). Effects of nitrate on the treatment of lead contaminated groundwater by nanoscale zerovalent iron. *Journal of Hazardous Materials, 280*, 504–513. https://doi.org/10.1016/j.jhazmat.2014.08.040.

Su, Y., Adeleye, A. S., Keller, A. A., Huang, Y., Dai, C., Zhou, X., & Zhang, Y. (2015). Magnetic sulfide-modified nanoscale zerovalent iron (S-NZVI) for dissolved metal ion removal. *Water Research, 74*, 47–57. https://doi.org/10.1016/j.watres.2015.02.004.

Su, Y., Adeleye, A. S., Huang, Y., Zhou, X., Keller, A. A., & Zhang, Y. (2016). Direct synthesis of novel and reactive sulfide-modified nano iron through nanoparticle seeding for improved cadmium-contaminated water treatment. *Scientific Reports, 6*, 24358. https://doi.org/10.1038/srep24358.

Suzuki, I. (2001). Microbial leaching of metals from sulfide minerals. *Biotechnology Advances, 19*, 119–132. https://doi.org/10.1016/S0734-9750(01)00053-2.

Tang, J., Tang, L., Feng, H., Zeng, G., Dong, H., Zhang, C., Huang, B., Deng, Y., Wang, J., & Zhou, Y. (2016). pH-dependent degradation of p-nitrophenol by sulfidated nanoscale zerovalent iron under aerobic or anoxic conditions. *Journal of Hazardous Materials, 320*, 581–590. https://doi.org/10.1016/j.jhazmat.2016.07.042.

Wang, J., & Farrell, J. (2003). Investigating the role of atomic hydrogen on chloroethene reactions with iron using Tafel analysis and electrochemical impedance spectroscopy. *Environmental Science & Technology, 37*, 3891–3896. https://doi.org/10.1021/es0264605.

Wang, L., Cao, M., Ai, Z., & Zhang, L. (2014). Dramatically enhanced aerobic atrazine degradation with Fe@Fe 2O3 core-shell nanowires by tetrapolyphosphate. *Environmental Science & Technology, 48*, 3354–3362. https://doi.org/10.1021/es404741x.

Xia, X., Ling, L., & Zhang, W. (2017). Genesis of pure Se(0) nano- and micro-structures in wastewater with nanoscale zero-valent iron (NZVI). *Environmental Science. Nano, 4*, 52–59. https://doi.org/10.1039/C6EN00231E.

Xu, L., & Wang, J. (2011). A heterogeneous Fenton-like system with nanoparticulate zero-valent iron for removal of 4-chloro-3-methyl phenol. *Journal of Hazardous Materials, 186*, 256–264. https://doi.org/10.1016/j.jhazmat.2010.10.116.

Xu, C., Zhang, B., Wang, Y., Shao, Q., Zhou, W., Fan, D., Bandstra, J. Z., Shi, Z., & Tratnyek, P. G. (2016). Effects of Sulfidation magnetization and oxygenation on azo dye reduction by zerovalent iron. *Environmental Science & Technology, 50*, 11879–11887.

Zhang, Y., Su, Y., Zhou, X., Dai, C., & Keller, A. A. (2013). A new insight on the core-shell structure of zerovalent iron nanoparticles and its application for Pb(II) sequestration. *Journal of Hazardous Materials, 263*, 685–693. https://doi.org/10.1016/j.jhazmat.2013.10.031.

Zhang, Y., Li, Y., Dai, C., Zhou, X., & Zhang, W. (2014). Sequestration of Cd(II) with nanoscale zero-valent iron (NZVI): Characterization and test in a two-stage system. *Chemical Engineering Journal, 244*, 218–226. https://doi.org/10.1016/j.cej.2014.01.061.

Zhang, Q., Guo, W., Yue, X., Liu, Z., & Li, X. (2016). Degradation of rhodamine B using FeS-coated zero-valent iron nanoparticles in the presence of dissolved oxygen. *Environmental Progress & Sustainable Energy, 35*, 1673–1678. https://doi.org/10.1002/ep.12412.

Chapter 10
Microbial Perspective of NZVI Applications

Panaya Kotchaplai, Eakalak Khan, and Alisa S. Vangnai

Abstract Nanoscale zero-valent iron (NZVI), due to its small size and high reactivity, is regarded as a promising alternative especially for in situ environmental remediation. There has already been a number of successful in situ contaminant removal/remediation using NZVI. In this context, interactions between NZVI and environmental microorganisms at the contaminated site are inevitable. The high reactivity of NZVI could potentially cause an adverse effect to microorganisms that are involved in environmental restoration. The interactions between NZVI and microorganism may in turn affect NZVI reactivity. Accordingly, it is important to understand the microbial aspects of NZVI applications. This chapter provides an overview of the consequent effect of the interactions between NZVI and microorganisms including the effect of NZVI on microorganisms as well as the effect of microorganisms on NZVI behavior. It specifically focuses on the reported effects of NZVI on microbial survival and activity, as well as several factors causing the complication of toxicity assessment. The prospects of NZVI-enhanced bioremediation is also discussed. Finally, this chapter presents future research needs in furtherance of successful NZVI applications.

P. Kotchaplai
International Program in Hazardous Substance and Environmental Management, Graduate School, Chulalongkorn University, Bangkok, Thailand

Center of Excellence on Hazardous Substance Management (HSM), Chulalongkorn University, Bangkok, Thailand

E. Khan
Department of Civil and Environmental Engineering, North Dakota State University, Fargo, ND, USA

A. S. Vangnai (✉)
Center of Excellence on Hazardous Substance Management (HSM), Chulalongkorn University, Bangkok, Thailand

Biocatalyst and Environmental Biotechnology Research Unit, Department of Biochemistry, Faculty of Science, Chulalongkorn University, Bangkok, Thailand
e-mail: alisa.v@chula.ac.th

T. Phenrat, G. V. Lowry (eds.), *Nanoscale Zerovalent Iron Particles for Environmental Restoration*, https://doi.org/10.1007/978-3-319-95340-3_10

Keywords Nanoscale zerovalent iron · Microbiology · Bioremediation · Combined remedies with NZVI

10.1 Role of Microorganisms in Environmental Restoration

Ecological systems consist of plants, animals, and microorganisms (i.e., bacteria, fungi, archaea, and viruses), of which microorganisms are the most abundant. The numbers, distribution, and diversity of microorganisms depend on their habitat or environment. For instance, microorganisms are most abundant in soil or sediment (10^7–10^{10} cell/g dry weight), with fewer in water or groundwater (10^2–10^6 cell/g dry weight) (Vorobyova et al. 1997). Many microorganisms play important roles in a range of biological processes or ecological functions in ecosystems, such as nutrient recycling, soil processing, transformation or detoxification of toxic contaminants, protecting plants from disease and abiotic stress, or supporting plant growth.

Because they are versatile and able to adapt, some microorganisms can survive and even thrive in a wide range of sometimes extreme environments including contaminated sites. Several indigenous bacteria, for example, *Pseudomonas, Burkholderia*, and *Acinetobacter*, are known for their metabolic versatility, which allows them to use soil or groundwater contaminants to support their growth and, at the same time, reduce the contaminant concentrations. Microorganisms can degrade or detoxify organic pollutants and can also immobilize contaminants (e.g., heavy metals) through intracellular accumulation or sorption/precipitation. Accordingly, biological treatment of pollutants at contaminated sites by microorganisms has been extensively studied. Environmental pollutants can be attenuated naturally by indigenous microorganisms, and/or by addition of nutrients (biostimulation) to support microbial growth/metabolism, and/or by addition of pollutant-degrading strains (bio-augmentation) to speed up the remediation of the contaminated sites, respectively.

Rhizospheric microorganisms reside in the rhizosphere, which is the region next to the plant roots in the soil. Soil microorganisms are attracted to the rhizosphere by plant root exudates, which contain a broad range of compounds including sugars, organic acids, and amino acids that are sources of carbon and nitrogen. Root exudates may also contain antimicrobials, phenolic compounds, and specialized signaling compounds that are selectively beneficial for certain groups of microorganisms. So, while root exudates are beneficial for these selected microorganisms, some rhizospheric microorganisms, known as plant growth-promoting rhizospheric microorganisms (PGPRs), are also in turn beneficial for plants by providing nutrients (i.e., phosphate solubilization) and phytohormones, accelerating mineralization processes, and reducing plant diseases and infections (Somers et al. 2004; Badri et al. 2009). *Bacillus*, *Burkholderia*, *Clostridium*, *Enterobacter*, and *Pseudomonas* are the most abundant PGPRs reported so far (Somers et al. 2004; Badri et al. 2009). Some of these rhizospheric microorganisms are capable of degrading pollutants of which the process is known as rhizoremediation (Kuiper et al. 2004). Rhizoremediation has been frequently used to reduce the contamination of pesticides and/or herbicides in agricultural soil and other organic pollutants in biosolids that have been used as fertilizer in agricultural areas (Silambarasan and Vangnai 2017).

The abundance and distribution of environmental microorganisms change with changes in their habitats and surroundings. Climate change, fertilizer, pesticide, and herbicide applications, and pollutant release are all known to affect bacterial communities directly or indirectly by affecting microbial cells or changing the environmental conditions, respectively. Fluctuations in the environment and the presence of toxic compounds may alter microbial communities such that their ecological functions are affected. This includes the addition of NZVI for remediation, which can alter redox conditions, provide an electron source (H_2), and can potentially be toxic to microorganisms. This chapter discusses aspects of the interactions between microorganisms and NZVI.

10.2 Impact of NZVI on Microorganisms

10.2.1 Toxicity of NZVI

Nanoscale zero-valent iron (NZVI) has been used for in situ environmental pollutant treatment, and thus large amounts of NZVI have purposely been released into the environment (Mueller et al. 2012). Consequently, concern about the environmental impact of these reactive iron particles has also increased. In general, iron is an essential trace nutrient for microorganisms, generally existing in the environment as soluble Fe(II) and insoluble Fe(III) depending on the environmental conditions, such as the oxygen concentration (ORP) or pH (Straub et al. 2001). Even though large amounts of natural iron already exists in the environment, substantial release of NZVI particles may cause unintended consequences due to their high reactivity and propensity to release Fe(II).

Information from the literature regarding NZVI toxicity is summarized in Table 10.1. Initially, NZVI was reported to be strongly toxic to bacteria (Auffan et al. 2008; Lee et al. 2008; Diao and Yao 2009) and was proposed as a potential antibacterial agent (Lee et al. 2008; Kim et al. 2010). While the practical dose of NZVI for in situ remediation applications ranges from 0.3 to 30 g/L (Kocur et al. 2016), NZVI can be noticeably toxic at concentrations as low as 0.009 g/L in anaerobic conditions (3.4 log inactivation) or 0.09 g/L in aerobic conditions (2.6 log inactivation) and may lead to a decrease in the number of viable bacteria cells (Lee et al. 2008). For example, Diao and Yao (2009) reported that, under aerobic conditions, the cell viability of *Pseudomonas fluorescens* and *Bacillus subtilis* decreased by almost two and one orders of magnitude, respectively, when exposed to NZVI at a concentration of 0.1 g/L, whereas a fungal strain, *Aspergillus versicolor*, was able to tolerate exposure to NZVI at concentrations of up to 10 g/L (Diao and Yao 2009). Shah et al. (2010) reported that iron-based

Table 10.1 NZVI toxicity on microorganisms

Microbial strains	NZVI characteristics	Experimental condition (NZVI concentration, medium, and the test conditions)	Findings	References
Agrobacterium sp. PH-08	Synthesized NZVI with surface modification	0.1–0.25 g/L, water, shaking under aerobic condition	Significant accumulation of NZVI onto cell surface Surface modification of NZVI by carboxymethyl cellulose (CMC) reduced NZVI toxicity	Zhou et al. (2014)
Agrobacterium sp. PH-08	NZVI (NaBond, China)	0.1–10 g/L, water, shaking under aerobic condition	No significant effect on cell viability and biodegradability	Le et al. (2014)
Alcaligenes eutrophus	Synthesized NZVI with surface modification	0.65 g/L, medium, shaking under anaerobic condition	The presence of NZVI decreased the total RNA content; however, the rebound in RNA content was observed after 2 days Surface modification affected nitrate removal rate and the generated product. The presence of surface modifier reduced NZVI toxicity	An et al. (2010)
B. cereus ATCC 14579	Commercial NZVI (NANOFER 25S, NANOIRON s.r.o., Czech Republic)	1–10 g/L, NaCl solution, non-shaking under aerobic condition	Proteomic analysis revealed the upregulation of thioredoxin, an oxidative stress response protein, and downregulation of membrane-bound proteins TEM analysis suggested an early sporulation of bacterial cells	Fajardo et al. (2013)
B. nealsonii	Commercial NZVI (NANOFER 25S, NANOIRON s.r.o., Czech Republic)	1–10 g/L, NaCl solution, non-shaking under aerobic condition	One-log cell inactivation was observed when cells were exposed to 10 g/L NZVI	Fajardo et al. (2012)

B. subtilis ATCC 9372	Synthesized NZVI	1–10 g/L, deionized water, shaking under aerobic and anaerobic condition	High toxicity (80–100% inactivation) was observed, especially under aerobic condition High accumulation of iron nanoparticles onto cell surface was observed	Diao and Yao (2009)
E. coli strains (*E. coli* strain Qc1301 and Qc2472 (*sodA-sodB* mutant))	Synthesized NZVI	0.007–0.7 g/L, ultrapure water, aerobic condition	Toxicity levels from iron species are as follows: NZVI > nMagnetite > nMaghemite Significant reduction in number of viable cells could be detected at 0.07 g/L of NZVI and 0.7 g/L of nMagnetite Significant effect of all three iron nanoparticles was observed in cells devoid of antioxidants enzymes	Auffan et al. (2008)
E. coli ATCC 8739	Synthesized NZVI	1.2–110 mg/L, 2 mM carbonate buffer, stirring under aerobic and anaerobic conditions	Anaerobically NZVI exposure resulted in higher cell inactivation, comparing to aerobic condition Iron oxide nanoparticles, soluble iron species, and Fe^0 powder showed insignificant toxicity under anaerobic condition	Lee et al. (2008))
E. coli ATCC 8739	Synthesized NZVI	1.2–110 mg/L, 3 mM carbonate buffer, mixing using magnetic stirrer under aerobic and anaerobic condition	Anaerobically NZVI exposure resulted in higher level of intracellular ROS, disturbed cell membrane integrity and respiratory activity, comparing to aerobic condition	Kim et al. (2010)
E. coli ATCC 33876			Anaerobically NZVI exposure resulted in higher cell inactivation,	Li et al. (2010b)

(continued)

Table 10.1 (continued)

Microbial strains	NZVI characteristics	Experimental condition (NZVI concentration, medium, and the test conditions)	Findings	References
	Commercial NZVI (Toda Kogyo Corp., Japan) with surface modification	0.001–2 g/L, 5 mM bicarbonate buffer, shaking under aerobic and anaerobic condition	comparing to aerobic condition Surface modification or the presence of natural organic matters reduced NZVI attachment to cells, thus reducing the toxicity	
E. coli strains (*E. coli* JM109 and *E. coli* BW25113)	Commercial NZVI (Toda Kogyo Corp., Japan)	1 g/L, 50 mM carbonate buffer, mixing using magnetic stirrer under aerobic condition	Bacterial cells at exponential and decline growth phases were more susceptible to NZVI exposure than those at lag and stationary phases. Toxic effect is strain-dependent	Chaithawiwat et al. (2016)
E. coli strains (*E. coli* BW25113 and its mutants)	Commercial NZVI (Toda Kogyo Corp., Japan)	1 g/L, 50 mM carbonate buffer, mixing using magnetic stirrer under aerobic condition	Bacterial strains lacking antioxidant enzyme or sigma S factor were more sensitive to NZVI exposure	Chaithawiwat et al. (2016)
Klebsiella planticola DSZ	Commercial NZVI (NANOFER 25S, NANOIRON s.r.o., Czech Republic)	1–10 g/L, NaCl solution, non-shaking under aerobic condition	Accumulation of NZVI onto cell surface was observed No significant toxicity was observed when cells were exposed to 1–10 g/L NZVI	Fajardo et al. (2012)
K. oxytoca K5 (isolated from NZVI-treated soil)	Commercial NZVI (NANOFER 25S, NANOIRON s.r.o., Czech Republic)	1–10 g/L, NaCl solution, non-shaking under aerobic condition	No significant NZVI toxicity was observed	Sacca et al. (2013)
Paracoccus sp. YF1	Synthesized NZVI	0.05–1 g/L, minimal medium, shaking under aerobic condition	NZVI at 50 mg/L promoted bacterial growth and nitrate removal. However, cell growth and activity were adversely affected when cells were exposed to high concentration of NZVI	Jiang et al. (2015)

P. fluorescens ATCC 13525	Synthesized NZVI	1–10 g/L, deionized water shaking under aerobic and anaerobic condition	Cell inactivation of more than 99% was observed in both aerobic and anaerobic NZVI exposure Exposure to oxidized iron nanoparticles did not affect cell survival	Diao and Yao (2009)
P. stutzeri ATCC 14405	Commercial NZVI (NANOFER 25S, NANOIRON s.r.o., Czech Republic)	1–10 g/L, NaCl solution, non-shaking under aerobic condition	Two-log reduction in cell viability was observed within 10 min of exposure to 1 and 5 g/L of NZVI. Exposure to 10 g/L NZVI resulted in lower toxicity (one-log reduction) due to the aggregation of NZVI The oxidative stress responses were induced as indicated by the upregulation of *katB* (encoding catalase) and *sodB* gene product Downregulation of membrane transport proteins upon NZVI attachment onto cell surface suggested NZVI-induced damage on bacterial membrane	Sacca et al. (2014b)
P. putida G7	Commercial NZVI (NANOFER STAR, NANOIRON s.r.o., Czech Republic)	0.1 μg/L – 1 g/L, phosphate buffer, aerobic condition	0.2-log cell inactivation was observed in cell exposed to 0.1–1 g/L Negative chemotactic response was detected when cells were exposed to subinhibitory concentration of NZVI	Ortega-Calvo et al. (2016)
P. putida F1	Commercial NZVI (NANOFER 25S, NANOIRON s.r.o., Czech Republic)	0.1–5 g/L, minimal medium, shaking under aerobic condition	2–4 log of cell inactivation within 1 h of NZVI exposure Rebound in number of viable cells	Kotchaplai et al. (2017)

(continued)

Table 10.1 (continued)

Microbial strains	NZVI characteristics	Experimental condition (NZVI concentration, medium, and the test conditions)	Findings	References
			was observed with prolonged exposure time Repetitive NZVI exposure resulted in the NZVI-persistent phenotype Membrane damage and membrane adaptation were reported	
P. aeruginosa PAO1 (planktonic and biofilm)	Commercial NZVI (Golder Associates, Canada) with palladization and surface modification	0.015 and 0.15 g/L, Luria-Bertani (LB) medium, shaking under aerobic condition	Adverse effect on bacterial growth was observed when cells were exposed to 0.15 g/L Pd-NZVI. While rhamnolipid-coated Pd-NZVI reduced biofilm formation, the CMC-coated Pd-NZVI showed no adverse effect on biofilm formation	Basnet et al. (2016)
Sphingomonas sp. PH-07	Synthesized NZVI with palladization	0.5–10 g/L, minimal medium, shaking under aerobic condition	*Sphingomonas* sp. PH-07 could grow in the presence of NZVI up to 2.5 g/L. Bacterial growth was completely inhibited when cells were exposed to NZVI at 7.5 g/L	Kim et al. (2012b)
MS2 coliphages	Synthesized NZVI	0.09–0.9 mM, 3 mM carbonate buffer, mixing using magnetic stirrer under aerobic and anaerobic condition	Higher NZVI toxicity to MS2 coliphages under aerated condition, than that in deaerated condition Direct interaction between NZVI and viral surface mainly contributes to the toxicity	Kim et al. (2011)
Aspergillus versicolor ATCC 26644	Synthesized NZVI	1–10 g/L, deionized water shaking under aerobic and anaerobic condition	No significant effect was observed	Diao and Yao (2009)

nanoparticles at a concentration of 0.1 mM or 5.58 mg/L had little effect on fungal growth (*Trametes versicolor*) but had adverse effects on the production of several lignocellulose-degrading enzymes (Shah et al. 2010). Shah et al. (2010), however, did not report either the characteristics of the iron nanoparticles (i.e., iron species, reactivity) (Shah et al. 2010) or the mechanisms behind the processes.

While most of the studies reviewed reported that there were obvious toxic effects on bacteria from NZVI, Fajardo et al. (2012) and Sacca et al. (2013) observed that it was only slightly toxic to *Klebsiella planticola* D5Z and *K. oxytoca* K5, a Gram-negative bacteria isolated from NZVI-treated soil, respectively, at 10 g/L (Fajardo et al. 2012; Sacca et al. 2013). Cell viability of *Sphingomonas* sp. strain PH-07 decreased by less than one order of magnitude at NZVI concentrations up to 5.0 g/L, showing tolerance at high concentrations (Kim et al. 2012b). It should be noted that these studies were conducted under aerobic conditions, in which the toxicity of NZVI is assumed to be low. Under aerobic conditions, the oxidation of NZVI occurs rapidly, so that less toxic oxidation products such as iron oxide are produced (Lee et al. 2008). However, these studies have not reported information about the NZVI reaction times or reactivity; thus it is difficult to distinguish the intrinsic strain-dependent tolerance of NZVI from the limited toxicity of oxidized NZVI. However, it does suggest that risks associated with NZVI are decreased when the material has become oxidized to the corresponding Fe oxide.

The aforementioned studies indicate that most bacteria are susceptible to NZVI exposure; however, in situ applications of NZVI do not always have deleterious effects on the abundance of indigenous microorganisms. Pawlett et al. (2013) reported that the total biomass of indigenous microorganisms in agricultural soil was not affected by additions of NZVI, except in soil that was supplemented with organic matter, in which the microbial biomass decreased by 29% (Pawlett et al. 2013). Examination of a microbial phospholipid profile indicated that there was little change in either the microbial community or an indicator of the abundance of fungal and bacterial membrane stress (*trans/cis* ratio of 16:1ω7) when exposed to NZVI (Pawlett et al. 2013), but that monounsaturated fatty acids of Gram-negative bacteria did change slightly upon exposure to NZVI (Pawlett et al. 2013). Nevertheless, there is still uncertainty as to whether phospholipid profiles can be used effectively to investigate alterations in microbial communities in stressful conditions because these profiles can change as bacteria adapt to a range of stresses. For instance, decreases in the sum, rather than the abundance (caused by higher vulnerability to NZVI), of monounsaturated fatty acids of Gram-negative bacteria may reflect how bacteria respond to membrane stress, as proposed in the study (Pawlett et al. 2013). Also, *cis* monounsaturated fatty acids may change to *trans* monounsaturated fatty acids as a short-term response mechanism to membrane damage in certain bacterial strains, so the *trans/cis* ratio of 16:1ω7 reported in this study for a microbial community 4 months after NZVI was added may not be valid.

Kirschling et al. (2010) investigated the effect of NZVI on microbial communities in groundwater containing aquifer material from three different trichloroethene (TCE)-contaminated sites (Kirschling et al. 2010). When NZVI (1.5 g/L) was added, the conditions were more reduced, the pH increased, and the activity of the sulfate-reducing population was stimulated by the H_2 generated by the NZVI (Kirschling et al. 2010). Further, polyaspartate, a biodegradable polymer that is used to modify the NZVI surface, supports the growth of microorganisms (Kirschling et al. 2010). In contrast to Kirschling et al. (2010), Kumar et al. (2014) reported that NZVI at a w/w ratio of 1% inhibited sulfate-reducing bacteria in an aquifer microcosm, whereas the granular and microscale zero-valent iron stimulated sulfate reduction (Kumar et al. 2014). The dose-dependent inhibition of bacterial sulfate reduction occurred when NZVI concentrations exceeded 0.5 g/L (Kumar et al. 2014).

The different routes by which NZVI is introduced to the environment may determine its potential toxicity on the recipient microorganisms. When NZVI is intentionally used in combination with bioremediation, microorganisms, especially those that reside in contaminated sites, are exposed to the reactive form of NZVI. On the other hand, NZVI in industrial discharges or sludge (which can be later used as biosolids) may end up in agricultural areas in an oxidized form. While several studies have shown that nanoparticles can have adverse effects on the abundance and activity of functional microorganisms in agricultural soils (Ben-Moshe et al. 2013; Chai et al. 2015; Xu et al. 2015; He et al. 2016b), iron oxide nanoparticles appear to have limited adverse effects. Li et al. (2007) used NZVI to react with sulfide compounds, thereby reducing the odor of the biosolids (Li et al. 2007). They also proposed that iron oxides and hydroxides, end products of NZVI, were similar to iron minerals that occurred naturally in the environment and that small amounts of NZVI could increase the bioavailable iron in soils (Li et al. 2007). He et al. (2011) reported increased activity of invertase and urease, enzymes involved in soil nutrient cycling, because Actinomycetes species were stimulated when iron oxide nanoparticles were added (He et al. 2011). Ben-Moshe et al. (2013) reported that soil microbial community alterations induced by iron oxide nanoparticles were linked to soil properties (Ben-Moshe et al. 2013).

Biofilms are complex communities of single or multiple microorganism species associated with surfaces that are held together by a matrix of extracellular polymeric substances (EPS) (Davey and O'Toole 2000). Microorganisms in groundwater aquifers, soil, marine environment, industrial effluents discharge, as well as those applied in bioreactors are mainly in the form of biofilms. The EPS matrix generally consists of carbohydrates and proteins, and environmental DNA protects microorganisms from many stresses, including nanoparticles (Miao et al. 2009; Dimkpa et al. 2011; Sudheer Khan et al. 2011; Thuptimdang et al. 2015; Wang et al. 2016a), by providing binding/sequestration sites (Khan et al. 2011; Ikuma et al. 2015; Wang et al. 2016a). So far, the interactions between NZVI and biofilms have been mainly assessed in terms of the retardation of NZVI transport or retention of NZVI in soil columns. Basnet et al. (2016) reported that Pd-NZVI had a limited effect on bacterial biofilm but had adverse effects on biofilm formation (Basnet et al. 2016). Since studies of the interactions between nanoparticles and biofilms are limited (Ikuma et al. 2015), more investigations are needed, particularly on the stages, components, and structure of biofilms, since these factors have been reported to affect biofilm susceptibility to stresses (Joshi et al. 2012; Thuptimdang et al. 2015; Thuptimdang et al. 2017).

10.2.2 Mechanisms of NZVI Toxicity and Microbial Responses

Figure 10.1 shows a simplified overview of NZVI-induced damage in microbial cells and microbial responses. Under aerobic conditions, bacteria normally encounter some deleterious effects from endogenous reactive oxygen species (ROS) generated during respiration. Reactive oxygen species, and in particular hydroxyl radicals (HO·), are highly reactive and attack bacterial biomolecules such as DNA, proteins, and lipids (Sevcu et al. 2011). Response mechanisms, including ROS-scavenging enzymes or damage-repair systems, are triggered to balance or reduce the levels of ROS generated, thereby alleviating the oxidative stress to cells. As highly redox-active materials, NZVI particles are expected to generate significant amounts of ROS. In the presence of oxygen, Fe^0 in NZVI particles can directly transfer electrons to oxygen molecules generating hydrogen peroxide (Auffan et al. 2008). The oxidation products of Fe^0, Fe(II), and Fe(III) have been known to catalyze Fenton and Haber-Weiss reactions, thereby generating ROS, particularly HO·. The abrupt increases in ROS following exposure to NZVI may exceed the response capability of bacteria, and may trigger oxidative stress, and subsequent cell death (Lefevre et al. 2016). So far, numerous studies have highlighted both the NZVI-mediated oxidative damage and the responses of microorganisms (Auffan et al. 2008; Lee et al. 2008; Kim et al. 2010; Fajardo et al. 2013; Sacca et al. 2014b; Chaithawiwat et al. 2016). A study of *P. stutzeri* showed that the *katB* expression that encodes for catalase, the antioxidant enzyme, increased by 570% after exposure to NZVI (Sacca et al. 2014b). *E. coli* that lack the genes to encode for superoxide dismutase, the antioxidant enzyme, are more susceptible to NZVI than the wild type, which suggests that the systems that defend bacteria from oxidative stress during NZVI exposure are important (Auffan et al. 2008; Chaithawiwat et al. 2016). The oxidation products of NZVI (e.g., maghemite, magnetite, and lepidocrocite) are much less toxic to bacteria (Auffan et al. 2008; Lee et al. 2008; Diao and Yao 2009).

Toxicity mediated by NZVI results in intracellular rather than extracellular oxidative stress, as scavenging the extracellular ROS (in bulk solution) did not influence bacterial inactivation (Kim et al. 2010), and may reflect the very short half-lives of ROS. For instance, the half-lives of HO· and H_2O_2 are about 10^{-9} and 10^{-3} s, respectively (D'Autreaux and Toledano 2007). When NZVI interacts with dissolved O_2, the generated ROS may dissipate before attacking microorganisms. The filtrate of NZVI was not toxic to bacteria (Diao and Yao 2009; Chaithawiwat et al. 2016), which confirms that NZVI toxicity decreased because of limited interactions between bacterial and NZVI particles, as reported previously (Li et al. 2010b; Chen et al. 2011). Recently, He et al. (2016a) reported that Fe(II) and H_2O_2

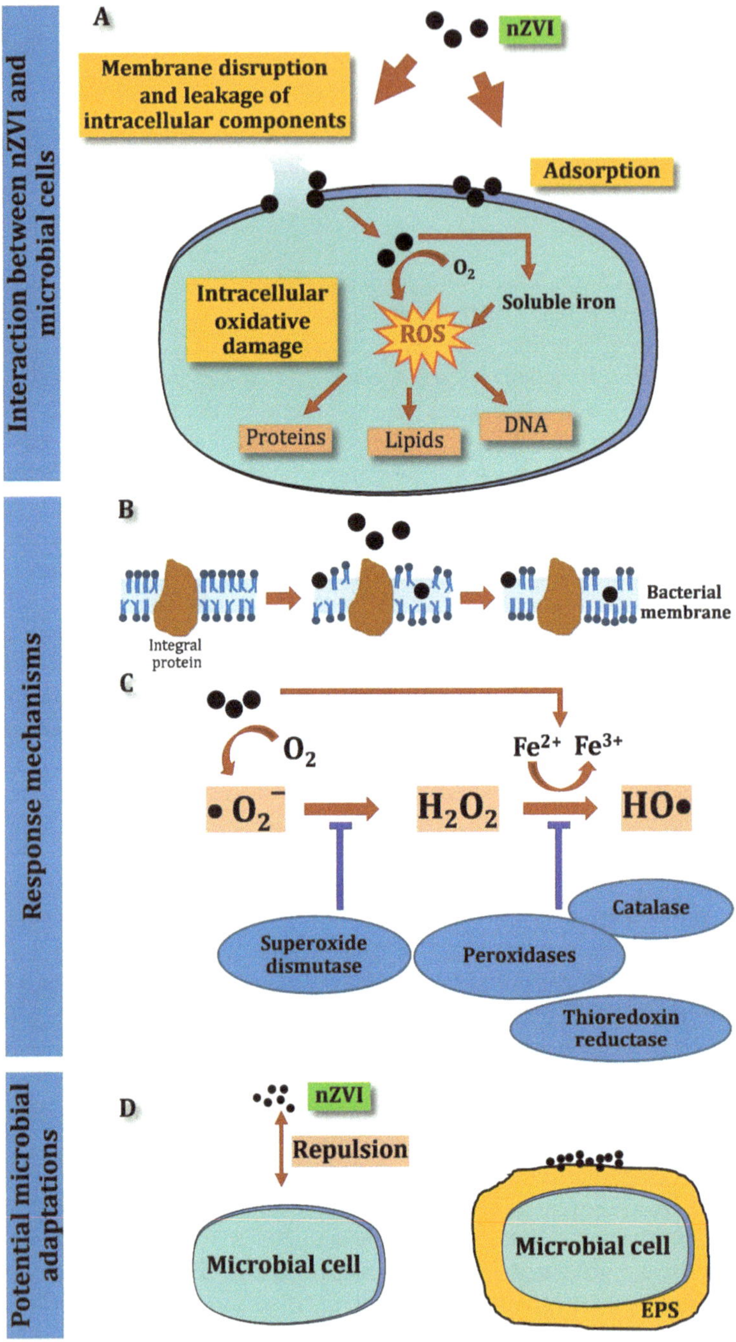

Fig. 10.1 (**a**) A simplified overview of NZVI-induced damage in microbial cells. Once NZVI particles attach to the cell surface, they may disturb and disrupt microbial membrane causing

generated under aerobic conditions were associated with the surface of NZVI particles (He et al. 2016a). Accordingly, the interactions between bacterial membranes and the iron particles contribute significantly to the toxicity of NZVI to microorganisms (Popova et al. 2008; Li et al. 2010b; Chen et al. 2011; Chaithawiwat et al. 2016; Lefevre et al. 2016). It appears that, during exposure to NZVI, iron nanoparticles tend to attach to bacterial membranes, disrupting membrane composition, interfering with membrane functionality, and promoting iron-induced intracellular oxidative stress (Lefevre et al. 2016). However, as discussed later in the chapter, there are mechanisms that limit the adhesion of the NZVI to microbial cell walls, thereby decreasing toxicity (Li et al. 2010b; Chen et al. 2011).

When used in situ, NZVI is mainly applied to subsurface environments where the conditions are anaerobic (few studies have demonstrated aerobic in situ applications). While it is known that NZVI is much more toxic in anaerobic conditions than aerobic conditions (Lee et al. 2008; Kim et al. 2010; Lv et al. 2017), few studies have examined the influence of oxygen on toxicity of NZVI in detail. Lv et al. (2017) recently explored different mechanisms of NZVI toxicity under different levels of dissolved oxygen (Lv et al. 2017). The inactivation mechanisms observed under anaerobic conditions were mainly attributed to physical disruption with a small contribution from soluble Fe(II), whereas oxidative stress and physical disruption both had major roles in inactivation mechanisms under aerobic conditions (Lv et al. 2017). These findings support the toxicity mechanisms of NZVI proposed earlier (Lee et al. 2008).

Even though it has been reported as highly toxic, the toxicity of NZVI appears to be transient, such that the activities or abundance of microorganisms can be recovered when NZVI is removed or has completely lost its reactivity (Kotchaplai et al. 2017). Accordingly, microbes must either adapt or respond to NZVI to withstand any damage that occurs so that they can survive the exposure. While NZVI toxicity

Fig. 10.1 (continued) intracellular component leakage. Reactive oxygen species (e.g., O_2^-, H_2O_2, or OH·) may be generated when electrons are transferred from NZVI to intracellular oxygen and thus damage cell components including proteins, DNA, and lipids. Several stress response mechanisms have been reported toward membrane damage and oxidative stress. (**b**) NZVI-induced membrane damage and microbial response. The interactions between NZVI and microbial cells fluidize the membrane phospholipid bilayer, which may in turn affect the functionality of membrane-bound proteins. In response to the NZVI-induced damage, certain bacteria may adjust their membrane rigidity by increasing the proportion of saturated fatty acids or by converting *cis*- to *trans*-unsaturated fatty acids, resulting in a more tightly packed and more rigid membrane (i.e., less membrane fluidity), as shown in (**b**). (**c**) NZVI-induced oxidative stress and microbial response. ROS species generated during NZVI exposure could attack intracellular biomolecules. Microbial antioxidant enzymes are generally induced and expressed as a responsive mechanism. In addition, sporulation has also been reported as microbial response to NZVI exposure. (**d**) Microbial adaptation to NZVI-induced damage. Since NZVI and the cell surface need to be attached for NZVI toxicity to be exerted, it has been proposed that microorganisms may induce modifications in cell surfaces or increase the production of EPS to reduce interactions between NZVI and cells. The presence of a polymer coating on the NZVI may also inhibit attachment of NZVI to the cell surfaces through steric or electrosteric repulsions is triggered by interactions with microbial surfaces, and cell surfaces are clearly disturbed when exposed to NZVI, currently there is limited information about how microbes respond to NZVI-induced membrane damage.

Bacterial cytoplasmic membranes, which are phospholipid bilayers embedded with protein, serve as protective layers for cells and provide sites for many cellular processes. Several studies have reported that bacterial cells can quantitatively and qualitatively modify their membrane composition in response to membrane disturbance caused by exposure to nanoparticles (Fang et al. 2007; Hachicho et al. 2014; Zhu et al. 2014). For example, Kotchaplai et al. (2017) recently reported that *P. putida* F1 rapidly increased the *trans*- to *cis*- monounsaturated fatty acid ratio in response to an NZVI-induced fluidized membrane (Kotchaplai et al. 2017). This energy-independent *cis-trans* isomerization of the membrane fatty acid means that bacteria can maintain a membrane state that allows their survival (Heipieper et al. 2003). Microorganisms that lack this mechanism must rely on energy-demanding biosynthesis of saturated fatty acids and so may not adapt so easily to NZVI as the strain studied. The same study also reported that *P. putida* F1 adapted when repeatedly exposed to NZVI by producing a phenotypic variant that was more tolerant to NZVI (Kotchaplai et al. 2017). While it is still not clear about the mechanisms involved, alterations to the surface of the adapted cells probably means that there is less interaction between NZVI and bacterial cells, thereby mitigating NZVI toxicity (Kotchaplai et al. 2017). Dhas et al. (2014) reported that, when repeatedly exposed to increasing concentrations of silver and zinc oxide nanoparticles, several bacterial strains adapted by producing more EPS and extracellular proteins, thereby reducing direct contact between nanoparticles and cells (Dhas et al. 2014).

Chemotaxis is a bacterial behavioral response in which flagellated cell can sense and move toward or repel in response to the presented gradient of both organic and inorganic compounds (Pandey and Jain 2002). The positive taxis or chemical attraction allows certain microorganisms to reach nutrients as well as metabolizable compounds including environmental contaminants, thus increasing bioavailability and potentially bioremediation efficiency (Vangnai et al. 2013). On the other hand, the negative taxis or chemical repellent allows bacterial cells to escape the toxic condition ensuring cell survival (Pandey and Jain 2002). Ortega-Calvo et al. (2016) reported the negative chemotactic response of *P. putida* G7 exposed to subinhibitory concentration (1–10 μg/L) of NZVI, whereas higher NZVI concentration resulted in the reduction of viable cells (Ortega-Calvo et al. 2016). The same study also reported that bacterial taxis became undetectable after aerobically aging the NZVI (Ortega-Calvo et al. 2016). It was proposed that the NZVI-induced negative effect is associated with the NZVI toxicity rather than the detection by bacterial specific chemoreceptors (Ortega-Calvo et al. 2016).

10.2.3 Factors Affecting Toxicity Assessments of NZVI

Exposure conditions are critical when determining and interpreting the environmental impacts of nanoparticles (Holden et al. 2016). Unlike an ex situ process that is conducted under controlled conditions, the exposure conditions during in situ applications of NZVI are diverse and site-dependent. Therefore, contaminated sites need to be accurately characterized so that the fate of NZVI and its interactions with, and potential impacts on, the surrounding species in the environment can be predicted. Most previous studies, especially those that demonstrate high NZVI toxicity, have been conducted in media that are very chemically simple, such as water (Diao and Yao 2009), a buffer (e.g., 2–5 mM carbonate buffer) (Lee et al. 2008; Chaithawiwat et al. 2016), or NaCl solution (Fajardo et al. 2013; Sacca et al. 2013, 2014a, b). The ions or organic matter present also play an important role in the characteristics of NZVI and its toxicity. Phosphate species can form an insoluble complex with Fe(III) that results in passivation of the NZVI surface, thereby reducing the redox reactivity of NZVI, whereas the presence of oxalate can generate soluble Fe(II)-oxalate, which reduces passivation of the NZVI surface (Lee et al. 2008). The adsorption of natural organic matter on the surface of NZVI decreases the interactions between the cell and NZVI particles and so reduces the toxicity (Chen et al. 2011; Hachicho et al. 2014). The different surface characteristics (e.g., adsorbed polymers or other surface treatments) of NZVI lead to different interactions between cell surfaces and NZVI, thereby altering the effects of NZVI on microorganisms. Ionic strength has also been reported to influence the stability and aggregation of NZVI particles (Keller et al. 2012; Adeleye et al. 2013).

It is still difficult to compare and define the impact of NZVI on environmental microorganisms because of a lack of consistency in experimental conditions and a lack of clarity about the characteristics of NZVI. The fact that the characteristics of contaminated sites vary means that it is difficult to accurately predict the impacts of NZVI on environmental microorganisms. As such, the environmental impacts of NZVI appear to be site-dependent. While most studies have focused on the characteristics of NZVI, very few studies have assessed the impact of the microbe properties on the interactions between NZVI and the organisms (Chaithawiwat et al. 2016). Interactions between microorganisms and abiotic surfaces are influenced by microbial surface components that are in turn affected by solution chemistry (Gaboriaud et al. 2008; Shephard et al. 2010; Jacobson et al. 2015). The compounds present can also affect the metabolic state of microorganisms and may alter their susceptibility to stress. The toxicity of NZVI also depends on the bacterial growth phase (Chaithawiwat et al. 2016). According to Chaithawiwat et al. (2016), bacteria in the exponential phase are much more susceptible to NZVI exposure than cells in the stationary phase (Chaithawiwat et al. 2016). Some organic contaminants can serve as growth substrates for microorganisms and may promote an exponential growth phase of metabolically capable strains. It should also be noted that that the microbial communities will vary in different areas of a contaminated site, and this spatial heterogeneity of microorganisms in contaminated sites is very important as microbial susceptibility to NZVI also appears to be strain-dependent (Chaithawiwat et al. 2016). Groundwater, however, is hydrologically connected meaning that, as attempts to increase the mobility and stability of NZVI in subsurface environments are made, there is the potential for unintended migration of NZVI to a down-gradient non-contaminated site where the abundances and structures of bacterial communities differ from those in the source area. So better understanding of the susceptibility of different organisms to NZVI is needed to fully assess the potential risks from exposure to NZVI.

Microorganisms are typically associated with aquifer materials, soils, or sediments and are present at much lower numbers in groundwater (Hazen et al. 1991; Holm et al. 1992). The initial concentration of microorganisms also affects the toxicity of NZVI. The NZVI particles must be in contact with the microbial surface for further NZVI-induced damage to occur. So, when the ratio between the microbial concentration and NZVI particles is low, there may be a greater chance of interactions between NZVI particles and cells, while a high ratio may indicate a greater number of uninjured cells that can continue to grow and function.

10.3 Impact of Microorganisms on NZVI

10.3.1 Impact of Microorganisms on Transformations, Reactivity, and Characteristics of NZVI

The rapid aging of NZVI is a constraint to NZVI environmental applications. Several methods to regenerate or reactivate NZVI particles have been proposed (Xie and Cwiertny 2010; Yang et al. 2017). Dissimilatory iron-reducing bacteria (DIRB) are capable of shuttling electrons from organic matter to a variety of terminal electron acceptors including ferric iron and play a significant role in anaerobic or reductive transformation of the contaminants. Fe(III) reduction can occur via direct interactions between the surface of DIRB and iron oxide or via indirect shuttling of electrons through the secreted electron shuttle. Yang et al. (2017) reported that abiotic TCE reduction of NZVI could be restored by adding *Shewanella putrefaciens* and that the restorative effect was much more pronounced in a micro NZVI system (Yang et al. 2017). Reduction of the iron oxide shell of aged NZVI results in sorption of Fe(II), which can serve as a reductant, while the dissolution of Fe(II) from aged NZVI may limit the growth of a passivating Fe-oxide layer over the Fe(0) core (Yang et al. 2017). As mentioned earlier, some DIRB can indirectly reduce Fe(III) via the secreted electron shuttle (e.g., flavins). Riboflavin has been reported to compete with organic contaminants for electrons, thereby decreasing the effectiveness of the NZVI treatment (Bae and Lee 2014). However, during NZVI aging, the reduced riboflavin can reduce and solubilize Fe(II) and allow access to the reactive Fe(0) core, which prolongs the reactivity of NZVI (Bae and Lee 2014).

While studies have shown that NZVI particles can effectively adsorb heavy metals, more recent studies have reported release of adsorbed heavy metals from aged NZVI particles (Su et al. 2014; Calderon and Fullana 2015). Ye et al. (2017) reported that *Pantoea* sp. strain IMH, a Gram-negative bacteria containing *arsC* that aerobically reduces soluble As(V) to As(III), could slow down the aging process of NZVI (Ye et al. 2017). While so far As mobilization has been negligible, researchers have suggested that mobilization of As may be case-dependent (Ye et al. 2017).

Some studies modify NZVI with polymer coatings to improve the stability and mobility of NZVI in a contaminated site (Phenrat et al. 2009a). The abundance of the microbial population in a TCE-degrading community increased when PAA-coated NZVI was added (Kirschling et al. 2010). Some of the selected polymers, for example, PAA and CMC, are known to be biodegradable (Sieger et al. 1995; Tabata et al. 1999), providing a carbon source for bacterial growth. The loss of coating on the NZVI may also increase its reactivity since it has been shown that the adsorbed polymer decreases NZVI reactivity compared to uncoated NZVI (Phenrat et al. 2009b, 2018).

10.3.2 Impact of Microorganisms on NZVI Transport

The effective transport of NZVI to the target site is one of the most important factors for successful remediation. Therefore, many approaches, such as modification of NZVI particles and the solution chemistry that affects NZVI mobility, have been extensively studied so that the performance of NZVI applications can be predicted and improved (Saleh et al. 2008; Phenrat et al. 2009a; Lin et al. 2010; Jiemvarangkul et al. 2011; Kim et al. 2012a; Laumann et al. 2014). An association between NZVI and biofilms is inevitable as biofilms are the most common form of microorganisms in the environment, especially in soils and aquifers. However, the interactions between NZVI particles and biofilm have received little attention to date. Lerner et al. (2012) reported that NZVI transport in a glass bead column that contained bacterial biofilm depended on the ionic strength (1 and 25 mM NaCl solution) (Lerner et al. 2012). Similarly, Basnet et al. (2016) reported that biofilm retarded the transport of Pd-NZVI in both monovalent (3–100 mM NaCl) and divalent (1–30 mM $CaCl_2$) solutions (Basnet et al. 2016). The particles tested did not significantly damage the biofilm (Basnet et al. 2016). However, Basnet et al. (2016) also reported that biofilm formation decreased in the presence of 0.15 g/L rhamnolipid-modified Pd-NZVI, bare Pd-NZVI, or surface modifiers (i.e., rhamnolipid and CMC) but increased considerably when exposed to CMC-modified Pd-NZVI (Basnet et al. 2016). Even though NZVI transport may improve when the surface is modified, a decrease in transport during in situ applications would be expected when biofilms are present (Basnet et al. 2016).

To date, few studies have reported any effects, including retardation, of biofilms on nanoparticle transport; Mitzel and Tufenkji (2014) reported that the retention time (i.e., faster travelling) of silver nanoparticles in $NaNO_3$ and $Ca(NO_3)_2$ decreased, presumably due to the repulsive electrosteric force between bacterial EPS and the polymer coating on the silver nanoparticles (Mitzel and Tufenkji 2014).

10.4 Synergistic Effects During Potential Nanotechnology-Bioremediation

One of the most promising aspects of NZVI has been its ability to stimulate biodegradation of contaminants. Therefore, the combined remedy of NZVI and concomitant bioremediation is an important but under-explored application. Researchers have reported that, when NZVI is injected, the oxidation-reduction potential (ORP) may decrease rapidly, causing the conditions to become more reduced and favor the reduction of chlorinated contaminants (Wei et al. 2010; Kocur et al. 2014). The large amount of dissolved H_2 that is generated when NZVI interacts with water, a predominant electron acceptor in groundwater, also contributes to an immediate decline in ORP. However, despite the sharp decline when NZVI is introduced, the ORP gradually increases or rebounds afterward because of migration of NZVI (Wei et al. 2010) or from the complete, or near complete, oxidation of the NZVI. The NZVI particles, dissolved Fe(II), and the H_2 generated can act as inorganic electron donors for several biological processes including denitrification, reductive dehalogenation, sulfate reduction, and methanogenesis (Shi et al. 2015; Wang et al. 2016b). Thus, the stimulation of microbial activity during in situ applications of NZVI, particularly after abiotic processes, or the reactive period of NZVI, is attractive.

In addition to the stimulation by NZVI, the coatings on NZVI or co-injected substrate may promote biodegradation. Modifiers, such as CMC, anionic polymers, surfactants, and oil-based emulsifiers, have been used to modify NZVI surfaces or have been co-injected to improve the stability, reactivity, and mobility of NZVI. Some of these substrates, such as CMC, act as electron donors for microbial activity, and selected organic compounds can be metabolized by fermentative microorganisms, and may provide H_2. Some polymers may also reduce the interactions between NZVI particles and cell surfaces, thereby mitigating NZVI toxicity. Several lab-scale studies have reported reduced or inhibited microbial activity when NZVI, particularly uncoated NZVI, is added; however, NZVI toxicity may be alleviated and microbial growth or activity stimulated when biodegradable polymers are added (Kirschling et al. 2010; Xiu et al. 2010a). Field demonstrations have shown that, when co-injected, these modifiers can enhance long-term microbial activity (e.g., reductive dechlorination) (He et al. 2010; Kocur et al. 2015). Kocur et al. (2015) investigated how, when co-injected with CMC, NZVI influenced in situ reductive dechlorination of chlorinated volatile organic compounds (cVOCs) (Kocur et al. 2015) and reported abiotic dechlorination of cVOCs followed by biotic dechlorination. The increased abundance of organohalide-respiring bacteria (i.e., *Dehalococcoides* spp.) and biological activity along the flowpath of the transported CMC and NZVI suggested that co-injection of CMC and NZVI could provide conditions that favor biotic reductive dichlorination, i.e., the CMC and H_2 generated could serve as electron donors (Kocur et al. 2015).

While large amounts of ammonia, an undesirable end product, may be generated by abiotic reduction of nitrate by NZVI, the process of biological denitrification, in which nitrates are reduced to N_2, could overcome this problem (Shin and Cha 2008;

An et al. 2009). Several studies have reported that biological denitrification was more efficient when NZVI was added and that, for example, ammonia generation was minimized (An et al. 2009) and nitrate removal was increased (Liu et al. 2014; Jiang et al. 2015). The enhanced nitrate reduction may be attributed to the increase in electron donors generated by NZVI, including H_2, Fe(II) (Shin and Cha 2008; An et al. 2009; Jiang et al. 2015), or Fe_3O_4 (Liu et al. 2014).

As well as the reductive and oxidative transformation mechanisms already discussed, the adsorption of contaminants onto the iron oxide shell of NZVI may be beneficial for the biodegradation process. In the presence of goethite (α-FeOOH), dichlorodiphenyltrichloroethane (DDT) may be dechlorinated faster by *Shewanella decolorationis* S12 than with biotic (absence of goethite) or abiotic (Fe(II)-goethite) systems alone (Li et al. 2010a). Dichlorodiphenyldichloroethane, the chlorination product of DDT, may not be further transformed by bacteria but may decrease slightly in bacteria-goethite or Fe(II)-goethite systems (Li et al. 2010a). DIRB can reduce contaminants and iron oxide at the same time, while Fe(III) reduction may support the growth of DIRB and contaminant transformation (Fig. 10.2).

10.5 Future Needs

While the NZVI-based treatment process has advantages of high reactivity, rapid degradation, and low cost, challenges associated with NZVI in situ applications still exist, such as incomplete dehalogenation, non-specific electron transfer (which reduces electron transfer to the contaminants), and low mobility (Wang et al. 2016b). In field treatment, target contaminants may not be completely removed by NZVI addition (Mace 2006). Thus, long-term attenuation of the contaminants by biodegradation may be required (Kocur et al. 2015). However, there is still limited information about the effects of NZVI on environmental microorganisms at the field scale, especially those involved in bioremediation processes. This prevents optimization of injection conditions to support bioremediation.

The ecological effects of NZVI appear to be site-specific and may be governed by several factors, including the physical (soil properties), chemical (e.g., pH, ionic strength, natural organic matter) and biological (abundance and species of indigenous microorganisms) characteristics of contaminated sites. While NZVI has been reported to be highly toxic, most field studies have demonstrated limited adverse effects on microorganisms. To ensure the ecological impacts of NZVI can be predicted accurately, we need a sound understanding of the interactions between NZVI and indigenous microbial species in contaminated sites or sites where NZVI may migrate to after injection.

The rapid generation of H_2 or soluble Fe(II) when NZVI is injected is expected to stimulate microbial growth and activity; however, this beneficial process may be hindered by the toxicity of NZVI. Kumar et al. (2014) reported that additions of granular (gZVI), microscale ZVI (mZVI), and NZVI could enhance the pH and

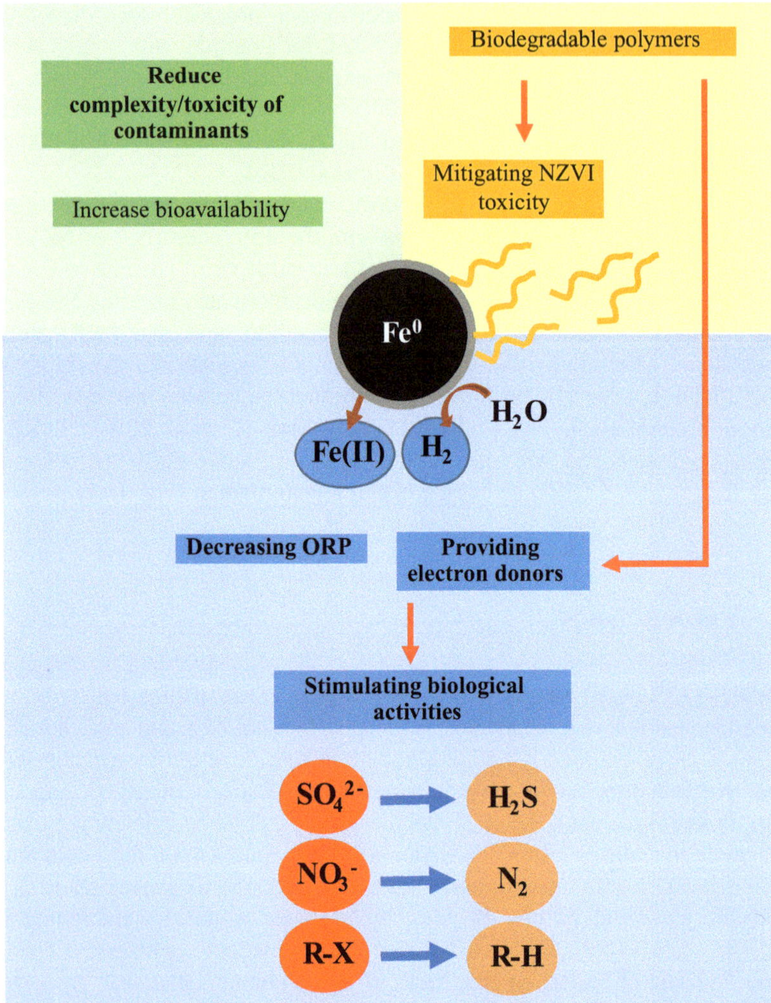

Fig. 10.2 Conceptual model of synergistic effects between NZVI nanotechnology and bioremediation. The high reactivity of NZVI could rapidly reduce the complexity and toxicity of the contaminants, thus increasing the bioavailability for further degradation. The H_2 generated when NZVI interacts with water, and the released Fe(II) could act as an electron donor for many biological processes. The decreased ORP also favors biotic dechlorination. NZVI toxicity may be alleviated and microbial growth or activity stimulated when co-injected with, or modified by, biodegradable polymers

decrease the ORP, thereby favoring microbial growth and activity (Kumar et al. 2014). However, one study reported that sulfate reduction was only stimulated when either gZVI or mZVI was added along with glycerol, a required organic carbon source for heterotrophic sulfate reducers, whereas additions of NZVI (both with and without glycerol) inhibited sulfate reduction because of the toxicity of the nanoparticles (Kumar et al. 2014). Thus, co-injection of a carbon source may be a necessary component of a mixed NZVI/bioremediation strategy.

For effective remediation, NZVI needs to be applied at appropriate concentrations since, at high concentrations, NZVI may inhibit the viability and activity of bacteria. For instance, Jiang et al. (2015) reported that, while additions of NZVI stimulated biodenitrification, biological processes were inhibited as the NZVI concentrations increased (Jiang et al. 2015). Similarly, Koenig et al. (2016) reported that the combination of abiotic treatment of chloroethane mixture by NZVI and biotic treatment by organochlorine respiring bacteria could resolve the issues with each technique (Koenig et al. 2016). However, microbial activity is inhibited by NZVI at higher concentrations, and the reactivity of NZVI is affected by nutrients that are needed during bacterial growth (Koenig et al. 2016). Accordingly, an understanding of the mobility of NZVI is essential because limited mobility may result in NZVI hotspots with high Fe^0 concentrations that are potentially detrimental to the surrounding microorganisms.

Concern remains about some aspects of biotic reductive chlorination even though it is a promising alternative for treating contaminants. For example, the amount or presence of functioning microorganisms and competition between other indigenous strains for growth may affect the bioremediation efficiency (Wang et al. 2016b). Directly and indirectly, NZVI additions may exert selective pressure on certain microbial species, thereby causing shifts in the bacterial communities of contaminated sites and competition between microbial strains (Kumar et al. 2014; Wang et al. 2016b). For instance, methanogens compete with dechlorinating bacteria for H_2, so NZVI additions are more beneficial for methanogenesis (Xiu et al. 2010b). Because the preferences for electron donors and optimum conditions are strain-dependent, information is needed about the microbial strains responsible for the desired biological processes and about the optimal concentrations and types of electron donor to achieve the desired biostimulation effects. The properties of the NZVI may also affect the ability to biodegrade TCE. For example, one study reported that bare NZVI down regulated the expression of the tceA gene responsible for TCE degradation, while polymer coated NZVI upregualted tceA expression (Xiu et al. 2010a).

Wang et al. (2016b) has provided an excellent overview of the integrated system between in situ reductive dehalogenation of halogenated contaminants by NZVI and bioremediation, in which advantages, disadvantages, challenges, and future perspectives are discussed (Wang et al. 2016b). Once introduced in situ, NZVI rapidly interacts with contaminants and water, thereby generating further H_2, with the result that biological processes, particularly biotic reductive dichlorination, are much slower (Wang et al. 2016b). This gap has become an important challenge for the effective integration of NZVI and biological treatments (Wang et al. 2016b). The reaction time of NZVI is also important for evaluating the potential long-term stimulation of biological processes. Surface modifiers, redox-active organic molecules, or ion species, when present, have been reported to impact the reactivity or reactive time of NZVI (Koenig et al. 2016). Information about the effects of these factors on the reactivity of NZVI and effectiveness of biostimulation, especially during in situ applications of NZVI, is required to ensure remediation attempts are successful.

Acknowledgments The authors thank Deborah Ballantine, PhD, from Edanz Group (www.edanzediting.com/ac) for editing a draft of this manuscript.

References

Adeleye, A. S., Keller, A. A., Miller, R. J., & Lenihan, H. S. (2013). Persistence of commercial nanoscaled zero-valent iron (nZVI) and by-products. *Journal of Nanoparticle Research, 15*, 1–18.

An, Y., Li, T., Jin, Z., Dong, M., Li, Q., & Wang, S. (2009). Decreasing ammonium generation using hydrogenotrophic bacteria in the process of nitrate reduction by nanoscale zero-valent iron. *The Science of the Total Environment, 407*, 5465–5470.

An, Y., Li, T., Jin, Z., Dong, M., Xia, H., & Wang, X. (2010). Effect of bimetallic and polymer-coated Fe nanoparticles on biological denitrification. *Bioresource Technology, 101*, 9825–9828.

Auffan, M., Achouak, W., Rose, J., Roncato, M. A., Chaneac, C., Waite, D. T., Masion, A., Woicik, J. C., Wiesner, M. R., & Bottero, J. Y. (2008). Relation between the redox state of iron-based nanoparticles and their cytotoxicity toward *Escherichia coli*. *Environmental Science & Technology, 42*, 6730–6735.

Badri, D. V., Weir, T. L., van der Lelie, D., & Vivanco, J. M. (2009). Rhizosphere chemical dialogues: Plant-microbe interactions. *Current Opinion in Biotechnology, 20*, 642–650.

Bae, S., & Lee, W. (2014). Influence of riboflavin on nanoscale zero-valent iron reactivity during the degradation of carbon tetrachloride. *Environmental Science & Technology, 48*, 2368–2376.

Basnet, M., Gershanov, A., Wilkinson, K. J., Ghoshal, S., & Tufenkji, N. (2016). Interaction between palladium-doped zerovalent iron nanoparticles and biofilm in granular porous media: Characterization, transport and viability. *Environmental Science-Nano, 3*, 127–137.

Ben-Moshe, T., Frenk, S., Dror, I., Minz, D., & Berkowitz, B. (2013). Effects of metal oxide nanoparticles on soil properties. *Chemosphere, 90*, 640–646.

Calderon, B., & Fullana, A. (2015). Heavy metal release due to aging effect during zero valent iron nanoparticles remediation. *Water Research, 83*, 1–9.

Chai, H., Yao, J., Sun, J., Zhang, C., Liu, W., Zhu, M., & Ceccanti, B. (2015). The effect of metal oxide nanoparticles on functional bacteria and metabolic profiles in agricultural soil. *Bulletin of Environmental Contamination and Toxicology, 94*, 490–495.

Chaithawiwat, K., Vangnai, A., McEvoy, J. M., Pruess, B., Krajangpan, S., & Khan, E. (2016). Role of oxidative stress in inactivation of *Escherichia coli* BW25113 by nanoscale zero-valent iron. *The Science of the Total Environment, 565*, 857–862.

Chen, J., Xiu, Z., Lowry, G. V., & Alvarez, P. J. (2011). Effect of natural organic matter on toxicity and reactivity of nano-scale zero-valent iron. *Water Research, 45*, 1995–2001.

D'Autreaux, B., & Toledano, M. B. (2007). ROS as signalling molecules: Mechanisms that generate specificity in ROS homeostasis. *Nature Reviews Molecular Cell Biology, 8*, 813–824.

Davey, M. E., & O'Toole, G. A. (2000). Microbial biofilms: From ecology to molecular genetics. *Microbiology and Molecular Biology Reviews: MMBR, 64*, 847–867.

Dhas, S. P., Shiny, P. J., Khan, S., Mukherjee, A., & Chandrasekaran, N. (2014). Toxic behavior of silver and zinc oxide nanoparticles on environmental microorganisms. *Journal of Basic Microbiology, 54*, 916–927.

Diao, M., & Yao, M. (2009). Use of zero-valent iron nanoparticles in inactivating microbes. *Water Research, 43*, 5243–5251.

Dimkpa, C. O., Calder, A., Gajjar, P., Merugu, S., Huang, W., Britt, D. W., McLean, J. E., Johnson, W. P., & Anderson, A. J. (2011). Interaction of silver nanoparticles with an environmentally

beneficial bacterium, Pseudomonas chlororaphis. *Journal of Hazardous Materials, 188*, 428–435.
Fajardo, C., Ortiz, L. T., Rodriguez-Membibre, M. L., Nande, M., Lobo, M. C., & Martin, M. (2012). Assessing the impact of zero-valent iron (ZVI) nanotechnology on soil microbial structure and functionality: A molecular approach. *Chemosphere, 86*, 802–808.
Fajardo, C., Sacca, M. L., Martinez-Gomariz, M., Costa, G., Nande, M., & Martin, M. (2013). Transcriptional and proteomic stress responses of a soil bacterium Bacillus cereus to nanosized zero-valent iron (nZVI) particles. *Chemosphere, 93*, 1077–1083.
Fang, J., Lyon, D. Y., Wiesner, M. R., Dong, J., & Alvarez, P. J. (2007). Effect of a fullerene water suspension on bacterial phospholipids and membrane phase behavior. *Environmental Science & Technology, 41*, 2636–2642.
Gaboriaud, F., Gee, M. L., Strugnell, R., & Duval, J. F. (2008). Coupled electrostatic, hydrodynamic, and mechanical properties of bacterial interfaces in aqueous media. *Langmuir: The ACS Journal of Surfaces and Colloids, 24*, 10988–10995.
Hachicho, N., Hoffmann, P., Ahlert, K., & Heipieper, H. J. (2014). Effect of silver nanoparticles and silver ions on growth and adaptive response mechanisms of *Pseudomonas putida* mt-2. *FEMS Microbiology Letters, 355*, 71–77.
Hazen, T. C., Jimenez, L., Lopez de Victoria, G., & Fliermans, C. B. (1991). Comparison of bacteria from deep subsurface sediment and adjacent groundwater. *Microbial Ecology, 22*, 293–304.
He, F., Zhao, D., & Paul, C. (2010). Field assessment of carboxymethyl cellulose stabilized iron nanoparticles for *in situ* destruction of chlorinated solvents in source zones. *Water Research, 44*, 2360–2370.
He, S. Y., Feng, Y. Z., Ren, H. X., Zhang, Y., Gu, N., & Lin, X. G. (2011). The impact of iron oxide magnetic nanoparticles on the soil bacterial community. *Journal of Soils and Sediments, 11*, 1408–1417.
He, D., Ma, J., Collins, R. N., & Waite, T. D. (2016a). Effect of structural transformation of nanoparticulate zero-valent iron on generation of reactive oxygen species. *Environmental Science & Technology, 50*, 3820–3828.
He, S., Feng, Y., Ni, J., Sun, Y., Xue, L., Feng, Y., Yu, Y., Lin, X., & Yang, L. (2016b). Different responses of soil microbial metabolic activity to silver and iron oxide nanoparticles. *Chemosphere, 147*, 195–202.
Heipieper, H. J., Meinhardt, F., & Segura, A. (2003). The cis-trans isomerase of unsaturated fatty acids in Pseudomonas and Vibrio: Biochemistry, molecular biology and physiological function of a unique stress adaptive mechanism. *FEMS Microbiology Letters, 229*, 1–7.
Holden, P. A., Gardea-Torresdey, J. L., Klaessig, F., Turco, R. F., Mortimer, M., Hund-Rinke, K., Cohen Hubal, E. A., Avery, D., Barcelo, D., Behra, R., Cohen, Y., Deydier-Stephan, L., Ferguson, P. L., Fernandes, T. F., Herr Harthorn, B., Henderson, W. M., Hoke, R. A., Hristozov, D., Johnston, J. M., Kane, A. B., Kapustka, L., Keller, A. A., Lenihan, H. S., Lovell, W., Murphy, C. J., Nisbet, R. M., Petersen, E. J., Salinas, E. R., Scheringer, M., Sharma, M., Speed, D. E., Sultan, Y., Westerhoff, P., White, J. C., Wiesner, M. R., Wong, E. M., Xing, B., Steele Horan, M., Godwin, H. A., & Nel, A. E. (2016). Considerations of environmentally relevant test conditions for improved evaluation of ecological hazards of engineered nanomaterials. *Environmental Science & Technology, 50*, 6124–6145.
Holm, P. E., Nielsen, P. H., Albrechtsen, H. J., & Christensen, T. H. (1992). Importance of unattached bacteria and bacteria attached to sediment in determining potentials for degradation of xenobiotic organic contaminants in an aerobic aquifer. *Applied and Environmental Microbiology, 58*, 3020–3026.
Ikuma, K., Decho, A. W., & Lau, B. L. (2015). When nanoparticles meet biofilms-interactions guiding the environmental fate and accumulation of nanoparticles. *Frontiers in Microbiology, 6*, 591.
Jacobson, K. H., Gunsolus, I. L., Kuech, T. R., Troiano, J. M., Melby, E. S., Lohse, S. E., Hu, D., Chrisler, W. B., Murphy, C. J., Orr, G., Geiger, F. M., Haynes, C. L., & Pedersen, J. A. (2015). Lipopolysaccharide density and structure govern the extent and distance of nanoparticle

interaction with actual and model bacterial outer membranes. *Environmental Science & Technology, 49*, 10642–10650.

Jiang, C., Xu, X., Megharaj, M., Naidu, R., & Chen, Z. (2015). Inhibition or promotion of biodegradation of nitrate by *Paracoccus* sp. in the presence of nanoscale zero-valent iron. *The Science of the Total Environment, 530–531*, 241–246.

Jiemvarangkul, P., Zhang, W. X., & Lien, H. L. (2011). Enhanced transport of polyelectrolyte stabilized nanoscale zero-valent iron (nZVI) in porous media. *Chemical Engineering Journal, 170*, 482–491.

Joshi, N., Ngwenya, B. T., & French, C. E. (2012). Enhanced resistance to nanoparticle toxicity is conferred by overproduction of extracellular polymeric substances. *Journal of Hazardous Materials, 241–242*, 363–370.

Keller, A. A., Garner, K., Miller, R. J., & Lenihan, H. S. (2012). Toxicity of nano-zero valent iron to freshwater and marine organisms. *PLoS One, 7*, e43983.

Khan, S. S., Srivatsan, P., Vaishnavi, N., Mukherjee, A., & Chandrasekaran, N. (2011). Interaction of silver nanoparticles (SNPs) with bacterial extracellular proteins (ECPs) and its adsorption isotherms and kinetics. *Journal of Hazardous Materials, 192*, 299–306.

Kim, J. Y., Park, H. J., Lee, C., Nelson, K. L., Sedlak, D. L., & Yoon, J. (2010). Inactivation of *Escherichia coli* by nanoparticulate zerovalent iron and ferrous ion. *Applied and Environmental Microbiology, 76*, 7668–7670.

Kim, J. Y., Lee, C., Love, D. C., Sedlak, D. L., Yoon, J., & Nelson, K. L. (2011). Inactivation of MS2 coliphage by ferrous ion and zero-valent iron nanoparticles. *Environmental Science & Technology, 45*, 6978–6984.

Kim, H. J., Phenrat, T., Tilton, R. D., & Lowry, G. V. (2012a). Effect of kaolinite, silica fines and pH on transport of polymer-modified zero valent iron nano-particles in heterogeneous porous media. *Journal of Colloid and Interface Science, 370*, 1–10.

Kim, Y. M., Murugesan, K., Chang, Y. Y., Kim, E. J., & Chang, Y. S. (2012b). Degradation of polybrominated diphenyl ethers by a sequential treatment with nanoscale zero valent iron and aerobic biodegradation. *Journal of Chemical Technology & Biotechnology, 87*, 216–224.

Kirschling, T. L., Gregory, K. B., Minkley, E. G., Jr., Lowry, G. V., & Tilton, R. D. (2010). Impact of nanoscale zero valent iron on geochemistry and microbial populations in trichloroethylene contaminated aquifer materials. *Environmental Science & Technology, 44*, 3474–3480.

Kocur, C. M., Chowdhury, A. I., Sakulchaicharoen, N., Boparai, H. K., Weber, K. P., Sharma, P., Krol, M. M., Austrins, L., Peace, C., Sleep, B. E., & O'Carroll, D. M. (2014). Characterization of nZVI mobility in a field scale test. *Environmental Science & Technology, 48*, 2862–2869.

Kocur, C. M., Lomheim, L., Boparai, H. K., Chowdhury, A. I., Weber, K. P., Austrins, L. M., Edwards, E. A., Sleep, B. E., & O'Carroll, D. M. (2015). Contributions of abiotic and biotic dechlorination following carboxymethyl cellulose stabilized nanoscale zero valent iron injection. *Environmental Science & Technology, 49*, 8648–8656.

Kocur, C. M., Lomheim, L., Molenda, O., Weber, K. P., Austrins, L. M., Sleep, B. E., Boparai, H. K., Edwards, E. A., & O'Carroll, D. M. (2016). Long-term field study of microbial community and dechlorinating activity following carboxymethyl cellulose-stabilized nanoscale zero-valent Iron injection. *Environmental Science & Technology, 50*, 7658–7670.

Koenig, J. C., Boparai, H. K., Lee, M. J., O'Carroll, D. M., Barnes, R. J., & Manefield, M. J. (2016). Particles and enzymes: Combining nanoscale zero valent iron and organochlorine respiring bacteria for the detoxification of chloroethane mixtures. *Journal of Hazardous Materials, 308*, 106–112.

Kotchaplai, P., Khan, E., & Vangnai, A. S. (2017). Membrane alterations in Pseudomonas putida F1 exposed to nanoscale zerovalent iron: Effects of short-term and repetitive nZVI exposure. *Environmental Science & Technology, 51*, 7804–7813.

Kuiper, I., Lagendijk, E. L., Bloemberg, G. V., & Lugtenberg, B. J. (2004). Rhizoremediation: A beneficial plant-microbe interaction. *Molecular Plant-Microbe Interactions: MPMI, 17*, 6–15.

Kumar, N., Omoregie, E. O., Rose, J., Masion, A., Lloyd, J. R., Diels, L., & Bastiaens, L. (2014). Inhibition of sulfate reducing bacteria in aquifer sediment by iron nanoparticles. *Water Research, 51*, 64–72.

Laumann, S., Micic, V., & Hofmann, T. (2014). Mobility enhancement of nanoscale zero-valent iron in carbonate porous media through co-injection of polyelectrolytes. *Water Research, 50*, 70–79.

Le, T. T., Murugesan, K., Kim, E. J., & Chang, Y. S. (2014). Effects of inorganic nanoparticles on viability and catabolic activities of agrobacterium sp. PH-08 during biodegradation of dibenzofuran. *Biodegradation, 25*, 655–668.

Lee, C., Kim, J. Y., Lee, W. I., Nelson, K. L., Yoon, J., & Sedlak, D. L. (2008). Bactericidal effect of zero-valent iron nanoparticles on Escherichia coli. *Environmental Science & Technology, 42*, 4927–4933.

Lefevre, E., Bossa, N., Wiesner, M. R., & Gunsch, C. K. (2016). A review of the environmental implications of *in situ* remediation by nanoscale zero valent iron (nZVI): Behavior, transport and impacts on microbial communities. *The Science of the Total Environment, 565*, 889–901.

Lerner, R. N., Lu, Q., Zeng, H., & Liu, Y. (2012). The effects of biofilm on the transport of stabilized zerovalent iron nanoparticles in saturated porous media. *Water Research, 46*, 975–985.

Li, X. Q., Brown, D. G., & Zhang, W. X. (2007). Stabilization of biosolids with nanoscale zero-valent iron (nZVI). *Journal of Nanoparticle Research, 9*, 233–243.

Li, F. B., Li, X. M., Zhou, S. G., Zhuang, L., Cao, F., Huang, D. Y., Xu, W., Liu, T. X., & Feng, C. H. (2010a). Enhanced reductive dechlorination of DDT in an anaerobic system of dissimilatory iron-reducing bacteria and iron oxide. *Environmental Pollution, 158*, 1733–1740.

Li, Z., Greden, K., Alvarez, P. J., Gregory, K. B., & Lowry, G. V. (2010b). Adsorbed polymer and NOM limits adhesion and toxicity of nano scale zerovalent iron to E. coli. *Environmental Science & Technology, 44*, 3462–3467.

Lin, Y. H., Tseng, H. H., Wey, M. Y., & Lin, M. D. (2010). Characteristics of two types of stabilized nano zero-valent iron and transport in porous media. *The Science of the Total Environment, 408*, 2260–2267.

Liu, Y., Li, S., Chen, Z., Megharaj, M., & Naidu, R. (2014). Influence of zero-valent iron nanoparticles on nitrate removal by Paracoccus sp. *Chemosphere, 108*, 426–432.

Lv, Y., Niu, Z., Chen, Y., & Hu, Y. (2017). Bacterial effects and interfacial inactivation mechanism of nZVI/Pd on Pseudomonas putida strain. *Water Research, 115*, 297–308.

Mace, C. (2006). Controlling groundwater VOCs. (cover story). *Pollution Engineering, 38*, 24–28.

Miao, A. J., Schwehr, K. A., Xu, C., Zhang, S. J., Luo, Z., Quigg, A., & Santschi, P. H. (2009). The algal toxicity of silver engineered nanoparticles and detoxification by exopolymeric substances. *Environmental Pollution, 157*, 3034–3041.

Mitzel, M. R., & Tufenkji, N. (2014). Transport of industrial PVP-stabilized silver nanoparticles in saturated quartz sand coated with Pseudomonas aeruginosa PAO1 biofilm of variable age. *Environmental Science & Technology, 48*, 2715–2723.

Mueller, N. C., Braun, J., Bruns, J., Cernik, M., Rissing, P., Rickerby, D., & Nowack, B. (2012). Application of nanoscale zero valent iron (NZVI) for groundwater remediation in Europe. *Environmental Science and Pollution Research International, 19*, 550–558.

Ortega-Calvo, J. J., Jimenez-Sanchez, C., Pratarolo, P., Pullin, H., Scott, T. B., & Thompson, I. P. (2016). Tactic response of bacteria to zero-valent iron nanoparticles. *Environmental Pollution, 213*, 438–445.

Pandey, G., & Jain, R. K. (2002). Bacterial chemotaxis toward environmental pollutants: Role in bioremediation. *Applied and Environmental Microbiology, 68*, 5789–5795.

Pawlett, M., Ritz, K., Dorey, R. A., Rocks, S., Ramsden, J., & Harris, J. A. (2013). The impact of zero-valent iron nanoparticles upon soil microbial communities is context dependent. *Environmental Science and Pollution Research International, 20*, 1041–1049.

Phenrat, T., Kim, H.-J., Fagerlund, F., Illangasekare, T., Tilton, R. D., & Lowry, G. V. (2009a). Particle size distribution, concentration, and magnetic attraction affect transport of polymer-modified Fe0 nanoparticles in sand columns. *Environmental Science & Technology, 43*, 5079–5085.

Phenrat, T., Liu, Y., Tilton, R. D., & Lowry, G. V. (2009b). Adsorbed polyelectrolyte coatings decrease Fe^0 nanoparticle reactivity with TCE in water: Conceptual model and mechanisms. *Environmental Science & Technology, 43*, 1507–1514.

Phenrat, T., Schoenfelder, D., Kirschling, T. L., Tilton, R. D., & Lowry, G. V. (2018). Adsorbed poly(aspartate) coating limits the adverse effects of dissolved groundwater solutes on Fe0 nanoparticle reactivity with trichloroethylene. *Environmental Science and Pollution Research International, 25*(8), 7157–7169.

Popova, O. B., Sanina, N. M., Likhatskaya, G. N., & Bezverbnaya, I. P. (2008). Effects of copper and cadmium ions on the physicochemical properties of lipids of the marine bacterium Pseudomonas putida IB28 at different growth temperatures. *Russian Journal of Marine Biology, 34*, 179–185.

Sacca, M. L., Fajardo, C., Nande, M., & Martin, M. (2013). Effects of nano zero-valent iron on Klebsiella oxytoca and stress response. *Microbial Ecology, 66*, 806–812.

Sacca, M. L., Fajardo, C., Costa, G., Lobo, C., Nande, M., & Martin, M. (2014a). Integrating classical and molecular approaches to evaluate the impact of nanosized zero-valent iron (nZVI) on soil organisms. *Chemosphere, 104*, 184–189.

Sacca, M. L., Fajardo, C., Martinez-Gomariz, M., Costa, G., Nande, M., & Martin, M. (2014b). Molecular stress responses to nano-sized zero-valent iron (nZVI) particles in the soil bacterium *Pseudomonas stutzeri. PLoS One, 9*, e89677.

Saleh, N., Kim, H. J., Phenrat, T., Matyjaszewski, K., Tilton, R. D., & Lowry, G. V. (2008). Ionic strength and composition affect the mobility of surface-modified Fe0 nanoparticles in water-saturated sand columns. *Environmental Science & Technology, 42*, 3349–3355.

Sevcu, A., El-Temsah, Y. S., Joner, E. J., & Cernik, M. (2011). Oxidative stress induced in microorganisms by zero-valent iron nanoparticles. *Microbes and Environments, 26*, 271–281.

Shah, V., Dobiasova, P., Baldrian, P., Nerud, F., Kumar, A., & Seal, S. (2010). Influence of iron and copper nanoparticle powder on the production of lignocellulose degrading enzymes in the fungus Trametes versicolor. *Journal of Hazardous Materials, 178*, 1141–1145.

Shephard, J. J., Savory, D. M., Bremer, P. J., & McQuillan, A. J. (2010). Salt modulates bacterial hydrophobicity and charge properties influencing adhesion of Pseudomonas aeruginosa (PA01) in aqueous suspensions. *Langmuir: The ACS Journal of Surfaces and Colloids, 26*, 8659–8665.

Shi, Z., Fan, D., Johnson, R. L., Tratnyek, P. G., Nurmi, J. T., Wu, Y., & Williams, K. H. (2015). Methods for characterizing the fate and effects of nano zerovalent iron during groundwater remediation. *Journal of Contaminant Hydrology, 181*, 17–35.

Shin, K. H., & Cha, D. K. (2008). Microbial reduction of nitrate in the presence of nanoscale zero-valent iron. *Chemosphere, 72*, 257–262.

Sieger, C. H. N., Kroon, A. G. M., Batelaan, J. G., & Vanginkel, C. G. (1995). Biodegradation of carboxymethyl celluloses by Agrobacterium Cm-1. *Carbohydrate Polymers, 27*, 137–143.

Silambarasan, S., & Vangnai, A. S. (2017). Plant-growth promoting Candida sp. AVGB4 with capability of 4-nitroaniline biodegradation under drought stress. *Ecotoxicology and Environmental Safety, 139*, 472–480.

Somers, E., Vanderleyden, J., & Srinivasan, M. (2004). Rhizosphere bacterial signalling: A love parade beneath our feet. *Critical Reviews in Microbiology, 30*, 205–240.

Straub, K. L., Benz, M., & Schink, B. (2001). Iron metabolism in anoxic environments at near neutral pH. *FEMS Microbiology Ecology, 34*, 181–186.

Su, Y., Adeleye, A. S., Zhou, X., Dai, C., Zhang, W., Keller, A. A., & Zhang, Y. (2014). Effects of nitrate on the treatment of lead contaminated groundwater by nanoscale zerovalent iron. *Journal of Hazardous Materials, 280*, 504–513.

Sudheer Khan, S., Bharath Kumar, E., Mukherjee, A., & Chandrasekaran, N. (2011). Bacterial tolerance to silver nanoparticles (SNPs): Aeromonas punctata isolated from sewage environment. *Journal of Basic Microbiology, 51*, 183–190.

Tabata, K., Kasuya, K. I., Abe, H., Masuda, K., & Doi, Y. (1999). Poly(aspartic acid) degradation by a Sphingomonas sp. isolated from freshwater. *Applied and Environmental Microbiology, 65*, 4268–4270.

Thuptimdang, P., Limpiyakorn, T., McEvoy, J., Pruss, B. M., & Khan, E. (2015). Effect of silver nanoparticles on Pseudomonas putida biofilms at different stages of maturity. *Journal of Hazardous Materials, 290*, 127–133.

Thuptimdang, P., Limpiyakorn, T., & Khan, E. (2017). Dependence of toxicity of silver nanoparticles on Pseudomonas putida biofilm structure. *Chemosphere, 188*, 199–207.

Vangnai, A. S., Takeuchi, K., Oku, S., Kataoka, N., Nitisakulkan, T., Tajima, T., & Kato, J. (2013). Identification of CtpL as a chromosomally encoded chemoreceptor for 4-chloroaniline and catechol in Pseudomonas aeruginosa PAO1. *Applied and Environmental Microbiology, 79*, 7241–7248.

Vorobyova, E., Soina, V., Gorlenko, M., Minkovskaya, N., Zalinova, N., Mamukelashvili, A., Gilichinsky, D., Rivkina, E., & Vishnivetskaya, T. (1997). The deep cold biosphere: Facts and hypothesis. *FEMS Microbiology Reviews, 20*, 277–290.

Wang, Q., Kang, F., Gao, Y., Mao, X., & Hu, X. (2016a). Sequestration of nanoparticles by an EPS matrix reduces the particle-specific bactericidal activity. *Scientific Reports, 6*, 21379.

Wang, S., Chen, S., Wang, Y., Low, A., Lu, Q., & Qiu, R. (2016b). Integration of organohalide-respiring bacteria and nanoscale zero-valent iron (bio-nZVI-RD): A perfect marriage for the remediation of organohalide pollutants? *Biotechnology Advances, 34*, 1384–1395.

Wei, Y. T., Wu, S. C., Chou, C. M., Che, C. H., Tsai, S. M., & Lien, H. L. (2010). Influence of nanoscale zero-valent iron on geochemical properties of groundwater and vinyl chloride degradation: A field case study. *Water Research, 44*, 131–140.

Xie, Y., & Cwiertny, D. M. (2010). Use of dithionite to extend the reactive lifetime of nanoscale zero-valent iron treatment systems. *Environmental Science & Technology, 44*, 8649–8655.

Xiu, Z. M., Gregory, K. B., Lowry, G. V., & Alvarez, P. J. (2010a). Effect of bare and coated nanoscale zerovalent iron on tceA and vcrA gene expression in Dehalococcoides spp. *Environmental Science & Technology, 44*, 7647–7651.

Xiu, Z. M., Jin, Z. H., Li, T. L., Mahendra, S., Lowry, G. V., & Alvarez, P. J. (2010b). Effects of nano-scale zero-valent iron particles on a mixed culture dechlorinating trichloroethylene. *Bioresource Technology, 101*, 1141–1146.

Xu, C., Peng, C., Sun, L. J., Zhang, S., Huang, H. M., Chen, Y. X., & Shi, J. Y. (2015). Distinctive effects of TiO2 and CuO nanoparticles on soil microbes and their community structures in flooded paddy soil. *Soil Biology and Biochemistry, 86*, 24–33.

Yang, Z., Wang, X.-l., Li, H., Yang, J., Zhou, L.-Y., & Liu, Y.-d. (2017). Re-activation of aged-ZVI by iron-reducing bacterium Shewanella putrefaciens for enhanced reductive dechlorination of trichloroethylene. *Journal of Chemical Technology and Biotechnology, 92*, 2642–2649.

Ye, L., Liu, W., Shi, Q., & Jing, C. (2017). Arsenic mobilization in spent nZVI waste residue: Effect of Pantoea sp. IMH. *Environmental Pollution, 230*, 1081–1089.

Zhou, L., Thanh, T. L., Gong, J., Kim, J. H., Kim, E. J., & Chang, Y. S. (2014). Carboxymethyl cellulose coating decreases toxicity and oxidizing capacity of nanoscale zerovalent iron. *Chemosphere, 104*, 155–161.

Zhu, B., Xia, X., Xia, N., Zhang, S., & Guo, X. (2014). Modification of fatty acids in membranes of bacteria: Implication for an adaptive mechanism to the toxicity of carbon nanotubes. *Environmental Science & Technology, 48*, 4086–4095.

Chapter 11
Electromagnetic Induction of Nanoscale Zerovalent Iron for Enhanced Thermal Dissolution/Desorption and Dechlorination of Chlorinated Volatile Organic Compounds

Tanapon Phenrat and Gregory V. Lowry

Abstract A major problem plaguing the success of in situ dechlorination using NZVI is the slow rate of dissolution of chlorinated volatile organic compounds (CVOCs) from dense nonaqueous phase liquid (DNAPL) or slow desorption of CVOCs from soil in the aqueous phase. This is because the dechlorination using NZVI is surface mediated; therefore, contaminants must be dissolved to transport to the NZVI surface. For this reason, any action to enhance the DNAPL dissolution or desorption of CVOCs from the soil and DNAPL can speed the reaction rate and improve the electron utilization efficiency of the remediation. This chapter summarizes the state of knowledge about using a low-frequency (LF) electromagnetic field (EMF) (150 kHz) with NZVI to enhance the CVOC degradation rate in a DNAPL system and in a soil and groundwater system via thermal-enhanced CVOC dissolution or desorption followed by enhanced dechlorination using NZVI. NZVI is a ferromagnetic particle capable of magnetic induction heating under an applied LF EMF. The heat generated can speed up the dechlorination reaction and can promote DNAPL dissolution or desorption of contaminants from soils. The most recent work on using this novel approach is summarized as a proof of concept. The CVOC degradation kinetics in groundwater and in soil with groundwater as well as in a DNAPL system by NZVI both with and without LF EMF were compared to quantify the benefits of using LF EMF for enhanced thermal dissolution and magnetically enhanced NZVI corrosion.

T. Phenrat (✉)
Department of Civil Engineering, Environmental Engineering Program, Naresuan University, Phitsanulok, Thailand

Center of Excellence for Sustainability of Health, Environment and Industry (SHEI), Faculty of Engineering, Naresuan University, Phitsanulok, Thailand

G. V. Lowry
Center for Environmental Implications of Nanotechnology (CEINT), Durham, NC, USA

Department of Civil & Environmental Engineering, Carnegie Mellon University, Pittsburgh, PA, USA

T. Phenrat, G. V. Lowry (eds.), *Nanoscale Zerovalent Iron Particles for Environmental Restoration*, https://doi.org/10.1007/978-3-319-95340-3_11

Keywords Nanoscale zerovalent iron · Electromagnetic field · Low frequency · Electromagnetic induction heating · Thermal enhanced dissolution · Thermal enhanced desorption · DNAPL · Ferromagnetic · Combined remedies with NZVI

11.1 Limitations of Chlorinated Organic Dechlorination by NZVI in the Presence of NAPL and Soil

As discussed in Chaps. 3 and 9, nanoscale zerovalent iron (NZVI) particles can reductively dechlorinate many chlorinated volatile organic compounds (CVOCs) (He et al. 2007; Johnson et al. 2013; Kocur et al. 2015; O'Carroll et al. 2013; Phenrat et al. 2009; Phenrat et al. 2015, 2010; Sakulchaicharoen et al. 2010; Tratnyek and Johnson 2006; Zhang et al. 1998; Zhao et al. 2016). Due to the small size and high reactivity, NZVI has been proposed for source-zone remediation with the aim of destroying dense nonaqueous phase liquids (DNAPLs) entrapped in the subsurface (Bishop et al. 2010; Fagerlund et al. 2012; He et al. 2010; Henn and Waddill 2006; Phenrat et al. 2011a, b; Quinn et al. 2005; Saleh et al. 2005, 2007; Su et al. 2012; Taghavy et al. 2010; Zhan et al. 2009). Removal of mass of the contaminant sources is designed to speed up site closure.

One conceptual model of active and selective DNAPL source-zone removal is the delivery of polymer-modified NZVI to target DNAPL sources and subsequently perform reductive dechlorination (see Chap. 5). However, delivering NZVI to the DNAPL/water interface appears to be insufficient for effective DNAPL source-zone removal because DNAPL dissolution is the rate-limiting step of dechlorination (Miller et al. 1990; Powers et al. 1992; Saba and Illangasekare 2000).

Even though conventional wisdom may suggest that the most effective way to remediate a DNAPL source is to place the NZVI as close as possible to or within the source, recent research suggests that DNAPL mass depletion will not be significantly enhanced by NZVI-mediated dechlorination by placing NZVI directly in the DNAPL source zone (Fagerlund et al. 2012; Taghavy et al. 2010). Fagerlund et al. (2012) conducted 2D flow-through experiments using a PCE-DNAPL pool (saturation = 70%) with 10 g/L of NZVI emplaced right next to the source zone. Two different groundwater flow rates were assessed (pore-water velocity = 81 and 9.4 cm/day for Experiments 1 and 2, respectively; Fig. 11.1a).

Over the duration of the experiment, approximately 30% of the initial PCE-DNAPL mass reacted to form ethene, 50% was eluted as dissolved-phase PCE, and 20% remained in the column as PCE-DNAPL. Interestingly, they modeled the experimental results assuming rate-limited DNAPL dissolution and NZVI-mediated dechlorination of PCE to its three main reaction by-products, allowing for partitioning these by-products back into the DNAPL. When comparing the results from these models with the average total mass-transfer rate during the quasi-steady-state periods, the dechlorination reactions only had a minor effect on the total mass-transfer rate. In Experiment 1, the dechlorination by NZVI is predicted to increase

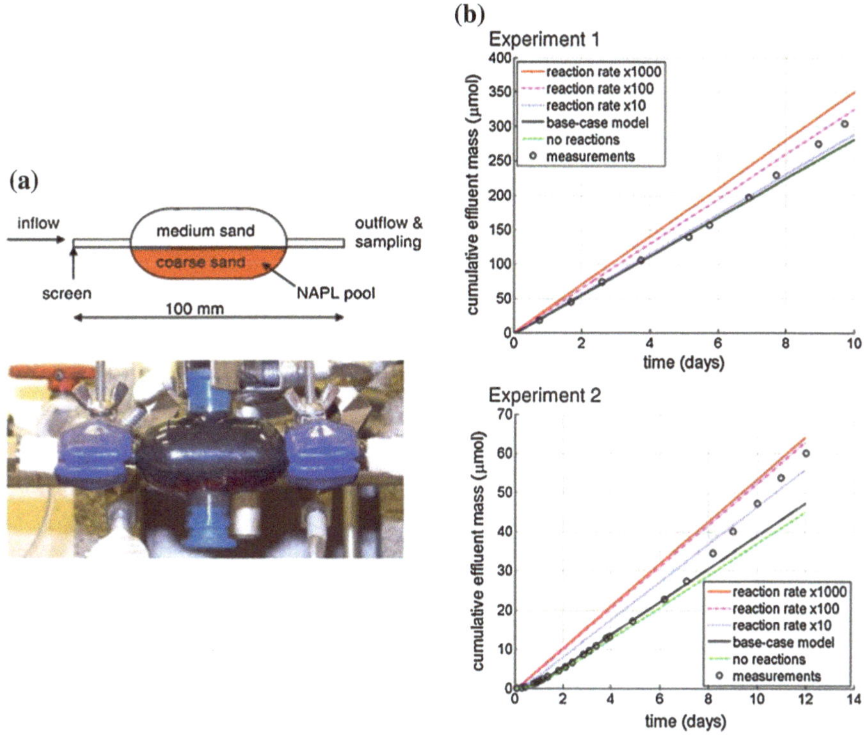

Fig. 11.1 (**a**) Schematic and photograph of the experimental setup taken after packing the cell and creating the DNAPL-PCE pool (red) and (**b**) cumulative effluent mass of PCE dechlorination using NZVI at various reaction rates. (Reprinted with permission from (Fagerlund et al. 2012). Copyright (2012) Elsevier)

the DNAPL dissolution rate by less than 0.5% in comparison with the mass-transfer rate without NZVI. For Experiment 2 (the slower groundwater flow rate), in which more than 25% of the total dissolved PCE reacted to non-chlorinated by-products, the predicted difference in the DNAPL dissolution rate due to NZVI reactions is still only 3.5% (4.2 μmol/day with reactions and 4.06 μmol/day without). Comparing Experiments 1 and 2, it can be concluded that, for slower groundwater flow through the DNAPL source zone (Experiment 2), there is more time for dechlorination reactions to occur, and the presence of NZVI can have a larger effect on the DNAPL mass-depletion rate compared to faster flow (Experiment 1). The cases considered here are within a scenario in which a high concentration (10 g/L) of NZVI has been effectively emplaced directly in the DNAPL source zone, everywhere surrounding the DNAPL-water interfaces. However, despite this, regarding what may be considered an "optimal" emplacement to target the source, very little enhancement of the source depletion rate was seen.

Furthermore, Fagerlund et al. (2012) investigated how higher reaction rates (corresponding to more emplaced particles or faster reacting particles) would influence the DNAPL mass removal and PCE-effluent concentrations from the source zone. To this end, they conducted a sensitivity study for the reaction rate by using rate constants that are 10, 100, and 1000 times higher than the ones measured in their batch experiments of the NZVI particles for Experiments 1 and 2 (Fig. 11.1b). Looking at the hypothetical model scenarios for higher reaction rates, they revealed that an 18% acceleration of the mass removal rate by NZVI is possible if the reaction rate is increased by a factor of 10 in the slower flowing Experiment 2 (Fig. 11.1b). However, the maximum enhanced acceleration of the mass removal was approximately 33% when the NZVI reaction rate was increased by 1000 times the rate measured in batch reactors. In this case, PCE is degraded much faster than it can dissolve, resulting in almost no PCE in the solution and therefore no possibilities for additional mass removal. Similarly, in Experiment 1, they reached 24% acceleration of mass removal at the rate of 1000 times that measured in the batch reactor, which should also be very close to the maximum limit because very little PCE remains in the effluent water. Noticeably, with 1000 times increase in NZVI reactivity, the DNAPL source depletion is only around 3.3 times faster. This finding is logical because the dechlorination reactions occur in the aqueous phase (as they require water to take place) and the mass transfer of PCE from DNAPL to the aqueous phase is generally rate limited. Thus, significant acceleration of the DNAPL mass-depletion rate is unlikely, even with highly reactive particles.

The mass transfer of CVOCs is also rate limiting for dechlorination of CVOCs sorbed to the soil, e.g., clay lenses. The CVOC sorbed to soil may behave as a long-term secondary source, gradually leaching dissolved CVOCs to contaminate the groundwater downgradient. Slow desorption of CVOCs from soil can result in retardation of reductive detoxification using NZVI. This is because the reaction with NZVI only occurs on the fraction of TCE that is in the water phase. Using trichloroethylene (TCE) as an example, TCE has an arithmetic mean of organic carbon partitioning coefficient (k_{oc}) of 86 (unitless) (ranging from 18.5 to 150). The soil-water partitioning coefficient (k_d) is the k_{oc} x fraction of organic carbon in soil (f_{oc}). This partitioning coefficient can be used to calculate the retardation factor as shown in Eq. 11.1. Subsequently, we can estimate the decrease in TCE dechlorination rate by NZVI in the soil-water system ($k_{\text{TCE-aq-Soil}}$) with R using Eq. 11.2 in comparison to the TCE dechlorination rate constant using NZVI in water (no sorption onto soil) ($k_{\text{TCE-aq}}$):

$$R = 1 + \frac{k_d{}^{*}\rho_b}{n} \tag{11.1}$$

$$k_{\text{TCE-aq-Soil}} = \frac{k_{\text{TCE-aq}}}{R}, \tag{11.2}$$

where ρ_b and n are the bulk density of soil and porosity, respectively. Under equilibrium with the partitioning coefficient (k_d) of 1.46 L/kg, R is calculated as 7.3 using Eq. 11.1. Thus, the TCE sorption into soil can decrease the TCE dechlorination rate constant in the soil-water system by 7.3 times in comparison to the system with groundwater alone.

This is problematic because, instead of using its reducing power to destroy contaminants, the NZVI reacts with the water to form H_2, which increases the amount of NZVI required for remediation (Berge and Ramsburg 2010; Liu et al. 2007). This is a mass-transfer limitation problem that cannot be solved by modifying NZVI to have greater reactivity, such as by doping with catalysts. Instead, any action to enhance the DNAPL dissolution or desorption of CVOCs from soil and DNAPL can speed the reaction rate and improve the electron utilization efficiency of the remediation. A possible solution is to use NZVI with thermal-enhanced CVOC dissolution or desorption. Increasing the temperature can increase the fraction of TCE in the water phase and increase the rate of reaction with the NZVI.

This chapter summarizes the state of knowledge about using a low-frequency (LF) electromagnetic field (EMF) (150 kHz) with NZVI to enhance the CVOC degradation rate in a DNAPL system and in a soil and groundwater system. The most recent pieces of work on using this novel approach are summarized as proof of concept. The TCE degradation kinetics in groundwater and in soil with groundwater as well as in a DNAPL system by NZVI both with and without LF EMF were compared to quantify the benefits of using LF EMF for enhanced thermal dissolution and magnetically enhanced NZVI corrosion.

11.2 Conceptual Model for Thermally Enhanced Dissolution/Desorption and Dechlorination Using NZVI with LF EMF

In addition to being a reducing agent, NZVI is ferromagnetic (Dalla Vecchia et al. 2009; Phenrat et al. 2007; Rosická and Šembera 2011). This magnetism gives NZVI the ability to enhance thermal dissolution and desorption of CVOCs when subjected to LF EMF. This is similar to magnetic nanoparticles used in medicine for thermal treatments, such as magnetic-assisted hyperthermia (Pablico-Lansigan et al. 2013), where functionalized magnetic nanoparticles target tumor cells and then are heated by applied LF EMF to kill the cancer cells (Bañobre-López et al. 2013). Similarly, NZVI should be able to induce heat under an applied LF EMF. This is caused by hysteresis loss, which is one of the three losses including eddy current loss and residual loss. Under applied LF EMF, upon the magnetization and demagnetization cycle, magnetic particles respond irreversibly, causing hysteresis and loss of energy as heat. The degree of irreversibility, ΔU, is related to the amount of energy dissipation upon the reversal of the field, which is material dependent. Under LF EMF, the reversal happens continuously and yields heat by energy dissipation from the particles. For a particular frequency (f) of LF EMF, the heat (P) generated due to the hysteresis loss is given by Eq. 11.3 (Li et al. 2010):

$$P = f\Delta U. \qquad (11.3)$$

Phenrat et al. (2016) characterized magnetic properties of NZVI (including NF25 from Nanoiron, Czech Republic, and MRNIP from Toda Corporation, Japan), ZVI materials (H200 and H150 from Hepure, USA), and magnetite nanoparticles using a vibrational sample magnetometer (VSM). They reported the hysteresis curves from the field of 1000 to -1000 G to illustrate the hysteresis loops of each ZVI or NZVI sample as shown in Fig. 11.2a. They revealed that the order of ΔU for ZVI and NZVI is as follows: H150 (12×10^4 emu G/g) > H200 (7×10^4 emu G/g) > NF25 (6.7×10^4 emu G/g) > MRNIP2 (5.5×10^4 emu G/g) > magnetite (4.4×10^4 emu G/g).

Unsurprisingly, under an applied LF EMF, ZVI and NZVI generated heat and raised the temperature of the suspension (Fig. 11.2b). In DI water, all ZVI and NZVI raised the temperature of the suspension above 80 °C in less than 15 min. However, their heat induction kinetics and rate constants depend on their magnetic properties. Their different heating kinetics agreed very well with their different ΔU values (Phenrat et al. 2016).

In addition, LF EMF may accelerate NZVI corrosion, resulting in the increase of intrinsic NZVI reactivity. The accelerated ZVI corrosion was explained by the magneto-convection of Fe^{2+}. According to this theory, the magnetic field causes an additional convective transfer of solutes to the surface of a ferromagnetic electrode (such as ZVI) due to the difference in the magnetic susceptibility of solutes and the electrode surface (Waskaas and Kharkats 1999) as well as the difference in the magnetic susceptibility of the solute and solvent molecules (Kitazawa et al. 2002). Guan's research group conducted a simulation to illustrate that the field gradient force can accumulate paramagnetic Fe^{2+} ions along the higher field gradient at the ZVI particle surface, which creates localized galvanic couples and electromagnetic forces that stimulate the migration of ions, the breakdown of passive film, and, eventually, the enhancement of localized corrosion (Jiang et al. 2015; Liang et al. 2014). This phenomenon results in the higher reduction rate of ZVI than that without DC MF. Similarly, the same field gradient force can act on paramagnetic metal ions, such as Cu^{2+}, to expedite Cu^{2+} adsorption on the ZVI surface (Jiang et al. 2015).

More quantitatively, the total enhancement factor (EF_{T}) (see Eq. 11.4) of CVOC degradation using NZVI with magnetic induction heating (MIH) by LF EMF can be attributed to three theoretical enhancement factors, namely, (1) the enhancement factor according to the Arrhenius equation (EF_{A}), (2) the thermally enhanced dissolution factor (EF_{D}), and (3) the accelerated corrosion enhancement factor (EF_{C}; see Eq. 11.3). Phenrat and Kumloet (2016) proposed that the total enhancement factor (E_{T}), as the ratio between the observed k_{SA} with MIH and the observed k_{SA} without MIH, can be expressed as:

$$EF_{\mathrm{T}} = EF_{\mathrm{A}} * EF_{\mathrm{C}} * EF_{\mathrm{D}}. \tag{11.4}$$

The theoretical EF_{A} can be calculated from Eq. 11.5:

$$\ln \frac{k_{\mathrm{MIH}}}{k_{\mathrm{No\ MIH}}} = \ln EF_{\mathrm{A}} = -\frac{E_a}{R}\left[\frac{1}{T_{\mathrm{MIH}}} - \frac{1}{T_{\mathrm{No\ MIH}}}\right]. \tag{11.5}$$

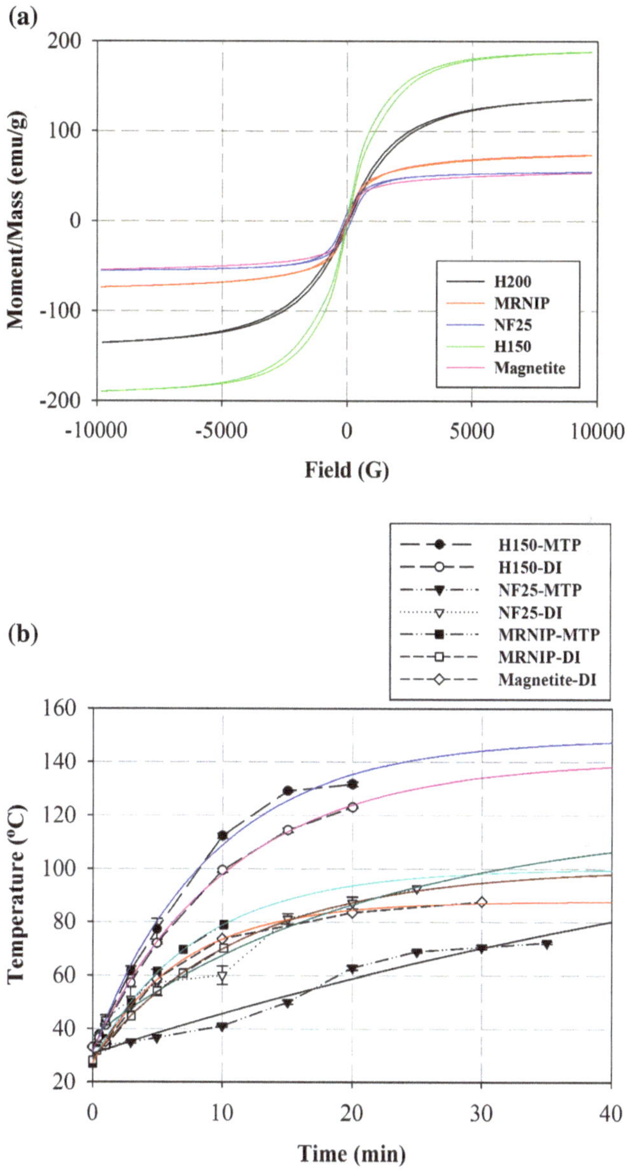

Fig. 11.2 (**a**) Hysteresis curves from the field of 1000 to −1000 G and (**b**) heat induction kinetics of various NZVI, ZVI, and magnetite nanoparticles (Phenrat et al. 2016). (Reprinted with permission from (Phenrat et al. 2016). Copyright (2016) American Chemical Society)

where k_{MIH} and $k_{\text{No MIH}}$ are the dechlorination rate constants with and without MIH, theoretically governed only by the effect of the Arrhenius equation, E_a is the activation energy of dechlorination, and R is the ideal gas constant (8.314 J/

K*mole). Moreover, T_{MIH} is the average temperature of the reactors as a result of MIH, and $T_{\mathrm{No\ MIH}}$ is the room temperature when the experiments without MIH were conducted.

We can theoretically estimate EF_{D} by comparing the solubility of CVOCs at T_{MIH} (S_{TMIH}) and $T_{\mathrm{No\ MIH}}$ ($S_{\mathrm{T\ No\ MIH}}$), as in Eq. 11.6:

$$EF_{\mathrm{D}} = \frac{S_{\mathrm{TMIH}}}{S_{\mathrm{T\ No\ MIH}}}. \tag{11.6}$$

The last theoretical EF is EF_{C}. Unlike the EF_{A} and EF_{D}, EF_{C} cannot be theoretically calculated due to the complexity and the current insufficient fundamental understanding of the phenomenon. Thus, we have to determine this by solving Eq. 11.4 using the calculated EF_{A} and EF_{D} and the measured EF_{T}.

Figure 11.3 illustrates the conceptual model of enhanced remediation of CVOC-sorbed soil using NZVI with LF EMF. The conceptual model of enhanced DNAPL source depletion is illustrated in Fig. 11.4. For these applications, NZVI must be

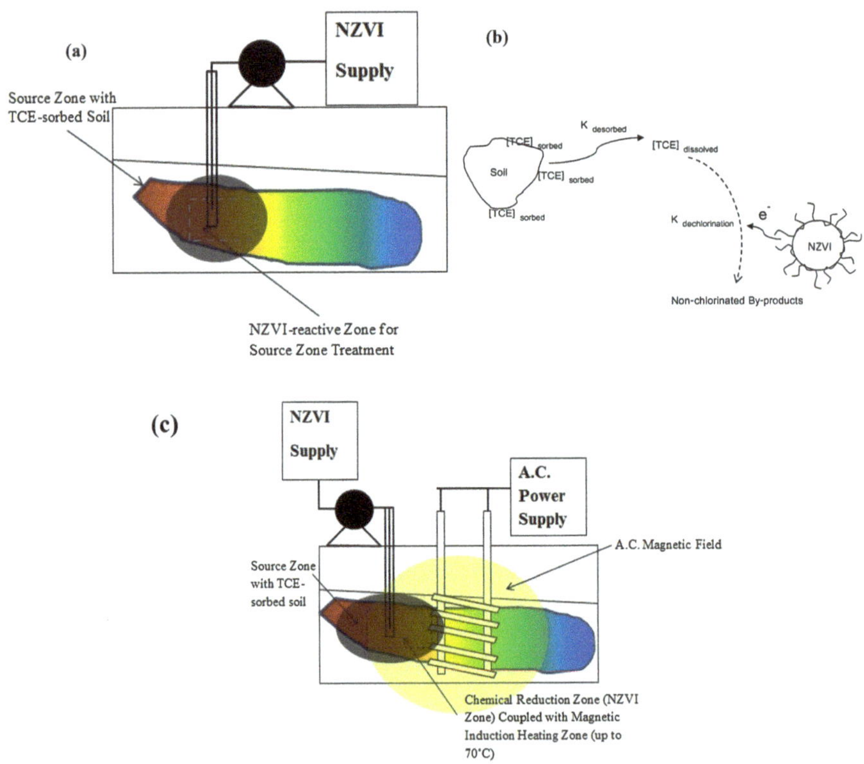

Fig. 11.3 Schematics of contaminated groundwater and soil remediation by (**a**) conventional NZVI, which relies on (**b**) two-step reaction including TCE desorption from soil and TCE dechlorination at NZVI surface and (**c**) combined remediation using NZVI with LF EMF

delivered into the subsurface to the source. Then, LF EMF will be applied to magnetically induce heat from the deposited NZVI and speed up the mass transfer as well as enhance the NZVI corrosion for higher intrinsic CVOC detoxification. The next sections will summarize the proof of concept of these novel applications in aqueous, soil-sorbed, and DNAPL phases.

11.3 Enhanced Dissolved TCE Dechlorination Using NZVI and LF EMF

Phenrat et al. (2016) conducted batch experiments to investigate the role of LF EMF on dissolved TCE dechlorination using NZVI. They prepared the reactors using 20 mL serum bottles capped with a Mininert™ closures. Each reactor contained 10 mL of headspace and 10 mL of TCE (50 mg/L) in DI or groundwater from Map Ta Phut (MTP) Industrial Estate, Thailand. Then, the TCE dechlorination kinetics were monitored using the headspace method for both ZVI (50 g/L of H150) and NZVI (10 g/L of NF25). The dechlorination was performed with and without LF EMF. For TCE dechlorination experiments under applied LF EMF, the experimental setup was modified by placing the insulated reactors into the center of the induction coil of the EMF generator that generated LF EMF at the current intensity of 15 A and the frequency of 150 kHz (see Fig. 11.5). For NF25 (see Table 11.1), five cycles of electromagnetic induction heating were applied over 10 h. Each cycle was 2 h with 60 min of LF EMF application, followed by 60 min of reaction without LF EMF (i.e., cooling down). The temperature was measured at the end of the heating cycle. During the 60 min of reaction without LF EMF for each cycle, the insulated reactors were rotated on an end-over-end rotator at 30 rpm to allow dechlorination to take place at a gradually decreasing temperature. A similar approach was used for H150 (ZVI). However, each cycle of reaction for H150 reactors consisted of 65 min with only 5 min of LF EMF application, followed by 60 min of reaction without LF EMF.

From the experiments discussed above, they revealed that applied LF EMF greatly enhanced the dechlorination of dissolved TCE in DI and MTP groundwater by both NZVI (NF25) and ZVI (H150) (see Fig. 11.6a and Table 11.2). Without LF EMF, the mass-normalized TCE dechlorination rate constants by NF25 in DI and MTP groundwater were $7.10 \pm 1.22 \times 10^{-3}$ and $5.43 \pm 0.47 \times 10^{-3}$ L/g/hr., respectively. For H150, the mass-normalized pseudo first-order rate constants were $0.84 \pm 0.09 \times 10^{-3}$ and $1.02 \pm 0.09 \times 10^{-3}$ L/g/hr. for DI and MTP groundwater, respectively. Without LF EMF, the rate constants of H150 in DI are around 5–10 times smaller than the NF25 because NZVI has a much larger specific surface area than ZVI, and dechlorination is a surface-mediated reaction.

However, the application of LF EMF-induced heat to 60 °C to 80 °C and increased the rate of TCE dechlorination compared to the absence of applied LF EMF. The TCE dechlorination afforded by NF25 with LF EMF yielded the rate constants of $35.20 \pm 4.23 \times 10^{-3}$ and $20.20 \pm 3.10 \times 10^{-3}$ L/g/hr. for DI and MTP

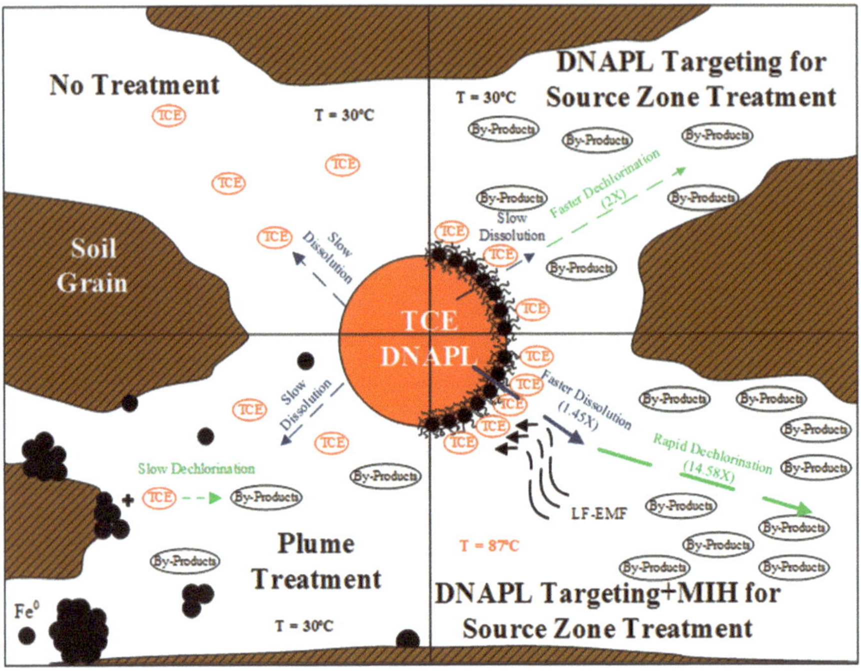

Fig. 11.4 Schematic of enhanced DNAPL source-zone depletion using polymer-modified to target DNAPL (Chap. 5) followed by applying LF EMF for speeding up the CVOC dissolution rate (Phenrat and Kumloet 2016). (Reprinted with permission from (Phenrat and Kumloet 2016). Copyright (2016) Elsevier)

Fig. 11.5 EMF generator and an induction coil (white coil) to hold vials for induction heating and LF EMF-enhanced dechlorination experiments

Table 11.1 Physicochemical properties of Nanofer 25 (NF25)

Physicochemical properties of Nanofer 25 (NF25)	
Chemical composition of Fe^0	Fe(core) FeO (shell)
Content of solid phase in dispersion by weight	20%
Content Fe^0 in solid phase Other ingredients in solid phase Content of Fe^0 in dispersion by weight	≈ 85% Fe_3O_4,FeO,C 17%
Crystalline structure of Fe^0	Alpha Fe
Particles morphology	Spherical
Average particle size	d50 < 50 nm
Particle-specific surface area	>25m^2/g
Dispersion color	Black
Dispersion density Fe^0 particles density Fe_3O_4 density	1210 kg/m3 7870 kg/m3 5700 kg/m3

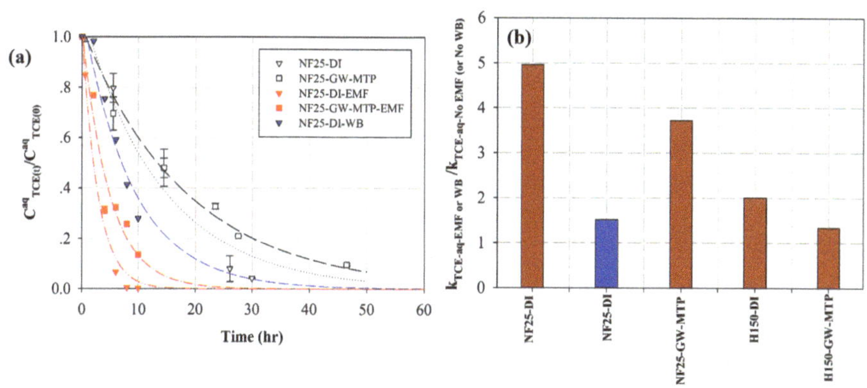

Fig. 11.6 (**a**) TCE dechlorination kinetics using NF25 in DI water (triangles) and MTP groundwater (squares) with AC EMF (red symbols) and without AC EMF (white symbols) as well as NF25 in DI water heated by water bath (WB; blue triangles). The dashed lines represent the pseudo first-order models. (**b**) Summary of enhanced dissolved TCE dechlorination by NZVI and ZVI with AC EMF and WB depicted as the ratio of the TCE dechlorination rate constant with AC EMF or WB ($k_{\text{TCE-aq-AC EMF or WB}}$) over the TCE dechlorination rate constant without AC EMF or WB ($k_{\text{TCE-aq-No AC EMF or No WB}}$) for each case. (Reprinted with permission from Phenrat et al. 2016. Copyright (2016) American Chemical Society)

groundwater, respectively. This is 5 and 3.7 times greater than without LF EMF for DI water and MTP groundwater, respectively (Fig. 11.6b). Similarly, when ZVI (H150) was placed under applied LF EMF for dissolved TCE dechlorination, it induced greater heat in every consecutive cycle. LF EMF increased the TCE degradation rate constants of ZVI (H150) in both in DI and MTP groundwater, although the increase was not as pronounced as the case of NZVI (NF25). The TCE dechlorination afforded by H150 with LF EMF yielded the rate constants of 2.01 and

Table 11.2 Mass-normalized pseudo first-order rate constants of TCE dechlorination for various experimental conditions in this study

NZVI or ZVI	Type of media	Heating condition	Mass-normalized dechlorination rate constant (10^{-3} L/g/hr)
N25	DI	None	7.10 ± 1.22
		Water bath	10.80 ± 1.74
		EMF	35.20 ± 4.23
N25	MTP	None	5.43 ± 0.47
		EMF	20.20 ± 3.10
N25	Soil + MTP	None	2.20 ± 0.68
		EMF	11.80 ± 1.13
H150	DI	None	0.84 ± 0.09
		EMF	1.69 ± 0.25
H150	MTP	None	1.02 ± 0.09
		EMF	1.37 ± 0.25

1.34 times greater than without EMF for DI water and MTP groundwater, respectively (Fig. 11.6b).

Intuitively, for the same NZVI, the enhanced reaction is a function of electromagnetic induction time (the longer the induction time, the faster the dechlorination) and the particle concentration under the induction (the greater the particle concentration, the faster the dechlorination). Figure 11.7a illustrates the linear correlation (with the slope of 0.9) between the enhanced dissolved TCE dechlorination rate constant due to EMF ($k_{\text{TCE-aq-EMF}}/k_{\text{TCE-aq-No EMF}}$) and the ratio of EMF application time (Σt_{EMF}) over the total dechlorination time ($t_{\text{Dechlorination}}$) multiplied by particle concentration ($C_{\text{ZVI or NZVI}}$) for all the four cases in this study. This correlation confirms that longer EMF application time and higher concentrations of ZVI or NZVI will yield greater dechlorination rates, as theoretically expected.

The enhanced dissolved TCE dechlorination rate is presumably due to the higher temperature (Arrhenius theory) and enhanced ZVI corrosion. To separate the two effects, they conducted another set of TCE dechlorination experiments using NF25 in a heated water bath to simulate the temperature profile for dechlorination obtained from the MIH study of NF25 with LF EMF. As shown in Table 11.2, the TCE dechlorination afforded by NF25 in DI with the heated water bath yielded the rate constant of $10.80 \pm 1.74 \times 10^{-3}$ L/g/hr. This is around 1.52 times greater than the TCE dechlorination rate constant of NF25 in DI without heating but still 3.26 times smaller than the TCE dechlorination rate constant of NF25 in DI with LF EMF. They proposed that, besides the temperature increase, the enhanced ZVI corrosion may also be enhancing the reaction with TCE. Figure 11.7b supports this hypothesis by illustrating the decrease of ORP for both NZVI reactors as a function of time under LF EMF. The ORP represents the reduction capability of the system. Increasingly negative ORP indicates a greater reduction capacity of the system (i.e., NZVI is more active in donating electrons, or other species in the water are making the system more reducing). Presumably, the applied LF EMF induces the field gradient force on

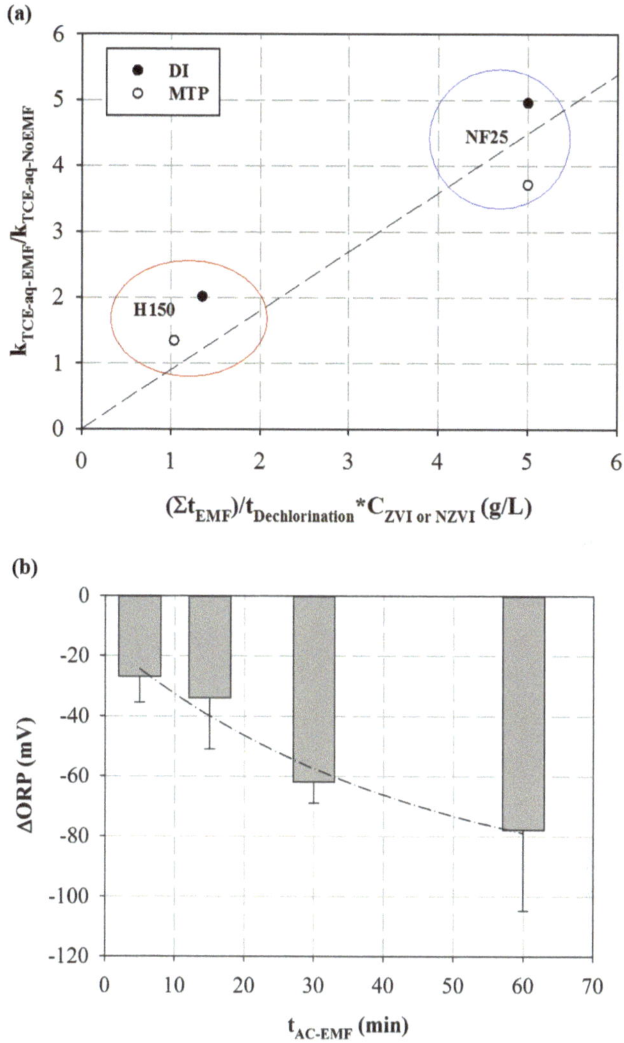

Fig. 11.7 (**a**) Linear correlation (with the slope of 0.9) between the enhanced dissolved TCE dechlorination rate constant due to AC EMF ($k_{\text{TCE-aq-AC EMF}}/k_{\text{TCE-aq-No AC EMF}}$) and the ratio of AC EMF application time ($\Sigma t_{\text{AC EMF}}$) over the total dechlorination time ($t_{\text{Dechlorination}}$) multiplied by particle concentration ($C_{\text{ZVI or NZVI}}$) for all four cases in this study (both NF25 and H150 in DI and MTP groundwater) and (**b**) the increasingly negative ORP magnitude (ΔORP) of NZVI as a function of AC EMF time ($t_{\text{AC EMF}}$)

the NZVI surface, which subsequently induces paramagnetic Fe^{2+} ions, the by-product of NZVI oxidation, to deplete from the NZVI surface where the field gradient is low and to accumulate on the NZVI surface where the field gradient is high. This Fe^{2+} accumulation creates localized galvanic couples, which promote the breakdown of the passive iron-oxide layer, which is known to decrease NZVI reactivity (Martin et al. 2008). This local galvanic couple activity eventually enhances localized ZVI corrosion (Jiang et al. 2015; Liang et al. 2014), resulting in increasingly negative ORP magnitude and the increase of the reduction power of the system.

11.4 Enhanced Dechlorination of TCE-Sorbed Soil Using NZVI and LF EMF

Phenrat et al. (2016) also investigated the influence of LF EMF on the dechlorination of TCE sorbed onto soil (i.e., in a TCE pre-sorbed soil-water system). To perform this experiment, they used the same batch experimental setup as for dissolved TCE dechlorination discussed previously, but instead of using TCE (50 mg/L) in DI or MTP groundwater, these experiments used TCE in a soil-water system. To prepare the TCE-contaminated soil, 20 mL reactors were filled with 8 mL of soil slurry (4:6 of MTP soil-to-groundwater ratio by volume) and 100 mg/L of TCE. Figure 11.8 illustrates TCE sorption kinetics and partitioning behavior in MTP soil and groundwater. The other steps of the experiments were the same as discussed previously.

They found that using LF EMF with NZVI resulted in a more rapid and complete TCE degradation in the soil-groundwater system. As shown in Fig. 11.9, the dissolved TCE dechlorination first-order rate constant without LF EMF was $2.20 \pm 0.68 \times 10^{-3}$ L/g/hr., around 2.5 times smaller than the rate constant in MTP groundwater without soil. The decreased TCE dechlorination rate constant in the presence of soil agrees very well with the theoretically retarded dechlorination rate constant in Eq. 11.2. Noticeably, without applied LF EMF, the dissolved TCE concentration did not reach zero even after 300 h of reaction. Instead, the dechlorination seemed to progress to the system where TCE in the water remained constant at around 30% of the initial dissolved TCE in the system. No further decrease of dissolved TCE in the system occurred from 150 h to 300 h. After the end of the

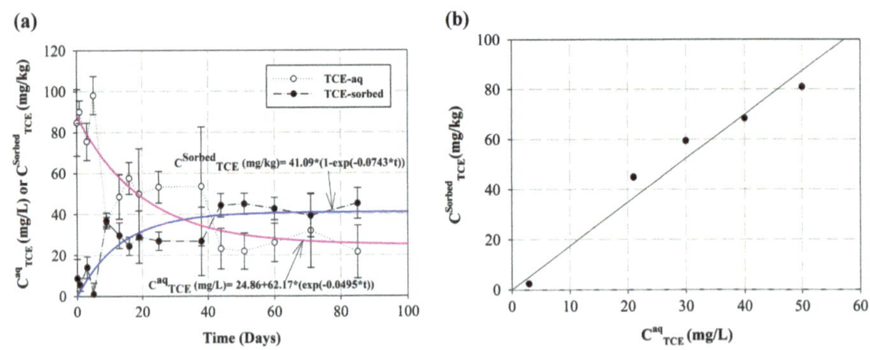

Fig. 11.8 (**a**) TCE sorption kinetics on MTP soil and (**b**) K_d of TCE on MTP soil as determined from the slope of the graph

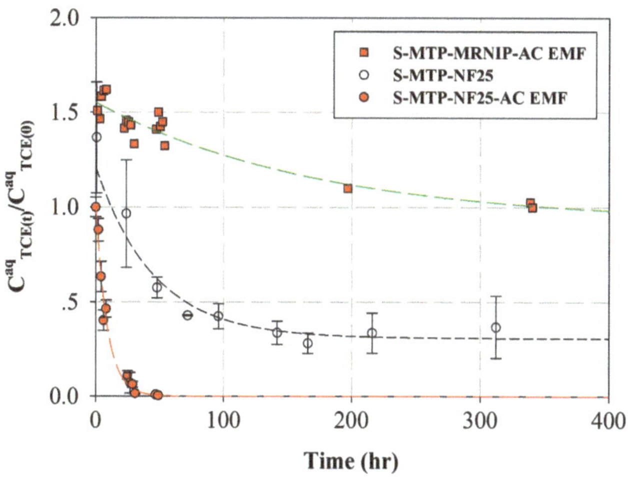

Fig. 11.9 Dissolved TCE dechlorination kinetics in a soil-water system with pre-adsorbed TCE. Unfilled circles represent TCE dechlorination by NF25 without LF EMF, red circles represent TCE dechlorination by NF25 with LF EMF, and red squares represent TCE desorption from soil by (unreactive) MRNIP with LF EMF. The dashed lines represent the pseudo first-order model fits. (Reprinted with permission from (Phenrat et al. 2016). Copyright (2016) American Chemical Society)

experiment, they recovered 35% by mass of the initial TCE from the reactor using hexane extraction. Additionally, 75% of the recovered TCE was sorbed into the soil, while 25% was dissolved into the MTP groundwater. This result illustrates the effect of the mass-transfer limitation due to TCE sorption and slow desorption on TCE dechlorination using NZVI (Phenrat et al. 2010; Zhang et al. 2011).

This problem was eliminated using LF EMF. As shown in Fig. 11.9, the TCE dechlorination on rate constant is 5.4 times higher with LF EMF, and unlike the absence of LF EMF, the dissolved TCE concentration goes to zero. The dissolved TCE was completely degraded after 31 h of reaction, and no dissolved TCE was observed over another 200 h, suggesting that the TCE degradation was complete, even for the adsorbed TCE. After the study, they performed hexane extraction to recover TCE sorbed onto soil. They reported only around 3% by mass of the initial TCE. This finding agrees with the conceptual model in that heat induced by electromagnetic induction of NF25 raised the temperature of the soil-water system, which enhanced TCE desorption from the soil (Fig. 11.3). Desorbed TCE was available for dechlorination and rapidly degraded by NF25. This thermally enhanced desorption hypothesis was demonstrated using the nonreactive (aged) NZVI, MRNIP. Over four MIH cycles, MRNIP generated heat and raised the temperature to around 65 °C. As a result, the dissolved TCE concentration increased to around 1.5 times the initial concentration (red squares in Fig. 11.10), indicating that TCE was desorbed from soil. However, since MRNIP was aged and not reactive anymore, allowing the system to cool to the original temperature slowly returned the TCE

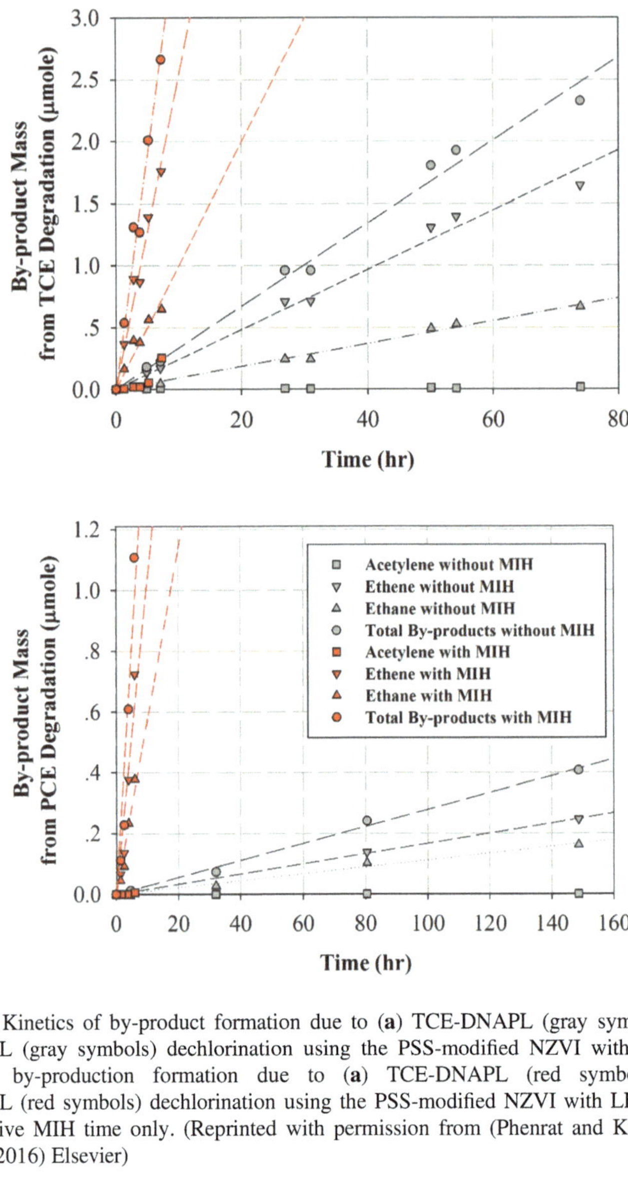

Fig. 11.10 Kinetics of by-product formation due to (**a**) TCE-DNAPL (gray symbols) and (**b**) PCE-DNAPL (gray symbols) dechlorination using the PSS-modified NZVI without LF EMF. Cumulative by-production formation due to (**a**) TCE-DNAPL (red symbols) and (**b**) PCE-DNAPL (red symbols) dechlorination using the PSS-modified NZVI with LF EMF during the cumulative MIH time only. (Reprinted with permission from (Phenrat and Kumloet 2016). Copyright (2016) Elsevier)

concentration back to its original value. The sorption rate constant was approximately 5.8×10^{-3} hr.$^{-1}$. For the case of NF25 with LF EMF, Phenrat et al. (2016) did not observe desorption and resorption of TCE because the desorbed TCE was rapidly dechlorinated at the elevated temperature.

11.5 Enhanced Dechlorination of TCE and PCE DNAPL Using NZVI and LF EMF

The LF EMF has the potential to be a successful DNAPL source-zone removal method if used with NAPL targeting by polymer-modified NZVI. Phenrat and Kumloet (2016) demonstrated this novel application using batch experiments similar to the experiments for dissolved and soil-sorbed TCE discussed previously except that they put the initial TCE and PCE DNAPL of 0.47 and 0.43 mL (corresponding to 0.1 g) into the reactor instead of dissolved or soil-sorbed TCE. In addition, instead of measuring TCE and PCE concentration, they tracked TCE- and PCE-DNAPL degradation by measuring the formation kinetics of dechlorination by-products including acetylene, ethene, and ethane. They applied the DNAPL targeting strategy (Fig. 11.4) for these DNAPL dechlorination experiments by preparing polystyrene sulfonate (PSS)-modified NZVI, which targeted TCE and PCE DNAPL via the Pickering emulsion protocol (see Chap. 5).

Without LF EMF, the PSS-modified NZVI (10 g/L) that attached to TCE DNAPL and PCE DNAPL transformed TCE and PCE to ethene and ethane (gray symbols in Fig. 11.10a, b). Neither acetylene nor other chlorinated organic intermediates were observed. The reaction followed pseudo zero-order reaction kinetics with the by-product formation rate constant (equal to the TCE- and PCE-DNAPL degradation rates) of 33.5×10^{-3} and 2.8×10^{-3} μmole/hr., for TCE DNAPL and PCE DNAPL, respectively. Conversely, PSS-modified NZVI with LF EMF greatly improved the degradation of TCE and PCE DNAPL relative to the absence of LF EMF. With four cycles of LF EMF applied, PSS-modified NZVI-induced heating elevated the temperature up to 87 °C and increased the pseudo zero-order by-product formation rate constants (equal to the TCE- and PCE-DNAPL degradation rates) to 490×10^{-3} and 160×10^{-3} μmole/hr. for TCE DNAPL and PCE DNAPL (red symbols in Fig. 11.10a, b), respectively. This means that the rate constants of TCE DNAPL and PCE DNAPL are up to 15-fold and 60-fold, respectively, versus the dechlorination rate without LF EMF (Phenrat and Kumloet 2016).

They performed an analysis of the contribution from the enhanced thermal dissolution from DNAPL (EF_D), of the effect of the increasing temperature on the rate constant (the Arrhenius equation) (EF_A), and of the accelerated NZVI corrosion caused by the applied LF EMF (EF_C) on enhanced TCE- and PCE-DNAPL dechlorination. They concluded that EF_A theoretically contributed to the enhanced TCE-DNAPL and PCE-DNAPL dechlorination by 4.2 and 16.8 times, respectively, while the EF_D theoretically contributed to the enhanced TCE-DNAPL and PCE-DNAPL dechlorination by 1.5 and 2.2 times, respectively. Based on solving Eq. 11.4, the EF_C theoretically contributed to the enhanced TCE-DNAPL and PCE-DNAPL dechlorination by 2.4 and 1.5 times, respectively.

Acknowledgments The authors are thankful for research funding from (1) the Thailand Research Fund (TRF) (MRG5680129); (2) the National Nanotechnology Center (Thailand), a member of the National Science and Technology Development Agency, through grant number P-11-00989; and (3) the National Research Council (R2556B070).

References

Bañobre-López, M., Teijeiro, A., & Rivas, J. (2013). Magnetic nanoparticle-based hyperthermia for cancer treatment. *Reports of Practical Oncology and Radiotherapy, 18*(6), 397–400.

Berge, N. D., & Ramsburg, C. A. (2010). Iron-mediated trichloroethene reduction within nonaqueous phase liquid. *Journal of Contaminant Hydrology, 118*(3–4), 105–116.

Bishop, E. J., Fowler, D. E., Skluzacek, J. M., Seibel, E., & Mallouk, T. E. (2010). Anionic homopolymers efficiently target zerovalent iron particles to hydrophobic contaminants in sand columns. *Environmental Science & Technology, 44*(23), 9069–9074.

Dalla Vecchia, E., Coisson, M., Appino, C., Vinai, F., & Sethi, R. (2009). Magnetic characterization and interaction modeling of zerovalent iron nanoparticles for the remediation of contaminated aquifers. *Journal of Nanoscience and Nanotechnology, 9*(5), 3210–3218.

Fagerlund, F., Illangasekare, T. H., Phenrat, T., Kim, H.-J., & Lowry, G. V. (2012). PCE dissolution and simultaneous dechlorination by nanoscale zero-valent iron particles in a DNAPL source zone. *Journal of Contaminant Hydrology, 131*(1–4), 9–28.

He, F., Zhao, D., Liu, J., & Roberts, C. B. (2007). Stabilization of Fe-Pd nanoparticles with sodium carboxymethyl cellulose for enhanced transport and dechlorination of trichloroethylene in soil and groundwater. *Industrial and Engineering Chemistry Research, 46*(1), 29–34.

He, F., Zhao, D., & Paul, C. (2010). Field assessment of carboxymethyl cellulose stabilized iron nanoparticles for *in situ* destruction of chlorinated solvents in source zones. *Water Research, 44* (7), 2360–2370.

Henn, K. W., & Waddill, D. W. (2006). Utilization of nanoscale zero-valent iron for source remediation - a case study. *Remediation Journal, 16*, 57–77.

Jiang, X., Qiao, J., Lo, I. M. C., Wang, L., Guan, X., Lu, Z., Zhou, G., & Xu, C. (2015). Enhanced paramagnetic Cu2+ ions removal by coupling a weak magnetic field with zerovalent iron. *Journal of Hazardous Materials, 283*, 880–887.

Johnson, R. L., Nurmi, J. T., O'Brien Johnson, G. S., Fan, D., O'Brien Johnson, R. L., Shi, Z., Salter-Blanc, A. J., Tratnyek, P. G., & Lowry, G. V. (2013). Field-scale transport and transformation of carboxymethylcellulose-stabilized nano zero-valent iron. *Environmental Science & Technology, 47*(3), 1573–1580.

Kitazawa, K., Hirota, N., Ikezoe, Y., Uetake, H., Kaihatsu, T., & Takayama, T. (2002). Magneto-convection processes observed in non-magnetic liquid–gas system. *Riken Review, 44*, 156–158.

Kocur, C. M., Lomheim, L., Boparai, H. K., Chowdhury, A. I., Weber, K. P., Austrins, L. M., Edwards, E. A., Sleep, B. E., & O'Carroll, D. M. (2015). Contributions of abiotic and biotic dechlorination following carboxymethyl cellulose stabilized nanoscale zero valent iron injection. *Environmental Science & Technology, 49*(14), 8648–8656.

Li, Z., Kawashita, M., Araki, N., Mitsumori, M., Hiraoka, M., & Doi, M. (2010). Magnetite nanoparticles with high heating efficiencies for application in the hyperthermia of cancer. *Materials Science and Engineering: C, 30*(7), 990–996.

Liang, L., Sun, W., Guan, X., Huang, Y., Choi, W., Bao, H., Li, L., & Jiang, Z. (2014). Weak magnetic field significantly enhances selenite removal kinetics by zero valent iron. *Water Research, 49*, 371–380.

Liu, Y., Phenrat, T., & Lowry, G. V. (2007). Effect of TCE concentration and dissolved groundwater solutes on NZVI-promoted TCE dechlorination and H2 evolution. *Environmental Science & Technology, 41*(22), 7881–7887.

Martin, J. E., Herzing, A. A., Yan, W., Li, X. Q., Koel, B. E., Kiely, C. J., & Zhang, W. X. (2008). Determination of the oxide layer thickness in core-shell zerovalent iron nanoparticles. *Langmuir, 24*(8), 4329–4334.

Miller, C. T., Poirer-McNeill, M. M., & Mayer, A. S. (1990). Dissolution of trapped nonaqueous phase liquids: Mass transfer characteristics. *Water Resources Research, 26*(11), 2783–2796.

O'Carroll, D. M., Sleep, B., Krol, M., Boparai, H., & Kocur, C. (2013). Nanoscale zero valent iron and bimetallic particles for contaminated site remediation. *Advances in Water Resources, 51*, 104–122.

Pablico-Lansigan, M. H., Situa, S. F., & Samia, A. C. S. (2013). Magnetic particle imaging: Advancements and perspectives for real-time in vivo monitoring and image-guided therapy. *Nanoscale, 5*, 4040–4055.

Phenrat, T., & Kumloet, I. (2016). Electromagnetic induction of nanoscale zerovalent iron particles accelerates the degradation of chlorinated dense non-aqueous phase liquid: Proof of concept. *Water Research, 107*, 19–28.

Phenrat, T., Saleh, N., Sirk, K., Tilton, R. D., & Lowry, G. V. (2007). Aggregation and sedimentation of aqueous nanoscale zerovalent iron dispersions. *Environmental Science & Technology, 41*(1), 284–290.

Phenrat, T., Liu, Y., Tilton, R. D., & Lowry, G. V. (2009). Adsorbed polyelectrolyte coatings decrease Fe^0 nanoparticle reactivity with TCE in water: Conceptual model and mechanisms. *Environmental Science & Technology, 43*(5), 1507–1514.

Phenrat, T., Schoenfelder, D., Losi, M., Yi, J., Peck, S. A., & Lowry, G. V. (2010). In C. L. Geiger & K. M. Carvalho-Knighton (Eds.), *Environmental applications of nanoscale and microscale reactive metal particles* (pp. 183–202). Washington, DC: American Chemical Society.

Phenrat, T., Crimi, M., Illanagasekare, T., & Lowry, G. V. (2011a). In M. Ram, E. S. Andreescu, & D. Hanming (Eds.), *Nanotechnology for environmental decontamination* (pp. 271–322). New York: McGraw-Hill Publisher.

Phenrat, T., Fagerlund, F., Illanagasekare, T., Lowry, G. V., & Tilton, R. D. (2011b). Polymer-modified Fe0 nanoparticles target entrapped NAPL in two dimensional porous media: Effect of particle concentration, NAPL saturation, and injection strategy. *Environmental Science & Technology, 45*(14), 6102–6109.

Phenrat, T., Schoenfelder, D., Kirschling, T. L., Tilton, R. D., & Lowry, G. V. (2015). Adsorbed poly(aspartate) coating limits the adverse effects of dissolved groundwater solutes on Fe0 nanoparticle reactivity with trichloroethylene. *Environmental Science and Pollution Research, 25*(8), 7157–7169.

Phenrat, T., Thongboot, T., & Lowry, G. V. (2016). Electromagnetic Induction of Zerovalent Iron (ZVI) powder and nanoscale Zerovalent Iron (NZVI) particles enhances Dechlorination of trichloroethylene in contaminated groundwater and soil: Proof of concept. *Environmental Science & Technology, 50*(2), 872–880.

Powers, S. E., Abriola, L. M., & Weber, W. J., Jr. (1992). An experimental investigation of nonaqueous phase liquid dissolution in saturated subsurface systems: Steady state mass transfer rates. *Water Resources Research, 28*(10), 2691–2705.

Quinn, J., Geiger, C., Clausen, C., Brooks, K., & Coon, C. (2005). Field demonstration of DNAPL dehalogenation using emulsified zero-valent iron. *Environmental Science & Technology, 39*(5), 1309–1318.

Rosická, D., & Šembera, J. (2011). Assessment of influence of magnetic forces on aggregation of zero-valent Iron nanoparticles. *Nanoscale Research Letters, 6*, 10.

Saba, T., & Illangasekare, T. H. (2000). Effect of groundwater flow dimensionality on mass transfer from entrapped nonaqueous phase liquid contaminants. *Water Resources Research, 36*(4), 971–980.

Sakulchaicharoen, N., O'Carroll, D. M., & Herrera, J. E. (2010). Enhanced stability and dechlorination activity of pre-synthesis stabilized nanoscale FePd particles. *Journal of Contaminant Hydrology, 118*(3–4), 117–127.

Saleh, N., Phenrat, T., Sirk, K., Dufour, B., Ok, J., Sarbu, T., Matyjaszewski, K., Tilton, R. D., & Lowry, G. V. (2005). Adsorbed triblock copolymers deliver reactive iron nanoparticles to the oil/water interface. *Nano Letters, 5*(12), 2489–2494.

Saleh, N., Sirk, K., Liu, Y., Phenrat, T., Dufour, B., Matyjaszewski, K., Tilton, R. D., & Lowry, G. V. (2007). Surface modifications enhance nanoiron transport and NAPL targeting in saturated porous media. *Environmental Engineering Science, 24*(1), 45–57.

Su, C., Puls, R. W., Krug, T. A., Watling, M. T., O'Hara, S. K., Quinn, J. W., & Ruiz, N. W. (2012). A two and half-year-performance evaluation of a field test on treatment of source zone

tetrachloroethene and its chlorinated daughter products using emulsified zero valent iron nanoparticles. *Water Research, 46*(16), 5071–5084.
Taghavy, A., Costanza, J., Pennell, K. D., & Abrio, L. M. (2010). Effectiveness of nanoscale zero-valent iron for treatment of a PCE–DNAPL source zone. *Journal of Contaminant Hydrology, 118*(3–4), 128–142.
Tratnyek, P. G., & Johnson, R. L. (2006). Nanotechnologies for environmental cleanup. *Nano Today, 1*(2), 44–48.
Waskaas, M., & Kharkats, Y. I. (1999). Magnetoconvection phenomena: A mechanism for influence of magnetic fields on electrochemical processes. *The Journal of Physical Chemistry. B, 103*, 4876–4883.
Zhan, J., Zheng, T., Piringer, G., Day, C., McPherson, G. L., Lu, Y., Papadopoulos, K., & John, V. T. (2009). Transport characteristics of nanoscale functional zerovalent iron/silica composites for in situ remediation of trichloroethylene. *Environmental Science & Technology, 42*(23), 8871–8876.
Zhang, W.-X., Wang, C.-B., & Lien, H.-L. (1998). Treatment of chlorinated organic contaminants with nanoscale bimetallic particles. *Catalysis Today, 40*, 387–395.
Zhang, M., He, F., Zhao, D., & Hao, X. (2011). Degradation of soil-sorbed trichloroethylene by stabilized zero valent iron nanoparticles: Effects of sorption, surfactants, and natural organic matter. *Water Research, 45*(7), 2401–2414.
Zhao, X., Liu, W., Cai, Z., Han, B., Qian, T., & Zhao, D. (2016). An overview of preparation and applications of stabilized zero-valent iron nanoparticles for soil and groundwater remediation. *Water Research, 100*, 245–266.

Chapter 12
Improving the Reactivity of ZVI and NZVI Toward Various Metals and Metal(loid)s with Weak Magnetic Field

Jinxiang Li, Yuankui Sun, Liping Liang, and Xiaohong Guan

Abstract This chapter provides an overview of employing weak magnetic field (WMF) and premagnetization to improve the reactivity of ZVI toward various metal (loid)s. The rate constants of metal(loid)s sequestration by ZVI were increased by 1.1–383.7- and 1.2–12.2-fold due to the application of WMF and premagnetization, respectively. The mechanisms of WMF-induced improvement in contaminant sequestration by ZVI are also summarized. Finally, this chapter identifies the current knowledge gaps and future research needs of WMF/ZVI system for environmental application.

Keywords Nanoscale Zerovalent Iron · Zerovalent Iron · Weak Magnetic Field · Metals · Metalloids · Sequestration

12.1 Limitations of ZVI Application

There is an increasing interest on employing zero-valent iron (ZVI) and nanoscale zero-valent iron (NZVI) for the removal of contaminants from groundwater and wastewater (Chaps. 3 and 4), as reflected by the exponentially increasing number of publications in the last two decades (see Chap. 1). Nevertheless, considering the intrinsic properties of ZVI and the reactions that occur in the process of contaminants sequestration by ZVI, employing granular iron, iron filings, microscale, or millimetric zero-valent iron (known collectively as ZVI) for contaminant removal has several intrinsic limitations, as schematically illustrated in Fig. 12.1.

ZVI has low reactivity since it is covered with an intrinsic passive layer generated in its manufacturing process and has a relatively low specific surface area (Ritter

J. Li · Y. Sun · X. Guan (✉)
Tongji University, Shanghai, People's Republic of China
e-mail: guanxh@tongji.edu.cn

L. Liang
Shaoxing University, Shaoxing, People's Republic of China

T. Phenrat, G. V. Lowry (eds.), *Nanoscale Zerovalent Iron Particles for Environmental Restoration*, https://doi.org/10.1007/978-3-319-95340-3_12

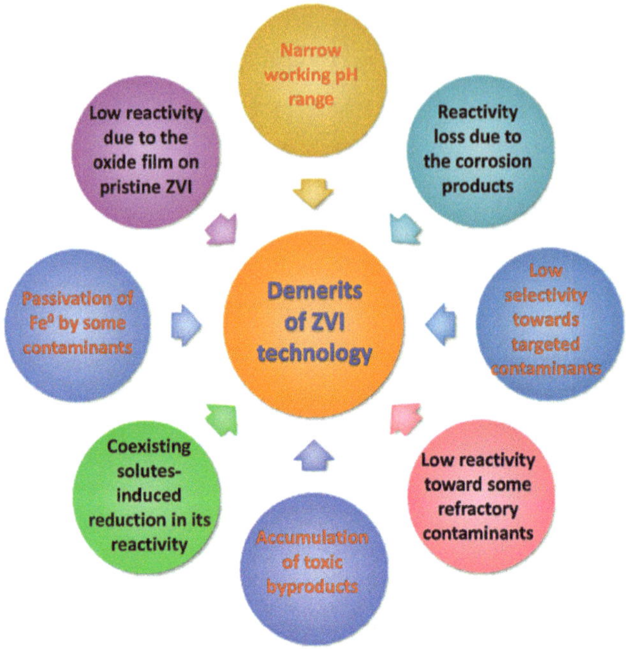

Fig. 12.1 Summary of the limitations of ZVI technology (Guan et al. 2015b)

et al. 2002). The reactivity of ZVI drops over time, due to the surface passivation caused by the corrosion products (e.g., the precipitation of metal hydroxides, such as $Fe(OH)_2$, $Fe(OH)_3$, and metal carbonates, such as $FeCO_3$ on the surface of ZVI). Moreover, some contaminants can passivate ZVI, leading to a drop in the reactivity of ZVI (Liang et al. 2014b):

(i) Due to the direct involvement of H^+ in the corrosion reactions and the mass transport limitations imposed by the precipitation of a passive film on the metal surface, the reactivity of ZVI drops sharply with increasing pH.
(ii) ZVI has low reactivity toward some refractory contaminants, e.g., Se(VI), and it has low adsorption capacity and selectivity for targeted contaminants, especially for the organic contaminants.

Some toxic intermediates may accumulate in the process of contaminant removal (especially the lightly chlorinated hydrocarbons) by ZVI due to the low reactivity of ZVI (Wang and Zhang 1997).

Although various countermeasures, including fabricating NZVI (Wang and Zhang 1997), to overcome or mitigate these limitations of ZVI technologies have been gradually developed in the past two decades, further research and development in this area are still necessary since each countermeasure has drawbacks (Guan et al. 2015b). In the past several years, our group explored the feasibility of employing weak magnetic field (WMF) to overcome some limitations of ZVI. The performance and mechanism of employing WMF to improve ZVI technology will be discussed in this chapter. In addition, the potential application of WMF to enhance the reactivity of NZVI will also be presented.

12.2 The Feasibility of Employing WMF to Overcome the Limitations of ZVI Technologies

It has been recognized that the adsorptive capability and reductive activity of ZVI toward various contaminants rely on the effectiveness of iron to serve as electron donor (Agrawal and Tratnyek 1996; Gillham and Ohannesin 1994; Matheson and Tratnyek 1994; Miehr et al. 2004). However, the granular ZVI, or iron filings, extensively examined in laboratory studies and field demonstrations, has low intrinsic reactivity toward contaminants due to its inherent passive film (Obiri-Nyarko et al. 2014; Prasad et al. 2011). Additionally, the formation of oxide and hydroxide layers results in passive layer on ZVI surface during contaminant removal reduces its reactivity (Ansaf et al. 2016; Guan et al. 2015b). For these reasons, since the publication of two full papers on the degradation of halogenated aliphatics by ZVI in 1994 (Gillham and Ohannesin 1994; Matheson and Tratnyek 1994), various countermeasures have been developed in the past two decades to overcome the limitations of ZVI-based technology. According to the commonalities and characteristics of these countermeasures, they can be divided into seven categories: pretreatment of pristine ZVI, NZVI, ZVI-based bimetals, physically enhanced ZVI technology, coupling ZVI with other materials, chemically enhanced ZVI technology, and methods to recover the reactivity of aged ZVI (Guan et al. 2015b). Even though the reactivity of ZVI can be unequivocally enhanced by those abovementioned alternatives, it should be notified that the complex procedures, extra costs, and potential ecotoxicity have limited their practical applications in improving the performance of ZVI (Guan et al. 2015b; Guo et al. 2016).

Contaminant removal, especially metals and metalloids, by ZVI is a complex interplay of adsorption onto ZVI corrosion products, coprecipitation or sequestration in the matrix of ZVI corrosion products, and reduction by Fe^0, Fe^{II}, or H_2/H (Noubactep 2008, 2009a). Thus, it is acknowledged that the corrosion rate of ZVI governs the rate of contaminant removal by ZVI (Triszcz et al. 2009). In our lab, it was accidently found that replacing the mechanical stirrer with a magnetic stirrer could considerably accelerate the corrosion of ZVI, leading to a more rapid release of Fe^{2+} and a more dramatic drop in ORP compared to the case with the mechanical stirrer (Liang et al. 2014a). Careful examination of the literatures implied that this phenomenon should be ascribed to the weak magnetic field (WMF) (<70 mT) supplied by the magnetic stirrer and the magnetic rotor. Although the study of magnetic field effects on electrochemical systems dates back more than 100 years with the research efforts of Faraday, very few studies have employed magnetic field to enhance contaminant removal by ZVI (Guan et al. 2015b). WMF can be generated by only two pieces of permanent magnets, which is thus promising and environmentally friendly. Therefore, the WMF-improved performance of ZVI for water decontamination has been extensively investigated in the past several years. It was found that the application of WMF could increase the reactivity of pristine ZVI, broaden the applicable pH range of ZVI, enhance the resistance of ZVI toward some passivating contaminants and solutes, improve the reactivity of ZVI toward refractory contaminants, and recover the reactivity of aged ZVI samples.

12.2.1 Increase the Reactivity of Pristine ZVI

As for the reactivity of pristine of ZVI toward contaminant, Guan's group pioneered that the application of WMF could significantly enhance the sequestration of Se(IV)/Se(VI) by ZVI (7.4-mic$^{\#}$) (Liang et al. 2014a, 2015), and the WMF induced greater enhancement in Se(IV) removal at lower initial Se(IV) concentrations. Applying a WMF could also substantially enhance the removal rates and efficiencies of As(V)/As(III) removal by the pristine ZVI (Jinshan$^{\#}$) (Sun et al. 2014). The WMF-induced improvement in As(V)/As(III) removal by ZVI is primarily associated with the accelerated ZVI corrosion, evidenced by the pH variation, Fe^{2+} release, and the formation of corrosion products characterized with X-ray absorption spectroscopy. Interactions of the pristine Jinshan$^{\#}$ ZVI with real arsenic-bearing groundwater revealed that WMF could greatly improve the reactivity of the pristine ZVI toward arsenic even in the presence of various cations and anions. WMF was also applied to enhance the sequestrations of Cu(II)/EDTA-Cu(II) and Cr(VI) by pristine Guoyao$^{\#}$ ZVI (Guan et al. 2015a; Jiang et al. 2015) and 33.1-mic$^{\#}$ ZVI (Feng et al. 2015), respectively. In order to better elucidate the promoting effects of WMF, a ratio (R_{WMF}) of rate constants for contaminant removal measured with (w/) and without (w/o) the WMF by the pristine ZVI was calculated, according to Eq. 12.1.

$$R_{\mathrm{WMF}} = \frac{k_{\mathrm{w/WMF\ obs}}}{k_{\mathrm{w/o\ WMF\ obs}}} \qquad (12.1)$$

where $k_{\mathrm{w/WMF\ obs}}$ and $k_{\mathrm{w/o\ WMF\ obs}}$ are the corresponding observed rate constants with and without WMF, respectively. Note that the resulting ratios are unitless. Figure 12.2 depicted that the rate constants of paramagnetic Cu(II) removal by ZVI with WMF were 10.8- to 383.7-fold greater than those without WMF. X-ray diffraction (XRD) and X-ray photoelectron spectroscopy (XPS) analyses revealed that applying a WMF enhanced both the Cu(II) adsorption to the ZVI surface and the transformation of Cu(II) to Cu(0) by ZVI (Jiang et al. 2015). The removal rate of Cr (VI) by ZVI was elevated by 1.1–5.9 times due to the application of a WMF, and there was a positive correlation between the WMF-induced promotion factor of Cr (VI) removal rate and that of Fe^{2+} release rate in the absence of Cr(VI) at pH 4.0–5.5 (Feng et al. 2015). In addition, the application of a WMF increased the rate constants of Sb(V) sequestration by 5.6–7.7 times at $[Sb(V)]_{ini}$ of 5.0–40.0 mg/L and enhanced the removal capacity of Sb(V) from 18.1 to 39.2 mg Sb(V)/g Fe (Li et al. 2015a). The effect of WMF on Sb(III) removal by ZVI was also determined in open, well-mixed batch reactors as a function of experimental factors, including addition of Fe(II), Sb dose, mixing rate, pH, initial concentrations of Sb(III), etc. (Xu et al. 2016a). It was found that the largest effect observed was the roughly 6.0–8.0-fold increase in Sb (III) removal rate due to the application of a WMF during the experiments.

Generally, WMF improved the sequestrations of contaminants by the pristine iron materials to different extents. Figure 12.2 shows that R_{WMF} generally falls in the range of 1.2–88.0, with only a few values outside that range (e.g., WMF induced a 330.4-fold improvement in Se(IV) removal by 7.4-mic# (Liang et al. 2014a), and WMF induced a 383.7-fold improvement in Cu(II) removal by Guoyao# (Jiang et al. 2015)). Overall, there are no trends in R_{WMF} with respect to metalloid type or iron material (symbol shape and color). Since it is known, or expected, that there will be significant variability in the solution chemistry for removal of these various metal (loid)s by ZVI from different origins, the performance of WMF for enhancing the sequestration of various metal(loid)s by ZVI was specified and discussed in the following parts. As for the mechanisms and rate-controlling processes for removal of these contaminants by ZVI, it will be further discussed in sections. 12.3 and 12.4, respectively.

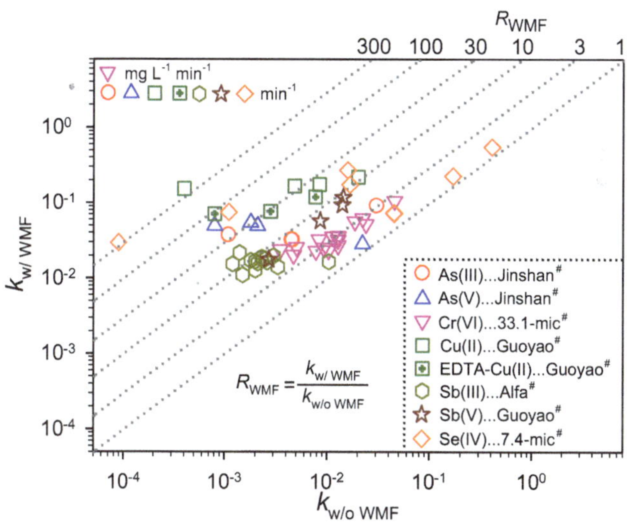

Fig. 12.2 Summary of the performances of WMF for enhancing various metal(loid)s sequestration by the pristine ZVI from different origins

12.2.2 Broaden the Applicable pH Range of ZVI

One limitation of ZVI technology is its narrow working pH, e.g., it is generally only effective under acidic conditions. Therefore, Liang et al. (2014a) and Jiang et al. (2015) investigated the influence of WMF on Se(IV) and Cu(II) removal by ZVI as a function of pH, respectively. Sodium acetate of 0.10 M, 2-(N-morpholino) ethanesulfonic acid (MES) of 0.10 M, and tris(hydroxymethyl)aminomethane (TRIS) of 0.20 M were employed as buffers for the experiments conducted at pH 4.0–5.0, pH 6.0, and pH 7.0–7.2, respectively, to maintain pH stable (±0.1) (Jiang et al. 2015; Liang et al. 2014a). As expected, the kinetics of Se(IV) removal was strongly dependent on pH, regardless of the presence of WMF, as shown in Fig. 12.3. In the absence of WMF, the removal of Se(IV) by ZVI exhibited self-accelerating characteristics and involved a lag period (defined as the time necessary for 10% Se(IV) removal in this study) before the initiation of a rapid removal period. The duration of lag period lasted 5 min, 10 min, and 10 min, respectively, at pH 4.0, 5.0, and 6.0. Beyond the lag period, Se(IV) was rapidly removed by ZVI at pH 4.0–6.0. However, Se(IV) removal by ZVI was completely inhibited at pH 7.0 in the absence of WMF at the time scale of our experiments. The application of WMF shortened the lag period and markedly accelerated Se(IV) removal by ZVI at pH 4.0–6.0. The lag period only lasted for 0 min, 3 min, and 5 min, respectively, at pH 4.0, 5.0, and 6.0, and it extended to 20 min and 240 min, respectively, at pH 7.0 and 7.2. Although negligible Se(IV) was removed by ZVI at pH 7.0 in 24 h or at 7.2 in 10 days without WMF, 97.5% of Se(IV) could be removed in 3 h at pH 7.0 and 70.5% of Se(IV) could be removed in 47 h at pH 7.2 in the presence of WMF. Therefore, the application of WMF could extend the working pH range of ZVI for Se (IV) removal from 4.0–6.0 to 4.0–7.2 (Liang et al. 2014a). Similar phenomenon was

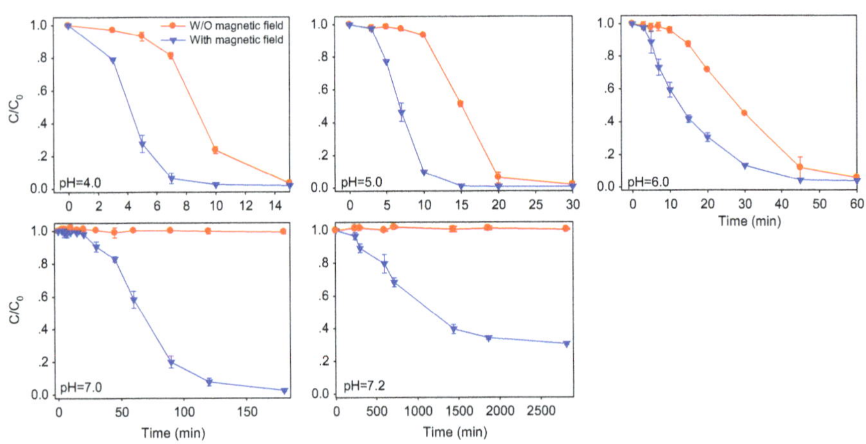

Fig. 12.3 Influence of WMF on Se(IV) removal by ZVI at different pH levels (Se (IV) = 40.0 mg L^{-1}, Fe^0 = 1.0 g L^{-1}). (Reprinted with permission from Liang et al. (2014a, b). Copyright (2014) Elsevier)

observed in the process of Cu(II) removal by ZVI. When pH was increased from 3.0 to 6.0, the removal efficiency of Cu(II) in 2 h dropped sharply from 97.5% to 4.5% without WMF (Jiang et al. 2015). However, over 99% of Cu(II) could be sequestered by ZVI in 30 min over the pH range of 3.0–6.0 upon the application of a WMF, implying that the application of a WMF could compensate the strong pH dependence of Cu(II) removal by ZVI (Jiang et al. 2015).

12.2.3 Enhance the Resistance of ZVI Toward Some Passivating Contaminants and Solutes

ZVI has been widely used as a reducing agent for the treatment of Cr(VI)-contaminated water. However, it is also well documented that the reduction of Cr (VI) by ZVI is a self-inhibiting process due to the formation of $(Cr_xFe_{1-x})(OH)_3$ or $Cr_xFe_{1-x}OOH$ layer on the ZVI surface. This passivating layer can increase the resistance for the electron transfer from ZVI to oxidants like Cr(VI) or oxygen and thus hinders further reduction of Cr(VI) by ZVI (Gheju 2011). Indeed, according to our previous study (Feng et al. 2015), the rate constants of Cr(VI) removal by ZVI could drop from 24.01×10^{-3} to 5.02×10^{-3} mg/(L · min) in the absence of WMF as the initial Cr(VI) concentration increased from 1.56 to 10.40 mg/L (see Fig. 12.4a). While, as demonstrated in Fig. 12.4a, imposing a WMF can remarkably improve Cr(VI) removal by ZVI at various Cr(VI) concentrations, the rate constants varied from 52.10×10^{-3} to 27.01×10^{-3} mg/(L · min), which were 1.2–4.1 times higher than without WMF. Thus, the inhibitory effect of passivation layers at a higher initial Cr(VI) concentration could be greatly alleviated by imposing a WMF.

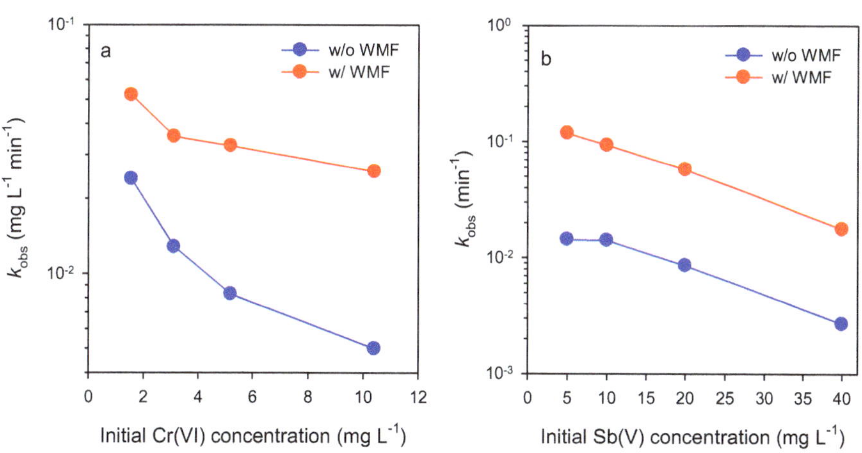

Fig. 12.4 WMF effects on contaminant removal rate by ZVI (**a**) at various initial Cr (VI) concentrations with a ZVI dosage of 0.1 g/L (pH = 4.0) and (**b**) at various initial Sb (V) concentrations with a ZVI dosage of 1.0 g/L (pH = 5.0)

Similarly, the removal rates of Sb(V) by ZVI without the presence of WMF dropped with increasing the Sb(V) concentrations from 5.0 to 40.0 mg/L (Fig. 12.4b), implying that Sb(V) could decrease the ZVI reactivity, especially at higher Sb(V) concentration. However, upon the application of a WMF, the rate constants of Sb(V) sequestration increased by 5.56–7.71 times compared to their counterparts without WMF, and the removal capacity of Sb(V) was enhanced from 18.1 to 39.2 mg Sb(V)/g Fe (Li et al. 2015a). The Fe *K*-edge X-ray absorption fine structure (XAFS) spectroscopy, XRD, and SEM results further confirmed that elevated Sb(V) concentrations could inhibit the corrosion of ZVI without WMF, while the application of WMF could alleviate the passivation of ZVI by Sb(V) and thus promote the removal of Sb(V) by corroded ZVI.

Silicate, commonly present in natural water, is a known corrosion inhibitor that could depress the ZVI performance through the formation of surface complexes at the particle-water interface. As shown in Fig. 12.5, for As(III) sequestration, the presence of 0.5 mM SiO_3^{2-} decreased the As(III) removal efficiencies from 74.9% to 21.3% without WMF at 5 h, or from 63.2% to 30.1% with WMF at 3 h, which indicated that ZVI corrosion was markedly depressed by silicate regardless of the application of WMF. However, a closer comparison of the Fe^0 consumption between the two scenarios with silicate (see the insert of Fig. 12.5) revealed that more Fe^0

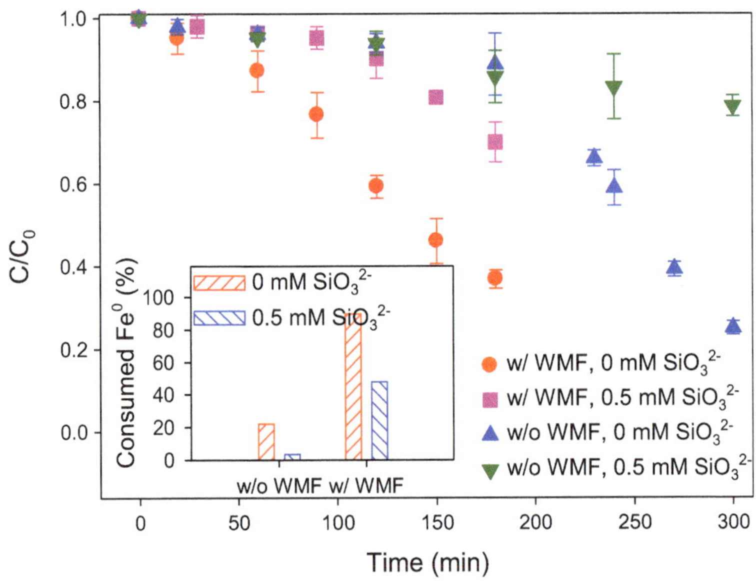

Fig. 12.5 Effect of SiO_3^{2-} (relative to 3 mM SO_4^{2-}) on As(III) removal by ZVI in the absence or presence of WMF (ZVI = 0.20 g/L, As(III) = 1000 μg/L, initial pH = 10.5, T = 25 °C). Insert represents the consumed Fe^0 fractions under different conditions which were determined by LCF of Fe k^3-weighted EXAFS spectra. (Reprinted with permission from Sun et al. (2017). Copyright (2017) American Chemical Society)

(48.2% at 3 h vs. 3.6% at 5 h) was reacted when WMF was introduced. This suggested that, from another perspective, the presence of WMF could significantly alleviate the adverse influences of silicate on ZVI corrosion.

In summary, employing WMF can also enhance the resistance of ZVI toward the passivating effects of some contaminants or solutes, which may count much for maintaining or improving the performance of ZVI-based technology when passivating contaminants like Cr(VI) and Sb(V) or coexisting solutes like silicate were present in water.

12.2.4 *Improve the Reactivity of ZVI Toward Refractory Contaminants*

Some contaminants such as Se(VI) are very refractory to be removed by ZVI. Several studies had been carried out to examine the performance of ZVI toward Se (VI) removal. However, the iron filings or microscale iron powder had low reactivity toward Se(VI) removal. To improve the removal rate of Se(VI) by ZVI, NZVI and Ni-Fe bimetal were used (Mondal et al. 2004). Moreover, it was proposed to apply Co^{2+}, Mn^{2+}, or Fe^{2+} to improve Se(VI) removal by ZVI (Tang et al. 2014). However, both methods bear some demerits. Although iron is inexpensive in bulk form, NZVI and nano-sized Ni-Fe bimetal were much more expensive because the costly precursor reagents and complicated processes are needed to synthesize them. Furthermore, the toxicity of nanomaterial has arisen much concern. The application of Co^{2+}, Mn^{2+}, or Fe^{2+} would generate extra sludge, and the residual Co^{2+} and Mn^{2+} would cause secondary pollution. Therefore, it is critical to explore an environmentally friendly method that can significantly improve the reactivity of ZVI to remove Se(VI).

Liang et al. examined the influence WMF on Se(VI) removal by ZVI at pH 6.0 buffered with 0.1 M MES (Liang et al. 2015). Figure 12.6a shows that negligible Se (VI) (<4%) was removed by ZVI without the application of WMF within 72 h. The presence of WMF dramatically enhanced Se(VI) sequestration by ZVI and complete removal of 10.0 mg L^{-1} Se(VI) by ZVI was achieved in 90 min. The removal efficiencies were 96.3%, 67.6%, and 36.9% after 12 h of reaction, respectively, when the initial Se(VI) concentrations were 20.0, 40.0, and 100.0 mg L^{-1}. Moreover, Se (VI) removal by ZVI with WMF was almost completed within 5 h when the initial Se (VI) concentration was in the range of 20.0–100.0 mg L^{-1} and negligible removal of Se(VI) was observed with prolonged reaction time. The main portion of each data set in Fig. 12.6b could be well described by the zero-order kinetics. The fitting results are presented with dashed lines in Fig. 12.6b. It was found that the zero-order rate constants (k_{obs}) were in the range of 0.094–0.141 mg L^{-1} min^{-1}, no obvious dependence on the initial concentration of Se(VI). Furthermore, the zero-order rate constants of Se(VI) removal increased progressively as the ZVI dosage increased. Compared to other methods reported in the literature, Se(VI) removal by ZVI coupling with WMF open to the air was a promising method.

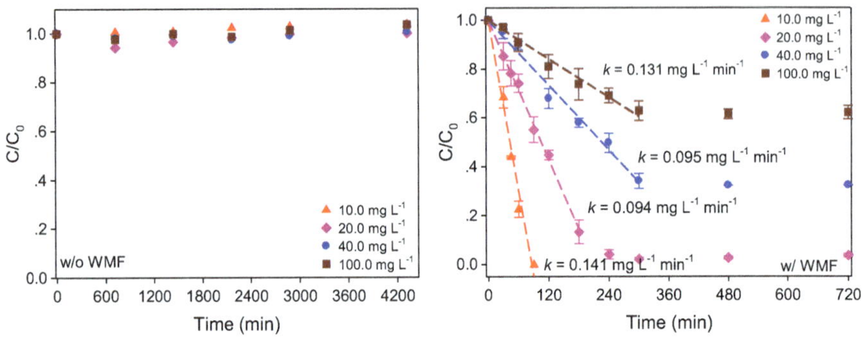

Fig. 12.6 Concentration profiles of dissolved Se(VI) in the absence of WMF and those in the presence of WMF. The dotted lines are the results of simulating the kinetics Se(VI) removal by ZVI in the presence of WMF with zero-order rate law ($[Se(VI)]_0 = 10.0–100.0\ mg\ L^{-1}$, $ZVI = 1.0\ g\ L^{-1}$, $NaCl = 0.01\ M$, $pH = 6.0$). (Reprinted with permission from Liang et al. (2015). Copyright (2015) Elsevier)

12.2.5 Recover the Reactivity of Aged ZVI Samples

The sequestration of metals in ZVI-based systems is generally favored by the growth and accumulation of iron oxides—and other iron-containing solid phases—that are formed by the Fe^{2+} released during corrosion of ZVI (Noubactep 2009a). However, the specific effects of these authigenic phases are multiple and variable, so the relative importance of specific effects is not always clear, despite many studies that have addressed aspects of this issue. For example, although thicker passive films (and the formation of secondary authigenic phases) increase the surface area for contaminant removal by non-reductive processes, breakdown of the passive film on ZVI generally favors contaminant reduction by exposing more strongly reducing surface area (Turcio-Ortega et al. 2012; Xie and Cwiertny 2010). When ZVI is deployed for water treatment under oxic conditions, the role of authigenic iron oxides is even more critical because the phase transformations are more dynamic and oxidative pathways for contaminant treatment become more significant (Mylon et al. 2010).

The process of generating corrosion products is referred to as “aging.” Aging effects are relevant to laboratory- and field-scale applications of ZVI for water treatment, so the increased attention to them in recent studies is an important development in the maturation of this field. Given the key role that oxide/passive films play in the short- and long-term performance of ZVI in remediation applications, it is inevitable that many strategies for maintaining or enhancing ZVI treatments involve manipulating the properties of the oxides. The earliest example of this was acid washing, which removes oxides and therefore increases the rates of contaminant reduction in laboratory tests (Agrawal and Tratnyek 1996; Lai and Lo 2008; Lin and Lo 2005; Matheson and Tratnyek 1994). Another example is ultrasonication, which favors contaminant degradation by physically disrupting the oxide coating on ZVI and may also dislodge oxide precipitates that obstruct pore channels in reactive barriers (Geiger et al. 2002; Hung and Hoffmann 1998; Liu et al. 2007). A third approach involves application of an electrical potential from an external source to reduce the oxides that comprise the passive film, thereby increasing reaction rates and possibility adding to the longevity of ZVI treatments (Chen et al. 2012; Lu et al. 2012).

We prepared aged ZVI (AZVI) for time periods from 0 to 96 hr, characterized the structure and composition of the resulting materials, and determined the feasibility of recovering the reactivity of AZVI toward Se(IV) with WMF (Liang et al. 2014b). Characterization of the ZVI used in this study by XRD confirmed that the major component of the pristine material was α-Fe^0 (JCPDS 99-0064). XRD peaks corresponding to magnetite (Fe_3O_4, JCPDS 88-0315) started to appear in AZVI-12 (ZVI aged for 12 h) and then increased in intensity progressively with further aging. Peak intensity for Fe^0 decreased with aging, disappearing completely in AZVI-96 (ZVI aged for 96 h) with concomitant appearance of two new peaks identified to be lepidocrocite (γ-FeOOH, JCPDS 74-1877).

As shown in Fig. 12.7, the sequestration of Se(IV) by zero-valent iron (ZVI) was strongly influenced by the coupled effects of aging ZVI and presence of a weak magnetic field (WMF). It is evident that WMF consistently gives faster selenium removal kinetics and that this enhancement is greatest for intermediate aging time. The defining features of Fig. 12.7 are threefold: (i) unaged ZVI (AZVI-0) gives the fastest Se(IV) removal rates, with little increase due to WMF; (ii) aging 80 or more hours gives the slowest Se(IV) rates, with little enhancement by WMF; and (iii) aging between 6 and 60 h gives nearly constant rates of Se(IV) removal but with rate constants that are 10- to 100-fold greater with WMF than without.

It has been elucidated that the constituents and structure of corrosion products were strongly dependent on the water chemistry and hydraulic conditions (Furukawa

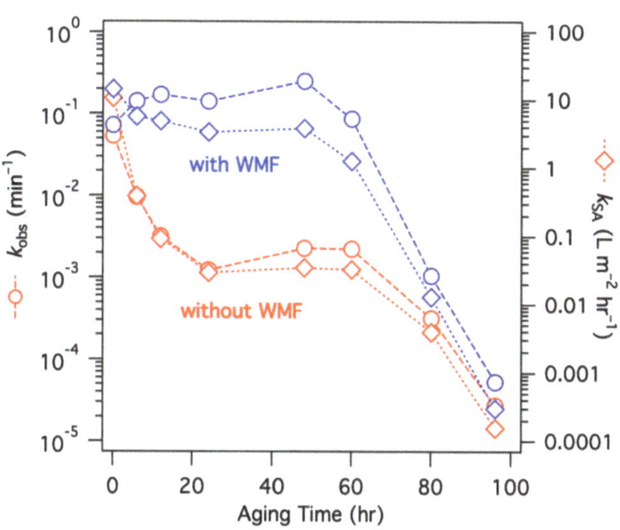

Fig. 12.7 Effect of aging and WMF on the kinetics of Se(IV) sequestration by ZVI

et al. 2002). One depassivation method effective for ZVI covered by one category of passive layer may be not efficient for ZVI sample with another category of iron oxide film. In other words, the performance of a depassivation method may be strongly associated with the amount, constituents, and in-depth distribution of passive layer coated on AZVI samples. However, none of the previous studies has systematically characterized the passive layer of AZVI prepared under various aging conditions and explored the feasibility of employing one depassivation approach to recover the reactivity of these AZVI samples. Therefore, we further fabricated AZVI samples under different conditions so as to get AZVI samples with passive layers of different depths and constituents, systematically characterized the structure and composition of passive films of these samples, and explored the feasibility of using uniform magnetic field (UMF) to depassivate these AZVI samples with passive layers of different constituents and depths (Xu et al. 2016b).

The AZVI samples prepared under different conditions exhibited the passive layers of different morphologies, amounts, and constituents, as demonstrated in Fig. 12.8. Owing to the accumulation of iron oxides on their surface, all the prepared AZVI samples were much less reactive than the pristine ZVI for Se(IV) removal. However, the reactivity of all AZVI samples toward Se(IV) sequestration could be significantly enhanced by applying a uniform magnetic field (UMF). Moreover, the flux intensity of UMF necessary to depassivate an AZVI sample was strongly dependent on the properties of its passive layer. The UMF of 1 mT was strong enough to restore the reactivity of the AZVI samples with Fe_3O_4 as the major constituent of the passive film or with a thin layer of α-Fe_2O_3 and γ-FeOOH in the external passive film. The flux intensity of UMF necessary to depassivate the AZVI samples would increase to 2 mT and even 5 mT if the AZVI samples were covered with passive films being thicker and denser and contained more γ-FeOOH and

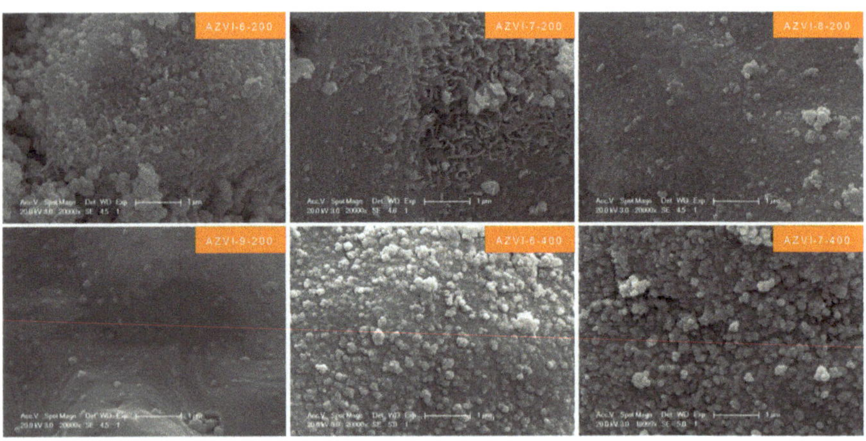

Fig. 12.8 SEM images of the AZVI samples synthesized in SGW under various conditions. (Reprinted with permission from Xu et al. (2016a, b). Copyright (2016) American Chemical Society)

α-Fe_2O_3. Furthermore, increasing the flux intensity of UMF facilitated the reduction of Se(IV) to Se(0) by AZVI samples.

The results of this work suggest that magnetization after deployment might provide a noninvasive way to maintain or restore the reactivity of ZVI even after there has been substantial accumulation of passivating oxides. This method might be most applicable to permeable reactive barriers (PRBs) emplaced in aerobic aquifers and in aboveground canisters or filter beds, where passivation by accumulation of iron oxides is most severe (Mackenzie et al. 1999).

12.3 Influence of WMF on the Mechanisms of Contaminant Removal by ZVI

All the above results showed that the application WMF greatly accelerated the removal of various contaminants by ZVI under various conditions. However, it kept unknown whether the application of WMF changed the removal mechanisms of contaminants. Therefore, in our lab, we examined the influence of WMF on the reductive removal, oxidative removal, and adsorptive removal of contaminants by ZVI taking Se(IV), As(III), and Sb(V) as the probe contaminants, respectively.

12.3.1 Influence of WMF on the Reductive Removal of Contaminants

To explore the influence of WMF on the mechanisms of Se(IV) removal by ZVI, the oxidation states of Se in the precipitates collected at different reaction time, pH, and initial Se(IV) concentration were analyzed with XAFS. The Se *K*-edge XANES data for two Se model compounds including Se(0) and Se(IV) were analyzed to establish the reference X-ray absorption *K*-edge energies (E_0). XANES spectra of Se(IV)-treated ZVI particles reacted with 40.0 mg L^{-1} Se(IV) at pH 6.0 open to air for 30 min and 24 h in the absence of WMF showed that selenium was present as Se(0). The application of WMF had no influence on the removal mechanisms of Se (IV) under this condition. The XANES spectra of Se(IV)-treated ZVI in the presence of WMF collected at different time at pH 7.0 demonstrated that the fraction of Se (0) increased with elapsed time and Se(IV) was completely transformed to Se(0) at the end of 24 h. Thus, Se(IV) dosed at 40.0 mg L^{-1} was removed by ZVI via adsorption followed by reduction at pH 7.0 in the presence of WMF but via reduction to Se(0) at pH 6.0 regardless of the presence of WMF.

Figure 12.9 illustrates the XANES spectra of ZVI particles reacted with Se(IV) of different concentrations open to air in the absence or presence of WMF at pH 6.0. In the presence of WMF, all the removed Se(IV) was present as Se(0) at the end of 1 h reaction, independent of the initial Se(IV) concentration. However, the initial Se (IV) concentration had great impact on the removal mechanisms of Se(IV) at pH 6.0

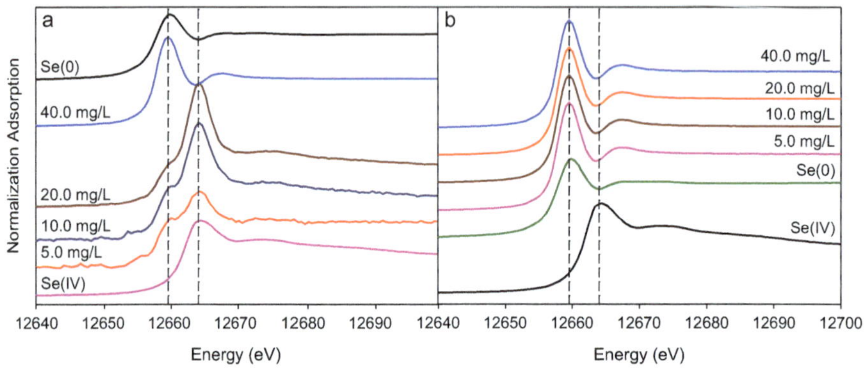

Fig. 12.9 XANES spectra of (**a**) ZVI particles reacted with Se(IV) of different concentrations open to air for 24 h in the absence of WMF at pH 6.0; (**b**) ZVI particles reacted with Se(IV) of different concentrations open to air for 1 h in the presence of WMF at pH 6.0 ($Fe^0 = 1.0$ g L^{-1}). (Reprinted with permission from Liang et al. (2014a, b). Copyright (2014) Elsevier)

in the absence of WMF. When the initial Se(IV) concentration was in the range of 5.0–20.0 mg L^{-1}, only a small fraction of removed Se(IV) was reduced to Se(0) even after 24 h. All the removed Se(IV) was reduced to Se(0) at pH 6.0 in the absence of WMF when Se(IV) was dosed at 40.0 mg L^{-1}. Combining the results of XANES analysis, the influence of WMF on Fe^{2+} release, and ORP variation, it could be concluded that the application of WMF did not change the mechanisms of Se (IV) removal by ZVI but accelerated the reduction of Se(IV) by ZVI.

12.3.2 *Influence of WMF on the Oxidative Removal of Contaminants*

The corrosion of ZVI by oxygen also produces strong oxidants capable of oxidizing various organic and inorganic compounds. Under acidic conditions, Fe^0 surfaces transfer two electrons to oxygen (Eq. 12.2) to produce hydrogen peroxide (H_2O_2). The interaction of H_2O_2 with Fe^{2+} will generate hydroxyl radical (•OH) following Eq. 12.3 at low pH or Fe(IV) at pH above 5.0 via Eq. 12.4 (Bataineh et al. 2012; Hug and Leupin 2003; Lee and Sedlak 2008).

$$Fe^0 + O_2 + 2H^+ \rightarrow Fe^{2+} + H_2O_2 \tag{12.2}$$

$$Fe^{2+} + H_2O_2 \rightarrow Fe^{3+} + \cdot OH + OH^- \tag{12.3}$$

$$Fe^{2+} + H_2O_2 \rightarrow Fe(IV) + H_2O \tag{12.4}$$

Several spectroscopic investigations of arsenic speciation after reaction with ZVI have evidenced the partial oxidation of As(III) to As(V) (Manning et al. 2002; Neumann et al. 2013; Su and Puls 2001). Therefore, As(III) was selected as a

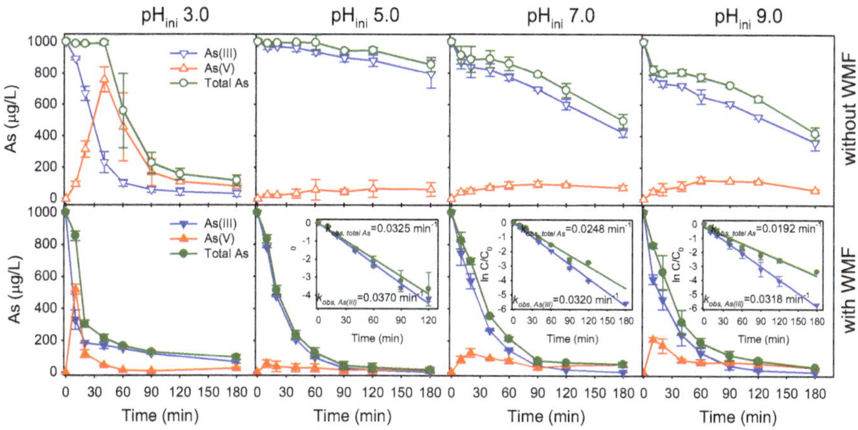

Fig. 12.10 Influence of WMF on the depletion of total arsenic and As(III) as well as the evolution of As(V) in the process of As(III) removal by ZVI at various initial pH values (ZVI = 0.10 g/L, initial As(III) = 1000 μg/L, $T = 25$ °C). Inset: simulation of the rates of total arsenic and As(III) removal in the presence of WMF with pseudo-first-order model

probe pollutant to test whether the oxidative capacity of a ZVI system could be affected by WMF.

Figure 12.10 shows the influence of WMF on the depletion of total As and As(III) as well as the evolution of As(V) in the process of As(III) removal by ZVI at different pH_{ini} levels. Without WMF, the removal efficiencies of total As in 3 h were 88.4%, 14.4%, 50.4%, and 58.1%, respectively, at pH_{ini} of 3.0, 5.0, 7.0, and 9.0. The removal of aqueous As(III) by ZVI at pH_{ini} 3.0 in the absence of WMF obviously consisted of two periods. During the first period (the first 40 min of reaction), the decrease in As (III) concentration was accompanied with an elevation of As(V), whereas the concentration of total As kept almost constant, indicating the quick transformation of As (III) to As(V) but negligible sequestration of As(V) involved during this period. During the second stage, the concentration of total As decreased progressively, mainly associated with the generation of iron (oxyhydr)oxides and the resultant As (V) removal. However, over the pH_{ini} range of 5.0–9.0, little As(V) was detected in solutions without WMF since the Fenton reaction rates were relatively low at high pH values. Therefore, As(III) was primarily removed by the oxide film coated on the pristine ZVI and/or the newly formed iron oxides without oxidizing to As(V) in the absence of WMF at pH_{ini} 5.0–9.0 (Noubactep 2009b; Pang et al. 2009).

The application of WMF remarkably accelerated the removal of both As(III) and total As at all tested pH values, as illustrated in Fig. 12.10. The removal efficiencies of total As with WMF were as high as 90.5–98.4%. Upon the introduction of WMF at pH_{ini} 3.0, it was observed that the lag period of total As removal disappeared, the depletion rate of As(III) increased, and the accumulation of As(V) was alleviated. Since As(III) has low affinity for iron (oxyhydr)oxides due to its electric neutrality, while As(V) is much more easily to be adsorbed or entrapped under acidic conditions (Dixit and Hering 2003; Guan et al. 2012), the acceleration of As(III) disappearance rate indicated the As(III) conversion to As(V) in solution was improved by WMF at pH_{ini} 3.0. The oxidizing •OH produced by Fenton reaction is generally believed to be responsible for the As(III) oxidation at low pH; thus WMF may favor the Fenton reaction (via Eq. 12.3) involved in the ZVI system (Katsoyiannis et al. 2008; Pang et al. 2011). Although As(III) was oxidized at a greater rate in the presence of WMF, a drop in As(V) accumulation compared to its counterpart in the absence of WMF was observed, which should be explained by the quick generation of iron (oxyhydr)oxides. These results provided strong evidence that As(III) was first oxidized to As(V) before As depletion at pH_{ini} 3.0 and the oxidation mainly occurred in solution (Katsoyiannis et al. 2008, 2009a, b).

As(III) oxidation could also occur at $pH_{ini} \geq 5.0$, since previous studies have suggested that Fenton reaction (via Eq. 12.4) could take place at neutral pH with Fe (IV) rather than •OH as the predominant products (Bataineh et al. 2012; Katsoyiannis et al. 2008). Indeed, the generation of As(V) was observed in the process of As(III) removal at pH_{ini} 5.0–9.0 regardless of the application of WMF (Fig. 12.10). Although the concentration of soluble As(V) after 3 h at pH_{ini} 5.0–9.0 with WMF was smaller than its counterpart without WMF, the appearance of As(V) in the first 10–20 min in the presence of WMF was more pronounced and increased with increasing pH_{ini}. The observed phenomena implied that the introduction of WMF facilitated the oxidation of As(III) to As(V) in aqueous phase at pH_{ini} 5.0–9.0. The rate of As(III) oxidation by Fe^0/O_2 in the presence of WMF exceeded that of As(V) adsorption by the corrosion products, resulting in the As(V) accumulation in the solution, as illustrated in Fig. 12.10. Thus, the WMF-induced enhancement in As(III) removal by ZVI at pH_{ini} 5.0–9.0 should be ascribed to the accelerated ZVI corrosion and the subsequent elevated As(III) oxidation rate.

In the past two decades, most studies on ZVI mainly focused on using its reducing capacity, while few studies on contaminant removal by the oxidants generated in the process of ZVI corrosion were carried out, possibly due to the low generation of Fenton reagents and the pH limitation (it only works well at acidic pH range) (Lee et al. 2007; Sedlak and Andren 1991). Given that the application of WMF accelerated the Fenton reaction favoring As(III) oxidation under both acidic and neutral pH, it can be expected that the oxidative removal of many organic contaminants by ZVI under oxic conditions may also be improved with the aid of WMF.

12.3.3 *Influence of WMF on the Adsorptive Removal of Contaminants*

The absorption edges of Sb in the precipitates collected in the presence or absence of WMF were similar to that of $K_2H_2Sb_2O_7 \cdot 4H_2O$, as illustrated in Fig. 12.11a, indicating that the oxidation state of Sb in the samples was predominantly Sb(V). Although the reduction of Sb(V) to more toxic Sb(III) was observed in the process of Sb(V) removal by NZVI (Dai et al. 2014; Dorjee et al. 2014), Sb(V) was not reduced to Sb(III) by micron-sized ZVI in our study (Li et al. 2015a), which was advantageous

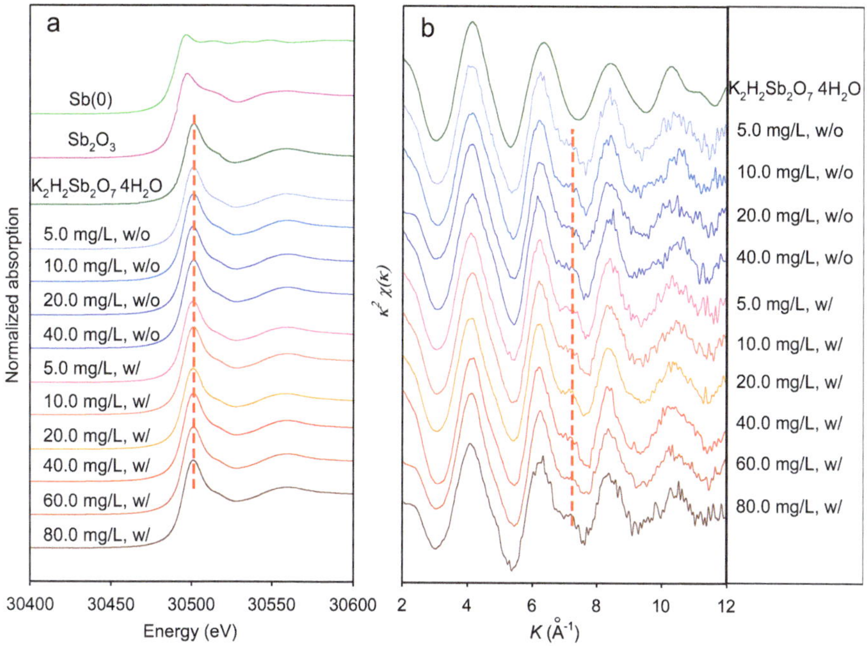

Fig. 12.11 Sb *K*-edge XANES spectra (**a**) and k^2-weighted EXAFS spectra (**b**) of corrosion samples with WMF as well as without WMF and reference materials (ZVI dosage 1.0 g/L, $[Sb(V)]_{ini} = 5.0–80.0$ mg/L, $pH_{ini} = 4.0$, reaction time: 3.0 h). (Reprinted with permission from Li et al. (2015a). Copyright (2015) Elsevier)

for Sb(V) removal since the transformation of removed Sb(III) to Sb(V) may lead to the release of Sb from the adsorbent (Leuz et al. 2006; Xi et al. 2011). The k^2-weighted Sb *K*-edge EXAFS spectra of Sb(V)-treated ZVI samples shown in Fig. 12.11b were essentially identical and independent of $[Sb(V)]_{ini}$ and WMF. A shoulder peak was observed at $k \approx 7$ Å^{-1} in all spectra (indicated by the dashed line in Fig. 12.11b), which was not found in spectrum of $K_2H_2Sb_2O_7 \cdot 4H_2O$ and could be due to the incorporation of Sb(V) into the iron (hydr)oxide structure (Mitsunobu et al. 2010).

The k^2-weighted Sb *K*-edge EXAFS spectra of Sb(V)-treated ZVI samples obtained at the $[Sb(V)]_{ini}$ of 5.0 mg/L without and with WMF and their FTs are shown in Fig. 12.12a, b, respectively. The FT of the EXAFS spectra isolates the contributions of different coordination shells, in which the peak positions correspond to the interatomic distances. However, the peak positions in Fig. 12.12b have not been corrected for the phase shift effects, and thus they deviate from the real distance by 0.3–0.5 Å. First-neighbor contributions were fitted with 6 oxygen atoms at 1.98 ± 0.02 Å, which corresponded to an $Sb(OH)_6^-$ octahedron, confirming that the geometry of Sb(V) was not changed by incorporating into the corroded ZVI. Only the fit with two Sb-Fe shells resulted in a good fit of the spectra and in physically meaningful values of distance and coordination number. The two Sb-Fe

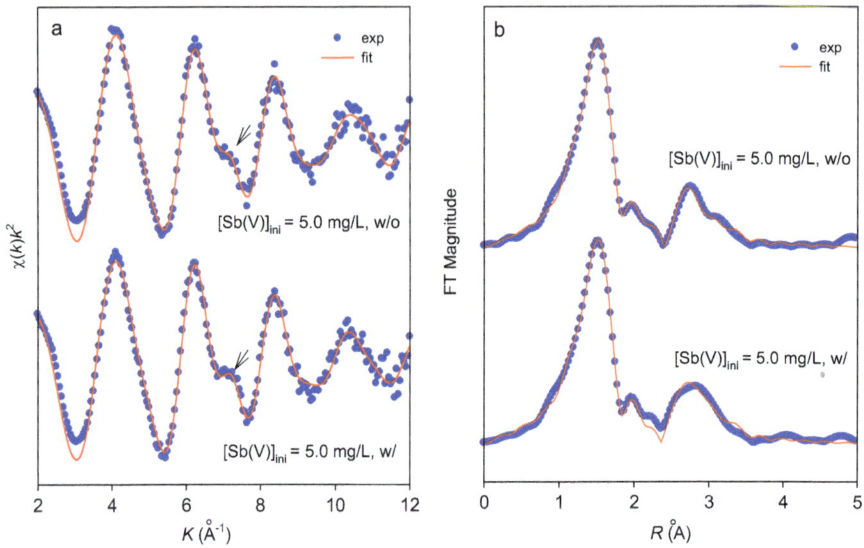

Fig. 12.12 (**a**) k^2-weighted Sb *K*-edge EXAFS spectra and (**b**) FT spectra of Sb(V)-treated ZVI samples with or without WMF. The arrows show the featured peak of the incorporated Sb(V) (ZVI dosage 1.0 g/L, $[Sb(V)]_{ini}$ = 5.0 mg/L, pH_{ini} = 4.0, reaction time: 3.0 h). (Reprinted with permission from Li et al. (2015a, b, c). Copyright (2015) Elsevier).

distances were in the ranges of 3.07–3.11 and 3.52–3.57 Å for the sample obtained either with or without WMF. The first value corresponds to the Sb-Fe distance along the edge-sharing linkage, and the second corresponds to the Sb-Fe distance along the corner-sharing link (Mitsunobu et al. 2010; Scheinost et al. 2006). The coordination numbers (CNs) of both bonds were 1.0–2.4 and 2.8–4.2, respectively, for the sample obtained without WMF, and 1.7–2.3 and 2.6–3.6, respectively, for the sample collected with WMF. The Sb-Fe distances determined in this study were very close to those reported in the literature, while the CNs were larger than those in the literature (Guo et al. 2014; Mitsunobu et al. 2010; Scheinost et al. 2006), further confirming that Sb(V) was incorporated into the structure of iron oxide. Moreover, the similarities in the Sb-Fe distances and CNs for the samples obtained with and without WMF suggested that the application of WMF only accelerated Sb (V) sequestration by ZVI but did not affect the removal mechanisms of Sb(V).

12.4 The Mechanisms of WMF-Induced Improvement on Contaminant Removal by ZVI

Although increasing research suggests that WMF can promote the reactivity of ZVI toward a variety of metal(loid)s, the mechanisms keep unclear. With regard to the WMF-induced improvement on contaminant removal by ZVI, it should be firstly addressed that the superimposed WMF had negligible influence on the removal mechanisms of contaminants but facilitated the processes of contaminant removal by ZVI based on the XAFS analysis (Liang et al. 2014a, b; Sun et al. 2014). Additionally, the superimposed WMF had negligible influence on the apparent activation energy of Cr(VI) removal by ZVI, indicating that WMF accelerated Cr(VI) removal by ZVI but did not change the removal mechanisms (Feng et al. 2015). Jiang et al. (2015) further showed that the mechanism of Cu(II) removal by ZVI was adsorption followed by reduction to metallic Cu^0 regardless of the presence or absence of WMF. The superimposed WMF seems to only enhance the mass transfer of the reactants including oxygen and H^+ to the ZVI surface. To further verify the aforementioned conclusions and explicit the contributors in the WMF-induced improvement, this section systematically summarized the findings identified by the electrochemical test, the induced MF around a ZVI grain in a WMF, and the WMF influences as functions of MF intensity and ZVI particle size.

12.4.1 Electrochemical Test

To further evaluate the role of MF in ZVI corrosion, potentiostatic polarization experiments were performed at different anode polarization potentials. A schematic diagram for lab-scaled configuration is shown in Fig. 12.13. Potentiostatic anodic

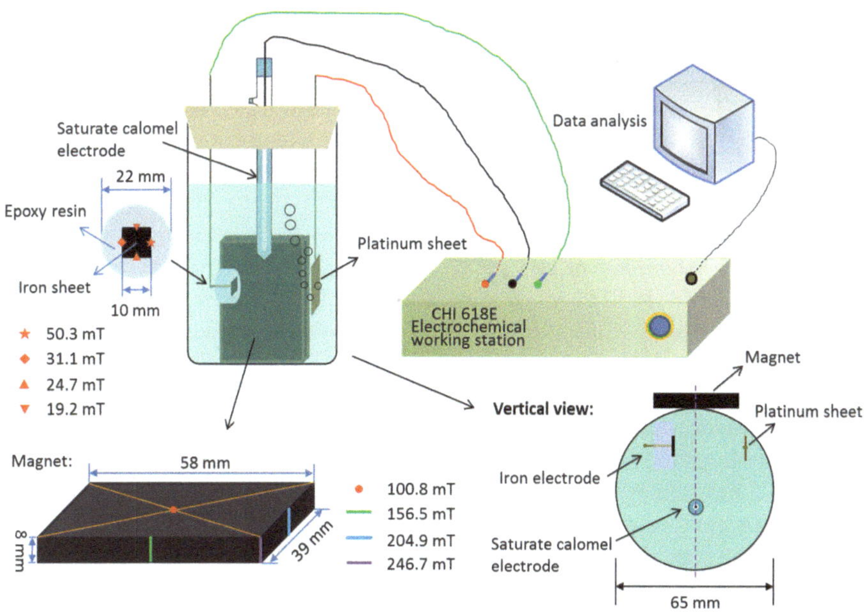

Fig. 12.13 Schematic diagram of lab-scaled setup for electrochemical measurements. (Reprinted with permission from Jiang et al. (2015). Copyright (2015) Elsevier)

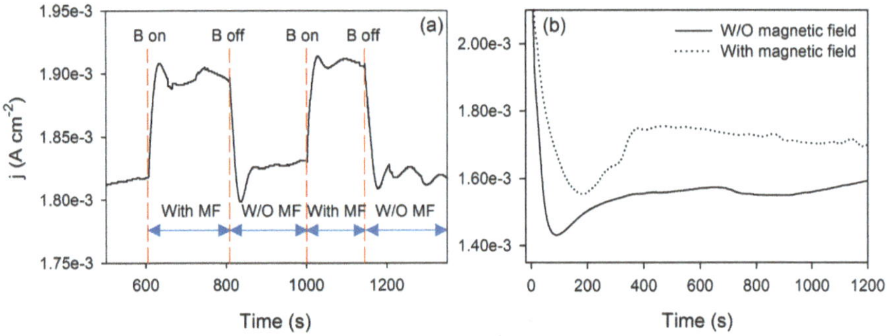

Fig. 12.14 Potentiostatically recorded current density transient (absolute value) of Fe in 0.01 M HCl (**a**) at +240 mV (versus SHE) and impact of applying and withdrawing a magnetic field, (**b**) at +200 mV (versus SHE) with or without an external magnetic field. (Reprinted with permission from Jiang et al. (2015). Copyright (2015) Elsevier)

polarization and potentiodynamic polarization measurements were performed using a CHI 618E electrochemical workstation to assess the effect of MF on ZVI corrosion. A three-electrode system, an iron sheet (1.0 cm^2 exposed to electrolyte) as the working electrode, a platinum sheet as the counter electrode, and a saturated calomel electrode (0.2412 V vs. SHE) as reference electrode, was used in this study. The magnetic field was provided by placing a magnet beside the electrolytic cell and the detailed information for configuration was illustrated in Fig. 12.13. Prior to each test, the iron sheet electrode was polished for a mirror face by abrasive paper of various sizes. After the open circuit potential (OCP) had stabilized, potentiodynamic polarization of the iron electrode was carried out from −0.5 V to +0.5 V with a scan rate of 0.1 mV/s. Moreover, the iron electrode was polarized at −280, +200, and + 240 mV (ca. +60, +540, and + 580 mV vs. OCP), respectively, and the current versus time was recorded. All potentials were reported with respect to the standard hydrogen electrode (SHE).

The potentiostatic polarization j-t curves recorded for iron electrode in 0.01 M HCl (pH = 2.0) are shown in Fig. 12.14. The applied polarization potential in Fig. 12.14a was set at +240 mV (ca. +580 mV vs. OCP), under which the current density transient was diffusion-controlled. This potential is located in the transition region of the potentiodynamic polarization curve, as shown in Fig. 12.15, in which the Tafel region was followed by a transition region entering into a diffusion-controlled region with almost constant current density (Sueptitz et al. 2009). After applying a MF (B on) at 605 s, the current density (absolute value) sharply increased from 1.82 mA cm^{-2} to 1.91 mA cm^{-2} within 30 s, then kept constant at ca. 1.90 mA cm^{-2}, indicating that the applied MF enhanced the mass transfer and thus accelerated anodic dissolution of iron. Once the MF was withdrawn (B off) at 808 s, the current density rapidly declined to ca. 1.80 mA cm^{-2}, followed by a slight increase. Similar phenomena could be reproduced by applying and withdrawing the MF during 1000~1150 s. Figure 12.14b shows the results of potentiostatic

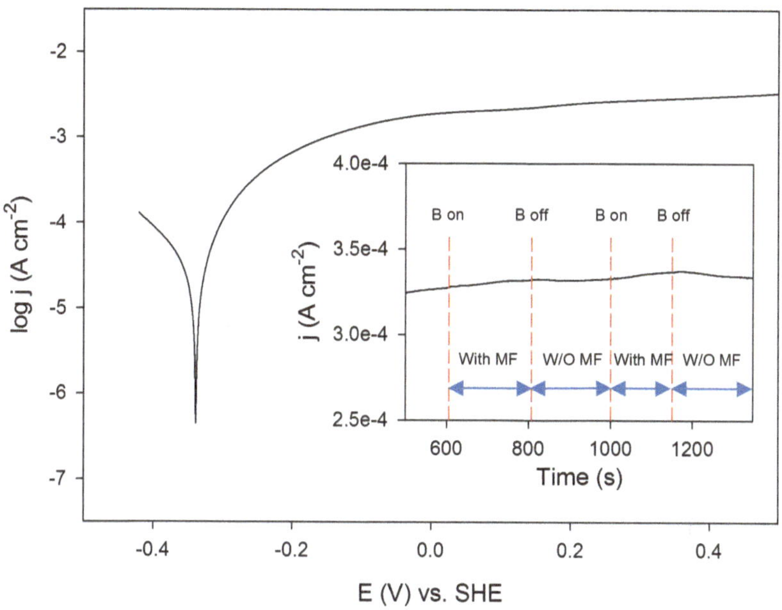

Fig. 12.15 Potentiodynamic polarization curve of iron in 0.01 M HCl (pH = 2.0). The insert shows the potentiostatically recorded current density transient (absolute value) of Fe in 0.01 M HCl at −280 mV (versus SHE, ~ +60 mV vs. OCP) and the negligible impact of applying and withdrawing a magnetic field. (Reprinted with permission from Jiang et al. (2015). Copyright (2015) Elsevier)

polarization at +200 mV (ca. +540 mV vs. OCP). The current density obtained in the presence of MF was always greater than that obtained in the absence of MF from the outset, further confirming that the application of a MF could accelerate the iron electrode corrosion.

However, negligible effects of applying or withdrawing the MF were observed when the polarization potential was set at −280 mV (ca. +60 mV vs. OCP, insert in Fig. 12.15), which is located in the Tafel region, and the current density transient was governed by electron transfer. Therefore, the acceleration of iron corrosion in the presence of MF arose from the MF-induced enhancement of mass transfer (Fe^{2+}, H^+).

12.4.2 The Induced MF Around a ZVI Grain in a WMF

Being ferromagnetic, ZVI is magnetized in a uniform WMF and an inhomogeneous MF is induced around its surface. To evaluate the flux density distribution near the ZVI sphere surface, the Mechanical APDL (ANSYS) 14.0 software was employed to numerical simulations, assuming a 10 μm diameter spherical particle of pure ZVI in

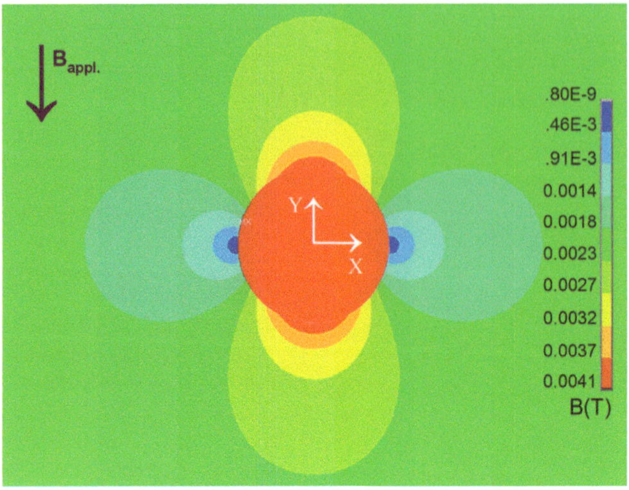

Fig. 12.16 Distribution of the MF induced density of the plane parallel to the applied MF and through the center of a ZVI sphere in a homogeneous MF with an applied flux intensity ($B_{appl.}$) of 2.0 mT. The color bar on its right side represents magnetic field strength changes. From top to bottom, the gradient colors indicate increased magnetic field strength. (Reprinted with permission from Li et al. (2017). Copyright (2017) Elsevier)

an externally applied magnetic field with homogenous flux density (B_{appl}) of 2.0 mT, as depicted in Fig. 12.16.

It was found that the magnetic field around the ZVI particle surface was stronger than the external imposed one and the magnetic field gradient at 1.0 μm away from the ZVI surface was as large as 7154 T m^{-1}. There are two forces including the magnetic gradient force ($F_{\Delta B}$) expressed with Eq. 12.5 and the Lorentz force (F_L) expressed with Eq. 12.6 that have been proposed to contribute to the WMF-induced improvement in the reactivity of ZVI (Liang et al. 2014b). The involvement of $F_{\Delta B}$, which acts on the paramagnetic ions to drive their movement toward the site with higher induced MF density (Ragsdale et al. 1998) in the process of contaminant removal by ZVI with WMF, had been qualitatively verified by the uneven distribution of the paramagnetic Cu^{2+} and iron corrosion products along the iron wire with the presence of WMF in our previous studies (Jiang et al. 2015; Liang et al. 2014b). It was reported that Lorentz force acting on the charged species cutting across the magnetic lines of force could narrow the diffusion layer and thus enhance mass transportation (Lioubashevski et al. 2004). However, we have not made effort to get direct evidence for the contribution of F_L. Consequently, the relative contribution of $F_{\Delta B}$ and F_L in the process of contaminant removal by ZVI with WMF is clarified in Sect. 12.4.3.

$$F_{\Delta B} = \left(\chi \times \frac{c}{\mu_0}\right) B(x) \frac{dB(x)}{dx} \tag{12.5}$$

$$F_{\mathrm{L}} = J \times B \tag{12.6}$$

where χ is the molar magnetic susceptibility ($m^3\ mol^{-1}$), μ_0 is the magnetic permeability of vacuum (equal to $4\pi \times 10^{-7}$ T m A^{-1}), $B(x)$ is the MF intensity at position x, J is the flux of charged species (coulombs $cm^{-2}\ s^{-1}$), and B is the flux intensity of external magnetic field (T).

12.4.3 The WMF Influences as Functions of MF Intensity and ZVI Particle Size

12.4.3.1 Influence of MF Flux Intensity on the Reactivity of ZVI Roward Cr(VI)

A very weak MF may be not enough to induce great improvement in contaminant removal by ZVI, but a strong MF may cause the aggregation of ZVI particles and thus deteriorate the performance of ZVI. Consequently, to address these issues, the kinetics of Cr(VI) removal by ZVI as a function of MF intensity was investigated, as shown in Fig. 12.17. Obviously, the application of a WMF with intensity ranging from 0.2 to 20 mT remarkably improved the reactivity of ZVI toward Cr(VI). Approximately 69.4% of Cr(VI) was removed by ZVI in 120 min in the absence of WMF, whereas more than 80.9% of Cr(VI) could be sequestrated by ZVI within 60 min with the presence of MF of 0.2 mT. The most rapid Cr(VI) removal was achieved in the MF of 2.0 mT, and a further increase in the MF intensity resulted in a drop in the Cr(VI) removal rate. To quantitatively describe the influence of MF intensity on the reactivity of ZVI toward Cr(VI), the pseudo-zero-order rate law via

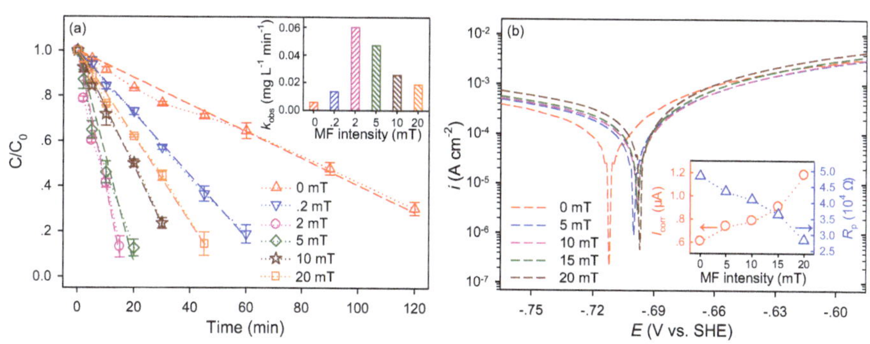

Fig. 12.17 (**a**) Influence of MF intensity on the kinetics of Cr(VI) removal by ZVI. Inset of (**a**): the pseudo-zero-order rate constants of Cr(VI) removal by ZVI as a function of MF intensity. Reaction conditions for (**a**): $[Cr(VI)]_0 = 4.0$ mg L^{-1}, $[Fe^0] = 0.50$ g L^{-1}. (**b**) Linear sweep voltammograms (LSVs) for the stationary PIPE as function of the MF intensity. Inset of (**b**): change of the corrosion current (I_{corr}, μA) and polarization resistance (R_p, Ω) derived from the LSVs with MF intensity. (Reprinted with permission from Li et al. (2017). Copyright (2017) Elsevier)

Eq. 12.7 was used to simulate the kinetics of Cr(VI) elimination (Gheju 2011), as illustrated by the dashed lines in Fig. 12.17a.

$$\frac{\mathrm{d}[\mathrm{Cr(VI)}]}{\mathrm{d}t} = -k_{\mathrm{obs}} \tag{12.7}$$

where k_{obs} is the pseudo-zero-order rate constant (mg L^{-1} min^{-1}) of Cr(VI) removal by ZVI. As depicted in the inset of Fig. 12.17a, the Cr(VI) sequestration rate constant was enhanced progressively from 0.0059 to 0.0603 mg L^{-1} min^{-1} with increasing the MF intensity from 0 to 2.0 mT and the application of a WMF of 2.0 mT induced a 10.2-fold enhancement in the rate of Cr(VI) removal. Although the rates of Cr (VI) removal by ZVI progressively decreased to 0.0470, 0.0256, and 0.0190 mg L^{-1} min^{-1}, respectively, by increasing the MF intensity of MF to 5.0, 10.0, and 20.0 mT, they were much larger than that by ZVI without WMF. The gradual drop in the Cr(VI) removal rate by increasing the intensity from 2.0 mT to 20.0 mT should be mainly ascribed to the aggregation of iron particles at higher MF intensity, which could decrease the surface sites necessary for the reaction to occur.

To further illustrate the WMF influence on the reactivity of ZVI but to exclude its influence on the aggregation of ZVI, electrochemical analysis was carried out. The Tafel polarization technique for a PIPE being similar to the shape of the LSVs of packed iron powder electrodes reported previously (Fig. 12.17b) (Nurmi et al. 2004; Nurmi and Tratnyek 2008; Turcio-Ortega et al. 2012) was applied to unravel the influence of MF flux intensity on the corrosion reactivity of ZVI. The resulting current density (i) vs. potential data (E) was used to determine the polarization resistance as defined by Eq. 12.8 (Stern and Geary 1957).

$$R_{\mathrm{p}} = \frac{b_{\mathrm{a}} b_{\mathrm{c}}}{2.303 I_{\mathrm{corr}} (b_{\mathrm{a}} + b_{\mathrm{c}})} \tag{12.8}$$

where R_{p} is the is the polarization resistance (Ω), I_{corr} is the corrosion current density (A cm^{-2}), and b_{a} and b_{c} are the anodic and cathodic Tafel constants (V), respectively.

As illustrated in the inset of Fig. 12.17b, R_{p} progressively dropped with increasing the MF flux intensity from 0 to 20.0 mT, which was accompanied with a gradual elevation in the I. Thus, the corrosion rate enhanced gradually with increasing the MF flux intensity. It should be additionally notified that the corrosion potential (E_{corr}) kept almost constant at various MF intensities, confirming that MF enhanced the mass-transfer controlled corrosion rather than the electron-transfer controlled iron corrosion (Sueptitz et al. 2009). The electrochemical results verified that the drop in the removal rate of Cr(VI) by ZVI as the MF flux intensity increased from 2.0 to 20.0 mT (shown in Fig. 12.17a) arose from the more obvious aggregation of iron particles at higher MF flux intensity.

As shown in Fig. 12.18, for a specific ZVI particle, the maximum flux intensity of induced MF and the MF gradient around the ZVI particle surface increased proportionally with the flux intensity of applied WMF. Thus, the contribution of $F_{\Delta \mathrm{B}}$ and

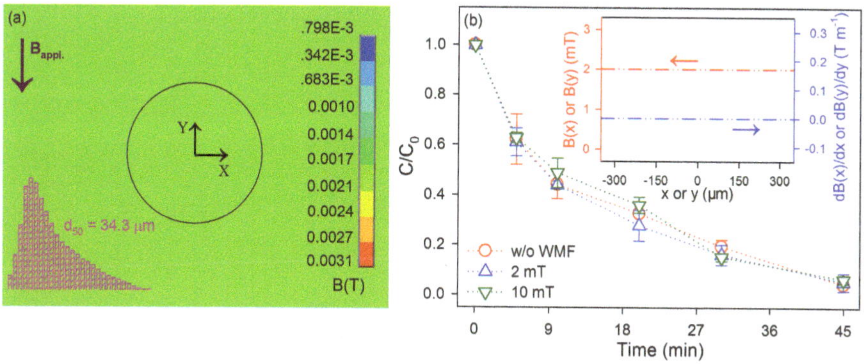

Fig. 12.18 (**a**) The MF strength distribution of the plane parallel to the applied MF with an applied flux density ($B_{appl.}$) of 2.0 mT through the center of a ZVZ sphere, inset: particle size distribution of the pristine ZVZ sample employed in this study. (**b**) Influence of MF flux intensity on the kinetics of Cr(VI) removal by ZVZ, inset: the magnetic field intensity distribution and the magnetic field gradient multiplied by magnetic field intensity along the coordinate X or Y of ZVZ particle. Reaction conditions: $[Cr(VI)]_0 = 4.0$ mg L^{-1}, $[Zn^0] = 0.50$ g L^{-1}. (Reprinted with permission from Li et al. (2017). Copyright (2017) Elsevier)

F_L to the WMF-induced improvement in Cr(VI) removal by ZVI could not be differentiated just based on the results of the influence of MF flux intensity on Cr (VI) removal by ZVI. Different from ZVI, zero-valent zinc (ZVZ) is not ferromagnetic and is not magnetized in a MF. The numerical simulation by the Mechanical APDL (ANSYS) 14.0 software shown in Fig. 12.18a validated that the MF across the ZVZ particle is uniform in a homogenous MF and the MF induced density across the ZVZ particle was identical to that of the superimposed MF. Accordingly, the charged ions that moved toward and away from ZVZ particles (along the X or Y coordinate) were only subject to F_L. Therefore, the influence of MF flux intensity on the kinetics of Cr(VI) of ZVZ was determined, as demonstrated in Fig. 12.18b. Obviously, the effect of MF with flux intensity of 2.0 or 10 mT on Cr(VI) removal by ZVZ was negligible, indicating that F_L in this case is too small to narrow the diffusion boundary layer and to accelerate mass transportation. Although the literatures declared that F_L could narrow the diffusion boundary layer and thus accelerated mass transport (Lioubashevski et al. 2004; Peipmann et al. 2010), the flux intensity of MF employed in their studies was much greater than that superimposed in our study. Therefore, the observed improving effect of WMF on Cr(VI) removal by ZVI should be mainly ascribed to the magnetic gradient force, $F_{\Delta B}$, which will be further verified in the following parts.

12.4.3.2 Effects of WMF on the Kinetics of Cr(VI) Sequestration by ZVI with Different Particle Sizes

In a constant flux intensity of MF, the induced MF gradient along a ZVI particle is expected to drop with increasing the ZVI particle size, while the flux intensity of the

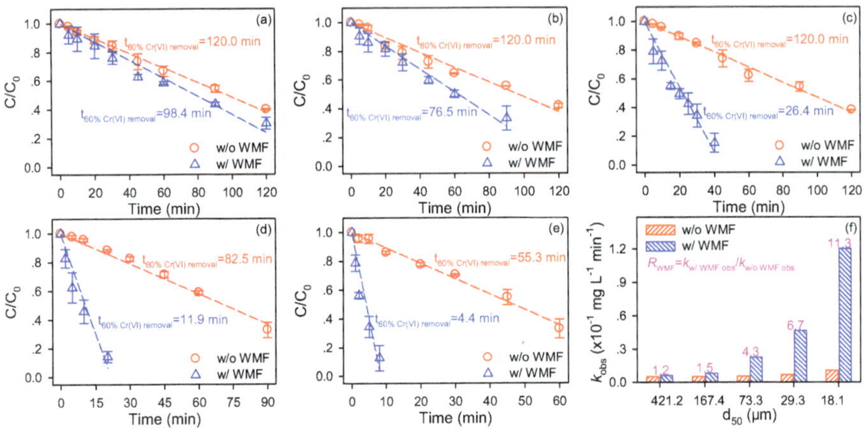

Fig. 12.19 Effects of WMF on the kinetics of Cr(IV) removal by ZVI-421.2 μm (**a**), ZVI-167.4 μm (**b**), ZVI-73.3 μm (**c**), ZVI-29.3 μm (**d**), and ZVI-18.1 μm (**e**), respectively. (**f**) Influence of WMF on the pseudo-zero-order rate constants of Cr(VI) removal by ZVI and R_{WMF} as a function of the iron particle size. Reaction conditions: $[Fe^0] = 0.50$ g L^{-1}, $[Cr(VI)]_0 = 4.0$ mg L^{-1}, pH = 4.0 (0.1 M HAc-NaAc), [flux intensity of MF] = 2.0 mT. (Reprinted with permission from Li et al. (2017). Copyright (2017) Elsevier)

induced MF is fixed being irrespective of the ZVI particle size (Fujiwara et al. 2006; Tanimoto et al. 1997). Thus, F_L is irrespective with the particle size of ZVI, while $F_{\Delta B}$ drops with the increase in ZVI particle size. Consequently, the influence of a uniform MF of 2.0 mT on the reactivity of ZVI with different particle sizes was studied, as demonstrated in Fig. 12.19, to identify the contribution of $F_{\Delta B}$ and F_L. It should be specified that the ZVI samples used in this part were produced in the same batch and they were divided to five parts with sieves of different pore sizes. Therefore, the properties, especially the passive layer covering on the surface of ZVI samples, of these ZVI samples are identical except their particle size.

Apparently, the WMF-induced improvement on Cr(VI) removal was more significant for ZVI with smaller particle size. In the absence of WMF, as illustrated by Fig. 12.19a–e, the durations necessary to achieve 60% of Cr(VI) removal ($t_{60\ \%Cr\ (VI)\ removal}$) were 120.0, 120.0, 120.0, 82.5, and 55.3 min for ZVI-421.2 μm, ZVI-167.4 μm, ZVI-73.3 μm, ZVI-29.3 μm, and ZVI-18.1 μm, respectively. Alternatively, the time to achieve 60% of Cr(VI) removal by these ZVI samples was dropped to 98.4, 76.5, 26.4, 11.9, and 4.4 min, respectively, with the presence of WMF. Therefore, the volume of the reactor with WMF could be reduced to ~8.0% of its counterpart without WMF to achieve 60% of Cr(VI) removal if the ZVI-18.1 sample was employed. The morphologies of the reacted iron samples collected at the time point, where 60% of Cr(VI) removal was achieved with or without WMF, were determined. To quantitatively compare the WMF influence on the disappearance kinetics of Cr(VI) by ZVI with different particle sizes, the kinetics of Cr(VI) removal was simulated with the pseudo-zero-order rate law and the obtained zero-order rate constants. As iron particle size decreased from 421.2 to 18.1 μm, the rate constants of

Cr(VI) removal by ZVI without WMF increased slightly from 0.0051 to 0.0107 mg L^{-1} min^{-1}, which should be largely ascribed to the increase in specific surface area and in the concentration of reactive sites with decreasing ZVI size (Johnson et al. 1996; Noubactep et al. 2005). Upon the application of WMF, the pseudo-zero-order rate constants for Cr(VI) removal by ZVI increased sharply from 0.0063 to 0.1204 min^{-1} with decreasing the iron particle size from 421.2 to 18.1 μm. To provide a generalized basis for the comparison among the ZVI samples of different sizes, we calculated the ratio (R_{WMF}) of kinetic constants measured with (w/) and without (w/o) the WMF, according to Eq. 12.1. Appreciably, R_{WMF} were 1.2, 1.5, 4.3, 6.7, and 11.3 by the ZVI samples with the particle size of 421.2, 167.4, 73.3, 29.3, and 18.1 μm, respectively (as shown in Fig. 12.19f), suggesting that the WMF effect was more significant for ZVI with smaller particle size. Figure 12.19 also demonstrates that WMF is a measure that can improve the reactivity of ZVI much more considerably than decreasing the size of ZVI particle, especially for the ZVI with particle size ≤73.3 μm.

The reactivity of ZVI of different particle sizes toward Cr(VI) removal was enhanced to different extents, which should be mainly ascribed to the different contribution of $F_{\Delta B}$ and F_L to the WMF-induced improvement in the reactivity of ZVI. The distributions of the flux intensity and the gradients of the induced MF of the plane parallel to the applied MF and through the center of a ZVI sphere with size varying from 18.1 to 421.2 μm were simulated with ANSYS 14.0. Since F_L and $F_{\Delta B}$ were proportional to the flux intensity and gradient of MF, respectively, the correlation between the MF flux intensity (or the MF gradient) and the value of R_{WMF} for ZVI samples with different particles sizes was built, as demonstrated in Fig. 12.20. It should be specified that the maximum flux density of MF ($B(x)$ or $B(y)$) and the maximum MF gradient ($B(x)\mathrm{d}B(x)/\mathrm{d}x$ or $B(y)\mathrm{d}B(y)/\mathrm{d}y$) along the x or y coordinate were used to build the correlation. R_{WMF} of ZVI samples with different sizes was

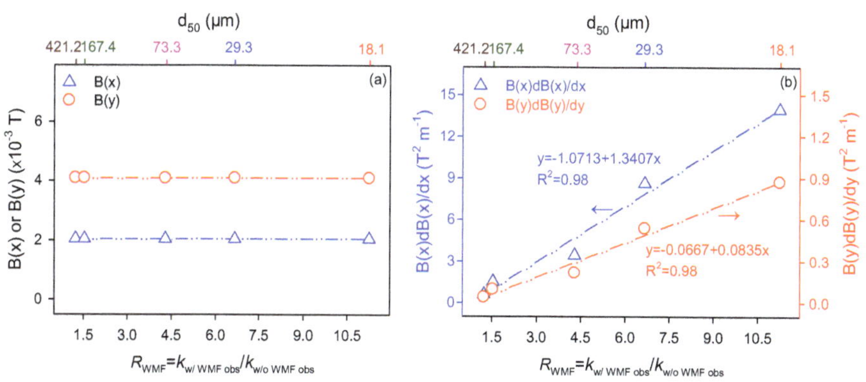

Fig. 12.20 (**a**) Correlation of R_{WMF} with the maximum MF induced density along the coordinate X/Y ($B(x)$ *or* $B(y)$, T) at the surface of ZVI of different sizes. (**b**) Correlation of R_{WMF} with the MF gradient multiplied by MF induced density at surface of ZVI of different sizes along the coordinate X/Y ($B(x)\mathrm{d}B(x)/\mathrm{d}x$ *or* $B(y)\mathrm{d}B(y)/\mathrm{d}y$, T^2 m^{-1}). (Reprinted with permission from Li et al. (2017). Copyright (2017) Elsevier)

irrelevant to $B(x)$ or $B(y)$ of these ZVI samples (Fig. 12.20a), but there was a strong positive correlation between R_{WMF} of ZVI samples with different sizes and ($B(x)dB(x)/dx$ or $B(y)dB(y)/dy$) of these ZVI samples (Fig. 12.20b). These results suggested that $F_{\Delta B}$ rather than F_L was the major driving force for the observed WMF effect on Cr(VI) by ZVI.

12.4.3.3 Direct Evidence for $F_{\Delta B}$-Derived Movement of Fe(II) Along the Surface of a ZVI Sphere

The dual-energy contrast images of the ZVI samples corroded in acetate buffer for 30 min under stationary state with and without WMF were collected with the synchrotron-based STXM (Liang et al. 2013), as illustrated in Fig. 12.21. It should be specified that the mic-7.4 ZVI sample employed in our previous study (Li et al. 2015b) was used in this part because of its regular and smooth sphere, which was beneficial for distinguishing the WMF-induced discrepancies in topography. The dual-energy contrast images explicitly showed that the corrosion products generated in the absence of WMF were almost evenly distributed around the ZVI sphere (Fig. 12.21a). Combining Figs. 12.16 and 12.21b, the corrosion products were concentrated at both ends of the ZVI sphere where the maximum flux intensity of the induced MF appeared with the presence of WMF. Being pulled by $F_{\Delta B}$, the paramagnetic Fe^{2+} tended to move along magnetic lines to the place with higher MF flux intensity (as directed by the arrows in the Fig. 12.21b) (Sueptitz et al. 2011), thereby resulting in the uneven distribution of Fe^{2+} and eventually localized

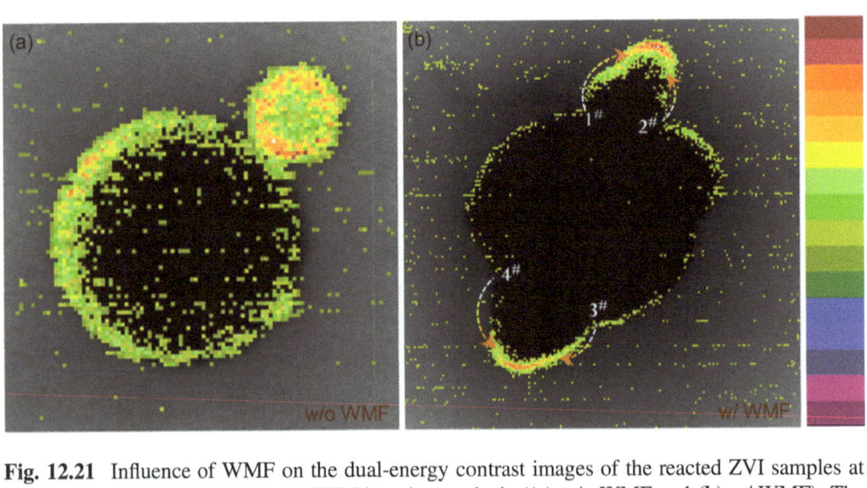

Fig. 12.21 Influence of WMF on the dual-energy contrast images of the reacted ZVI samples at 30 min with the iron element by STXM imaging analysis ((**a**) w/o WMF and (**b**) w/ WMF). The color bar on the right side stands for the relative content changes. From bottom to top, the gradient colors indicate an increased iron content. Reaction conditions for preparing the reacted ZVI samples: $[Fe^0] = 0.50$ g L^{-1}, [flux intensity of MF] = 2.0 mT. (Reprinted with permission from Li et al. (2017). Copyright (2017) Elsevier)

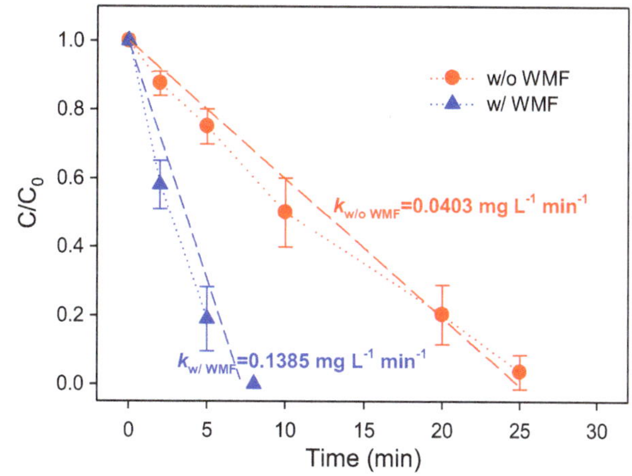

Fig. 12.22 Influence of WMF on the kinetics of Cr (VI) removal by NZVI. Reaction conditions: $[Cr(VI)]_0 = 4.0\ mg\ L^{-1}$, $[nFe^0] = 0.50\ g\ L^{-1}$, pH = 4.0 (0.1 M HAc-NaAc)

distribution of corrosion products in the vicinity of the ZVI sphere. This phenomenon was consistent with the uneven distribution of corrosion products and Cu along the iron wire in the uneven MF observed in previous studies (Jiang et al. 2015; Liang et al. 2014b). Furthermore, in the presence of a MF, where a concentration gradient of paramagnetic ions (Fe^{2+} in this study) exists, an additional driving force acting on the reaction mixture will arise. This force, which has the same direction as the gradient of the paramagnetic ions, will cause a redistribution of velocities in the diffusion layer. As a result, an additional convective transport of all the components of solution will be generated, which is referred to as a magnetoconvection phenomenon (Waskaas and Kharkats 1999, 2001).

In addition, the kinetics of Cr(VI) removal by NZVI purchased from CNC Material company with d_{50} of 500 nm could be improved because of the accelerated NZVI corrosion in the presence of WMF. As shown in Fig. 12.22, the rate constant of Cr(VI) removal by the NZVI/WMF system was 3.4-fold greater than that without WMF effect. It should be noted that the WMF-induced improvement on the reactivity of this commercial NZVI sample is even smaller than that on the reactivity of ZVI with particle sizes of 29.3 or 18.1 μm. This is because the WMF effects are strongly dependent on the properties of the ZVI particles' passive layer besides the size of ZVI particles. Anyway, the application of WMF can improve the reactivity of both micro-sized ZVI and NZVI. The possibility of employing WMF to overcome the limitations of NZVI technologies will be explored in the near future.

12.5 Employing Premagnetization to Enhance the Reactivity of ZVI and the Involved Mechanisms

Although employing WMF to enhance various metal(loid)s removal by ZVI from different origins is a promising and environmentally friendly method since it does not need extra energy and costly reagents, it may be a great challenge in practice to supply a WMF around a ZVI-based filter or PRB. Therefore, it is highly desirable to optimize this technology so that it can be applied in real practice.

Applying a WMF generated by permanent magnets to improve the reactivity of ZVI is efficient, chemical-free, and environmentally friendly. Nevertheless, it may be a great challenge in practice to supply a WMF, generated either by permanent magnets or by electromagnets, for a large treatment unit, which would increase the costs for construction and operation, respectively. ZVI is ferromagnetic, and it can be magnetized by an external magnetic field and remain magnetized even after the external field is removed (Aziz et al. 2014; Ghosh et al. 2012). Thus, taking advantage of the magnetic memory of ZVI, premagnetization may be employed to improve the reactivity of ZVI. However, it keeps unknown whether it is possible to take advantage of the magnetic remanence of ZVI to achieve enhanced contaminant removal by ZVI. Recently, premagnetization of iron by taking advantage of its magnetic memory was employed to improve the ZVI performance (Li et al. 2015c). It was surprisingly observed that, compared to the pristine ZVI, a greater As(III) elimination rate was obtained by the premagnetized ZVI over the initial pH range of 4.0–9.0, and the reaction rate increased progressively with increasing the intensity of the magnetic field for premagnetization (Fig. 12.23).

Moreover, from a broad perspective, as illustrated in Fig. 12.24, premagnetization could enhance the performances of multiple ZVI samples for the sequestration of various oxidative contaminants (including amaranth (AR27), Pb(II), Cu(II), Se(IV), Ag(I), and Cr(VI)) with rate constants that were 1.2- to 12.2-fold greater than those by pristine ZVI under well-controlled experimental conditions (Li et al. 2015b). Further investigation indicated that the improved performance of premagnetized ZVI was not induced by the physical squeezing effect of the ZVI grains during magnetization, but more likely ascribed to the remanence kept by the magnetized ZVI. The remanence kept in the ZVI particles after premagnetization makes each magnetized

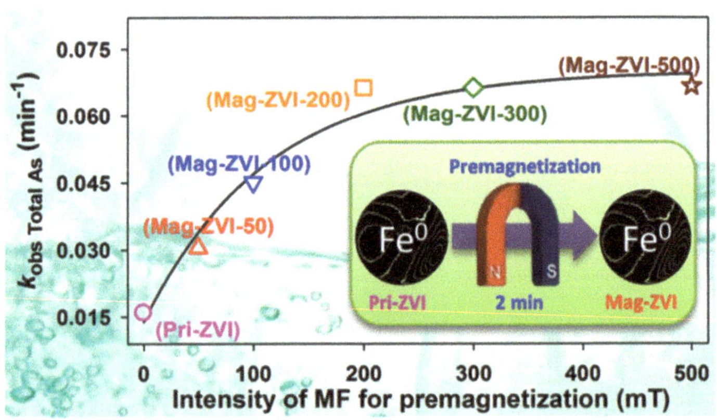

Fig. 12.23 Schematic illustration of the improved reactivity of ZVI toward As(III) by taking advantage of its magnetic memory. (Reprinted with permission from Li et al. (2015a, b, c). Copyright (2015) American Chemical Society)

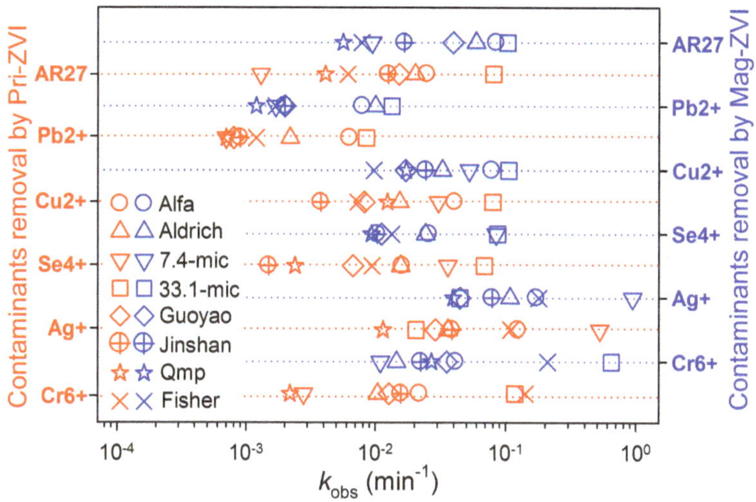

Fig. 12.24 Rate constants (k_{obs}, min^{-1}) for Cr6+, Ag+, Se4+, Cu2+, Pb2+, and AR27 sequestration by pristine ZVI (Pri-ZVI-left ordinate) and premagnetized ZVI (Mag-ZVI-right ordinate). (Reprinted with permission from Li et al. (2015a, b, c). Copyright (2015) American Chemical Society)

ZVI particle a small magnet, which can generate an inhomogeneous MF around the surface of magnetized ZVI particle. Thus, being analogous to applying an external WMF, an enhancing effect of premagnetization on ZVI performance was observed.

12.6 Summary

Increasing research suggests that WMF promotes the reactions of ZVI with a variety of metal(loid)s in water, whereas it should be notified that the WMF effect is associated with the MF flux intensity and the iron particle size, which was of practical and fundamental importance to the environmental decontamination by the ZVI/WMF system. It is impractical to employ strong MF to accelerate contaminants sequestration by ZVI since the aggregation of ZVI particles will occur in a strong MF. Our group is designing the effective magnetic reactors and making efforts to apply the WMF-assisted ZVI technology in real wastewater treatment. Moreover, to further explore the interaction mechanisms of WMF-induced enhancement, efforts are also being made to investigate the synergetic effect of paramagnetic ions (e.g., Fe^{2+}, Mn^{2+}, Cu^{2+}, Co^{2+}, etc.) and WMF on the performance of ZVI toward various contaminants.

Given that employing premagnetization possesses many other merits such as chemical-free, ease of operation, and environmental friendliness, this method will definitely count much in the field of ZVI-based technology. While the WMF effect appears to have great promise for enhancement of water treatment by ZVI, many aspects of this strategy require further investigation, such as whether the residual effect of magnetization ("remanence") is sufficient to be useful, the significance of the WMF enhancement with other classes of contaminants (e.g., metal cations and organics), and the trade-off between accelerated contaminant removal and consumption of the ZVI.

References

Agrawal, A., & Tratnyek, P. G. (1996). Reduction of nitro aromatic compounds by zero-valent iron metal. *Environmental Science & Technology, 30*(1), 153–160.

Ansaf, K. V. K., Ambika, S., & Nambi, I. M. (2016). Performance enhancement of zero valent iron based systems using depassivators: Optimization and kinetic mechanisms. *Water Research, 102*, 436–444.

Aziz, F., Pandey, P., Chandra, M., Khare, A., Rana, D. S., & Mavani, K. R. (2014). Surface morphology, ferromagnetic domains and magnetic anisotropy in BaFeO3−δ thin films: Correlated structure and magnetism. *Journal of Magnetism and Magnetic Materials, 356*(0), 98–102.

Bataineh, H., Pestovsky, O., & Bakac, A. (2012). pH-induced mechanistic changeover from hydroxyl radicals to iron(IV) in the Fenton reaction. *Chemical Science, 3*(5), 1594–1599.

Chen, L., Jin, S., Fallgren, P. H., Swoboda-Colberg, N. G., Liu, F., & Colberg, P. J. S. (2012). Electrochemical depassivation of zero-valent iron for trichloroethene reduction. *Journal of Hazardous Materials, 239–240*(0), 265–269.

Dai, C. M., Zhou, Z., Zhou, X. F., & Zhang, Y. L. (2014). Removal of Sb(III) and Sb(V) from aqueous solutions using nZVI. *Water, Air, and Soil Pollution, 225*(1), 12.

Dixit, S., & Hering, J. G. (2003). Comparison of arsenic(V) and arsenic(III) sorption onto iron oxide minerals: Implications for arsenic mobility. *Environmental Science & Technology, 37*(18), 4182–4189.

Dorjee, P., Amarasiriwardena, D., & Xing, B. S. (2014). Antimony adsorption by zero-valent iron nanoparticles (nZVI): Ion chromatography-inductively coupled plasma mass spectrometry (IC-ICP-MS) study. *Microchemical Journal, 116*, 15–23.

Feng, P., Guan, X. H., Sun, Y. K., Choi, W. Y., Qin, H. J., Wang, J. M., Qiao, J. L., & Li, L. N. (2015). Weak magnetic field accelerates chromate removal by zero-valent iron. *Journal of Environmental Sciences (China), 31*, 175–183.

Fujiwara, M., Mitsuda, K., & Tanimoto, Y. (2006). Movement and diffusion of paramagnetic ions in a magnetic field. *Journal of Physical Chemistry B, 110*(28), 13965–13969.

Furukawa, Y., Kim, J. W., Watkins, J., & Wilkin, R. T. (2002). Formation of ferrihydrite and associated iron corrosion products in permeable reactive barriers of zero-valent iron. *Environmental Science & Technology, 36*(24), 5469–5475.

Geiger, C. L., Ruiz, N. E., Clausen, C. A., Reinhart, D. R., & Quinn, J. W. (2002). Ultrasound pretreatment of elemental iron: Kinetic studies of dehalogenation reaction enhancement and surface effects. *Water Research, 36*, 1342–1350.

Gheju, M. (2011). Hexavalent chromium reduction with zero-valent iron (ZVI) in aquatic systems. *Water, Air, and Soil Pollution, 222*(1-4), 103–148.

Ghosh, N., Mandal, B. K., & Mohan Kumar, K. (2012). Magnetic memory effect in chelated zero valent iron nanoparticles. *Journal of Magnetism and Magnetic Materials, 324*(22), 3839–3841.

Gillham, R. W., & Ohannesin, S. F. (1994). Enhanced degradation of halogenated aliphatics by zero-valent iron. *Ground Water, 32*(6), 958–967.

Guan, X. H., Du, J. S., Meng, X. G., Sun, Y. K., Sun, B., & Hu, Q. H. (2012). Application of titanium dioxide in arsenic removal from water: A review. *Journal of Hazardous Materials, 215*, 1–16.

Guan, X., Jiang, X., Qiao, J., & Zhou, G. (2015a). Decomplexation and subsequent reductive removal of EDTA-chelated CuII by zero-valent iron coupled with a weak magnetic field: Performances and mechanisms. *Journal of Hazardous Materials, 300*, 688–694.

Guan, X. H., Sun, Y. K., Qin, H. J., Li, J. X., Lo, I. M., He, D., & Dong, H. R. (2015b). The limitations of applying zero-valent iron technology in contaminants sequestration and the corresponding countermeasures: The development in zero-valent iron technology in the last two decades (1994-2014). *Water Research, 75*, 224–248.

Guo, X. J., Wu, Z. J., He, M. C., Meng, X. G., Jin, X., Qiu, N., & Zhang, J. (2014). Adsorption of antimony onto iron oxyhydroxides: Adsorption behavior and surface structure. *Journal of Hazardous Materials, 276*, 339–345.

Guo, X., Yang, Z., Dong, H., Guan, X., Ren, Q., Lv, X., & Jin, X. (2016). Simple combination of oxidants with zero-valent-iron (ZVI) achieved very rapid and highly efficient removal of heavy metals from water. *Water Research, 88*, 671–680.

Hug, S. J., & Leupin, O. (2003). Iron-catalyzed oxidation of arsenic(III) by oxygen and by hydrogen peroxide: pH-dependent formation of oxidants in the Fenton reaction. *Environmental Science & Technology, 37*(12), 2734–2742.

Hung, H. M., & Hoffmann, M. R. (1998). Kinetics and mechanism of the enhanced reductive degradation of CCl_4 by elemental iron in the presence of ultrasound. *Environmental Science & Technology, 32*(19), 3011–3016.

Jiang, X., Qiao, J., Lo, I. M. C., Wang, L., Guan, X., Lu, Z., Zhou, G., & Xu, C. (2015). Enhanced paramagnetic Cu2+ ions removal by coupling a weak magnetic field with zero valent iron. *Journal of Hazardous Materials, 283*(0), 880–887.

Johnson, T. L., Scherer, M. M., & Tratnyek, P. G. (1996). Kinetics of halogenated organic compound degradation by iron metal. *Environmental Science & Technology, 30*(8), 2634–2640.

Katsoyiannis, I. A., Ruettimann, T., & Hug, S. J. (2008). pH dependence of Fenton reagent generation and As(III) oxidation and removal by corrosion of zero valent iron in aerated water. *Environmental Science & Technology, 42*(19), 7424–7430.

Katsoyiannis, I. A., Ruettimann, T., & Hug, S. I. (2009a). Response to comment on "pH dependence of Fenton reagent generation and As(III) oxidation and removal by corrosion of zero valent iron in aerated water". *Environmental Science & Technology, 43*(10), 3980–3981.

Katsoyiannis, I. A., Ruettimann, T., & Hug, S. J. (2009b). Response to comment on "pH dependence of Fenton reagent generation and As(III) oxidation and removal by corrosion of zerovalent iron in aerated water". *Environmental Science & Technology, 43*(1), 234–234.

Lai, K. C. K., & Lo, I. M. C. (2008). Removal of chromium (VI) by acid-washed zero-valent iron under various groundwater geochemistry conditions. *Environmental Science and Technology, 42*(4), 1238–1244.

Lee, C., & Sedlak, D. L. (2008). Enhanced formation of oxidants from bimetallic nickel-iron nanoparticles in the presence of oxygen. *Environmental Science & Technology, 42*(22), 8528–8533.

Lee, J., Kim, J., & Choi, W. (2007). Oxidation on zerovalent iron promoted by polyoxometalate as an electron shuttle. *Environmental Science & Technology, 41*(9), 3335–3340.

Leuz, A. K., Monch, H., & Johnson, C. A. (2006). Sorption of Sb(III) and Sb(V) to goethite: Influence on Sb(III) oxidation and mobilization. *Environmental Science & Technology, 40*(23), 7277–7282.

Li, J., Bao, H., Xiong, X., Sun, Y., & Guan, X. (2015a). Effective Sb(V) immobilization from water by zero-valent iron with weak magnetic field. *Separation and Purification Technology, 151*, 276–283.

Li, J., Qin, H., & Guan, X. (2015b). Premagnetization for enhancing the reactivity of multiple zerovalent iron samples toward various contaminants. *Environmental Science & Technology, 49* (24), 14401–14408.

Li, J. X., Shi, Z., Ma, B., Zhang, P. P., Jiang, X., Xiao, Z. J., & Guan, X. H. (2015c). Improving the reactivity of zerovalent iron by taking advantage of its magnetic memory: Implications for arsenite removal. *Environmental Science & Technology, 49*(17), 10581–10588.

Liang, L., Yang, W., Guan, X., Li, J., Xu, Z., Wu, J., Huang, Y., & Zhang, X. (2013). Kinetics and mechanisms of pH-dependent selenite removal by zero valent iron. *Water Research, 47*(15), 5846–5855.

Liang, L., Sun, W., Guan, X., Huang, Y., Choi, W., Bao, H., Li, L., & Jiang, Z. (2014a). Weak magnetic field significantly enhances selenite removal kinetics by zero valent iron. *Water Research, 49*, 371–380.

Liang, L. P., Guan, X. H., Shi, Z., Li, J. L., Wu, Y. N., & Tratnyek, P. G. (2014b). Coupled effects of aging and weak magnetic fields on sequestration of selenite by zero-valent iron. *Environmental Science & Technology, 48*(11), 6326–6334.

Liang, L., Guan, X., Huang, Y., Ma, J., Sun, X., Qiao, J., & Zhou, G. (2015). Efficient selenate removal by zero-valent iron in the presence of weak magnetic field. *Separation and Purification Technology, 156*(Part 3), 1064–1072.

Lin, C. J., & Lo, S. L. (2005). Effects of iron surface pretreatment on sorption and reduction kinetics of trichloroethylene in a closed batch system. *Water Research, 39*(6), 1037–1046.

Lioubashevski, O., Katz, E., & Willner, I. (2004). Magnetic field effects on electrochemical processes: A theoretical hydrodynamic model. *Journal of Physical Chemistry B, 108*(18), 5778–5784.

Liu, H., Li, G., Qu, J., & Liu, H. (2007). Degradation of azo dye Acid Orange 7 in water by Fe^0/granular activated carbon system in the presence of ultrasound. *Journal of Hazardous Materials, 144*(1-2), 180–186.

Lu, X., Li, M., Tang, C., Feng, C., & Liu, X. (2012). Electrochemical depassivation for recovering Fe^0 reactivity by Cr(VI) removal with a permeable reactive barrier system. *Journal of Hazardous Materials, 213–214*(0), 355–360.

Mackenzie, P. D., Horney, D. P., & Sivavec, T. M. (1999). Mineral precipitation and porosity losses in granular iron columns. *Journal of Hazardous Materials, 68*(1-2), 1–17.

Manning, B. A., Hunt, M. L., Amrhein, C., & Yarmoff, J. A. (2002). Arsenic(III) and arsenic (V) reactions with zerovalent iron corrosion products. *Environmental Science & Technology, 36* (24), 5455–5461.

Matheson, L. J., & Tratnyek, P. G. (1994). Reductive dehalogenation of chlorinated methanes by iron metal. *Environmental Science & Technology, 28*(12), 2045–2053.

Miehr, R., Tratnyek, P. G., Bandstra, J. Z., Scherer, M. M., Alowitz, M. J., & Bylaska, E. J. (2004). Diversity of contaminant reduction reactions by zerovalent iron: Role of the reductate. *Environmental Science & Technology, 38*(1), 139–147.

Mitsunobu, S., Takahashi, Y., Terada, Y., & Sakata, M. (2010). Antimony(V) incorporation into synthetic ferrihydrite, goethite, and natural iron oxyhydroxides. *Environmental Science & Technology, 44*(10), 3712–3718.

Mondal, K., Jegadeesan, G., & Lalvani, S. B. (2004). Removal of selenate by Fe and NiFe nanosized particles. *Industrial & Engineering Chemistry Research, 43*(16), 4922–4934.

Mylon, S. E., Sun, Q. A., & Waite, T. D. (2010). Process optimization in use of zero valent iron nanoparticles for oxidative transformations. *Chemosphere, 81*(1), 127–131.

Neumann, A., Kaegi, R., Voegelin, A., Hussam, A., Munir, A. K. M., & Hug, S. J. (2013). Arsenic removal with composite iron matrix filters in Bangladesh: A field and laboratory study. *Environmental Science & Technology, 47*(9), 4544–4554.

Noubactep, C. (2008). A critical review on the process of contaminant removal in Fe-0-H2O systems. *Environmental Technology, 29*(8), 909–920.

Noubactep, C. (2009a). An analysis of the evolution of reactive species in Fe-0/H2O systems. *Journal of Hazardous Materials, 168*(2-3), 1626–1631.

Noubactep, C. (2009b). Comment on "pH dependence of Fenton reagent generation and As(III) oxidation and removal by corrosion of zero valent iron in aerated water". *Environmental Science & Technology, 43*(1), 233–233.

Noubactep, C., Meinrath, G., Dietrich, P., Sauter, M., & Merkel, B. J. (2005). Testing the suitability of zerovalent iron materials for reactive walls. *Environmental Chemistry, 2*(1), 71–76.
Nurmi, J. T., & Tratnyek, P. G. (2008). Electrochemical studies of packed iron powder electrodes: Effects of common constituents of natural waters on corrosion potential. *Corrosion Science, 50*(1), 144–154.
Nurmi, J. T., Bandstra, J. Z., & Tratnyek, P. G. (2004). Packed powder electrodes for characterizing the reactivity of granular iron in borate solutions. *Journal of the Electrochemical Society, 151*(6), B347–B353.
Obiri-Nyarko, F., Grajales-Mesa, S. J., & Malina, G. (2014). An overview of permeable reactive barriers for in situ sustainable groundwater remediation. *Chemosphere, 111*, 243–259.
Pang, S. Y., Jiang, J., Ma, J., Pang, S. Y., & Ouyang, F. (2009). New insight into the oxidation of arsenite by the reaction of zerovalent iron and oxygen. Comment on "pH dependence of Fenton reagent generation and As(III) oxidation and removal by corrosion of zero valent iron in aerated water". *Environmental Science & Technology, 43*(10), 3978–3979.
Pang, S. Y., Jiang, J., & Ma, J. (2011). Oxidation of sulfoxides and arsenic(III) in corrosion of nanoscale zero valent iron by oxygen: Evidence against ferryl ions (Fe(IV)) as active intermediates in Fenton reaction. *Environmental Science & Technology, 45*(1), 307–312.
Peipmann, R., Lange, R., Kubeil, C., Mutschke, G., & Bund, A. (2010). Magnetic field effects on the mass transport at small electrodes studied by voltammetry and magnetohydrodynamic impedance measurements. *Electrochimica Acta, 56*(1), 133–138.
Prasad, P., Das, C., & Golder, A. K. (2011). Reduction of Cr (VI) to Cr (III) and removal of total chromium from wastewater using scrap iron in the form of zerovalent iron (ZVI): Batch and column studies. *The Canadian Journal of Chemical Engineering, 89*(6), 1575–1582.
Ragsdale, S. R., Grant, K. M., & White, H. S. (1998). Electrochemically generated magnetic forces. Enhanced transport of a paramagnetic redox species in large, nonuniform magnetic fields. *Journal of the American Chemical Society, 120*(51), 13461–13468.
Ritter, K., Odziemkowski, M. S., & Gillham, R. W. (2002). An in situ study of the role of surface films on granular iron in the permeable iron wall technology. *Journal of Contaminant Hydrology, 55*(1-2), 87–111.
Scheinost, A. C., Rossberg, A., Vantelon, D., Xifra, I., Kretzschmar, R., Leuz, A. K., Funke, H., & Johnson, C. A. (2006). Quantitative antimony speciation in shooting-range soils by EXAFS spectroscopy. *Geochimica et Cosmochimica Acta, 70*(13), 3299–3312.
Sedlak, D. L., & Andren, A. W. (1991). Oxidation of chlorobenzene with Fenton's reagent. *Environmental Science & Technology, 25*(4), 777–782.
Stern, M., & Geary, A. L. (1957). Electrochemical polarization I. A theoretical analysis of the shape of polarization curves. *Journal of the Electrochemical Society, 104*(1), 56–63.
Su, C. M., & Puls, R. W. (2001). Arsenate and arsenite removal by zerovalent iron: Effects of phosphate, silicate, carbonate, borate, sulfate, chromate, molybdate, and nitrate, relative to chloride. *Environmental Science & Technology, 35*(22), 4562–4568.
Sueptitz, R., Koza, J., Uhlemann, M., Gebert, A., & Schultz, L. (2009). Magnetic field effect on the anodic behaviour of a ferromagnetic electrode in acidic solutions. *Electrochimica Acta, 54*(8), 2229–2233.
Sueptitz, R., Tschulik, K., Uhlemann, M., Schultz, L., & Gebert, A. (2011). Effect of high gradient magnetic fields on the anodic behaviour and localized corrosion of iron in sulphuric acid solutions. *Corrosion Science, 53*(10), 3222–3230.
Sun, Y. K., Guan, X. H., Wang, J. M., Meng, X. G., Xu, C. H., & Zhou, G. M. (2014). Effect of weak magnetic field on arsenate and arsenite removal from water by zerovalent iron: An XAFS investigation. *Environmental Science & Technology, 48*(12), 6850–6858.
Sun, Y., Hu, Y., Huang, T., Li, J., Qin, H., & Guan, X. (2017). Combined Effect of Weak Magnetic Fields and Anions on Arsenite Sequestration by Zerovalent Iron: Kinetics and Mechanisms. Environ. Sci. Technol., *51*(7), 3742–3750.
Tang, C., Huang, Y. H., Zeng, H., & Zhang, Z. (2014). Promotion effect of Mn 2+ and Co 2+ on selenate reduction by zero-valent iron. *Chemical Engineering Journal, 244*, 97–104.

Tanimoto, Y., Katsuki, A., Yano, H., & Watanabe, S. (1997). Effect of high magnetic field on the silver deposition from its aqueous solution. *Journal of Physical Chemistry A, 101*(40), 7359–7363.

Triszcz, J. M., Port, A., & Einschlag, F. S. G. (2009). Effect of operating conditions on iron corrosion rates in zero-valent iron systems for arsenic removal. *Chemical Engineering Journal, 150*(2-3), 431–439.

Turcio-Ortega, D., Fan, D. M., Tratnyek, P. G., Kim, E. J., & Chang, Y. S. (2012). Reactivity of Fe/FeS nanoparticles: Electrolyte composition effects on corrosion electrochemistry. *Environmental Science & Technology, 46*(22), 12484–12492.

Wang, C., & Zhang, W. (1997). Synthesizing nanoscale iron particles for rapid and complete dechlorination of TCE and PCBs. *Environmental Science & Technology, 94*(18), 9602–9607.

Waskaas, M., & Kharkats, Y. I. (1999). Magnetoconvection phenomena: A mechanism for influence of magnetic fields on electrochemical processes. *Journal of Physical Chemistry B, 103*(23), 4876–4883.

Waskaas, M., & Kharkats, Y. I. (2001). Effect of magnetic fields on convection in solutions containing paramagnetic ions. *Journal of Electroanalytical Chemistry, 502*(1-2), 51–57.

Xi, J. H., He, M. C., & Lin, C. Y. (2011). Adsorption of antimony(III) and antimony(V) on bentonite: Kinetics, thermodynamics and anion competition. *Microchemical Journal, 97*(1), 85–91.

Xie, Y., & Cwiertny, D. M. (2010). Use of dithionite to extend the reactive lifetime of nanoscale zero-valent iron treatment systems. *Environmental Science & Technology, 44*(22), 8649–8655.

Xu, C., Zhang, B., Zhu, L., Lin, S., Sun, X., Jiang, Z., & Tratnyek, P. G. (2016a). Sequestration of antimonite by zerovalent iron: Using weak magnetic field effects to enhance performance and characterize reaction mechanisms. *Environmental Science & Technology, 50*(3), 1483–1491.

Xu, H., Sun, Y., Li, J., Li, F., & Guan, X. (2016b). Aging of zerovalent iron in synthetic groundwater: X-ray photoelectron spectroscopy depth profiling characterization and depassivation with uniform magnetic field. *Environmental Science & Technology, 50*(15), 8214–8222.

Chapter 13
Vadose Zone Remediation of Dense Nonaqueous Phase Liquid Residuals Using Foam-Based Nanoscale Zerovalent Iron Particles with Low-Frequency Electromagnetic Field

Tanapon Phenrat and Gregory V. Lowry

Abstract This chapter presents a novel combined remedy using foam-based NZVI (F-NZVI) for vadose zone remediation of volatile organic compound (VOC) contamination. Conceptually, F-NZVI serves two remedial actions. First, - F-NZVI can flush the VOC or nonaqueous phase liquid (NAPL) from the soil. Second, in addition to flushing NAPL from the soil, the NZVI that is deposited on the soil grain, if electromagnetically induced by a low-frequency (LF) electromagnetic field (EMF) (Chap. 11), should generate heat and speed up VOC removal in the vadose zone via thermally enhanced volatilization when used with soil vapor extraction (SVE). This chapter reviews the use of various surfactants to produce foam and F-NZVI for soil flushing. Moreover, characterization and transport experiments of foam and F-NZVI in unsaturated porous media demonstrate the thermally enhanced evaporation of VOCs using F-NZVI and LF EMF (up to 40 times enhanced evaporation of trichloroethylene). The feasibility of this novel approach is compared with a thermally enhanced SVE using radio-frequency heating (RFH) without F-NZVI. The chapter points out that using F-NZVI with LF EMF could theoretically be an alternative to RFH because it does not require as high of an irradiation frequency as RFH and should lead to lower capital and operational costs versus RFH.

T. Phenrat (✉)
Department of Civil Engineering, Environmental Engineering Program, Naresuan University, Phitsanulok, Thailand

Center of Excellence for Sustainability of Health, Environment and Industry (SHEI), Faculty of Engineering, Naresuan University, Phitsanulok, Thailand

G. V. Lowry
Center for Environmental Implications of Nanotechnology (CEINT), Durham, NC, USA

Department of Civil & Environmental Engineering, Carnegie Mellon University, Pittsburgh, PA, USA

T. Phenrat, G. V. Lowry (eds.), *Nanoscale Zerovalent Iron Particles for Environmental Restoration*, https://doi.org/10.1007/978-3-319-95340-3_13

Keywords Nanoscale Zerovalent Iron · Vadose Zone · Foam · Thermal Enhanced Soil Vapor Extraction · Combined Remedies with NZVI · Electromagnetic · Low Frequency · Electromagnetic Induction Heating

13.1 Limitations of Aqueous NZVI Dispersion for Vadose Zone Remediation

Vadose zone contamination with chlorinated volatile organic compounds (CVOCs) is a persistent environmental problem, jeopardizing environmental quality and public health. The average half-life for CVOC abiotic transformations ranges from 2 months to greater than 10^{10} years (Barbee 2007). The CVOCs may be entrapped as dense nonaqueous phase liquid (DNAPL) residuals in pores in the vadose zone (Fig. 13.1) and behave as long-term sources of vapor intrusion problems (Abreu and Johnson 2005; Brusseau et al. 2013; Ronen et al. 2004) and of groundwater contamination by discharging CVOCs to underlying groundwater via infiltration and percolation (Fagerlund et al. 2007a, b). Furthermore, CVOCs can also sorb into soil and subsequently discharge both vapor and dissolved CVOCs to contaminate soil gas and underlying groundwater (Brusseau et al. 2013; Ronen et al. 2004). While cleanup of a CVOC-contaminated saturated zone has several remediation alternatives, remediation techniques for CVOCs for the vadose zone are limited to excavation, soil vapor extraction (SVE) (Yang et al. 2001; Yoon et al. 2003), thermal-enhanced SVE (Truex et al. 2009; Nakamura et al. 2000; Poppendieck et al. 1999a, b), and bioventing (Boulding and Ginn 2003; Brusseau et al. 2013).

As discussed throughout this book, in situ chemical reduction (ISCR) using nanoscale zerovalent iron (NZVI) is a promising groundwater remediation alternative (Babakhani et al. 2015; Fagerlund et al. 2012; Phenrat et al. 2010b, c; 2011, 2018; Tratnyek and Johnson 2006; Zhang et al. 1998; Zhang 2003). Evidently, 70 pilot-scale and field-scale NZVI remediation tests in saturated aquifers had been conducted (Bardos et al. 2015; Karn et al. 2009; Mueller et al. 2012) (see also

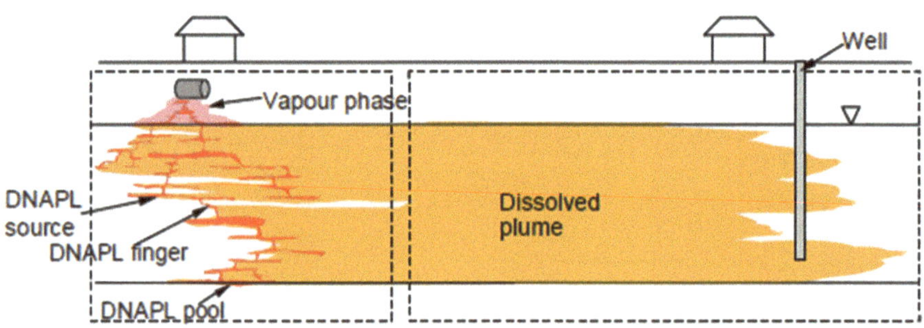

Fig. 13.1 Schematics of dense nonaqueous phase liquid (DNAPL) contamination in the vadose zone that acts as a long-term source of groundwater contamination and vapor intrusion

Chaps. 1 and 9). However, water-based deliver methods for NZVI are not applicable for dechlorination of CVOCs in the vadose zone. For example, gravity may preferentially induce the migration of water-based NZVI dispersion in the vertical direction, limiting lateral transport and the radius of influence (ROI) (Ding et al. 2013). In addition, flushing water-based NZVI dispersion in the vadose zone may cause unintended CVOC dissolution and migration to the underlying aquifers, leading to more severe contamination.

Consequently, Zhong and Li's group from the Pacific Northwest National Laboratory proposed using foam-assisted delivery of NZVI in the vadose zone to overcome these technical difficulties (Ding et al. 2013; Shen et al. 2011; Zhong et al. 2010). They revealed that sodium lauryl ether sulfate (SLES)-stabilized foam could transport NZVI through an unsaturated sand-packed column much better than a water-based NZVI dispersion (Ding et al. 2013). Even though the difficulties in particle delivery are solved by the foam-assisted delivery option, NZVI still faces technical difficulty in CVOC degradation, as reductive dechlorination by NZVI requires water molecules for the corrosion of iron and subsequent transformation of CVOCs to more benign by-products (see Chap. 3) (Chen et al. 2014; Fagerlund et al. 2012; Liu et al. 2007; Phenrat et al. 2009; Sakulchaicharoen et al. 2010; Saleh et al. 2005; Tratnyek and Johnson 2006; Wei et al. 2012; Zhang 2003). It is expected that the vadose zone may not have enough water and may not have sufficient CVOC dissolution into pore water for substantial reductive dechlorination by NZVI. Thus, the reactivity of NZVI may not be as useful to vadose zone cleanup as it is for the saturated zone.

In this chapter, we first describe a possible strategy of using foam-based NZVI (F-NZVI) with low-frequency (LF) electromagnetic field (EMF) (Chap. 11) to assist in thermal-enhanced SVE for vadose zone restoration. In detail, we describe the synthesis and characterization of F-NZVI and assess its stability and mobility in an unsaturated porous medium. The TCE volatilization kinetics, both with and without LF EMF, are compared to quantify the benefits of using LF EMF with F-NZVI in enhanced thermal treatment. Finally, we theoretically analyze the feasibility of this LF EMF and F-NZVI combined remediation concept in comparison with a conventional radio-frequency heating (RFH) without F-NZVI.

13.2 Conceptual Model for Soil Flushing Followed by Thermal-Enhanced SVE Using Foam-Based NZVI

The most commonly used vadose zone remediation technique for volatile organic compounds (VOCs), SVE, relies on extracting VOCs in the soil vapor phase, as the name implies. Nevertheless, at in situ natural temperature, its removal efficacy is limited by mass transfer due to slow desorption and slow diffusion of VOCs from the soil matrix, especially if low permeable layers exist (Heron et al. 1998; Werth and Reinhard 1997). Thermal-enhanced options (Truex et al. 2009; Nakamura et al. 2000;

Poppendieck et al. 1999a, b; Price et al. 1999; Roland et al. 2008; Vermeulen and McGee 2000; Werth and Reinhard 1997) have received increasing attention, as they can increase vapor pressure to overcome the mass transfer limitations of SVE. Electrical heating, such as electrical resistance heating (ERH), and electromagnetic heating, such as RFH and microwave heating (MWH), are novel thermal alternatives overcoming technical problems associated with the low heat capacity of air as well as the poor accessibility and heat transfer of low-permeable soil layers if steam or hot air injection is used as a thermal-enhanced option (Heron et al. 1998; Price et al. 1999; Vermeulen and McGee 2000).

Moreover, ERH delivers an electric current to the subsurface and converts energy from electrical to thermal due to the resistance of the subsurface material to the flow of electricity, i.e., Ohmic heating. The amount of power dissipated by resistance (P_{ERH}) (watts) can be expressed using Ohm's law (Rizzoni 2004):

$$P_{\mathrm{ERH}} = RI^2 \tag{13.1}$$

$$\text{or } P_{\mathrm{ERH}} = VI, \tag{13.2}$$

where V is the applied electrical potential (V), R is the resistance of the material between the electrodes (Ω), and I is the current passing through the heating zone (A). Resistance is related to the material's effective electrical conductivity (σ (S/m)), length of the material between the electrodes (L (m)), and cross-sectional area of the material (A (m^2)) by Rizzoni (2004):

$$R = \frac{L}{\sigma A'} \tag{13.3}$$

For vadose zones with partially saturated, nonconducting grains, such as silica sand, the effective electrical conductivity can be described by Archie's law (Jackson et al. 2008):

$$\sigma = \sigma_W \varnothing^M, \tag{13.4}$$

where ∅ is the water-filled porosity, M is the cementation exponent (unitless) depending on type of materials, and σ_W is the electrical conductivity of water (S/m).

Conversely, electromagnetic heating, such as RFH and MWH, converts electromagnetic energy into thermal energy through the interaction between the EMF and the atoms or molecules present in the irradiated material. Thus, this interaction is governed by the nature of the irradiated material and the radiation frequency. Both RFH and MWH use frequencies as high as 500 kHz to 500 MHz (Lowe et al. 1999; Price et al. 1999) and 500 MHz to 500 GHz (Falciglia et al. 2013; Lowe et al. 1999), respectively. Heating of nonmagnetic aquifer materials arises from conduction losses and dielectric losses as follows:

$$Q = 2\pi f \varepsilon_0 \varepsilon''_{\mathrm{eff}} |E|^2, \tag{13.5}$$

where Q is the heat generated in a unit volume of the irradiated media, $|E|$ is the magnitude of the applied electric field, f is the irradiated frequency, ε_0 is the permittivity of free space (8.85×10^{-12} F/m), and ε''_{eff} is an effective imaginary permittivity, which can be described as:

$$\varepsilon''_{\text{eff}} = \varepsilon'' + \frac{\sigma}{2\pi f \varepsilon_0}. \quad (13.6)$$

Both ERH and electromagnetic heating have been used in field-scale applications for oil recovery (Bera and Babadagli 2015; Vermeulen and McGee 2000) and soil remediation (Beyke and Fleming 2005; Gavaskar et al. 2007; Price et al. 1999; Vermeulen and McGee 2000). Moreover, ERH was reported to remove as much as 98% of entrapped trichloroethylene (TCE) in a water-saturated soil after 175 days of a field-scale test (Beyke and Fleming 2005). RFH was reported to achieve the same removal efficiency as ERH in a water-saturated soil for a field-scale test (Smith and Hinchee 1993). Based on 40 sites remediated by ERH, Beyke and Fleming (2005) estimated that the average capital cost of ERH was \$200,000, while the operational cost was between \$52 and \$91 per cubic meter, claiming that ERH is a proven and cost-effective remediation alternative (Beyke and Fleming 2005). On the other hand, Bientinesi et al. (2015) reported that the capital cost of RFH for soil remediation was \$1.1 million, while the operational cost was \$140 per cubic meter (Bientinesi et al. 2015). Importantly, both capital and operational costs of RFH are higher than that of ERH due to the fact that the electrical transmission losses and capital cost of RFH equipment increase with the frequency of EMFs (Vermeulen and McGee 2000).

While conduction and dielectric losses are the majority of losses in RFH of soil, magnetic loss is another heating mechanism capable of operating at lower frequency and thus theoretically capable of decreasing the transmission losses and the capital cost of the EMF generation equipment. The magnetic induction heating (MIH) is through hysteresis loss of magnetic materials due to the irreversible magnetization in an EMF. At a particular frequency (f) of the EMF, the amount of power dissipated by MIH (P_{MIH}) (W) can be expressed as follows (Li et al. 2010):

$$P_{\text{MIH}} = f\Delta U, \quad (13.7)$$

where ΔU is the area under the hysteresis curves, governed by the characteristics of the magnetic materials. The degree of irreversibility, ΔU, is related to the amount of energy dissipation upon the reversal of the field. The MIH in a LF EMF (30–300 kHz) is well studied in medicine for thermal treatments, such as magnetic-assisted hyperthermia (Lim et al. 2007; Pablico-Lansigan et al. 2013), where functionalized magnetic nanoparticles, such as magnetite, target tumor cells and then are heated by applied LF EMF to kill the cancer cells (Bañobre-López et al. 2013; Lim et al. 2007). For this reason, ferromagnetic nanoparticles, such as NZVI, can be coupled with the LF EMF to serve as a combined remediation technique for increasing the rate and completeness of an in situ cleanup of CVOCs in the subsurface (See Chap. 11).

Conceptually, foam-based NZVI (F-NZVI) can perform two remediation processes. First, F-NZVI can flush the CVOC from the soil. See examples of soil remediation using foam flushing in Table 13.1 and Fig. 13.2. Foaming surfactant has recently gained a lot of attention as a foam-based soil-flushing technique (Zhao et al. 2016) to remove pesticides (Wang et al. 2015), polychlorinated biphenyl (Wang and Chen 2012), and CVOCs (Jeong et al. 2000; Maire and Fatin-Rouge 2017) from unsaturated porous media. Foam flushing mechanistically enhances removal of NAPL by direct and indirect displacements, increased solubility, and NAPL blob snap-off (Jeong et al. 2000). Interestingly, the foam-flushing surfactant achieved high DNAPL removal efficiency without lowering the interfacial tension to a critically low value, which may cause vertical migration (Jeong et al. 2000; Maire and Fatin-Rouge 2017). Thus, unintended vertical migration of DNAPL as a result of foam flushing should not be of serious concern when applied properly in a medium with appropriate permeability.

Second, in addition to flushing NAPL from the soil, the NZVI that is deposited on the soil grain, if electromagnetically induced by LF EMF, should generate heat and speed up CVOC removal in the vadose zone via thermally enhanced volatilization when used with SVE (Fig. 13.3). Raising the chemical vapor pressures by heating soil in situ can decrease the remediation time as RFH does, but the lower frequency of LF EMF should result in a lower-cost system and operation.

13.3 Generation and Characterization of F-NZVI

The F-NZVI generation requires ingredients including aqueous NZVI dispersion, gas (typically N_2 to prevent NZVI oxidation), and surfactant (as a foam stabilizing agent) with a foam generation column. Figure 13.4 illustrates the schematic diagram of F-NZVI generation proposed by Zhong and Li's group from the Pacific Northwest National Laboratory (Ding et al. 2013; Shen et al. 2011; Zhong et al. 2010). This is a typical foam generation unit for surfactant foam flushing, except that, for F-NZVI generation, NZVI is added into the surfactant solution (Jeong et al. 2000; Lv et al. 2017; Maire and Fatin-Rouge 2017). The F-NZVI generation unit is where N_2 gas and NZVI in surfactant dispersion (pumped by a peristaltic pump) meet at the inlet of the column to help create F-NZVI by passing through a 60-mesh stainless steel screen (Srirattana et al. 2017). Foam quality and stability are important properties for using F-NZVI as a remediation agent. They are affected by surfactant performance (type and concentration), NZVI concentration, and N_2 flow rate. The foam quality (F (%)) is defined by Eq. 13.8 (Mulligan and Eftekhari 2003):

$$F = \frac{Vg}{V\,l + Vg} \times 100, \qquad (13.8)$$

Table 13.1 Examples of surfactant foam-enhanced soil remediation

Experimental condition	Surfactant	Performance	References
DNAPL recovery using foam flushing followed by surfactant flushing (column)	Dihexyl sulfosuccinate as a foaming agent and Tergitol as a solubilizing agent	At pressure gradient <90 kPa/m, 34% to 60% DNAPL was recovered using foam flushing, and 95% removal was achieved using foam flushing followed by 2 pore volumes (PVs) of Tergitol flushing	Maire and Fatin-Rouge (2017)
DDT desorption from unsaturated soil using four different surfactants (column)	TX-100, tween 80, Brij-35, SDS	At surfactant solution flow rate of 3 mL/min and N_2 flow rate of 57 mL/min, the maximum DDT concentrations flushed out by surfactant foam were 4.06, 1.46, 2.10, and 1.67 mg/L for TX-100, SDS, tween 80, and Brij-35, respectively. The desorption efficiency of the surfactants followed the order TX-100 > tween 80 > SDS > Brij-35. TX-100 showed the highest DDT concentration in effluent due to its relatively high foamability, stability, and high ability to enhance solubility	Lv et al. (2017)
DDT desorption from soil using four different surfactants (batch)	Tween 80, TX-100, SDS, Brij-35	The solubilizing ability for DDT followed the order of tween 80 > TX-100 > SDS > Brij-35, while the order of desorption efficiency for DDT was TX-100 (60%) > tween 80 (40%) > Brij-35 (12%) > SDS (10%) at surfactant concentration of 5 g/L	Wang et al. (2015)
Remediation of PCP-contaminated soil (column)	TX-100, JBR 425	TX-100 (1%) foam removed 85% and 84% of PCP from fine sand and sandy silt, respectively, while JBR 425 (1%) removed 60% and 61% of PCP from fine sand and sandy silt, respectively.	Mulligan and Eftekhari (2003)

(continued)

Table 13.1 (continued)

Experimental condition	Surfactant	Performance	References
		Comparison of liquid and foam injection of 1% TX-100 indicated that the foam removed more than twice as much PCP in all cases than the liquid surfactant solution	
Removal of n-pentadecane (column)	Triton SP series	Over 74% of the n-pentadecane was removed at a gas-liquid ratio of 10:1. The recovery of n-pentadecane using foam flushing was increased significantly compared with that obtained by surfactant solution flushing, probably due to the reduced channeling flow effect	Huang and Chang (2000)
Removal of residual TCE (micromodel)	Sodium $C_{14–16}$ olefin sulfonate 2%-(w)	The surfactant foam flushing removed 99% of the residual TCE, while surfactant flushing removed 41% with 25 PV of the same surfactant solution	Jeong et al. (2000)
Surfactant foam/ bioaugmentation technology of TCE DNAPL (batch)	STEOL CS-330	Injecting the foam in a pulsed operation removed 75% of the contaminant, and adding the microbes resulted in 95%–99% degradation of the residual	Rothmel et al. (1998)
HCB removal from the soil by soil flushing using colloidal gas aphron (CGA) suspensions generated from a plant-based surfactant (column)	Natural surfactant from *Sapindus mukorossi* (soapnut)	The CGA suspension recovered 6701 g of HCB in 12 PV compared to 81 g by water flooding	Kommalapati et al. (1998)
Surfactant/foam process for removal of mixture of TCE, TCA, and PCE DNAPL at hill air Force Base in Utah (field)	Sodium dihexyl sulfosuccinate	The average DNAPL saturation of the swept volume was reduced to 0.03%	Hirasaki et al. (1997)
Remediation of PAH-contaminated soils using foams	Biosurfactants +50% ethanol	Desorption and removal of PAHs at pressure of 33.9 kPa/m (1.5 psi/ft) or less	Kilbane II et al. (1997)

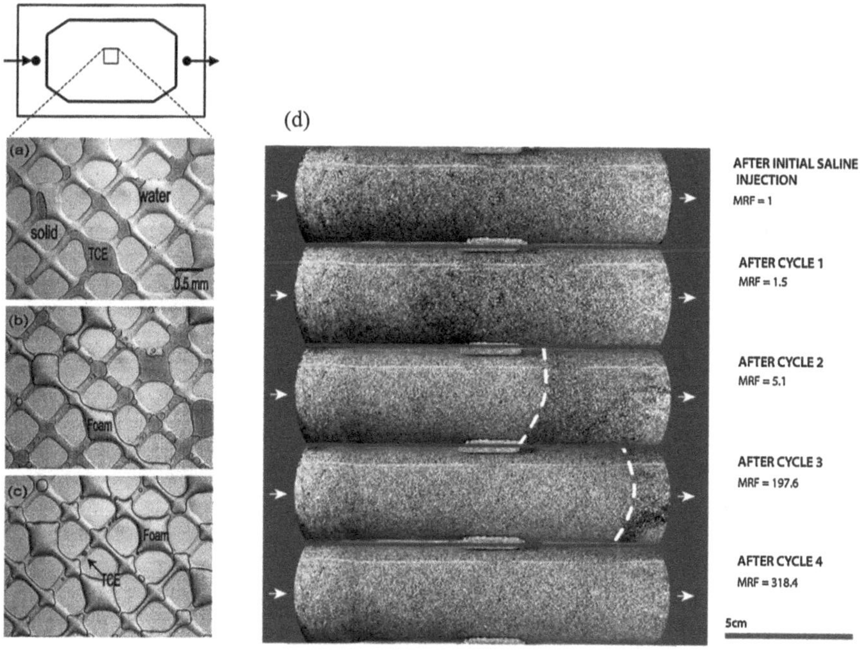

Fig. 13.2 (**a–c**) Images from a micromodel during surfactant foam flooding at a capillary number (N_{ca}) of 1.1 × 1^{-5}; 66% gas fraction: (**a**) initial TCE distribution in the micromodel, (**b**) after surfactant foam injection of 5 pore volume (PV), (**c**) final image after surfactant foam flooding at 25 PV (Jeong et al. 2000). (Reprinted with permission from (Jeong et al. 2000). Copyright (2000) the American Chemical Society), and (**d**) foam propagation and black dense nonaqueous phase liquid (DNAPL) displacement (from left to right) after each cycle of injecting DHSS surfactant and nitrogen in alternation in soil (SAG process). Set point = 70 kPa m^{-1}. When clearly visible, the foam front was emphasized by a white dotted line (Maire and Fatin-Rouge 2017). Mobility reduction factor: MRF. See more details in the work by (Maire and Fatin-Rouge 2017). (Reprinted with permission from (Maire and Fatin-Rouge 2017). Copyright (2017) Elsevier)

where V_l and V_g are the volume of the liquid and gas components of the foam, respectively. To quantify the foam quality, a known volume of foam sample is collected. Then, methanol is used as a defoaming agent for each foam sample (Shen et al. 2011; Zhong et al. 2010). The volumes of the defoamed solutions are measured using a graduated cylinder to determine V_l. The difference between the total foam volume and the liquid volume (V_l) is the gas volume (V_g). The amount of NZVI in the liquid phase of each F-NZVI sample can be quantified by separating NZVI out of the defoamed solution using magnetic separation followed by weighting of the separated NZVI after drying in an oven at 105 °C for around 2 h. Foam stability is defined as the time required by F-NZVI to reach half of its initial volume. The longer the time, the greater the stability of F-NZVI. The bubble size distribution of F-NZVI is also another foam property that is easily measured by a light microscope with appropriate magnification.

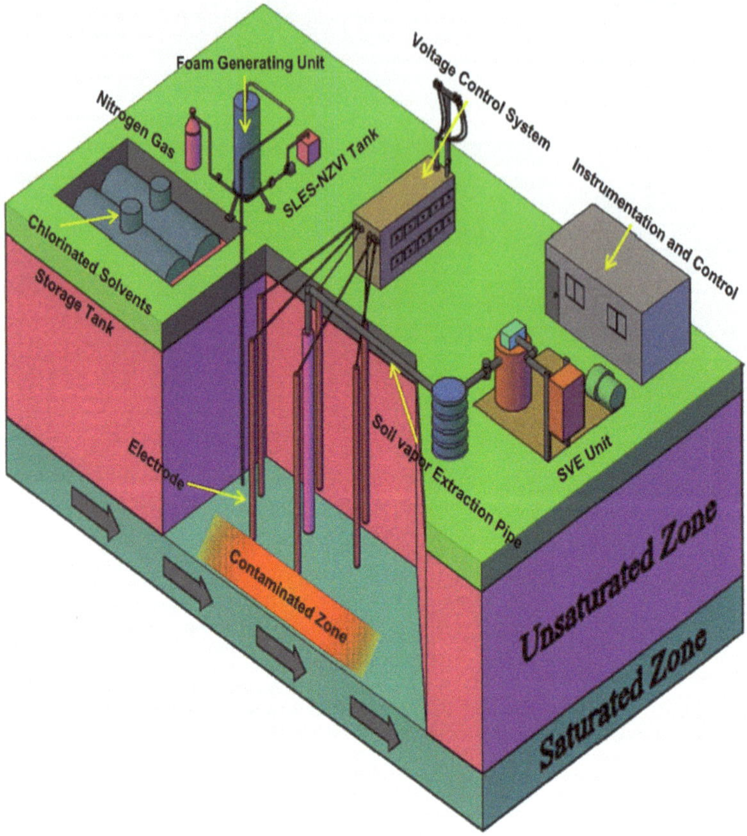

Fig. 13.3 Soil vapor extraction (SVE) with thermally enhanced volatilization by foam-based NZVI with LF EMF

The type of surfactant greatly affects F-NZVI characteristics. We evaluated the ability of five commercially available surfactants including PEG-20 sorbitan monostearate (Tween 60), sorbitan monooleate (Span 80), Vertex Type 1 (proprietary foaming agent for lightweight cellular concrete), Vertex Type 2 (proprietary foaming agent for lightweight cellular concrete), and SLES to stabilize F-NZVI. We found that only three types of surfactant including SLES, Vertex Type 1, and Vertex Type 2 successfully generated foam and F-NZVI, while Span 80 and Tween 60 appeared to be too sparingly soluble in water to assist foam formation (Srirattana et al. 2017). The quality of foam was very high (99.60–99.69%) both with and without NZVI for all three surfactants that produced foam. At 1% (w/w) surfactant without NZVI and a N_2 flow rate of 125 mL/min, SLES appeared to be the most stable (foam half-life ($t_{1/2}$) = 143 min) followed by Vertex Type 1 ($t_{1/2}$ = 130 min) and Vertex Type 2 ($t_{1/2}$ = 56 min) (Fig. 13.5a). The bubble size distribution of foam

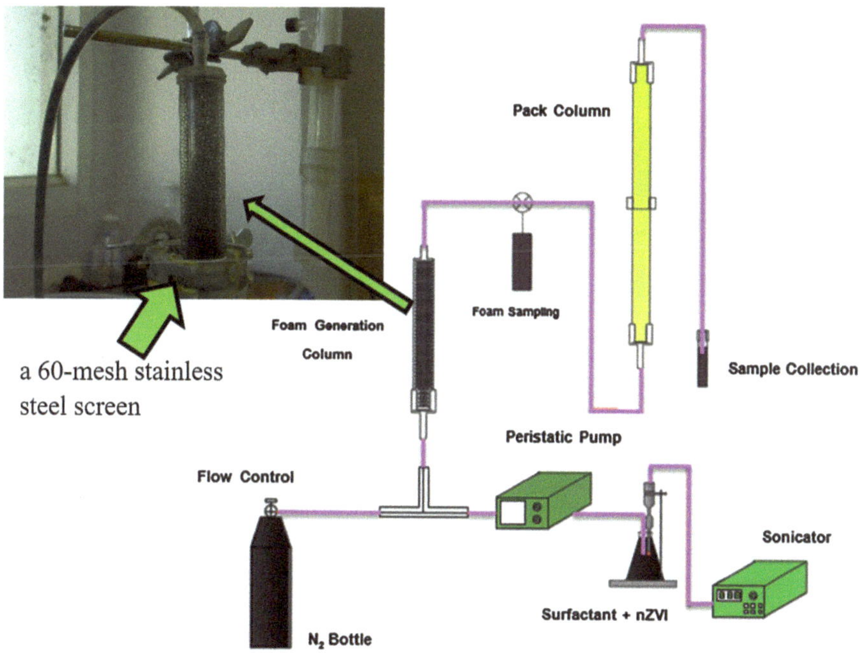

Fig. 13.4 F-NZVI generation unit connected to an unsaturated sand-packed column for F-NZVI delivery

without NZVI was from 5 to 35 μm (Fig. 13.5b). Most of the bubble sizes for foam stabilized by Vertex Type 2 (see Fig. 13.5c) and SLES were 6–10 μm, while most of the bubble sizes for foam stabilized by Vertex Type 1 were slightly bigger (i.e., 11–15 μm). Interestingly, adding NZVI into the foam did not seriously alter the foam stability for both SLES-F-NZVI (NZVI foam stabilized by SLES) ($t_{1/2} = 140$ min) and Vertex Type 1 F-NZVI ($t_{1/2} = 131$ min), while NZVI substantially enhanced the stability of the foam stabilized by Vertex Type 2, for which $t_{1/2}$ became 112 min (Fig. 13.5a), lasting twice as long as without NZVI. This similar synergetic effect of nanoparticles (silica nanoparticles) and surfactant in foam stabilization has previously been reported (Binks et al. 2008; Lv et al. 2015). Figure 13.5d illustrates a microscopic demonstration of surfactant-modified NZVI adsorption into the foam bubble. Noticeably, black aggregates of surface-modified NZVI are located on the contact line of the two phases (gas and liquid). The accumulation of surfactant-modified NZVI on the foam surface is similar to interfacial targeting of the NAPL source zone by amphiphilic polymer-modified NZVI in saturated porous media reported in recent studies (Phenrat et al. 2011; Saleh et al. 2005). The film of attached surfactant-modified NZVI on the foam bubble may increase the surface elasticity of the foam, reduce the coarsening tendency, and enhance foam stability (Binks et al.

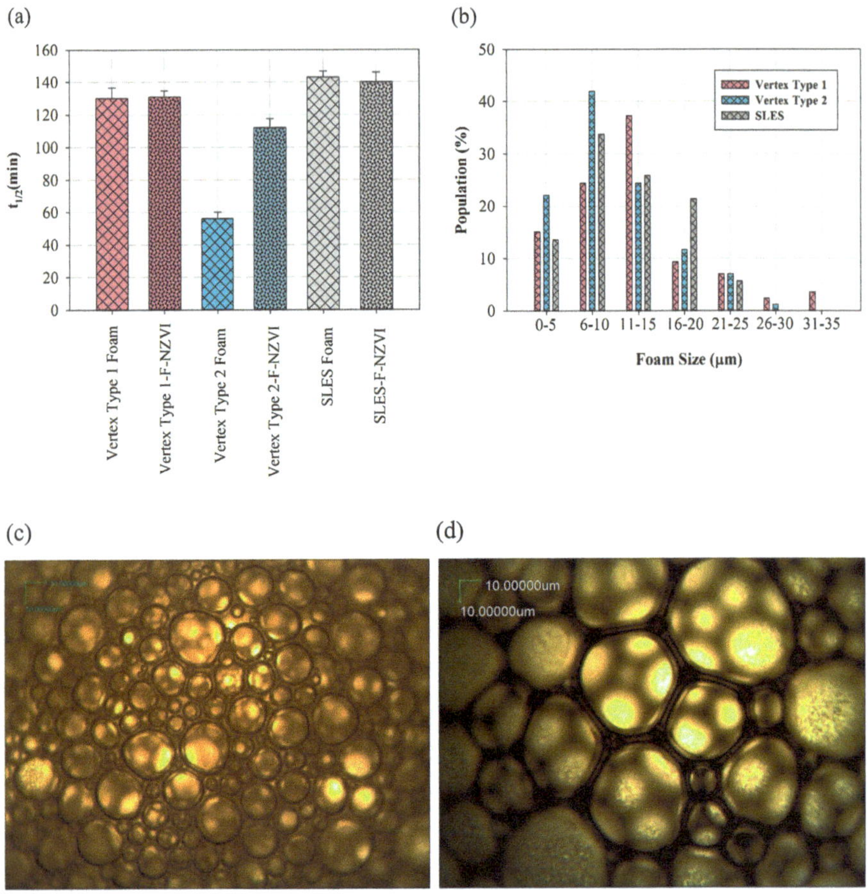

Fig. 13.5 (**a**) Half-life ($t_{1/2}$) of foam and F-NZVI formed by three different surfactants, (**b**) bubble size distribution of foams made from three different surfactants, and micrograph of (**c**) Vertex Type 2 foam and (**d**) Vertex Type 2-F-NZVI

2008). This could explain the increased stability of Vertex Type 2-F-NZVI, mentioned previously.

Surfactant concentration and N_2 flow rate also play a substantial role on F-NZVI stability (Srirattana et al. 2017). As shown in Fig. 13.6a, at an N_2 flow rate of 125 mL/min, SLES-F-NZVI at a surfactant concentration of 3% (w/w) performed the best in terms of foam stability ($t_{1/2} = 173$ min) among the SLES concentration from 1 to 9% (w/w). Similarly, as shown in Fig. 13.6b, at SLES concentration of 3% (w/w), the greater the N_2 flow rate, the greater the concentration of NZVI in foam. At a N_2 flow rate of 500 mL/min, SLES-F-NZVI contained NZVI as much as 33.5 ± 3.7 g/L when started with an initial concentration of 50 g/L. For SLES-F-NZVI at 3% (w/w) SLES and 500 mL/min N_2 on NZVI, the carrying capacity was 67% of the NZVI stock solution.

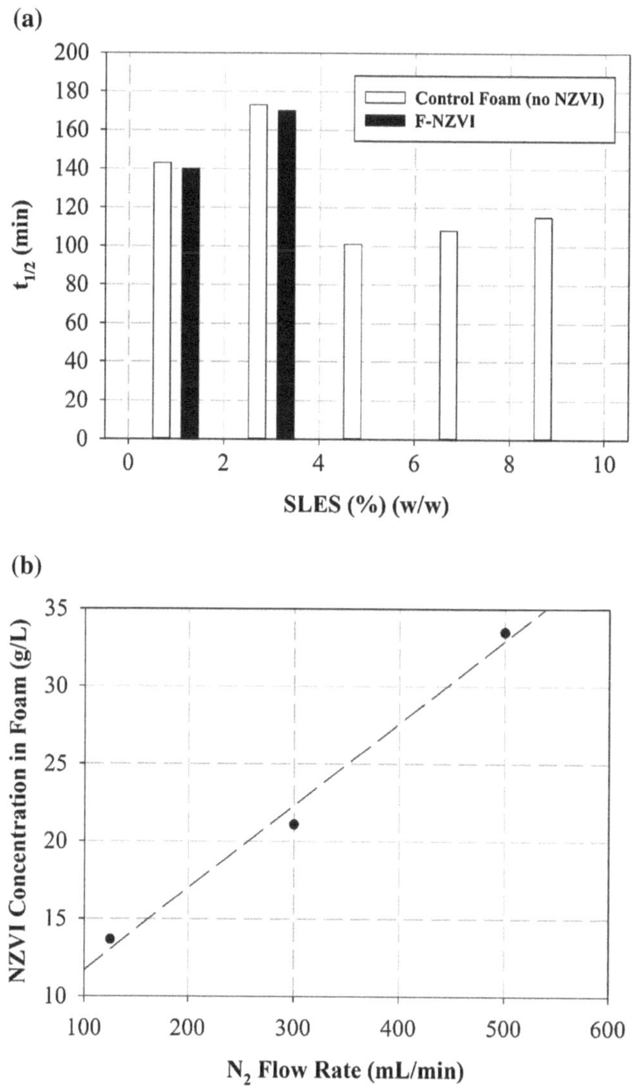

Fig. 13.6 (**a**) Effect of SLES concentration on foam stability and (**b**) effect of N_2 flow rate on capability to carry NZVI in foam (the initial concentration of NZVI in stock solution = 50 g/L)

13.4 Delivery of F-NZVI in Unsaturated Porous Media

As mentioned earlier, F-NZVI is supposed to perform two remediation tasks, first by flushing CVOC from the soil and second by leaving behind NZVI for electromagnetic induction heating that will thermally enhance SVE. Thus, understanding

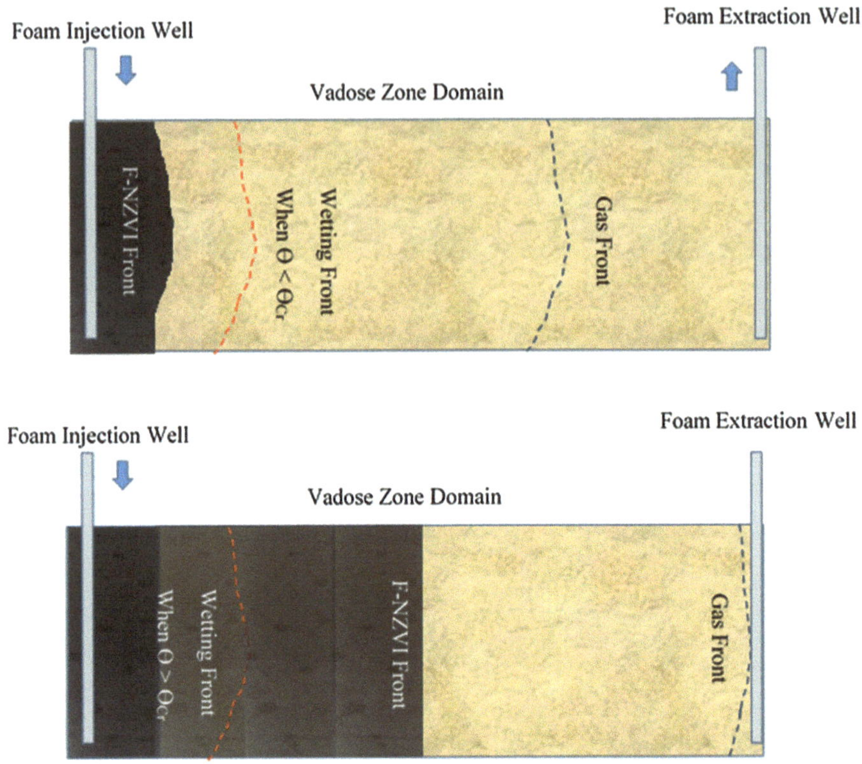

Fig. 13.7 Schematics of F-NZVI transport in the vadose zone at two different moisture contents (θ)

F-NZVI delivery and resulting NZVI emplacement is critical for the design of efficient F-NZVI injection in a vadose zone. Figure 13.7 illustrates the F-NZVI transport as well as the concept of the separation of foam, wetting, and gas fronts as well as the NZVI deposition during foam transport in the vadose zone (Srirattana et al. 2017; Zhong et al. 2010).

Initially, foam bubbles rupture at the foam flow front in the unsaturated soil, releasing the foaming gas and depositing liquid as well as NZVI into the sediment. Bubble rupture is inevitable when foam is initially injected into a porous medium with relatively low water saturation because the initial capillary pressure in the medium is much larger than the critical capillary pressure of foam destruction, $P_{\text{cr-foam}}$, which is defined as the limiting capillary pressure at which foam bubbles can flow through a porous media (Khristov et al. 2002).

By continuing to deliver F-NZVI, the wetting front moves forward, and the length of the wetted soil increases. The liquid content in the wetted soil increases with time until a certain saturation is reached, and the flow of foam bubbles is initiated at the upstream end of the wetted sediment section. This is when the liquid saturation is

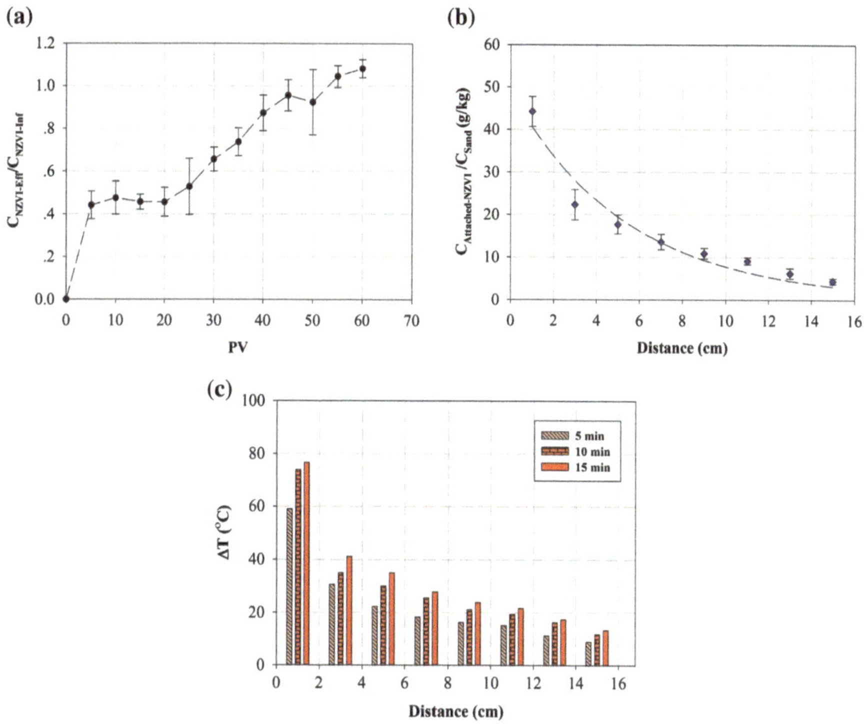

Fig. 13.8 (**a**) Breakthrough curve of SLES-F-NZVI over 60 PVs delivered through an unsaturated sand-packed column, (**b**) concentration of deposited NZVI in unsaturated porous media as a function of distance from the inlet, and (**c**) the ΔT for SLES-F-NZVI deposited onto unsaturated sand at each distance from the inlet according to the NZVI emplacement profile in (**b**) (Srirattana et al. 2017). (Reprinted with permission from (Srirattana et al. 2017). Copyright (2017) Elsevier)

built up, and the surfactant concentration in the liquid is increased. Consequently, the capillary pressure of the porous medium decreases to below $P_{\text{cr-foam}}$ so that foam bubbles may stay unruptured or even be regenerated from the pore liquid and be transported. This moisture content of soil is called the critical moisture content (θ_{cr}) (Saiers and Lenhart 2003).

This foam transport phenomena can be illustrated through a breakthrough curve of F-NZVI shown in Fig. 13.8a (Srirattana et al. 2017). This breakthrough curve was generated at 50 g/L NZVI stock solution in 3% (w/w) SLES at a 500 mL/min N_2 flow rate. The liquid portion of the foam (0.17%) carried 41.3 g/L of NZVI. The transport experiments were performed using a cylindrical acrylic column, 16 cm in length and 2 cm in inner diameter, packed with quartz sand with an average size of 0.85 mm. Sixty pore volumes (PVs) of F-NZVI was delivered into the unsaturated sand-packed column. Notably, from the fifth PV to the 20th PV, the effluent NZVI concentration seemed to reach a steady state at around 45% of the influent concentration. The unsaturated medium filtered around half of the NZVI carried by the foam. This finding is in good agreement with a recent study that reported that NZVI (3 g/L) carried by foam generated by SLES (0.5% (w/w)) was filtered at approximately 20% by an unsaturated sand-packed column (grain diameter of 0.8–1.25 mm) (Ding et al. 2013). Presumably, Srirattana's study generated foam at a higher NZVI concentration than Ding's study, which caused more agglomeration of NZVI in the foam and thus was more easily filtered out by the unsaturated porous medium. Nevertheless, from the 20th to 60th PV, the effluent NZVI concentration gradually increased to 100% through the unsaturated packed bed. Presumably, after the 20th PV, the moisture content in unsaturated porous media exceeded the critical moisture content (θ_{cr}) where immobilized NZVI started to be released at a rate proportional to the product of the pore water velocity and the attached particle concentration as discussed earlier (Saiers and Lenhart 2003). Nevertheless, it is worth highlighting that, typically, the critical moisture content is from 10% to 20% of the PV (Saiers and Lenhart 2003). Consequently, the accumulation of liquid to reach the critical moisture content at the location close to the foam injection point should not significantly result in any unintended vertical percolation of DNAPL to the underlying groundwater (Jeong et al. 2000).

Figure 13.8b illustrates the NZVI attached to the sand in the unsaturated packed column after 60 PVs of F-NZVI containing 100 g/L NZVI were flushed through the column. The concentration of NZVI deposited on sand decreased exponentially with the distance from the inlet, in good agreement with typical infiltration behavior of colloidal and nanoparticles in an unsaturated porous medium (Ding et al. 2013). The NZVI emplacement in an unsaturated porous medium to achieve high particle concentrations is desirable because the greater the emplaced NZVI, the more effective the thermally enhanced evaporation of TCE promoted by NZVI under LF EMF. In summary, F-NZVI can be delivered and can leave behind a significant amount of NZVI for thermally enhanced remediation.

13.5 Magnetic Induction Heating of F-NZVI

The F-NZVI can induce heat under LF EMF. The NZVI is predominantly Fe^0 and Fe_3O_4, both of which are magnetic, as discussed in Chap. 11. Noticeably, upon the magnetization and demagnetization cycle, NZVI responded irreversibly, causing hysteresis and loss of energy as heat. The degree of irreversibility, ΔU, is related to the amount of energy dissipation upon the reversal of the field. Based on Fig. 11.2 (Chap. 11), the ΔU for NZVI is 6.7×10^4 emu G/g. Under applied LF EMF at a current density of 13 A and a frequency of 150 kHz, the DI, SLES solution, and SLES foam control samples without NZVI could not generate any heat. However, DI and SLES solutions with NZVI as well as F-NZVI generated heat and substantially raised the temperature of the suspension, as expected (Fig. 13.9). The increase in temperature was proportional to the amount of deposited NZVI. Their heat induction rate constants (k_{HI}) (determined from Eq. 13.9) and maximum induced temperature

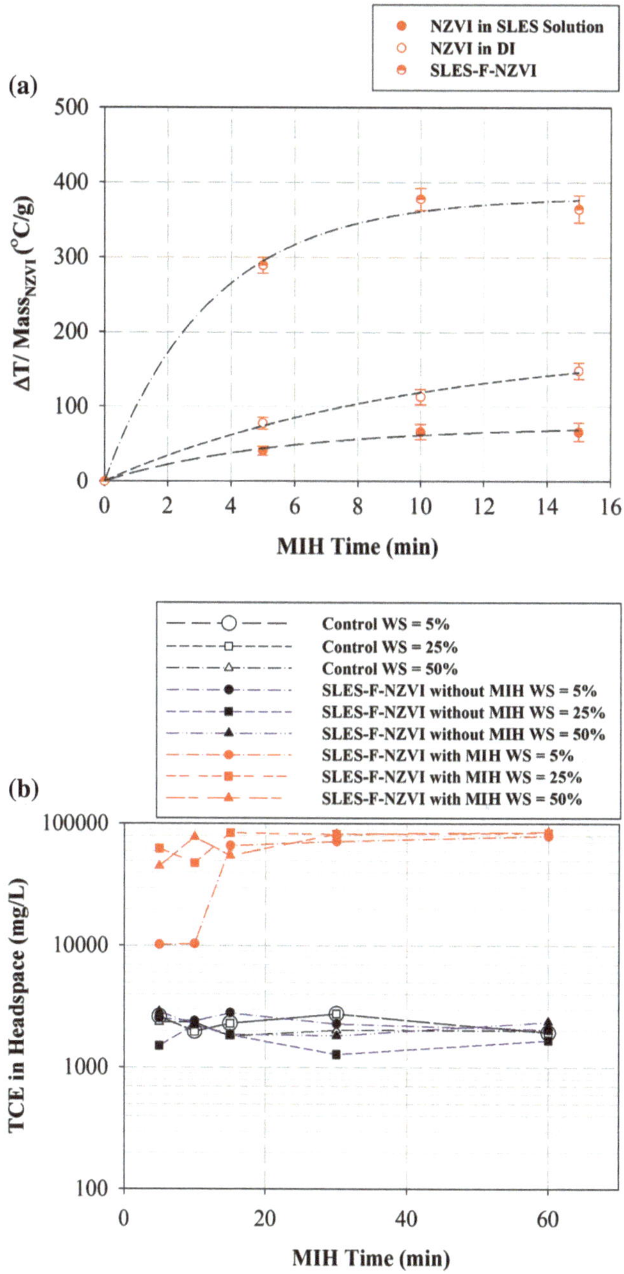

Fig. 13.9 (**a**) MIH kinetics of SLES-F-NZVI in comparison to NZVI in DI water and SLES concentration (3% (w/w)) and (**b**) kinetics of TCE partitioning to headspace with and without applied LF EMF and as a function of MIH time (Srirattana et al. 2017). (Reprinted with permission from (Srirattana et al. 2017). Copyright (2017) Elsevier)

increases (ΔT_{max}/$Mass_{NZVI}$) significantly depended on their vehicles of delivery (foam-based or aqueous-based vehicles):

$$\frac{\Delta T}{Mass_{NZVI}} = \frac{\Delta T_{Max}}{Mass_{NZVI}}\left(1 - e^{-k_{HI}t}\right). \quad (13.9)$$

Interestingly, Srirattana et al. (2017) found that SLES-F-NZVI ($K_{HI} = 29.94 \times 10^{-2}$ min^{-1} and ΔT_{max}/$Mass_{NZVI}$ = 380 °C/g) was the fastest in terms of heating kinetics, followed by NZVI in DI water (K_{HI} = 10.14 × 10^{-2} min^{-1} and ΔT_{max}/ $Mass_{NZVI}$ = 186 °C/g) and NZVI in SLES solution (K_{HI} = 17.88 × 10^{-2} min^{-1} and ΔT_{max}/$Mass_{NZVI}$ = 74 °C/g). The difference between the heat induction rate constants (K_{HI}) and ΔT_{max}/$Mass_{NZVI}$ generated by NZVI in the three different vehicles may be attributed to the specific heat capacity (C_P) of each carrying medium. Since the specific heat capacity of air (1.012 J/g/C at 25 °C) is lower than the specific heat capacity of DI water (4.1813 J/g/C at 25 °C) by a factor of around 4, it is comprehensible that the heat induction rate constant (K_{HI}) of SLES-F-NZVI was 3 times greater than that of the NZVI in DI water (Srirattana et al. 2017).

As expected, SLES-F-NZVI emplaced onto unsaturated sand could also generate heat and raise the temperature under an applied LF EMF, but the increase of temperature was lower than that of free SLES-F-NZVI based on the same foam-generating conditions. For example, Fig. 13.8c shows the ΔT for SLES-F-NZVI deposited into unsaturated sand at each distance from the inlet. The maximum NZVI deposition into the sand was 77 g/kg at 0–2 cm from the inlet. With this NZVI emplacement profile in each segment, the induced ΔT substantially increased with time from 5 to 15 min (Fig. 13.8c). Intuitively, the greater the emplaced NZVI, the higher the increased ΔT at a particular MIH time. Promisingly, the ΔT values were 41 °C and 77 °C for emplaced NZVI concentrations of 22 g/kg and 45 g/kg for the unsaturated sand segment of 2–4 cm and 0–2 cm, respectively, after 15 min of MIH. This finding suggests that, to make SLES-F-NZVI a viable alternative for an induced thermal remediation agent for the vadose zone, a significant amount of NZVI, at least at a concentration of 22 g/kg to 44 g/kg, must be emplaced in the unsaturated porous media. This could be accomplished easily by delivering F-NZVI using NZVI dispersion at the initial concentration of 100 g/L (Srirattana et al. 2017).

13.6 Enhanced TCE Evaporation by F-NZVI Under LF EMF

The second goal of using F-NZVI in the vadose zone is to thermally enhance SVE performance. Thus, enhanced TCE evaporation by F-NZVI under LF EMF is an effective key indicator of the concept. Srirattana et al. (2017) did a simple batch test of enhanced TCE evaporation. They used a 25-ml glass vial containing 2 mL of SLES-F-NZVI (generated using 100 g/L of NZVI in 3% (w/w) SLES and 500 mL/

min of N_2) and 17.20 g of quartz sand with three different water saturations (WS) (5%, 25%, and 50%) and 0.25 mL (or 3.5% saturation) of pure-phase TCE promptly capped with Teflon Mininert™. Control reactors including TCE in sand without SLES-F-NZVI were made under the same setup with the same three WS. The vial was insulated and placed into the center of the induction coil of the EMF generator at 13 A and 150 kHz for MIH study for 5, 10, 15, 30, and 60 min. The TCE evaporation due to SLES-F-NZVI coupled with MIH was quantified by gas measurement using gas chromatography.

The control reactors (i.e., unsaturated sand with WS = 5%, 25%, and 50% and TCE (pure-phase) saturation = 3.5% without SLES-F-NZVI) and the reactors with SLES-F-NZVI but without MIH cannot generate any heat, and the temperature remains at 25 °C for 60 min of the study. However, under an applied LF EMF, deposited SLES-F-NZVI generated heat to achieve a temperature from 95 °C to 110 °C (ΔT from 70 °C to 85 °C) in 15 min. The amount of WS in the unsaturated porous medium did not obviously affect the heating kinetics of the SLES-F-NZVI reactors in LF EMF. Thus, the deposited SLES-F-NZVI concentration is the main factor controlling the heating. As theoretically suggested, the heat generated by SLES-F-NZVI under LF EMF substantially enhanced TCE evaporation in batch reactors. As shown in Fig. 13.9b (presented in log scale), the TCE concentrations in the headspace of the control reactors and SLES-F-NZVI reactors without LF EMF application were relatively constant over 60 min of the study (i.e., 2236 ± 329 mg/L and 2114 ± 465 mg/L, respectively). The degree of WS did not appear to play any obvious role in TCE evaporation in these reactors. After around 15 min of MIH, the TCE concentration in the headspace of all MIH reactors reached a similar maximum concentration of 81,504 ± 4751 mg/L, regardless of the degree of WS. Consequently, the MIH of SLES-F-NZVI deposited in unsaturated porous media at the NZVI concentration of 45 g/kg increased TCE evaporation by 39.5 ± 6.6 times in comparison to the reactors without MIH. This proof-of-concept experiment strengthens the VOC remedial feasibility using SVE thermally enhanced by F-NZVI and LF EMF.

13.7 Comparative Analysis of the Theoretical Performance of MIH with F-NZVI and RFH Without F-NZVI

The benefit of F-NZVI with LF EMF may be more pronounced once we theoretically compare the performance of LF EMF with F-NZVI to that of LF electromagnetic heating and RFH without F-NZVI. Srirattana et al. (2017) performed this task in their recent study. Both LF electromagnetic heating and RFH are forms of electromagnetic heating; the only difference is their frequencies. For the application of F-NZVI with LF EMF in the unsaturated porous medium discussed previously, the ΔT was 77 °C. They revealed that, given that the specific heat capacity of air is 1.006 kJ/kg K, air density is 1.184 kg/m^3, the reactor volume was 1.5×10^{-5} m^3, and the heating time was 5 min, the theoretical amount of power dissipated by MIH (P_{MIH}) of F-NZVI was estimated to be 4.6×10^{-3} W.

However, under LF EMF of 150 kHz and $I = 13$ A, causing $|E|$ to be 79.9 V/m, even without F-NZVI, EMF would interact with the sand grain to create heat via conduction and dielectric loss, but not magnetic loss. They estimated the theoretical amount of power dissipated by electromagnetic heating, P_{EM} (W) without F-NZVI, using the following:

$$P_{EM} = V 2\pi f \varepsilon_0 \varepsilon''_{eff} |E|^2, \tag{13.10}$$

where V is the volume of the reactor, $|E|$ is the magnitude of the applied electric field (V/m), f is the irradiated frequency (1/s), ε_0 is the permittivity of free space (8.85×10^{-12} F/m), and ε''_{eff} is an effective imaginary permittivity (unitless) that is a function of f (Revil 2013). Given that $|E| = 79.9$ V/m, as measured from LF EMF, $\varepsilon''_{eff} = 39.8$ for $f = 150$ kHz (Revil 2013), and the reactor volume was 1.5×10^{-5} m^3 ; therefore, the irradiated frequency of 150 kHz yielded $P_{EM} = 3.18 \times 10^{-5}$ W. Thus, using F-NZVI with LF EMF can induce heat 145-fold more efficiently than that via LF EMF without F-NZVI.

Similarly, they theoretically calculated the frequency of RFH necessary to produce the same theoretical amount of power dissipation obtained from P_{MIH}. The RFH is a form of electromagnetic heating that can be estimated using Eq. 13.10 (referred to as the theoretical amount of power dissipated by RFH (P_{RFH})). Given $\varepsilon''_{eff} = 15.9$ for $f = 54.5$ MHz (Revil 2013), the irradiated frequency of RFH to yield $P_{RFH} = 4.6 \times 10^{-3}$ W, equivalent to the P_{MIH}, is 54.5 MHz. This is in line with the irradiation frequency typically used in RFH remediation (500 kHz to 500 MHz (Lowe et al. 1999, Price et al. 1999)). Notably, Srirattana et al. (2017) reported that the irradiated frequency of RFH was approximately 360-fold greater than that required by LF EMF when applied with F-NZVI. They also pointed out that using F-NZVI with LF EMF could theoretically be an alternative to RFH because it does not require as high of an irradiation frequency as RFH and should still lead to both lower capital and operational costs versus RFH (Vermeulen and McGee 2000).

Using F-NZVI for this thermally enhanced remediation requires an extra cost for the NZVI in comparison to LF electromagnetic heating or RFH. However, the additional material cost is in the same range as using NZVI for saturated zone remediation. Srirattana et al. (2017) estimated that for 60 PVs of F-NZVI delivery to deposit enough NZVI to generate heat under LF EMF to 77 °C, given a porosity of 0.33 and foam quality of 99.84% (i.e., 99.84% air and 0.16% NZVI dispersion), only 3.16 kg of NZVI is necessary. This concurs with the NZVI mass required for saturated zone remediation. In general, the NZVI concentration emplaced for saturated zone remediation is from 1 to 20 g/L (Phenrat et al. 2010a). Thus, for one PV injection (1 cubic meter), the mass for NZVI injection is between 0.33 and 6.6 kg (for a NZVI concentration of 1 to 20 g/L). As a result, the material cost of F-NZVI for this application should not be prohibitive.

Acknowledgments The authors are thankful for the research funding from (1) the Thailand Research Fund (TRF) (MRG5680129); (2) the National Nanotechnology Center (Thailand), a member of the National Science and Technology Development Agency, through grant number P-11-00989; and (3) the National Research Council (R2556B070).

References

Abreu, L. D. V., & Johnson, P. C. (2005). Effect of vapor source–building separation and building construction on soil vapor intrusion as studied with a three-dimensional numerical model. *Environmental Science & Technology, 39*(12), 4550–4561.

Babakhani, P., Fagerlund, F., Shamsai, A., Lowry, G. V., & Phenrat, T. (2015). Modified MODFLOW-based model for simulating the agglomeration and transport of polymer-modified Fe0 nanoparticles in saturated porous media. *Environmental Science and Pollution Research, 25* (8), 7180–7199.

Bañobre-López, M., Teijeiro, A., & Rivas, J. (2013). Magnetic nanoparticle-based hyperthermia for cancer treatment. *Reports of Practical Oncology and Radiotherapy, 18*(6), 397–400.

Barbee, G. C. (2007). Fate of chlorinated aliphatic hydrocarbons in the vadose zone and ground water. *Ground Water Monitoring and Remediation, 14*(1), 129–140.

Bardos, P., Bone, B., Černík, M., Elliott, D. W., Jones, S., & Merly, C. (2015). Nanoremediation and international environmental restoration markets. *Remediation Journal, 25*(2), 83–94.

Bera, A., & Babadagli, T. (2015). Status of electromagnetic heating for enhanced heavy oil/bitumen recovery and future prospects: A review. *Applied Energy, 151*, 206–226.

Beyke, G., & Fleming, D. (2005). In situ thermal remediation of DNAPL and LNAPL using electrical resistance heating. *Remediation Journal, Summer*, 5–22.

Bientinesi, M., Scali, C., & Petarca, L. (2015). *Radio frequency heating for oil recovery and soil remediation* (pp. 1199–1204). Whistler, BC, Canada: The International Federation of Automatic Control.

Binks, B. P., Kirkland, M., & Rodrigues, J. A. (2008). Origin of stabilisation of aqueous foams in nanoparticle–surfactant mixtures. *Soft Matter, 4*, 2373–2382.

Boulding, J. R., & Ginn, J. S. (2003). *Practical handbook of soil, vadose zone, and ground-water contamination: Assessment, prevention, and remediation*. New York: CRC Press.

Brusseau, M. L., Carroll, K. C., Truex, M. J., & Becker, D. J. (2013). Characterization and remediation of chlorinated volatile organic contaminants in the vadose zone. *Vadose Zone Journal, 12*(4). https://doi.org/10.2136/vzj2012.0137.

Chen, M. Y., Su, Y. F., & Shih, Y. H. (2014). Effect of geochemical properties on degradation of trichloroethylene by stabilized zerovalent iron nanoparticle with Na-acrylic copolymer. *Journal of Environmental Management, 144*, 88–92.

Ding, Y., Liu, B., Shen, X., Zhong, L., & Li, X. (2013). Foam-assisted delivery of nanoscale zero valent iron in porous media. *Journal of Environmental Engineering, 139*(9), 1206–1212.

Fagerlund, F., Illanagasekare, T., & Niemi, A. (2007a). Nonaqueous-phase liquid infiltration and immobilization in heterogeneous media: 2. Application to stochastically heterogeneous formations. *Vadose Zone Journal, 6*(3), 483–495.

Fagerlund, F., Illangasekare, T., & Niemi, A. (2007b). Nonaqueous-phase liquid infiltration and immobilization in heterogeneous media: 1. Experimental methods and two-layered reference case. *Vadose Zone Journal, 6*(3), 471–482.

Fagerlund, F., Illangasekare, T. H., Phenrat, T., Kim, H.-J., & Lowry, G. V. (2012). PCE dissolution and simultaneous dechlorination by nanoscale zero-valent iron particles in a DNAPL source zone. *Journal of Contaminant Hydrology, 131*(1–4), 9–28.

Falciglia, P. P., Urso, G., & Vagliasindi, F. G. A. (2013). Microwave heating remediation of soils contaminated with diesel fuel. *Journal of Soils and Sediments, 13*, 1396–1407.

Gavaskar, A., Bhargava, M., & Condit, W. (2007). *Cost and performance review of Electrical Resistance Heating (ERH) for source treatment* (p. 121). Port Hueneme, CA: Naval Facilities Engineering Service Center.

Heron, G., Van Zutphen, M., Christensen, T. H., & Enfield, G. C. (1998). Soil heating for enhanced remediation of chlorinated solvents: A laboratory study on resistive heating and vapor extraction in a silty, low-permeable soil contaminated with trichloroethylene. *Environmental Science & Technology, 32*(10), 1474–1481.

Hirasaki, G. J., Miller, C. A., Szafranski, R., Lawson, J. B., Tanzil, D., Jackson, R. E., Londergan, J., & Meinardus, H. (1997). *Field demonstration of the surfactant/foam process for aquifer remediation* (p. SPE 39393). San Antonio, TX: Society of Petroleum Engineers. https://www.onepetro.org/conference-paper/SPE-39292-MS

Huang, C.-W., & Chang, C.-H. (2000). A laboratory study on foam-enhanced surfactant solution flooding in removing n-pentadecane from contaminated columns. *Colloids and Surfaces A: Physicochemical and Engineering Aspects, 173*(1–3), 171–179.

Jackson, P. D., Williams, J. F., Lovell, M. A., Camps, A., Rochelle, C., & Milodowski, A. E. (2008). *An investigation of the exponent in Archie's equation: Comparing numerical modeling with laboratory data: Towards characterising disturbed samples from the Cascadia margin:-Iodp expedition 311* (pp. 1–11). Austin, TX: Society of Petrophysicists and Well-Log Analysts.

Jeong, S.-W., Corapcioglu, M. Y., & Roosevelt, E. W. (2000). Micromodel study of surfactant foam remediation of residual trichloroethylene. *Environmental Science & Technology, 34*(16), 3456–3461.

Karn, B., Kuiken, T., & Otto, M. (2009). Nanotechnology and in situ remediation: A review of the benefits and potential risks. *Environmental Health Perspectives, 117*(12), 1823–1831.

Khristov, K., Exerowa, D., & Minkov, G. (2002). Critical capillary pressure for destruction of single foam films and foam: Effect of foam film size. *Colloids and Surfaces A: Physicochemical and Engineering Aspects, 210*, 159–166.

Kilbane, J. J., II, Chowdiah, P., Kayser, K. J., Misra, B., Jackowski, K. A., Srivastava, V. J., Sethu, G. N., Nikolov, A. D., Wasan, D. T., & Hayes, T. D. (1997). Remediation of contaminated soils using foams. *Land Contamination & Reclamation, 5*, 41–54.

Kommalapati, R. R., Valsaraj, K. T., Constant, W. D., & Roy, D. (1998). Soil flushing using colloid gas Aphron suspensions generated from a plant-based surfactant. *Journal of Hazardous Materials, 60*, 73–87.

Li, Z., Kawashita, M., Araki, N., Mitsumori, M., Hiraoka, M., & Doi, M. (2010). Magnetite nanoparticles with high heating efficiencies for application in the hyperthermia of cancer. *Materials Science and Engineering: C, 30*(7), 990–996.

Lim, J. K., Tilton, R. D., Eggman, A., & Majetich, S. A. (2007). Design and synthesis of plasmonic magnetic nanoparticles. *Journal of Magnetism and Magnetic Materials, 311*, 78–83.

Liu, Y., Phenrat, T., & Lowry, G. V. (2007). Effect of TCE concentration and dissolved groundwater solutes on NZVI-promoted TCE dechlorination and H2 evolution. *Environmental Science & Technology, 41*(22), 7881–7887.

Lowe, D. F., Oubre, C. L., & Ward, C. H. (1999). *Soil vapor extraction using radio frequency heating: Resource manual and technology demonstration*. New York: CRC Press.

Lv, Q., Li, Z., Li, B., Li, S., & Sun, Q. (2015). Study of nanoparticle–surfactant-stabilized foam as a fracturing fluid. *Industrial and Engineering Chemistry Research, 54*(38), 9468–9477.

Lv, C., Chen, J., & Wang, X. (2017). Evaluation of surfactant performance in *in situ* foam flushing for remediation of dichlorodiphenyltrichloroethane-contaminated soil. *International journal of Environmental Science and Technology, 14*, 631–638.

Maire, J., & Fatin-Rouge, N. (2017). Surfactant foam flushing for *in situ* removal of DNAPLs in shallow soils. *Journal of Hazardous Materials, 321*, 247–255.

Mueller, N. C., Braun, J., Bruns, J., Černík, M., Rissing, P., Rickerby, D., & Nowack, B. (2012). Application of nanoscale zero valent iron (NZVI) for groundwater remediation in Europe. *Environmental Science and Pollution Research, 19*, 550–558.

Mulligan, C. N., & Eftekhari, F. (2003). Remediation with surfactant foam of PCP-contaminated soil. *Engineering Geology, 70*(3–4), 269–279.

Nakamura, T., Senior, C. L., Burns, E. G., & Bell, M. D. (2000). Solar-powered soil vapor extraction for removal of dense nonaqueous phase organics from soil. *Journal of Environmental Science and Health. Part A, Toxic/Hazardous Substances & Environmental Engineering, 35*(6), 795–816.

Pablico-Lansigan, M. H., Situa, S. F., & Samia, A. C. S. (2013). Magnetic particle imaging: Advancements and perspectives for real-time in vivo monitoring and image-guided therapy. *Nanoscale, 5*, 4040–4055.

Phenrat, T., Liu, Y., Tilton, R. D., & Lowry, G. V. (2009). Adsorbed polyelectrolyte coatings decrease Fe0 nanoparticle reactivity with TCE in water: Conceptual model and mechanisms. *Environmental Science & Technology, 43*(5), 1507–1514.

Phenrat, T., Kim, H.-J., Fagerlund, F., Illangasekare, T., & Lowry, G. V. (2010a). Empirical correlations to estimate agglomerate size and deposition during injection of a polyelectrolyte-modified Fe0 nanoparticle at high particle concentration in saturated sand. *Journal of Contaminant Hydrology, 118*(3–4), 152–164.

Phenrat, T., Kim, H.-J., Fagerlund, F., Illangasekare, T., & Lowry, G. V. (2010b). Empirical correlations to estimate agglomerate size and deposition during injection of a polyelectrolyte-modified Fe0 nanoparticle at high particle concentration in saturated sand. *Journal of Contaminant Hydrology, 118*(3–4), 152–164.

Phenrat, T., Song, J. E., Cisneros, C. M., Schoenfelder, D. P., Tilton, R. D., & Lowry, G. V. (2010c). Estimating attachment of nano- and submicrometer-particles coated with organic macromolecules in porous media: Development of an empirical model. *Environmental Science & Technology, 44*(12), 4531–4538.

Phenrat, T., Fagerlund, F., Illanagasekare, T., Lowry, G. V., & Tilton, R. D. (2011). Polymer-modified Fe0 nanoparticles target entrapped NAPL in two dimensional porous media: Effect of particle concentration, NAPL saturation, and injection strategy. *Environmental Science & Technology, 45*(14), 6102–6109.

Phenrat, T., Schoenfelder, D., Kirschling, T. L., Tilton, R. D., & Lowry, G. V. (2018). Adsorbed poly(aspartate) coating limits the adverse effects of dissolved groundwater solutes on Fe^0 nanoparticle reactivity with trichloroethylene. *Environmental Science & Pollution Research, 25*(8), 7157–7169.

Poppendieck, D. G., Loehr, R. C., & Webster, M. T. (1999a). Predicting hydrocarbon removal from thermally enhanced soil vapor extraction systems. 1. Laboratory studies. *Journal of Hazardous Materials, 69*(1), 81–93.

Poppendieck, D. G., Loehr, R. C., & Webster, M. T. (1999b). Predicting hydrocarbon removal from thermally enhanced soil vapor extraction systems: 2. Field study. *Journal of Hazardous Materials, 69*(11), 95–109.

Price, S. L., Kasevich, R. S., Johnson, M. A., Wiberg, D., & Marley, M. C. (1999). Radio frequency heating for soil remediation. *Journal of the Air & Waste Management Association, 49*(2), 136–145.

Revil, A. (2013). Effective conductivity and permittivity of unsaturated porous materials in the frequency range 1 mHz–1GHz. *Water Resources Research, 49*, 306–327.

Rizzoni, G. (2004). *Principles and applications of electrical engineering*. New York: McGraw-Hill Inc.

Roland, U., Buchenhorst, D., Holzer, F., & Kopinke, F. D. (2008). Engineering aspects of radio-wave heating for soil remediation and compatibility with biodegradation. *Environmental Science & Technology, 42*(4), 1232–1237.

Ronen, D., Graber, E. R., & Laor, Y. (2004). Volatile organic compounds in the saturated–unsaturated Interface region of a contaminated phreatic aquifer. *Vadose Zone Journal, 4*(2), 337–344.

Rothmel, R. K., Peters, R. W., St. Martin, E., & DeFlaun, M. F. (1998). Surfactant foam/bioaugmentation technology for in situ treatment of TCE-DNAPLs. *Environmental Science & Technology, 32*, 1667–1675.

Saiers, J. E., & Lenhart, J. J. (2003). Colloid mobilization and transport within unsaturated porous media under transient-flow conditions. *Water Resources Research, 39*(1), 1019.

Sakulchaicharoen, N., O'Carroll, D. M., & Herrera, J. E. (2010). Enhanced stability and dechlorination activity of pre-synthesis stabilized nanoscale FePd particles. *Journal of Contaminant Hydrology, 118*(3–4), 117–127.
Saleh, N., Phenrat, T., Sirk, K., Dufour, B., Ok, J., Sarbu, T., Matyjaszewski, K., Tilton, R. D., & Lowry, G. V. (2005). Adsorbed triblock copolymers deliver reactive iron nanoparticles to the oil/water interface. *Nano Letters, 5*(12), 2489–2494.
Shen, X., Zhao, L., Ding, Y., Liu, B., Zeng, H., Zhong, L., & Li, X. (2011). Foam, a promising vehicle to deliver nanoparticles for vadose zone remediation. *Journal of Hazardous Materials, 186*(2–3), 1773–1780.
Smith, L. A., & Hinchee, R. E. (1993). *In situ thermal technologies for site remediation*. Boca Raton: CRC Press.
Srirattana, S., Piaowan, K., Lowry, G. V., & Phenrat, T. (2017). Electromagnetic induction of foam-based nanoscale Zerovalent Iron (NZVI) particles to thermally enhance non-aqueous phase liquid (NAPL) volatilization in unsaturated porous media: Proof of concept. *Chemosphere, 183*, 323–331.
Tratnyek, P. G., & Johnson, R. L. (2006). Nanotechnologies for environmental cleanup. *Nano Today, 1*(2), 44–48.
Truex, M. J., Gillie, J. M., Powers, J. G., & Lynch, K. P. (2009). Assessment of *in situ* thermal treatment for chlorinated organic source zones. *Remediation Journal, 19*(2), 7–17.
Vermeulen, F., & McGee, B. (2000). In-situ electromagnetic heating for hydrocarbon recovery and environmental remediation. *Journal of Canadian Petroleum Technology, 39*(8), 25–29.
Wang, H., & Chen, J. (2012). Enhanced flushing of polychlorinated biphenyls contaminated sands using surfactant foam: Effect of partition coefficient and sweep efficiency. *Journal of Environmental Sciences (China), 24*(7), 1270–1277.
Wang, X., Chen, J., & Lv, C. (2015). Evaluation of foam surfactant for foam-flushing technique in remediation of DDT-contaminated soil using data envelopment analysis method. *Environmental Science and Pollution Research, 22*(4), 2994–3003.
Wei, Y. T., Wu, S. C., Yang, S. W., Che, C. H., Lien, H. L., & Huang, D. H. (2012). Biodegradable surfactant stabilized nanoscale zero-valent iron for *in situ* treatment of vinyl chloride and 1,2-dichloroethane. *Journal of Hazardous Materials, 211–212*, 373–380.
Werth, C. J., & Reinhard, M. (1997). Effects of temperature on trichloroethylene desorption from silica gel and natural sediments. 2. Kinetics. *Environmental Science & Technology, 31*, 697–703.
Yang, J.-W., Cho, H.-J., Choi, G.-Y., & Lee, S.-H. (2001). Cost-effective monitoring for a soil vapor extraction (SVE) system. A simplified modeling and gas sensor test. *Environmental Monitoring and Assessment, 70*(1), 201–210.
Yoon, H., Valocchi, A. J., & Werth, C. J. (2003). Modeling the influence of water content on soil vapor extraction. *Vadose Zone Journal, 2*, 368–381.
Zhang, W. (2003). Nanoscale iron particles for environmental remediation: An overview. *Journal of Nanoparticle Research, 5*, 323–332.
Zhang, W.-X., Wang, C.-B., & Lien, H.-L. (1998). Treatment of chlorinated organic contaminants with nanoscale bimetallic particles. *Catalysis Today, 40*, 387–395.
Zhao, Y. S., Su, Y., Lian, J. R., Wang, H. F., Li, L. L., & Qin, C. Y. (2016). Insights on flow behavior of foam in unsaturated porous media during soil Flushing. *Water Environment Research, 88*(11), 2132–2141.
Zhong, L., Szecsody, J. E., Zhang, F., & Mattigod, S. V. (2010). Foam delivery of amendments for vadose zone remediation: Propagation performance in unsaturated sediments. *Vadose Zone Journal, 9*, 757–767.

Chapter 14
Carbothermal Synthesis of Aerosol-Based Iron-Carbon Nanocomposites for Adsorption and Reduction of Cr(VI)

Jiawei He, Ling Ai, Yiyan Wang, Yuan Long, Chaoliang Wei, and Jingjing Zhan

Abstract Spherical iron-carbon nanocomposites were synthesized through a facile aerosol-based process and a subsequent carbothermal reduction. Carbothermal treatment reduces iron species to zero-valent iron rather than using expensive sodium borohydride. In addition, the high porosity of iron-carbon composites allows the entry of contaminants to reactive sites. These composites with nanoscale zero-valent iron particles incorporated in the carbon matrix exhibit synergistic adsorption and reaction for more efficient removal of Cr(VI) in water. Under identical experimental conditions, aerosol-assisted iron-carbon composites showed the highest removal efficiency compared to other materials including nanoscale zero-valent iron particles, aerosol-assisted carbon, and their physical mixture. Meanwhile, X-ray photoelectron spectroscopy analysis proved as-prepared iron-carbon composites could effectively transform Cr(VI) to much less toxic Cr(III). These iron-carbon composites can be designed at low cost, the process is amenable to scale-up for in situ application, and the materials are intrinsically benign to the environment.

Keywords Iron-carbon nanocomposite · Adsorption · Reduction · Hexavalent chromium · Aerosol-based synthesis

14.1 Introduction

Hexavalent chromium (Cr(VI)) is a frequently detected groundwater contaminant originating from various industrial processes such as leather tanning, pigment synthesis, electroplating, and metal finishing (Krishnani and Ayyappan 2006; Owlad et al. 2009). Because it is a potential carcinogen or mutagen to most

J. He · L. Ai · Y. Wang · Y. Long · C. Wei · J. Zhan (✉)
School of Food and Environment, Dalian University of Technology, Panjin, People's Republic of China
e-mail: jingjingzhan@dlut.edu.cn

T. Phenrat, G. V. Lowry (eds.), *Nanoscale Zerovalent Iron Particles for Environmental Restoration*, https://doi.org/10.1007/978-3-319-95340-3_14

organisms, Cr(VI) has been listed as one of priority pollutants for groundwater contamination, and its concentration in drinking water has been regulated by many countries. Therefore, eliminating the risks from Cr(VI) contamination has received significant attention from both academics and industry. During the past few decades, extensive efforts have been carried out to develop and synthesize nanomaterials with unique reactivity and functional characteristics for environmental applications (Wang and Zhang 1997; Al-abed and Chen 2001; Nyer and Vance 2001; Zhang 2003; Gatmiri and Hosseini 2004; Liu et al. 2005a, 2005b; He and Zhao 2007; Phenrat et al. 2007; Li et al. 2008). For example, nanoscale zero-valent iron (NZVI) particles are a promising approach in the treatment of Cr(VI) because of its strong reducing ability and fast reaction kinetics. Here, NZVI serves as an electron donor, converting Cr(VI) to its lower oxidation state trivalent chromium (Cr(III)) (Li et al. 2008; Xie and Cwiertny 2012; Huang et al. 2013). Compared to Cr(VI), Cr(III) is 300–500 times less toxic (Miretzky and Cirelli 2010). Furthermore, Cr(III) is only sparingly soluble $Cr(OH)_3$ ($K_{sp} = 6.3 \times 10^{-31}$) at the high pH that results from Fe (0)/H_2O buffering and thus is immobilized in situ or could be completely removed as solid waste.

However, the intrinsic ferromagnetism of NZVI particles leads to rapid aggregation, decreasing the reactivity and limiting the mobility of NZVI in the subsurface. Prior studies have shown that the aggregation of NZVI particles can be inhibited dramatically by adsorption of hydrophilic or amphiphilic polymers such as guar gum (Tiraferri et al. 2008), poly(acrylic acid) (PAA) (Schrick et al. 2004), starch (He and Zhao 2005), carboxymethyl cellulose (CMC) (He and Zhao 2007; He et al. 2007), and triblock copolymers (PMAA-PMMA-PSS) (Saleh et al. 2005) on the NZVI particle surface (Chap. 5). These adsorbed polymer molecules inhibit NZVI aggregation and enhance solution stability through steric hindrance and/or electrostatic repulsion, while they may decrease the reactivity by blocking reactive surface sites (Phenrat et al. 2009) (Chap. 3). Another option is to distribute and immobilize NZVI particles onto/into a support, generating NZVI-support composites (Chap. 2). In addition to prevent the aggregation of NZVI particles, the main feature of these composite particles is that the utilized supports such as functionalized silica (Zhan et al. 2008; Zheng et al. 2008), commercial available activation carbon (Choi et al. 2008, 2009), carbon black (Schrick et al. 2004; Hoch et al. 2008), and uniform spherical pyrolysis carbon (Zhan et al. 2009; Sunkara et al. 2010) are effective in adsorbing dissolved contaminants, thereby increasing local concentrations of contaminants in the vicinity of the NZVI particles and facilitating the removal and degradation of contaminants. Although these composite materials are effective and feasible, the techniques to synthesize these composites are constrained by one or more of the following limitations: costly precursors (Zhan et al. 2008, 2009; Zheng et al. 2008), multiple steps in the process (Schrick et al. 2004; Choi et al. 2008, 2009; Hoch et al. 2008; Sunkara et al. 2010), low iron loading percentage (Choi et al. 2008; Zhan et al. 2009; Sunkara et al. 2010), and inappropriate size range for transport through the subsurface (Choi et al. 2008, 2009).

In recent work from our laboratory, we demonstrated that an aerosol-based process followed by liquid-phase reduction can be employed to prepare Fe-C composites, where nanoscale zero-valent iron particles were supported on the surface of carbon particles (Zhan et al. 2011). During the aerosol process, the "chemistry in a droplet" leads to submicrometer-sized Fe-C composites (100 nm–1 μm), which are in the optimal size range for effective transport through a porous medium like soil (Schrick et al. 2004; Zhan et al. 2008). Importantly, aerosol-based carbon spheres serve as strong adsorbents for contaminants increasing local concentrations at the NZVI reaction sites thereby enhancing the driving force of reaction. One current disadvantages of this process is that NZVI only can be decorated on the carbon particle surface, which may limit the loading percentage of NZVI. Moreover, the sodium borohydride reductant is costly, and it generates large amounts of hydrogen during reduction, impeding the safe scale-up of the process (Hoch et al. 2008).

In this chapter, we retain the merits of our earlier work to use the aerosol-based technology but with carbothermal treatment to produce iron-carbon composites attempting to distribute NZVI particles throughout the carbon matrix and avoid the use of sodium borohydride. In the obtained iron-carbon composites, NZVI particles have been well separated on the surface of carbon and inside the carbon spheres as well, making high loading of NZVI possible. Meanwhile, carbothermal treatment introduces high porosity in the iron-carbon composites, allowing access of contaminants to internal reactive sites. In the application, these materials provide dual functions of adsorption coupling with reaction in the removal Cr(VI) species from the solution. Characterization of the obtained composites was carried out by SEM, TEM, XRD, and BET. X-ray photoelectron spectroscopy (XPS) analysis was applied to determine the predominant mechanism of the removal of Cr(VI), and the influencing factors include solution pH; the initial contaminant concentration and composite (NZVI) dosage were also investigated in detail. To confirm a synergistic effect of the combined carbon/NZVI material, parallel experiments using NZVI, carbon, and a mixture of separated NZVI and carbon, NZVI and carbon were also tested.

14.2 Synthesis of Aerosol-Based Iron-Carbon Nanocomposites

Figure 14.1a shows the schematic of the aerosol apparatus, consisting of an atomizer, a heating zone, and a filter. Starting with a homogenous aqueous solution containing sucrose and iron (II) sulfate, a commercial atomizer atomizes the solution into droplets that undergo a heating and drying step, generating particles that are collected on a filter. Figure 14.1b is a representation of the formation route of Fe-C composites. First, continuous streams of the precursor solution are broken into a chain of aerosol droplets after atomization. When these aerosol droplets pass through the heating zone, the precipitation and phase transition of iron species are concomitant with the dehydration and carbonization of sucrose under high-temperature

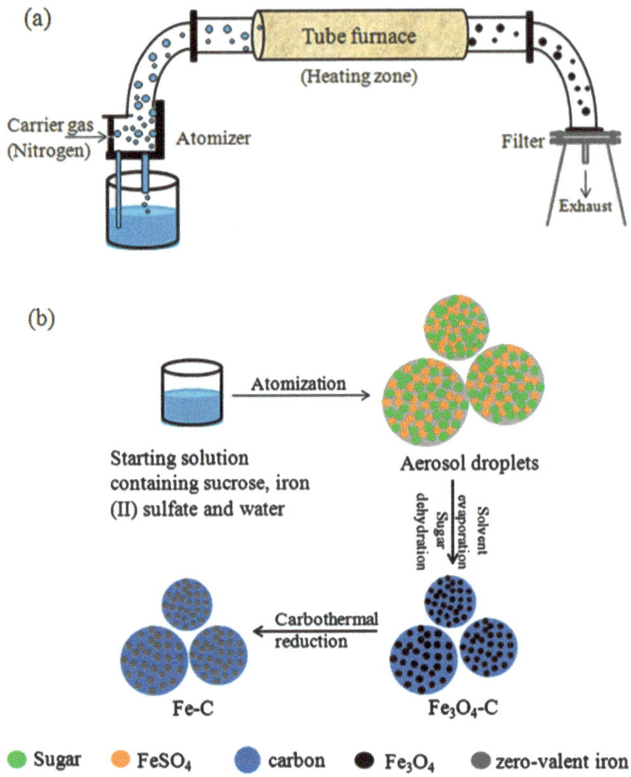

Fig. 14.1 (**a**) Schematic of the aerosol-based process and (**b**) schematic of the synthesis route for iron-carbon composites

conditions, fabricating Fe_3O_4-C composites with Fe_3O_4 nanoparticles incorporated in the carbon matrix. In the following carbothermal reduction process, a "self-redox" reaction occurs in the Fe_3O_4-C composites, where entrapped Fe_3O_4 particles are thermally reduced by the carbon frameworks (present in excess), resulting in the formation of final product Fe-C composites. The final weight percentage of zero-valent iron in the Fe-C composites is approximately 40%. This content was determined by weighing the residual solid (Fe_2O_3) of a known mass of Fe-C composites after calcination under 500 °C in the air for 4 h. It is noteworthy that Fe-C composites can also be obtained by reduction of Fe_3O_4-C using hydrogen at 500 °C for 5 h, which is very similar to the reduction process used for production of RNIP (Toda Kogyo Corp.) (Uegami 2007). In both cases, we have found Fe-C composites with the same requisite characteristics.

14.3 Characterization of Aerosol-Based Iron-Carbon Nanocomposites

The morphology and microstructure of as-made particles were analyzed through scanning and transmission electron microscopy. Figure 14.2 shows representative SEM and TEM images of Fe_3O_4-C composites obtained from the aerosol process. Apparently, Fe_3O_4-C composites are spherical with sizes ranging from 100 nm to 1 μm, characteristic of particles synthesized by the aerosol-based technique. The rough surface as indicated in SEMs (Fig. 14.2a, b) shows some Fe_3O_4 particles are embedded on the surface of carbon matrix. Meanwhile, TEM images as displayed in Fig. 14.2c, d suggest that Fe_3O_4 particles also can be found in the inside of carbon spheres, where Fe_3O_4 are visualized clearly with dark spots due to high electron density. Figure 14.3 shows typical images of the final product Fe-C composites obtained after carbothermal reduction process. As shown in SEM images (Fig. 14.3a, b), the presence of individual zero-valent iron particles embedded in carbons implies the distribution of nanoiron throughout the carbon without aggregation. TEMs in Fig. 14.3c, d further confirmed the fact that zero-valent iron particles were incorporated in porous carbon spheres, where zero-valent iron particles with higher electron contrast are observed clearly. It is worth noting that Fe-C

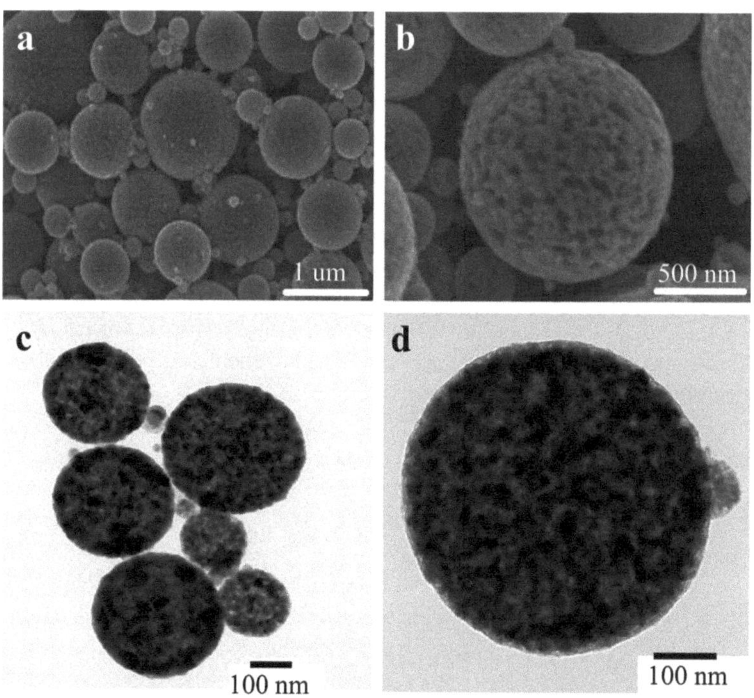

Fig. 14.2 (**a**, **b**) SEM and (**c**, **d**) TEM images show the morphology of aerosol-based Fe_3O_4-C composites

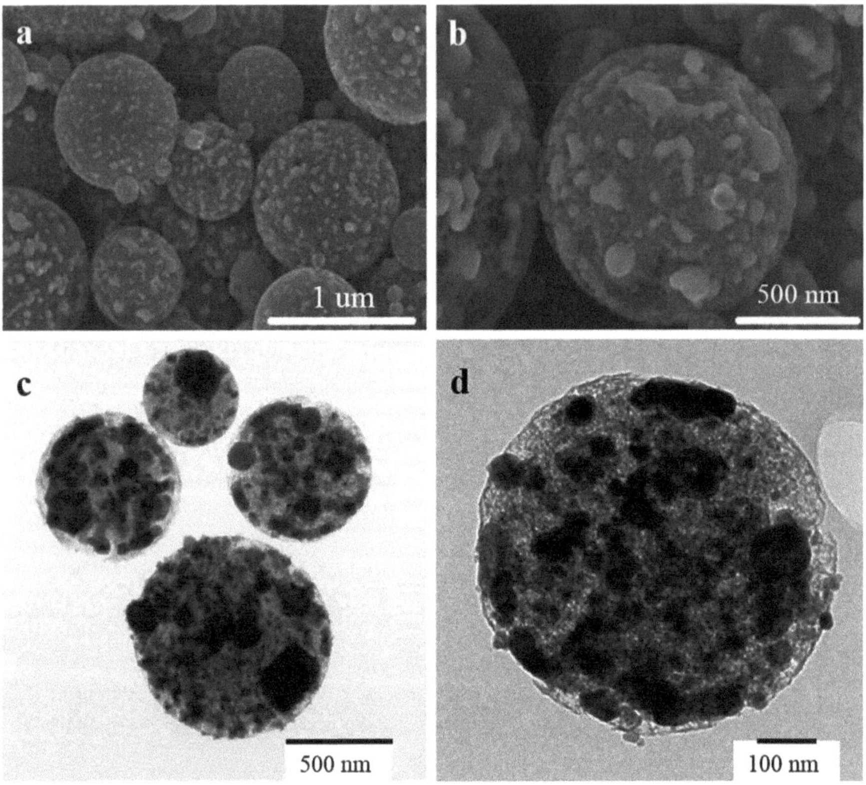

Fig 14.3 (**a**, **b**) SEM and (**c**, **d**) TEM images show the morphology of Fe-C composites

composites appear to be more porous compared with Fe_3O_4-C composites. We consider that the enlargement of pore structure is attributed to three aspects: carbon consumption by carbothermal reduction, persistent carbonization under high temperature, and Fe_3O_4 or Fe sintering. Sintering is also responsible for the observation that the average size of zero-valent iron particles has increased in contrast to that of Fe_3O_4 particles after longer carbothermal treatment times.

Figure 14.4a shows the N_2 adsorption isotherms obtained for these composites. Surface areas at 77 K were calculated based on Brunauer-Emmett-Teller (BET) method. For Fe_3O_4-C and Fe-C composites, the BET surface areas were found to be 130 m^2/g and 224 m^2/g, respectively, and the corresponding Barrett-Joyner-Halenda (BJH) desorption pore volumes were determined to be 0.282 cm^3/g and 0.4 cm^3/g. Both samples exhibit a IV-type isotherm with H_3-type hysteresis loop on the basis of the IUPAC (International Union of Pure and Applied Chemistry) classification (Sing 1985). The type-IV isotherm is associated with the capillary condensation taking place in mesopores, and the H_3-type hysteresis loop is attributed to asymmetric slit-shape mesopores. The type of isotherm indicates an interconnected porous system and high pore connectivity according to the percolation theory (Seaton 1991; Liu

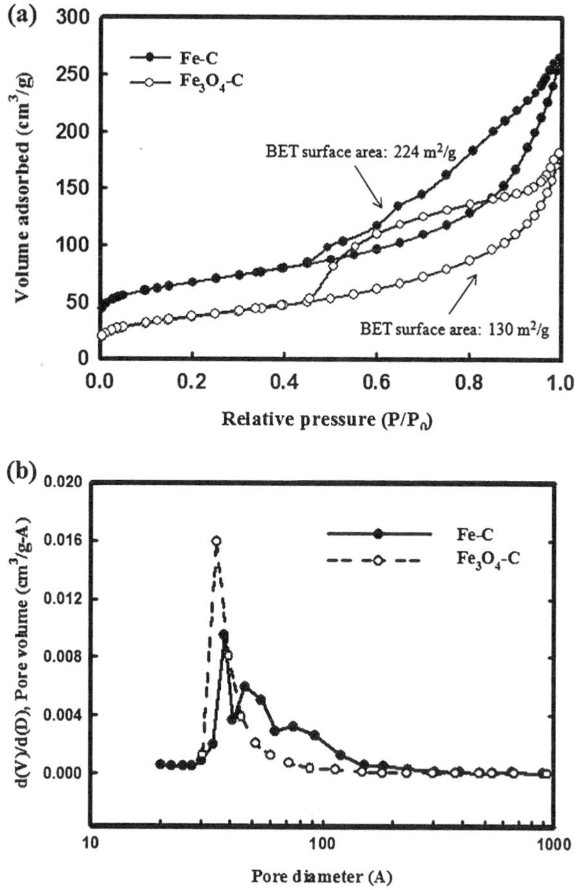

Fig. 14.4 (**a**) Nitrogen adsorption-desorption isotherms for Fe-C and Fe_3O_4-C composite microspheres; (**b**) the BJH pore size distribution derived from the desorption branch of the isotherm of composite microspheres

et al. 2005c). In agreement with TEM images of Figs. 14.2 and 14.3, the evolution of pore structure from Fe_3O_4-C composites to Fe-C composites is confirmed from the BJH pore size distribution, illustrated in Fig. 14.4b. Clearly, a narrow pore size distribution at the mean value of 3.4 nm is validated for Fe_3O_4-C composites. In contrast, the pores in Fe-C composites are larger and various with the peaks in the distribution centered at 3.8 nm, 4.6 nm, and 7.5 nm. The high surface area and porosity of Fe-C composites are desired and will promote entry of contaminants into the particles to contact the iron species within porous carbon matrixes.

The phases of iron species are verified by the XRD patterns shown in Fig. 14.5. For particles collected directly from the aerosol process prior to the carbothermal treatment, magnetite peaks dominate the pattern indicating that the starting material $FeSO_4$ transformed primarily to Fe_3O_4 during aerosolization. The transition of iron oxides to element iron occurred in the carbothermal step and is temperature-dependent. When the temperature was increased from 400 to 700 °C, partial

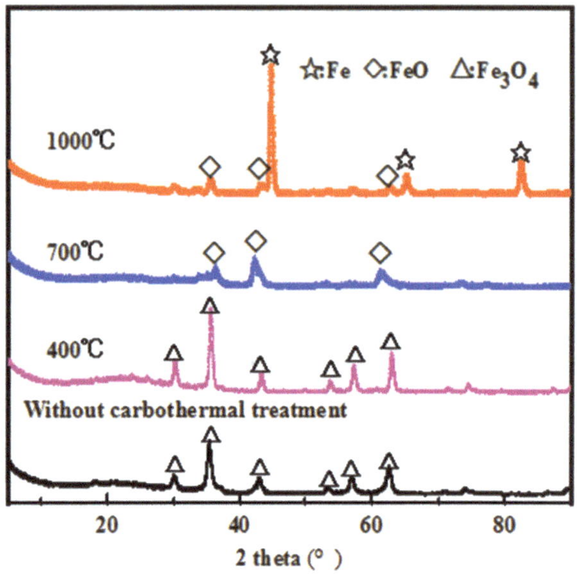

Fig. 14.5 XRD patterns of particles without and with carbothermal treatment under various temperatures (400 °C, 700 °C, and 1000 °C)

reduction of magnetite took place, and FeO was the existing form of iron. Furthermore, the desired zero-valent iron was obtained at 1000 °C indicated by strong diffraction peaks at 2θ of 45°, 65°, and 82°.

14.4 Removal of Cr(VI) in Water by Aerosol-Assisted Fe-C Composites: Performance and Proposed Mechanism

14.4.1 Performance of Aerosol-Assisted Fe-C Composites in the Removal of Cr(VI)

The performance of aerosol-assisted Fe-C composites in the removal of Cr(VI) is presented in Fig. 14.6. Performance of the composite material is compared to a physical mixture of NZVI and aerosol-assisted carbon (Fe + C), just NZVI, and just aerosol-assisted carbon alone under identical experimental conditions. In the experiments, either 10 mg of the Fe-C composites (40% is Fe and 60% is carbon), 10 mg of Fe + C mixture (Fe/C = 4:6), 4 mg of NZVI, or 6 mg of aerosol-assisted C was added to water containing 10 mg L^{-1} of Cr(VI). These amounts were chosen to keep the dosages of iron and carbon same. All of the materials were effective in the removal of Cr(VI) to some extent, but the Fe-C composites afforded the best removal. Within the first 5 min, 91% of Cr(VI) ions were removed by Fe-C composites, and the final removal efficiency over a 2 h reaction time was 99%, which meant that the concentration of Cr(VI) has been reduced to 0.1 mg L^{-1}.

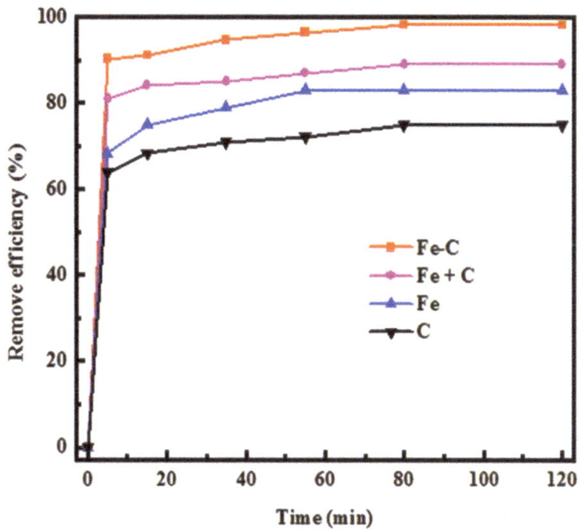

Fig. 14.6 A comparison of Cr(VI) removal efficiencies by Fe-C, Fe + C, Fe, and C at pH = 5 and an initial Cr (VI) concentration of 10 mg L^{-1} (10 mg Fe-C, 10 mg Fe + C, 4 mg Fe, and 6 mg C)

In contrast, for carbon and NZVI alone, the removal efficiencies were only 63% and 68% in the first 5 min due to adsorption and reduction, respectively, and the residual concentrations of Cr(VI) were 2.5 mg L^{-1} and 1.7 mg L^{-1}, respectively, after 2 h. It is worth noting that the Fe + C mixture showed a greater capacity for removal of Cr (VI) than individual NZVI or carbon, and the removal efficiency reached 81% and 89% in 5 min and 2 h, respectively, but was lower than the aerosol-assisted Fe-C. This is indicative of the synergistic effects of adsorption together with reaction for Fe-C composites in the removal of Cr(VI), where adsorption concentrates Cr (VI) in the vicinity of the NZVI particles and thus facilitates the reaction. This finding is in agreement with our previous work in the remediation of trichloroethylene (TCE) (Zhan et al. 2011). The above improvements in removal efficiency indicated that Fe-C composites have potential advantages in Cr(VI) removal.

14.4.2 Synergetic Removal Mechanism

The speciation of adsorbed chromium on the Fe-C composites was determined by X-ray photoelectron spectroscopy. For the chromium spectra shown in the Fig. 14.7, the photoelectron peaks for the chromium 2p3/2 and 2p1/2 centered at 577.1 and 587.6 eV, respectively, and both were similar to the binding energies of Cr(III)-containing materials (Tang et al. 2014; Zhou et al. 2015). This data demonstrated that adsorbed Cr(VI) anions were reduced to Cr(III) after exposure to aerosol-assisted Fe-C composites. However, bands of Cr(III) in the XPS spectra are weak even though batch experiments had shown that most of the Cr(VI) was removed after contact with aerosol-assisted Fe-C composites. This discrepancy is probably because

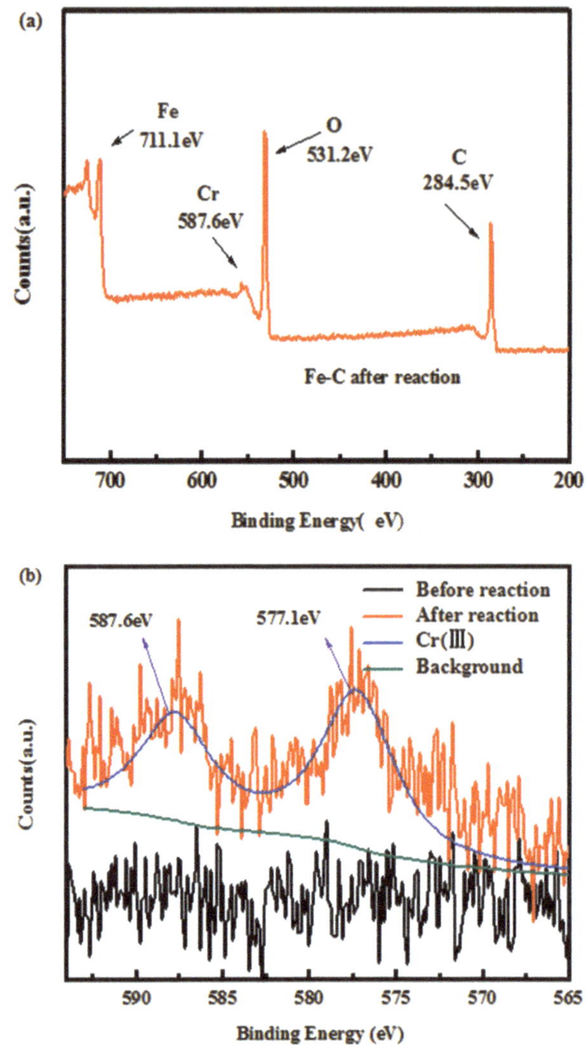

Fig. 14.7 XPS full survey (**a**) and Cr 2p spectrum (**b**) for Fe-C composites after Cr (VI) removal. For comparison, the Cr 2p spectrum of before reaction is shown

the reduction of Cr(VI) occurs on the surface of the iron particles, which are mainly entrapped in the interior of the composites. However, XPS only provides information about chromium on the exterior surfaces of the particles. In addition, an excessive dosage of Fenton's reagent was used to re-oxidize the supernatant to confirm if any reduced Cr(III) remained in the solution phase according to a previous method (Cao et al. 2006). The fact that no Cr(VI) was detected after re-oxidation further validates the immobilization efficiency of Cr(III) by Fe-C composites.

14.4.3 *Effect of pH*

The initial pH of contaminated water is an important controlling parameter in the removal of Cr(VI). Considering the pH value of typical natural groundwater that ranges from 5.0 to 9.0 (Lv et al. 2011), we chose the range of pH 5–10 to evaluate the removal efficiency of Fe-C composites. The results in Fig. 14.8a show that pH has a

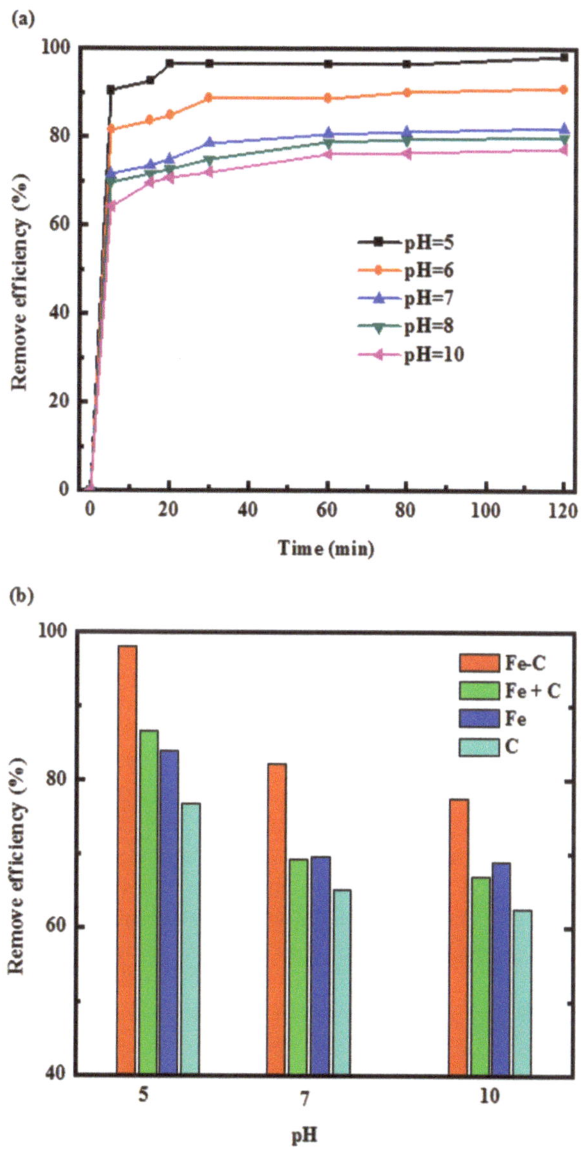

Fig. 14.8 (**a**) Effects of the initial pH on Cr(VI) removal efficiency of Fe-C composites (initial Cr(VI) concentration, 10 mg L^{-1}; Fe-C dosage, 10 mg) and (**b**) Cr(VI) removal efficiencies at different solution pH for various materials including Fe-C, Fe + C, Fe, and C

direct effect on the Cr(VI) treatment, where the removal efficiency decreased from 99% to 82% with pH increasing from 5.0 to 7.0 and then continued to drop to 77% with pH continuously increasing to 10.0. The fact that Fe-C composites are more effective under acidic conditions could be attributed to two aspects. On the one hand, the pH impact on Cr(VI) removal is largely associated with the Cr(VI) chemistry in the solution and the surface properties of the composites. Although the existing forms of Cr(VI) ion in aqueous solutions vary with pH, all of them are anions including $HCrO_4^-$, CrO_4^{2-}, and $Cr_2O_7^{2-}$. Hence, at lower pH, the Fe-C composites were positively charged leading to the electrostatic attraction between Cr(VI) and the composites. A large quantity of H^+ in acidic conditions could also promote the reaction rate of iron, accelerating the reduction of Cr(VI) to Cr(III):

$$2HCrO_4^- + 3Fe^0 + 14H^+ \rightarrow 3Fe^{2+} + 2Cr^{3+} + 8H_2O \tag{14.1}$$

$$HCrO_4^- + 3Fe^{2+} + 7H^+ \rightarrow 3Fe^{3+} + Cr^{3+} + 4H_2O \tag{14.2}$$

However, the removal efficiency did not significantly differ under neutral and alkaline conditions (pH = 7–10) compared to the acidic condition (pH = 5–7). It may be because H^+ is the limiting factor in the removal of Cr(VI) for Fe-C composites, while competition between chromium ions and OH^- plays a relatively small role. In addition, the effect of pH on Cr(VI) removal for other types of materials were also investigated. As shown in Fig. 14.8b, all removal efficiencies depend on pH, and all of them were higher at lower pH. It is worth noting that the lowest removal efficiency for Fe-C composites is still 77% at pH 10, compared to that of bare aerosol-assisted carbon under the optimal condition, showing its potential application in heavy metal removal over a wide pH range.

14.4.4 *Effect of Fe-C Dose and Initial Cr(VI) Concentration*

Figure 14.9a compares the removal efficiencies in the treatment of Cr(VI) when different amounts of Fe-C composites were employed; the initial concentration of Cr (VI) was 25 mg L^{-1}. Although the initial rate of uptake was not affected by the Fe-C concentration, the final adsorbed amount was indeed affected by the Fe-C dose. The removal efficiency for the first 5 min decreased from 98% to 94%, 91%, 85%, and finally 82% after 2 h of contact time when the dosage of Fe-C composites was reduced from 2 g L^{-1} to 1.5, 1.0, 0.75, and finally 0.5 g L^{-1}, respectively, implying a finite number of reaction sites for Fe-C composites. Meanwhile, the effect of initial Cr(VI) concentration on the removal efficiency was investigated in the range of 10–50 mg L^{-1} as shown in Fig. 14.9b. The removal percentage of Cr(VI) decreased as the initial Cr(VI) concentration increased. The Cr(VI) removal efficiency was 99% at 10 mg L^{-1} and decreased to 90%, 89%, and 81% when the initial Cr (VI) concentration reached 15, 25, and 50 mg L^{-1}, respectively. The data presented in Fig. 14.9 could not be fit well to a pseudo-first-order model because the overall

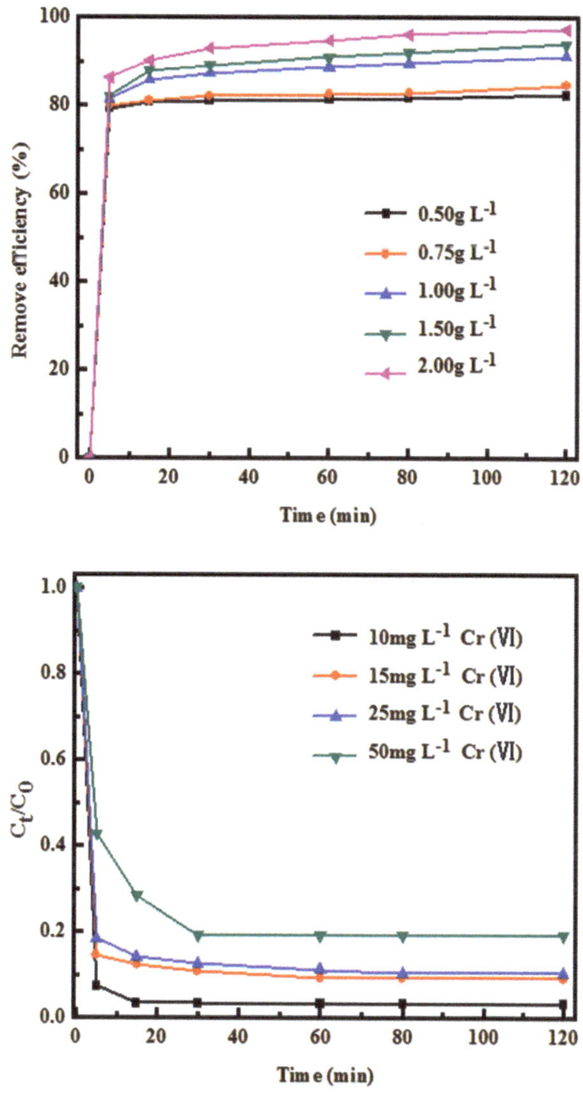

Fig. 14.9 Effects of Fe-C composites dosage (**a**) and the initial Cr (VI) concentration (**b**) on Cr (VI) removal efficiency

mechanism was more complicated than a simple chemical reaction or adsorption. While it is not clear that Cr ions are expected to preferentially interact with the carbon or NZVI components of the Fe-C composites in our experiments, we suppose all the Cr(VI) ions were reduced to Cr(III) on the iron surface eventually, which was supported by aforementioned XPS analysis.

14.5 Conclusion

Iron-carbon nanocomposites with a spherical mesoporous structure were fabricated through a simple aerosol-assisted process followed by an inexpensive carbothermal reduction process with only common sucrose and ferrous sulfate as starting materials. These composites exhibit dual functionalities in the removal of Cr(VI), where carbon matrices allow adsorption of Cr(VI) and immobilization of iron nanoparticles prevents zero-valent nanoiron aggregation, thereby providing enhanced reactivity compared to NZVI alone. XPS analysis proved Fe-C composites could effectively transform Cr(VI) to Cr(III), which is much less toxic. Meanwhile, these composites display high efficiency compared to other materials including a physical mixture of separated NZVI and aerosol-assisted carbon, NZVI alone, and aerosol-assisted carbon alone, which indicates a synergistic effect for the removal of Cr(VI) due to the supporting the iron on the carbon surface. These iron-carbon composites can be produced at low cost, and the process is amenable to scale-up to large reactors and continuous processing, showing significant potential for the remediation of groundwater and sediment contaminated with heavy metals such as Cr(VI).

Acknowledgements We wish to thank Dr. Yanqiang Huang at Dalian Institute of Chemical Physics for his assistance with the XPS analysis. Funding from the Fundamental Research Funds for the Central Universities is gratefully acknowledged.

References

Al-abed, S. R., & Chen, J. (2001). Transport of trichloroethylene (TCE) in natural soil by electroosmosis. In: Smith, J. A. & Burns, S. E. (eds.), *Physicochemical groundwater remediation* (pp. 91–114). New York: Springer.

Cao, J., & Zhang, W. (2006). Stabilization of chromium ore processing residue (COPR) with nanoscale iron particles. *Journal of Hazardous Materials*, *132*(2–3), 213–219.

Choi, H., Al-Abed, S. R., Agarwal, S., & Dionysiou, D. D. (2008). Synthesis of reactive Nano-Fe/Pd bimetallic system-impregnated activated carbon for the simultaneous adsorption and dechlorination of PCBs. *Chemistry of Materials*, *20*, 3649–3655.

Choi, H., Agarwal, S., & Al-Abed, S. R. (2009). Adsorption and simultaneous dechlorination of PCBs on GAC/Fe/Pd: mechanistic aspects and reactive capping barrier concept. *Environmental Science & Technology*, *43*, 488–493.

Gatmiri, B., & Hosseini, A. H. (2004). Conceptual model and mathematical formulation of NAPL transport in unsaturated porous media. In: Thomas, H. R. & Young, R. N. (eds.), *Geoenvironmental engineering: Integrated management of groundwater and contaminated land* (pp. 67–75). London, UK: Thomas Telford Publishing.

He, F., & Zhao, D. (2005). Preparation and characterization of a new class of starch-stabilized bimetallic nanoparticles for degradation of chlorinated hydrocarbons in water. *Environmental Science & Technology*, *39*, 3314–3320.

He, F., & Zhao, D. (2007). Manipulating the size and dispersibility of zerovalent iron nanoparticles by use of carboxymethyl cellulose stabilizers. *Environmental Science & Technology*, *41*, 6216–6221.

He, F., Zhao, D., Liu, J., & Roberts, C. B. (2007). Stabilization of Fe–Pd nanoparticles with sodium carboxymethyl cellulose for enhanced transport and dechlorination of trichloroethylene in soil and groundwater. *Industrial & Engineering Chemistry Research, 46*, 29–34.

Hoch, L. B., Mack, E. J., Hydutsky, B. W., Hershman, J. M., Skluzacek, J. M., & Mallouk, T. E. (2008). Carbothermal synthesis of carbon-supported nanoscale zero-valent iron particles for the remediation of hexavalent chromium. *Environmental Science & Technology, 47*, 2600–2605.

Huang, P., Ye, Z., Xie, W., Chen, Q., Li, J., Xu, Z., & Yao, M. (2013). Rapid magnetic removal of aqueous heavy metals and their relevant mechanisms using nanoscale zero valent iron (nZVI). *Water Research, 47*, 4050–4058.

Krishnani, K., & Ayyappan, S. (2006). Heavy metals remediation of water using plants and lignocellulosic agrowastes. In: Ware, G. W., Whitacre, D. M., Albert, L. A., de Voogt, P., Gerba, C. P., Hutzinger, O., Knaak, J. B., Mayer, F. L., Morgan, D. P., Park, D. L., Tjeerdema, R. S., Yang, R. S. H., Gunther, F. A. (eds.), *Reviews of environmental contamination and toxicology* (Vol. 188, pp. 59–84).

Li, X., Cao, J., & Zhang, W. (2008). Stoichiometry of Cr(VI) immobilization using nanoscale zerovalent iron (nZVI): A study with high-resolution X-ray photoelectron spectroscopy (HR-XPS). *Industrial & Engineering Chemistry Research, 47*, 2131–2139.

Liu, Y., Choi, H., Dionysiou, D., & Lowry, G. V. (2005a). Trichloroethene hydrodechlorination in water by highly disordered monometallic nanoiron. *Chemistry of Materials, 17*, 5315–5322.

Liu, Y., Majetich, S. A., Tilton, R. D., Sholl, D. S., & Lowry, G. V. (2005b). TCE dechlorination rates, pathways, and efficiency of nanoscale iron particles with different properties. *Environmental Science & Technology, 39*, 1338–1345.

Liu, Z., Fan, T., Zhang, W., & Zhang, D. (2005c). The synthesis of hierarchical porous iron oxide with wood templates. *Microporous and Mesoporous Materials, 85*, 82–88.

Lv, X., Xu, J., Jiang, G., & Xu, X. (2011). Removal of chromium(VI) from wastewater by nanoscale zero-valent iron particles supported on multiwalled carbon nanotubes. *Chemosphere, 85*, 1204–1209.

Miretzky, P., & Cirelli, A. F. (2010). Cr(VI) and Cr(III) removal from aqueous solution by raw and modified lignocellulosic materials: A review. *Journal of Hazardous Materials, 180*, 1–19.

Nyer, E. K., & Vance, D. B. (2001). Nano-scale iron for dehalogenation. *Groundwater Monitoring & Remediation, 21*, 41–46.

Owlad, M., Aroua, M. K., Daud, W. A. W., & Baroutian, S. (2009). Removal of hexavalent chromium-contaminated water and wastewater: A review. *Water, Air, and Soil Pollution, 200*, 59–77.

Phenrat, T., Saleh, N., Sirk, K., Tilton, R. D., & Lowry, G. V. (2007). Aggregation and sedimentation of aqueous nanoscale zerovalent iron dispersions. *Environmental Science & Technology, 41*, 284–290.

Phenrat, T., Liu, Y., Tilton, R. D., & Lowry, G. V. (2009). Adsorbed polyelectrolyte coatings decrease Fe^0 nanoparticle reactivity with TCE in water: Conceptual model and mechanisms. *Environmental Science & Technology, 43*, 1507–1514.

Saleh, N., Phenrat, T., Sirk, K., Dufour, B., Ok, J., Sarbu, T., Matyjaszewski, K., Tilton, R. D., & Lowry, G. V. (2005). Adsorbed triblock copolymers deliver reactive iron nanoparticles to the oil/water interface. *Nano Letters, 5*, 2489–2494.

Schrick, B., Hydutsky, B. W., Blough, J. L., & Mallouk, T. E. (2004). Delivery vehicles for zerovalent metal nanoparticles in soil and groundwater. *Chemistry of Materials, 16*, 2187–2193.

Seaton, N. A. (1991). Determination of the connectivity of porous solids from nitrogen sorption measurements. *Chemical Engineering Science, 46*, 1895–1909.

Sing, K. S. W. (1985). Reporting physisorption data for gas/solid systems with special reference to the determination of surface area and porosity (recommendations 1984). *Pure and Applied Chemistry, 57*, 603–619.

Sunkara, B., Zhan, J., He, J., McPherson, G. L., Piringer, G., & John, V. T. (2010). Nanoscale zerovalent iron supported on uniform carbon microspheres for the in situ remediation of chlorinated hydrocarbons. *ACS Applied Materials & Interfaces, 2*, 2854–2862.

Tang, L., Yang, G., Zeng, G., Cai, Y., Li, S., Zhou, Y., Pang, Y., Liu, Y., Zhang, Y., & Luna, B. (2014). Synergistic effect of iron doped ordered mesoporous carbon on adsorption-coupled reduction of hexavalent chromium and the relative mechanism study. *Chemical Engineering Journal, 239*, 114–122.

Tiraferri, A., Chen, K., Sethi, R., & Elimelech, M. (2008). Reduced aggregation and sedimentation of zero-valent iron nanoparticles in the presence of guar gum. *Journal of Colloid and Interface Science, 324*, 71–79.

Uegami M., Kawano J., Okita T., Fujii Y., Okinaka K., Kayuka K., & Yatagi S. (2006). Iron particles for purifying contaminated soil or groundwater. US Patent 7,022,256, Apr. 4, 2006.

Wang, C., & Zhang, W. (1997). Synthesizing nanoscale iron particles for rapid and complete dechlorination of TCE and PCBs. *Environmental Science & Technology, 31*, 2154–2156.

Xie, Y., & Cwiertny, D. M. (2012). Influence of anionic cosolutes and pH on nanoscale zerovalent iron longevity: Time scales and mechanisms of reactivity loss toward 1,1,1,2-tetrachloroethane and Cr(VI). *Environmental Science & Technology, 46*, 8365–8373.

Zhan, J., Zheng, T., Piringer, G., Day, C., McPherson, G. L., Lu, Y., Papadopoulos, K., & John, V. T. (2008). Transport characteristics of nanoscale functional zerovalent iron/silica composites for in situ remediation of trichloroethylene. *Environmental Science & Technology, 42*, 8871–8876.

Zhan, J., Sunkara, B., Le, L., John, V. T., He, J., McPherson, G. L., Piringer, G., & Lu, Y. (2009). Multifunctional colloidal particles for in situ remediation of chlorinated hydrocarbons. *Environmental Science & Technology, 43*, 8616–8621.

Zhan, J., Kolesnichenko, I., Sunkara, B., He, J., McPherson, G. L., Piringer, G., & John, V. T. (2011). Multifunctional iron–carbon nanocomposites through an aerosol-based process for the in situ remediation of chlorinated hydrocarbons. *Environmental Science & Technology, 45*(5), 1949–1954.

Zhang, W. (2003). Nanoscale iron particles for environmental remediation: An overview. *Journal of Nanoparticle Research, 5*, 323–332.

Zheng, T., Zhan, J., He, J., Day, C., Lu, Y., McPherson, G. L., Piringer, G., & John, V. T. (2008). Reactivity characteristics of nanoscale zerovalent iron–silica composites for trichloroethylene remediation. *Environmental Science & Technology, 42*, 4494–4499.

Zhou, X,. Lv, B., Zhou, Z., Li, W., & Jing, G. (2015). Evaluation of highly active nanoscale zero-valent iron coupled with ultrasound for chromium(VI) removal. *Chemical Engineering Journal, 281*, 155–163.

Chapter 15
Sustainable Environmental Remediation Using NZVI by Managing Benefit-Risk Trade-Offs

Khara Grieger, Rune Hjorth, Alexis Wells Carpenter, Frederick Klaessig, Emilie Lefevre, Claudia Gunsch, Kullapa Soratana, Amy E. Landis, Fern Wickson, Danail Hristozov, and Igor Linkov

Abstract Ensuring the sustainable development and use of NZVI for in situ remediation requires the incorporation of a multitude of factors and criteria, including those related to technology performance, cost, potential impacts to the environment and human health, as well as ethical, social, and legal concerns. This chapter

---

K. Grieger (✉)
RTI International, Health and Environmental Risk Analysis Program, Research Triangle Park, NC, USA
e-mail: kgrieger@rti.org

R. Hjorth
Department of Environmental Engineering, Technical University of Denmark, Kongens Lyngby, Denmark

A. W. Carpenter · E. Lefevre · C. Gunsch
Duke University, Department of Civil and Environmental Engineering, Durham, NC, USA

F. Klaessig
Pennsylvania Bio Nano Systems, LLC, Philadelphia, PA, USA

K. Soratana
School of Logistics and Supply Chain, Naresuan University, Phitsanulok, Thailand

School of Sustainable Engineering and the Built Environment, Arizona State University, Tempe, AZ, USA

A. E. Landis
Department of Civil and Environmental Engineering, Colorado School of Mines, Golden, CO, USA

F. Wickson
GenØk Centre for Biosafety, Forskningsparken, Tromsø, Norway

D. Hristozov
University Ca'Foscari Venice, Venice, Italy

I. Linkov
Environmental Laboratory, Engineer Research and Development Center, US Army Corps of Engineers, Vicksburg, MS, USA

T. Phenrat, G. V. Lowry (eds.), *Nanoscale Zerovalent Iron Particles for Environmental Restoration*, https://doi.org/10.1007/978-3-319-95340-3_15

provides an overview of these factors in order to help guide the sustainable development of NZVI. Among other main results, we find that while there are promising findings regarding its performance and effectiveness as a remediation technique, there are also growing concerns regarding its impacts to the environment and health. To date, most of this research has focused on the potential (eco)toxicological effects of NZVI with limited research on broader issues such as social or ethical implications. In fact, the social implications of NZVI, including the ability for a range of stakeholders and members of the public to be active participants in decision-making processes, have either been minimal or nonexistent. We also find that marketplace limitations appear to be serious obstacles to ensuring the sustainable development and use of NZVI as an environmental remediation technology, including questions pertaining to the validity of its cost-competitiveness. In order to balance the potential benefits, risks, and uncertainty characteristics of NZVI, there are a number of decision support frameworks and risk analysis tools which may be applied, including multi-criteria decision analysis, life cycle assessment, as well as diverse risk characterization or screening tools (e.g., NanoRiskCat). While several of these frameworks and tools may be suited for NZVI in theory, very few of them have been applied to NZVI in practice. In conclusion, these results indicate that while NZVI has potential to reduce environmental contaminants through in situ remediation, its development and use, particularly at field-scale sites, has not proceeded in the most sustainable manner possible thus far. In light of this, we provide specific recommendations to help ensure the sustainable development and use of NZVI, including recommendations specific for diverse stakeholder groups such as researchers, academics, industry, and government officials.

Keywords Nanoscale zerovalent iron · Sustainability · Benefit-risk tradeoffs · Human health impact · Ecological impact · Decision support frameworks

15.1 Introduction

Environmental remediation involves the removal or degradation of contaminants in polluted soil and/or groundwater using a variety of methods and techniques. The overall goal of environmental remediation is therefore to reduce an environmental – and in some cases also potential health – risk that is present within a contaminated site. Specific treatment methods and approaches to reduce an environmental risk primarily involve the degradation of contaminants through the application of different treatment technologies, such as ex situ treatment options (e.g., pump-and-treat) or in situ techniques (e.g., chemical oxidation, use of zero valent iron particles, etc.). As described in previous chapters, the use of NZVI as an in situ environmental remediation technique has increased in popularity and attention in recent years as it has frequently been cited to be a potentially cost-effective alternative to conventional treatment technologies, which are often very time- and resource-intensive. In fact, NZVI is the most widely used engineered nanomaterial in the USA and Europe for soil and water remediation to date.

Since NZVI is a new and novel technology, many questions remain related to its potential to cause adverse effects on the environment and human health, as is the case with other emerging technologies employing engineered nanomaterials. For example, it is not yet confirmed if NZVI and its various formulations pose a hazard to organisms living in the environment, such as microbial populations in the subsurface that are essential for ecosystem functioning. Moreover, there is at least the potential that the use of NZVI could result in long-term issues of persistency and bioaccumulation that may not be observed until well after the technology has been deployed. Furthermore, there are also societal implications that need to be considered in order to ensure the sustainable use and deployment of NZVI. For example, there are ethical, social, and potentially legal concerns to injecting engineered nanoparticles such as NZVI into the subsurface especially given the uncertainty regarding its potential to cause unanticipated, adverse effects. Moreover, public participation and involvement in decision-making regarding the use of NZVI for remediation purposes has either been minimal or absent thus far (Grieger et al. 2012a), potentially leading to greater public controversies and debates regarding its use. While dedicated research and time is needed to answer many of these questions and concerns (e.g., Grieger et al. 2010b), many of the broader societal issues should arguably be incorporated in the near term while this technology is still in development. Meanwhile, site remediators, researchers, decision-makers, and regulators all need cost-effective strategies to remediate contaminated soil and ground waters in the near term, before many uncertainties surrounding the use of NZVI can be decreased. Therefore, diverse stakeholder groups, including government agencies, academia, research institutes, and industry, are interested in ways in which NZVI may be developed and used in a sustainable manner that incorporates not only effectiveness and economic feasibility but also additional concerns related to environmental and human health and societal impacts (Fig. 15.1). In other words, in order for the potential benefits of NZVI to be realized as a competitive remediation technology to clean up contaminated sites, the environmental, health, societal, ethical, as well as economic considerations need to be included in the development and deployment of NZVI along with decision-making aspects throughout the entire value chain of this novel material.

This chapter provides an overview of the key factors involved and guidance for decision support in order to promote the sustainable development and use of NZVI for environmental remediation. More specifically, Sect. 15.2 provides a brief overview of the use of NZVI; Sect. 15.3 details the benefits expected from its use; Sect. 15.4 provides an overview of the main concerns regarding the potential ecological and health risks and limitations of NZVI, as well as ethical, legal, and societal concerns; Sect. 15.5 provides an overview of how the potential benefits and risks of NZVI may be balanced using structured decision support tools; and finally Sect. 15.6 provides overall conclusions for the chapter and specific recommendations to promote the sustainable use of NZVI. We hope that the information presented herein will ultimately help promulgate the pursuit of a sustainable remediation technology

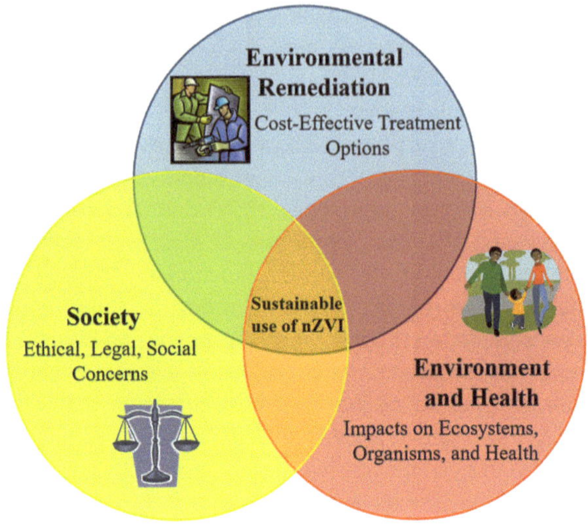

Fig. 15.1 The sustainable development and use of NZVI for environmental remediation requires the integration of cost-effective remediation techniques, mitigation of potential risks related to the environment and human health, and consideration of societal factors

that incorporates not only traditional factors relevant for environmental remediation, such as cost and effectiveness of treatment options, but also broader aspects related to environmental, health, and societal factors.

15.2 Current State of NZVI Use: Brief Overview

To date, the use of NZVI in pilot testing or field trials has been mainly in the USA and Europe (Chap. 1). In the USA, NZVI has been used in a large number of field-scale applications including private, residential, military, and manufacturing sites, particularly within the southwest and eastern states (Grieger et al. 2010a; Woodrow Wilson International Center for Scholars 2014). In Europe, there have been over a dozen pilot tests using NZVI although only three full-scale field applications: one in Germany and two in the Czech Republic (Mueller et al. 2012). The Project for Emerging Nanotechnology has recorded more than 40 sites worldwide using NZVI for nanoremediation in seven different countries (Woodrow Wilson International Center for Scholars 2014). However, the project's list is not all-inclusive nor regularly updated, and new sites are being established regularly.

Of the recorded sites, chlorinated compounds including trichloroethylene (TCE), dichloroethylene (DCE), and tetrachloroethylene (PCE) are the primary targets for NZVI treatment. Through the use of stabilized NZVI and multiple injection sites, near-complete remediation of field sites with chlorinated contaminants has been demonstrated. For example, an abandoned metal processing plant in Alabama with initial PCE and TCE concentrations of 4133 and 3710 μg/L, respectively, showed ca. 75% removal only 7 days after injection of a carboxymethylcellulose (CMC)-stabilized NZVI (He et al. 2010). The success of NZVI for dechlorination has inspired current research to focus on developing NZVI for remediation of other halogenated compounds, such as brominated flame retardants and fluorinated compounds (Keum and Li 2005; Ghauch 2008). Moreover, NZVI's proven sorption capabilities are being explored for removal of heavy metals and radioactive complexes such as Cr(VI), As(III), As(V), and U(VI) (Yan et al. 2013).

Currently, the most significant obstacle to using NZVI in the field is its low particle mobility, deliverability, and contaminant selectivity, which is important to ensure a good contact between NZVI nanoparticles and the source of contamination. This is particularly the case for uncoated (or bare) NZVI nanoparticles that tend to aggregate and adhere to soil particles. As described in previous chapters of this book, most field studies indicate that an average migration distance of less than 1 m can be expected with bare NZVI (Tratnyek and Johnson 2006; Saleh et al. 2008; Müller and Nowack 2010). To address this immobility hurdle, researchers have engineered coated or stabilized NZVI for greater migration (Kim et al. 2009). Furthermore, at least one study also documented that certain stabilizers allow NZVI to remain mobile for up to 8 months under some hydrogeological conditions (Kim et al. 2009). Currently, polymer-stabilized NZVI (e.g., carboxymethyl cellulose, poly(styrene sulfonate), poly(aspartate)), emulsified NZVI, and bimetallic NZVI are most commonly employed for field-scale studies. Of course, one drawback of employing a stabilizer is a concomitant decrease in reactivity as the surface of NZVI is covered. Moreover, hydrophilic stabilizers that aid in mobility decrease interaction between NZVI and hydrophobic target compounds. To address this, Saleh et al. (2005) proposed using NZVI modified by an amphiphilic triblock copolymer that contained polyanionic blocks to stabilize NZVI in suspension and a hydrophobic poly(methyl methacrylate) block to drive nanoparticle adsorption to the NAPL/water interface. They demonstrated the potential for NAPL targeting by demonstrating that polymer-modified NZVI adsorbed strongly to NAPL/water interfaces. NZVI surface activity was confirmed by the ability of triblock copolymer-modified NZVI to emulsify trichloroethylene (DNAPL) or dodecane (NAPL) in water. The high emulsification effectiveness of polymer-modified NZVI indicated that the potential for NAPL interfacial targeting via NZVI adsorption would be enhanced by amphiphilic polymer coatings on NZVI. Similarly, Phenrat et al. (2011) successfully utilized NZVI modified by olefin-maleic acid copolymer to target dense nonaqueous phase liquid (DNAPL) source zone in a flow-through porous media model. They reported that the amount of olefin-maleic acid copolymer-modified NZVI that targets a NAPL source depends on NAPL saturation (Sn), presumably because the saturation controls the available NAPL/water interfaces for targeting and the flow field through NAPL source. Furthermore, Edmiston et al. (2011) report on the development of Iron-Osorb®, a composite material where NZVI is embedded in a hydrophobic organosilane matrix, where the matrix serves to selectively sorb hydrophobic compounds (e.g., TCE) and increase interaction with NZVI surfaces. A surfactant was co-injected with Iron-Osorb® to allow proper mixing with groundwater. Along with other improvements in injection engineering, the NZVI-silica composite was found to have an improved sphere of influence up to 1472 m^3. Clearly, the success of NZVI for in situ remediation will not only require complex formulations but also careful engineering of proper drilling and injection methods.

Results from recent field studies also indicate that while the high reactivity of NZVI serves as an advantage for quick reduction of target compounds, NZVI's short lifetime also serves a major obstacle for its ultimate success. These nanoparticles are often synthesized at the site and/or are stored under inert atmospheric conditions to maximize its reactivity prior to use. After injection, NZVI's nonselective reductive reactivity, not only with target contaminants but also any other reducible species, limits its lifetime. As such, contamination levels may increase back to pre-injection levels following exhaustion of NZVI. Improving mobility, enhancing selectivity for targets, and exploring effective injection techniques (i.e., concentrations, composition, number of injections, etc.) are current areas of development in order to establish this emerging material as a successful in situ remediation technique. One example of this approach is described in Chap. 9 on the synthesis, reactivity, and lifetime of sulfidized NZVI.

15.3 Potential Benefits of Using NZVI

Given the increased attention on the use of NZVI as a promising remediation technique in recent years, it is not surprising that there are clear expected benefits to its employment. These benefits primarily relate to environmental benefits, attributed to the reduction of contaminants in soil and groundwater, as well as market (economic) benefits from its development and deployment. These aspects, including the mechanisms behind NZVI's ability to degrade environmental contaminants as well as comparison to other commonly used and other in situ techniques, are described in more detail in the subsequent sections.

15.3.1 Environmental Benefits

As an environmental remediation option, the expected environmental benefits from the use of NZVI are clearly related to the reduction of contaminants, both in terms of their concentrations as well as the size of contaminant plumes. Since the overall objective of using NZVI for environmental remediation is to improve environmental quality, its inherent use is designed to enhance environmental health and sustainability. Since NZVI is one of several remediation treatment options from which a site remediator or project manager must select, a review of these different options including both conventional treatment techniques as well as new or emerging in situ options for remediation is provided below.

Remediating contaminated soil and groundwater continues to be an expensive process. A US Government Accountability Office (GAO) report from 2010 estimated that the US government spent approx. \$335–681 million on remediating contaminated sites between 2010 and 2014 (Stephenson 2010). It has also been estimated that there are 235,000–355,000 sites that still require cleanup in the USA, with an estimated price tag of $174–253 billion using conventional remediation strategies (US EPA 2005). In another estimate, the cost of NZVI was significantly lower than that of pump-and-treat and reactive barrier technologies: $450,000 for NZVI, $2.2 million for reactive barrier technology, and $4.16 million for pump and treat. Based on these estimates and other similar comparisons, the use of NZVI for in situ remediation therefore has often been cited as a cost-effective technique to clean up contaminated soil/groundwater, which may be particularly attractive for federal agencies with limited available funds for remediation projects. However, it should also be mentioned here that a full validation of NZVI's cost-competitiveness is suspected to require more long-term studies and in-depth reviews of costs related to the production and use of NZVI across its entire life cycle (e.g., Crane and Scott 2012). In addition to cost, soil excavations and treatments of contaminated sites deep below the surface also cause large environmental disturbances. Therefore in situ techniques like NZVI have an advantage of avoiding these large-scale environmental disturbances by remediating the contaminants in the subsurface.

Overall, the development of stabilized NZVI has yielded systems with increased mobility, allowing transport of NZVI particles along and through contaminated soils and groundwater for more efficient remediation than stationary methods. The use of NZVI may also provide faster remediation, with timelines over the course of days to months compared to pump-and-treat methods that can span years to decades (Mueller et al. 2010). Other authors have, however, also suggested that longer-term trials may be required to fully document the complete remediation capacity of NZVI since rebound of contamination has also been observed in some cases (O'Carroll et al. 2013).

As described in previous chapters of this book, contaminants can be degraded into benign by-products by NZVI through reductive mechanisms. For example, dehalogenation can occur to transform chlorinated compounds (TCE) to ethene. Moreover, toxic metals such as Cr(VI), As(V), As(III), Pb(II), Zn(II), Cd(II), Ni(II), and Cu(II) can be treated either by reduction or sorption/complexation (Li and Zhang 2007). However, complete degradation does not always occur, and some instances result in the generation of compounds of equal or greater toxicity than the initial target contaminant. For example, laboratory studies have proven NZVI is capable of debrominating BDE-209, a toxic flame retardant with extreme hydrophobicity and low bioavailability. However, the reaction yields an array of lower bromo congeners that are not only just as toxic but also exhibit greater bioavailability due to decreased hydrophobicity (Keum and Li 2005). The production of by-products is also relevant for other remediation strategies, including thermal and chemical oxidation, and therefore although a potential problem requiring consideration, this issue is not completely unique to NZVI (Müller and Nowack 2010). In light of this, NZVI-based technologies with longer reactivity times should be developed to help ensure complete degradation to benign products. Some researchers suggest that the reduction of contaminants into congeners with greater bioavailability may enhance the opportunity for subsequent biodegradation to occur (Jeon et al. 2013). This interplay between nano- and bioremediation may result in an overall more efficient treatment strategy (see also Chaps. 7 and 10).

In terms of conventional techniques to which NZVI may be compared, pump-and-treat and capping methods are the most widely used remediation methods. In addition to being expensive, however, these methods do not address the ultimate need for complete removal of the contaminant. With pump and treat, there still remains a large amount of contaminated mass that must be treated following excavation. Since capping methods do not actually remove contaminants from the site, close monitoring must continue indefinitely to ensure issues do not reappear over time. It is likely that ex situ remediation techniques will be phased out in the coming decade (Karn et al. 2009), and therefore in situ remediation options are likely the way forward to remediate contaminated soil and groundwater. Permeable barriers provide a slight advantage over ex situ approaches in that they can capture the contaminants for removal. While some permeable barriers are composed of bulk zero valent iron (ZVI) such that degradation can occur, others are composed of clays and zeolites where sorption serves as the main remediation strategy. The issue of disposal arises for these types of passive permeable barriers though since the contaminant is removed from the groundwater but not degraded. Furthermore, permeable barriers often require appropriate placement and sufficient residence time in order to achieve complete contaminant removal/degradation. Other drawbacks include costly excavation efforts and the inability to treat areas other than shallow aquifers.

Given this information regarding ex situ techniques, it seems clear that in situ methods can potentially represent a more sustainable alternative to ex situ methods, primarily due to lower anticipated remediation costs and decreased environmental disturbances inherent to the in situ processes. In addition to NZVI treatment, other currently practiced in situ techniques include chemical oxidation, thermal treatment, purging/flushing methods, and bioremediation (US EPA 2013). In situ chemical oxidation involves the use of an oxidative chemical agent that is injected into the subsurface matrices for contaminant degradation. This process uses relatively harsh chemical reagents that raise concerns about downstream environmental effects such as geochemical changes (e.g., mobilization of metals) and negative impacts on microbial communities (Crimi and Siegrist 2003). Thermal remediation treatments employ high temperatures to mobilize and volatilize contaminants that are then captured in collection wells. While effective, high-energy consumption and infrastructure demands result in high costs (Müller and Nowack 2010). Variability in soil and aquifers can also result in nonuniform heating and uncertainty in verifying the success of the treatment. Similar to in situ chemical oxidation treatment, effects of the high temperatures on microbial communities remain a major environmental concern (Friis et al. 2006). Air sparging and soil vapor extraction provide an in situ method with lower-energy requirements but are also restricted to use with volatile contaminants. In situ flushing is another technique for mobilizing contaminants for capture and removal. Flushing solutions range from acidic or basic solutions to aqueous solutions that contain chelating agents, cosolvents, or surfactants depending on the identity of the target for removal. Field studies employing flushing techniques are few due to difficulties in distribution in the subsurface (US EPA 2013).

Bioremediation and phytoremediation are also popular in the literature as potentially sustainable techniques. Bioremediation involves using microbes or fungi to biotically capture and degrade contaminants. For example, strains of bacteria with known dehalogenation enzymes have been used to degrade TCE (Suttinun et al. 2013). These microbes can either be naturally occurring or can be introduced to the contaminated site through a process called bioaugmentation. In both cases, long-term effects on the natural floral are a concern. Phytoremediation, where vegetation is used to treat contaminants, has proven effective in removing petroleum hydrocarbons, crude oil, halogenated compounds, pesticides, and explosives as well as inorganic compounds. For example, rice grass plants are effective at removing arsenic from contaminated soils and ground waters (Ampiah-Bonney et al. 2007). Phytoremediation is, however, limited in its ability to treat areas far below the surface, and the issue of disposal is a concern. As with other remediation strategies, the generation of equally or more toxic intermediates and by-products (e.g., production of vinyl chloride from biotic degradation of TCE) also remains a major hindrance especially considering the longer timelines required for bioremediation techniques (Mueller et al. 2010).

15.3.2 Market Benefits

In order for NZVI to be a truly sustainable remediation option, it must be competitive with other treatment techniques. A part of this competition is in terms of its marketplace performance, cost-competitiveness, in addition to issues of performance and reactivity. Translating laboratory advances into commercial products is a challenge that crosses many product types and technologies. Furthermore, the commercial success of utilizing NZVI chemistry for in situ remediation is not automatic nor is it based solely on technical performance. In many cases, as with other in situ remediation options, there are existing markets with established norms of practice as well as profitability and community acceptance issues. While early claims of performance and economic value may not always be confirmed in later trials or revenues, as discussed later, the active probing of technical and economic considerations is an inherent component to successful innovation.

In general, the market benefits of developing and employing NZVI for environmental remediation include the NZVI market itself (i.e., developing NZVI, functional coatings, etc.) and trickle-down benefits from this market (e.g., remediation equipment based on NZVI, trained staff, etc.). As mentioned in the previous section, NZVI is often cited as a potentially cost-effective remediation option characterized by being a fraction of the cost compared to conventional technologies. In this way, the potential cost-effectiveness of NZVI would help its market mature and become established as a robust treatment option. For example, some studies have compared the overall costs of different remediation technologies, such as the 2004 field trial of bimetallic NZVI at the Klockner Road site of PSE&G led PARS Environmental (Karn et al. 2009). As mentioned above, this study resulted in the following estimates across three different remediation technologies: $450,000 for NZVI, $2.2 million for reactive barrier technology, and $4.16 million for pump and treat, demonstrating that NZVI was extremely cost-effective in this case. It should be noted here, however, that some authors have questioned the long-term suitability of these estimates when calculating the overall marketplace benefits of NZVI (Müller and Nowack 2010; Crane and Scott 2012; Mueller et al. 2012; O'Carroll et al. 2013).

In addition to cost comparisons, the US EPA also performed field trials on NZVI from 2000 to 2010 to demonstrate a proof-of-concept project under field conditions and to investigate potential market-oriented benefits of this emerging technology (US EPA OSWER 2013). These demonstration projects involving NZVI exemplify the fact that the NZVI market is being heavily considered by governmental agencies as a feasible remediation strategy worthy of consideration. Related to the successful employment of NZVI and post-remediation, future remediation costs may also be avoided not only in regard to site cleanup but also as property values are not subject to further losses due to proximity to contaminated sites. After site remediation, the site may be used for other purposes that may have increased market value, such as conversion to a recreational area and/or generally increasing the flexibility of the overall land use. While there are potential market-related benefits to using NZVI as a remediation option, we have identified only a few firms that advertise that they specifically use NZVI (e.g., NANO IRON®), while many firms and suppliers advertise that they use larger-sized iron particles (ZVI). In order to ensure marketplace success and achieve acceptance of NZVI technology, a remediation firm that is selling or employing NZVI will likely need "brand recognition" whereby their name is associated with successful remediation projects that span a broad spectrum of remediation trials over heterogeneous sites.

15.3.3 Conclusion on Benefits of NZVI as Environmental Remediation Treatment Option

Overall, there are numerous potential benefits of using NZVI for in situ remediation. For instance, NZVI is characterized by a high and broad-spectrum reactivity as a result of increased surface area per volume ratio and the large oxidation potential of iron in its zero valent state (Lin et al. 2008; Chang and Kang 2009). Indeed, NZVI has been shown to be effective against a wide range of contaminants, including metals, nonmetallic inorganics, halogenated aliphatics and aromatics, as well as other organic compounds (Yan et al. 2013). Thus, NZVI treatment could ideally be used as a single strategy to remediate sites containing multiple contaminants (Lien et al. 2007). Another benefit of NZVI is that the iron oxide-based by-products are generally considered inert, and the short reactive lifetimes of NZVI minimize potential long-term effects from the active material, i.e., Fe^0 (Müller and Nowack

2010). This is a significant advantage over other remediation strategies that rely upon more persistent nanoremediation strategies, such as photocatalytic titanium dioxide nanoparticles and sorbent carbon nanotubes. NZVI also has several additional advantages in terms of its ease of synthesis, limited cleanup/processing prior to use (see Chap. 2), as well as ability to be applied underneath buildings and other infrastructures similar to other in situ techniques (Elliott and Zhang 2001).

In addition to performance and ease of use benefits, NZVI is also often cited as being significantly cost-effective compared to conventional technologies, in terms of both labor and energy (Karn et al. 2009). In this way, these anticipated cost-savings has made in situ NZVI treatment a very attractive option for risk managers or site remediators. In addition, the successful development and deployment of NZVI may also lead to other market benefits, including the development of a specific NZVI market and associated impacts from its use, and may lead to cost-savings for remediators (such as federal agencies) with limited available funding for remediation projects. It should be noted, however, that the potential cost-effectiveness of NZVI has been questioned by several authors, and therefore future studies are needed to validate its full, long-term cost-effective potential as a remediation option.

15.4 Potential Risks and Limitations of Using NZVI

While the previous section described the potential benefits of using NZVI as an in situ remediation technique, this section describes some of the main concerns regarding its development and use. As detailed below, there has been an increasing amount of ecotoxicological and toxicological studies on the potential impacts of NZVI published in the past 5–10 years which raise concerns regarding the potential environmental and health impacts of NZVI. These studies are extremely relevant for risk assessment purposes and for the design of clear regulatory guidelines regarding emerging remediation techniques such as NZVI. In addition to potential impacts on ecological and human health, other concerns exist in terms of broader social, ethical, and legal aspects, as discussed in the latter part of this section. These potential risks and limitations of using NZVI as an in situ remediation method are therefore described in more detail below.

15.4.1 Ecological Risks

Since NZVI is directly injected into soil and groundwater through its intended use as an in situ remediation technique, one of the primary concerns regarding its use and deployment pertains to the potential impacts to ecological systems and organisms. If NZVI were to be engineered to be more mobile in the subsurface, potential exposures may occur from migration of the injected particles to surface waters. Alternatively, the upward migration of NZVI during injection could place the particles into the rhizosphere where interactions may occur with important soil organisms such as rhizobia, fungi, or plants. To date, there are various ecotoxicological studies published which relate to the potential impacts of NZVI on selected ecological receptors, and these are grouped below according to research investigations on (i) microbial species and fungi based on in vitro models, (ii) complex microbial communities, and (iii) macroorganisms.

15.4.1.1 In Vitro Toxicity Studies on Microbial Species and Fungi

A relatively large number of studies have been conducted in order to assess the toxicity of NZVI on individual microbial species and fungi (Auffan et al. 2008; Lee et al. 2008; Diao and Yao 2009; Li et al. 2010; Shah et al. 2010; Chen et al. 2011b; Kim et al. 2011; Fajardo et al. 2012; Keller et al. 2012; Marsalek et al. 2012; Fajardo et al. 2013; Otero-González et al. 2013; Sacca et al. 2013b; Zhou et al. 2013). These studies primarily utilize standard in vitro toxicity assays, monitoring cell viability (e.g., counting plate method, microscopy), metabolic activity (e.g., O_2 consumption, protein transcriptional level, enzymatic activities), or cell integrity (e.g., electron microscopy and enzymatic assays). Overall, the effect of NZVI on prokaryotes seems to be highly species dependent. While no toxic effect was detected on *Klebsiella* spp. even at NZVI concentrations up to 10 g/L (Fajardo et al. 2012; Sacca et al. 2013b) and sporulation seems to be used as a defense mechanism in *Bacillus* species in response to NZVI exposure (Fajardo et al. 2012, 2013), severe toxic effects have been documented for *Escherichia coli* (Auffan et al. 2008; Lee et al. 2008; Li et al. 2010; Chen et al. 2011b), *Pseudomonas fluorescens* (Diao and Yao 2009), and *Agrobacterium* spp. (Zhou et al. 2013) for concentration below 1 g/L (i.e., 1/10th the concentration of NZVI slurries injected into the ground), and these generally followed a standard dose-response relationship. Electron microscopic observations showed that NZVI adsorbed on the bacterial cell wall, leading to physical disruption of cell membranes and leakage of cellular content. The generation of reactive oxygen species (ROS) by NZVI particles (i.e., through the Fenton reaction) is the most frequently suggested mechanism behind NZVI toxicity as extracellular ROS can induce oxidative stress response and, in the worst-case scenario, irreversible cell damage (Fajardo et al. 2013). Because of its acute toxicity toward viral particles (Kim et al. 2011) and the toxic cyanobacteria *Microcystis aeruginosa* (Marsalek et al. 2012), NZVI has been suggested as a potential water disinfectant and a treatment to prevent the formation of cyanobacterial bloom. In contrast, fungi seem to display a relatively low sensitivity toward NZVI toxicity (Diao and Yao 2009; Shah et al. 2010; Otero-González et al. 2013). Eukaryotic microalgae, however, present some degree of sensitivity toward NZVI (Keller et al. 2012; Marsalek et al. 2012), although less pronounced than that observed for prokaryotic species.

As discussed above, the tendency of bare NZVI to agglomerate and oxidize rapidly once released in the environment causes a decrease in both its transport and reactivity, reducing its effectiveness for contaminant remediation. NZVI coated with synthetic polymers or surfactants has been used as an alternative to bare NZVI in order to enhance mobility and slow down aging (i.e., oxidation). The toxic effects of coated and aging (i.e., oxidized) NZVI particles have also been tested, and results show a significant decrease in toxicity compared to bare NZVI (An et al. 2010; Li et al. 2010; Zhou et al. 2013). Presumably, macromolecule surface coating of NZVI prevents attachment of NZVI to the bacterial cell wall eliminating physical disruption of cell membranes and leakage of cellular content as well as ROS generation.

15.4.1.2 Ecotoxicity Studies on Complex Microbial Communities

An increasing number of studies have also been conducted on investigating the toxicity of NZVI for indigenous complex microbial communities in various natural or engineered environments (Kirschling et al. 2010; Xiu et al. 2010; Cullen et al. 2011; Fajardo et al. 2012; Barnes et al. 2010; Kumar et al. 2013; Pawlett et al. 2013; Sacca et al. 2013a; Tislton et al. 2013; Wu et al. 2013; Yang et al. 2013). These studies have integrated traditional microbial ecology as well as molecular techniques but more importantly used a microcosms approach. This approach has enabled these studies to take into account the interaction of the microbial components and the environmental matrix, which is arguably more relevant than in vitro toxicity assays given that the characteristics of the environmental matrix are likely modulate the potential toxic effects of NZVI (Chen et al. 2011b; Pawlett et al. 2013).

In the explored studies, community composition was determined using techniques such as fluorescence in situ hybridization, denaturing gradient gel electrophoresis, phospholipid fatty acid analysis profiles, environmental cloning, and pyrosequencing. Results of these studies are confounding, and sometimes contradictory, as some do not detect any change in microbial community composition while others do. One possible explanation for this could be that the different methods used to assess community composition display different degrees of taxonomic resolution. As discussed above, in vitro toxicity studies showed that NZVI toxicity was species dependent. Therefore, NZVI might be toxic to certain species and promote more tolerant and competitive species. Furthermore, upon its addition into an anaerobic system, NZVI decreases the redox potential, increases the pH, and produces cathodic H_2, creating environmental conditions which could select for certain bacterial species over others and cause a shift in microbial community composition (Xiu et al. 2010; Kumar et al. 2013). Another explanation for differences in effects observed between studies is that the soil properties and initial bacterial populations were different prior to adding the NZVI. More studies would be needed to elucidate the impact of environmental variables on the susceptibility of microbial communities to NZVI.

Microbial communities play critical functions in global nutrient cycling. Therefore, only looking at changes in community species composition gives an incomplete picture of the impact of NZVI on ecosystem functioning. Indeed, some bacterial species might be functionally redundant and a shift in microbial species might not necessarily result in functionality loss. Some of the published studies

(Kirschling et al. 2010; Fajardo et al. 2012; Sacca et al. 2013a, b; Yang et al. 2013) have used more functional approaches involving various molecular and enzymatic techniques to look either at specific functional communities of interest (e.g., methanogens and sulfate reducers) or specific metabolic activities (e.g., oxidative stress response and ammonia oxidation potential). Fajardo et al. (2012), who quantified the expression of functional denitrifying genes in an agricultural soil amended with 34 g/kg NZVI, did not detect any detrimental effect of NZVI. Although no change in community composition was detected in a soil amended with 17 g/kg NZVI, Sacca et al. (2013a, b) found that catalase was overexpressed, indicating that soil bacteria were able to counteract the oxidative stress induced by NZVI. Other studies have focused on particular functional microbial groups such as methanogens or sulfate-reducing bacteria (Kirschling et al. 2010; Yang et al. 2013). Results from these studies, again, show confounding findings. While Kirschling et al. (2010) found that sulfate-reducing bacteria and methanogen populations of aquifer sediment increased after addition of 1.5 g/L NZVI, Yang et al. (2013) and Kumar et al. (2013) found that methanogenesis and sulfate reduction were inhibited in digested sludge and aquifer sediment at 1 g/L NZVI, respectively. These disparate findings are likely related to the fact that NZVI toxicity is largely mediated by initial environmental conditions (e.g., organic material content, physicochemical characteristics such as pH, ionic strength, porosity) (Chen et al. 2011b; Pawlett et al. 2013) and initial microbial community structure in the different environments studied.

In situ NZVI deployment may also negatively affect indigenous microbial species whose biodegradative potential could contribute to contaminant remediation. Therefore looking at the impact of NZVI on these particular communities is extremely relevant to assess long-term impacts. Xiu et al. (2010) showed that targeted trichloroethylene-degrading communities were temporarily inactivated by 1 g/L NZVI but recovered their functionality as NZVI aged. Evidence that dechlorination was predominantly the result of bacterial activity in this particular system was also provided. Therefore the addition of NZVI was, in this case, detrimental to trichloroethylene bioremediation. In line with these results, Tilston et al. (2013) showed that a soil chloroaromatic mineralizing community was inhibited by the addition of 10 g/kg NZVI but that the community fully recovered functionality after 7 days.

15.4.1.3 Toxicity Studies on Macroorganisms

Toxicity studies on animals and plants have also revealed that NZVI could have a strong negative impact (Li et al. 2009; Chen et al. 2011a; El-Temsah and Joner 2012a, b, 2013; Keller et al. 2012; Jiamjitrpanich et al. 2012, Marsalek et al. 2012, Chen et al. 2013; Kadar et al. 2013; Ma et al. 2013; Sacca et al. 2013a). All these studies employed standard toxicity assays or adapted assay protocols to organisms of common occurrence in the biota considered.

The primary biological parameters measured in these studies were reproduction success, embryological development, and adult and larval survival and growth. All of the surveyed studies showed a severe effect of NZVI on the organisms tested, especially at the highest concentrations (i.e., up to 10 g/L or 17 g/kg in assays conducted in solution and soil, respectively). Using such toxicity assays, marked decreases in growth, reproductive success, and survival were observed for earthworms, collembola, and the nematode *C. elegans* (El-Temsah and Joner 2013; Sacca et al. 2013a). However, Sacca et al. (2013a), who conducted a toxicity assay directly in soil, showed that growth and reproduction of *C. elegans* was enhanced even at high NZVI concentrations (i.e., 17 g/kg). This surprising result indicates, again, that NZVI toxicity is highly dependent on the environmental matrix (Chen et al. 2011a, b; Pawlett et al. 2013). As standard toxicity assays are not carried out under environmental conditions, these methods are inadequate to accurately assess the in situ toxic effects of NZVI under realistic environmental conditions. Severe effects of NZVI on the survival of planktonic species (Keller et al. 2012; Marsalek et al. 2012; El-Temsah and Joner 2013) as well as for the embryologic development and fertilization success of marine organisms (sea squirt, mussel; Kadar et al. 2013) have also been reported. Studies conducted on medaka fish showed an acute toxicity on juveniles accompanied by an increase in larval mortality and embryologic developmental delay or failure (Li et al. 2009; Chen et al. 2011a, b, 2013). Suggested mechanisms explaining the delay or failure in development are again related to the capacity of NZVI to catalyze the formation of ROS. Interestingly, Li et al. (2009) showed that although adult fish displayed some severe histopathological and morphological alterations of the gills and intestine, they were able to recover over time as NZVI aged.

In addition to concerns on the adverse effects of NZVI to microbial populations in the environment, concerns have also been raised in terms of NZVI's effects on terrestrial plants. Since plants and other autotrophic organisms serve as the primary producers in food webs, adverse impacts and potential for bioaccumulation may have consequences for food webs and organisms in higher trophic levels. Although direct comparisons between studies are complicated by methodological differences, the toxic effect of NZVI on plants seems to vary greatly across species (El-Temsah and Joner 2012a; Jiamjitrpanich et al. 2012; Marsalek et al. 2012; Ma et al. 2013). Several studies have demonstrated some negative impacts of NZVI in early developmental stages with germination failure and reduction of shoot growth but also in mature plants with a reduction of growth rate. Electron microscopic observations revealed the formation of an iron coating on plant root surfaces and occasionally internalization of iron particles in root cells (Ma et al. 2013). These findings suggest that iron particles could interfere with the uptake of water and nutrients by blocking root membrane pores, thus potentially explaining a reduction of growth rate over time.

15.4.1.4 Potential for Bioaccumulation

The potential for bioaccumulation in the food web is often one of the most significant parameters when discussing the potential health and environmental risks of a substance or material. This is because the ability for a substance such as a chemical or metal to accumulate to higher trophic levels (e.g., predators or humans) can cause greater toxicity for these organisms. To the best of the authors' knowledge, there have been no published studies that have investigated the bioaccumulation potential of NZVI. Therefore, the ability for NZVI to bioaccumulate in the environment and within the food web remains largely unknown at this time.

15.4.1.5 Potential for Migration in the Environment

In addition to the studies documenting the hazard and toxicity potential of NZVI on a variety of test organisms, other studies have started to shed some light on the potential for NZVI to migrate in the environment. The migration of NZVI in the environment would thereby increase the potential for direct exposure of NZVI for ecological receptors as well as indirect human exposures if substantial environmental migration (e.g., to natural water bodies) were to occur. To date, most research related to the migration of NZVI has been framed within the context of remediation and the ability of NZVI to migrate within a contaminated plume, which is essential for sufficient degradation of contaminants to occur. The average migration distance for uncoated or bare NZVI has been documented to be approximately one meter, although other studies have also cited migration to no more than a few centimeters under most environmental conditions (Saleh et al. 2008) (Chap. 6). This is due to contact with the surrounding environmental media, such as groundwater aquifers, as well as other processes that may limit NZVI's migration ability (e.g., retention due to mineral concentrations (Ca, Mg) or ionic strength conditions; Theron et al. 2008). However, the application of coatings, or the use of supports for the NZVI (e.g., carbon), many of which are designed to help migration of NZVI in the environment, may also increase migration distances (Chaps. 5 and 6). For example, one study documented that some coatings may allow NZVI at low particle concentration to migrate tens or even hundreds of meters in consolidated sandy aquifers (Saleh et al. 2008). Other studies have shown that NZVI may remain mobile for up to 8 months with the application of certain coatings and under some hydrogeological conditions (Kim et al. 2009). These studies suggest the potential for low concentrations of NZVI to continue to migrate in a porous medium for large distances. However, the limited reactive lifetime of those materials suggests that exposures would likely be to Fe-oxide reaction products rather than to reactive NZVI (Chap. 3).

15.4.2 Human Health Risks

The rapid development of nanoremediation and deployment of NZVI in the field has naturally raised concerns regarding human health. However, because NZVI is directly injected underground, direct exposure to NZVI is unlikely to happen outside of inhalation from improper handling or from accidental spillage during nanoparticle production, transport, and field application. Exposure from groundwater is also possible, but this would likely be exposure to the oxidized product. Nevertheless, a few toxicity studies using human and rodent cell cultures have been conducted to study potential hazards of NZVI in the case of skin exposure, inhalation, or ingestion. Keenan et al. (2009) showed a decrease of human bronchial epithelial cell viability exposed to NZVI in vitro, suggesting that inhalation of NZVI could potentially result in lung irritation. Phenrat et al. (2009) who assessed the neurotoxicity of NZVI using cultured rodent microglia and neurons revealed evidence of mitochondrial swelling and apoptosis in microglia as well as internalization of nanoparticles in neurons. However, the oxidized product was much less toxic than the reactive NZVI containing Fe^0. Nadagouda et al. (2010), using human keratinocyte cells as model for skin exposure, found that NZVI was damaging cell membrane integrity.

Although sufficient literature currently does not exist to assess the potential for transport of NZVI into groundwater or surface waters used for drinking water or irrigation for agricultural, it could be theorized that indirect human exposures may occur through the ingestion of food or water containing NZVI nanoparticles. This would of course assume that NZVI would have the ability to migrate from the contaminated site to drinking water or groundwater sources and not be filtered through drinking water processes. While there are some studies documenting the potential for NZVI to migrate up to 100 m in some cases when applied with certain coatings, there are currently no data to support the migration of NZVI in the subsurface to the extent that it could penetrate current drinking water systems. It should also be noted that the migration of NZVI to distances up to 100 m would likely involve transformed NZVI to iron oxide nanoparticles. Therefore, toxicity investigations involving iron oxide nanoparticles may be more relevant than bare or pristine NZVI nanoparticles at these distances.

15.4.3 Other Considerations: Ethical, Legal, and Social Aspects

Assessing the sustainability of a new technology necessitates the adoption of a broader framework for assessment than simply evaluating the potential risks it may pose to human health and the environment. This is because the requirements for sustainability extend beyond simply avoiding potential harm, to actively creating and facilitating the development of socio-ecological systems that can be maintained in a healthy state over generations and which are robust and resilient in the face of change. Sustainability is now widely accepted as involving a “triple bottom line” (Elkington 1997) in environmental, economic, and social dimensions should be incorporated simultaneously (See Fig. 15.1). As some of the environmental and economic dimensions of NZVI development and use are addressed above, this section will highlight some of the social questions that arise and need to be considered, including some of the ethical and legal issues at stake.

15.4.3.1 Social Implications and Public Participation

The social dimension of sustainability has traditionally received less attention and conceptual development than economic and environmental aspects (Dillard et al. 2009). Work in this field is, however, increasing both theoretically and practically, and articulations of the types of issues that are relevant to consider in any assessment of social sustainability are emerging. These include issues and questions concerning distributive justice, equitable inclusion, collective agency, participatory opportunity, political accountability, and institutional sufficiency. The generation of public debate, community resistance, and social controversy around previous emerging technologies, such as genetically modified organisms (GMOs) and nuclear power, has led to a strong call for enhanced democratization of science and innovation and enhanced public engagement in decision-making on the development and use of new technologies. Access to participate in decision-making processes and giving people the agency to be involved in decisions that affect them is also an element of social sustainability. For nanotechnology generally, there has been an extensive effort to facilitate what has been called "upstream public engagement" in the innovation and decision-making process, although this has not been without problems (e.g., Delgado et al. 2011), and it has represented a significant shift in the conceptualization of science, technology, and innovation policy.

This effort to facilitate greater public involvement in nanotechnology has, however, failed to extend to the case of nanoremediation. One recent study highlights that although NZVI is being seriously considered, and in some cases implemented as a viable remediation strategy in several countries, there has been very little specific public engagement undertaken to ensure that all stakeholders are involved in decision-making regarding this technology (Grieger et al. 2012a). While the case has been discussed in a couple of events focused on nanotechnology broadly, the actors that are actively developing and deploying the technology do not appear to have made any substantial effort to advance participatory opportunities, equitable inclusion, or collective agency in this case. Within the broader context of a clear commitment to public engagement in nanotechnology, the lack of apparent stakeholder and public participation in the development and use of NZVI for remediation seems particularly noteworthy and of concern, especially since nanoremediation may represent one of the largest point source releases of free nanoparticles into the environment and seems to have potential for societal and environmental implications.

As highlighted in other sections of this chapter, there are serious gaps and uncertainties in the available knowledge on questions relating to health and environmental risk. While such uncertainties may be a concern on their own for assessing the sustainability of NZVI, they take on enhanced significance in light of the lack of public engagement. When the scientific basis from which an assessment of potential risks could be conducted is highly uncertain and still in development, then the role for various stakeholders and publics in the evaluation of the technology becomes particularly crucial. It is important to provide a critical evaluation of how the available knowledge has been framed, generated, and interpreted and to canvas a range of different perspectives and values for informing decision-making processes. In the case of NZVI where there seems to be challenges of translating the relevance of the available results generated under controlled laboratory conditions into real world contexts in which the technology is to be deployed, having a decision-making process that is open, transparent, democratic, and empowering citizen engagement is a crucial element of the technology's social sustainability.

15.4.3.2 Ethical and Regulatory Implications

In addition to these concerns relating to the opportunities and mechanisms available for inclusive and participatory decision-making, there are also potential issues related to the impacts of the technology on social structures and organization, as well as on resource and risk distribution concerns. As with the development of any technology, it is important to consider the types of actors, arrangements, and infrastructure involved and required to make the technology function (e.g., including those not only related to the production of NZVI particles but also to their transport, storage, and injection into sites); how the required system could lead to potential concentrations of knowledge, money, and power; and whether the same people that stand to benefit from the technology's development and use are also those that will bear the potential risks. An assessment of such factors may need to be done on a case-by-case basis specific to the contexts within which the technology will be deployed. However, one of the common general concerns may relate to the ability of NZVI to actually address the root cause of the problem. That is, concerns about whether NZVI is actually able to treat and reduce the source of the pollution or whether use of the technology may create ongoing relations of dependence.

There may also be grounds for concerns regarding NZVI on factors of institutional sufficiency and political accountability, particularly in terms of the regulatory frameworks that apply. NZVI currently has an unclear status as a regulated material. That is, there is no agreed international position on whether the nano-component of ZVI as a remediation technology is recognized as a new material or requires any novel regulatory consideration. Since ZVI has been approved for use and deployed in the field for extended periods of time now, NZVI may not always be viewed by either innovation companies or regulatory agencies as a "new" material or technology that may require investment in new knowledge or additional regulatory scrutiny. This lack of clarity can be seen as an institutional insufficiency potentially affecting social sustainability, particularly if interpretations of what is required vary across national contexts.

Around the world, different nations have adopted different positions concerning the permissibility of nanoremediation. For example, the USA has adopted a very liberal position and already engaged in a range of commercial scale field applications (and indeed is the country with the largest number of field applications). The Czech Republic has also used full-scale field applications in at least two sites (Mueller et al.

2012). Germany has run limited and controlled field trials, seeking to gather more information and knowledge before proceeding to widespread use (DEFRA 2011), while the UK has effectively adopted a precautionary stance, with no use of nanoremediation techniques being encouraged before more knowledge is available on potential risks (The Royal Society and The Royal Academy of Engineering 2004). However, the proposed EU definition of a nanomaterial, i.e., that 50% of the material by particle number have at least one dimension that is less than 100 nm, makes NZVI a "nanomaterial" that requires special attention during product registration and use.

While the use of all remediation technologies must comply with local and national environmental legislation and operators will typically need to obtain local or regional permits to practice, under which an assessment of the safety of the technology will be included, political accountability for the use of NZVI can be of concern. This is because frameworks for regulating chemicals are typically developed at a national level (or at the European level for member states of the European Commission), but for nanomaterials, specific regulatory frameworks and requirements are still in development and contested. This creates a situation in which the use of NZVI or other nanoremediation technologies may be assessed at a local level, while there may be no clear or specific position on the safety testing requirements for nanomaterials available at a national or federal level. Social sustainability would require that clearer lines of political accountability be established, particularly concerning the regulatory status of nanoparticles and the authority responsible for their assessment, as well as appropriate methods for safety testing and implementation of monitoring requirements.

15.4.3.3 Market Limitations

While the overall purpose of NZVI as an environmental remediation option is to improve environmental quality, the market success and commercialization of NZVI does not rely solely on its technical performance. Other factors are also at play, including profitability, cost-effectiveness, and community acceptance of the novel technology. In fact, developing and using novel technologies and materials inherently requires some uncertainty in terms of its ultimate success, particularly on a long-term basis. In this way, innovation often tests a company's or firm's resilience in rebounding from potential setbacks or also in a firm's available resources to pursue these novel innovations. Past experience has also shown that when innovative technologies intrude on existing marketplace relationships, these relationships change as the technology itself matures (Utterback 1987; Utterback and Suarez 1993).

From the standpoint of NZVI innovation for in situ remediation, there currently is a heterogeneous landscape, in that individual contaminated sites may pose specific, unique challenges due to differences in environmental contaminants, hydrogeological characteristics, etc. There is currently no dominant NZVI product or application technique on the market and hence is still emerging as a viable remediation technology. Decision-makers are also typically from diverse stakeholder groups such as corporate landowners and government agencies and may not necessarily have a deep and interconnected knowledge of each other or with the site in question. Unlike a market defined by a commercial product, the soil and groundwater remediation market does not respond to the typical "supply and demand" forces, and the number or state of the contaminated sites do not change in light of available budgets, for example. Furthermore, environmental remediation also inherently involves public health considerations regarding drinking water quality and agricultural water use, among other issues. This heterogeneous landscape can be particularly challenging for NZVI as an emerging technology.

There are also unique challenges to NZVI's marketplace sustainability related to its novel properties as well as use in the subsurface. For instance, NZVI's novel chemistry employed, particularly if applied with various coatings, may create new supply chains as the marketplace emerges. These novel properties may also play a large role in terms gaining public trust and acceptance of this emerging technology, particularly if they lead to adverse impacts to health or the environment. Furthermore, subsurface chemical reactions may be difficult to monitor and which impact its market mature, although this is similar to other in situ treatment options as well. Finally, and perhaps more importantly, is the issue of whether NZVI is in fact a cost-effective environmental remediation option. While there have been some early reports documenting the potential cost-effectiveness of NZVI (e.g., Karn et al. 2009), other authors have also questioned whether more long-term studies which investigate the remediation capacity of NZVI may be needed as well as the inclusion of all costs related to its development and deployment. For instance, Müller and Nowack (2010) and Crane and Scott (2012) estimate that in order for NZVI to be cost-competitive with other in situ remediation options such as chemical oxidation, the price of NZVI should be reduced by a fifth to a tenth of current prices on a per kg basis. Others have also suggested that the full costs of NZVI development and deployment need to be considered such as the inclusion of the costly premiums when deploying NZVI as well as expensive manufacturing costs (Crane and Scott 2012). Based on cost-benefit analyses of NZVI, the German Federal Institute for Geosciences and Natural Resources concluded that the NZVI market was not yet ready for large-scale applications (Mueller et al. 2012). Therefore, these marketplace limitations appear to be serious obstacles to ensuring the sustainable development and use of NZVI as an environmental remediation technology.

15.4.4 Synopsis on NZVI Potential Risks and Limitations

Overall most research on the potential risks of NZVI have focused on performing (eco)toxicological studies (i.e., hazard potential) of NZVI as opposed to studies which have investigated potential exposures or ethical, societal, or legal implications. Looking across these studies which have been published in the last decade and particularly within the past 5 years, it is clear that there have been documented impacts on various microorganisms (microbes, fungi) and macroorganisms (plants, fish, aquatic and terrestrial species) as well as impacts on microbial communities and environmental systems. Most of these studies have shown that NZVI toxicity is species dependent and dose-response patterns have been documented in many cases. In addition to impacts on various test organisms, NZVI was also found to alter environmental conditions, such as redox potential, pH, etc., which have direct implications for surrounding microbial community compositions, some of which are critical to help aid the degradation of environmental contaminants. Also, the primary mechanism for toxicity based on the current literature is the development of ROS. At the same time, however, research in this field (similar to other engineered nanomaterials) is still evolving, and several of the studies produced confounding, and in some cases contradictory, results. This may be due to the use of different test methods or varying environmental conditions or initial microbial community structures between the studies. In terms of promoting the sustainable development of NZVI, several studies documented that coated or oxidized (aged) NZVI often reduced toxicity to various organisms.

It is expected that uncoated (bare) NZVI will only migrate a short distance in the environment which would reduce potential environmental exposures. However, other research also indicates that the application of certain coatings and under some hydrogeological conditions may increase the migration distances of NZVI up to tens or even hundreds of meters. If these migration distances would occur, this would increase the exposure potential of NZVI to a variety of ecological receptors as well as increase potentials for indirect human exposures. Most human exposures, however, are expected to derive from exposures during manufacturing, handling, as well as unintended use patterns (e.g., accidental spillage).

Implications of the development and use of NZVI also go beyond just the potential risks to human health and the environment. Social, legal, and ethical concerns must also be taken into account in order to ensure that NZVI is a sustainable technology for remediation. The social implications of NZVI, including the ability for a range of stakeholders and members of the public to be active participants in decision-making processes, have either been minimal or nonexistent. Within the broader context of a clear commitment to public engagement in nanotechnology, the lack of apparent stakeholder and public participation in the development and use of NZVI for remediation seems particularly of concern, especially since nanoremediation may represent one of the largest point source releases of free nanoparticles into the environment and seems to have potential for societal and environmental implications. In the case of NZVI, similar to other engineered nanomaterials characterized by high degrees of uncertainty, a decision-making process for NZVI development and deployment should be open, transparent, democratic, and empower citizen engagement in order for the technology to be socially sustainable.

In addition to these social concerns, questions of institutional sufficiency and political accountability are also relevant for NZVI. Currently, the regulatory status of NZVI is unclear, and there are no agreed international positions on whether the nano-component of NZVI as a remediation technology is recognized as a "new" material or requires novel regulatory considerations. Also in terms of regulatory implications, NZVI and other nanoremediation technologies seem to be assessed at local levels, while there are no clear or specific positions on the safety testing requirements for nanomaterials in general at national and federal levels. In order for NZVI to be developed and used in the most sustainable manner, clearer lines of political accountability should be established, particularly regarding the regulatory status of engineered nanomaterials, the responsible regulatory body(ies), as well as specific safety testing and monitoring requirements. Finally, marketplace limitations, including questions of its true cost-competitiveness, appear to be serious obstacles to ensuring the sustainable development and use of NZVI as an environmental remediation technology.

15.5 Balancing Benefits and Risks Using Structured Decision Support Frameworks

Given the information presented in the preceding sections on the potential benefits, risks, and uncertainties of NZVI as an environmental remediation technique, the work of many decision-makers may not be necessarily straightforward regarding the development and/or deployment of NZVI. This is similar to the use of other engineered nanomaterials, in which completing standard risk assessments is hampered by the significant degrees of both quantitative and qualitative forms of uncertainty (e.g., Grieger et al. 2009; Wickson et al. 2010). Therefore, decision-makers may not necessarily have the best tools available when deciding upon which remediation strategy to use in a given site if decisions are to be made on traditional risk assessment frameworks or other conventional strategies like cost-benefit analysis (Grieger et al. 2012b). This is especially true for regulatory agencies that operate on limited funding and need to demonstrate clear, transparent decision-making processes.

Decision-making in environmental remediation is a complex task involving trade-offs among economic and environmental benefits and risks. Some of the key questions an environmental engineer or site manager may ask are the following: Which remediation technology is best suited for the given site? What is the best method to decide which tool to use that is both transparent and defensible? How can decisions be made in a structured, transparent manner that incorporates diverse stakeholder preferences, if needed? Two of the main approaches for decision support when selecting between alternative remediation technologies are cost-benefit analysis and various types of risk analyses. Using these approaches would typically involve the evaluation of various criteria such as remediation time, efficiency of remediation, economic evaluations, as well as other parameters not easily translated into a monetary value (e.g., environmental benefits). The balancing of these different (and potentially competing) criteria complicates comparisons among alternative remediation technologies in order to select the most suitable remediation strategy or potential strategies. One method to balance diverse criteria may be to convert them in a manner which makes comparisons easier, such as converting them to a common "currency" in order to compare similar items. However, the incorporation of stakeholder preferences may also be important in many cases since remediation decisions may potentially involve social or ethical considerations as described in previous sections (e.g., public participation).

At the moment, there is currently no single assessment framework that takes a holistic view of the sustainability of remedial actions particularly involving engineered nanomaterials such as NZVI. Therefore, there is a strong demand for methods that allow for the consideration of the potential benefits and risks of remediation (including impacts at local and regional levels) and incorporates the range of uncertainties and potential stakeholder values. These assessment methods should also support broad-based participation in a robust and transparent way, as well as be able to transfer potential risks as expenses accrued through improvement in risk management procedures, which may be offset by the savings in insurance/risk transfer costs (Grieger et al. 2012a). In other words, decision support tools which help balance the potential risks, benefits, and underlying uncertainties and which take stakeholder preferences into account in a transparent manner are needed for remediation strategies like NZVI. These methods may be based on existing approaches such as multi-criteria decision analysis (MCDA) (Linkov and Moberg 2012), value of information (Keisler et al. 2014), Bayesian networks (e.g., Wiesner and Bottero 2011), or other methods including other risk screening tools, as described below in more detail.

15.5.1 Multi-criteria Decision Analysis (MCDA)

MCDA involves a large group of methods (Belton and Stewart 2011), designed to ensure that the synthesis of multiple sources of information is documented and directed toward a pre-defined goal (Linkov et al. 2011). Some important examples are MAUT/MAVT (multi-attribute utility/value theory), outranking, interactive, goal aspiration, AHP (Analytic Hierarchy Process), ELECTRE (Elimination and Choice Expressing Reality), PROMETHEE (Preference Ranking Organization Method for Enrichment Evaluations), and TOPSIS (Technique for Order Preference by Similarity to Ideal Solution) (Linkov and Moberg 2012). The MCDA methods have the ability to integrate the opinions of multiple experts, thus decreasing the bias of subjective judgments (Linkov et al. 2007, 2011). In this context MCDA can be modified for single- or multiple-person perspectives, such as diverse stakeholder groups. The multiple-person type is based on the group decision theory and involves multiple experts or decision-makers which provide various perspectives on the decision problem to reach objective conclusions. In this case, the MCDA algorithms have to include consensus measures showing how much the group of decision-makers agree or disagree on the results (Carlsson et al. 1992).

There are several reviews of decision-analytical methods used in environmental management of contaminated sites, including the use of MCDA (Kiker et al. 2005; Linkov et al. 2006a, b; Onwubuya et al. 2009; Huang et al. 2011). While Kiker et al. (2005) review applications of MCDA in real remediation projects, Onwubuya et al. (2009) provide an overview of the available decision support methods used to select less invasive, alternative remediation options such as phytoremediation, in situ immobilization etc., which, among others, cover also MCDA. Onwubuya et al. (2009) also provide an overview of the software-based decision support systems/tools used across Europe (e.g., DESYRE, PRESTO, CARO, ROSA), including information on the evaluation criteria they use (e.g., risk, cost, sustainability, and socioeconomic factors). These authors conclude that the only MCDA-based decision support tool that covers all criteria is DESYRE (Onwubuya et al. 2009). Therefore, DESYRE is a decision support system (DSS) for selecting remediation technologies for rehabilitation of contaminated sites, which can easily be applied to check if NZVI is most suitable for a certain remediation project. The methodology behind DESYRE involves three steps: the first step selects the range of suitable remediation technologies within a database according to their applicability to site-specific conditions; the second step sets comparative criteria, while the third step develops a ranking MCDA algorithm that compares and classifies the selected technologies (Critto et al. 2006). The following paragraphs describe this process in more detail while Fig. 15.2 provides an overview.

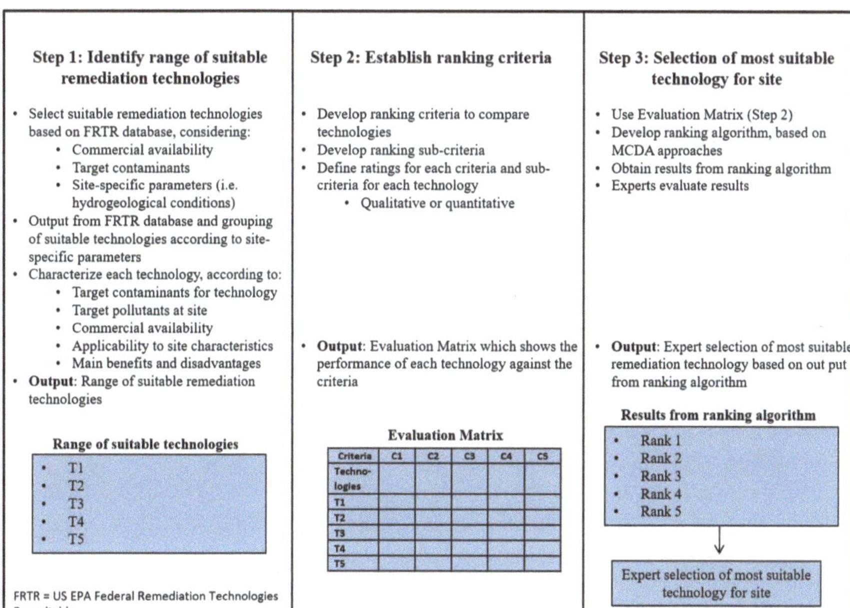

Fig. 15.2 Overview of DESYRE decision support system to select suitable remediation technologies, based on Critto et al. (2006)

Step 1 The goal of the first step is to provide a pool of remediation technologies suitable for a contaminated site. To select the remediation technologies, two subsequent selection filters are applied to the input database of technologies provided by the US EPA Federal Remediation Technologies Roundtable (FRTR). The first filter is based on (1) commercial availability of the considered technology and (2) its target contaminants overlapping those found in the investigated site. The second filter is concerned with site-specific parameters (i.e., hydrogeological and physico-chemical characteristics of the investigated environmental matrix) affecting the feasibility of remediation technologies. This selection procedure includes three interactive tables (A, B, C) and a characterization database, which logically drive the expert in the selection. The A table includes the input database of technologies and groups the cleanup technologies according to the treated contaminated matrix (i.e., soil, surface water, or groundwater, emitted off-gas). Each technology is characterized by (1) target contaminants, (2) the target pollutants found at the site, (3) commercial availability, (4) applicability to site characteristics, and (5) main benefits and disadvantages of application. The last column of the table includes a synthetic expert judgment or evaluation. The overall process results in a pool of all remediation technologies applicable to the study site. The B table includes only the technologies selected by the expert in the previous Table A after application of the first filter. It is only descriptive and provides the characterization of the selected technologies according to a set of additional criteria, including extraction, removal and retrieval, biodegradation, immobilization, destruction, and chemical transformation, specific cleanup effectiveness, and capability to be included in train technology treatments. The C table lists site-specific parameters that affect the applicability of the selected technologies. Those are either related to the treated matrix (e.g., pH, total organic carbon, hydraulic conductivity) or related to the target contaminants (e.g., vapor pressure, solubility) (Critto et al. 2006).

Step 2 The goal of the second step is to establish a set of comparative criteria. Critto et al. (2006) used the following six comparative criteria: reliability, course of action (i.e., intervention condition), hazardousness, community acceptability/impacts, effectiveness, and cost. Then the evaluation sub-criteria are associated with each criterion. Some of those are strictly correlated with technical aspects (i.e., costs, cleanup time, performance, reliability and maintenance, technology development status, cleanup operation locations, train technology, hazardous reagents use, contaminated matrix removal, and residuals production), whereas others refer to the potential effects on human health and the environment (i.e., dust and volatile substances emission, effects on water, and consequences to soil and community acceptability). For each evaluation sub-criterion, a qualitative or quantitative rating is defined. This rating scheme allowed the so-called evaluation matrix to be obtained. The matrix reports evaluation criteria in the columns and the selected technologies in the rows. The evaluation matrix provides the input data for the third step in the analysis.

Step 3 The goal of the third step is to decide upon the most suitable remediation technology by experts. In this step, the results from the evaluation matrix are used to make a comparative ranking of the remediation technologies by developing a ranking algorithm, which is based on a MCDA ranking algorithm. The results of the rankings will be evaluated by experts, whereby the experts select the most suitable remediation technology based on the ranking results.

The above approach demonstrates the ease and analytical rigor of this decision-analytical process and the many benefits derived from quantitative assessments. This displays how quantitative MCDA analysis can be integrated into a larger decision framework and proposes a formal methodology for selection of technologies in a remediation strategy, which can incorporate expert judgment. The application of such approaches is promising, since decision-making can be altered for stakeholder preferences through weighing of certain criteria more over other criteria. This could help decision-makers decide which environmental remediation technology or which form of NZVI may be "best" for the remediation site at hand, for example.

15.5.2 Value of Information and Bayesian Logic

Remediation of contaminated land is associated with high costs. In order to design remediation programs that are cost-effective, it is relevant to use value of information (VOI) analysis. VOI compares the benefit at the present state of knowledge with the benefit from acquiring new information from site investigations (Back et al. 2007). Bayesian networks (BN) can be used to calculate the expected change, i.e., the value of the investigation (Back et al. 2007).

BN methods can combine mechanistic and empirical information with expert judgment to build probabilistic forecasting models that can incorporate uncertainty and be easily updated with new information (Wiesner and Bottero 2011). They can integrate multiple lines of evidence to aid in decisions of whether to invest more resources in field investigations or not. BN models are adaptable because the context in which they are applied is always different in terms of unique site specificity, priorities, objectives, and influence of various stakeholders. Therefore, they are flexible in incorporating a wide range of variables and operating at different scales.

Dakins et al. (1996) were among the first to illustrate a methodology that utilizes Bayesian Monte Carlo analysis in combination with VOI analysis to predict the value of future data collection programs in environmental remediation. The authors applied the methodology to the specific case study of polychlorinated biphenyl (PCB) sediment contamination. It calculated the expected value of sample information (EVSI) as the difference between the expected loss of the optimal decision based on an outdated versus an updated set of data (Dakins et al. 1996). The application of the model showed that the EVSI is most sensitive to the unit cost of remediation and is rather insensitive to the penalty cost associated with under-remediation. The work of Dakins et al. (1996) was followed by several other publications discussing applications of VOI and BN approaches to environmental remediation (Freeze et al. 1992; James and Gorelick 1994; Abbaspour et al. 1996; Cox 1999; Back 2007).

15.5.3 Risk Screening Tools

NanoRiskCat is one risk screening tool to identify, categorize, and rank exposures and effects of engineered nanomaterials in consumer products based on current information (Danish Ministry of the Environment 2011; Hansen et al. 2014). The overall goal of this screening tool is to assist manufacturers, downstream end users, regulators, and other stakeholder groups in evaluating, ranking, and finally communicating the potential exposures and effects of select nanomaterials through a tiered approach, based on information related to the conventional form of a chemical as well as existing data on the nanoform. Potential exposures are related to professional end users, consumers, as well as the environment, and potential hazard effects are related to both humans and the environment. The final outcome of the screening for a given nanomaterial in a specific application is essentially a short title that describes the intended use of the nanomaterial (e.g., TiO_2 in sunscreen or NZVI as in situ remediation technique) together with a five-color-coded dot display, using a traffic light color-code model (i.e., red, yellow, green, and grey to indicate level of potential risk for each criterion; whereby grey denotes a lack of available information). In terms of potential exposure to the selected nanomaterial, information based on the "location" of the nanomaterial in the application (i.e., bulk, on the surface, in a liquid or air matrix) as well as the potential for exposure related to the description and explanation of the process or product category (e.g., manufacture method) is taken into consideration. In terms of the potential to cause a hazard to health or the environment, factors such as toxicity data (e.g., genotoxic, mutagenic, carcinogenic, etc.) as well as other factors related to more long-term concerns (i.e., persistency, bioaccumulation, potential for irreversible harm to the environment, readily dispersed, and novel) are taken into account (Hansen et al. 2014).

One of the main strengths of NanoRiskCat is that it can be used in conditions of extreme uncertainty, such as is the case with NZVI (Hansen et al. 2014). Since traditional assessment tools (such as chemical risk assessment paradigm) significantly struggle in cases of uncertainty, NanoRiskCat may be advantageous in this regard. The results from using this screening tool may also be easily communicated to interested parties, using the five-colored dot system and traffic light color coding. At the same time, one of the weaknesses of this tool is that the cutoff values used in the environmental hazard evaluation are based on dose by mass and it is likely that this may not be valid for all nanomaterials (Danish Ministry of the Environment 2011). Another potential weakness of this tool is that a holistic assessment based on expert judgment is used to evaluate the hazard potential for human health, whereby, e.g., carcinogenicity, mutagenicity, etc., are grouped together in the evaluation (Danish Ministry of the Environment 2011).

While the NanoRiskCat tool has been applied to ten individual case studies thus far involving other engineered nanomaterials (e.g., C_{60} in a lubricant and TiO_2 in sunscreen (Danish Ministry of Environment 2011), nanosilver in NANOVERTM cleansing soap, C_{60} in a synthetic oil, and carbon nanotubes in a tennis racket (Hansen et al. 2014)), NanoRiskCat has not yet been applied to NZVI for remediation purposes. Based on the information presented in preceding sections of this chapter related to potential hazards and exposures of NZVI, using NanoRiskCat in the future would seem appropriate to generate and present useful risk screening information to a site manager, environmental engineer, regulator, or other decision-makers when deciding upon the use of NZVI at a particular contaminated site. The output of applying this tool to NZVI would be a short title (e.g., NZVI for in situ remediation of chlorinated organic compounds) followed by five dots, corresponding to the potential for health and environmental exposures and hazard potentials based on available data.

Another risk screening approach that may be relevant for NZVI is the Nano Risk Framework developed through a collaboration between the Environmental Defense Fund and Dupont (ED and Dupont 2007a). This approach focuses primarily on risk assessment and risk management aspects related to an engineered nanomaterial across its product life cycle. The Nano Risk Framework is a structured guide that risk assessors and managers within an organization can use to evaluate and make decisions related to the potential risks of a nanomaterial or the product in which it is contained. There are six main steps to completing this approach: (1) describe the material and application, (2) describe the product life cycle(s), (3) evaluate the potential risks to health and environment, (4) assess the risk management options, (5) formulate decisions regarding risk management strategies, and (6) review and adapt decisions in light of new information.

The Nano Risk Framework was actually applied to at least three documented case studies, all of which were DuPont potential products: surface treated high-rutile phase TiO_2 in the DuPont Light Stabilizer 210 (ED and DuPont 2007b), single- and multi-walled carbon nanotubes in polymer nanocomposites by melt processing (ED and DuPont 2007c), and NZVI as used in in situ environmental remediation (ED and DuPont 2007d). In regard to the case study involving NZVI, DuPont concluded that it had "no immediate plans to implement this technology at any DuPont site" (ED and DuPont 2007d). This was largely due to the uncertainties related to the end products of the remediation reactions following in situ injection or following a spill with NZVI. In addition, DuPont noted questions related to the persistency of the nanoparticles and whether NZVI may be converted into a soluble iron salt or hydroxide.

15.5.4 Life Cycle Assessment

Because there has been a general consensus that the potential environmental, health, and safety risks of engineered nanomaterials should be evaluated over their life

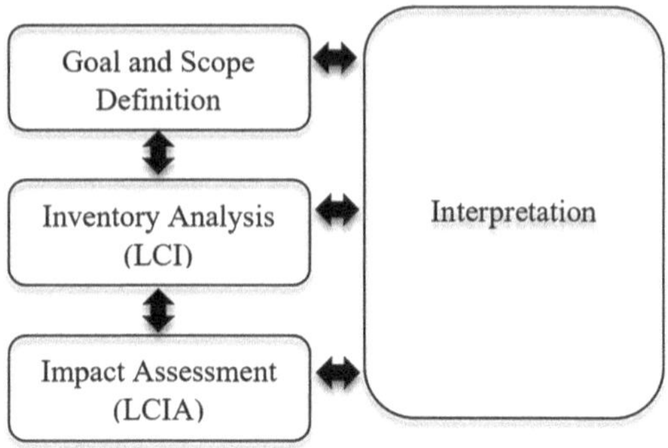

Fig. 15.3 Overview of Life cycle Assessment framework. (Source: International Organization for Standardization (2006))

cycle, Life Cycle Assessment (LCA) has naturally been one decision-support tool proposed to this end. LCA is a standardized decision support tool that assesses the potential impacts of a product, technology, or service throughout its various life cycle stages, spanning from "cradle" (i.e., raw material acquisition) to "grave" (i.e., end of life) (Hauschild 2005; Grieger et al. 2012c). Essentially, LCA is an approach to quantify life cycle resource consumption and environmental impacts from a product or services. The final results from performing LCA on a product or material is a quantitative assessment regarding a range of environmental impacts, including climate change, resource depletion, and human toxicity. These results are presented on a relative basis in order to be able to compare different alternatives, which ultimately aid in decision-making. LCA is structured into four main steps, as shown in Fig. 15.3 below (International Organization for Standardization 2006): (1) goal and scope definition, (2) inventory analysis, (3) impact assessment, and (4) interpretation of the results.

Step 1: Goal and scope definition The objectives of LCA are defined. For example, objectives of LCA can be a process LCA to quantify environmental footprints of a product or a comparative LCA to compare environmental footprints among different types of product or process. A system boundary and a functional unit are also identified in this LCA step.

Step 2: Life cycle inventory analysis (LCI) Inventory data is collected from experiments, literature, government reports, and LCA databases, such as USLCI, ecoinvent, ETH-ESU, and IdeMat (Design for Sustainability Program 2001; Frischknecht and Jungbluth 2004; National Renewable Energy Laboratory 2012; The Centre for Life Cycle Inventories 2014). Inventory data is collected according to the defined system boundaries and the necessity to achieve the defined goal. This LCA step is crucial since the results rely on quality of LCI.

Step 3: Life cycle impact assessment (LCIA) The LCI data are converted to quantifiable environmental impacts, such as global warming, eutrophication, acidification, and smog formation potentials. There are two main groups of LCIA tool (Bare et al. 2000). One is where contaminants are calculated on a midpoint basis, which can avoid estimation in LCIA, such as Tool for Reduction and Assessment of Chemicals and other environmental Impacts (TRACI), CML2001, and EDIP97 (Bare et al. 2003; Dreyer et al. 2003). Environmental impacts on a midpoint basis are reported in kg equivalent (eq) of a reference material (e.g., kg CO_2 eq for global warming potential). Another group is where midpoint impact categories are allocated into one or more damage categories or endpoint impacts, such as IMPACT 2002+ and Eco-indicator 99 (Dreyer et al. 2003; Jolliet et al. 2003).

Step 4: Interpretation of the results In this step, LCI and LCIA results are correlated and interpreted for more meaningful information to enable a consistent decision-making process and provide results and explain limitations of a product or service to the industries and decision-makers.

While these four steps outline the main approach taken in conducting LCAs, there are other methods to consider potential impacts of a product in different life cycles stages. For example, the term life cycle "thinking" refers to the life cycle stages spanning from cradle to grave, cradle to gate, or cradle to cradle. Similarly, a "cradle-to-grave" design includes all life cycle stages, while a "cradle-to-gate" design includes up to manufacturing of life cycle stages or where the products are produced and ready to be shipped. A "cradle-to-cradle" design includes all life cycle stages and is a model within which wastes generated over the life cycle are recirculated and used as a raw material in the production process.

15.5.4.1 LCA Applied to NZVI

While LCA has been applied to a number of engineered nanomaterials (Lloyd et al. 2005; Osterwalder et al. 2006; Khanna et al. 2008; Linkov et al. 2009; Lemming et al. 2010), to the best of the authors' knowledge, it has not been applied to NZVI for environmental remediation specifically. Nonetheless, we provide some preliminary information below that may be used by researchers or practitioners interested in conducting an LCA for NZVI.

Step 1: Goal and scope definition Generally, a goal and scope definition of LCA applied to NZVI for environmental remediation would be to quantify environmental impacts and/or trade-offs within the process or among different remediation technologies. LCA can be categorized into either a "process LCA," where only environmental impacts related to NZVI environmental remediation would be quantified, or a "comparative LCA," where environmental impacts from NZVI would be quantified and compared to the impacts from other environmental remediation technologies, such as thermal treatment, in situ chemical oxidation, or surfactant cosolvent flushing, in a comparable system boundary. Comparing different scenarios of NZVI use, varying in materials, production processes, and/or application methods can also be referred to as a comparative LCA.

A system boundary must be clearly defined in the goal and scope definition step. System boundaries can vary depending on the goal of the study. For example, if the goal is to evaluate the environmental sustainability of different NZVI particles, a cradle-to-gate LCA must be conducted to quantify environmental impacts from the acquisition of raw materials (e.g., iron, catalyst, nonmetallic carrier material, and energy) through transportation and production processes, based on the designed NZVI particles. For a cradle-to-grave LCA of NZVI, application methods (use phase) and end of life must be included in the system boundary, as illustrated in Fig. 15.4. With literature review, experiments, and knowledge on NZVI environmental remediation technology, researchers or practitioners should be able to identify details, e.g., types and quantities of resources and processes, to be included in the system boundary. Other factors, such as information on NZVI particles, the contaminated sites, application technologies, and follow-up process of the remediation, which have an influence over the life cycle of NZVI, should also be considered.

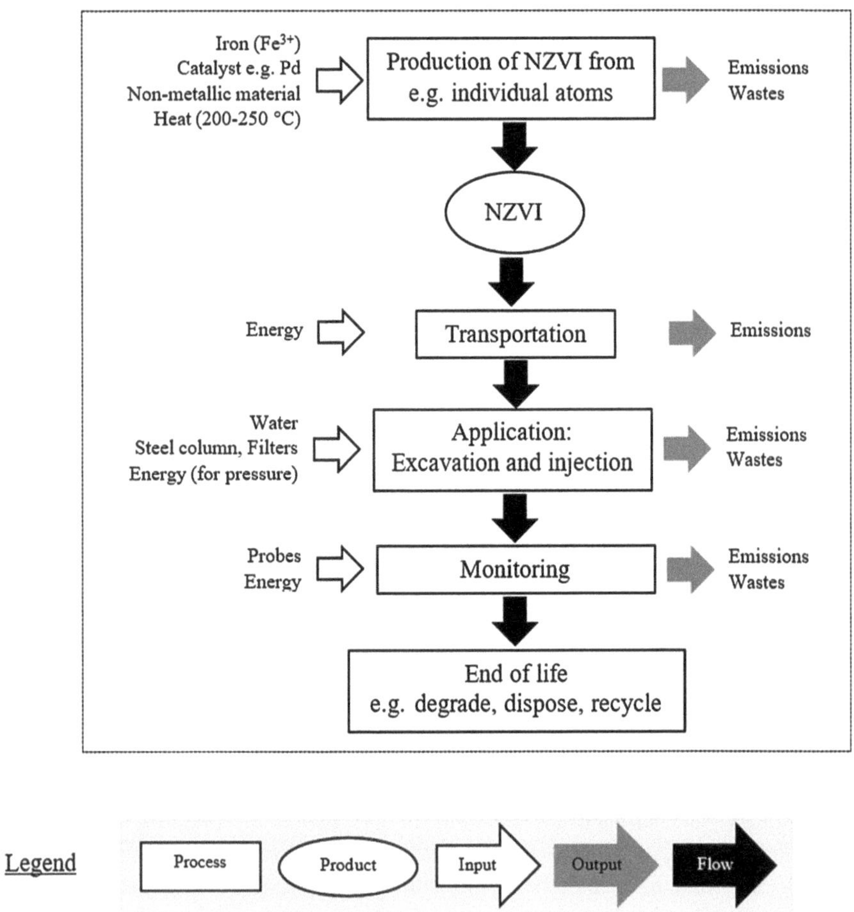

Fig. 15.4 Example of a system boundary of NZVI for in situ environmental remediation. The dashed line defines the system boundary of this LCA

For a fair comparison among different scenarios, processes, or technologies, inventories of inputs (e.g., resources and energy) and outputs (e.g., emissions and wastes) must be converted and examined on the same functional unit basis. A functional unit is a unit to quantitatively measure the performance of products or services in relation to their inputs and outputs. Based on the description, a functional unit for NZVI might be kg of NZVI particles required to remediate one km^3 of PCE-contaminated soil. The impact results could be reported in kg CO_2 eq of global warming potential or MJ of energy consumed per kg of NZVI used to remediate the contaminated soil.

Step 2: Life cycle inventory analysis (LCI) Inventories are input and output data related to the product or service that are collected and evaluated in order to meet the defined goal and scope. Input data is the quantity of resources required in each process over a life cycle of a product or service. Output data is the quantity of emissions and wastes generated from each process over a life cycle of a product or service. Inventories can be collected from experiments or literature, such as peer-reviewed publications, government reports and industrial standards, and/or life cycle databases. Several life cycle databases, such as USLCI, ecoinvent, IdeMat, and BUWAL, are partially available on their websites and fully available via LCA software, namely, SimaPro and GaBi (PE International 2011; PRé Consultants 2013). To date, inventory for the production of NZVI or ZVI does not exist in any life cycle database; however, the production of high-iron content cast iron was used to model the impact results from ZVI (Higgins and Olson 2009). It is important to convert all inventories from their original unit to per functional unit for a fair comparison. For example, the unit of the inventories can be converted to kg of Fe^{3+}/functional unit or kWh of electricity/functional unit.

In the case where the production system results in several products, one product is defined as a primary product, while the other remaining products are defined as coproducts. The original system boundary is then expanded to include the production of a displaced coproduct. The process is called system expansion. The avoided environmental impacts from the production of the coproducts are credited to the primary product as its environmental benefits. With system expansion, issues on impact allocation either on a mass or cost basis between the primary product and its coproducts can be avoided.

It is useful to keep track of the inventories, for example, in a table format or an Excel spreadsheet with original value and the value per functional unit. Moreover, it is important to record how the inventories are obtained to ensure that they are based on the same system boundary or under different conditions. Typically, more inventories may be collected as the study is being developed to answer the goal and scope defined and to improve the quality of the study. An example of a LCI table is given in Table 15.1. Functional unit is, e.g., 5.5 m^3 of NZVI particle or 10 ha of contaminated site.

Some life cycle studies may omit the LCIA results and end in this LCI step. It is acceptable to quantify and report only total inputs and outputs of the products or processes. However, they will not be considered as a LCA study but rather a LCI study.

Table 15.1 Example of a life cycle inventory table for life cycle inventory analysis (LCI) step

Parameter	Original value (unit)	Value (unit) per functional unit	Reference (experiment, database, etc.)	Note
Iron required during NZVI production	A (kg/10 kg of NZVI)	B (kg)	...	...
NZVI medium longevity	I (year/5 ha of the site)	J (year)		
Energy consumption during the injection (operation)	P (MJ/100 tons of NZVI)	Q (kWh)	...	...
Greenhouse gas emissions from the production of Pb	X (g CO^2 eq[a]/ 50 kg of Pb)	Y (kg CO_2 eq[a])	...	...

Functional unit is, e.g., 5.5 m^3 of NZVI particle or 10 ha of contaminated site
[a]*eq* equivalent

Step 3: Life cycle impact assessment (LCIA) As listed in previous sections, there are several LCIA methods developed by institutes in European countries and the USA. The tools developed by the European institutes, e.g., CML 2002, Eco-indicator 99, EDIP 2003, IMPACT 2002+, and ReCiPe, were based on European databases for European countries or globally (PRé Consultants 2000; Jolliet et al. 2003; Technical University of Denmark 2003; The National Institute for Public Health and the Environment (RIVM) et al. 2008; Institute of Environmental Sciences (CML) 2012). ReCiPe was developed based on eco-indicator 99 and CML 2002 methods. The method reported both midpoint and endpoint impact categories, which were combined in a consistent environmental framework (The National Institute for Public Health and the Environment (RIVM) et al. 2008). The LCIA method developed by the US agency is TRACI, which was developed particularly for the USA (Bare et al. 2003). Different LCIA methods have different impact categories (see Table 15.2).

LCIA results are normally presented in a bar graph format with an error bar representing the possible range of results. The figure can report actual values of the results or normalize the results to the impact from the process with the highest impact contribution in each category. Impact results can be positive or negative. A process with positive results indicates that the process makes a positive impact on the environment or human health, while a process with negative results indicates that the process can reduce or mitigate impacts. The range or uncertainty of the results can be obtained from Monte Carlo Analysis (MCA). The results are in a distribution curve. Generally, MCA software, such as Crystal Ball and @Risk, not only provides uncertainty of the results but also provides sensitivity analysis results via tornado correlation coefficient (Oracle 2008; Palisade Corporation 2010). The impact results can be used to identify the process with a major environmental impact contribution for area of improvements, while the results from tornado correlation coefficient provide further information on which parameter is causing the major impact contribution in that process. For example, assuming that NZVI particle production has the highest impact contribution among all the processes, based on the tornado correlation coefficient, we might find that surface area of NZVI particle is the main factor for the high-impact contribution.

Table 15.2 List of impact categories of various LCIA methods

LCIA method	Impact categories	Reference
TRACI	*Midpoint*: Acidification, global warming, carcinogenic, non-carcinogenic, respiratory effect, ecotoxicity, eutrophication, ozone depletion, and smog formation	Bare et al. (2003)
CML 2002	*Midpoint*: Abiotic depletion, acidification, ecotoxicity – (1) freshwater aquatic, (2) freshwater sediment, (3) marine aquatic, (4) marine sediment, and (5) terrestrial, eutrophication, global warming, human toxicity, incremental reactivity, land competition, malodorous air, ozone depletion, photochemical oxidation, radiation, and smog	Institute of Environmental Sciences (CML) (2012)
Eco-indicator 99	*Endpoint*: Acidification/eutrophication, carcinogenics, climate change, ecotoxicity, fossil fuels, land use, mineral extraction, ozone layer, radiation, respiratory inorganics, and respiratory organics	PRé Consultants (2000)
EDIP 2003	*Endpoint*: Acidification, bulk waste, ecotoxicity, (1) soil chronic, (2) water acute, and (3) water chronic; eutrophication, (1) aquatic (N), (2) aquatic (P), and (3) terrestrial, global warming, and hazardous waste; human toxicity, (1) air, (2) soil, and (3) water, ozone depletion, ozone depletion (human), ozone depletion (vegetation), radioactive waste, resources (all), and slags/ashes	Technical University of Denmark (2003)
ReCiPe	*Midpoint*: Climate change, ozone depletion, human toxicity, photochemical oxidant formation, particulate matter formation, ionizing radiation, terrestrial acidification, eutrophication, (1) freshwater and (2) marine; ecotoxicity, (1) terrestrial, (2) freshwater, and (3) marine; occupation, (1) agricultural land and (2) urban land and natural land transformation *Endpoint*: Human health, ecosystem, and resources	The National Institute for Public Health and the Environment (RIVM) et al. (2008)
IMPACT 2002+	*Endpoint*: Acid/nutri(terrestrial), acidification (aquatic), carcinogens, ecotoxicity (aquatic), ecotoxicity (terrestrial), eutrophication (aquatic), global warming, land occupation, mineral extraction, non-carcinogenics, nonrenewable energy, ozone depletion, radiation (ionizing), respiratory inorganics, and respiratory organics	Jolliet et al. (2003)

The following section is developed with an intention only to provide examples on how the impact results are normally presented. The results used in Figs. 15.5 and 15.6 are all theoretical and should not be used to provide comparisons with other studies. Figure 15.5 illustrates how LCIA results from a process LCA can be

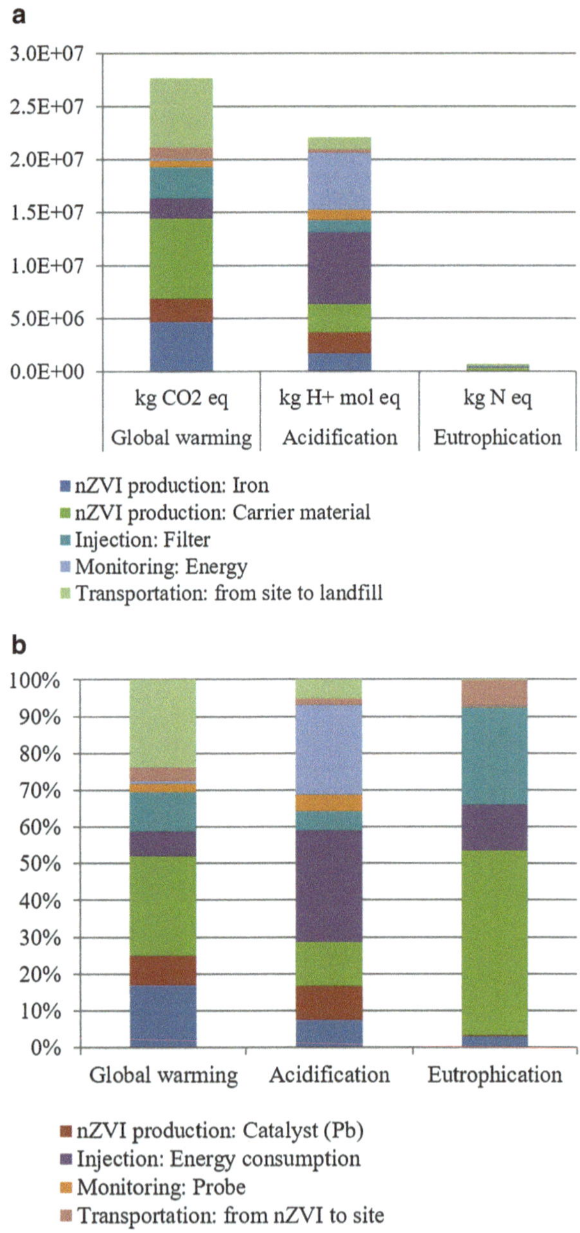

Fig. 15.5 Examples of LCIA results from a process LCA: (**a**) actual values of the LCIA results and (**b**) normalized LCIA results. All results are theoretical and should not be used for comparison with other studies

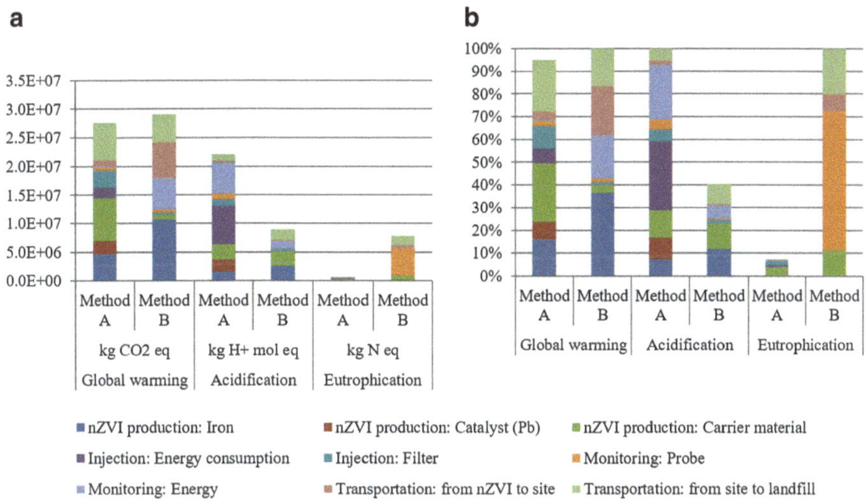

Fig. 15.6 Examples of LCIA results from a comparative LCA: (**a**) actual values and (**b**) normalized LCIA results. All results are theoretical and should not be used for comparison with other studies

presented. This figure presents global warming, acidification, and eutrophication potential results per functional unit (e.g., 100 m^3 of contaminated soil). Figure 15.6 also indicates which process is the major impact contribution or the area of improvement. However, by presenting actual values of the results, some impact results may overshadow other results. In this case, the global warming and acidification potential results overshadow the eutrophication potential result. The process with the major eutrophication potential contribution is the most obvious result as shown in Fig. 15.6b. Therefore, it is common in LCA to present normalized results. Impacts from each process are normalized to the total impact of each impact category. Normalized results can also be used to help a company protect their proprietary data and product formula since there is no possibility to track back from the actual results without any additional information provided.

A comparative LCA is normally conducted to quantify environmental impacts and/or identify trade-offs among different products or methods. The results can aid in decision-making processes on which products or methods to use. The LCIA results in Fig. 15.6 not only indicate a process with a major impact contribution but also show trade-offs between methods A and B. The results indicated that method A contributes lower global warming potential but higher acidification and eutrophication potentials compared to method B. Based on such results, a process with the major global warming potential contribution in method A is a carrier material in NZVI production process, while a process with the major global warming potential contribution in method B is iron in the NZVI production process. Acidification and eutrophication potential results can be discussed in a similar format as the global warming potential results.

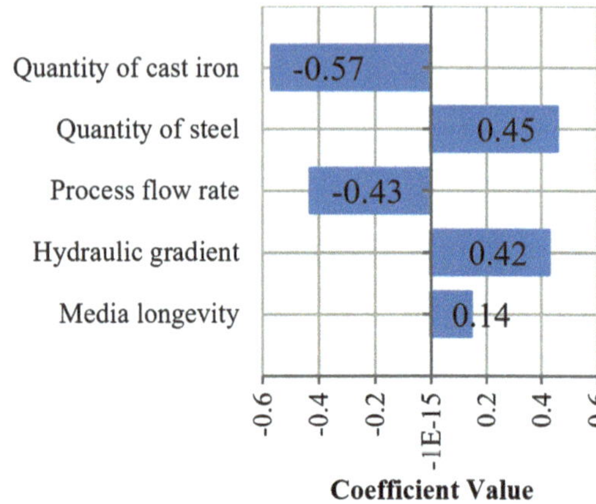

Fig. 15.7 Tornado correlation coefficient. All results are theoretical and should not be used for comparison with other studies

Monte Carlo Analysis is a statistical analysis where input values are randomly selected from parameter distributions to construct an output distribution (Woller 1996). Monte Carlo simulation has been used in conjunction with LCA to evaluate the range of possible values for a system (Soratana and Marriott 2010; Soratana et al. 2012, 2013). Figure 15.7 illustrates the resultant probability distribution for global warming potential. The assumed results indicate that the environmental impact would change significantly given changes to the quantity of cast iron, quantity of steel, process flow rate, and hydraulic gradient. In this case, parameters, which are quantity of cast iron and process flow rate, have an inverse effect, while quantity of steel and hydraulic gradient have a direct effect on the impact. It can also be stated that the quantity of iron is the most sensitive parameter, with an inverse effect on the impact, and that the media longevity is the least sensitive parameter with a direct effect on the impact.

Step 4: Interpretation of the results Interpretation of the results has to be conducted at any step of the LCA study, as depicted in Fig. 15.3. The main objective of the interpretation step is to provide meaningful and understandable results, conclusions, and/or recommendations. Examples of how LCI and LCIA can be discussed and interpreted, consistent with their goal and scope, are already provided in Step 2 and Step 3. Typical interpretations of the results that can be made based on LCI and LCIA are:

- Present mass/energy flow of the process to identify loss
- Processes with the highest and/or lowest impact contribution for area of improvement
- Trade-offs among different technologies/products to support decision-making processes
- The most sensitive parameter in each impact category to identify the cause of the major impact contribution process

15.5.4.2 Future Steps for Applying LCA to NZVI

Traditional LCA can be applied to any generic product or technology. However, it might not be the most suitable approach for an emerging technology such as NZVI due to the lack of inventory data. Partially in response to this, some authors have started to define the functional unit for LCA slightly different, to take into account these limitations when working with this uncertainty. For instance, Lemming et al. (2012) propose that "the management of the target treatment zone which leads to a 99% removal of the contaminant mass" could serve as the functional unit for LCA. Moreover, multi-stakeholder collaborations are needed in order to connect, share, and access databases which may be used to advance inventory data required by the LCA process. Currently, the lack of inventory has led to a prospective LCA, which examines environmental trade-offs from different scenarios of the technology, resulting in a high uncertainty associated with the results. One proposed approach for such emerging technologies may be an anticipatory LCA, which integrates environmental, societal, and technological assessments. At the same time, knowledge generated from anticipatory LCAs may be fed back into the technology development process, also incorporating stakeholder feedback, in order to develop NZVI in the most sustainable manner possible (Wender 2013).

15.5.5 Other Decision Support Tools

The previously described tools and frameworks were included in this chapter since they are regarded as the most mature in their stage of development or have received substantial recognition in terms of decision support. In addition to these tools though, there are also other decision support techniques that may be relevant to NZVI but perhaps on a broader scale, such as for risk governance or to provide risk screening approaches. For example, Comprehensive Environmental Assessment was developed by US EPA and applied to other case studies involving engineered nanomaterials (US EPA 2010, 2012a, b) and may be an interesting approach in order to elicit expert opinion in light of significant data gaps. Due to its broad focus, it has been considered to be more of a risk governance framework than a risk screening tool or a tool intended, for example, for risk managers for industry (Grieger et al. 2012b). In addition, control banding has been applied to several engineered nanomaterials and is considered a promising technique to estimate potential risks of occupational workers combining known data on exposure and hazard (Brouwer 2012; Riediker et al. 2012). In addition, the precautionary matrix (Höck 2011), Stoffenmanager Nano Risk Banding Tool (van Duuren-Stuurman et al. 2011, 2012), and NanoLCRA (Shatkin 2008, 2009a, b) have all been developed specifically for engineered nanomaterials and may be particularly useful for screening level analyses particularly for workplace health and safety analyses. A comprehensive review of available risk analysis and decision support tools for engineered nanomaterials in general is provided in Eisenberg et al. (2015) and Hoehener and Hoeck (2013). While these aforementioned decision support tools may be applicable and relevant to NZVI, none of them have as yet been applied to the case of NZVI for remediation purposes.

15.5.6 Decision Support for Diverse Stakeholders

It is clear that different decision problems are likely to be faced by different stakeholder groups and actors. For example, remediation site managers will likely evaluate which available remediation option may be the most cost-effective, whereby cost and performance are among the top decision criteria. Other important factors may also include the maturity of the technology (i.e., whether it is "tried and tested" and hence uncertainty reduced in terms of its performance), effectiveness (i.e., how well will the technology clean up the contaminated site, reducing performance uncertainty), and, to some extent, potential stakeholder views and preferences depending on the relationships between the decision-maker and different stakeholder groups. It is clear that public officials and government regulators responsible for remediating contaminated sites are likely to consider public opinion and stakeholder perspectives in decision-making (including the choice of remediation options), while managers of a private environmental consulting firm, for example, may not have mechanisms to readily incorporate public opinion or stakeholder views.

Similarly, different decisions may be best answered or supported by the use of different decision support tools as shown in the preceding sections. As shown, MCDA could help decision-makers choose between competing alternatives based on select criteria. For example, MCDA may help decision-makers choose which remediation treatment technology may be preferred based on select criteria such as cost, effectiveness, treatment time, potential toxicity to health, potential toxicity to the environment, as well as potential for human and environmental exposures, in which NZVI is one of several in situ remediation technologies. Other decision support tools may help a site manager or regulatory official better understand the potential risks of NZVI either to health or the environment through the application of diverse risk screening tools, including NanoRiskCat and the Nano Risk Framework. The application of LCA can also help evaluate the potential environmental impacts of a nanomaterial or nano-product (such as NZVI or NZVI-slurry) across its life cycle stages and covers a broad range of impacts including climate change, resource depletion, and toxicity which are not covered in the other risk analysis tools presented here. In this way, not all of the decision support tools and frameworks presented have been developed for the same purpose and for the same intention; and hence the selection of the "right" tool(s) for the decision at hand is of the utmost importance.

In order to promote the sustainable development and use of NZVI, using a decision support tool that includes criteria relevant for not only potential risks to health and the environment but also societal, ethical, and economic terms is fundamental. This is due to the foundational nature of the concept of sustainability, in which three "pillars" are deemed important for inclusion: protecting health and the environment, consideration of social concerns, and ensuring economic feasibility (Adams 2006). In the case of NZVI, choosing a decision support methodology or approach which inherently includes these factors will help formulate decisions to ensure the sustainable development and use of NZVI. While there is not yet an established method or tool that integrates the fields of risk assessment and sustainability formally, several attempts at doing so have been published (see Sexton and Linder (2014) for a review). None of these, however, have been in the field of engineered nanomaterials or have been applied to NZVI.

Nevertheless, the application of MCDA is promising for NZVI since site remediators, regulators, or other decision-makers can use this established methodology along with select criteria considered important for the decision on hand (e.g., potential for toxicity, persistency, cost, etc.) in order to decide which treatment may be the most sustainable option or which type of NZVI may be more sustainable than others (e.g., emulsified NZVI, bare NZVI, etc.). Stakeholder preferences, including public opinions on nanotechnology or contaminated soil and groundwater, may also be factored into these analyses. To date, public participation on nanoremediation options has been extremely limited to nonexistent (Grieger et al. 2012a), and therefore it is advisable to increase public participation in decision-making processes involving NZVI as this will likely lead to more sustainable outcomes. Meanwhile, risk analysis approaches including NanoRiskCat and the Nano Risk Framework may be coupled with LCA assessments, in order to include not only traditional parameters in risk (e.g., toxicity, exposure, etc.) but also other dimensions important for sustainability such as greenhouse gas emissions and resource depletions. These methods do not typically incorporate broader societal dimensions including public opinion; thus other governance mechanisms may also need to be applied when relying on risk analysis or LCA methodologies for decision support.

15.6 Conclusions and Recommendations

Ensuring the sustainable development and use of NZVI for in situ remediation requires the incorporation of a multitude of factors and criteria, including those related to technology performance, cost, potential impacts to the environment and human health, as well as ethical, social, and legal concerns. There is a growing body of literature on the use and development of NZVI for remediation purposes, with promising findings regarding its performance and effectiveness as a remediation technique with a broad range of application. However, studies on the impacts of NZVI are still very much within early stages. Most research on the potential implications of NZVI have focused on the potential (eco)toxicological effects, and very little work has focused on its social or ethical implications. As seen with other emerging technologies, the inclusion of diverse stakeholders and public participation can play a large role in the development of socially robust technologies. Therefore, it is strongly recommended that a greater emphasis is placed on ensuring an open, transparent, and democratic process that encourages citizen engagement. Moreover, the political accountability of NZVI's development and use is questionable given the fact that there are still no agreed-upon international positions on whether the nano-component of NZVI as a remediation technology is "new" and thereby requires novel regulatory considerations. Finally, marketplace limitations appear to be serious obstacles to ensuring the sustainable development and use of NZVI as an environmental remediation technology, including questions pertaining to the validity of its cost-competitiveness. These findings indicate that while NZVI has potential to reduce environmental contaminants through in situ deployments, its development and use, particularly at field-scale sites, has not proceeded in the most sustainable manncr possible.

When deciding upon the selection of suitable remediation technologies, be that forms of NZVI or other engineered nanomaterials to remediate a given site, traditional approaches to decision support may be hampered by the extensive uncertainties presented by nanotechnologies and nanomaterials. Alternative decision support tools such as MCDA, VOI, and Bayesian logic may offer advantages to traditional approaches. One type of MCDA tool (i.e., DESYRE) seems to be one of the most promising decision support systems for selecting remediation technologies, including the consideration of NZVI. The output from DESYRE includes the selection of the most suitable remediation technology based on expert review from ranking algorithm output. One of the main advantages of this approach is the incorporation of stakeholder preferences through the weighting of some criteria over others.

For decisions based on the potential risks of NZVI or to perform risk screening analyses, NanoRiskCat, Nano Risk Framework, Precautionary Matrix, Stoffenmanager Nano Risk Banding Tool, and Life Cycle Assessment may all be suitable for NZVI. To date, however, only the Nano Risk Framework has been applied to NZVI in documented reports. Life Cycle Assessment has great potential in order to assess the potential impacts of NZVI along its entire life cycle chain, although no published studies have documented its application to NZVI to date. In light of this, we have provided some preliminary information on how LCA may be applied to NZVI in order to aid other researchers or practitioners applying LCA in full for NZVI. As shown, one of the most challenging obstacles to completing a LCA for NZVI is, however, the lack of inventory data.

In conclusion, this book chapter has outlined the main factors which are important for inclusion in order to develop and use NZVI in the most sustainable manner. More specifically, environmental and health impacts are fundamental to incorporate not only into decision-making processes but also in the development and deployment of NZVI. Based on published studies, there is a clear potential to engineer or develop NZVI using the principles of "green chemistry" including the application of various coatings to reduce toxicity. At the same, it is clear that societal issues including public participation and engagement are essential for NZVI to be used most sustainably, particularly at field sites. While there are a number of decision support approaches and risk analysis tools available, very few of them have been tested on

NZVI to date. These tools and frameworks all have their advantages and disadvantages and may be better suited for some decisions at hand compared to others, including, e.g., the use of MCDA by site remediators or project managers to determine which remediation technology is best for a given contaminated site. Finally, while NZVI's cost-effectiveness has often been cited as one of its main advantages for in situ remediation, it appears that more studies should be conducted in order to fully evaluate this, particularly for long-term time frames.

In light of these findings and the information presented in this chapter, the following recommendations are made in order to ensure the sustainable development and use of NZVI for environmental remediation.

- First, given the relatively limited number of studies which have investigated the (eco)toxicity potential of NZVI, further research is needed in order to understand NZVI's impacts on a variety of test organisms. This should include (eco)toxicity tests on NZVI with surface coatings and functionalizations that are most relevant for field-scale applications. It should also include testing of fully oxidized NZVI as this is the more likely form of the material for ecological exposures. Some of these surface coatings and functionalizations have been shown to increase migration distances in the environment as well as potentially decrease NZVI's toxicity. Furthermore, exposure and ecotoxicity testing should ideally be conducted in ecosystem-like settings given the fact that NZVI is directly injected into environmental matrices under normal use conditions. Therefore, further work is needed to better understand realistic exposure scenarios of NZVI including the application of different surface coatings and within ecosystem-like settings.
- Second, more research is needed on NZVI's potential for persistency and bioaccumulation in the environment. As there are different forms of NZVI developed and deployed, including bimetallic NZVI, a better understanding of the potential for these various forms of NZVI to persist and potentially bioaccumulate in the environment is critical to understanding the long-term effects of this emerging technology. To date, very little is known regarding NZVI's potential for persistency and bioaccumulation based on publically available literature.
- Third, basic research on NZVI nanoparticles should be supplemented by translation research that establishes simulation, modeling, standard test methodologies, reference materials, and other enabling capabilities. The Center for the Environmental Implication on Nanotechnology (CEINT) housed at Duke University provides an interesting model for this translational research, in that established mesocosm studies have been set up in order to provide more realistic exposure scenarios to understand the fate, transformation, and behavior of engineered nanomaterials in the environment which may be coupled with ongoing risk modeling analyses. In addition, CEINT's sister organization CEIN (Center for Environmental Implications of Nanotechnology, housed at University of California) as well as international research consortia such as NanoReg (http://nanoreg.eu/) are significant sources of information regarding, e.g., production volumes,

reference materials, toxicity test protocols, and related aspects regarding environmental and regulatory testing of engineered nanomaterials.

- The continued support of databases and data sources such as the Nanomaterial Registry (https://www.nanomaterialregistry.org/) is, therefore, also extremely important in order to ensure that large data sets are maintained and expanded upon, given the fact that several assessment and decision analysis frameworks and tools are hampered by the lack of information pertaining to engineered nanomaterials such as NZVI. Therefore, it is recommended that the continued support and commitment to maintaining large data sets and data sources relevant to understanding NZVI's physicochemical properties, its behavior in the environment, as well as potential impacts on health and ecosystem receptors is ensured.
- Finally, and perhaps most importantly, it is recommended that public participation should play a greater role in ensuring that NZVI is developed and used in the most sustainable manner. It has been documented that despite the fact that NZVI has already been deployed in several countries, there has been relatively little or no public participation or citizen engagement activities to date. It also well known that public acceptance of technology is a significant factor in order to ensure sustainable outcomes, and therefore it is recommended that the public and other stakeholder groups are more actively involved in decision-making regarding the development and use of NZVI as an environmental remediation option.

References

Abbaspour, K., Schulin, R., Schläppi, E., & Flühler, H. (1996). A Bayesian approach for incorporating uncertainty and data worth in environmental projects. *Environmental Modeling & Assessment, 1*, 151–158.

Adams WM (2006) *The future of sustainability: Re-thinking environment and development in the twenty-first century*. Report of the IUCN Renowned Thinkers Meeting, 29–31 January 2006.

Ampiah-Bonney, R. J., Tyson, J. F., & Lanza, G. R. (2007). Phytoextraction of arsenic from soil by Leersia Oryzoides. *International Journal of Phytoremediation, 9*, 31–40.

An, Y., Li, T., Jin, Z., Dong, M., Xia, H., & Wang, X. (2010). Effect of bimetallic and polymer-coated Fe nanoparticles on biological denitrification. *Bioresource Technology, 101*, 9825–9828.

Auffan, M., Achouak, W., Rose, J., Roncato, M. A., Chanéac, C., Waite, D. T., Masion, A., Woicik, J. C., Wiesner, M. R., & Bottero, J. Y. (2008). Relation between the redox state of iron-based nanoparticles and their cytotoxicity toward Escherichia coli. *Environmental Science & Technology, 42*, 6730–6735.

Back, P. E. (2007). A model for estimating the value of sampling programs and the optimal number of samples for contaminated soil. *Environmental Geology, 52*, 573–585.

Back, P., Rosén, L., & Norberg, T. (2007). Value of information analysis in remedial investigations. *Ambio, 36*, 486–493.

Bardos, P., Nathanail, J., & Pope, B. (2002). General principles for remedial approach selection. *Land Contamination and Reclamation, 10*, 137–160.

Bare, J., Hofstetter, P., Pennington, D., & Haes, H. U. (2000). Midpoints versus endpoints: The sacrifices and benefits. *The International Journal of Life Cycle Assessment, 5*(6), 319–326.

Bare, J. C., Norris, G. A., Pennington, D. W., & McKone, T. (2003). TRACI: The tool for the reduction and assessment of chemical and other environmental impacts. *Journal of Industrial Ecology, 6*(3–4), 49–78.

Barnes, R. J., van der Gast, C. J., Riba, O., Lehtovirta, L. E., Prosser, J. I., Dobson, P. J., & Thompson, I. P. (2010). The impact of zero-valent iron nanoparticles on a river water bacterial community. *Journal of Hazardous Materials, 184*, 73–80.

Belton, V., & Stewart, T. (2011). *Multiple criteria decision analysis: An integrated approach.* Dordrecht: Kluwer, Academic Publishers.

Brouwer, D. H. (2012). Control banding approaches for nanomaterials. *Annals of Occupational Hygiene, 56*, 506–514.

Carlsson, C., Ehrenberg, D., Eklund, P., Fedrizzi, M., Gustafsson, P., Lindholm, P., Merkuryeva, G., Riissanen, T., & Ventre, A. (1992). Consensus in distributed soft environments. *European Journal of Operational Research, 61*, 165–185.

Chang, M., & Kang, H. (2009). Remediation of pyrene-contaminated soil by synthesized nanoscale zero-valent iron particles. *Journal of Environmental Science and Health, 44*, 576–582.

Chen, P. J., Su, C. H., Tseng, C. Y., Tan, S. W., & Cheng, C. H. (2011a). Toxicity assessments of nanoscale zerovalent iron and its oxidation products in medaka (*Oryzias latipes*) fish. *Marine Pollution Bulletin, 63*, 339–346.

Chen, J., Xiu, Z., Lowry, G. C., & Alvarez, P. J. J. (2011b). Effect of natural organic matter on toxicity and reactivity of nano-scale zero-valent iron. *Water Research, 45*, 1995–2001.

Chen, P. J., Wu, W. L., & Wu, K. C. (2013). The zerovalent iron nanoparticle causes higher developmental toxicity than its oxidation products in early life stages of medaka fish. *Water Research, 47*, 3899–3909.

Cox, L. (1999). Adaptive spatial sampling of contaminated soil. *Risk Analysis, 19*, 1059–1069.

Crane, R. A., & Scott, T. B. (2012). Nanoscale zero-valent iron: Future prospects for an emerging water treatment technology. *Journal of Hazardous Materials, 211-212*, 112–125.

Crimi, M. L., & Siegrist, R. L. (2003). Geochemical effects on metals following permanganate oxidation of DNAPLs. *Ground Water, 41*, 458–469.

Critto, A., Cantarella, L., Carlon, C., Giove, S., Petruzze, G., & Marcomini, A. (2006). Decision support-oriented selection of remediation technologies to rehabilitate contaminated sites. *Integrated Environmental Assessment and Management, 2*, 273–285.

Cullen, L. G., Tilston, E. L., Mitchell, G. R., Collins, C. D., & Shaw, L. J. (2011). Assessing the impact of nano- and micro-scale zerovalent iron particles on soil microbial activities: Particle reactivity interferes with assay conditions and interpretation of genuine microbial effects. *Chemosphere, 82*, 1675–1682.

Dakins, M. E., Toll, J. E., Small, M. J., & Brand, K. P. (1996). Risk-based environmental remediation: Bayesian Monte Carlo analysis and the expected value of sample information. *Risk Analysis, 16*, 67–79.

Danish Ministry of the Environment. (2011). NanoRiskCat-a conceptual decision support tool for nanomaterials. Copenhagen, Denmark, 269 pp.

Davis, M., Long, T. C., Shatkin, J. A., Wang, A., Graham, J. A., Gwinn, M., & Ranalli, B. (2010). *Comprehensive environmental assessment. Nanomaterial case studies: Nanoscale titanium dioxide in water treatment and in topical sunscreen.* U.S. Environmental Protection Agency (USEPA).

DEFRA. (2011). *A risk/benefit approach to the application of iron nanoparticles*. U.K. Department for Environment, Food and Rural Affairs.

Delgado, A., Kjølberg, K. L., & Wickson, F. (2011). Public engagement coming of age: From theory to practice in STS encounters with nanotechnology. *Public Understanding of Science, 20* (6), 826–845.

Design for Sustainability Program. (2001). *IdeMat Online*. Delft: Delft University of Technology.

Diao, M., & Yao, M. (2009). Use of zero-valent iron nanoparticles in inactivating microbes. *Water Research, 43*, 5243–5251.

Dillard, J., Dujon, V., & King, M. C. (2009). *Understanding the social dimension of sustainability.* New York: Routledge.

Dreyer, L. C., Niemann, A. L., & Hauschild, M. Z. (2003). Comparison of three different LCIA methods: EDIP97, CML2001 and eco-indicator 99. *International Journal of Life Cycle Analysis, 8*(4), 191–200.
Duuren-Stuurman, B., Vink, S., Brouwer, D., Kroese, D., Heussen, H., Verbist, K., Telemans, E., Niftrik, M. V., & Fransman, W. (2011). *Stoffenmanager nano: Description of the conceptual control banding model.* Zeist: Netherlands Organisation for Applied Scientific Research (TNO).
Edmiston, P. L., Osborne, C., Reinbold, K. P., Pickett, D. C., & Underwood, L. A. (2011). Pilot scale testing composite swellable organosilica nanoscale zero-valent iron—Iron-Osorb®—For in situ remediation of trichloroethylene. *Remediation Winter, 22*, 105–123.
Eisenberg, D., Grieger, K. D., Hristozov, D., Bates, M., & Linkov, I. (2015). Risk assessment, life cycle assessment, and decision methods for nanomaterials. In *Nanomaterials in the Environment*. Reston: American Society of Civil Engineers.
Elkington, J. (1997). *Cannibals with forks: The triple bottom line of 21st century business*. Oxford: Capstone Publishing.
Elliott, D. W., & Zhang, W. X. (2001). Field assessment of nanoscale bimetallic particles for groundwater treatment. *Environmental Science & Technology, 35*, 4922–4926.
El-Temsah, Y. S., & Joner, E. J. (2012a). Ecotoxicological effects on earthworms of fresh and aged nano-sized zero-valent iron (NZVI) in soil. *Chemosphere, 89*, 76–82.
El-Temsah, Y. S., & Joner, E. J. (2012b). Impact of Fe and Ag nanoparticles on seed germination and differences in bioavailability during exposure in aqueous suspension and soil. *Environmental Toxicology, 27*, 42–49.
El-Temsah, Y. S., & Joner, E. J. (2013). Effects of nano-sized zero-valent iron (NZVI) on DDT degradation in soil and its toxicity to collembola and ostracods. *Chemosphere, 92*, 131–137.
Environmental Defense (ED) and Dupont. (2007a). *Nano risk framework*. Washington DC: Environmental Defense – Dupont Nano Partnership.
Environmental Defense (ED) and Dupont. (2007b). *Nanomaterial risk assessment worksheet: DuPont light stabilizer for use as a polymer additive*. Washington DC: Environmental Defense – Dupont Nano Partnership.
Environmental Defense (ED) and Dupont. (2007c). *Nanomaterial risk assessment worksheet: Incorporation of single and multiwalled carbon nano tubes (CNTs) into polymer nanocomposites by melt processing*. Washington DC: Environmental Defense – Dupont Nano Partnership.
Environmental Defense (ED) and Dupont. (2007d). *Nanomaterial risk assessment worksheet: Zero valent nano sized iron nanoparticles (NZVI) for environmental remediation*. Washington DC: Environmental Defense – Dupont Nano Partnership.
Fajardo, C., Ortíz, L. T., Rodríguez-Membibre, M. L., Nande, M., Lobo, M. C., & Martin, M. (2012). Assessing the impact of zero-valent iron (ZVI) nanotechnology on soil microbial structure and functionality: A molecular approach. *Chemosphere, 86*, 802–808.
Fajardo, C., Saccà, M. L., Martinez-Gomariz, M., Costa, G., Nande, M., & Martin, M. (2013). Transcriptional and proteomic stress responses of a soil bacterium *Bacillus cereus* to nanosized zero-valent iron (NZVI) particles. *Chemosphere, 93*, 1077–1083.
Freeze, R. A., James, B., Massmann, J., Sperling, T., & Smith, L. (1992). Hydrogeological decision analysis: 4. The concept of data worth and its use in the development of site investigation strategies. *Ground Water, 30*, 574–588.
Friis, A. K., Heron, G., Albrechtsen, H. J., Udell, K. S., & Bjerg, P. L. (2006). Anaerobic dechlorination and redox activities after full-scale electrical resistance heating (ERH) of a TCE-contaminated aquifer. *Journal of Contaminant Hydrology, 88*, 219–234.
Frischknecht, R., & Jungbluth, N. (2004). *SimaPro database manual.* The ETH-ESU 96 Libraries version 2.1. ESU-services.
Ghauch, A. (2008). Rapid removal of flutriafol in water by zero-valent iron powder. *Chemosphere, 71*, 816–826.
Giove, S., Brancia, A., Satterstrom, F. K., & Linkov, I. (2009). Decision support systems and environment: Role of MCDA. In A. Marcomini, G. W. Suter II, & A. Critto (Eds.), *Decision support systems for risk-based management of contaminated sites*. Boston: Springer, US.

Grieger, K., Hansen, S. F., & Baun, A. (2009). The known unknowns of nanomaterials: Describing and characterizing uncertainty within environmental, health and safety risks. *Nanotoxicology, 3* (3), 1–12.

Grieger, K., Wickson, F., Andersen, H. B., & Renn, O. (2012a). Improving risk governance of emerging technologies through public engagement: The neglected case of nano-remediation? *International Journal of Emerging Technologies and Society, 10*, 61–78.

Grieger, K., Linkov, I., Hansen, S. F., & Baun, A. (2012b). Environmental risk analysis for nanomaterials: Review and evaluation of frameworks. *Nanotoxicology, 6*(2), 196–212.

Grieger, K., Laurent, A., Miseljic, M., Christensen, F., Baun, A., & Olsen, S. (2012c). Analysis of current research addressing complementary use of life-cycle assessment and risk assessment for engineered nanomaterials: Have lessons been learned from previous experience with chemicals? *Journal of Nanoparticle Research, 14*(7), 1–23.

Grieger, K., Fjordbøge, A., Hartmann, N. B., Eriksson, E., Bjerg, P. L., & Baun, A. (2010a). Environmental benefits and risks of zero-valent iron nanoparticles (NZVI) for in situ remediation: Risk mitigation or trade-off? *Journal of Contaminant Hydrology, 118*, 165–183.

Grieger, K., Baun, A., & Owen, R. (2010b). Redefining risk research priorities for nanomaterials. *Journal of Nanoparticle Research, 2*(2), 383–392.

Hansen, S. F., Jensen, K. A., & Baun, A. (2014). NanoRiskCat: A conceptual tool for categorization and communication of exposure potentials and hazard of nanomaterials in consumer products. *Journal of Nanoparticle Research, 16*, 2195.

Hauschild, M. Z. (2005). Assessing environmental impacts in a life-cycle perspective. *Environmental Science & Technology, 39*(4), 81A–88A.

He, F., Zhao, D., & Paul, C. (2010). Field assessment of carboxymethyl cellulose stabilized iron nanoparticles for in situ destruction of chlorinated solvents in source zones. *Water Research, 44*, 2360–2370.

Higgins, M. R., & Olson, T. M. (2009). Life-cycle case study comparison of permeable reactive barrier versus pump-and-treat remediation. *Environmental Science & Technology, 43*(24), 9432–9438.

Höck, J., Epprecht, T., Hofmann, H., Höhner, K., Krug, H., Lorenz, C., Limbach, L., Gehr, P., Nowack, B., Riediker, M., Schirmer, K., Schmid, B., Som, C., Stark, W., Studer, C., Ulrich, A., Götz, N. V., Wengert, S., & Wick, P. (2010). *Guidelines on the precautionary matrix for synthetic nanomaterials*. Federal Office of Public Health and Federal Office for the Environment.

Höck J., Behra R., Bergamin L., Bourqui-Pittet M., Bosshard C., Epprecht T., Furrer V., Frey S., Gautschi M., Hofmann H., Höhener K., Hungerbühler K., Knauer K., Krug H., Limbach L., Gehr P., Nowack B., Riediker M., Schirmer K., Schmid K., Som C., Stark W., Suarez Merino B., Ulrich A., von Götz N., Walser T., Wengert S., Wick P., Studer C.: Guidelines on the Precautionary Matrix for Syn-thetic Nanomaterials. Federal Office of Public Health and Federal Office for the Environment, Berne 2018, Version 3.1

Hoehener, K., & Hoeck, J. (2013). *Deliverable D2.6 draft (m30) consolidated framework for EHS of manufactured nanomaterials*. ERA-NET SIINN; safe implementation of innovative nanoscience and nanotechnology.

Huang, I. B., Keisler, J., & Linkov, I. (2011). Multi-criteria decision analysis in environmental sciences: Ten years of applications and trends. *Science of the Total Environment, 409*, 3578–3594.

Institute of Environmental Sciences. (2012). CMLCA software program. Leiden University, RA Leiden, The Netherlands

International Organization for Standardization (ISO). (2006). ISO 14040:2006, Environmental management, life cycle assessment – Principles and framework.

James, B. R., & Gorelick, S. M. (1994). When enough is enough: The worth of monitoring data in aquifer remediation design. *Water Resources Research, 30*, 3499–3513.

Jeon, J. R., Murugesan, K., Nam, I. H., & Chang, Y. S. (2013). Coupling microbial catabolic actions with abiotic redox processes: A new recipe for persistent organic pollutant (POP) removal. *Biotechnology Advances, 31*, 246–256.

Jiamjitrpanich, W., Parkpian, P., Polprasert, C., Laurent, F., & Kosanlavit, R. (2012). The tolerance efficiency of *Panicum maximum* and *Helianthus annuus* in TNT-contaminated soil and NZVI-contaminated soil. *Journal of Environmental Science and Health, 47*, 1506–1513.

Jolliet, O., Margni, M., Charles, R., Humbert, S., Payet, J., Rebitzer, G., & Rosenbaum, R. (2003). IMPACT 2002+: A new life cycle impact assessment methodology. *The International Journal of Life Cycle Assessment, 8*(6), 324–330.

Kadar, E., Dyson, O., Handy, R. D., & Al-Subiai, S. N. (2013). Are reproduction impairments of free spawning marine invertebrates exposed to zero-valent nano-iron associated with dissolution of nanoparticles? *Nanotoxicology, 7*, 135–143.

Karn, B., Kuiken, T., & Otto, M. (2009). Nanotechnology and in situ remediation: A review of the benefits and potential risks. *Environmental Health Perspectives, 17*(12), 1823–1831.

Keenan, C. R., Goth-Goldstein, R., Lucas, D., & Sedlak, D. L. (2009). Oxidative stress induced by zero-valent iron nanoparticles and Fe(II) in human bronchial epithelial cells. *Environmental Science & Technology, 43*, 4555–4560.

Keller, A. A., Garner, K., Miller, R. J., & Lenihan, H. S. (2012). Toxicity of nano-zero valent iron to freshwater and marine organisms. *PLoS One, 7*(8), e43983. https://doi.org/10.1371/journal.pone.0043983.

Keum, Y. S., & Li, Q. X. (2005). Reductive debromination of polybrominated diphenyl ethers by zerovalent iron. *Environmental Science & Technology, 39*, 2280–2286.

Khanna, V., Bakshi, B. R., & Lee, L. J. (2008). Carbon nanofiber production. *Journal of Industrial Ecology, 12*(3), 394–410.

Kiker, G. A., Bridges, T. S., Varghese, A., Seager, T. P., & Linkov, I. (2005). Application of multicriteria decision analysis in environmental decision making. *Integrated Environmental Assessment and Management, 1*, 95–108.

Kim, L. Y., Changha, L., Love, D. C., Sedlak, D. L., Yoon, J., & Nelson, K. L. (2011). Inactivation of MS2 coliphage by ferrous ion and zero-valent iron nanoparticles. *Environmental Science & Technology, 45*, 6978–6984.

Kim, H. J., Phenrat, T., Tilton, R. D., & Lowry, G. V. (2009). Fe0 nanoparticles remain mobile in porous media after aging due to slow desorption of polymeric surface modifiers. *Environmental Science & Technology, 43*, 3824–3830.

Kirschling, T., Gregory, K., Minkley, N., Lowry, G., & Tilton, R. (2010). Impact of nanoscale zero valent iron on geochemistry and microbial populations. *Environmental Science & Technology, 44*, 3474–3480.

Kumar, N., Omoregie, E. O., Rose, J., Masion, A., Lloyd, J. R., Diels, L., & Bastiaens, L. (2013). Inhibition of sulfate reducing bacteria in aquifer sediment by iron nanoparticles. *Water Research, 51*, 64–72.

Lee, C., Kim, J. Y., Lee, W. I., Nelson, K. L., Yoon, J., & Sedlak, D. L. (2008). Bactericidal effect of zero-valent iron nanoparticles on *Escherichia coli*. *Environmental Science & Technology, 42*, 4927–4933.

Lemming, G., Hauschild, M., & Bjerg, P. (2010). Life cycle assessment of soil and groundwater remediation technologies: Literature review. *The International Journal of Life Cycle Assessment, 15*(1), 115–127.

Lemming, G., Chambon, J. C., Binning, P. B., & Bjerg, P. L. (2012). Is there an environmental benefit from remediation of a contaminated site? Combined assessments of the risk reduction and life cycle impact of remediation. *Journal of Environmental Management, 112*, 392–403.

Li, X. Q., & Zhang, W. X. (2007). Sequestration of metal cations with zerovalent iron nanoparticles-a study with high resolution x-ray photoelectron spectroscopy (HR-XPS). *Journal of Physical Chemistry C, 111*, 6939–6946.

Li, H., Zhou, Q., Wu, Y., Fu, J., Wang, T., & Jiang, G. (2009). Effects of waterborne nano-iron on medaka (*Oryzias latipes*): Antioxidant enzymatic activity, lipid peroxidation and histopathology. *Ecotoxicology and Environmental Safety, 72*, 684–692.

Li, Z., Greden, K., Alvarez, P., Gregory, K., & Lowry, G. (2010). Adsorbed polymer and NOM limits adhesion and toxicity of nano scale zero-valent iron (NZVI) to *E. coli*. *Environmental Science & Technology, 44*, 3462–3467.

Lien, H. L., Jhuo, Y. S., & Chen, L. H. (2007). Effect of heavy metals on dechlorination of carbon tetrachloride by iron nanoparticles. *Environmental Engineering Science, 24*, 21–30.

Lin, K., Chang, N., & Chuang, T. (2008). Fine structure characterization of zerovalent iron nanoparticles for decontamination of nitrites and nitrates in wastewater and groundwater. *Science and Technology of Advanced Materials, 9*, 025015.

Linkov, I., & Moberg, E. (2012). *Multi-criteria decision analysis: Environmental applications and case studies*. Boca Raton: CRC Press.

Linkov, I., Satterstrom, F. K., Kiker, G., Batchelor, C., Bridges, T., & Ferguson, E. (2006a). From comparative risk assessment to multi-criteria decision analysis and adaptive management: Recent developments and applications. *Environment International, 32*, 1072–1093.

Linkov, I., Satterstrom, F. K., Kiker, G., Seager, T. P., Bridges, T., Gardner, K. H., Rogers, S. H., Belluck, D. A., & Meyer, A. (2006b). Multicriteria decision analysis: A comprehensive decision approach for management of contaminated sediments. *Risk Analysis, 26*, 61–78.

Linkov, I., Satterstrom, F., Steevens, J., Ferguson, E., & Pleus, R. (2007). Multi-criteria decision analysis and environmental risk assessment for nanomaterials. *Journal of Nanoparticle Research, 9*, 543–554.

Linkov, I., Loney, D., Cormier, S., Satterstrom, F. K., & Bridges, T. (2009). Weight-of-evidence evaluation in environmental assessment: Review of qualitative and quantitative approaches. *Science of the Total Environment, 407*(19), 5199–5205.

Linkov, I., Welle, P., Loney, D., Tkachuk, A., Canis, L., Kim, J. B., & Bridges, T. (2011). Use of multicriteria decision analysis to support weight of evidence evaluation. *Risk Analysis, 31*, 1211–1225.

Lloyd, S. M., Lave, L. B., & Matthews, H. S. (2005). Life cycle benefits of using nanotechnology to stabilize platinum-group metal particles in automotive catalysts. *Environmental Science & Technology, 39*(5), 1384–1392.

Ma, X., Gurung, A., & Deng, Y. (2013). Phytotoxicity and uptake of nanoscale zero-valent iron (NZVI) by two plant species. *Science of the Total Environment, 443*, 844–849.

Marsalek, B., Jancula, D., Marsalkova, E., Mashlan, M., Safarova, K., Tucek, J., & Zboril, R. (2012). Multimodal action and selective toxicity of zerovalent iron nanoparticles against cyanobacteria. *Environmental Science & Technology, 46*, 2316–2323.

Mueller, N. C., Braun, J., Bruns, J., Černík, R. P., Rickerby, D., & Nowack, B. (2012). Application of nanoscale zero valent iron (NZVI) for groundwater remediation in Europe. *Environmental Science and Pollution Research, 19*, 550–558.

Müller, N.C., & Nowack, B. (2010). *Nano zero valent iron – THE solution for water and soil remediation?* ObservatoryNANO focus report.

Nadagouda, M. N., Castle, A. B., Murdock, R. C., Hussain, S. M., & Varma, R. S. (2010). In vitro biocompatibility of nanoscale zerovalent iron particles (NZVI) synthesized using tea polyphenols. *Green Chemistry, 12*, 114–122.

National Renewable Energy Laboratory. (2012). U.S. life cycle inventory database. Golden, CO, USA

O'Carroll, D. M., Sleep, B. E., Karol, M., Boparai, H. K., & Kocur, C. (2013). Nanoscale zero valent iron and bimetallic particles for contaminated site remediation. *Advances in Water Resources, 51*, 104–122.

Onwubuya, K., Cundy, A., Puschenreiter, M., Kumpiene, J., Bone, B., Greaves, J., Teasdale, P., Mench, M., Tlustos, P., Mikhalovsky, S., Waite, S., Friesl-Hanl, W., Marschner, B., & Müller, I. (2009). Developing decision support tools for the selection of "gentle" remediation approaches. *Science of the Total Environment, 407*, 6132–6142.

Oracle. (2008). *Oracle crystal ball*. The Decision Table Tool.

Osterwalder, N., Capello, C., Hungerbühler, K., & Stark, W. (2006). Energy consumption during nanoparticle production: How economic is dry synthesis? *Journal of Nanoparticle Research, 8*(1), 1–9.

Ostiguy, C., Riediker, M., Triolet, J., Troisfontaines, P., & Vernez, D. (2010). *Development of a specific control banding tool for nanomaterials*. French Agency for Food, Environmental and Occupational Health & Safety.

Otero-González, L., García-Saucedo, C., Field, J. A., & Sierra-Álvarez, R. (2013). Toxicity of TiO_2, ZrO_2, Fe^0, Fe_2O_3, and Mn_2O_3 nanoparticles to the yeast, *Saccharomyces cerevisiae*. *Chemosphere, 93*, 1201–1206.

Paik, S. Y., Zalk, D. M., & Swuste, P. (2008). Application of a pilot control banding tool for risk level assessment and control of nanoparticle exposures. *Annals of Occupational Hygiene, 52*, 419–428.

Palisade Corporation. (2010). @Risk Industrial. Risk analysis software. ITHACA, NY, USA

Pawlett, M., Ritz, K., Dorey, R. A., Rocks, S., Ramsden, J., & Harris, J. A. (2013). The impact of zero-valent iron nanoparticles upon soil microbial communities is context dependent. *Environmental Science and Pollution Research International, 20*, 1041–1049.

PE International. (2011). GaBi 5. Life cycle assessment modeling software. Leinfelden-Echterdingen, Germany

Phenrat, T., Long, T. C., Lowry, G. V., & Veronesi, B. (2009). Partial oxidation ("aging") and surface modification decrease the toxicity of nanosized zerovalent iron. *Environmental Science & Technology, 43*, 195–200.

Phenrat, T., Fagerlund, F., Illanagasekare, T., Lowry, G. V., & Tilton, R. D. (2011). Polymer-modified Fe0 nanoparticles target entrapped NAPL in two dimensional porous media: Effect ofparticle concentration, NAPL saturation, and injection strategy. *Environmental Science & Technology, 45*, 6102–6109.

PRé Consultants. (2000). Eco-indicator 99. A damage oriented method for life cycle impact assessment.

PRé Consultants. (2013). SimaPro 8. LCA software.

Riediker, M., Ostiguy, C., Triolet, J., Troisfontaine, P., Vernez, D., Bourdel, G., Thieriet, N., & Cadene, A. (2012). Development of a control banding tool for nanomaterials. *Journal of Nanomaterials, 2012*, 8.

Sacca, M. L., Fajardo, C., Costa, G., Lobo, C., Nande, M., & Martin, M. (2013a). Integrating classical and molecular approaches to evaluate the impact of nanosized zero-valent iron (NZVI) on soil organisms. *Chemosphere, 104*, 184–189.

Sacca, M. L., Fajardo, C., Nande, M., & Martin, M. (2013b). Effects of nano zero-valent iron on *klebsiella oxytoca* and stress response. *Environmental Microbiology, 66*, 806–812.

Saleh, N., Kim, H., Phenrat, T., Matyjaszewksi, K., Lowry, G. V., & Tilton, R. D. (2008). Ionic strength and composition affect the mobility of surface-modified Fe0 nanoparticles in water-saturated sand columns. *Environmental Science & Technology, 42*, 3349–3355.

Saleh, N., Phenrat, T., Sirk, K., Dufour, B., Ok, J., Sarbu, T., Matyjaszewski, K., Tilton, R. D., & Lowry, G. V. (2005). Adsorbed triblock copolymers deliver reactive iron nanoparticles to the oil/water interface. *Nano Letters, 5*, 2489–2494.

Sexton, K., & Linder, S. H. (2014). Integrated assessment of risk and sustainability in the context of regulatory decision making. *Environmental Science & Technology, 48*, 1409–1418.

Shah, V., Dobiasova, P., Baldrian, P., Nerud, F., Kumar, A., & Seal, S. (2010). Influence of iron and copper NP powder on the production of lignocellulose degrading enzymes in the fungus *Trametes versicolor*. *Journal of Hazardous Materials, 178*, 1141–1145.

Shatkin, J. A. (2008). *Nanotechnology: Health and environmental risks*. Boca Raton: Taylor & Francis.

Shatkin, J. A. (2009a). *Investigating the life-cycle risks of a nanomaterial in a coating using nano LCRA*. Society for risk analysis annual meeting. Symposium M4-I, Baltimore.

Shatkin, J. A. (2009b). *Risk analysis for nanotechnology: State of the science and implications*. Washington, DC: US Department of Agriculture.

Soratana, K., & Marriott, J. (2010). Increasing innovation in home energy efficiency: Monte Carlo simulation of potential improvements. *Energy and Buildings, 42*(6), 828–833.

Soratana, K., Harper, W. F., & Landis, A. E. (2012). Microalgal biodiesel and the renewable fuel standard's greenhouse gas requirement. *Energy Policy, 46*(0), 498–510.

Soratana, K., Khanna, V., & Landis, A. E. (2013). Re-envisioning the renewable fuel standard to minimize unintended consequences: A comparison of microalgal diesel with other biodiesels. *Applied Energy, 112*(0), 194–204.

Stephenson, J. B. (2010). *EPA's estimated costs to remediate existing sites exceed current funding levels, and more sites are expected to be added to the National Priorities List*. US Government Accountability Office. http://www.gao.gov/products/GAO-10-380

Suttinun, O., Luepromchai, E., & Müller, R. (2013). Cometabolism of trichloroethylene: Concepts, limitations and available strategies for sustained biodegradation. *Reviews in Environmental Science and Biotechnology, 12*, 99–114.

Technical University of Denmark. (2003). Environmental design of industrial products (EDIP) 2003. Lyngby, Denmark

The Centre for Life Cycle Inventories. (2014). *Swiss center for life cycle inventories*. Ecoinvent Version 3.

The National Institute for Public Health and the Environment (RIVM), Institute of Environmental Sciences (CML), PRé Consultants, Nijmegen, R. U. (2008). ReCiPe. Life cycle impact assessment methodology.

The Royal Society and The Royal Academy of Engineering. (2004). *Nanoscience and nanotechnologies: Opportunities and uncertainties- two year review of progress on government actions*. Joint academies' response to the council for science and technology's call for evidence, London.

Theron, J., Walker, J. A., & Cloete, T. E. (2008). Nanotechnology and water treatment: Applications and emerging opportunities. *Critical Reviews in Microbiology, 34*, 43–69.

Tilston, E. L., Collins, C. D., Mitchell, G. R., Princivalle, J., & Shaw, L. J. (2013). Nanoscale zerovalent iron alters soil bacterial community structure and inhibits chloroaromatic biodegradation potential in Aroclor 1242-contaminated soil. *Environmental Pollution, 173*, 38–46.

Tratnyek, P. G., & Johnson, R. L. (2006). Nanotechnologies for environmental cleanup. *Nano Today, 1*, 44–48.

United States Environmental Protection Agency (US EPA). (2005). *US EPA workshop on nanotechnology for site remediation*. Washington, DC: US EPA.

US EPA. (2010). *Nanomaterial case studies. Nanoscale titanium dioxide in water treatment and topical sunscreen (final)*, Research Triangle Park.

US EPA. (2012a). *Nanomaterial case study. A comparison of multiwalled carbon nanotube and decabromodiphenyl ether flame-retardant coatings applied to upholstery textiles (draft)*, Research Triangle Park.

US EPA. (2012b). *Nanomaterial case study: Nanoscale silver in disinfectant spray (final report)*, Washington, DC.

US EPA. (2013). *Technology innovation and field services division*. http://www.epa.gov/superfund/partners/osrti/tifsd.htm

US EPA Office of Solid Waste and Emergency Response (OSWER). (2013). *The project on emerging nanotechnologies: Selected sites using or testing nanoparticles for remediation*. www.cluin.org/download/remed/nano-site-list.pdf

Utterback, J. M. (1987). Innovation and industrial evolution in manufacturing industries. In B. R. Guile & H. Brooks (Eds.), *Technology and global industry: Companies and nations in the world economy* (pp. 16–48). Washington: National Academic Press.

Utterback, J. M., & Suarez, F. F. (1993). Innovation, competition, and industry structure. *Research Policy, 22*(1), 1–21.

van Duuren-Stuurman, B., Vink, S. R., Verbist, K. J. M., Heussen, H. G. A., Brouwker, D. H., Lroese, D. E. D., van Niftrik, M. F. J., Tielemans, E., & Fransman, W. (2011). *Stoffenmanager nano: Description of the conceptual control banding model*. Zeist: Netherlands Organisation for Applied Scientific Research (TNO).

van Duuren-Stuurman, B., Vink, S. R., Verbist, K. J. M., Heussen, H. G. A., Brouwker, D. H., Lroese, D. E. D., van Niftrik, M. F. J., Tielemans, E., & Fransman, W. (2012). Stoffenmanager

nano version 1.0: Web-based tool for risk prioritization of airborne manufactured nano objects. *The Annals of Occupational Hygiene, 56*(5), 525–541.
Vegter, J., Lowe, J., & Kasamas, H. (2002). *Sustainable management of contaminated land: An overview*. Austrian Federal Environment Agency on behalf of CLARINET.
Wender, B. (2013). LCA and responsible innovation of nanotechnology. In *School of Sustainable Engineering and the Built Environment, Master of science*. Tempe: Arizona State University.
Wickson, F., Gillund, F., & Myhr, A. (2010). Treating nanoparticles with precaution: Recognising qualitative uncertainty in scientific risk assessment. In K. Kjølberg & F. Wickson (Eds.), *Nano meets macro* (pp. 445–472). Singapore: Pan Stanford Publishing.
Wiesner, M. R., & Bottero, J. Y. (2011). A risk forecasting process for nanostructured materials, and nanomanufacturing. *Comptes Rendus Physique, 12*, 659–668.
Woller, J. (1996). *The basic of Monte Carlo simulations*. Lincoln: University of Nebraska-Lincoln Physical Chemistry Lab.
Woodrow Wilson International Center for Scholars. (2014). *Project on emerging nanotechnologies*. http://www.nanotechproject.org/inventories/remediation_map/
Wu, D., Shen, Y., Ding, A., Mahmood, Q., Liu, S., & Tu, Q. (2013). Effects of nanoscale zero-valent iron particles on biological nitrogen and phosphorus removal and microorganisms in activated sludge. *Environmental Technology, 34*, 2663–2669.
Xiu, Z., Jin, Z., Li, T., Mahendra, S., Lowry, G. V., & Alvarez, P. J. J. (2010). Effects of nano-scale zero-valent iron particles on a mixed culture dechlorinating trichloroethylene. *Bioresource Technology, 101*, 1141–1146.
Yan, W., Lien, H. L., Koel, B. E., & Zhang, W. X. (2013). Iron nanoparticles for environmental clean-up: Recent developments and future outlook. Environ Sci. *Processes Impacts, 15*, 63–77.
Yang, Y., Guo, J., & Hu, Z. (2013). Impact of nano zero valent iron (NZVI) on methanogenic activity and population dynamics in anaerobic digestion. *Water Research, 47*, 6790–6800.
Zhou, L., Thanh, T. L., Gong, J., Kim, J. H., Kim, E. J., & Chang, Y. S. (2013). Carboxymethyl cellulose coating decreases toxicity and oxidizing capacity of nanoscale zerovalent iron. *Chemosphere, 104*, 155–161.

Chapter 16
State of Knowledge and Future Needs for NZVI Applications in Subsurface Remediation

Gregory V. Lowry and Tanapon Phenrat

Abstract This final chapter summarizes the key lessons learned from a large number of laboratory-scale, intermediate-scale, and well-documented field-scale experiments using NZVI for in situ remediation presented in this book. The goal is to provide guidance to site managers and remediation professionals who are considering the use of NZVI for in situ treatment of a contaminant source zone, particularly those containing dense nonaqueous phase liquids or reducible heavy metals, as these are the most prevalent types of sites for application of NZVI. It also provides guidance on site conditions where NZVI treatment may be a viable option in terms of reactivity with contaminants of concern, reactive lifetime, and particle deliverability. The chapter also highlights the importance of NZVI characteristics to provide longer reactive lifetimes, viable injection methods for emplacing NZVI in the desired location in the subsurface, and detailed site characterizations as well as suitable laboratory feasibility studies. Future research needs for more effective NZVI applications in subsurface remediation, including a better understanding of combined NZVI-monitored natural recovery or bioremediation strategies and more detailed performance evaluations, are also discussed.

Keywords Nanoscale zerovalent iron · Lesson learned · Future need · Guidance · Field implementation

G. V. Lowry (✉)
Center for Environmental Implications of Nanotechnology (CEINT), Carnegie Mellon University, Pittsburgh, PA, USA

Department of Civil & Environmental Engineering, Carnegie Mellon University, Pittsburgh, PA, USA
e-mail: glowry@cmu.edu

T. Phenrat
Department of Civil Engineering, Environmental Engineering Program, Naresuan University, Phitsanulok, Thailand

Center of Excellence for Sustainability of Health, Environment and Industry (SHEI), Faculty of Engineering, Naresuan University, Phitsanulok, Thailand

T. Phenrat, G. V. Lowry (eds.), *Nanoscale Zerovalent Iron Particles for Environmental Restoration*, https://doi.org/10.1007/978-3-319-95340-3_16

16.1 Introduction

The application of NZVI for in situ remediation of groundwater pollutants is becoming a mature remediation technology. There are ~77 documented pilot and full-scale demonstration sites where NZVI was used as part of the remediation technology, primarily for chlorinated solvents (see Chap. 1) (Bardos et al. 2015; NanoRem 2018). Significant progress has been made in understanding how environmental conditions affect reactivity (Chaps. 3 and 4), how to emplace NZVI in situ where it is needed (Chap. 6), and how to leverage the complex, coupled interactions between NZVI and bioremediation (Chap. 10). Performance expectations are becoming clear, depending on the type of contaminant, site conditions, and injection methods (Chaps. 7 and 8). Despite these advances, the "windows of opportunity" for NZVI in a remediation plan are not always clear. This chapter summarizes some of the key lessons learned from a large number of laboratory, intermediate-scale, and well-documented field-scale experiments using NZVI for in situ remediation that are summarized in this book. Figure 16.1 illustrates the relationship between various factors discussed in this chapter as well as the windows of opportunity for successful in situ remediation using NZVI. The goal is to provide guidance to site managers and remediation professionals who are considering the use of NZVI for in situ treatment of a contaminant source zone, particularly those containing dense nonaqueous phase liquids (DNAPL) or reducible heavy metals as these are the most prevalent types of sites for application of NZVI. It also provides guidance on site conditions where NZVI treatment may be a viable option and highlights the need for NZVI having sufficient selectivity for contaminants over water to provide longer reactive lifetimes, viable injection methods for emplacing NZVI in the desired location in the subsurface, and additional research that is needed to develop better combined NZVI-monitored natural recovery (MNR) strategies. The health and safety as well as the regulatory implications of using NZVI at a site were discussed in detail in Chap. 15 and are not elaborated further in this chapter.

16.2 Conceptual Model for Subsurface Remediation Using NZVI

The basic conceptual model for remediation using NZVI is shown in Fig. 16.2 and described in more detail in Chap. 7. Here we consider an injected mass of NZVI that is distributed in the subsurface within and downgradient of the source. The injected NZVI lowers the redox potential in and directly downgradient of the injected NZVI. The NZVI reduces the contaminants or other reducible compounds that adsorb to the NZVI surface. Both dissolved Fe^{2+} and H_2 are released into the pore water surrounding the NZVI.

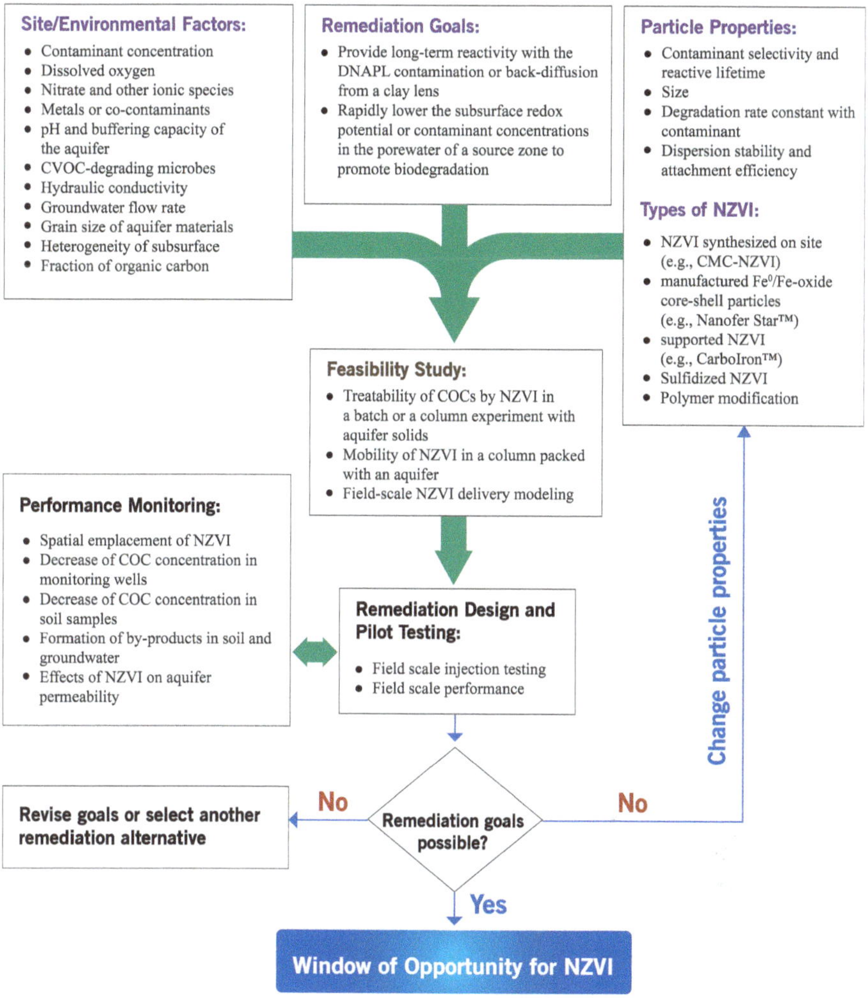

Fig. 16.1 Relationships between various factors discussed in this chapter as well as the windows of opportunity for successful in situ remediation using NZVI

16.3 Considerations About the Types of NZVI Available for Remediation

An important consideration of any project is what type of NZVI to use. While different methods have been proposed to synthesize NZVI at scales and costs required for viable commercial application (Chap. 2), the most dominant categories of NZVI include (1) those synthesized on site using borohydride or other suitable reductants and injected immediately; (2) those that are manufactured as Fe^0/Fe-oxide

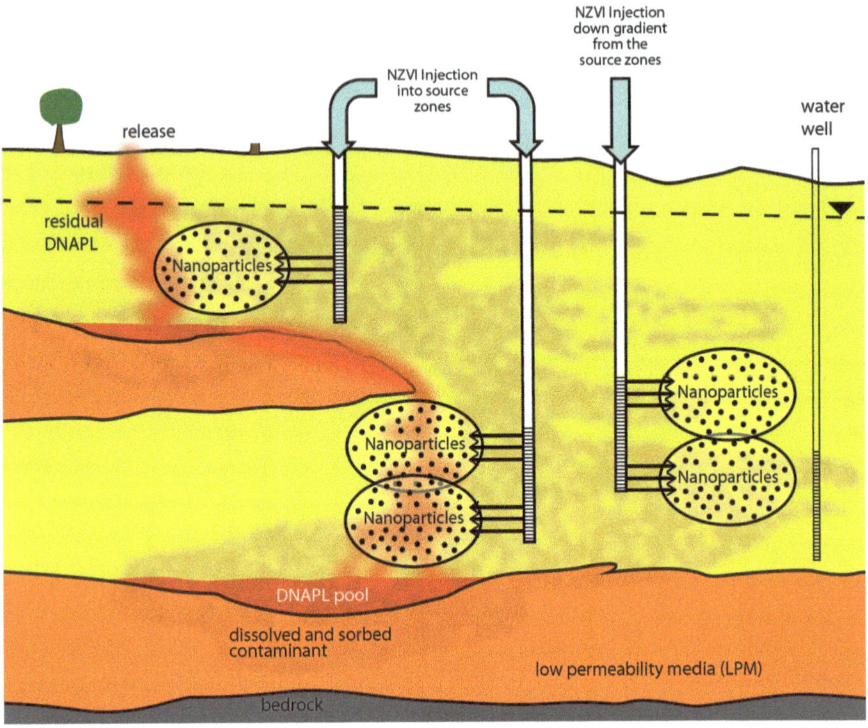

Fig. 16.2 Conceptual model for NZVI injection into and just downgradient from a DNAPL source area. Injected NZVI forms a reactive barrier that degrades the organic compound dissolved in the pore water. A lower mass of contaminant in the source, lower pore water concentrations, and stimulated biodegradation can all lead to lower concentration in downgradient wells

core-shell particles, shipped to the remediation site, and dispersed in water prior to injection (e.g., Nanofer Star™); and (3) supported NZVI products, typically supported on a carbon material (e.g., Carbo-Iron™). The supported NZVI products are engineered to be more mobile in the subsurface, and pilot field-scale studies with these materials indicate that they are indeed more easily injected into the subsurface, but transport distances in the field are still on the order of only a few meters (see summary of transport distances in Chap. 1) (Bardos et al. 2015; Mackenzie et al. 2016). Noticeably, polymers, such as carboxymethyl cellulose (CMC), are required for enhanced deliverability of all three kinds of NZVI. There are newer types of NZVI products being developed, e.g., sulfidized NZVI that may potentially overcome the limitation of NZVI's poor selectivity for the contaminant over water. However, the benefits of this approach have primarily been observed in laboratory studies, and more field trials are needed to quantify the benefits of this approach for increasing the reactive lifetime of the materials. Moreover, it is unclear if sulfidation increases transport and injectability in the subsurface. The sulfidized NZVI was thoroughly discussed in Chap. 9. Thus, we focus our attention here on these three main categories of NZVI.

The desired reactive properties of the NZVI product used will depend on the goals of the remediation. If the *remediation goal is to provide long-term reactivity with the DNAPL contamination as it dissolves or as it back-diffuses from a clay lens or other low hydraulic conductivity sources*, then high selectivity for the contaminant and a long reactive lifetime are desirable because both back-diffusion and DNAPL dissolution are slow processes. A higher selectivity and reactive lifetime will decrease the number of NZVI injections required to achieve the remediation goals. A longer reactive lifetime could be accomplished using a NZVI that has a much lower reactivity with the contaminant of interest than many of the NZVI products that are discussed in the literature or that are commercially available (as discussed in Chap. 1, the average abiotic reactive lifetime of NZVI in field-scale applications is 211 ± 134 days). Many studies have aimed at increasing the reaction rates of NZVI with contaminants, but this is not likely to be essential for good performance in the field where groundwater flows are slow or where the degradation rate of the contaminant is controlled by the rate of dissolution of DNAPL. In this regard, it may be desirable to use larger NZVI particles (e.g., several hundred nanometers in size) rather than the very small and highly reactive NZVI that is typically produced and studied (see good examples using Nanofer 25S vs. Nanofer Star (Stejskal et al. 2017) or using NZVI with micron-sized ZVI (Mueller et al. 2012) in Chap. 1). There is not likely to be a "one-size-fits-all" solution. Rather, the optimal NZVI particle size will ultimately be a trade-off between the required rate of reaction, the material's reactive lifetime, and ability to emplace the NZVI into the subsurface at the site.

If the *remediation goal is to rapidly lower the redox potential in the subsurface or to relatively quickly lower contaminant concentrations in the pore water of a source zone to promote biodegradation*, then injecting small, highly reactive NZVI would likely achieve this goal. It should be noted that there is not yet any compelling field evidence that NZVI injected into a source area containing free-phase DANPL results in mixing of NZVI with DNAPL and rapid degradation of entrapped DNAPL. It is more likely that the injected NZVI is serving as a reactive barrier for contaminants that are dissolving into the pore water. Regardless of the conceptual model, it is important to recognize that the reactive lifetime of the smallest NZVI particles will be as short as a few hours, days, or potentially weeks depending on the concentration injected and the influx of oxidant in the groundwater (target contaminant or other oxidants like dissolved O_2 or nitrate) entering the reactive NZVI zone. For achieving long-term reducing conditions in the subsurface, it may be desirable to inject a mixture of NZVI particle sizes such that you rapidly achieve reducing conditions with the smallest, most reactive particles and then provide long-term reducing conditions and reactivity with contaminants with the larger, less reactive particles in the mixture (see an example using Nanofer 25S and Nanofer Star (Stejskal et al. 2017) in Chap. 1). Co-injection of a mixture of particle sizes or consecutive injections of small and then large particles will depend on the properties of the mixed dispersion, the aquifer source zone, and the costs associated with multiple injections vs. a single injection. There is limited data available on the trade-offs between the different methods of injection and the effect of NZVI particle size distribution on injection and movement in the subsurface at the field scale. This currently prevents optimization of injection strategies that maximize remediation performance and minimize costs other than those conduced on site in pilot- and full-scale demonstrations.

16.4 Conditions Favorable to an NZVI-Based Remedy

NZVI is typically considered a source-zone treatment strategy. However, injection of NZVI into a DNAPL source effectively creates a reactive barrier where contaminants must dissolve (if DNAPL is present) or desorb (if back-diffusion is controlling the release) prior to reacting with the NZVI. The emplacement of NZVI into or near a DNAPL source did not enhance NAPL dissolution in laboratory studies (Fagerlund et al. 2012). Similarly, emplacement of NZVI around a tight source area, e.g., DNAPL or metals entrapped in a clay lens, would not be expected to significantly enhance mass transfer from that source into the groundwater. Practically speaking, in conditions where the rate of flux of contaminants away from the source area by groundwater flow is faster than the dissolution rate, the emplacement of NZVI at that interface would not be expected to increase the diffusion rate out of the low hydraulic conductivity lens. In laboratory studies, the NZVI directly at the DNAPL-water interface was rapidly oxidized, while the remaining NZVI that is emplaced farther away from the DNAPL-NZVI interface, even a few cm, served as a reactive barrier for the dissolved DNAPL components being released from the DNAPL (Fagerlund et al. 2012). This means that performance expectations for emplaced NZVI in an unconsolidated porous medium can, in part, be informed from lessons learned with in situ reactive barrier technologies (e.g., PRBs with ZVI). Keeping this conceptual model in mind, some guidance can be provided for identifying those sites that are favorable for NZVI treatment and those that are not. Favorable sites are those with relatively high contaminant concentrations (hence their use in source areas rather than for treating contaminant plumes), with limited intrusion of competing oxidant in upgradient water, with limited buffer capacity and neutral to slightly basic pH, and that are easily biostimulated through a reduction in redox potential. Rationales for these guidelines are discussed below.

Higher concentrations of contaminants, i.e., source zones, can be more efficient with respect to utilizing the reducing power of the NZVI for reduction of the contaminant over water. Given the relatively low selectivity of NZVI toward the contaminants, most of the reducing power (electrons) are used to reduce constituents other than the contaminants, primarily water to form H_2. Laboratory studies using TCE clearly indicated that higher contaminant concentrations favored the reaction with TCE over water (Liu et al. 2007). Strategies around the problem have included the use of carbon supports (such as Carbo-Iron colloids) to concentrate the contaminants near the NZVI or to make the NZVI more selective by doping it with sulfur (Fan et al. 2013, 2016). While these improvements are important recent advances, they do not yet provide sufficient selectivity to make NZVI work effectively as a long-lived reactive in situ barrier, so use in source zones remains the most attractive choice.

It is important to remember that NZVI is a *reactant* that is consumed in the reaction and not a catalyst. Thus, reactions with water and the continued influx of water with reductant demands will significantly impact the reactive lifetime of the emplaced NZVI. In field tests with Nanofer Star, reactive lifetimes for the NZVI ranged from 1 to 2 months (88% and 92% of the Fe^0 consumed, respectively). In one case, ~30% of the Fe^0 remained after 5 months (Bardos et al. 2015). Sites with significant recharge of aerated water or with NO_3 will dramatically shorten the reactive lifetime of the injected NZVI. Both dissolved O_2 and NO_3^- have been shown to be highly reactive with NZVI, and both require a significant number of electrons to reach their fully reduced products (H_2O requires 4e$^-$/mol and NH_3 requires 8e$^-$/mol) (Liu et al. 2007). This can be more than is required for TCE reduction to acetylene (4e$^-$/mol) and ethane (6e$^-$/mol). Significant influx of oxidants that react with NZVI or that are not reactive but would prevent the pore water from achieving strong reducing conditions would negatively impact NZVI performance. Thus, the site hydrology should be assessed to address any perennial inflow of oxidants to yield the best remediation performance. In particular, deep aquifers in arid regions with thick vadose zones or fractured media with significant and rapid surface water recharge will likely require a means of controlling these inflows of oxidant.

The water pH can also greatly affect the reactive lifetime of NZVI (Liu et al. 2007). Lower-pH waters lead to aggressive corrosion and formation of H_2. The will decrease the reactive lifetime of the emplaced NZVI. However, the corrosion of NZVI also leads to formation of OH^- and a resulting increase in pH of the pore water over time. If the groundwater is relatively poorly buffered, the pH of the pore water in the reactive zone relatively quickly becomes buffered by the Fe^0/H_20 redox couple which is ~10. This slows the corrosion rate of the NZVI with water but does not appreciably slow the desired reaction with contaminants like PCE (Kim et al. 2017). Thus, waters that are neutral pH or higher and that are poorly buffered are good candidates for NZVI applications.

16.5 Strategies for Injection

The injection of the NZVI and its final distribution in the source zone will determine the success of the remediation strategy. Unfortunately, one of the most difficult aspects of an NZVI-based subsurface remediation strategy is designing the injection strategy. Therefore, the other criterion for a favorable site for NZVI is that the porous medium must be amenable to injection of NZVI.

The ultimate distribution of NZVI into a specific formation cannot easily be predicted using bench-scale and intermediate-scale testing of injections, even using site-specific soils. Stability against aggregation, i.e., a highly stable NZVI suspension, does not guarantee good mobility in porous media. Simple filtration models are not good indicators for the distribution of NZVI in the subsurface. More complex models have been developed (see Chap. 6) (Phenrat et al. 2010a;

Babakhani et al. 2015; Bardos et al. 2015; Chowdhury et al. 2015), but these models must be calibrated for each site and each NZVI type and must take into account the heterogeneity of the site and potential confounding factors (e.g., the presence of fine – e.g., clay – particles). Methods need to be developed to determine the effective parameters of these models at the application scales. Evaluations of injectability have to be piloted at the site of interest (Bennett et al. 2010; Johnson et al. 2013; Busch et al. 2015; Chowdhury et al. 2015).

Most field injections have used either gravity-fed injections into wells or pressurized (direct push) injections using a Geoprobe® (see Chaps. 1, 7, 8). These injection strategies are appropriate for unconsolidated media with reasonable hydraulic conductivity (e.g., sands or sandy loams), fractured media, or consolidated media that must first be fractured to introduce the NZVI in a controlled manner. However, a porous medium is an excellent filter, and all field cases have shown that injected NZVI particles are readily filtered near the injection location. Based on the field data collected thus far (Table 1.6 in Chap. 1), the average travel distance of NZVI in pilot- or field-scale applications is 2.84 ± 1.60 m. For this reason, expectations of travel distances using current injection methods should be kept modest, less than 1–2 m (Bardos et al. 2015). Note that the "extent" of the transport is often reported using different metrics. The concentration of NZVI away from the injection source declines exponentially. So a transport distance where the injection concentration has decreased to 1/2 of the injected amount may be ~1 m, but the "extent" of the plume of injected NZVI may be much farther if one defines the transport distance where the NZVI concentration decreases to 0.05 or 0.01 of 0.001 of the injected NZVI concentration. The latter does not represent the location where the bulk of the iron (10–15 g/L) has been emplaced, which is the location where the required reducing conditions and reactive barriers are actually emplaced. Thus, the lack of effective injection strategies is still a significant barrier to effective implementation of NZVI.

16.6 Characterization of NZVI Emplacement

There are limited characterization data for the extent of the injected NZVI plume in the field, making it difficult to know what has been achieved in field sites. Although geochemical analysis of soil and water samples and visual observation of black NZVI dispersion in such samples can be conducted to directly determine the extent of NZVI emplacement, the spatial resolution of such data is very limited and cannot provide a complete picture of the subsurface distribution of NZVI. Moreover, sample collection and chemical analysis are time-consuming and not suitable for real-time monitoring of field activities like NZVI injection. Similarly, although in situ magnetic susceptibility measurements can be used to detect NZVI at ~0.5 g/L (Bardos et al. 2015), these must be deployed in selected monitoring wells, making them unable to deliver a complete subsurface distribution of NZVI.

For this reason, a geophysical technique, such as induced-polarization (IP) imaging (also known as complex electrical conductivity (CC) imaging), appears to be a good candidate for subsurface delineation of NZVI distribution. Furthermore, this nonintrusive technique can monitor continuous change in the subsurface during the NZVI injection. This approach uses four electrode impedance measurements in a time-domain manner, using a square wave with on and off times. Two electrodes are used to inject an electrical current while the other two to measure the resulting voltage. The IP measurement is interpreted using the ratio of the recorded voltage to the injected current amplitude (resistance) and the measurement of the voltage decay, once the current is switched off (chargeability).

The IP result can be reported in terms of the magnitude of complex electrical conductivity ($|\sigma|$) and the phase of complex electrical conductivity (ϕ). The magnitude ($|\sigma|$) of the complex electrical conductivity is primarily controlled by the electrical conductivity of the fluid (σ_f), the porosity, and the connectivity of the pore space. On the other hand, the phase of the complex electrical conductivity (ϕ) is solely related to surface conductivity (i.e., interface polarization effects) governed by electrochemical processes at the interface between mineral grains and pore fluid. Consequently, the distribution of the injected NZVI affects the surface conductivity of the subsurface, which is detected by the phase of the complex electrical conductivity (ϕ). In the laboratory, Joyce et al. (2012) demonstrated that the increase of ZVI concentration in a packed sand column contributed to the increase of ϕ. Similarly, for injected guar gum-coated micron-sized ZVI (MZVI) in a pilot-scale demonstration, Flores Orozco et al. (2015) reported the change of ϕ (25% change) as a result of the ZVI emplacement. Interestingly, they found this change is less than bare ZVI, presumably due to the presence of an adsorbed guar gum layer on MZVI, which hinders electronic conduction and electrode polarization processes (Flores Orozco et al. 2015). Nevertheless, this IP approach can still sufficiently provide a spatial distribution of injected MZVN and NZVI (Fig. 16.3) for remediation monitoring purposes.

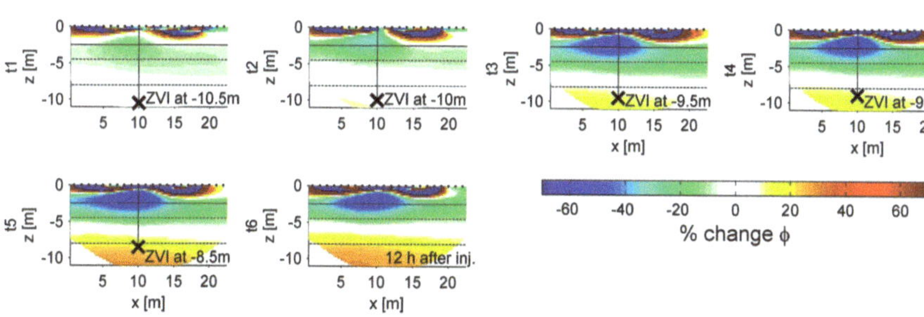

Fig. 16.3 Relative change of the phase of complex electrical conductivity (ϕ) at different times during and after GG-MZVI injection (the baseline (*t*0) images and those obtained for data collected during (*t*1−*t*5) and after (*t*6) the injection). The vertical line and black crosses indicate the respective positions of the injection point; black dots represent the electrodes placed at the surface; the solid horizontal line indicates the position of the groundwater level, and the dashed horizontal lines indicate the upper and lower limits of the aquitard (clayey-sandy materials). (Reprinted with permission from Flores Orozco et al. (2015). Copyright (2015) American Chemical Society)

16.7 Measuring the Remediation Performance of NZVI in the Field and Managing Expectations

Assessing the performance of NZVI in the field is challenging, but it is necessary to assess its potential as an effective remediation tool. Most performance data are based on the ability of NZVI treatment to decrease contaminant concentrations in monitoring wells downgradient from the source. Measurements of the reduction of mass of contaminant in the source area cannot be easily determined. Better tools and additional performance monitoring could provide the lines of evidence required to understand why NZVI treatment is successful or unsuccessful at a remediation site. These tools will have to differentiate between the desired effects from a lower contaminant mass in the source zone and contaminant degradation in a reactive barrier, with other possible factors that can influence contaminant concentrations in downgradient well. Degradation may also be a result of enhanced biodegradation occurring in the anaerobic or anoxic zone created by the NZVI. However, the lower mass flux from the source may also be due to changes in the formation permeability in the "reactive zone." If injection results in a lower hydraulic conductivity than the surrounding area, water may flow around the source and reactive zones. This lower water flux into the source zone can lead to a lower mass emission of contaminants from the source to the downgradient monitoring wells. While this is not the intention of the injected NZVI, it may lead to the desired effect of lower downgradient concentrations of contaminants. This has been shown to happen in quasi-2D flow experiments (Fig. 16.4), but the impacts of injection on the subsurface flow field and mass emission from the source are not commonly determined in field testing.

The recent NanoREM project published performance results from a series of NZVI injections (and other Fe-based materials). These can be found in a series of bulletins that have been published (http://www.nanorem.eu/Displaynews.aspx?ID=938) (NanoRem 2018). These tests included methods to monitor the flow of injected water (LiCl tracer) and monitoring wells to measure NZVI movement and redox levels over time (Stejskal et al. 2017). While the injected tracer is a good indicator of where the injected water migrated as dilution effects, it did little to indicate where the NZVI had migrated. Extensive monitoring of the concentration of chlorinated solvents and degradation by-products was also measured over time in monitoring wells nearby the NZVI injection locations (within 4–6 m). Contaminant concentrations decreased for a while (277 days) but then rebounded (Chap. 1). Part of this decrease is attributed to dilution from the injection water, but the formation of degradation products clearly indicates some degradation is also occurring. Collection and analysis of the NZVI a year after injection was used to assess the remaining Fe^0 content.

This suite of performance testing for in situ NZVI applications is impressive and informative and is unusual in its thoroughness. But this level of effort is required for many more sites to gather performance information to meter expectations of NZVI performance. However, the question remains as to whether or not these results (and

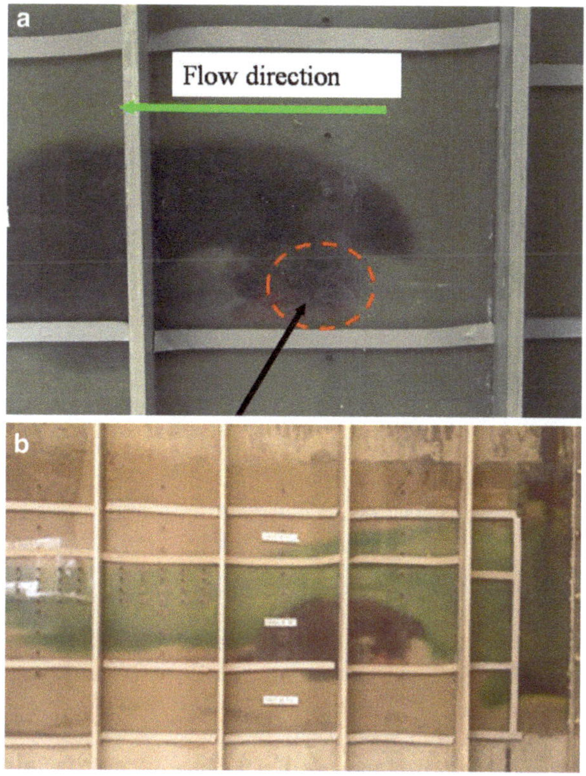

Fig. 16.4 (**a**) NZVI injected into and behind a PCE DNAPL source zone emplaced in large 2D tank experiment. (**b**) Flow field indicated by a green dye shows that NZVI emplaced into the source zone lowers the hydraulic conductivity in the source zone so that water flows above the source rather than through it. The impact of NZVI injection on the flow field in pilot-scale field studies is not generally reported, but may be impacted similarly in the field as in these laboratory tank studies

similar ones at other NanoREM sites as well as other published sites (see Chap. 1)) represent a "successful" application of NZVI. Concentrations of chlorinated solvents in monitoring wells remained low for ~277 days, and a significant part of that decrease was due to reaction with the NZVI. But concentrations then rebounded when the NZVI reactivity has been exhausted and the redox potential became oxidizing again. Could additional NZVI be injected into that same location each year until the source was exhausted and rebound was no longer observed? This would depend on how the injected NZVI (now oxidized) affected the aquifer permeability. Performance results from field trials obtained in several NanoREM pilot-scale tests set a good bar for expectations of performance in the field. Managing the performance expectations of NZVI at a particular site will require a clear definition of the metrics of success. A greater number of well-documented NZVI applications and corresponding remediation performance are needed to continue to refine these metrics.

16.7.1 Aquifer Characterization Needs and Limits of Treatability Studies

As noted in Chaps. 7 and 8, one of the most important factors for successful application of an NZVI-based remediation strategy is extensive site characterization data. The water quality (oxidant demand), contaminant distribution, and hydraulic conductivity field must be well characterized to devise a successful injection strategy. Heterogeneity in the hydraulic conductivity field also significantly impacts NZVI transport and emplacement. Without sufficient aquifer characterization, it is difficult to determine how much NZVI to inject, where to inject it, and how often to inject it or to accurately characterize its remediation performance. The required site characterization, especially for heterogeneous sites, can gain benefit from advanced real-time characterization techniques, including the membrane interface probe (MIP) for source-zone delineation and hydraulic profiling tools (HPT) for estimation of spatial hydraulic conductivity (Fig. 16.5a). The MIP has a halogen-specific detector (XSD), which specifically detects chlorinated volatile organic compounds (CVOCs) (Fig. 16.5b). The detector responses from MIP can be calibrated with field-contamination data, such as the CVOC concentration in the soil (Fig. 16.5c). Such a calibrated relationship can be used to convert the MIP detector response (in mV) in other stations to the contaminant concentration in the soil, which can subsequently be used to develop a three-dimensional (3D) distribution of CVOCs in the soil for a site. Similarly, HPT can quantify special hydraulic conductivity down to 0.03 m/day. The results from multiple HPT stations can be used to develop a 3D distribution of hydraulic conductivity for a site. These detailed site characterization data can be combined to develop a conceptual site model, which will be further converted to a mathematical model for predicting the transport of CVOCs and other nontargeting oxidants in the subsurface (Fig. 16.5d) at a particular time. This will help design the location, dimensions, and particle concentration of an appropriate treatment zone (for source reduction or plume interception) using NZVI.

Laboratory-based treatability studies are often performed to determine the potential for NZVI to work at a given site. However, a typical treatability study has some limitations with respect to what it can indicate for field-scale performance. First, the treatability study is often a batch reactivity study performed in groundwater collected from the site. A batch system is not representative of a flow through a system with perennial inputs of nontarget oxidants or other compounds in groundwater that may adsorb to or react with the Fe-oxide shell and change its reactivity on longer time scales. Often the pH increases to 10 in a batch system, and that may not be achieved in the aquifer under flow conditions. Despite these shortcomings, the laboratory treatability study at least provides some indication of the reaction rates and degradation products that may be expected at the field site. The uncertainty in these estimates should be considered in the design of a pilot-scale system.

Many times aquifer solids are often not included in the batch studies with site groundwater. This is an important oversight that may affect performance in the field relative to studies in groundwater alone. Aquifer solids, especially those containing

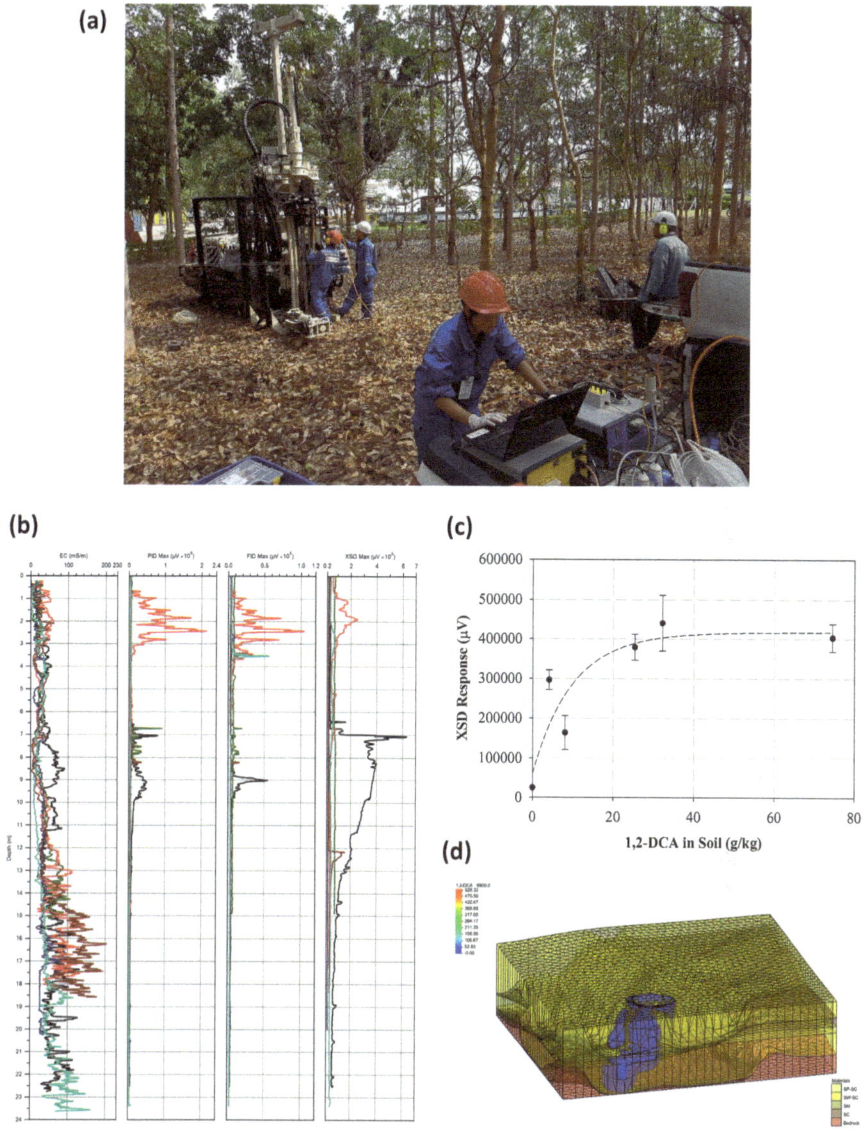

Fig. 16.5 (**a**) Hydraulic profiling tools (HPT) in action at a site. (**b**) Results of a membrane interface probe (MIP) from four different sensors, electrical conductivity sensor, photo ionization detector (PID), flame ionization detector (FID), and halogen-specific detector (XSD) (from left to right); Y-axis is the depth of investigation below ground surface, while the X-axis represents sensor or detector responses. For PID, FID, and XSD, the larger the response, the higher the concentration of different kinds of VOCs. (**c**) A relationship between XSD response (μV) and the 1,2-DCA concentration in the soil (mg/kg). (**d**) An example of modeling 1,2-DCA transport in the subsurface (blue plume) in a hydrogeological domain based on detailed site characterization

calcite, will often buffer the pH of the system to a much lower value that would be achieved in the batch test, even at 15 g/L of NZVI. This may affect the estimate of reactivity and reactive lifetime of the NZVI (Phenrat et al. 2010b). It will also neglect the potential for microbial degradation of contaminants and formation of unwanted by-products once a strong reducing environment is achieved. Using water, aquifer materials, and NZVI loadings anticipated in the field in batch tests will afford the best estimates for reactivity and potential for biostimulation or (unwanted) passivation, albeit still potentially different than for a flow through a system. For systems with higher groundwater flow rates, flow through column systems can be useful to approximate the length of reactive zone needed to achieve a desired reduction in concentration. While imperfect, these studies can at least provide some initial indication of the potential of NZVI to work at the site of interest; e.g., if a 20 m-long barrier is required to achieve remediation targets, then NZVI is probably not a good barrier technology.

16.7.2 Determination of Dose for Injection

Estimating the dose of NZVI required to achieve a remediation goal is difficult. This is because estimating the mass of the target contaminant, the reductant demand from other constituents in the water and targeting a specific redox potential goal for achieving the remediation targets are all challenging. Reductant demand will be site-specific and must be determined. All of these inputs, of course, depend on the flow rate of the groundwater and the retention time of the groundwater in the reactive zone. This retention time and the specific reactivity of the contaminant with NZVI in the system of interest can provide some guidance on the length of a reactive barrier that is needed to achieve remediation targets (Kim et al. 2017). This barrier length, along with a good estimate of the depth of the target zone can provide an estimate of the treatment volume required. There is a general "rule of thumb" that 10–15 g/L of NZVI in the pore water should provide the decrease in redox potential and adequate reactivity for in situ remediation. However, this is also site dependent, and it may not be possible to inject this much NZVI into the pore space. Thus, from the treatment volume, the porosity of the system, and this rule of thumb, one can estimate an NZVI mass required for injection to meet remediation targets. Using a NZVI to TCE mass ratios much greater than the stoichiometric amount (as much as 20:1 for one site evaluated) has been proposed (Lowry et al. 2012). This is all based on rough estimates, and more well-documented field trials are still required to better estimate the NZVI doses required.

16.7.3 Deployment of Combined Remedies

It may be advantageous to deploy NZVI as part of a combined remedy remediation strategy. One promising combined remedy is NZVI-promoted bioremediation for compounds that are amenable to reductive degradation, e.g., CVOCs or metals that are immobilized upon reduction. NZVI can quickly provide reducing conditions and maintain them in the field for extended periods under the right aquifer conditions. The limited number of studies available regarding NZVI-promoted bioremediation suggests that NZVI can promote the biodegradation of chlorinated solvents in situ (Wei et al. 2012; Kocur et al. 2015, 2016). They also indicate that NZVI will not have any long-lasting negative impacts on the subsurface microbial ecology and therefore its use should not preclude monitored natural attenuation (MNA) as part of the remediation strategy. Long-term field-pilot studies are still needed to elucidate the factors that promote long-term MNA after NZVI injection (e.g., injection doses, the use of biodegradable coatings or co-injected materials, spatial distribution requirements for enhancing MNA, impacts of NZVI-MNA on aquifer permeability and groundwater flow). Conditions that shift microbial communities toward those favorable for contaminant degradation must still be identified. Finally, the efficacy of treatment and the cost of deployment must be determined relative to other alternatives available, e.g., sugar substrates.

Furthermore, recent development in electromagnetic induction heating of NZVI may also provide a novel, in situ, combined remedy of thermally enhanced desorption and reductive dechlorination using NZVI in the future (Phenrat and Kumloet 2016; Phenrat et al. 2016; Srirattana et al. 2017). This should help alleviate the mass transfer limitation problem from source zone (dense nonaqueous phase liquid) dissolution (Fagerlund et al. 2012) and desorption of CVOCs from organic matter in the soil (Zhang et al. 2011), which adversely affects the electron utilization efficiency of NZVI, time to site closure, and cost-effectiveness of the in situ remediation using NZVI.

16.7.4 Closing Remarks

The potential use of NZVI for remediating a wide range of groundwater contaminants continues to be investigated around the world. New versions of NZVI are currently being synthesized and tested, e.g., supported NZVI and sulfidized NZVI. These materials may provide the materials with higher selectivity and improved injectability needed to make NZVI treatment more versatile and effective. Pilot-scale studies aimed at improving the injection methods and performance monitoring are still needed to identify the most promising injection methods for different aquifer conditions and for demonstrating long-term effectiveness of the method. Finally, leveraging NZVI's reducing capacity in combined remedies, particularly with bioremediation, is likely to afford the most versatile and cost-effective NZVI-based remediation approach. The great versatility of NZVI treatment and its low toxicity and potential for undesirable side effects make it an important remediation tool. Greater experience and more successful field demonstrations will increase its application as a remediation technology.

References

Babakhani, P., Fagerlund, F., Shamsai, A., Lowry, G. V., & Phenrat, T. (2015). Modified MODFLOW-based model for simulating the agglomeration and transport of polymer-modified Fe nanoparticles in saturated porous media. *Environmental Science and Pollution Research International, 25*(8), 7180–7199.

Bardos, P., Bone, B., Daly, P., Elliott, D., Jones, S., Lowry, G. V., & Merly, C. (2015). A risk/benefit appraisal for the application of nano-scale zero valent iron (nZVI) for the remediation of contaminated sites. *Taking nanotechnological remediation processes from lab scale to end user applications for the restoration of a clean environment.* European Union Seventh Framework Programme (FP7/2007-2013), Project Nr.: 309517, p. 76.

Bennett, P., He, F., Zhao, D., Aiken, B., & Feldman, L. (2010). In situ testing of metallic iron nanoparticle mobility and reactivity in a shallow granular aquifer. *Journal of Contaminant Hydrology, 116*, 35–46.

Busch, J., Meißner, T., Potthoff, A., Bleyl, S., Georgi, A., Mackenzie, K., Trabitzsch, R., Werban, U., & Oswald, S. E. (2015). A field investigation on transport of carbon-supported nanoscale zero-valent iron (nZVI) in groundwater. *Journal of Contaminant Hydrology, 181*, 59–68.

Chowdhury, A. I., Krol, M. M., Kocur, C. M., Boparai, H. K., Weber, K. P., Sleep, B. E., & O'Carroll, D. M. (2015). nZVI injection into variably saturated soils: Field and modeling study. *Journal of Contaminant Hydrology, 183*, 16–28.

Fagerlund, F., Illangasekare, T. H., Phenrat, T., Kim, H.-J., & Lowry, G. V. (2012). PCE dissolution and simultaneous dechlorination by nanoscale zero-valent iron particles in a DNAPL source zone. *Journal of Contaminant Hydrology, 131*, 9–28.

Fan, D., Anitori, R. P., Tebo, B. M., Tratnyek, P. G., Lezama Pacheco, J. S., Kukkadapu, R. K., Engelhard, M. H., Bowden, M. E., Kovarik, L., & Arey, B. W. (2013). Reductive sequestration of pertechnetate ($^{99}TcO_4^-$) by nano zero-valent iron (nZVI) transformed by abiotic sulfide. *Environmental Science & Technology, 47*, 5302–5310.

Fan, D., O'Brien Johnson, G., Tratnyek, P. G., & Johnson, R. L. (2016). Sulfidation of nano zerovalent iron (nZVI) for improved selectivity during in-situ chemical reduction (ISCR). *Environmental Science & Technology, 50*, 9558–9565.

Flores Orozco, A., Velimirovic, M., Tosco, T., Kemna, A., Sapion, H., Klaas, N., Sethi, R., & Bastiaens, L. (2015). Monitoring the injection of microscale zerovalent iron particles for groundwater remediation by means of complex electrical conductivity imaging. *Environmental Science & Technology, 49*, 5593–5600.

Johnson, R. L., Nurmi, J. T., O'Brien Johnson, G. S., Fan, D., O'Brien Johnson, R. L., Shi, Z., Salter-Blanc, A. L., Tratnyek, P. G., & Lowry, G. V. (2013). Field-scale transport and transformation of carboxymethylcellulose-stabilized nano zero-valent iron. *Environmental Science & Technology, 47*, 1573–1580.

Joyce, R. A., Glaser Ii, D. R., Werkema, D. D., Jr., & Atekwana, E. A. (2012). Spectral induced polarization response to nanoparticles in a saturated sand matrix. *Journal of Applied Geophysics, 77*, 63–71.

Kim, H. J., Leitch, M., Naknakorn, B., Tilton, R. D., & Lowry, G. V. (2017). Effect of emplaced nZVI mass and groundwater velocity on PCEdechlorination and hydrogen evolution in water-saturated sand. *Journal of Hazardous Materials, 322*, 136–144.

Kocur, C. M., Lomheim, L., Boparai, H. K., Chowdhury, A. I., Weber, K. P., Austrins, L. M., Edwards, E. A., Sleep, B. E., & O'Carroll, D. M. (2015). Contributions of abiotic and biotic dechlorination following carboxymethyl cellulose stabilized nanoscale zero valent iron injection. *Environmental Science & Technology, 49*, 8648–8656.

Kocur, C. M. D., Lomheim, L., Molenda, O., Weber, K. P., Austrins, L. M., Sleep, B. E., Boparai, H. K., Edwards, E. A., & O'Carroll, D. M. (2016). Long-term field study of microbial community and dechlorinating activity following carboxymethyl cellulose-stabilized nanoscale zero-valent iron injection. *Environmental Science & Technology, 50*, 7658–7670.

Liu, Y., Phenrat, T., & Lowry, G. V. (2007). Effect of TCE concentration and dissolved groundwater solutes on NZVI-promoted TCE dechlorination and H2 evolution. *Environmental Science & Technology, 41*, 7881–7887.

Lowry, G. V., Phenrat, T., Fagerlund, F., Illanagasekare, T., Tratnyek, P. G., & Johnson, R. L. (2012). *Fundamental study of the delivery of nanoiron to DNAPL source zones in naturally heterogeneous field systems.* SERDP Project ER-1485, p. 133.

Mackenzie, K., Bleyl, S., Kopinke, F. D., Doose, H., & Bruns, J. (2016). Carbo-Iron as improvement of the nanoiron technology: From laboratory design to the field test. *Science of the Total Environment, 563–564*, 641–648.

Mueller, N. C., Braun, J., Bruns, J., Černík, M., Rissing, P., Rickerby, D., & Nowack, B. (2012). Application of nanoscale zero valent iron (NZVI) for groundwater remediation in Europe. *Environmental Science and Pollution Research International, 19*, 550–558.

NanoRem. (2018). *Taking nanotechnological remediation processes from lab scale to end user applications for the restoration of a clean environment.* NanoRem.

Phenrat, T., & Kumloet, I. (2016). Electromagnetic induction of nanoscale zerovalent iron particles accelerates the degradation of chlorinated dense non-aqueous phase liquid: Proof of concept. *Water Research, 107*, 19–28.

Phenrat, T., Kim, H.-J., Fagerlund, F., Illangasekare, T., & Lowry, G. V. (2010a). Empirical correlations to estimate agglomerate size and deposition during injection of a polyelectrolyte-modified Fe^0 nanoparticle at high particle concentration in saturated sand. *Journal of Contaminant Hydrology, 118*, 152–164.

Phenrat, T., Schoenfelder, D., Losi, M., Yi, J., Peck, S. A., & Lowry, G. V. (2010b). Treatability study for a TCE contaminated area using nanoscale- and microscale-zerovalent iron particles: Reactivity and reactive life time. In C. L. Geiger (Ed.), *Environmental applications of nanoscale and microscale reactive metal particles* (pp. 183–202). Washington, DC: Oxford University Press.

Phenrat, T., Thongboot, T., & Lowry, G. V. (2016). Electromagnetic induction of zerovalent iron (ZVI) powder and nanoscale zerovalent iron (NZVI) particles enhances dechlorination of trichloroethylene in contaminated groundwater and soil: Proof of concept. *Environmental Science & Technology, 50*, 872–880.

Srirattana, S., Piaowan, K., Lowry, G. V., & Phenrat, T. (2017). Electromagnetic induction of foam-based nanoscale zerovalent iron (NZVI) particles to thermally enhance non-aqueous phase liquid (NAPL) volatilization in unsaturated porous media: Proof of concept. *Chemosphere, 183*, 323–331.

Stejskal, V., Lederer, T., Kvapil, P., Slunsky, J., & Skácelová, P. (2017). NanoRem Pilot Site – Spolchemie I, Czech Republic: Nanoscale zero-valent iron remediation of chlorinated hydrocarbons. *NanoRem bulletin*. CL:AIRE, UK, pp. 1–8.

Wei, Y. T., Wu, S. C., Yang, S. W., Che, C.-H., Lien, H.-L., & Huang, D. H. (2012). Biodegradable surfactant stabilized nanoscale zero-valent iron for in situ treatment of vinyl chloride and 1,2-dichloroethane. *Journal of Hazardous Materials, 211–212*, 373–380.

Zhang, M., He, F., Zhao, D., & Hao, X. (2011). Degradation of soil-sorbed trichloroethylene by stabilized zero valent iron nanoparticles: Effects of sorption, surfactants, and natural organic matter. *Water Research, 45*, 2401–2414.

Index

T. Phenrat, G. V. Lowry (eds.), *Nanoscale Zerovalent Iron Particles for Environmental Restoration*, https://doi.org/10.1007/978-3-319-95340-3

E

N

FSC
www.fsc.org
MIX
Papier | Fördert
gute Waldnutzung
FSC® C083411

Kolibri 360 GmbH
Ferdinand-Jühlke-Straße 7
99095 Erfurt, Deutschland
produktsicherheit@kolibri360.de